Combinatorial Chemistry

Combinatorial Chemistry

Edited by **Jina Redlin**

WILLFORD PRESS

New York

Published by Willford Press,
118-35 Queens Blvd., Suite 400,
Forest Hills, NY 11375, USA
www.willfordpress.com

Combinatorial Chemistry
Edited by Jina Redlin

International Standard Book Number: 978-1-68285-124-1 (Hardback)

Printed in the United States of America.

Contents

Preface

This book was inspired by the evolution of our times; to answer the curiosity of inquisitive minds. Many developments have occurred across the globe in the recent past which has transformed the progress in the field.

Combinatorial chemistry has become a prominent field due to the rapid technological advancements in chemistry and industrial manufacturing. The production of large numbers of synthetic molecules at surprisingly high rate is possible with combinatorial chemistry. It has played a significant role in materials science for exploration and development of novel materials. This book presents the complex techniques in the most comprehensible and easy to understand language. The aim of this book is to present researches that have transformed this discipline and aided its advancement. It will provide an in-depth knowledge about the discovery and optimization of biologically active compounds, and methodologies widely used in combinatorial chemistry. Those in search of information to further their knowledge will be greatly assisted by this book.

This book was developed from a mere concept to drafts to chapters and finally compiled together as a complete text to benefit the readers across all nations. To ensure the quality of the content we instilled two significant steps in our procedure. The first was to appoint an editorial team that would verify the data and statistics provided in the book and also select the most appropriate and valuable contributions from the plentiful contributions we received from authors worldwide. The next step was to appoint an expert of the topic as the Editor-in-Chief, who would head the project and finally make the necessary amendments and modifications to make the text reader-friendly. I was then commissioned to examine all the material to present the topics in the most comprehensible and productive format.

I would like to take this opportunity to thank all the contributing authors who were supportive enough to contribute their time and knowledge to this project. I also wish to convey my regards to my family who have been extremely supportive during the entire project.

Editor

An overview of crumb rubber modified asphalt

Nuha S. Mashaan*, Asim Hassan Ali, Mohamed Rehan Karim and Mahrez Abdelaziz

Center for Transportation Research, Faculty of Engineering, University of Malaya, Kuala Lumpur 50603, Malaysia.

Roadways are considered one of the most important elements of infrastructure and they play an essential role in our daily lives. In road pavement construction, the use of crumb rubber in the modification of bitumen binder is considered as a smart solution for sustainable development by reusing waste materials. It is believed that crumb rubber modifier (CRM) could be one of the alternative polymer materials in improving bitumen binder performance properties of hot mix asphalt. This study aims to present and discuss the findings from some of the studies, on the use of crumb rubber in asphalt pavement.

Key words: Rubberised bitumen, crumb rubber modifier, rheology, rutting, fatigue cracking.

INTRODUCTION

Over the years, road structures have deteriorated more rapidly due to increases in service traffic density, axle loading and low maintenance services. To minimise the damage of pavement surface and increase durability of flexible pavement, the conventional bitumen needs to be improved with regards to performance related properties, such as resistance to permanent deformation (rutting) and fatigue cracking. The modification of bituminous binder has been explored over the past years in order to improve road pavement performance properties.

There are many modification processes and additives that are currently used in bitumen modifications, such as styrene butadiene styrene (SBS), styrene-butadiene rubber (SBR), ethylene vinyl acetate (EVA) and crumb rubber modifier (CRM). The use of commercial polymers, such as SBS and SBR in road and pavement construction will increase the construction cost as they are highly expensive materials. However, with the use of alternative materials, such as CRM, will definitely be environmentally beneficial, and not only it can improve the bitumen binder properties and durability, but it also has a potential to be cost effective (Hamed, 2010).

In recent times, a serious problem that leads to environment pollution is the abundance and the increase of waste tyre disposal. Large amounts of rubbers are used as tyres for cars and trucks. Despite the long run in

service, these tyres are not discarded. Although, the rubber as a polymer is a thermosetting material cross linked to processing and moulding, however, it cannot be softened or remoulded by re-heating unlike other types of thermoplastics polymer which can be softened and reshaped when heated. The major approach to solve this issue is the recycle and reuse of waste tyre rubber and the reclaim of rubber raw materials (Adhilkarri et al., 2000). In recent years, researches on applications of rubberised bitumen binders have reported many adventages. These advantages include improved bitumen resistance to rutting due to high viscosity, high softening point and better resilience, improved bitumen resistance to surface initiated cracks, the reduction of fatigue/ reflection cracking, the reduction of temperature susceptibility, improved durability as well as the reduction in road pavement maintenance costs (Liu et al., 2009).

HISTORY OF USING CRUMB RUBBER AS A MODIFIER IN BITUMINOUS PAVEMENT

The earliest experiments date back to the 1840s, which involved incorporating natural rubber into bitumen to increase its engineering performance properties. The process of bitumen modification involving natural and synthetic rubber was introduced as early as 1843 (Thompson, 1979). Then, in 1923, natural and synthetic rubber modifications in bitumen were further improved (Isacsson and Lu 1999; Yildrim, 2007). According to the study of Yildrim (2007), the development of rubber-bitumen

---

*Corresponding author. E-mail: asim@siswa.um.edu.my.

materials being used as joint sealers, patches and membranes began in the late 1930s. In 1950, the use of scrap tyre in asphalt pavement was reported (Hanson et al., 1994).

In 1960, Charlie Mac Donald working as head material engineer in Phoenix, Arizona, used ground tyre rubber as an additive in bitumen binder modification. He found that after completing the mixing of crumb rubber with the conventional bitumen and allowing it to blend for mix duration of 45 to 60 min, there were new material properties produced, which resulted in swelling in the size of the rubber particles at higher temperatures allowing for higher concentrations of liquid bitumen contents in pavement mixes (Huffman, 1980). In the mid-1980s, the Europeans began the development of newer polymers and additives for use in bitumen binder modification (Brule, 1996). In recent years, the use of crumb rubber has gained interest in pavement modification and has shown that crumb tyre rubber can improve the bitumen performance properties (Brown et al., 1997).

CRUMB TYRE RUBBER GRINDING PROCESS

The crumb rubber is made by shredding scrap tyre, which is a particular material free of fibre and steel. The rubber particle is graded and found in many sizes and shapes. The crumb rubber is described or measured by the mesh screen or sieve size through which it passes during the production process. To produce crumb rubber, generally, it is important to reduce the size of the tyres. There are two techniques to produce crumb rubber: ambient grinding and the cryogenic process (Becker et al., 2001). The ambient grinding process can be divided into two methods: granulation and cracker mills. The ambient describes the temperature when the waste tyres rubber as its size is reduced. The material is loaded inside the crack mill or granulator at ambient temperature. The cryogenic grinding is a cleaner, slightly faster operation resulting in production of fine mesh size. The high cost of this process is a disadvantage due to the added cost of liquid nitrogen.

PERFORMANCE OF CRUMB TYRE RUBBER IN BITUMINOUS MATERIALS

There are two rather different methods in usage of tyre rubber in bitumen binders. Firstly, crumb rubber in the bitumen is dissolved as binder modifier. Second, is by substituting a portion of fine aggregates with ground rubber that does not completely react with bitumen (Huang et al., 2007).

Numerous factors can influence the modification effects which consist of base bitumen constituents, blending time and temperature, the percentage of rubber, the gradation of crumb rubber, the type of mixing (wet or dry) and the

grinding process method (Huang et al., 2007; Airey et al., 2003; Jeong et al., 2010). It observed that during the bitumen-rubber blending, due to higher stiffness and tensile strength at elevated temperatures, the mixture had decreased rutting capability (Palit et al., 2004). The design method for conventional bitumen mixture can be used for bitumen-rubber mixture as the mix stability being the primary factor. Also, standard paving machinery can be used for placement of bitumen-rubber mixture. However, a pneumatic tyre roller is not suitable as asphalt rubber will stick onto the roller tyres (Huang et al., 2007). Rubber pavement association found that using tyre rubber in open-graded mixture binder could decrease tyre noise by approximately 50%. In addition, in spray applications, rubber particles of multiple sizes had a better sound absorbing. Moreover, another advantage of using asphalt rubber is to increase the life-span of the pavement. However, recommendations were made to assess the cost effectiveness of asphalt rubber (Huang et al., 2007).

PHYSICAL AND RHEOLOGICAL PROPERTIES OF RUBBERISED BITUMEN

Penetration properties

The penetration is a measure of hardness or softness of bitumen binder which shows an effect by adding crumb rubber to bitumen binder; it decreases as rubber content is increased. The penetration shows lower values as rubber content increases at different mix conditions of rubberised bitumen binder, indicating that the binder becomes stiff and more viscous (Mashaan et al., 2011a). Mahrez (1999) investigated the properties of rubberised bitumen prepared by physical blending of bitumen 80/100 penetration grade with different crumb rubber content and various aging phases. The results of penetration values decreased over the aging as well as before aging by increasing the rubber content in the mix. Also, the modified binders have lower penetration values than unmodified binders.

Elastic recovery properties

The elastic recovery or elasticity describes the ability of a bitumen binder to elongate when the tension is applied and to recover its original shape when the tension is released. The degree of elastic recovery was used as an indicator of permanent deformation in pavement materials (Yildirim, 2007).

According to the study of Jensen and Abdelrahman (2006), the elastic recovery property is very important in both fatigue and rutting resistance selection and evaluation. The elastic recovery is a property that indicates the quality of polymer components in bitumen binders. Oliver

(1981) concluded from his study, that the elastic recovery of rubberised bitumen binders leads to an increase as the rubber particle size decreases. Modified bitumen binders showed a significant enhancement on the elastic recovery, and, in contrast, the ductility decreased with respect to unmodified binders (Mashaan et al., 2011a, b).

Ductility properties

The ductility is a distinct strength of bitumen, allowing it to undergo notable deformation or elongation. The ductility is defined as the distance in centimetre, to which a standard sample or briquette of the material will be elongated without breaking.

The studies of Mashaan et al. (2011a, b) concluded that finer rubber particles resulted in higher ductility elongation and also, that toughness would increase as rubber content increases. A combined effect of both time and temperature was noted with minimum elastic recovery value improved at maximum time and maximum temperature of two hours and 240°C, respectively (Jensen and Abdelrahman, 2006). The bitumen-rubber modification resulted in a better rutting resistance and higher ductility. However, the modified binder was susceptible to decomposition and oxygen absorption. There were problems of low compatibility, because of the high molecular weight. Furthermore, the recycled tyre rubber decreases reflective cracking, which in turn increases durability. In using waste tyre rubber, there are however, several practical and experimental issues, such as it requires an elevated composite of temperatures and extended digestion time during the mixing process for it to be diffused in the bitumen (Yildrim, 2007).

Viscosity and softening point properties

The viscosity refers to the fluid property of the bitumen, and it is a gauge of flow-resistance. At the application temperature, viscosity greatly influences the potential of the resulting paving mixes. During compaction or mixing, the low or high viscosity has been observed to result in lower stability values. The softening point refers to the temperature at which the bitumen attains a particular degree of softening. The use of crumb rubber in bitumen modification leads to an increase in the softening point and viscosity as rubber crumb content increases (Mahrez, 1999; Mashaan et al., 2011a). Mahrez and Rehan (2003) claimed that there is a consistent relationship between viscosity and softening point at different aging phases of rubberised bitumen binder.

Also, it is reported that the higher crumb rubber content leads to higher viscosity and softening point. The viscosity is a continuously increasing non linear function of rubber content and the relative increase is a factor related to the application of temperature (Bahia and Davies, 1995).

BITUMEN - RUBBER BLENDS INTERACTION AND ABSORPTION

Modified bitumen using crumb rubber showed an improvement in the performance of pavements over the base binders as a result of the interaction of crumb rubber with base binders. Due to this interaction, there are noticeable changes in the viscosity, physical and rheological properties of the rubberised bitumen binder (Airey et al., 2003; Bahia and Davies, 1995), leading to high resistance of rutting of pavements (Huang et al., 2007).

From the aforementioned review of literature, the primary mechanism of the interaction is swelling of the rubber particles caused by the absorption of light fractions into these particles and stiffening of the residual binder phase (Abedlrahman and Carpenter, 1999; Airey et al., 2003). The rubber particles are constricted in their movement into the binder matrix to move about due to the swelling process which limits the free space between the rubber particles. Compared to the coarser particles, the finer particles swell easily thus, developing higher binder modification (Abedlrahman and Carpenter, 1999). The swelling capacity of rubber particle is linked to the penetration grade of the binder, crude source and the nature of the crumb rubber modifier (Airey et al., 2003). According to the study of Shen et al. (2009), the factors which affect the digestion process of the bitumen and rubbers blends are rubber content, rubber size, binder viscosity and blending conditions.

Rubber content

According to a study conducted by Lee et al. (2008), the higher crumb rubber content produced increased viscosity at 135°C and improved the rutting properties. It was also observed that the increased crumb rubber amount (fine crumb rubber) produced rubberised bitumen with higher viscosity and lower resilience. However, optimum crumb rubber content still needs to be determined for each crumb rubber size and asphalt binder. It is believed that a physicochemical interaction that occurs between the asphalt and the crumb rubber alters the effective size and physical properties of the rubber particle, thus influencing pavement performance (Huang et al., 2007).

Becker et al. (2001) claimed that blend properties will be influenced by the amount of crumb rubber added to the bitumen. Higher amounts indicated significant changes in the blend properties. As rubber content generally increases, it leads to increased viscosity, increased resilience, increased softening point and decreases penetration at 25°C.

Rubber particle size

The study of Souza and Weissman (1994) using a binder with 15% rubber content (size of 0.2, 0.4 and 0.6 mm) in dense-graded bitumen.

The mixture showed improved performance in dynamic stability, 48 h residual stability, flexural strength and strain value. Asphalt containing 0.2 and 0.4 mm size rubber indicated the best laboratory results (Souza and Weissman, 1994). The particles size disruption of crumb rubber influenced the physical properties of bitumen-rubber blend. In general, small difference in the particles size has no significant effects on blend properties. However, the crumb rubber size can certainly make a big difference. According to a study of Shen et al. (2009), the particle size effects of CRM on high temperature properties of rubberised bitumen binders was an influential factor on visco-elastic properties. The coarser rubber produced a modified binder with high shear modules and an increased content of the crumb rubber decreased the creep stiffness which in turn showed significant thermal cracking resistance.

Blending conditions

In general, bitumen binder and ground tyre rubber are mixed together and blended at elevated temperatures for differing periods of time prior to using them as a paving binder. These two factors work together to evaluate the performance properties of rubberised bitumen binder through blending process of bitumen-rubber interaction. This variation in mixing time and temperature that results due to the normal activities are related to bitumen paving construction. Nevertheless, the consistency of rubberised bitumen rubber can be affected by the time and temperature used to combine the components and thus, must be cautiously used for it optimum potential to be achieved. Xiao et al. (2006) studied the effects of reaction time on the permanent changes of crumb rubber after mixing with bitumen.

The size reduction of rubber particle increased with the blend duration and with decreasing crumb rubber size. Lee et al. (2006) reported that a longer reaction time was ineffective in increasing the high temperature viscosity of the control binder. In a joined experiment, Mashaan et al. (2011b) also found that the blending time has an insignificant effect in the case of 30 and 60 min on rheological properties of rubberised asphalt. Paulo and Jorge (2008) investigated the effect of blending conditions (time and temperature) on penetration, softening point and resilience modulus. The results showed that there was no significant effect for three blending time (45, 60 and 90 min) on modified binder properties, although, there was a tendency for blending time effect to become constant between 60 and 90 min.

PAVEMENT DISTRESS

Rutting performance

Rutting defined as longitudinal depression in wheel paths

as a result of continued densification by traffic load. According to Sousa and Weissman (1994) study, rutting in bitumen pavement develops as load applications increases. The rutting seems as longitudinal depressions in the wheel paths with small upheavals on the sides. These are due to a combination of densification and shear deformation.

The rutting is a primary measure of the performance of pavement in several pavement design methods. The rutting can occur as a result of problematic sub grade, unbound base course. Rutting failures are consequence of heavy truckloads with high tyre pressures and high pavement temperatures. Hence, the considerable selection of bitumen binder and aggregate combination will boost in providing optimum performing asphalt pavements (Sousa and Weissman, 1994). Brown and Cross (1992) reported that permanent deformation in bituminous mixture is caused by consolidation and/or lateral movement of the mixture under traffic. Shear failure (lateral movement) of the bituminous mixture courses generally occurs in the top 100 mm of the pavement surface. However, it can run deeper if proper materials are not used. Moreover, it was evident that rutting is caused mainly by deformation flow rather than volume change.

Tayfur et al. (2007) claimed that after the initial densification, the permanent deformation of the bituminous mixture happens due to shear loads which take place close to the pavement surface which in fact is the contact area between the tyre and the pavement. These efforts increase without the volume variations in the bituminous mixture. They are the primary mechanisms in the development of rutting during the life span of the pavement design.

Fatigue cracking

Fatigue is one of most important distresses in asphalt pavement structure due to repeated load of heavy traffic services which occur at intermediate and low temperatures. Aflaki and Memarzadeh (2011) used different shear methods at low and intermediate temperature to study the effect of rheological properties of crumb rubber on fatigue cracking. The results showed that the high shear blending has more effect on improvement at low temperatures than the low shear blend. The use of crumb rubber modified with bitumen binder seems to enhance the fatigue resistance, as illustrated in a number of studies (Raad and Saboundjian, 1998; Soleymani et al., 2004; McGennis, 1995). The improved performance of bitumen rubber pavements when compared with conventional bitumen pavements has partly resulted from improved rheological properties of the rubberised bitumen binder.

CONCLUSION

This review study presented the application of crumb

rubber modifier in the asphalt modification of flexible pavement. From the results of previous studies, it aspires to consider crumb rubber modifier in hot mix asphalt to improve resistance to rutting and produce pavements with better durability by minimising the distresses caused in hot mix asphalt pavement. Hence, road users would be ensured of safer and smoother roads. Furthermore, the use of crumb rubber modifier as an additive in bitumen modified binder would reduce pollution problems and protect our environment as well.

REFERENCES

Abdelrahman MA, Carpenter SH (1999). Mechanism of interaction of asphalt cement with crumb rubber modifier. Transportation Research Record: J. Transport. Res. Board, 1661: 106-113.

Adhikari B, De D, Maiti S (2000). Reclamation and recycling of waste rubber. Progress in Polymer Science, 25: 909- 948.

Aflaki S, Memarzadeh M (2011). Using two-way ANOVA and hypothesis test in evaluating crumb rubber modification (CRM) agitation effects on rheological properties of bitumen. Construction and Building Materials, 25: 2094-2106.

Airey GD, Rahman MM, Collop AC (2003). Absorption of bitumen into crumb rubber using the basket drainage method. Int. J. Pavement Eng., 4(2): 105-119.

Bahia HU, Davies R (1995). Factors controlling the effect of crumb rubber on critical properties of asphalt binders. J. Assoc. Asphalt Paving. Technol., 64: 130-162.

Becker Y, Mendez MP, Rodriguez Y (2001). Polymer modified asphalt. Vision Tecnologica, 9(1): 39-50.

Brown DR, Jared D, Jones C, Watson D (1997). Georgia's experience with crumb rubber in hot – mix asphalt. Transportation Research Record: J. Transport. Res. Board, 1583: 45-51.

Brown DR, Cross SA (1992). A National Study of Rutting in Asphalt Pavement. J. Assoc. Asphalt. Paving. Technol., 61: 535-583.

Brule B (1996). Polymer-modified asphalt contents used in the road construction industry: Basic principles. Transportation Research Record: J. Transport. Res. Board, 1535: 48-53.

Hamed FK (2010). *Evaluation* of fatigue resistance for modified asphalt concrete mixture based on dissipate energy concept. Ph.D. Thesis, Technische University Darmstadt, Germany.

Hanson DI, Foo KY, Brown ER, Denson R (1994). Evaluation and characterization of a rubber-modified hot mix asphalt pavement. Transportation Research Record: J. Transport. Res. Board, 1436: 98-107.

Huang Y, Bird RN, Heidrich O (2007). A review of the use of recycled solid waste materials in asphalt pavements. Resources, Conservation and Recycling, 52: 58-73.

Huffman JE (1980). Sahuaro Concept of Asphalt-Rubber Binders. Presentation at the First Asphalt Rubber User Producer Workshop, Arizona.

Isacsson U, Lu X (1999). Characterization of bitumens modified with SEBS, EVA and EBA polymers. J.Mater. Sci., 34(15): 3737-3745.

Jensen W, Abdelrahman M (2006). Crumb rubber in performance-graded asphalt binder. Final Report SPR-01.Nebraska Department of Roads University of Nebraska-Lincoln, 05: 585.

Jeong KD, Lee SJ, Amirkhanian, SN, Kim KW (2010). Interaction effect of crumb rubber modified asphalt binder. Construction and Building Materials, 24: 824-831.

Lee SJ, Akisetty CK, Amirkhanian SN (2008). The effect of crumb rubber modifier (CRM) on the performance properties of rubberized binder in HMA pavements. Construction and Building Material, 22: 1368-1376.

Lee SJ, Amirkhanian SN, Shatanawi K (2006). Effect of reaction time on physical and chemical properties of rubber-modified binders. International Rubber Conference. Compendium of Papers CDRom, Lyon, France.

Liu S, Cao W, Fang J, Shang S (2009). Variance analysis and performance evaluation of different crumb rubber modified (CRM) asphalt. Construct. Build. Mater., 23: 2701-2708.

McGennis RB (1995). Evaluation of physical properties of fine crumb rubber – modified asphalt binder. Transportation Research Record: J. Transport. Res. Board, 1488: 62-71.

Mahrez A (1999). Properties of rubberised bitumen binder and its effect on the bituminous mix. M.Sc. Dissertation, Faculty of Engineering, University of Malaya, Malaysia.

Mahrez A, Rehan M (2003). Rheological evaluation of aging properties of crumb rubber-modified bitumen. Journal of the Eastern Asia Society for Transportation Studied, 5: 820-833.

Mashaan NS, Ali AH, Karim MR, Mahrez A (2011a). Effect of crumb rubber concentration on the physical and rheological properties of rubberised bitumen binders. Int. J. Phys. Sci., 6(4): 684-690.

Mashaan NS, Ali AH, Karim MR, Mahrez A (2011b). Effect of blending time and crumb rubber content on properties of crumb rubber modified asphalt binder. Int. J. Phys. Sci., 6(9): 2189-2193.

Oliver J (1981). Modification of paving asphalts by digestion with scrape rubber. J. Transport. Res. Board, 821: 37-44.

Palit SK, Sudhakar RK, Pandey BB (2004). Laboratory evaluation of crumb rubber modified asphalt mixes. J. Mater. Civil Eng., 16(1): 45-53.

Paul AP, Jorge CP (2008). Laboratory optimization of continuous blend asphalt rubber. Proceedings of 3rd European Pavement and Asset Management EPAM 3. Coimbra.

Raad L, Saboundjian S (1998). Fatigue behavior of rubber – modified pavements. Transportation Research Record. J. Transport. Res. Board, 1639: 73-82.

Shen J, Amirkhanian SN, Xiao F, Tang B (2009). Influence of surface area and size of crumb rubber on high temperature properties of crumb rubber and modified binders. J. Construct. Build. Mater., 23: 304-310.

Soleymani HR, Zhai H, Bahia H (2004). Role of modified binders in rheology and damage resistance behavior of asphalt mixture. Transportation Research Record: Transport. Res. Board, 1875: 70-79.

Sousa JB, Weissman SL (1994). Modeling permanent deformation of asphalt-aggregate mixes. J. Assoc. Asphalt. Paving. Technol., 63: 224-257.

Tayfur S, Ozen H, Aksoy A (2007). Investigation of rutting performance of asphalt mixtures containing polymer modifiers. Construct. Build. Mater., 21: 328–337.

Thompson DC, Hoiberg AJ (Eds) (1979). Bituminous Materials: Asphalt Tars and Pitches. New York: Krieger Publishing Co.

Xiao F, Ameirkhanian SN, Putman BJ (2006). Laboratory investigation of dimensional changes of crumb rubber reacting with asphalt binder. Asphalt Rubber 2006 Conference, Palms spring, California.

Yildrim Y (2007). Polymer modified asphalt binder. J. Build. Construct. Mater., 21: 66-72.

Potential reduction of concrete deterioration through controlled DEF in hydrated concrete

Samo Lubej[1] and Milan Radosavljevic[2]

[1]Faculty of Civil Engineering, University of Maribor, Smetanova 17, 2000 Maribor, Slovenia.
[2]School of Construction Management and Engineering, University of Reading, Whiteknights, P. O. Box 219, Reading, RG6 & AW, United Kingdom.

Delayed ettringite formation (DEF) is a chemical reaction with proven damaging effects on hydrated concrete. Ettringite crystals can cause cracks and their widening due to pressure on cracked walls caused by the positive volume difference in the reaction. Concrete may show improvements in strength at early ages but further growth of cracks causes widening and spreading through the concrete structure. In this study, finely dispersed crystallization nuclei achieved by adding air-entraining agent (AEA) and short vibration of specimens is presented as the main prerequisite for reducing DEF-induced deterioration of hydrated concrete. The study presents the method and mechanism for obtaining the required nucleation. Controlling long-term DEF by providing AEA-induced crystallisation nuclei, prevented excessive and rapid initial strength improvements, and resulted in a slight increase of compressive strength of fine grained concrete with only marginally lower density.

Key words: Delayed ettringite formation (DEF), aerated concrete, strength improvement.

INTRODUCTION

Delayed ettringite formation in cementitious materials is considered a harmful chemical reaction leading to a variety of damages (Diamond, 1996; Thomas, 2001; Barbarulo et al., 2005). The volume of the formed DEF crystals in the hardened concrete is larger than the volume of reactants and the main results are forces from the growing crystals acting upon walls of the crack. As a consequence, DEF cracks continue getting wider and spread through the concrete structure (Sahu and Thaulow, 2004; Thomas et al., 2008). In the study of DEF on railroad ties, Sahu and Thaulow (2004) found massive ettringite deposits at the interface between the paste and aggregate but with no signs of alkali-silica reaction, DEF was found to be the sole reason for map cracking. However, recent research has led to a better understanding of the mechanisms of DEF. It is believed that internal sulphate attack (ISA), which occurs in an environment free from external sources of sulphate, is the main mechanism that leads to DEF, particularly for heat-cured concrete structures (Collepardi, 2003).

In general, it is acknowledged that DEF is a result of a number of factors and conditions including excessive temperatures of above 70°C, the presence of sulfates, existing cracks, moist conditions etc (Taylor, 1990; Lawrence, 1995a, b; Ekolu et al., 2007a). Ekolu et al. (2007b) summarise various control measures that could be used for prevention of DEF, including the use of chemical additives. However, preventative measures and improvements in general durability require further attention.

In practice concrete and mortar mixes are normally based on Portland cement clinker, where the chemical process of hydration of clinker minerals yields hydrates and hydroxides. Because of the presence of gypsum, the chemical reaction between tricalcium aluminate (C_3A), gypsum ($CaSO_4$, $2H_2O$) and water, forms ettringite crystals ($3CaO$, Al_2O_3, $3CaOSO_4$, $31H_2O$). The volume difference in this reaction is positive and ettringite crystals grows faster, quickly growing on the unhydrated cement particles, which can slow down the hydration (Swaddiwudhipong et al., 2002). The presence of ettringite in a liquid cementitious system is unproblematic but its formation or re-formation in already hydrated concrete can lead to extensive damages (Collepardi, 2003). Due to the resulting volume difference particularly in the presence of sulfates, an expansive force within concrete can cause its disintegration (sulfate corrosion). It is well known that cements with low C_3A content are more resilient to sulfate corrosion although this also depends on the form of C_3A (Mather, 1968). For instance, crystalline C_3A is more reactive than its amorphous version.

The positive volume difference as a result of early ettringite formation (EEF) in cementitious materials rich with $C_4A_3\bar{S}$ calcium aluminate sulphate (expansive cement) can be used to compensate for the shrinkage during drying (Collepardi, 2001). In this case $C_4A_3\bar{S}$ hydrates within a few hours or days producing uniform distribution of ettringite and homogeneous expansion of hardened concrete at early stages. However, it is less known that ettringite could be formed in hardened cementitious materials without causing the well documented damages, which could potentially lead to their controlled strengthening.

The formation of a new phase characterized by volume expansion for the purpose of strength improvement is well known in the mainstream material science literature (Mishnaevsky, 2007). Such strengthening is based upon the creation of the internal compressive stresses on the contact between the existing matrix and the new phase particles, and depends on their shape, size and overall dispersion (Clifton and Ponnersheim, 1994). The newly formed particles should be small, spherical and located sufficiently apart from each other to avoid overlapping stress fields. Strength improvement of the Al_2O_3 ceramics with finely dispersed ZrO_2 particles is one example of this kind of strengthening (Cutler et al., 1987; Marshall et al., 1991). Internal stresses in the Al_2O_3 matrix created by applying the external force trigger polymorphic transformation of a tetragonal crystalline structure of ZrO_2 into a monoclinic crystalline structure. Increased volume creates substantial compressive stresses in the matrix surrounding the transformed particles leading to a several fold increase in compressive strength as well as resistance to the spreading of cracks. Studies that apply mechanisms of this kind for strength improvement of concrete are rare (Sobolev et al., 2006).

The creation of internal compressive stresses around particles or nuclei that have been transformed is the key requirement for such transformational strengthening. Its intensity depends on particle morphology, size and dispersion. Ideally, the particles should be small, spherical and uniformly dispersed to avoid overlapping their stress fields.

MATERIALS AND METHODS

In this study, the authors provide an investigation of a type of controlled DEF that may prevent deterioration or perhaps improve mechanical properties of hydrated fine grained concrete. A microstructure where crystallization nuclei are formed as a result of adding finely dispersed air-entraining agent (AEA) is the key prerequisite for such improvements. It is assumed that this would lead to localized and controlled formation of ettringite crystals, which could prevent DEF-induced deterioration. Their size and distribution depends on the level of dispersion of added AEA. The formation of localized ettringite crystals in the nuclei, particularly those adjacent to cracks, creates beneficial internal compressive stresses as the ettringite fills the nuclei. Similar to strength improvement of the Al_2O_3 ceramics described above, these internal compressive stresses lead to closure of nearby cracks preventing the growth of ettringite crystals. Furthermore, Ryu and Otsuki, (2002) show that the closure itself is beneficial since it decreases concrete permeability.

On the whole the experiments were based on fine grained concrete mixes using one type of cement, one type of AEA, fly ash and sand according to EN 196-1. Water-cement ratio, quantities of additives and hydration conditions were determined through laboratory testing. In addition, the experiments were based on specific climatic conditions necessary to achieve a controlled DEF in hydrated concrete. The investigation was based on four Portland cement (PC) CEM I 42, 5 R fine grained concrete mixes presented in Table 1.

The objective of this study is to investigate whether controlled DEF could lead to the reduction of deterioration and potential strengthening of hydrated fine grained concrete. To achieve this objective the study involves the following methods:

(1) Exposing hardened fine grained concrete prisms from the mix BI to the Duggan's test (Grabowski et al., 1992) (that is, prisms BI-DT); the test is essentially a cyclical interplay of heating and humidity and consists of a number of phases: prisms were first immersed in demineralised water for 72 h at 20 ± 2°C followed by 24 hours of drying in a drying chamber at 81 ± 2°C; they were then again immersed in the demineralised water for 24 h at 20 ± 2°C before being subject to 24 h drying in the chamber at 81 ± 2°C; in the last phase, prisms were immersed in the demineralised water for 24 h at 20 ± 2°C and then dried in the chamber for 72 h at 81 ± 2°C; the prisms were laboratory conditioned for 48 h in desiccators in between each of the above phases, and were once again immersed in the demineralised water for 24 h in order to fill the capillaries and voids with water,
(2) Strength comparison between hardened concrete prisms from mixes A, B, BI and prisms from mix BI that were exposed to the Duggan's test,
(3) Measuring ettringite formation by monitoring the length change (expansion) with a digital micrometer,
(4) microstructure comparison between hardened fine grained concrete samples from mixes A and AI where ettringite crystals were not detected with samples that were exposed to Duggan's test using electronic microscopy (that is, samples B-DT and BI-DT),
(5) Chemical analysis of ettringite crystals from samples B-DT and BI-DT in order to determine mass and atomic quantities of individual elements.

Four different mixes (A and AI, B and BI) were used for

Table 1. The four fine grained concrete mixes under investigation.

Parameter	Mix A (g)	Mix AI (g)	Mix B (g)	Mix BI (g)
Water	250	218,2	225	218,2
PC	450	450	310	310
Fly ash	-	-	140	140
AEA	-	6,8	-	6,8
Standard sand EN 196-1	1350	1350	1350	1350

Table 2. Results of the laboratory analysis for fly ash components.

Component part	Content (%)
Loss on ignition	0.41
Insoluble residue	16.67
SiO_2 impure	13.08
SiO_2 pure	47.62
SiO_2 soluble	0.64
SiO_2 total	48.26
SiO_2 active	35.18
Cao active	7.56
SO_3	1.88
CaO free	2.00

Figure 1. Apparatus for the measurement of length change of hardened concrete according to ASTM C490-86 placed in a climatic chamber.

comparative purposes because variations carried on:

(1) AEA content (A and B without AEA, versus AI and BI with AEA)
(2) Fly ash content (A and AI without fly ash versus B and BI with fly ash).

The comparison is necessary to identify the potential impact AEA and fly ash may have on mechanical properties before the prisms are exposed to Duggan's test (e.g. undesirable loss of compressive strength).

The above fine grained concrete mixes were prepared using a laboratory mixer according to EN 196-1. The conformity of fly ash for concrete was tested according to EN 450-1. The laboratory analysis of components presented in Table 2 confirmed that the fly ash used fulfils the criteria set in EN 197-1. Prisms were cast using 40 × 40 × 160 mm steel moulds. The samples from the mix B without AEA were then vibrated for 120 seconds with a standard low frequency of 50Hz and amplitude of 0.75 mm. The AEA-based samples from mixes AI and BI were vibrated for only 5 seconds under the same standard vibrating conditions as mix B samples in order to keep the volume of entrained air at an acceptable level (that is, preventing excessive reduction of strength). Past research shows that AEA-based samples should be vibrated with caution. For instance, Crawley (1953) found that high-frequency vibration causes more rapid loss of entrained air than moderate or low frequency vibration, and Hover (2001) reports that excessive vibration may lead to a complete loss of entrained air.

On the other hand, it has been found that short vibration cycles can improve compressive strength of concrete (Ozyldirim and Lane, 2003). A much shorter low-frequency vibration was therefore adopted to avoid a complete loss of finely dispersed crystallisation nuclei required for the controlled DEF. Controlled DEF is achieved by providing space for growth of ettringite crystals in the AEA-induced nuclei without any harm to hardened concrete. Specific climatic conditions were achieved by curing all prisms for 28 days in

a climatic chamber at a temperature of 20±2°C and relative humidity of 98±2%. After a required 28-day curing period six prisms from each of the mixes B and BI were exposed to Duggan's test in order to achieve the accelerated ettringite formation. The prisms were then placed into a standard apparatus for the determination of length change of hardened cement paste, mortar and concrete, constructed according to ASTM C490-86. During these measurements the apparatus itself was placed in the climatic chamber with a constant temperature of 20±2°C and relative humidity of 98±2% shown in Figure 1.

Ettringite formation was then monitored by measuring length change (expansion) with Mahr's MarCator 1080/12.5/0.005 mm digital micrometer. The results were recorded with an analogue/digital converter connected to a workstation. Developing expansion was measured regularly in 15 m intervals with a measurement accuracy of 0.005 mm, although intervals could well be longer considering the slow pace of DEF.

Density of hardened fine grained concrete (ρ), its compressive (f_c) and flexural strength (f_m) were measured on 10 additional prisms for each of the three mixes after standard 7, 14 and 28 days, and additionally after 56 and 121 days when compressive and flexural strengths should reach a plateau. Mechanical properties of fine grained concrete were examined with a universal dynamometer and a method according to EN 196-1. Optical microscopy using OLYMPUS SZX stereo microscope and QUANTA 200 3D electronic microscope was used to monitor the microstructure development in the hydrated fine grained concrete. The use of the latter enables longer low-vacuum observations without gold or carbon coating of samples. Chemical analysis of reactants in the hydrated paste was performed with Line Scan Microscopy (LSM) using SIRION 400 NC scanning electron microscope that works at high-vacuum but

Table 3. Densities and mechanical properties of hardened fine grained concrete (mix A).

Time interval (days)	ρ (kg/m^3)	f_m (MPa)	f_c (MPa)
121	1894	6.1	26.4

Table 4. Densities and mechanical properties of hardened fine grained concrete (mix AI).

Time interval (days)	ρ (kg/m^3)	f_m (MPa)	f_c (MPa)
7	1876	4.0	18.0
14	1904	4.4	21.8
28	1883	7.2	23.3
56	1885	7.3	23.5
121	1887	7.4	25.5

Table 5. Densities and mechanical properties of hardened fine grained concrete (mix B).

Time interval (days)	ρ (kg/m^3)	f_m (MPa)	f_c (MPa)
7	1860	3.2	11.9
14	1857	3.5	14.2
28	1868	3.9	17.4
56	1872	4.1	19.3
121	1866	5.6	22.1

Table 6. Densities and mechanical properties of hardened fine grained concrete (mix BI).

Time interval (days)	ρ (kg/m^3)	f_m (MPa)	f_c (MPa)
7	1801	2.7	11.6
14	1807	5.2	14.3
28	1803	4.1	17.7
56	1805	4.1	17.8
121	1818	5.6	21.0

specimens have to be dry and treated (carbon treatment in this case). The ettringite crystals were characterised with the Energy Dispersive X-Ray (EDX) spectroscopy using JEOL JSM 5610 scanning electron microscope operated at 20kV for single-point 100 s long EDX measurements.

RESULTS AND DISCUSSION

Tables 3, 4, 5 and 6 show measured densities, flexural f_m and compressive strength f_c of hardened fine grained concrete prisms for mixes A, AI, B and BI that were not exposed to Duggan's test. The compressive strength of AI (BI) prisms after 121 days is only 3.4 % (5 %) lower than that of prisms A (B), which indicates that loss of strength due to added AEA can be prevented with short low-frequency vibration. These results reveal that 5 s

vibration of BI prisms was sufficient to achieve a minimum loss of mechanical properties, which is normally expected from added AEA. On the other hand, the short vibration interval prevents the loss of crystallisation nuclei that can be seen later in the text.

Tables 7 and 8 show measured densities and mechanical properties of hardened fine grained concrete prisms for mortar mixes B and BI, but this time after 121 days, and exposed to Duggan's test after 28 days. Flexural strength of prisms B-DT that were exposed to Duggan's test has almost doubled to 11.2MPa, and compressive strength has more than doubled as well. These substantial increases in strength were not expected but to some extent there are some parallels with a study by Zhang et al. (2008). The study show that growth of ettringite crystals in a limited space of microvoids

Table 7. Density and mechanical properties of hardened fine grained concrete (B-DT: mix B exposed to Duggan's test).

Time (days)	ρ (kg/m^3)	f_m (MPa)	f_c (MPa)
121	2211	11.2	60.4

Table 8. Density and mechanical properties of hardened mortar prisms (BI-DT: mix BI with Duggan's test).

Time (days)	ρ (kg/m^3)	f_m (MPa)	f_c (MPa)
121	1810	6.0	22.4

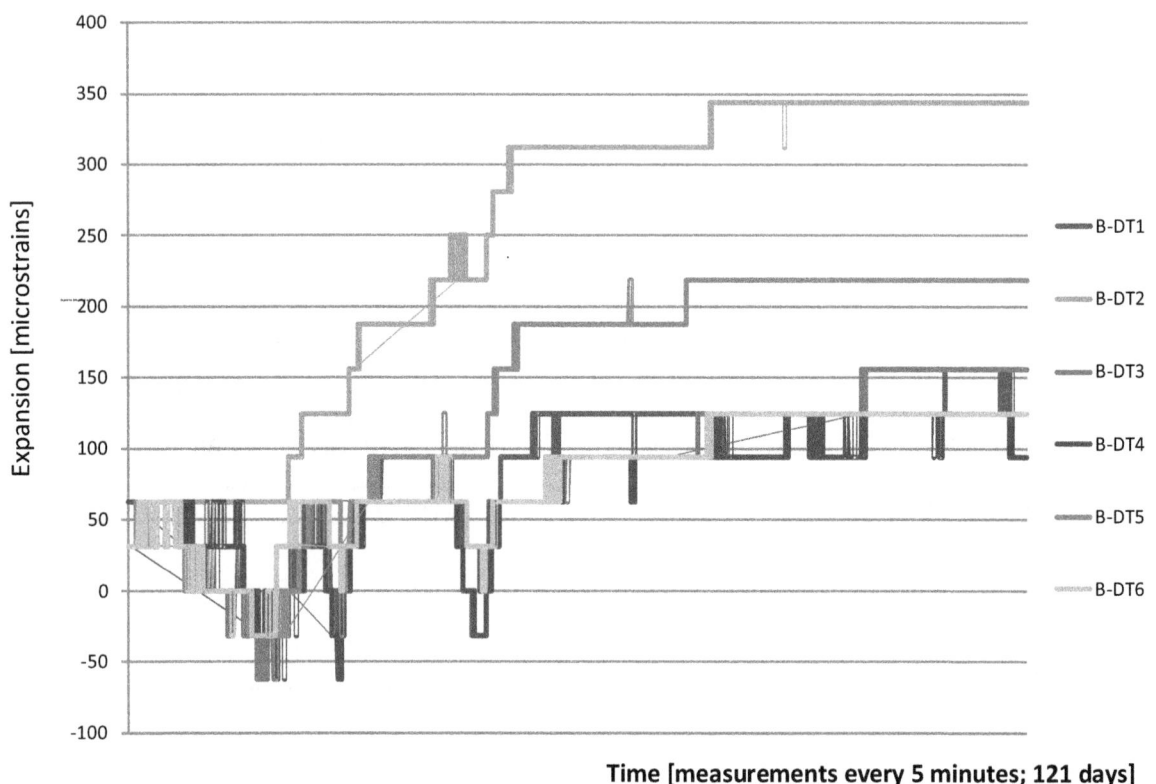

Figure 2. Change of length of six B-DT mortar prisms after the Duggan's test.

or preformed microcracks may lead to further evolution of localised microcracks. This process decreases the strength of prisms in early stages but ettringite crystals penetrate into the newly formed microcracks leading to partially recovered flexural strengths at later stages. One of the reasons for the increases could be associated with the increased ductility as a result of DEF-related expansion reported by Brunetaud et al. (2008). The observed causal relationship between tensile ductility and compressive strength has been observed in more detail by Bortolotti (1994). However, such increases are short-lived because the observed growth of ettringite crystals

leads to the expansion at later stages, causing further cracking and deterioration of concrete.

Figures 2 and 3 show change of length for the six fine grained concrete prisms from the mixes B and BI that were exposed to Duggan's test and a final 24 hour immersion in demineralised water. The change of length of prisms was recorded daily and stopped after 93 days when measurements did not show any further expansion.

Comparing microstructures of the specimens from fine grained concrete mixes A, AI, and BI presented in Figure 4a, b and c demonstrate that, as expected, similar AEA-induced nuclei exist in AI and BI specimens.

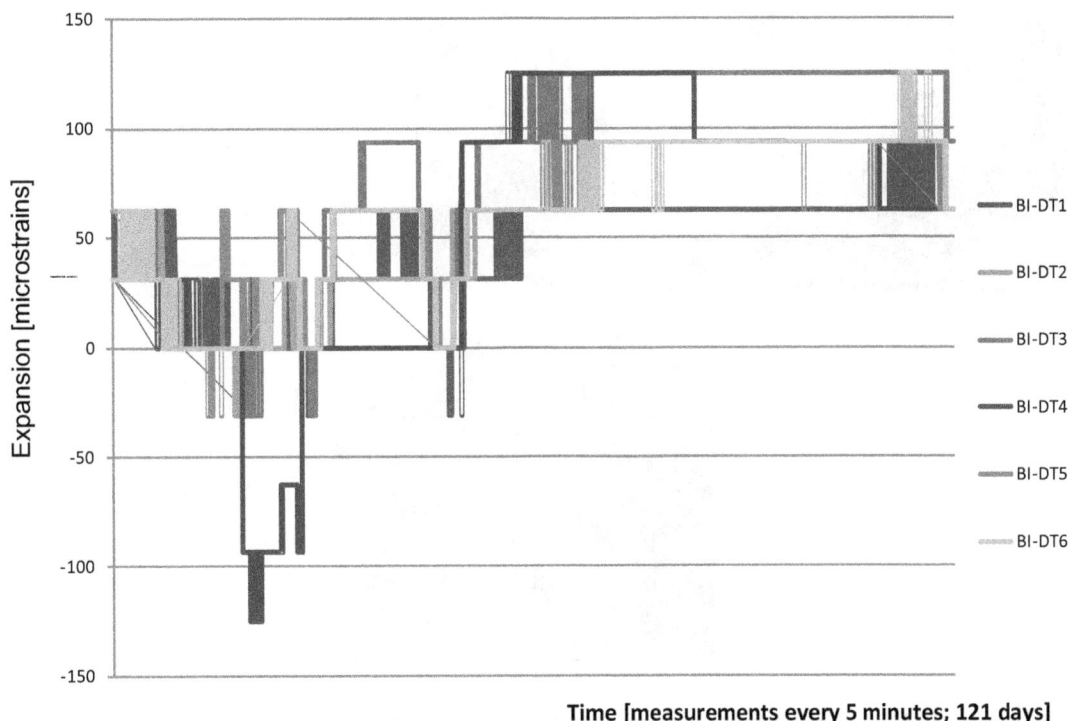

Figure 3. Change of length of six BI-DT fine grained concrete after the Duggan's test.

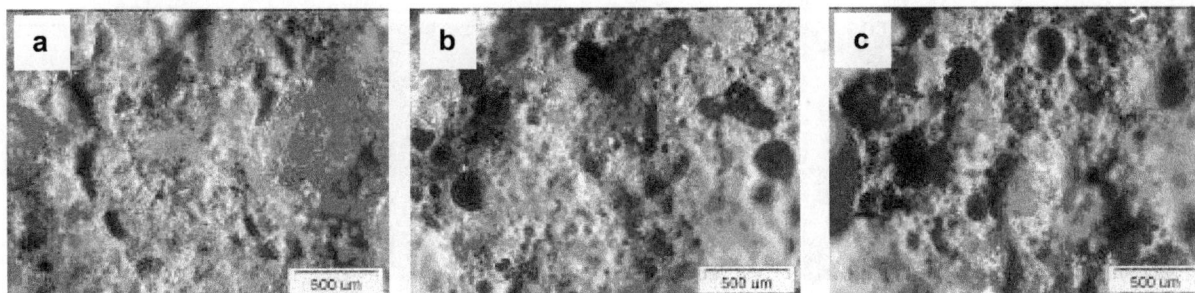

Figure 4 (a, b, c). Microstructures of fine grained concrete A, AI and BI after 121 days (500x magnification in all three cases; images obtained with OLYMPUS SZX stereo microscope).

Although ettringite crystals can be found in concretes produced by using pure Portland cement, no visible ettringite crystals were detected in any of the large number of prisms from the mix A. Ettringite crystals did appear in specimens of all other fine grained concrete mixes. Microstructures of specimen from the fine grained concrete mix AI show no visible ettringite crystals in AEA nuclei themselves displayed in Figure 5a, but they were detected in microcracks as seen in Figure 5b and c. Similar to Myneni et al. (1997) these crystals have thin, needle-shaped morphology and are approximately 2 µm in length revealing rapid growth. Fly ash in fine grained concrete mix BI may well be a source of soluble calcium for ettringite formation (Solem and McCarthy, 1992;

Zhang and Reardon, 2003; Chrysochoou and Dermatas, 2006), because its crystals were found in greater quantities in microcracks and within the AEA-induced nuclei. Figure 6a, b, c and 7a, b, c, d, e, f show that ettringite crystals have thin, needle-shaped morphology but those found in microcracks are only approximately 2µm in length shown in Figure 7f, as opposed to 10 µm crystals found in the nuclei in Figure 6b.

The microcrack that appeared on the surface of a nucleous in Figure 7f can be associated with the shrinkage of the matrix during hydration (Stang, 1996). The comparison between various different BI specimen shows that ettringite crystals start growing wherever there is enough space for growth before further damaging

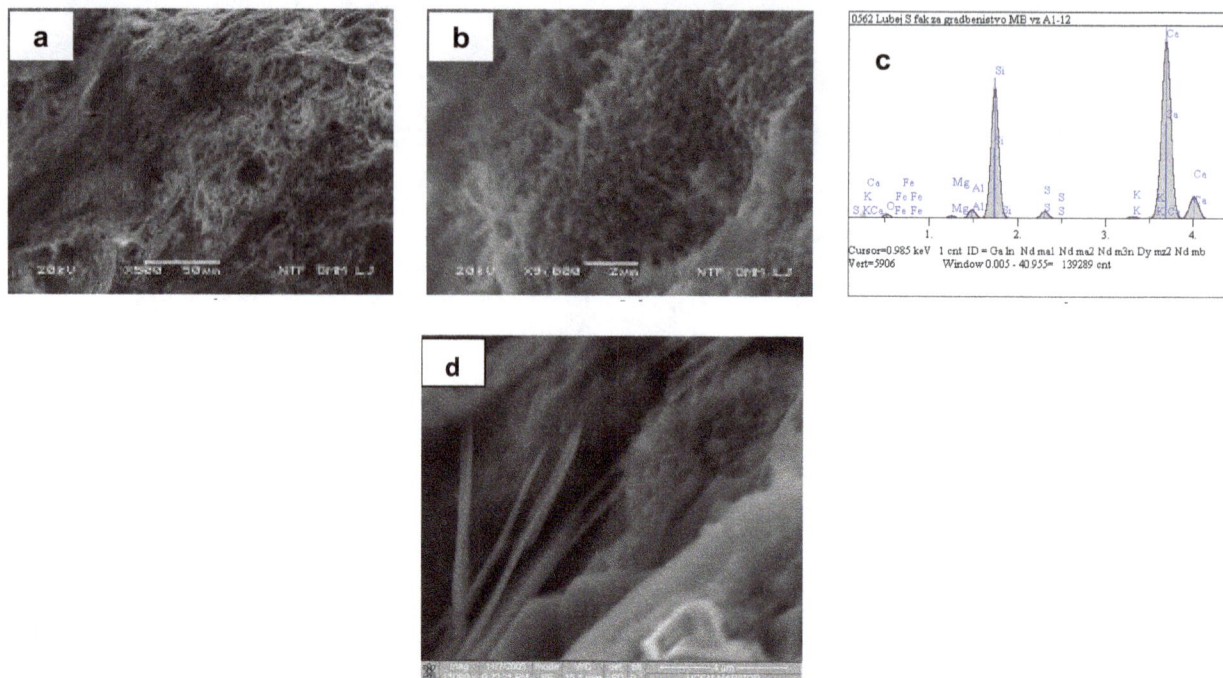

Figure 5 (a, b, c, d). Microstructure of fine grained concrete AI after 121 days (500x magnification in Figure 5a, and 9000x magnification of an area with thin approximately 2μm long ettringite crystals in figure 5b and 11000x magnification with QUANTA 200 3D electronic microscope in figure 5d; the results of the EDX analysis of a crystal from this area is in Figure 5c.

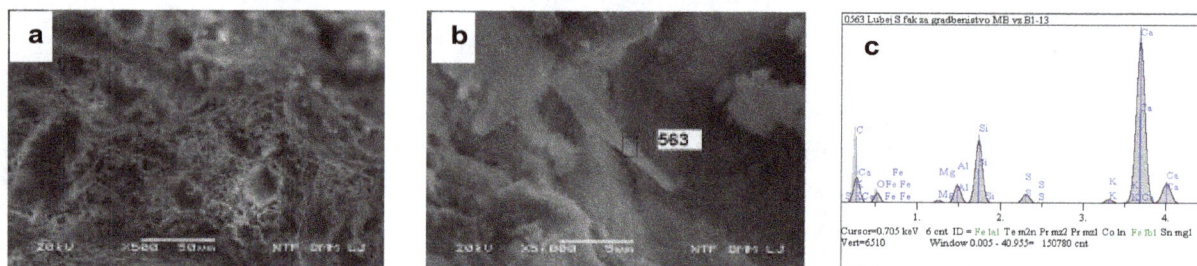

Figure 6 (a,b,c). Microstructure of fine grained concrete BI after 121 days (500x magnification in Figure 6a; the first studied area under 500x magnification of a nucleus with thin ettringite crystals with needle-shaped morphology and approximately 10um in length in figures 6b with the results of the EDX analysis of crystal 563 in Figure 6c).

concrete which enables further growth. AEA-induced nuclei may therefore act as relief reservoirs enabling the growth of substances like ettringite crystals in hardened concrete with minimum or no damaging effects. Hime, (1996) recommends air-entrainment as a way to prevent DEF and reports only a single incident where air-entrained concrete suffered from DEF.

Mix B has only been used to compare the microstructure and mechanical properties of specimen where fly ash has been added and the prisms were exposed to Duggan's test. Figure 8a reveals a microcrack in the B-DT sample at 500x magnification and Figure 8b clearly shows a cluster of up to 3 μm long ettringite crystals in the microcrack at 5000x magnification. Mehta

(1983) reports that such type II crystals with the length of 1 to 2 μm and thickness of 0.1 to 0.2 μm are known to be more expansive and potentially more damaging than much longer - type I crystals. Because there are no AEA-induced nuclei DEF in B-DT samples increased microcracking giving space to further growth of ettringite crystals, which inevitably damage the hardened concrete. However, expansion of B-DT samples was not dissimilar to expansion of other samples. This may be assigned to a relatively short time interval since expansion was monitored for only 93 days after the Duggan's test.

Ettringite crystals growing in microcracks fill the empty space and press against their walls causing the DEF-induced expansion of concrete. As discussed earlier, this

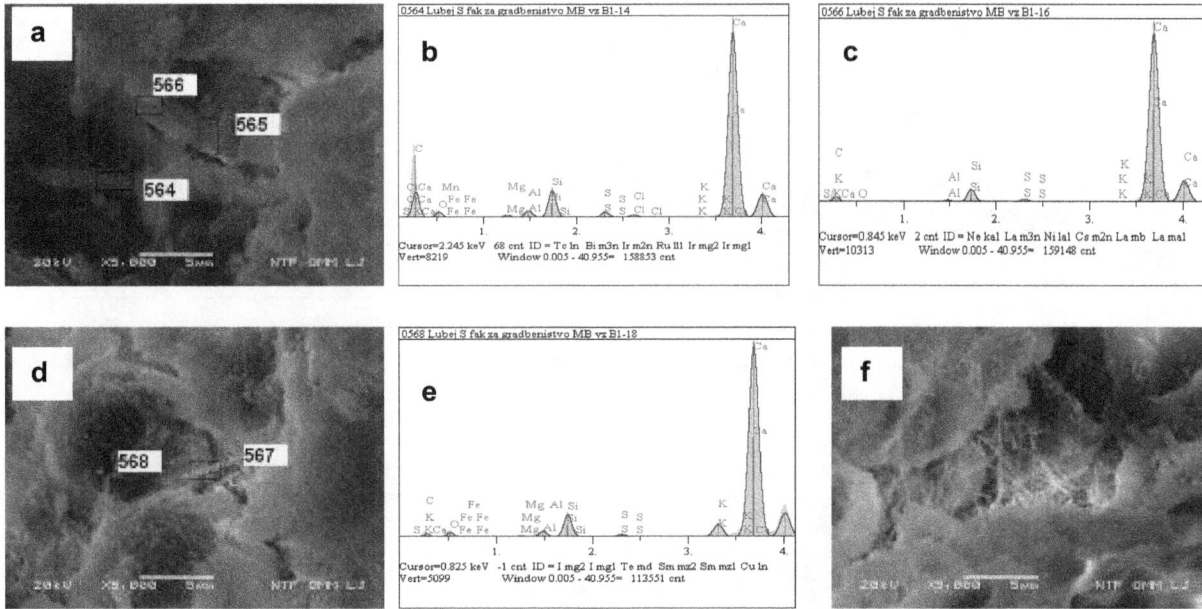

Figure 7 (a,b,c,d,e,f). Microstructure of fine grained concrete BI after 121 days (the second studied area from figure 6a under 5000x magnification in Figure 7a,d and f; the results of the EDX analysis of crystal 564, 566 and 568 are in Figure 7b,c and e respectively).

Figure 8 (a, b). Microstructure of fine grained concrete BI after 121 days (500x magnification in figure 8a and 5000x magnification in Figure 8b; this mortar mix was used for comparative purposes only).

may in turn lead to initially improved mechanical properties before they are significantly reduced due to further evolution of growth-related microcracking. Compressive and flexural strength figures for B-DT samples after 121 days confirm this with compressive strength more than double the compressive strength of all other specimen. This improvement of mechanical properties is probably caused by ettringite crystals filling the empty space in existing microcracks before further evolution of localised microcracks that would significantly decrease the mechanical properties and damage concrete.

If DEF is limited to growth in microcrasks then specimens BI-DT that were exposed to Duggan's test should exhibit similar properties to B-DT specimens.

Microstructure of the hardened fine grained concrete BI-DT after 121 days reveals visible ettringite crystals in AEA-induced nuclei in Figures 9b and 10b at 1,000 x and 5,000 x magnification respectively. These crystals are longer 10 μm type I crystals that are reportedly less damaging. The studied sections where AEA nuclei were present exposed some localised microcracking within the nuclei and presence of short ettringite crystals with needle-shaped morphology in the microcracks seen in Figure 11a and b. However, the majority of ettringite crystals were found within the nuclei themselves so the effect of their growth was assumed to be less damaging. Minimum expansion as shown in Figure 3 and comparable mechanical properties of specimens from all mixes, confirm this assumption.

Figure 9 (a, b, c). Microstructure of fine grained concrete BI after 121 days (500x magnification in Figure 8a with 5000x magnification of group crystals in the first studied area in figure 9b; the results of the EDX analysis of crystal 583 are in Figures 9c).

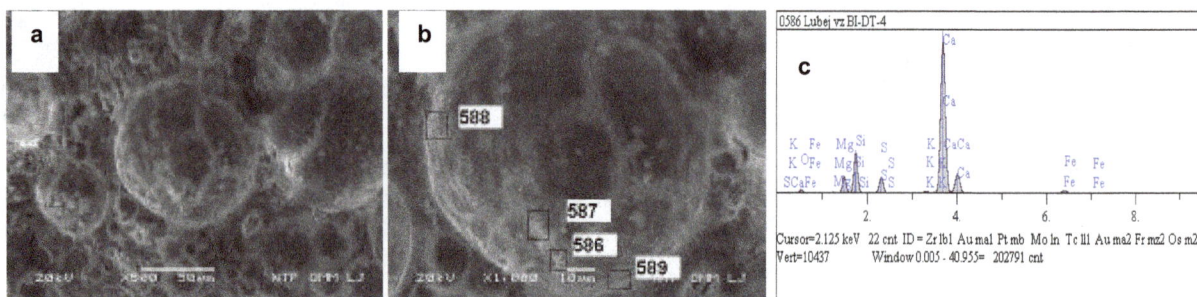

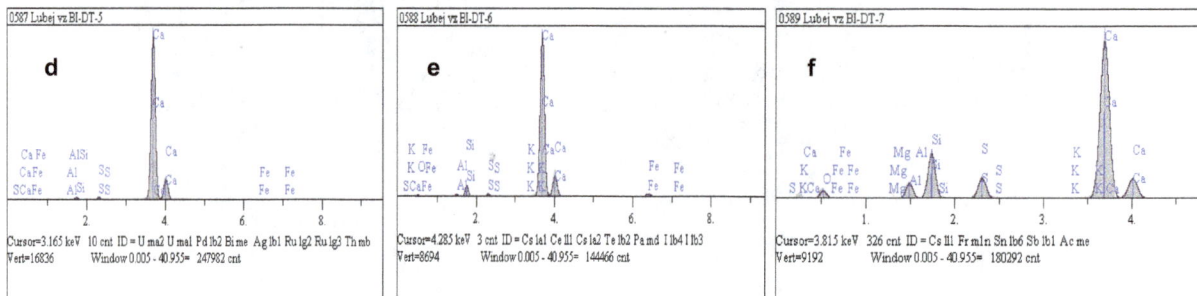

Figure 10 (a, b, c, d, e, f). Microstructure of fine grained concrete BI-DT after 121 days (500x magnification of a group of AEA-induced air voids in Figure 10a and 1000x magnification of an air void in Figure 10b; the results of EDX analysis of approximately 5 μm long crystals 586 and 587, and approximately 2 μm long crystals 588 and 589 are in Figures 7c, d, e and f respectively).

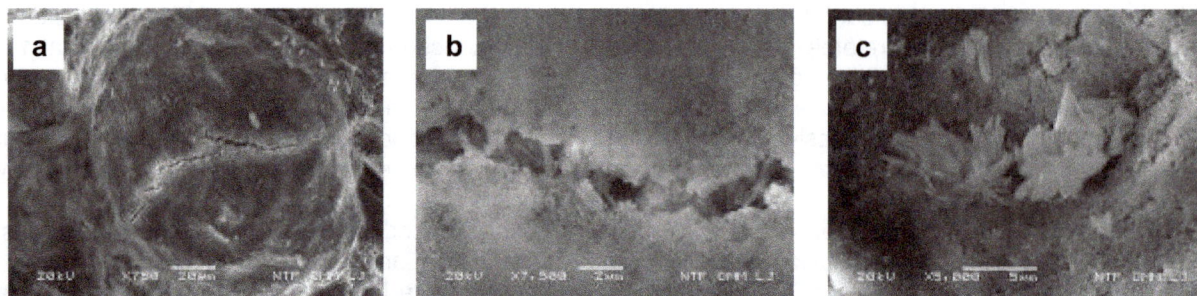

Figure 11 (a, b, c). Microstructure of fine grained concrete BI-DT after 121 days (750x magnification of an AEA nucleus with a microcrack in Figure 11a, 7,500x magnification of the microcrack in Figure 11b and 5000x magnification of the edge of the nucleus with a group of ettringite crystals in Figure 11c).

Table 9. Density and mechanical properties of hardened fine grained concrete without added AEA after 121 days for (prisms A, B and B-DT).

Variables	A	B	B-DT
ρ (kg/m^3)	1894	1866	2211
f_m (MPa)	6.1	5.6	11.2
f_c (MPa)	26.4	22.1	60.4

Table 10. Density and mechanical properties of hardened fine grained concrete with added AEA after 121 days for (prisms AI, BI and BI-DT).

Variables	AI	BI	BI-DT
ρ (kg/m^3)	1887	1818	1805
f_m (MPa)	5.6	5.6	6.0
f_c (MPa)	25.5	21.0	22.4

Table 11. EDX analysis results of the ettringite crystal at point 562 from sample AI in Figure 5c after 121 days.

Element	Mass %	Atomic %
Al	1.851	2.189
Si	26.567	30.187
S	1.740	1.732
Ca	55.105	43.878

The achieved compressive and flexural strengths of BI-DT specimens presented in Table 8 were similar to those of specimens B and BI in Tables 5 and 6 respectively and only 12% (15%) lower than those of specimens A(AI). The predominantly nuclei-based growth of ettringite crystals prevented excessive growth-related pressure on the walls of microcracks so no unusual initial improvements of mechanical properties were detected in the studied period. With a diameter of approximately 100 μm the nuclei offer sufficient space for ettrigite crystals to grow without causing progressive cracking and damaging the concrete.

The EDX analysis was performed on a number of identified crystals to examine the presence of ettringite, which was confirmed in all cases. Tables 9-15 show compositions of studied ettringite crystals from Figures 5-11. The cement paste close to fly ash particles often contains CH crystals (Xu et al., 1993). It is therefore not unusual to see variations in Al/Ca and S/Ca ratios between different crystals that are presented in the tables. Wherever CH crystals are present Al/Ca and S/Ca ratios remain high (0.1-0.18 and 0.05-0.24 respectively). On the other hand, Al/Ca and S/Ca ratios are low at points where CH crystals are absent (0.01-0.05 and 0.01-0.04 respectively). Relatively high content of Si at some points follows high Al/Ca and S/Ca ratios in a similar way

Table 12. EDX analysis results of the ettringite crystal from the first studied area at point 563 from sample BI in Figure 6c after 121 days.

Element	Mass %	Atomic %
Al	2.975	2.280
Si	10.235	7.536
S	1.595	1.029
Ca	38.594	19.915

as shown by Richardson (2000). This could be associated with the availability of active SiO_2 in the fly ash although the values remain high even for specimens AI that do not contain fly ash. This, on the other hand, could also be associated with increasing concentrations of Si in the cement paste during hydration as reported by Rothstein et al. (2002).

Ettringite crystals have formed as a result of volumetrically expansive sulfate reaction in porous areas predominantly within AEA-induced nuclei, microcracks and microvoids in general. Because the solid product volume in this reaction is greater than the solid reactant volume, this has led to the creation of internal stresses in the ettringite growth areas. Their growth within the AEA-

Table 13. EDX analysis results of the ettringite crystal from the second studied area at points 564, 566 and 568 from sample BI in Figure 7b, c and e after 121 days.

Point	564		566		568	
Element	Mass %	Atomic %	Mass %	Atomic %	Mass %	Atomic %
Al	1.464	1.212	0.702	0.825	1.329	1.736
Si	5.077	4.038	4.347	4.907	4.696	5.896
S	1.034	0.720	0.836	0.827	0.467	0.513
Ca	49.763	27.735	80.577	63.744	50.309	44.269

Table 14. EDX analysis results of the ettringite crystal from the first studied area at point 583 from sample BI-DT in Figure 9c after 121 days.

Element	Mass %	Atomic %
Al	5.970	6.509
Si	15.727	16.473
S	9.340	8.568
Ca	47.938	35.185

Table 15. EDX analysis results of the ettringite crystal from the second studied area at points 586, 587, 588 and 589 from sample BI-DT in Figures 10c, 10d, 10e and 10f after 121 days.

Point	586		587		588		589	
Element	Mass %	Atomic %	Mass %	Atomic %	Mass %	Atomic %	Mass %	Atomic %
Al	4.875	5.783	0.225	0.328	0.831	1.109	3.958	4.260
Si	11.808	13.457	1.148	1.607	4.107	5.264	11.331	11.718
S	4.657	4.648	0.960	1.177	0.897	1.006	5.384	4.876
Ca	62.473	49.892	95.707	93.929	82.188	73.815	57.537	41.696

induced nuclei created compressive stresses within the concrete matrix that then led to strength improvements as suggested by Springenschmid and Breitenbucher (1998).

Conclusion

DEF has been detected in all fine grained concrete prisms that were exposed to Duggan's test (B-DT and BI-DT). In addition, non-aerated fine grained concrete prisms B-DT show rapid and excessive increases in compressive strength, which leads to known harmful damages of hardened, concrete. Controlling DEF by using AEA as a nucleation agent results in a slight increase of compressive strength of fine grained concrete with only marginally lower density (comparison between BI-DT and BI). The detected ettringite crystals within the crystallisation nuclei prevented initial improvements of mechanical properties, which were observed in fine grained concrete B-DT without AEA as a result of excessive growth-related pressure on the walls of microcracks. This could be one of the reasons why air-entrained concrete normally does not suffer from DEF as

reported by Hime (1996). The investigation shows that AEA-induced nucleation can prevent deterioration of hardened concrete and may even offer a mechanism for improvement of mechanical properties but further investigations over longer test periods are required to establish whether the deterioration can be completely avoided and whether such improvements can be achieved and sustained.

REFERENCES

ASTM C490-86, Standard Specifications for Apparatus for Use in Measurement of Length Change Of Hardenet Cement Paste, Mortar, and Concrete, American Society of Testing and Materials, 1986.

Barbarulo R, Peycelon H, Prené S, Marchand J (2005) Delayed ettringite formation symptoms on mortars induced by high temperature due to cement heat of hydration or late thermal cycle, Cem. Conc. Res. 35(1):125-131.

Bortolotti L (1994). Influence of concrete tensile ductility on compressive strength of confined columns, J. mater. civil eng. 6(4):542-564.

Brunetaud X, Divet L, Damidot D (2008). Impact of unrestrained Delayed Ettringite Formation-induced expansion on concrete mechanical properties, Cem. Conc. Res. 38(11):1343-1348.

Chrysochoou M, Dermatas D (2006). Evaluation of ettringite and hydrocalumite formation for heavy metal immobilization: Literature review and experimental study, J. Hazardous Mater. 136(1):20-33.

Clifton JR, Ponnersheim JM (1994). Sulfate attack of cementitious materials: Volumetric relations and expansions. *NISTIR 5390*, US Department of Commerce.

Crawley WO (1953). Effect of Vibration on Air Content of Mass Concrete, ACI J. Proc. 49(6):909-920.

Collepardi M (2003). A state-of-the-art review on delayed ettringite attack on concrete, Cem. Conc. Composites 25(4-5):401-407.

Collepardi M (2001). Ettringite formation and sulfate attack on concrete. In: Proceedings of the 5th CANMET/ACI International Conference on Recent Advances in Concrete Technology, Singapore, pp. 27-39.

Cutler RA, Bright JD, Virkar AV, Shetty DK (1987). Strength improvement in transformation-toughened alumina by selective phase transformation, J. Am. Ceramic Soc. 70(10):714-718.

Diamond S (1996). Delayed ettringite formation - Processes and problems, Cem. Conc. Composites, 18(3):205-215.

Ekolu SO, Thomas MDA, Hooton RD (2007a). Dual effectiveness of lithium salt in controlling both delayed ettringite formation and ASR in concretes, Cem. Conc. Res. 37(6):942-947.

Ekolu SO, Thomas MDA, Hooton RD (2007b). Implications of pre-formed microcracking in relation to the theories of DEF mechanism, Cem. Conc. Res. 37(2):161-165.

EN 196-1:1999 Cement and concrete technology, Cements, Cement mortar, Compressive strength, Flexural strength, Compression testing, Bend testing, Compaction tests, Test equipment, Test specimens, Acceptance inspection, Conformity, Approval testing.

EN 197-1:2002 Cement – Part 1: Composition, specification and conformity criteria for common cements.

EN 450-1:2005 Fly ash, Concretes, Construction materials, Ashes, Chemical composition, Fineness, Density, Expansion (deformation), Marking, Quality control, Conformity, Sampling methods, Size classification, Defects, Statistical quality control, Power station waste aggregates.

Grabowski E, Czarnecki B, Gillott JE, Duggan CR, Scott JF (1992). Rapid test of concrete expansivity due to internal sulfate attack, ACI Mater. J. 89(5):469-480.

Hime WG (1996). Delayed ettringite formation – A concern for precast concrete? PCI Journal, July-August, 26-30.

Hover KC (2001). Vibration Tune-up, Conc. Int. 23(9):30-35.

Lawrence CD (1995a). Mortar expansions due to delayed ettringite formation. Effects of curing period and temperature, Cem. Conc. Res. 25(4):903-914.

Lawrence CD (1995b). Delayed ettringite formation: an issue? *Materials Science of Concrete*, vol. IV, American Ceramic Society, Ohio, USA, pp. 113-154.

Marshall DB, Ratto JJ, Lange FF (1991). Enhanced Fracture Toughness in Layered Microcomposites of Ce-ZrO_2 and Al_2O_3, J. Am. Ceramic Soc. 74(12):2979-2987.

Mather B (1968). Field and Laboratory Studies of the Sulphate Resistance of Concrete, Performance of Concrete-Resistance of Concrete to Sulphate and Other Environmental Conditions, Thorvaldson Symposium Proceedings, University of Toronto Press, Toronto, pp. 66-76.

Mehta PK (1983). Mechanism of sulfate attack on Portland cement concrete-another look, Cem. Conc. Res. 13(3):401-406.

Mishnaevsky L Jr (2007). Computational Mesomechanics of Composites, John Wiley. 74(12):2979-2987.

Myneni SCB, Traina SJ, Logan TJ, Waychunas GA (1997). Oxyanion behavior in alkaline environments: Sorption and desorption of arsenate in ettringite, Environ. Sci. Tech. 31:1761–1768.

Richardson IG (2000). The nature of the hydration products in hardened cement pastes, Cem. Conc. Composites, 22(2):97-113.

Rothstein D, Thomas JJ, Christensen BJ, Jennings HM (2002). Solubility behaviour of Ca-, S-, Al-, and Si-bearing solid phases in Portland cement pore solutions as a function of hydration time, Cem. Conc. Res. 32(10):1663-1671.

Ryu JS, Otsuki N (2002). Crack closure of reinforced concrete by electrodeposition technique, Cem. Conc. Res. 32(1):159-164.

Sahu S, Thaulow N (2004). Delayed ettringite formation in Swedish concrete railroad ties, Cem. Conc. Res. 34(9):1675-1681.

Sobolev K, Flores I, Hermosillo R, Torres-Martinez LM (2006). Nanomaterials and nanotechnology for high-performance cement composites, Proceedings of ACI Session on "Nanotechnology of Concrete: Recent Developments and Future perspectives", November 7, Denver, USA.

Springenschmid R, Breitenbucher R (1998). Influence of constituents, mix proportions and temperature on cracking sensitivity of concrete. In Springenschmid R (1998). Prevention of thermal cracking in concrete at early stages. RILEM Report 15, E & FN Spon.

Swaddiwudhipong S, Chen D, Zhang MH (2002). Simulation of the exothermic hydration process of Portland cement, Adv. Cem. Res. 14(2):61-69.

Taylor HFW (1990). Cement Chemistry, Academic Press, London, England.

Thomas MDA (2001). Delayed ettringite formation: Recent developments and future directions, in: S. Mindess, J. Skalny (Eds.), Materials Science of Concrete VI, American Ceramics Society, Westerville, OH, 2001, pp. 435–482.

Thomas M, Folliard K, Drimalas T, Ramlochan T (2008). Diagnosing delayed ettringite formation in concrete structures, Cem. Conc. Res. 38:841-847.

Ozyldirim C, Lane DS (2003). Investigation of self-consolidating concrete, Transportation Research Board Annual Meeting, National Research Council.

Solem JK, McCarthy GJ (1992). Hydration reactions and ettringite formation in selected cementitious coal conversion byproducts, Mater. Res. Soc. Symposium Proc. 245:71-79.

Stang H (1996). Significance of shrinkage – induced clamping pressure in fiber matrix bonding in cementitious materials, Adv. Cem. Based Mater. 4(3):106-115

Zhang M, Jiang M, Chen J (2008). Variation of flexural strength of cement mortar attacked by sulfate ions. Eng. Fracture Mechan. 75(17):4948-4957.

Zhang M, Reardon EJ (2003). Removal of B, Cr, Mo, and Se from wastewater by incorporation into hydrocalumite and ettringite, Environ. Sci. Technol. 37(15):2947-2952.

Xu A, Sarkar SL, Nilsson LO (1993). Effect of fly ash on the microstructure of cement mortar, Mater. Structures 26(7):414-424.

3

Quantum chemical calculations on molecular structures and solvents effect on 4-nitropicolinic and 4-methoxypicolinic acid

Semire Banjo and Adeoye Idowu Olatunbosun

Department of Pure and Applied Chemistry, Faculty of Pure and Applied Sciences, Ladoke Akintola University of Technology, P. M. B. 4000, Ogbomoso, Oyo State, Nigeria.

The density functional theory (DFT) (B3LYP) was used to study the solvents effect on electronic properties of 4-methoxypicolinic acid (4MOPIC) and 4-nitropicolinic acid (4NPIC). The calculated vibration frequencies at DFT/6-311++G** were compared to that of un-substituted picolinic acid to know the effect of donor/acceptor substituent on the molecules. Five solvents namely acetone, ethanol, diethyl ether, N, N-dimethylformamide (DMF) and tetrahydrofuran (THF) were used to study solvents effect. The methoxyl (OCH$_3$) group in 4MPIC pushed electrons into the picolinic acid ring thereby resulted in upfield resonance as compared to 4NPIC in which nitro (NO$_2$) group brought about downfield in ^{1}HNMR. The solvents increased the minimum energy required to remove an electron for 4MOPIC whereas it was lower in 4NPIC. The HOMO and LUMO energies calculated in the solvents revealed that both HOMOs and LUMOs experienced stabilization in 4NPIC but LUMOs were destabilized in 4MOPIC as compared to gas phase.

Key words: 4-methoxypicolinic acid, 4-nitropicolinic acid, solvents effect, density functional theory (DFT).

INTRODUCTION

Picolinic acid and its substituents are of interest to many researchers mainly because of their usefulness as ligands; the availability of some donor atoms on the compounds which can serve as binding sites to various metals in forming various monomeric and polymeric complexes (Yurovskaya et al., 1998). Picolinic acid is the body's prime natural chelator of vital trace elements such as chromium, zinc, manganese, copper, iron and molybdenum (Evan and Johnson, 1980). Picolinic acid is biosynthesized in the liver and kidneys from the amino acid tryptophan, and stored in the pancreas during digestion, secreted into the intestine(Fernandez-Poi et al., 2001).

The vibrational frequencies, UV-Visible, ^{1}H and ^{13}C NMR spectra of piconilic acid with its metal complexes of Fe(III), Ni(II), Cu(II), Zn(II) and Ag(I) were recently reported (Kalinowska et al., 2007; Goher and Abu-Youssef, 1996). The 3-methylpicolinic acid and 6-methylpicolinic acid with their cobalt complexes were synthesized and characterized by spectroscopic techniques and thermal characterization in the solid state (Kukovec et al., 2009).

In the past the chemical behaviour of ambident ligands such as mercaptobenzothiazole and mercaptobenzimidazole ligands has been investigated using density functional theory (AlHokbany and AlJammaz, 2011). This is with a view of understanding the coordination chemistry of the metal complexes to be formed. Density functional theory has been employed over the years to obtain thermochemical data, molecular structure force fields frequencies assignment of nuclear magnetic resonance (NMR), photoelectron, erythrocyte

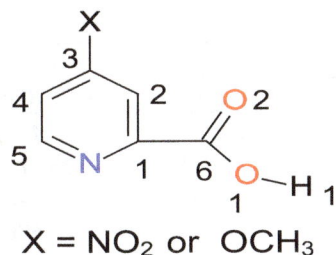

X = NO₂ or OCH₃

Figure 1. Schematic structures of the studied molecules: 4-nitropicolinic acid for X = NO₂ and 4-methoxypicolinic acid for X = OCH₃.

sedimentation rate (ESR) and ultra violet (UV) spectra, activation barriers, dipole moments and other one electron properties (Koczon et al., 2003; Parajón-Costal et al., 2004; Ziegler, 1991).

The effect of protic and aprotic solvents on the reactivity of picolinic, nicotinic and isonicotinic acid as well as of some substituted nicotinic acids with diazodiphenylmethane has been investigated (Dimitrijević et al., 1974; Drmanic et al., 2012; Marinkovi, 2005). In one of our recent work, DFT/B3LYP with various basis sets (6–31G*, 6-31G**, 6-311G**, 6-311+G**, 6-311++G**) was used to study solvents effect on geometry and electronic properties of picolinic acid. The results revealed that solvent-molecule interactions are more prominent around the heteroatoms in the case of the polar solvents. Therefore, this was a greater shielding/de-shielding on carbon nuclei around nitrogen and oxygen atoms of picolinic acid in the case of the polar solvents (Adeoye and Semire, 2013). The aim of this study is to employ density functional theory to study the effect of solvents on molecular structures and electronic properties of two 4-substituted picolinic acid namely 4-nitropicolinic acid and 4-methoxypicolinic acid as shown in Figure 1. The solvents employed are acetone, Ethanol, Diethyl ether, N,N-Dimethylformamide (DMF) and Tetrahydrofuran (THF) (Figure 1).

COMPUTATIONAL METHODS

The 4MOPIC and 4NPIC were optimized without symmetry constraint using density functional theory Beckes's three-parameter hybrid functional (Becke, 1988) employing the Lee, Yang and Parr correlation functional B3LYP (Lee et al., 1988) with 6–31G*, 6-31G** and 6-311G** basis sets. The vibration frequencies, electronic properties and NMR were calculated. In the calculation of vibration frequency, no imaginary frequency modes were obtained for the two molecules, therefore it was believed that true minimum on the potential energy surface were found for the molecules. The absorption transitions were calculated from the optimized geometry in the ground state S₀ using configuration interaction (CI) theory using density functional theory (DFT with 6-31G* in five different solvents namely acetone, ethanol, diethyl ether, N,N- Dimethylformamide (DMF) and tetrahydrofuran (THF). The convergence criteria for the energy calculations and geometry

optimizations used in the density functional methods were default parameters in the Spartan 06 program (Table 1).

RESULTS AND DISCUSSION

Chemical shifts and solvents effect

The molecular structure of the optimized 4MOPIC and 4NPIC using B3LYP method with 6-31G* basis set is used to calculate ¹³C and ¹H chemical shift calculations as shown in Table 1. It has been reported that ¹³C and ¹H NMR calculated using DFT are in good agreement with the experimental values (Teimouri et al., 2008; Karakurt et al., 2012; Cheeseman et al., 1996). Therefore, in the absence of experimental results the calculated ¹³C and ¹H NMR using DFT method can provide reasonable information that could assist in structural elucidation; thus the chemical shifts of the two isomers are compared. The results in Table 1 showed that ¹³C NMR chemical shifts for the two molecules are greater than 100 ppm which is the typical ¹³C NMR chemical shift for organic molecule (Kalinowski et al., 1988; Pihlaja and Kleinpeter, 1994). It has been reported that calculations based on the averaging of chemical shifts are in better agreement with the experimental values than plain calculations using optimized geometry (Stare et al., 2004). The chemical shifts of C2 and C4 are 109.373 and 115.541 ppm for 4MOPIC and 119.611 and 118.174 pmm for 4NPIC, therefore C2 and C4 experienced shielding effect because of its position in the molecules whereas other carbon atoms are de-shielded. C2 and C4 are more shielded in 4MOPIC because of higher electron density as a result of electron donating effect of CH₃ substituent. The chemical shifts of C3 are 165.388 and 155.204 ppm for 4MOPIC and 4NPIC respectively. The shielding/de-shielding of the substituent has no effect on C1, C5 and C6 of 4MOPIC as compared to un-substituted picolinic acid (Adeoye and Semire, 2013), although C1 and C5 in 4NPIC are slightly de-shielded. This is in agreement that the presence of electronegative atom attracts all electron clouds of carbon atoms towards itself, which in turn leads to de-shielding of carbon atoms and results in increase in chemical shift values (Varsanyi and Sohar, 1972). This de-shielding effect is noticed in 4NPIC molecule because of the attachment of NO₂ group.

The aromatic proton signals are observed at 5 to 7 ppm (Kalinowski et al., 1988; Pihlaja and Kleinpeter, 1994) and it has been found that presence of electrons on aromatic ring, double bonded atoms, and triple bonded atoms has been found to de-shield attached hydrogen (Varsanyi and Sohar, 1972). Its chemical shifts would be more susceptible to intermolecular interactions as compared to that for other heavier atoms. The methoxyl (OCH₃) group in 4MPIC pushed electrons into the picolinic acid ring thereby results in upfield resonance as compared to 4NPIC in which nitro (NO₂) group brought about downfield (Table 1). For instance, chemical shifts

Table 1. ^{13}C NMR and ^{1}H NMR (ppm) calculated at DFT/B3LYP level with various basis.

Atom	4-methoxylpicolinic acid				4-nitropicolinic acid			
	6-31G*	6-31G**	6-311G**	Aveg.	6-31G*	6-31G**	6-311G**	Aveg.
C1	152.432	145.375	154.018	150.608	154.234	147.115	155.407	152.252
C2	110.751	105.608	111.760	109.373	120.212	115.441	123.179	119.611
C3	164.568	158.971	172.625	165.388	153.907	150.061	161.643	155.204
C4	116.013	110.919	119.691	115.541	118.620	114.036	121.866	118.174
C5	153.926	145.624	155.270	151.607	155.851	147.514	156.685	153.350
C6	164.063	154.569	165.406	161.346	162.813	153.348	164.105	160.089
H1	5.583	6.757	5.511	5.950	5.970	7.166	5.868	6.335
H2	7.274	8.616	7.404	7.765	8.701	10.015	8.826	9.181
H4	6.600	7.867	6.862	7.110	7.892	9.198	8.015	8.368
H5	8.573	9.733	8.747	9.018	9.089	10.261	9.247	9.532

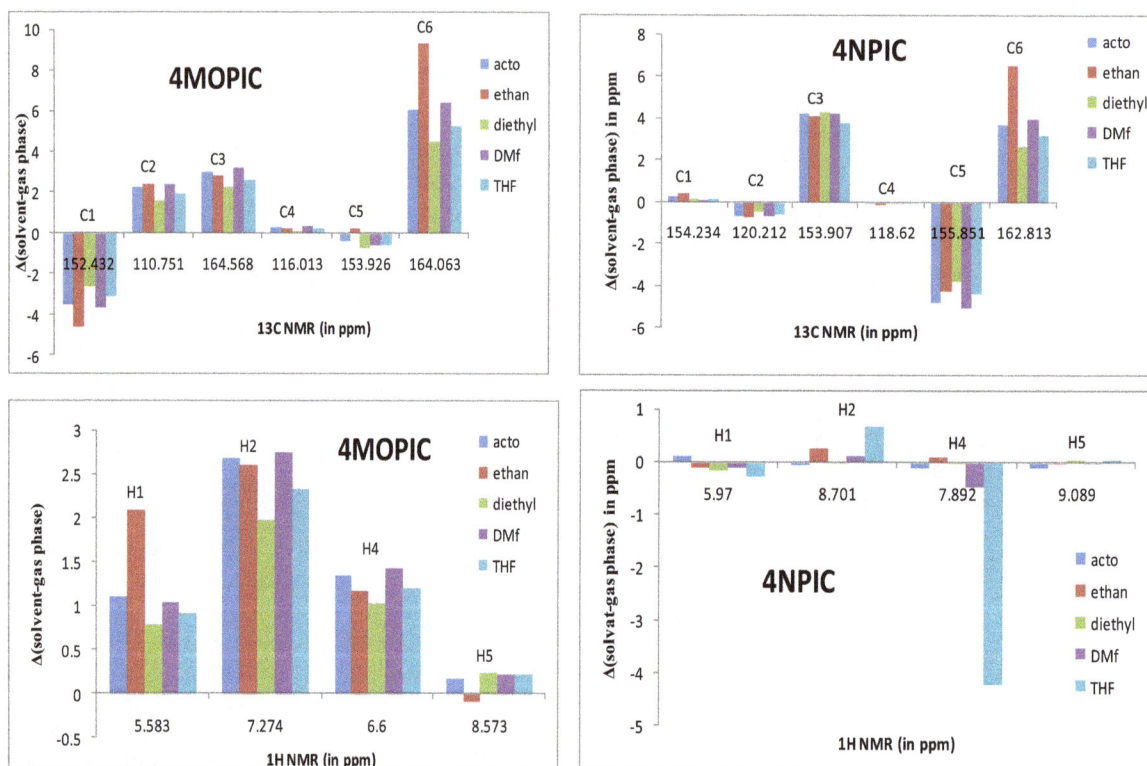

Figure 2. Solvents effect on ^{13}C NMR and ^{1}H NMR in ppm calculated at DFT/6-31G* level.

for H2 and H4 are 7.765 and 9.018 pmm for 4MOPIC and 9.181 and 9.9532 ppm for 4NPIC respectively. The hydroxyl oxygen atom shows electronegative property hereby contributed to hydroxyl hydrogen atom (H) downfield resonance as reflected in the two molecules. The calculated chemical shifts of hydroxyl hydrogen atom are 5.950 and 6.355 pmm for 4MOPIC and 4NPIC respectively which shows that NO_2 substituent increased downfield resonance experienced by hydroxyl hydrogen atom of 4NPIC (Table 1).

Theoretically, the solvent effect is estimated by comparing calculated chemical shifts for solution and gas phase at a particular level of calculation. In this work, the ^{13}CNMR and ^{1}HNMR calculated at DFT/6-31G* are used to estimate effect of solvents on chemical shifts as shown in Figure 2. The effects of solvents on ^{13}C NMR chemical shifts for 4MOPIC are pronounced on C1, C2, C3 and C6. For instance, the differences calculated on C6 for4MOPIC as compared to gas phase are 6.086, 9.301, 4.503, 6.379 and 5.256 ppm for acetone, ethanol, diethyl

Table 2. Selected vibrational frequencies calculated at B3LYP/6-311++G**.

4-methoxylpicolinic acid			4-nitropicolinic acid		
Cal.	Int.	Assign.	Cal.	Int.	Assign.
3773	96.50	vOH	3773	115.25	vOH
3242			3246		
3204		vCH	3229		vCH
3160			3171		
1821	353.81	vC=O	1828	346.81	vC=O
1623	244.52		1651	40.40	vC=C +vC=N
1606	21.09				
1516		vC=C +vC=N	1604	28.18	vC=C +vC=N
1427	5.27	C=C +πCH	1595	251.34	vC=C +vC=N
1355	110.82	C-O + πOH	1495		vC=N +πCH
1334	96.35	vC=N +C-OCH₃	1415		vC=C +πCH
1317	30.00	vC-C + C-N	1375	199.19	vNO₂
1285	71.94	vC-OCH₃ + πOH	1360	79.29	C-O + πOH
1177	182.74	πCH + πOH	1311		C-N + πOH+ πCH
1134	138.25	πCH + C-O	1301	30.91	vC=C-C=N+ πOH
1083	99.00	πCH + C-O	1206	87.52	vC-O + πOH +vC-NO₂
1058	7.26	vC-OCH₃	1139	203.08	vC-O+πCH
1008	58.98	δRing	1087	71.93	vC-O+πCH
998, 890, 860		σCH	1012		ωRing
805		π(C=C-C)	1001, 948, 880		σCH
613		σOH	734		π(C=C-C)
572		ωRing	660		φ (COOH)
262		γ CH₃	608		σOH
207		γ CH₃	276		ωRing
186		ω Ring	345		δRing
102		γ O-CH₃	146		γRing
36		γ COOH	37		γ (NO₂ + COOH)

v, stretching; π, in-plane bending; σ, out-of-plane bending; τ, torsion, δ, breathing; ω, wagging; γ, rocking; φ, scissor.

ether, DMF and THF respectively. The effect of ethanol is felt more on C6 and C1 for 4MOPIC molecule (Figure 2). In case of 4NPIC, the solvent effects are pronounced on C3, C5 and C6. For C6, the differences calculated are 3.713, 6.562, 2.681, 3.961 and 3.169 ppm for acetone, ethanol, diethyl ether, DMF and THF respectively as compared to results from gas phase calculations.

The effects of solvents on ^{1}HNMR chemical shifts for 4MOPIC are felt by all hydrogen atoms although more on H1, H2 and H4. For 4NPIC, the only solvent that has much effect on the molecule is THF as it is reflected in ^{1}HNMR on H2 and H4. The differences in chemical as compared to gas phase calculation could be explained in terms of changes in free energy hypersurface of the nuclei (Jernej et al., 2004).

Vibration frequencies

The vibration frequencies calculated at B3LYP/6-

311++G** are shown in Table 2. The ring carbon hydrogen stretching vibrations calculated at B3LYP/6-311++G** are in the region 3242-3160 cm[-1] for 4MOPIC and 3246-3171 cm[-1] for 4NPIC. The C=O stretching vibration for 4MOPIC and 4NPIC are calculated to be 1821 and 1826 cm[-1] with 353.81 and 346.81 intensities respectively. However, C=O stretching vibration for un-substituted picolinic acid was calculated to be 1820 cm[-1] (Adeoye and Semire, 2013) and experimentally observed at 1717 cm[-1] (Varsanyi and Sohar, 1972) and 1719 cm[-1] (Gfeller et al., 1976). The OH stretching vibration is calculated for both molecules at 3773 cm[-1] with ≈ 20% higher intensity in 4NPIC. This has been calculated at 3770 cm[-1] and experimentally observed at 3437 and 3464 cm[-1] for un-substituted picolinic acid (Varsanyi and Sohar, 1972; Gfeller et al., 1976). The C=C stretching coupled with C=N are calculated at 1623, 1606 and 1516 cm[-1]for 4MOPIC and 1651, 1604 and 1595 cm[-1] for 4NPIC. The C=C stretching coupled with C-H in plane deformation (πCH) are calculated to be 1427 and 1415 cm[-1] for

4MOPIC and 4NPIC respectively, this has been calculated at 1462 cm^{-1} and observed at 1439 cm^{-1} (Varsanyi and Sohar, 1972) and 1443 and 1719 cm^{-1} (Gfeller et al., 1976) for the un-substituted picolinic acid. The C-O stretching coupled with OH in plane deformation (πOH) is calculated at 1355 and 1360 cm^{-1} for 4MOPIC and 4NPIC respectively. Also, this has been calculated at 1358 cm^{-1} and observed experimentally at 1370 and 1347 cm^{-1} (Adeoye and Semire, 2013; Varsanyi and Sohar, 1972; Gfeller et al., 1976) un-substituted picolinic acid. The in-plane bending of C=C-C is calculated at 805 cm^{-1} for 4MOPIC and 734 cm^{-1} for 4NPIC.

The C-H out-of-plane deformation (σCH) for 4MOPIC is calculated at 998, 890 and 860 cm^{-1} whereas this is calculated at 1001, 948 and 880 cm^{-1} for 4NPIC. The OH out-of-plane deformation (σOH) is calculated to be 613 and 608 cm^{-1} for 4MOPIC and 4NPIC respectively. The COOH rocking vibration was at 36 cm^{-1} for 4MOPIC and 37 cm^{-1} for 4NPIC which is coupled with γNO$_2$.

Global electrophilicity and Electronic properties

The electronic properties of the 4MOPIC and 4NPIC are calculated from the total energies and the Koopmans' theorem. The ionization potential (IP) is determined from the energy difference between the energy of the compound derived from electron transfer which is approximated; IP ≈ -E$_{HOMO}$ while the electron affinity (EA) is given as; EA ≈ -E$_{LUMO}$, respectively. The chemical potential (μ), chemical hardness (η), electrophilicity index (ω) and softness (1/η) of a molecule are deduced form IP and EA values (Takusagawwa and Shimada, 1976; Zhou and Navangul, 1990; Chamizo et al., 1993; Koopmans, 1734; Parr et al., 1999) as shown in the following Equations 1, 2 and 3.

$$\mu = \left(\frac{\partial E}{\partial N}\right)v(r) \approx -\left[\frac{IP+EA}{2}\right] \approx -\left[\frac{E_{HOMO}+E_{LUMO}}{2}\right] \quad (1)$$

$$\eta = \left(\frac{\partial^2 E}{\partial N^2}\right)v(r) \approx \left[\frac{IP-EA}{2}\right] \approx [E_{HOMO} - E_{LUMO}] \quad (2)$$

$$\omega = \frac{\mu^2}{2\eta} \quad (3)$$

It is well known that when IP is small and EA is large and positive the molecule should be soft, therefore soft molecules are often more chemically reactive than hard molecules. Also, the electrophilicity index has being a useful structural depictor of the analysis of the chemical reactivity of molecules (Pearson, 1993; Bird, 1997; Chattaraj et al., 2003; Semire and Odunola, 2013; Semire, 2013). According to the definition, electrophilicity index measures the propensy of a species to accept electrons. As Domingo et al (Domingo et al., 2002) proposed the high nucleophility and electrophility of

heterocycles corresponds to opposite extremes of the scale of global reactivity indexes. A good, more reactive, nucleophile is characterized by a lower value of μ, ω and in opposite a good electrophile is characterized by a high value of μ, ω. The electronegativity and hardness are of course used extensively to make predictions about chemical behavior and these are used to explain aromaticity in organic compounds (De Proft and Geerlings, 2001). A hard molecule has a large HOMO–LUMO gap and a soft molecule has a small HOMO–LUMO. The LUMO represents electron(s) accepting ability and HOMO as electron donating ability of a molecule.

The HOMO, LUMO, energy band gap, dipole moment, energy of solvation, chemical potential, softness, electrophilicity/nucleophilicity index and UV-Vis adsorption maximum calculated are displaced in Table 3. The values of chemical hardness, chemical potential, softness, and electrophilicity index in gas phase for 4MOPIC are 5.46, -2.73, 0.183 and 0.6825 eV respectively and that of 4NPIC are 4.79, -2.40, 0.209 and 0.6013 eV respectively. Therefore, 4NPIC should be a better molecule to be involved in the interactions with electrophiles than for 4MOPIC. The dipole moment in a molecule is one of the important electronic properties when considering the interactions of molecules in solvents. The higher the value of dipole moment the stronger the intermolecular interactions would be expected, however the orientation of the dipole moment vector is also an important parameter. The calculated dipole moment values for 4MOPIC and 4NPIC in gas phase are 4.75 and 0.41Debye respectively.

The HOMO and LUMO energies calculated in solvents revealed that both HOMOs and LUMOs experienced stabilization in 4NPIC but LUMOs are destabilized in 4MOPIC as compared to gas phase. In line with our recent report (Adeoye and Semire, 2013), the magnitude of solvation energies reflect the degree of polarity of the solvents used in these calculations (that is, Ethanol > acetone ≈ DMF > THF ≈ Diethyl ether). The absorption maxima calculated in solvents are shifted to longer wavelengths as compared to gas phase (Table 3). The absorption maxima calculated at DFT (B3LYP) level with 6-31G*, 6-31G**, 6-311G**, 6-311+G** and 6-311++G** basis sets are shown in Table 4. The absorption maxima are 218.79, 218.82, 219.81, 225.53 and 225.57 nm for 4MOPIC and 276.06, 276.01, 274.52, 281.45 and 281.47 nm for 4NPIC at 6-31G*, 6-31G**, 6-311G** and 6-311+G** basis sets respectively. Another indicator of electrophilic attraction apart of the electrostatic potential is provided by the local ionization potential energy surface which is an overlaying of the energy of electron removal (ionization) onto the electron density. The regions with red color represent regions in the molecular surface where electron removal goes (with minimal energy) most easily, therefore easy of electron removal. The local ionization potential energy for 4MOPIC and

Table 3. HOMO, LUMO, energy band gap, dipole moment, solvation energy, global electrophilicity index and UV-Vis adsorption maximum calculated in gas phase and in solvents at DFT/6-31G* level.

	G. phase	Acetone	Ethanol	Diethyl ether	DMF	THF
4MOPIC						
HOMO (eV)	-6.96	-6.79	-6.88	-6.83	-6.78	-6.81
LUMO (eV)	-1.50	-1.52	-1.61	-1.51	-1.53	-1.51
Δ(H-L)	5.46	5.27	5.27	5.32	5.25	5.30
Sol. E (kJ/mol)	-43.47	-51.68	-55.54	-43.32	-51.69	-46.18
D.M	4.73	6.18	6.47	5.78	6.16	5.97
UV	218.79	219.43	219.50	219.34	219.70	217.98
H	5.46	5.27	5.27	5.32	5.25	5.30
M	-2.73	-2.69	-2.64	-2.66	-2.63	-2.65
(1/2 η)	0.183	0.190	0.190	0.188	0.190	0.189
Ω	0.6825	0.6865	0.6613	0.6650	0.6588	0.6625
4NPIC						
HOMO (eV)	-7.96	-7.70	-7.75	-7.76	-7.69	-7.73
LUMO (eV)	-3.17	-2.99	-3.11	-3.03	-2.98	-3.00
Δ(H-L)	4.79	4.71	4.64	4.73	4.71	4.73
Sol. E (kJ/mol)	-32.36	-50.38	-53.94	-43.41	-50.53	-45.93
D.M	0.41	0.55	0.39	0.49	0.64	0.55
UV	276.06	288.72	295.19	285.31	288.65	286.82
H	4.79	4.71	4.64	4.73	4.71	4.73
M	-2.40	-2.36	-2.32	-2.37	-2.36	-2.37
(1/η)	0.209	0.212	0.216	0.211	0.212	0.212
Ω	0.6013	0.5913	0.5800	0.5938	0.5913	0.5938

Table 4. λmax (in nm) calculated for 4-methoxyl and 4-nitropicolinic acid at DFT with various basis sets.

DFT/	4-methoxylpicolinic acid		4-nitropicolinic acid	
	UV	Intensity	UV	Intensity
6-31G*	218.79	0.16	276.06	0.035
6-31G**	218.82	0.16	276.01	0.035
6-311G**	219.81	0.16	274.52	0.038
6-311+G**	225.53	0.17	281.45	0.039
6-311++G**	225.57	0.17	281.47	0.039

4NPIC are calculated in different solvents as displayed in Figure 3. The local ionization potential energy shows that solvents affect ionization potential of the molecules. The solvents increased the minimum energy required to remove an electron for 4MOPIC whereas the minimum energy required was lower in 4NPIC as compared to gas phase. Therefore, the minimum energy required to remove an electron in different solvents could be arranged as ethanol > acetone $\approx$ DMF > THF > Diethyl ether > gas phase for 4MOPIC and gas phase > ethanol > diethyl ether > THF > acetone > DMF for 4NPIC. It could be suggested that increasing in electron density by electron donor substituent lowers the minimum energy required to remove an electron while decreasing in

electron density as a result of electron abstractor substituent raises the minimum energy required to remove an electron (Figure 3 and Table 5).

Thermodynamic properties

The standard enthalpy (H°), standard entropy (S°) and standard heat capacity at constant pressure ($C°_p$) are 378.00, 372.77 and 138.63.94 J/mol for 4MOPIC; and 299.25, 373.34 and 136.75 J/mol for 4NPIC respectively. The thermodynamic functions such as heat capacity ($C_{p,m}°$), entropy ($S_m°$) and enthalpy ($H_m°$) for 4MOPIC and 4NPIC are obtained from the theoretical harmonic frequencies as listed in Table 5. All the $C_{p,m}°$, $S_m°$ and $H_m°$

4-methoxylpicolinic acid

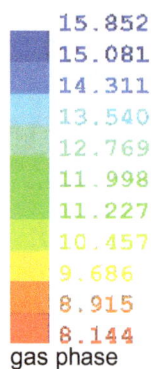

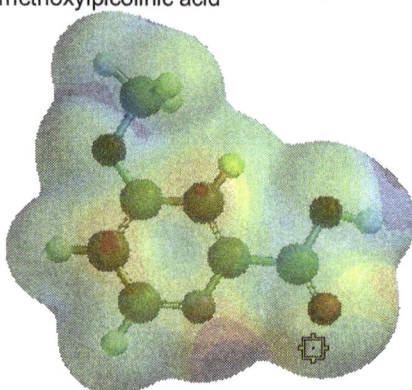

| 15.852 |
| 15.081 |
| 14.311 |
| 13.540 |
| 12.769 |
| 11.998 |
| 11.227 |
| 10.457 |
| 9.686 |
| 8.915 |
| 8.144 |

gas phase

4-nitropicolinic acid

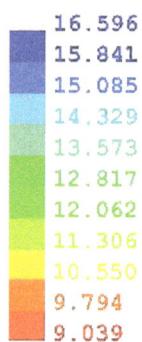

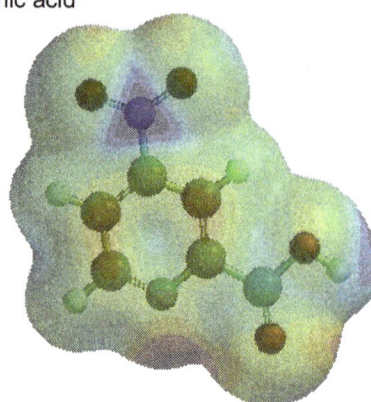

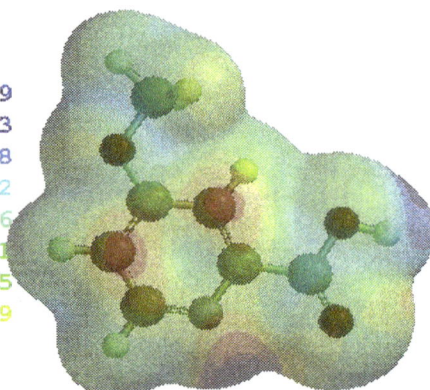

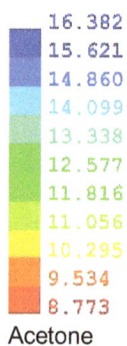

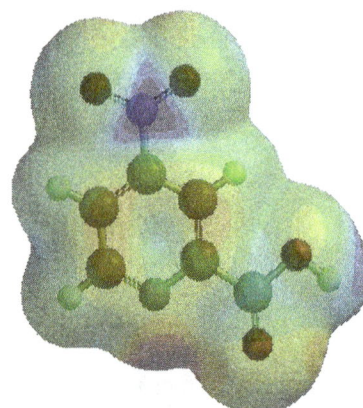

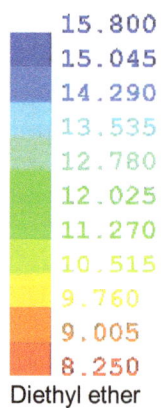

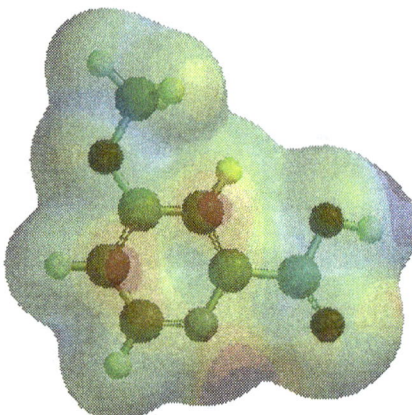

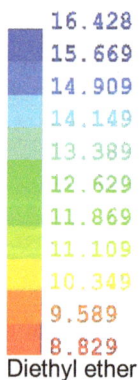

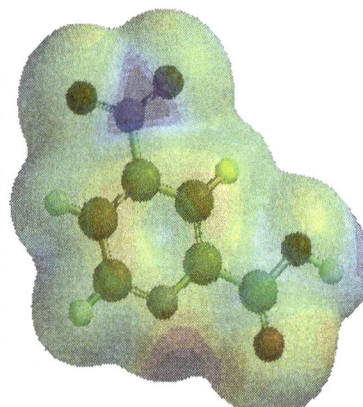

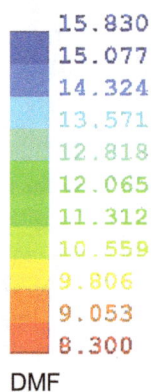

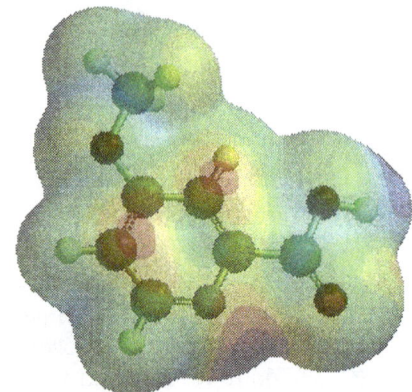

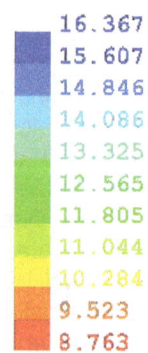

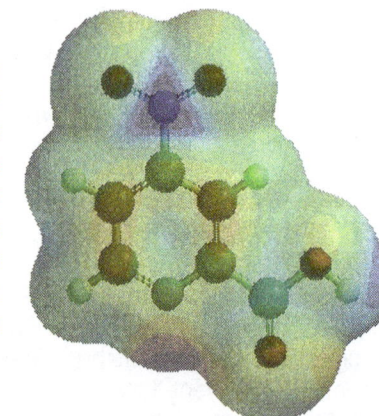

| 16.596 |
| 15.841 |
| 15.085 |
| 14.329 |
| 13.573 |
| 12.817 |
| 12.062 |
| 11.306 |
| 10.550 |
| 9.794 |
| 9.039 |

gas phase

| 15.859 |
| 15.103 |
| 14.348 |
| 13.592 |
| 12.836 |
| 12.081 |
| 11.325 |
| 10.569 |
| 9.814 |
| 9.058 |
| 8.302 |

Acetone

| 16.382 |
| 15.621 |
| 14.860 |
| 14.099 |
| 13.338 |
| 12.577 |
| 11.816 |
| 11.056 |
| 10.295 |
| 9.534 |
| 8.773 |

Acetone

| 15.800 |
| 15.045 |
| 14.290 |
| 13.535 |
| 12.780 |
| 12.025 |
| 11.270 |
| 10.515 |
| 9.760 |
| 9.005 |
| 8.250 |

Diethyl ether

| 16.428 |
| 15.669 |
| 14.909 |
| 14.149 |
| 13.389 |
| 12.629 |
| 11.869 |
| 11.109 |
| 10.349 |
| 9.589 |
| 8.829 |

Diethyl ether

| 15.830 |
| 15.077 |
| 14.324 |
| 13.571 |
| 12.818 |
| 12.065 |
| 11.312 |
| 10.559 |
| 9.806 |
| 9.053 |
| 8.300 |

DMF

| 16.367 |
| 15.607 |
| 14.846 |
| 14.086 |
| 13.325 |
| 12.565 |
| 11.805 |
| 11.044 |
| 10.284 |
| 9.523 |
| 8.763 |

DMF

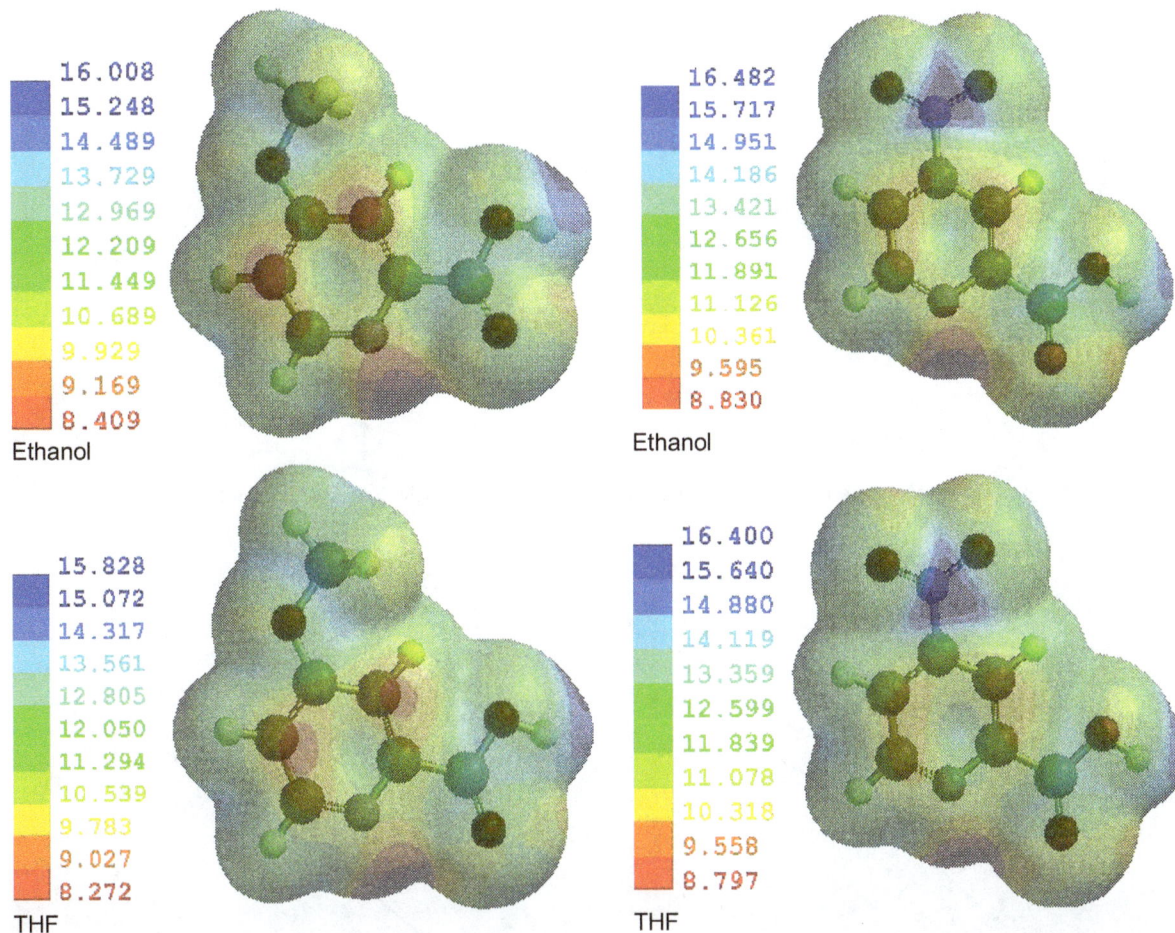

Figure 3. Local ionization energy surfaces on the molecular surfaces of 4MOPIC and 4NPIC with B3LYP/6-31G*. Color ranges, in kJ/mol.

Table 5. The thermodynamic properties obtained at different temperature for the 4-methoxyl- and 4-nitropicolinic acid at B3LYP/6-31++G** level.

Temp (K)	4MPIC			4NPIC		
	$H^{\circ}m$ (kcal/mol)	$S^{\circ}m$ (cal/mol)	C_p (cal/mol)	$H^{\circ}m$ (kcal/mol)	$S^{\circ}m$ (cal/mol)	C_p (cal/mol)
100	85.95	68.61	15.39	67.13	70.02	15.46
273	90.34	89.09	33.13	71.52	89.23	32.68
373	93.48	98.59	43.28	74.66	98.49	42.32
473	97.87	107.19	52.13	79.05	107.23	50.38
573	102.89	115.50	59.38	83.44	115.22	56.78
673	107.91	122.95	65.24	88.46	122.23	61.81
773	113.56	129.61	70.01	94.11	128.40	65.79
873	119.83	135.53	73.95	99.76	133.95	68.99
973	126.11	140.89	77.24	105.40	138.92	71.60

increases with the increase in temperature from 100 to 973K; this is due to the enhancement of molecular vibrations while temperature increases at constant pressure (1 atm). The correlations between these thermodynamic parameters and temperature (T) are plotted and fitted by quadratic equations as shown in Figures 4, 5 and 6. The fitting factor (R^2) for these parameters for 4MOPIC and 4NPIC is found to be 0.999

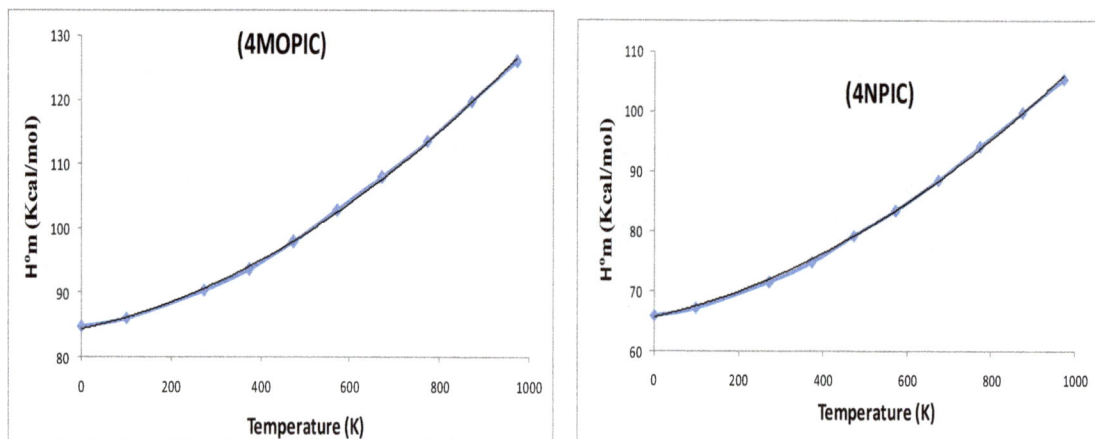

Figure 4. Correlation graph of enthalpy and temperature for 4-methoxylpicolinic and 4-nitropicolinic.

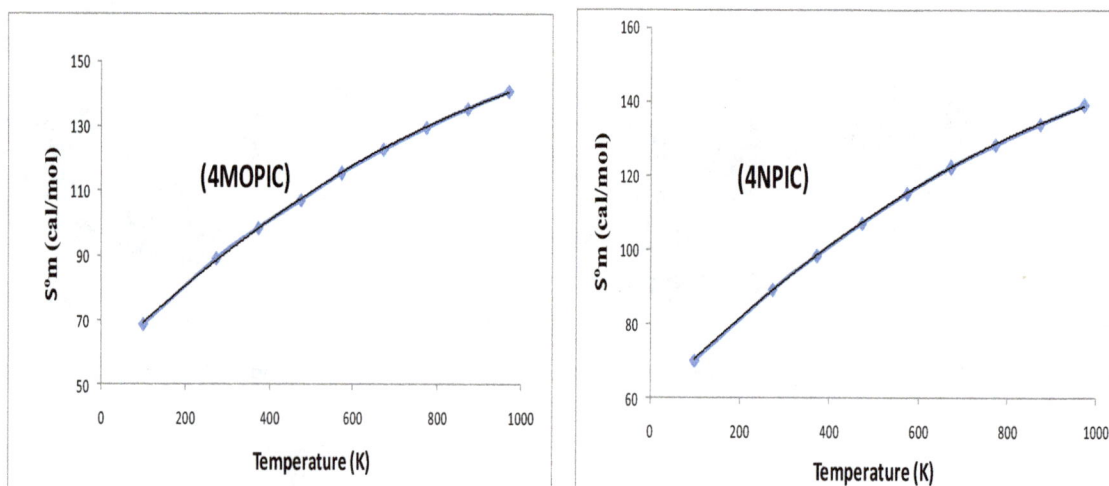

Figure 5. Correlation graph of entropy and temperature for 4-methoxylpicolinic and 4-nitropicolinic.

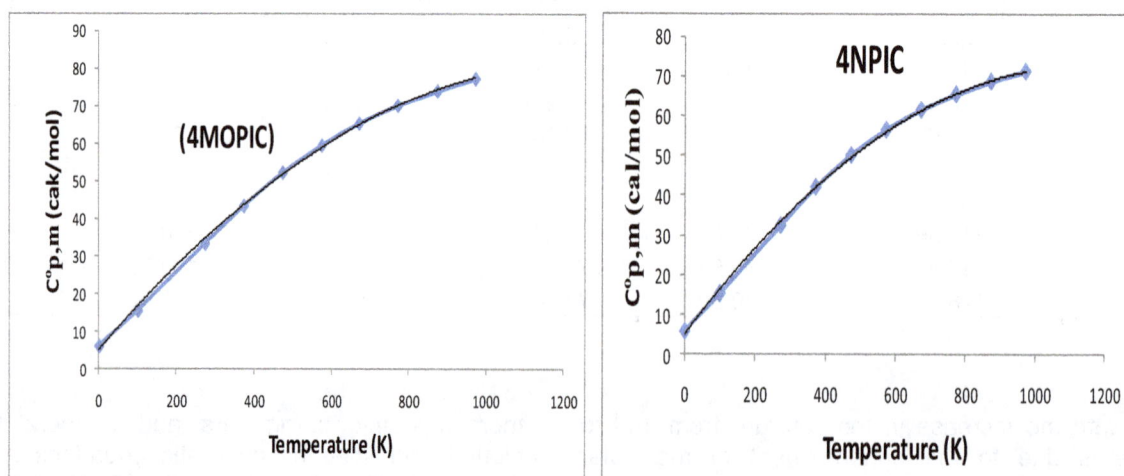

Figure 6. Correlation graph of heat capacity and temperature for 4-methoxylpicolinic and 4-nitropicolinic.

Table 6. The fitting factor (R^2) for parameters for 4MOPIC and 4NPIC.

4MOPIC	4NPIC
$C_{p,m}^\circ = 4.729 + 0.121T - 5.00T^2 \times 10^{-5}$; ($R^2 = 0.999$)	$C_{p,m}^\circ = 4.98 + 0.119T - 5.00T^2 \times 10^{-5}$; ($R^2 = 0.999$)
$H_m^\circ = 84.32 + 0.015T + 3.00T^2 \times 10^{-5}$ ($R^2 = 0.999$)	$H_m^\circ = 65.44 + 0.016T + 3.00T^2 \times 10^{-5}$ ($R^2 = 0.999$)
$S_m^\circ = 56.88 + 0.126T - 4.00 \times 10^{-5}T^2$ ($R^2 = 0.999$)	$S_m^\circ = 58.37 + 0.123T - 4.00 \times 10^{-5}T^2$ ($R^2 = 0.999$)

for heat capacity, enthalpy and entropy as shown in Table 6.

All the thermodynamic calculations are performed in the gas phase; therefore scale factors have been recommended for better accurate prediction (Zhang et al., 2010). All these thermodynamic data would be helpful in providing information for further study of the two isomers which can be useful to determine the directions of chemical reactions according to the second law of thermodynamics (Yazıcı et al., 2011; Nataraj et al., 2013; Govindarajanaet al., 2012).

Conclusion

In this work, we have performed the quantum chemical calculations on geometry, vibrational frequency and electronic properties of 4MOPIC and 4NPIC using B3LYP with various basis sets. Solvents effect on the molecules are studied by using five solvents namely ethanol, acetone, diethyl ether, DMF and THF. The results show that the HOMO and LUMO energies experienced stabilization in the solvents for 4NPIC but LUMO is destabilized in 4MOPIC as compared to gas phase. The HOMO, LUMO, electrophilicity index and softness revealed that 4NPIC would be a better molecule to be involved in the interactions with electrophiles than for 4MOPIC. The absorption maxima calculated using B3LYP/6-31G*/CIS are shifted to longer wavelengths in solvents as compared to gas phase. The minimum energy required to remove an electron in different solvents could be arranged as ethanol > acetone ≈ DMF > THF > Diethyl ether > gas phase for 4MOPIC and gas phase > ethanol > diethyl ether > THF > acetone > DMF for 4NPIC.

REFERENCES

Adeoye IO, Semire B, (2013). Solvents effect geometries and electronic properties of picolinic acid: A theoretical study. Int. J. Eng. Sci.13:101-107.

AlHokbany N, AlJammaz I (2011). Spectrocopic investigation and density functional theory calculations of mercaptobenzothiazole and mercaptobenzimidazole ligands and their rhenium complexes. Open J. Inorg. Chem. 1:23-32.

Becke AD (1988). Density-functional exchange-energy approximation with correct asymptotic behavior". Phys. Rev. A 38:3098–3100.

Bird CW (1997). Heteroaromaticity. 10. The Direct Calculation of Resonance Energies of Azines and Azoles from Molecular Dimensions. Tetrahedron. 53:2497-2501.

Chamizo JA, Morgado J, Sosa O (1993). Organometallic Aromaticity, Organometalliks 12:5005-5007.

Chattaraj PK, Maiti B, Sarkar U (2003). Philicity: A unified treatment of chemical reactivity and selectivity. J. Phys. Chem. A 107:4973–4975.

Cheeseman JR, Trucks GW, Keith TA, Frisch MJ, (1996) A comparison of models for calculating nuclear magnetic resonance shielding tensors", J. Chem. Phys. 104:5497-5509.

De Proft F, Geerlings P (2001). Conceptual and computational DFT in the study of aromaticity. Chem. Rev. 101:1451-1464.

Dimitrijević DM, Tadić ŽD, Mišić-Vuković MM, Muškatriović M (1974). Investigation of electronic effects in the reaction of diazodiphenylmethane with pyridine and pyridine N-oxide carboxylic acids, J. Chem. Soc. Perkin Trans. 2:1051-1055.

Domingo LR, Aurell M, Contreras M, Perez P (2002). Quantitative Characterization of the Local Electrophilicity of Organic Molecules. Understanding the Regioselectivity on Diels–Alder Reactions. J. Phys. Chem. A. 106:6871-6876.

Drmanic SZ, Nikolic JB, Marinkovic AD, Jovanovic BZ (2012). A comparative study of the linear salvation energy relationship for the reactivity of pyridine carboxylic acids with diazodiphenylmethane in protic and aprotic solvents. J. Serb. Chem. Soc. 77:1311-1338.

Evan GW, Johnson PE (1980). Characterization and quatititation of a zinc binding ligand in human milk. Peiatr. Res. 14:876-880.

Fernandez-Poi JA, Hamilton PD, Klos DJ (2001). Essential viral and cellular zinc and iron containing mettalloproteins as targets for prevention and therapy of viral diseases and cancer. Anticancer Res. 21:931-56.

Gfeller M, Furrer G, Schullin R (1976). Use of completely mixed flow-through systems: hydrolysis of phenyl picolinate. Environ. Sci. Technol. 31:3692-3701.

Goher MAS, Abu-Youssef MAM (1996). Synthesis, Spectral and Structural Characterization of a Monomeric Chloro Complex of Zinc(II) with Picolinic Acid; [Zn(C5H4NCO2H)(C5H4NCO2)Cl]. Polyhedron 15:453-457.

Govindarajana M, Karabacak M, Periandy S, Tanuja D (2012). Spectroscopic (FT-IR, FT-Raman, UV and NMR) investigation and NLO, HOMO–LUMO, NBO analysis of organic 2,4,5-trichloroaniline. Spectrochim Acta. A. 97:321-245.

Kalinowska M, Borawska M, Swislocka R, Piekut J, Lewandowski W (2007). Spectroscopic (IR, Raman, UV, ^{1}H and ^{13}C NMR) and microbiological studies of Fe(III), Ni(II), Cu(II), Zn(II) and Ag(I) picolinates. J. Mol. Struct. 834-836:419-415.

Kalinowski HO, Berger S, Braun S (1988). Carbon-13 NMR spectroscopy, John Wiley and Sons, Chichester.

Karakurt T, Dinçer M, Çukurovalı A, Yılmaz I (2012). Ab initio and semi-empirical computational studies on 5-hydroxy-5,6-di-pyridin-2-yl-4,5-dihidro-2H-[1,2,4]triazine-3-thione J. Mol. Struct. 1024:176–188.

Koczon P, Dobrowolskic J.Cz, Lewandowsk W, Mazurek AP (2003). Experimental and theoretical IR and Raman spectra of picolinic, nicotinic and isonicotinic acids. J. Mol. Struct. 655:89-95.

Koopmans T (1734). Ordering of wave functions and eigenvalues to the individual electrons of an atom, Physica. 1:104-113.

Kukovec BM, Popovic Z, Komorsky-Lovric Š, Vojkovic V, Vinkovic M (2009). Synthesis, structural, spectroscopic and thermal characterization of cobalt complexes with 3- and 6-methylpicolinic acid. Voltammetricand spectrophotometric study in solution Inorganica Chimica Acta 362:2704–2714.

Lee CT, Yang WT, Parr RG (1988). Development of the Colle-Salvetti correlation-energy formula into a functional of the electron density, Parr, Phys. Rev. B 37:785-589.

Marinkovi AD (2005). Investigations of the reactivity of pyridine carboxylic acids with diazodiphenylmethane in protic and aprotic solvents. Part I. Pyridine mono-carboxylic acids. J. Serb. Chem. Soc. 70:557–567.

Nataraj A, Balachandran V, Karthick T (2013). Molecular orbital studies (hardness, chemical potential, electrophilicity,and first electron excitation), vibrational investigation and theoretical NBOanalysis of 2-hydroxy-5-bromobenzaldehyde by density functional method. J. Mol. Struct. 2031:221-233.

Parajón-Costa BS, Wagner CC, Baran EJ (2004). Vibrational spectra and electrochemical behavior of Bispicolinate copper(II). J. Argentine Chem. Soc. 92:109-117.

Parr RG, Szentpaly L, Liu S (1999). Electrophilicity Index, J. Am. Chem. Soc. 121:1922-1924.

Pearson RG (1993). The principle of maximum hardness, Acc. Chem. Res. 26:250-255.

Pihlaja K, Kleinpeter E (1994). Carbon-13 Chemical Shifts in Structural and Stero Chemical Analysis, VCH Publishers, Deerfield Beach.

Semire B (2013). Density Functional Theory studies on electronic properties of thiophene S-oxides as aromatic dienophiles for reactivity prediction in Diels-Alder reactions. Pakistan J. Sci. Ind. Res. A 56:14-18.

Semire B, Odunola OA (2013). Theoretical Study on nucleophilic behaviour of 3,4-Dioxa-7-thia-cylopenta[a]pentalene and 3,7-Dioxa-4-thia-cyclopenta[a]pentalene using ab intio and DFT based reactivity descriptors. Int. J. Chem. Mod. 4:87-96.

Stare J, Jezierska A, Ambrožič G, Košir IJ, Kidrič J, Koll A, Mavri J, Hadž D (2004). Density Functional Calculation of the 2D Potential Surface and Deuterium Isotope Effect on ^{13}C Chemical Shifts in Picolinic Acid N-Oxide. Comparison with Experiment. J. Am. Chem. Soc. 126:4437.

Takusagawwa F, Shimada A (1976). Isonicotinic Acid. Acta Cryst B 32:1925-1927.

Teimouri A, Chermahini AN, Emami M (2008). Synthesis, characterization, and DFT studies of a novel azo dye derived from racemic or optically active binaphthol. Tetrahedron 64:11776–11782.

Varsanyi G, Sohar P (1972). Infrared spectra of 1,2,3,5-tetra-. substituted benzene derivativesActa Chim. Acad. Sci. Hung. 74:315-333.

Yazıcı S, Albayrak C, Gümrükçüoglu I, Senel I, Büyükgüngör O (2011). Experimental and density functional theory (DFT) studies on (E)-2-Acetyl-4-(4-nitrophenyldiazenyl) phenol. J. Mol. Struct. 985:292-298.

Yurovskaya MA, Mitkin OD, Zaitsera FV (1998). Functionalization of pyridines 2. Synthesis of Acylpyridines, pyridinecarboxylic acids and their derivatives. Review Chemistry of Heterocylic compounds 34:871- 879.

Zhang R, Dub B, Sun G, Sun Y (2010). Experimental and theoretical studies on o-, m- and p-chlorobenzylideneaminoantipyrines. Spectrochim Acta, 75A:1115-1124.

Zhou Z, Navangul HV (1990). Absolute hardness and aromatisity: MNDO study of benzenoid hydrocarbons, J. Phys. Org. Chem. 3:784-788.

Ziegler T (1991) Approximate density functional theory as apractical tool in molecular energetic and dynamics. Chem. Rev. 91:651-667.

4

Novel nano photocatalyst for the degradation of sky blue 5b textile dye

Zulfiqar Ali[1,2], Syed Tajammul Hussain[2], Muhammad Nawaz Chaudhry[1], Syeda Adila Batool[3], and Tariq Mahmood[2]

[1]College of Earth and Environmental Sciences, University of the Punjab, Lahore, Pakistan.
[2]Nano Science and Catalysis Division, National Centre for Physics, Islamabad, Pakistan.
[3]Department of Space Science Punjab University Lahore, Pakistan.

A new Nano photocatalyst, $CuO-Al_2O_3-ZrO_2-TiO_2$ was synthesized by co-precipitation for the removal of textile dye under ultra violet and direct sunlight. The synthesized catalyst was characterized by using X-rays diffraction, X-rays fluorescent, thermo gravimetric analysis, DRS, scanning electron microscopy and EDX. The dopants reduced the recombination of e_{cb}^- and h_{vb}^+ and decreased the band gap of TiO_2 from 3.25eV to 1.38eV. Due to this change, light absorption increased under sun light. The photocatalytic activity of the degraded samples of direct sky blue5B was analyzed by UV/Visible spectrophotometer. The results showed an enhanced photocatalytic activity under sun light about 96.8% in time duration of 100 min. It was also found that the degraded samples followed the pseudo first order kinetic model. This technique is cheap, easy and novel for the treatment of textile industrial waste water. In present energy crises of the world it can replace old expensive technologies.

Key words: Nano photocatalysis, direct sky blue5B, photocatalytic degradation, textile dyes, nano catalytic characterization.

INTRODUCTION

The extensive disposal of textile and other industrial waste water that contains organic dyes creates a severe contamination throughout the world. About 1 to 20% of the total dyes produced globally are lost during dyeing and other processes. These dyes are released into water as textile effluent (Weber and Stickney, 1993; Rafols and Barcelo, 1997; Houas et al., 2001; Kansal et al., 2009). These organic dyes weathered through oxidation; hydrolysis and other chemical reactions taking place in the wastewater phase can produce toxic metabolites (Bianco-Prevot et al., 2001; Neppolian et al., 2002; Pagga and Brown, 1986; Saquib and Muneer, 2003). These products produce unfavorable effects on human and animal health (Davydov et al., 2001) which requires a suitable treatment of wastewater for environmental

contamination prevention. After the development of novel heterogeneous photocatalytic, oxidation process in the 1970s is of special interest particularly under solar irradiation (Malato et al., 2002). In these processes hydroxyl radicals (OH⋅) and superoxide anion ($O_2^{⋅-}$), are produced due to absorption of radiation by semiconductor catalyst in contact with water and oxygen (Herrmann, 1999). Amongst a variety of catalysts, semiconductor photocatalyst containing titanium dioxide (TiO_2) is widely used due to its strong oxidizing power, non-toxicity and long-term photo stability (Pirkanniemi and Sillanpaa, 2002).

On the other hand, the photocatalytic efficiency of TiO_2 regarding the degradation of dyes reduces significantly due to the high recombination ratio of photo-induced

Synthesis of Catalysts using Co-precipitation Method

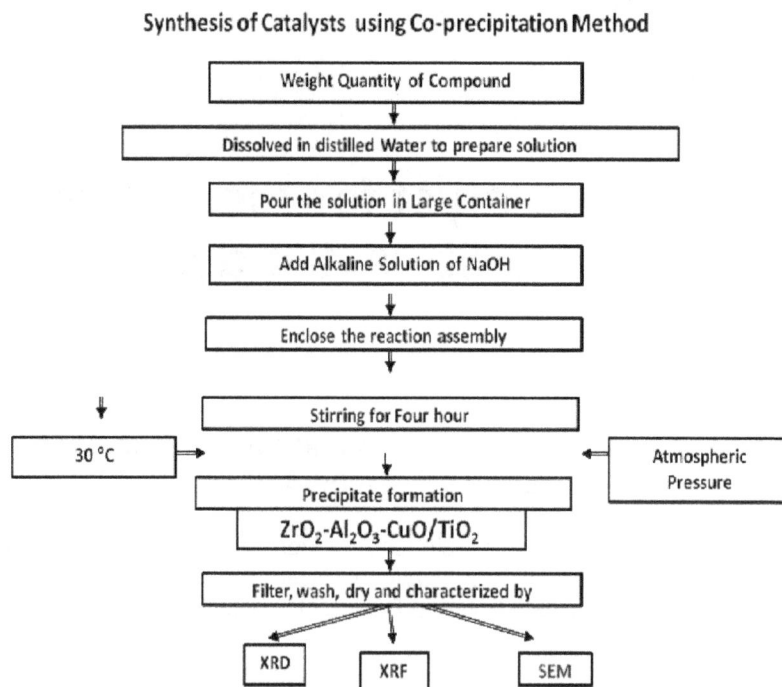

Figure 1. Summary of synthesis of photocatalyst Cuo-Zro$_2$-Al$_2$O$_3$-TiO$_2$.

electrons (e$^-$) and holes (h$^+$), which are produced due to irradiation under ultraviolet (UV) light (Bao et al., 2004; Malato et al., 2003). For TiO$_2$ the ideal wavelength regarding the band-gap energy is of 3.2 eV, which is at near-ultraviolet radiation (300nm) (Adewuyi, 2005; Meena and Pachwarya, 2009; Poulios and Aetopoulou, 2013). Thus the industrial efficacy of this process is bound to limited applications due to the need of ultra violet excitation energy source. As the solar light contains only 5% UV spectrum, hence further research in future is needed for the development of TiO$_2$ materials which is able to capture more amount of solar energy spectrum (Sonawane et al., 2004). Therefore the development of solar light active modified TiO$_2$ photo-catalyst with high catalytic efficiency in the visible region for the treatment of waste water is always of great interest and demand in both academic and commercial sectors. This paper aims at organic dyes degradation in wastewater treatment by modified TiO$_2$ photocatalytic systems. In particular, it focuses on enhancing the degradation efficiency of textile dyes by the use of visible light region, and improving the retrieval and reuse of TiO$_2$ photo-catalysts.

MATERIALS AND METHODS

TiO$_2$ (Degussa 25) was purchased from Sigma-Aldrich. Copper Nitrate, Zirconium Nitrate and Aluminum Nitrate (99.9% pure) were purchased from Merck Germany. The dye direct sky blue5B was purchased from Sigma-Aldrich. These chemicals were AR grade and there was no need of further purification before use.

The catalyst was prepared by a facile co-precipitation method as shown in Figure 1. CuO-ZrO$_2$-Al$_2$O$_3$-TiO$_2$ composite was synthesized by co-precipitation sedimentation method using ammonium carbonate as precipitating agent (Adewuyi, 2005). In this synthesis, the appropriate amount of TiO$_2$ (70 g) was suspended in 500 cm^3 of deionized water and homogenized for 30 min by stirring. Then equimolar copper nitrate, zirconium nitrate and aluminum nitrate each 10 g were mixed and dissolved in 100 cm^3 deionized water. The solution was homogenized by standard method. The mixture of these nitrates was then slowly mixed with TiO$_2$ solution and homogenized. The suspension was precipitated against drop wise addition of (NH$_4$)$_2$CO$_3$. The precipitate was then dried overnight with continuous stirring at room temperature. The resultant precipitate was then washed and filtered until the pH of the solution reached near 7 and then dried at 100°C overnight. It was then calcinated at 400°C for 4 h in Muffle furnace. The resultant material was grinded and sieved. The prepared sample was characterized by XRD, SEM (Mahmood, 2011), XRF, EDX, TGA and RBS.

The band gap of synthesized catalyst was found by UV/Vis/NIR Spectrometer Lambda 950 diffused reflectance spectroscopy. The Powder XRD of the catalyst was investigated by using Scintag XDS 2000 diffractometer with Cu Kα radiation source. The XRD analysis was performed from 0° to 90° (2θ) for the confirmation of structure and crystalline phases. By applying Scherer's equation, the crystallite size was evaluated. For the investigation of morphology of synthesized catalyst, Scanning Electron Microscope (JEOL, JSM 6490-A) was used. The X-Ray fluorescence (JEOL, Model JSX 3202M)) was used for the elemental analysis. The TGA was performed by Perkin Elmer Diamond Series, USA. The photo-degradation of the solution of dye direct sky blue5B was investigated by UV/Vis Spectrometer (Perkin Elmer Lambda 25).

The solution containing different concentration of dye and dosage of catalyst was exposed to UV and solar irradiation. The experimental setup under solar light is shown in Figure 2. The dye

Figure 2. Photocatalytic degradation of direct sky blue5B dye experimental set up.

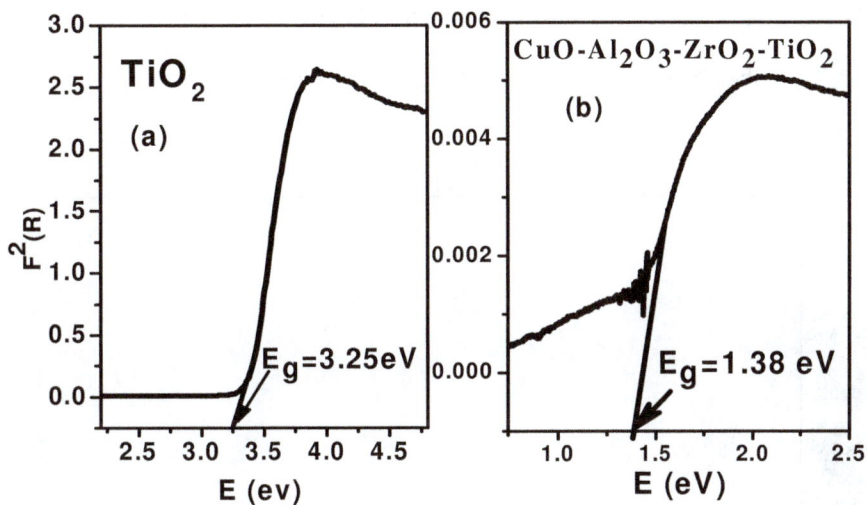

Figure 3. Band gap of TiO_2 and CuO-ZrO_2-Al_2O_3-TiO_2 nano composite.

solution was taken in a beaker with a magnetic stirrer for stirring and was exposed to the solar irradiation. 4 mirrors were used for focusing the sunlight to enhance the intensity of light. The degraded sample solution of 5 ml was taken after each 10 min interval up to 100 min. After filtering the sample solution, analyzed by UV/V is spectrometer. The removal percent of the dye direct sky blue5B samples were calculated by applying the formula:

Degradation/Decolonization (%) = (C_t-C_o) / C_o X 100 = (A_t-A_o) / A_o X 100

Where C_o and C_t are the concentrations while A_o and A_t are the absorbance values of the dye solution before and after the exposure to sunlight respectively.

RESULTS AND DISCUSSION

The band gap of the synthesized powder was evaluated by UV-Visible Diffuse Reflectance Spectroscopy (DRS) in the solid phase. The band gap analysis plots of Kubelka-Munk function versus energy for TiO_2 and the synthesized sample is shown in Figure 3. The band gap of TiO_2 is 3.25 eV which shows its anatase character. When TiO_2 is modified by CuO-Al_2O_3-ZrO_2 its band gap is decreased from 3.25 eV to 1.38 eV which is clear in Figure 3.

The shifting of the band gap from higher to lower wavelength is probably due to enhanced d-d and charge transfer transitions in the solid phase.

A comparison of XRD pattern of the synthesized composite photocatalyst is presented in Figure 4, where it can be observed that the reflections arising from TiO_2 are dominant in the XRD pattern composite. The XRD patterns were matched with the standard pattern for the TiO_2 (JCPDS - 71-1168). The reflections at 25.195°, 47.772°, 53.704°, 54.851° and 62.521° represent TiO_2

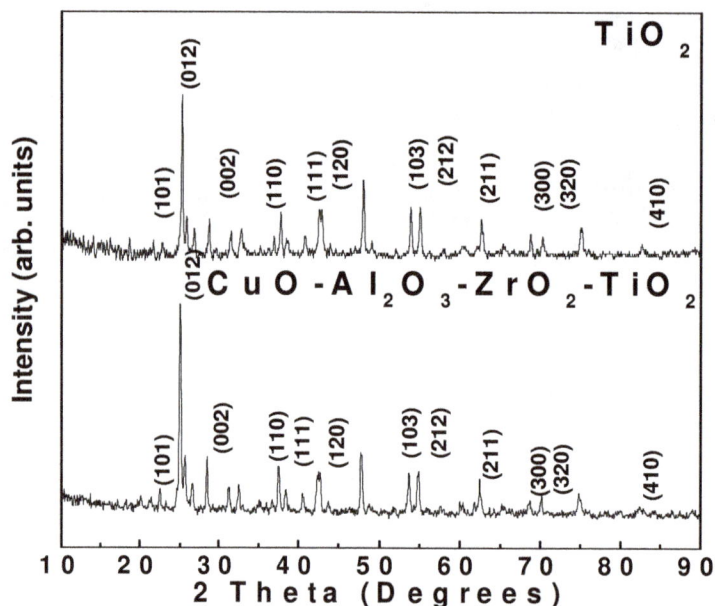

Figure 4. XRD Spectrum of TiO_2 and CuO-ZrO_2-Al_2O_3-TiO_2 nano composite.

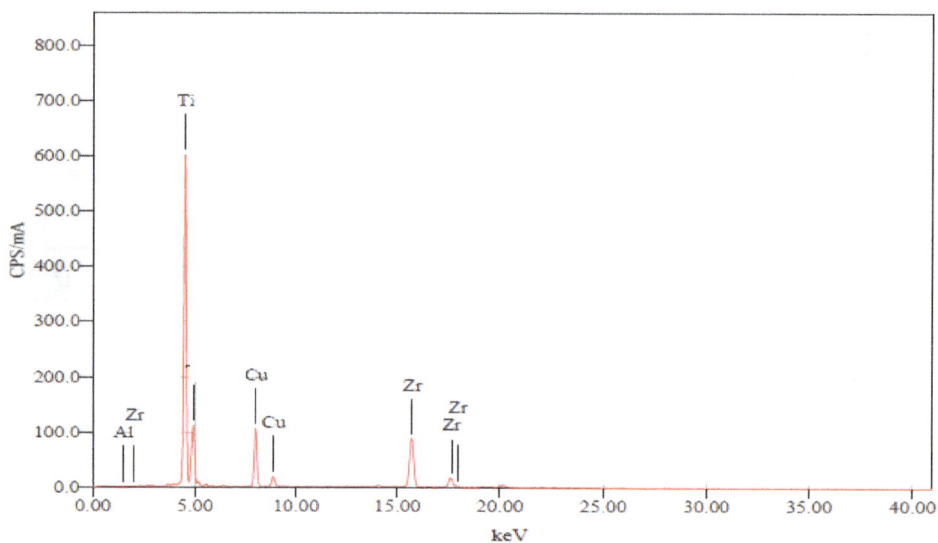

Figure 5. XRF peaks of photocatalyst CuO-Al_2O_3-ZrO_2-TiO_2.

(JCPDS -71 - 1168) whereas the reflections at 26.634°, 32.281° and 74.398° confirmed the presence of CuO phase (JCPDS - 34 - 1354). Similarly, the reflections at 2θ positions of 35.190° and 67.883° were matched with the Al_2O_3 (JCPDS - 34 - 0493). Low intensity peaks at the 2θ positions of 38.362°, 60.05° and 70.369° were identical with the standard pattern of ZrO_2 (JCPDS - 17 - 0385). The average crystallite sizes as calculated by Scherrer's equation using FWHM values of most intense peaks, was 29 nm.

The elemental composition of the synthesized sample CuO-Al_2O_3-ZrO_2/TiO_2 was confirmed by the X-ray Fluorescence analysis shown in Figure 5. It is found that all elements of the prepared catalyst were present and the major peak of TiO_2 shows that it is present as a base element in the catalyst. It is also confirmed from the analysis that no other peak rather than these elements is present in the catalyst, which shows its purity.

The thermo gravimetric analysis curve shows the weight loss against temperature rise as shown in Figure 6.

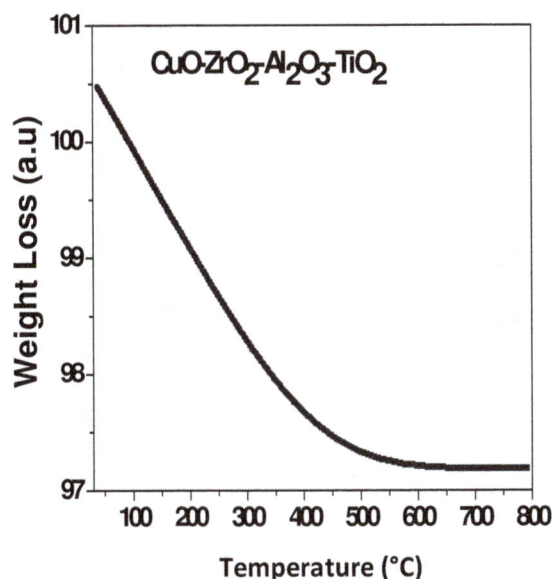

Figure 6. TGA of Photocatalyst (CuO-Al₂O₃-ZrO₂-TiO₂).

Figure 8. EDX peaks of Photocatalyst (CuO-Al₂O₃-ZrO₂-TiO₂).

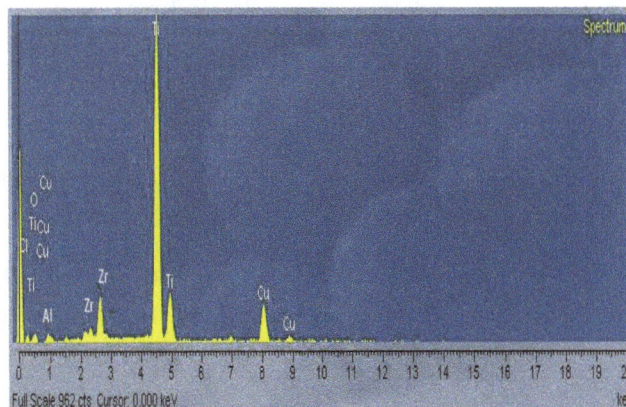

Figure 7. SEM of Photocatalyst (CuO-Al₂O₃-ZrO₂-TiO₂).

It was observed in the TGA curve of synthesized catalyst CuO-Al₂O₃-ZrO₂-TiO₂, there was a slight weight loss of about 2.49% in the temperature range of 100 °C to 600 °C, which corresponds to the removal of adsorbed water from the catalyst. The weight loss from 600 to 800 °C is about 0.134% which shows that the catalyst is thermally stable and suitable for photo degradation investigation.

The morphology of synthesized catalyst CuO-Al₂O₃-ZrO₂-TiO₂ was evaluated by SEM as presented in Figure 7. The porous nano-aggregates can be clearly seen in micrograph. It was observed that the metal loading, in the form of oxides, have no significant effect on the morphology of TiO_2, however a meek increase in the particle size is observed due to doping which is in accordance with XRD graph. Metal oxides can amend the surface of the catalyst by increasing the dispersion of active site. The sample composition is determined by the energy dispersive X-ray utility of SEM. The compositions obtained by EDX (Figure 8) and XRF (Figure 5) were in accordance with the theoretical values and within the limits of experimental errors.

The proficient degradation by a photocatalyst in aqueous medium depends upon its ability of absorbing light and its less rate of recombination of electron-hole pair. This ability promotes the generation of oxidizing hydroxyl (OH⁻) radicals that requires suitable band gap energy compared to that of incident photons. To enhance the generation of hydroxyl radicals is dependent on charge carrier recombination rate. The drawback of TiO_2 is that it has high recombination rate suffers, as the life time of excited states is of the order of 10^{-9} to 10^{-12} s, which in turns suppresses the formation of oxidizing species (Hoffmann et al., 1995). The other drawback is that, it has a wide band gap of 3.2eV, which lies in near-UV radiation (300 nm) (Adewuyi, 2005). This reduces its ability to absorb the photons in the visible spectrum that is, $E_{photon} \leq 380$ nm limits its use in the sunlight having major portion above 380 nm. The matter of recombination inhibition and light harvesting in the visible region can be made effective through the modification of the surface of TiO_2 by composite formation. This will induce the charge separation through the interfacial charge transfer between the allowed energy states. The possible mechanisms of recombination inhibition through mutual charge transfer between the interfacial energy states of the base material, TiO_2 and the components of the synthesized composites that is, CuO, Al₂O₃ and ZrO₂.

It is clear in Figure 9a that, when 100ppm solution of the dye was irradiated by a UV source (8 W lamp) by using TiO_2 (5 mg/L) as a catalyst, the degradation rate was about 65%, but under the same concentration of dye and catalyst under solar irradiation, the degradation rate

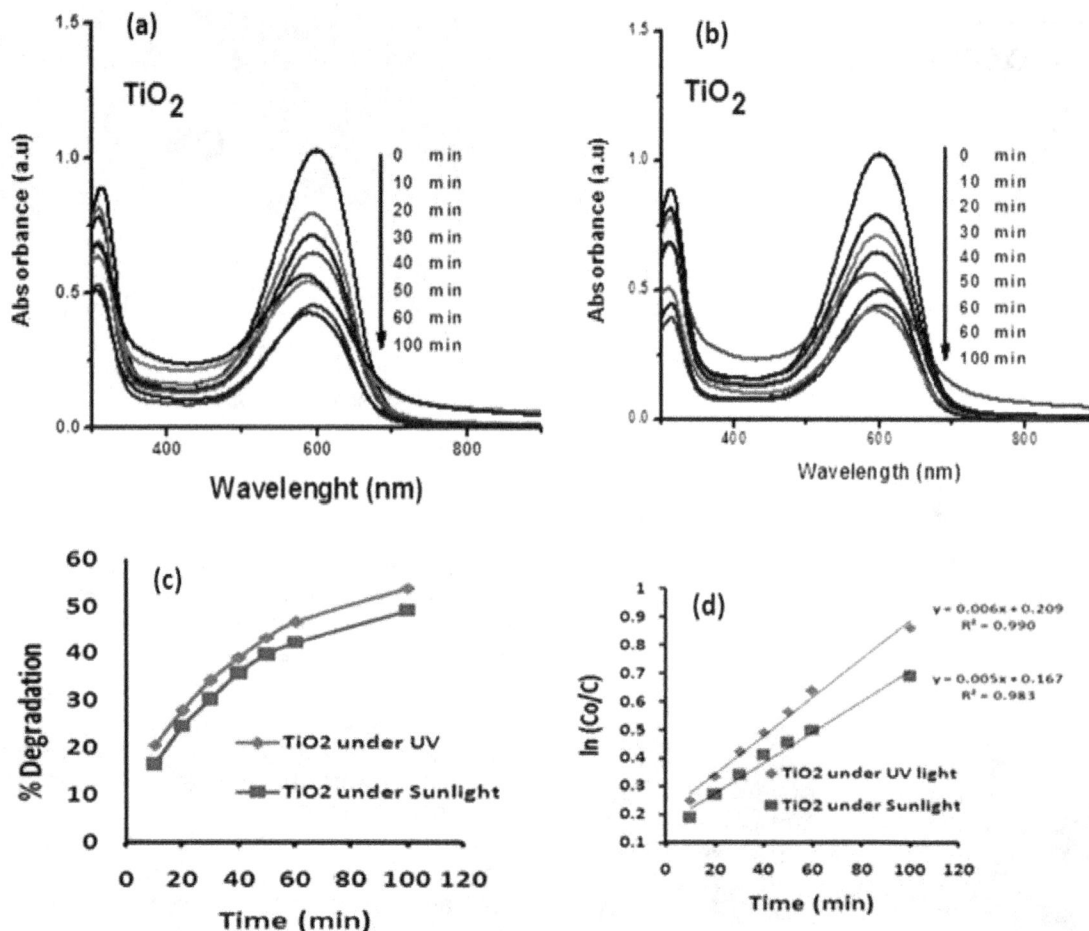

Figure 9. Graphical presentation of TiO$_2$ based degradation of Textile dye (Direct sky blue5B Dye).

Table 1. Comparisons between degradation rates of Direct sky blue5B Dye by TiO$_2$ and synthesized novel nano catalyst (CuO-Al$_2$O$_3$-ZrO$_2$-TiO$_2$) under UV & Sunlight.

Catalyst	Degradation rate (%) under UV light (100 min)	Degradation rate (%) under solar light (100 min)
TiO$_2$	65	58
CuO-Al$_2$O$_3$- ZrO$_2$-TiO$_2$	82.1	96.8

decreases by 58% (Figure 9b). This is due to the fact that the band gap of TiO$_2$ (3.2 eV) lies in the UV region (Table. 1).

It is clear from Table 1, the degradation rate of TiO$_2$ is 65% under UV light for 100 min duration, but the degradation rate of TiO$_2$ is 58% for the same duration of time. The less degradation rate under sunlight is due to band gap of TiO$_2$ (3.25 eV) which lies in the UV region. But when the TiO$_2$ is modified by CuO-Al$_2$O$_3$-ZrO$_2$ the band gap was reduced from 3.25 eV to 1.38 eV, which lies in the visible spectrum. From the result in Table 1, it is clear that the catalytic degradation efficiency enhanced up to 96.8% under solar irradiation, rather than 82.1%

under UV light (8W lamp) (Figure. 10).

The comparison of degradation profiles of direct sky blue5B in the presence of TiO$_2$ and the synthesized catalyst CuO-Al$_2$O$_3$-ZrO$_2$-TiO$_2$ shows that as the dye direct sky blue5B is an azo dye with two absorption bands at λ = 598 nm and λ = 313 nm. The high intensity band with λ_{max} = 598 nm wraps a region of 700 - 500 nm while the low intensity band at λ_{max} = 313 nm covers a region of 370 - 260 nm. The slow response of TiO$_2$ under solar irradiation is mainly due to its ability to utilize only 5% of solar spectrum ability with additional hindrance due to the low intensity band of dye in 370 - 260 nm, which absorbs a significant number of photons making theses

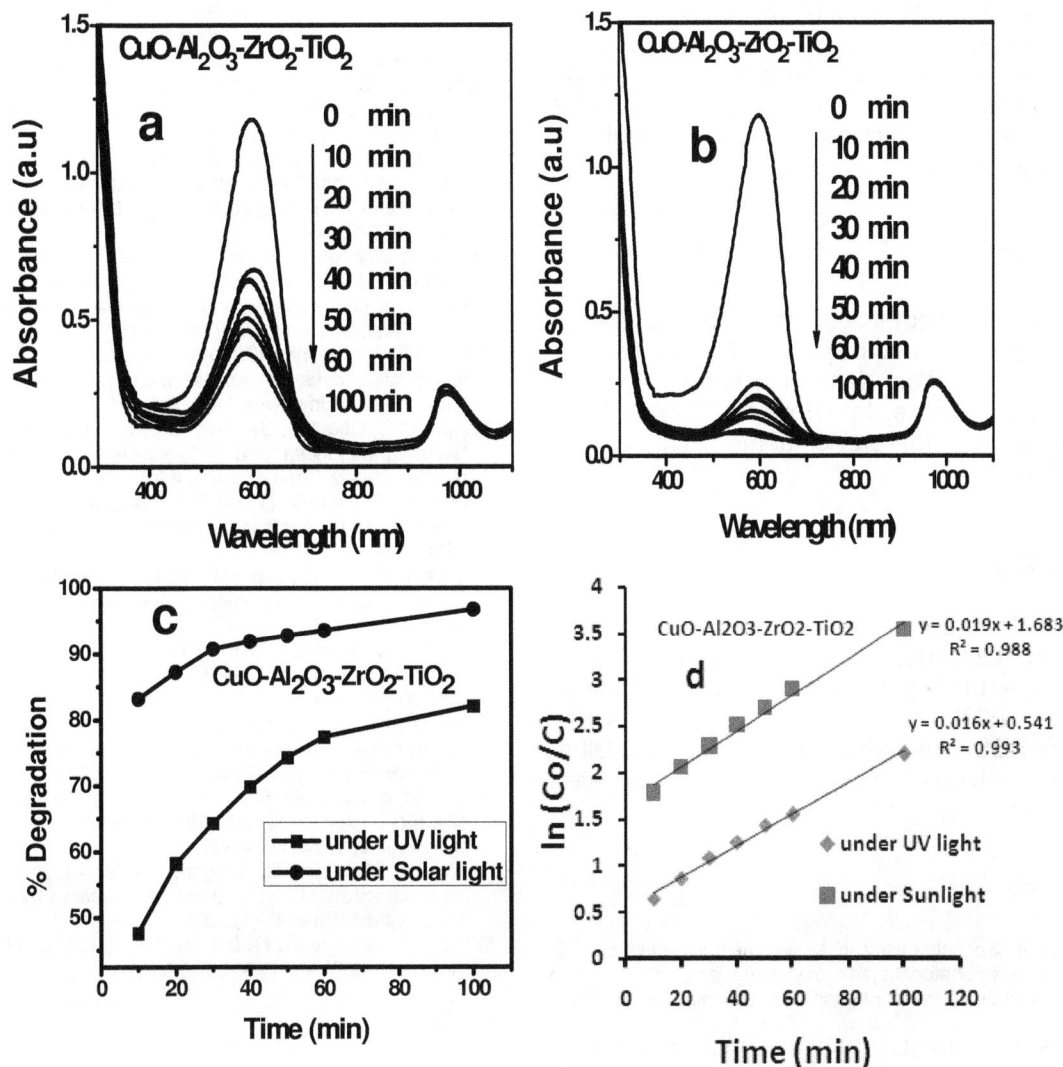

Figure 10. Graphical presentation of nano photocatalyst (CuO-Al_2O_3-ZrO_2-TiO_2) based degradation of textile dye (direct sky blue 5B dye).

inaccessible to TiO_2 thus causing a significant decrease in the degrading ability. As presented in Figure 9d, the kinetics of dye removal both in UV and under sunlight irradiation was evaluated by plotting ln (Co/C) against time. It is clear from the plots, the degradation of direct sky blue5B follows the pseudo-first order kinetics and the degradation rate depends on the concentration of dye with time.

Conclusion

From this study it is concluded that photocatalytic activity of TiO_2 can be enhanced by its suitable modification of the absorbing surface. However the best possible efficiency of the multiple photocatalyst can be obtained by selecting a suitable modifier with well-matched structure, morphological phase and chemical characteristics. It is also possible the photon initiated interfacial electron transfer between the allowed states of the semiconductors involved in the modified catalyst. Analysis of the synthesized catalyst pointed out that the dopants are very well dispersed in the anatase TiO_2. Optical properties enhancement, greater surface area and high quantum efficiency resulted due to adopted methodology TiO_2 modification. The highly efficient CuO-Al_2O_3-ZrO_2-TiO_2 nanoparticles of photocatalyst have extraordinary high activity in the degradation of direct sky blue5B under visible light irradiation. The use of visible light receptive nano photocatalysts are appropriate novel technique for the degradation of toxic organic pollutants and a cost effective way by using sunlight which is free of cost. Among other appropriate visible light active photocatalysts, CuO based systems are found to be a better choice for the modification of TiO_2 along with Al_2O_3 and ZrO_2. Nano photocatalysts, due to their low-price, simple synthesis process, high stability, high activity towards photo-induced redox reactions and reducing power are the best choice for degradation of environmental pollutants.

ACKNOWLEDGEMENTS

Author would like to acknowledge and extend his heartfelt gratitude to the persons who helped me for this research work. I also acknowledge Nanoscience and Catalysis Division, National Centre for Physics, Quaid-i-Azam University Islamabad, for providing me technical and analytical facilities.

ABBREVIATIONS

XRD, X-Rays Diffraction; **XRF,** X-Rays Fluorescent; **TGA,** Thermo Gravimetric Analysis; **DRS,** Diffuse Reflectance Spectroscopy; **SEM,** Scanning Electron Microscope; **EDX,** Energy-Dispersive X-Ray Spectroscopy; **RBS,** Rutherford Back Scattering Ultra-Violet/Visible/Near-Infra Red Spectrophotometer) UV/Vis/NIR.

REFERENCES

Adewuyi YG (2005). Sonochemistry in environmental remediation 2. Heterogeneous sonophotocatalytic oxidation process for the treatment of pollutants in water. Environ. Sci. Technol. 39(22):8557–8570.

Bao N, Feng X, Yang Z, Shen L, Lu X (2004). Highly efficient liquid-phase photooxidation of an azo dye methyl orange over novel nanostructured porous titanate-based fiber of self-supported radially aligned $H_2Ti_8O_{17}$ x $1.5H_2O$ nanorods.Environ. Sci. Technol. 38(9):2729-36.

Bianco-Prevot AB, Baiocchi C, Brussino MC, Pramauro E, Savarino P, Augugliaro V, Marcì G, Palmisano L (2001). Photocatalytic degradation of acid blue 80 in aqueous solutions containing TiO_2 suspensions. Environ. Sci. Technol. 35(5):971-976.

Davydov L, Reddy EP, France P, Smirniotis PG (2001). Sonophotocatalytic destruction of organic contaminants in aqueous systems on TiO_2 powders, Appl. Catal. B: Environ. 32(1):95-105(11).

Herrmann JM (1999). Heterogeneous photocatalysis: fundamentals and applications to the removal of various types of aqueous pollutants, Catal. Today 53:115–129.

Hoffmann MR, Martin ST, Choi W, Bahemannt DW (1995). Environmental Applications of Semiconductor Photocatalysis, Chem. Rev. 95(1):69-96.

Houas A, Lachheb H, Ksibi M, Elaloiu E, Guillard C, Herrmann JM (2001). Photocatalytic degradation pathway of methylene blue in water, Appl. Catal. B: Environ. 31:145-157.

Kansal SK, Kaur N, Sing S (2009). Photocatalytic degradation of commercial reactive dyes in aqueous phase using Nanophotocatalysts, Nanoscale Res. Lett. 4:709-716.

Mahmood T (2011). Metallic Phytoremediation and Nanobiotechnology of water hyacinth, PhD Thesis, Department of Biochemistry, Quid-i-Azam University Islamabad Pakistan.

Malato S, Blanco J, Caceres J, Fernandez AR, Aguera A, Rodriguez A (2002). Photocatalytic treatment of water soluble pesticides by photo-Fenton and TiO_2 using solar energy. Catalysis Today 76(2):209-220. ISSN 0920-5861.

Malato S, Blanco J, Campos A, Ca´eres J, Guillard C, Herrmann JM, Ferna´dez-Alba AR (2003). Effect of operating parameters on the testing of new industrial titania catalysts at solar pilot plant scale, Appl. Catal. B: Environ. 42(4):349–357.

Meena RC, Pachwarya RB (2009). Photo catalytic degradation of model textile azo dyes in textile wastewater using methylene blue immobilized resin dowex-11, J. Sci. Ind. Res. 68:730-734.

Neppolian B, Choi HC, Sakthivel S, Arabindoo B, Murugesan V (2002). Solar light induced and TiO_2 assisted degradation of textile dye reactive blue 4, Chemosphere, 46(8): 1173–1181.

Pagga U, Brown D (1986).The degradation of dyestuffs: Part II Behaviour of dyestuffs in aerobic biodegradation tests, Chemosphere 15(4):479–491.

Pirkanniemi K, Sillanpaa M (2002). Heterogeneous water phase catalysis as an environmental application: a review,Chemosphere, 48(10):1047–1060.

Poulios I, Aetopoulou I (2013). Photocatalytic degradation of the textile dye reactive Orange 16 in the presence of TiO_2 suspensions, Environ. Technol. 20:479-487.

Rafols C, Barcelo D (1997). Determination of mono- and disulphonated azo dyes by liquid chromatography–atmospheric pressure ionization mass spectrometry, J. Chromatogr. A. 777(1): 177–192.

Saquib M, Muneer M (2003). TiO_2-mediated photocatalytic degradation of a triphenylmethane dye (gentian violet), in aqueous suspensions, Dyes Pigments, 56(1):37–49.

Sonawane RS, Kale BB, Dongare MK (2004). Preparation and photo-catalytic activity of $FeTiO_2$ thin films prepared by sol–gel dip coating, Mater. Chem. Phys. 85(1):52–57.

Weber EJ, Stickney VC (1993). Hydrolysis kinetics of Reactive Blue 19-Vinyl Sulfone, Water Res. 27(1):63–67.

5

Synthesis, structural and optical characterizations of cadmium oxide (CdO) thin films by chemical bath deposition (CBD) technique

B. A. Ezekoye[1] , V. A. Ezekoye[1], P. O. Offor[2] and S. C. Utazi[1]

[1]Crystal Growth and Characterization Laboratory, Department of Physics and Astronomy, University of Nigeria, Nsukka, Enugu State, Nigeria.
[2]Department of Metallurgical and Materials Engineering, University of Nigeria, Nsukka, Enugu State, Nigeria.

Thin films of cadmium oxide (CdO) were synthesized and characterized using chemical bath deposition (CBD) technique. The XRD studies revealed amorphous CdO thin films which upon annealing at 623K transformed to polycrystalline structure. The optical studies showed that the CdO films have high average transmittance over 60% in the visible region and direct optical bandgaps of 2.02 ± 0.05 eV, 2.03 ± 0.05eV and 2.05 ± 0.05eV for samples X, Y, Z respectively. These characteristics make them good candidates for applications in photodiodes, phototransistors, photovoltaics, transparent electrodes, liquid crystal displays, IR detectors and anti-reflection coatings.

Key words: Thin films, chemical bath method, cadmium oxide, characterizations, bandgap.

INTRODUCTION

Metallic oxides are becoming prominent and important group of materials due to their versatile physiochemical, structural, and optical characteristics which include high temperature superconductivity, ferroelectricity, ferromagnetism, piezoelectricity, semiconductivity, optical, opto-electronic, magnetic, electric, thermal, electrochemical, catalytic and sensor properties (Jia et al., 2004; Vayssieres, 2004). Other attractive optical properties are low bandgap, high transmission, coefficient, invisible spectral domain, remarkable and luminescence characteristics (Ortega et al., 1999; Rusu and Rusu, 2005). The diversity emanates from the more complex crystal and electronic structures of metal oxides in comparison to other classes of materials. The elegancy of the metal oxides are found in the oxidation states, coordination numbers, symmetry, crystal-field stabilization, density, stoichiometry and acid-base surface properties that they exhibit. These characteristics made

them to find applications in photodiodes, phototransistors, photovoltaics, transparent electrodes, liquid crystal displays, IR detectors and anti-reflection coatings (Dhawale et al., 2008).

The cadmium oxide (CdO) thin films are n-type semiconductor that exhibits rock salt structure (FCC) with wide optical bandgap of 2.2 eV, high conductivity and high transmission in VIS region (Ortega et al., 1999; Henriguez et al., 2008). The thin films have attracted attention in recent years due to their attractive properties and wide range of technical applications in transparent electrodes, photovoltaics and display devices, saucers and others (Dhawale et al., 2008). The films also have been used as transparent contact in CuInSe2 and Si Solar Cells (Ortega et al., 1999).

Different techniques have been used by researchers in the past ten years for synthesizing the thin films such as spray pyrolysis, sputtering, pulsed laser deposition, sol-gel spin coating, electrochemical, activated reactive evaporation, metal organic chemical vapour deposition (MOCVD) and chemical bath method or solution growth methods (Henriguez et al., 2008; Caglar and Yakuphanoglu, 2009). In this study chemical bath method (CBD) was used in synthesizing cadmium oxide (CdO) and the synthesized samples were characterized using XRD, SEM and UV-VIS spectrophotometric techniques in order to find their possible areas of applications.

MATERIALS AND METHODS

The starting chemicals used in this work without further purification include cadmium chloride ($CdCl_2.4H_2O$) (AR chemicals grade), ammonia ($NH_3.H_2O$) and double distilled water (DDW). PVA and PVP cadmium chloride ($CdCl_2.4H_2O$) was used as precursor for the preparation of CdO thin films as a Cd^{2+} ionic source.

A detail of chemical bath deposition (CBD) has been discussed elsewhere (Ezekoye and Okeke, 2006; Ezema and Osuji, 2008). Stock solution (5 ml) of $CdCl_2$ was poured in a beaker followed by addition of 4 ml of 30% NH_3, with slight shake gave a whitish solution which is odourless. More of 30% NH_3 was poured into the set-up until it was clear by a total of 4 ml of NH_3. Double distilled water (H_2O, 34 ml) was added to the set-up which gave a fair whitish colour. The set-up was kept into an open conical flask where a good amount of oxygen was sufficiently supplied to it. A whitish film was deposited on the glass slide after 45 h, which turned clearly whitish when rinsed with distilled water.

Sample Y was post-treated at 200°C for 1 h which gave a whitish colour after the annealing. Sample X was post-treated at 400°C for 1 h, forming a brownish colour after the annealing. Sample Z was as-deposited, that is, no treatment given.

For the complex formation, an excess ammonium hydroxide solution (NH_3+H_2O) was added (30%) till a clear solution was obtained. The clear solution was kept under unstirred condition and glass substrate was dipped in it for 45 h. Whitish films due to the $Cd(OH)_2$ were formed on glass substrate. The CdO films were annealed in oxygen air-tight container at 473 and 673K for 1 h which generally facilitates decrease in dislocations, stresses, and inhomogeneities (Dhawale et al., 2008). The change in colour from whitish to brown films during annealing confirmed the formation of CdO. The equations for the reactions are as follows:

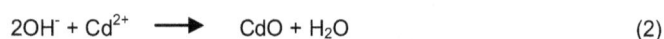

$$CdCl_2 + NH_3 + H_2O \longrightarrow [Cd(NH_3)_4]^{2+} + 2OH^- + 2Cl^- \qquad (1)$$

$$2OH^- + Cd^{2+} \longrightarrow CdO + H_2O \qquad (2)$$

RESULTS AND DISCUSSION

Thickness measurement

Film thickness (d) was determined by gravimetric weight difference method using high precision electronic balance given by the relation (Ezekoye and Okeke, 2006):

$$d = \frac{M}{(\rho \; x \; A)} \qquad (3)$$

where M is the mass of the film deposited on the substrate in gram, A is the area of the deposited film and

ρ, the density of the deposited material. The maximum thickness obtained for CdO thin film was 180 nm. This thickness is greater than the nano range classification of materials.

Optical studies

Optical properties of cadmium oxide (CdO) thin films were investigated by UV–VIS spectrophotometric technique. The absorption coefficient(α) is related to incident photons by the relation $\alpha hv = A(hv - E_g)^n$ (Caglar and Yakuphanoglu, 2009; Ezema and Osuji, 2008; Mohamed and Ali, 2008), where A is a constant and n is an index that characterizes the optical absorption process and is theoretically equal to 1/2, 2, 3/2 and 3 for direct allowed, indirect allowed, direct forbidden and indirect forbidden transitions, respectively. Since CdO is a direct band gap semiconductor, the $(\alpha hv)^2$ versus the hv diagram is depicted in Figure 1. The straight line on the curve at horizontal axis shows the energy band gap of the CdO thin films and is corresponding to 2.02 ±0.05eV (Sample Z) for the as-grown film. By annealing at 200 and 400°C, the band gap increased to 2.03 ± 0.05eV (Sample Y) and 2.05 ± 0.05eV (Sample X), respectively.

Figure 2 shows absorption spectra of CdO. The variation of optical absorbance (α) with wavelength (λ) of CdO film is shown in Figure 2. These spectra reveal that as-grown CdO films have low absorbance in the visible region. Figure 3 shows the transmittance spectra for CdO films. The increase in transmittance with increasing wavelength in ultra-violet (UV) region is fairly sharp. The absorption coefficient (α) was determined from the relation, $\alpha = \left(\frac{1}{d}\right)\ln\left(\frac{1}{T}\right)$, where d is the thickness of the film and T is the transmittance (Caglar and Yakuphanoglu, 2009; Mohamed and Ali, 2008) (Figure 3). This indicates that the absorption band gap transitions, which is characteristic of CdO and the fundamental absorption, which corresponds to electron excitation from the valence band to conduction band can be used to determine the nature and value of the optical band gap. The optical studies showed that the CdO films have high average transmittance over 60% in the visible region.

Film structure

Figures 4 (X), 4 (Y) and 4 (Z) show the XRD spectra of the grown CdO thin films. The film structure was studied by X-ray diffraction (XRD) technique using the CuK_α a radiation (λ = 0.1790 nm) and the grain size or crystallite size D was obtained using the Debye Scherrer relation (Barman et al., 2005):

$$D_{hkl} = \frac{K\lambda}{\beta\cos\theta}, \qquad (4)$$

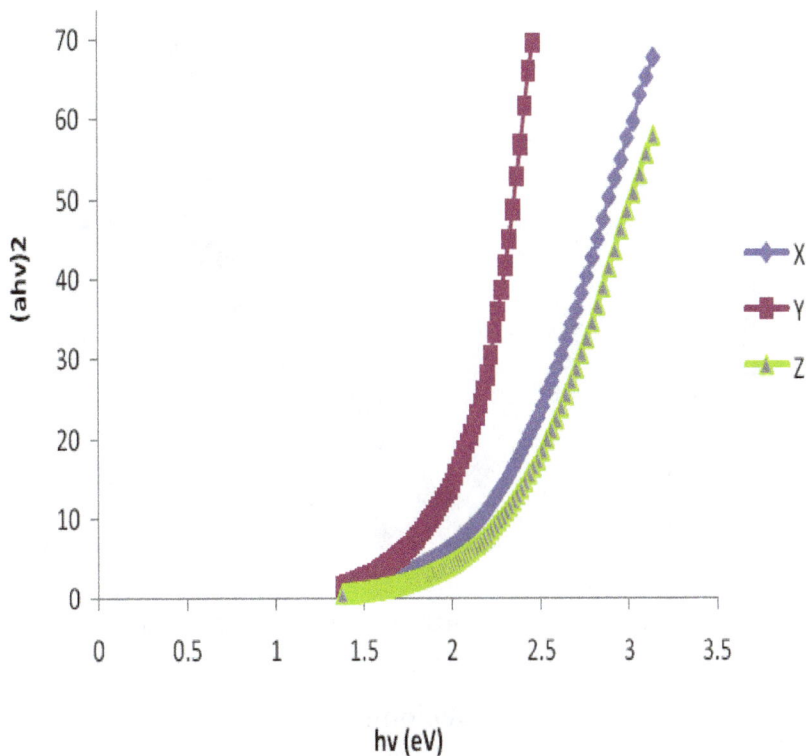

Figure 1. Plot of $(\alpha h\nu)^2$ against photon energy $h\nu$ (eV).

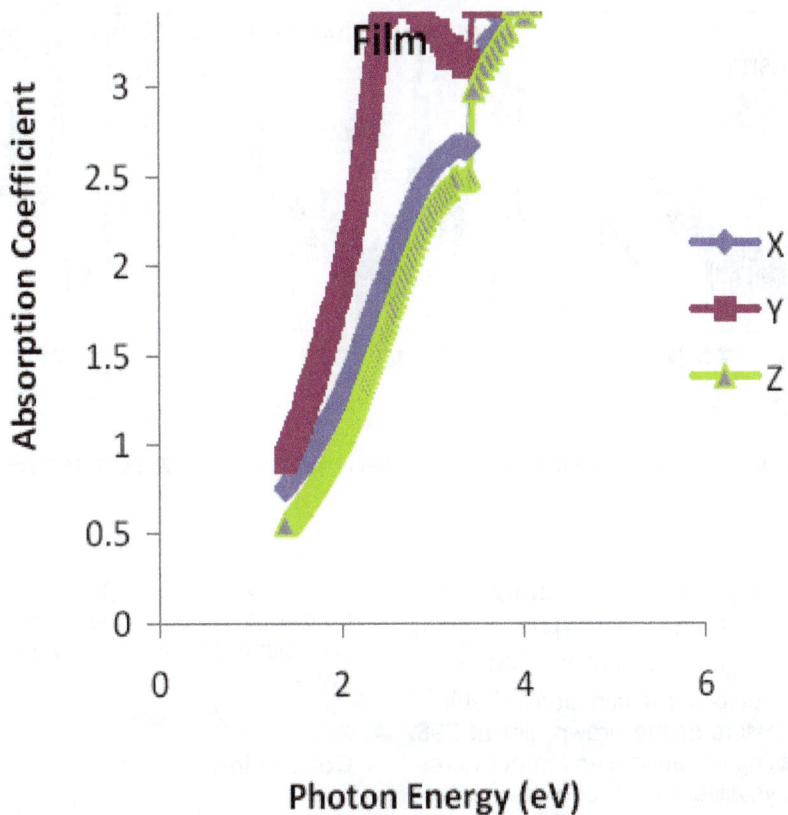

Figure 2. Plot of absorption coefficient against photon energy hv (eV).

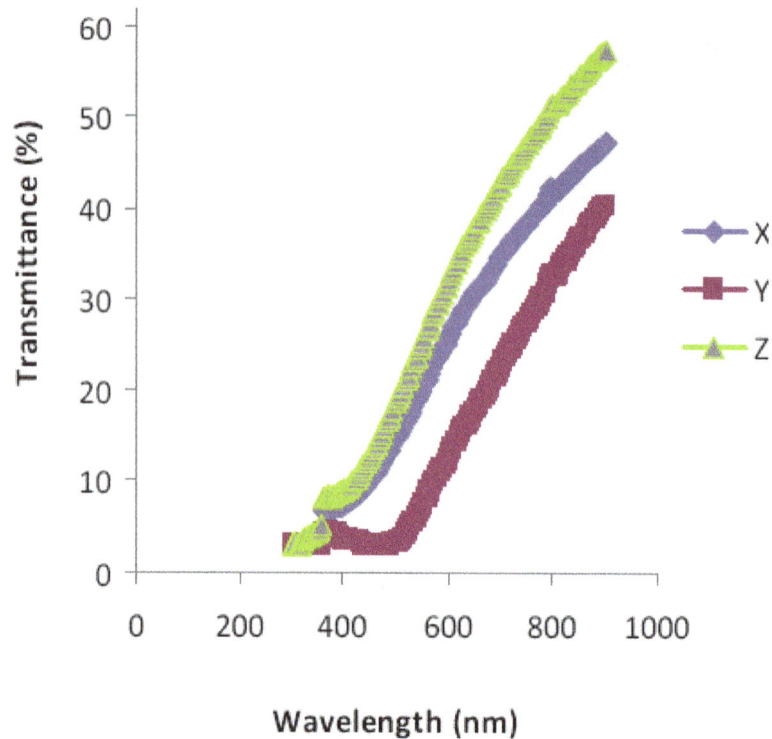

Figure 3. Plot of Transmittance (%) against photon energy h V (eV).

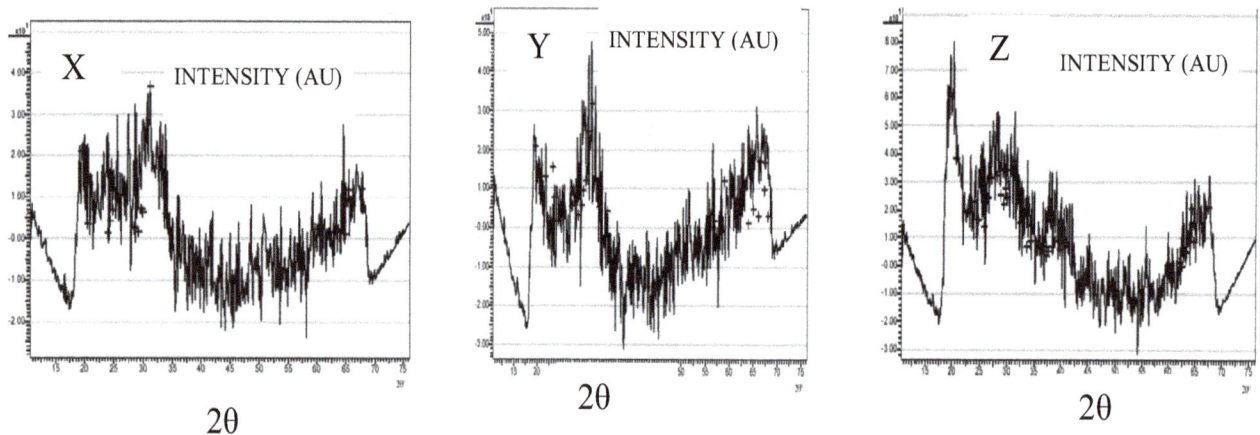

Figure 4. XRD spectra of the cadmium oxide thin film (X) annealed at 400°C (Y) annealed at 200°C and (Z) as-grown.

where K = 0.94, is the shape factor, $\lambda = 15408\text{Å}$, θ is the diffraction peak angle (Bragg's angle) in degrees and β is the full width at half maximum (FWHM) in radians, of the corresponding diffraction peak. Table 1 shows the various parameters of the grown film at 298, 473 and 673K. By increasing the annealing temperatures, the grain size of the crystallite was found to increase which is in agreement to Barman et al. (2005). Figure 5 shows the micrographs of the as-grown films (Z), and the

annealed at 200°C (Y) and annealed at 400°C (X). The as-grown is amorphous while the annealed became crystalline as can be seen by increase in grain-size in Table 1.

Conclusion

Thin films of cadmium Oxide (CdO) were synthesized and characterized using CBD technique. The XRD

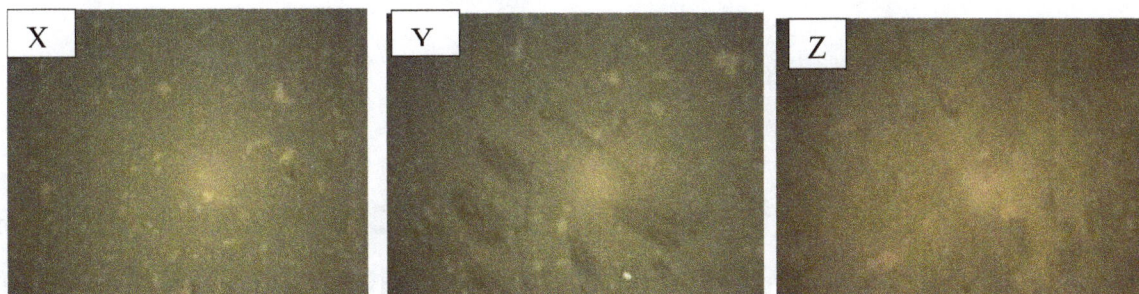

Figure 5. SEM image of samples X, Y, Z of the cadmium oxide thin films.

Table 1. XRD Parameters of the grown cadmium oxide (CdO) thin films.

Sample	Peak type	Peak position	Grain size (nm)	FWHM (°)	Bandgap (eV)
Z(278K)	(111)	31.19	28.675	1.4259	2.05 ± 0.05
Y(473K)	(111)	30.11	29.6749	0.96616	2.03 ± 0.05
X(673K)	(111)	20.64	43.0241	0.78548	2.02 ± 0.05

studies revealed amorphous CdO thin films which upon annealing at 623K transformed to polycrystalline structure. The optical studies showed that the CdO film has a high average transmittance of about 60% in the visible region with presence of direct bandgaps values of 2.02 ± 0.05eV, 2.03 ± 0.05eV and 2.05 ± 0.05eV for samples X, Y, Z, respectively. These characteristics make them good candidates for applications in photodiodes, phototransistors, photovoltaics, transparent electrodes, liquid crystal displays, IR detectors and anti-reflection coatings.

REFERENCES

Barman J, Borah JP, Sarma KC (2005). Synthesis and Characterization of CdS Nanoparticles by Chemical Growth. Optoelectronic Adv. Mater. Rapid Commun. 2(12):770-774.

Caglar M, Yakuphanoglu F (2009). Fabrication and Electrical Characterization of Flower-like CdO/p-Si Heterojunction Diode. J. Phys D: Appl. Phys. 42(045102):5.

Dhawale DS, More AM, Latthe SS, Rajpure KY, Lokhande CD (2008). Room Temperature Synthesis and Characterization of CdO Nanowires by Chemical Bath Deposition (CBD) Method. Appl. Surf. Sci. 254:3269-3273.

Ezekoye BA, Okeke CE (2006). Optical Properties in PbHgS Ternary Thin Films Deposited by Solution Growth Method. Pac. J. Sci. Technol. 7(2):108-113.

Ezema FI, Osuji RU (2008). Preparation and Optical Properties of Chemical Bath Deposited MnCdS2 Thin Films. FIZIKA A (Zagreb). 16(2):107-116.

Henriguez R, Grez P, Munoz E, Lincot D, Dalchiele EA, Marotti R, Gomez H (2008). One-step Pentiodynamic Synthesis of Polycrystalline Cadmium Oxide (CdO) Thin films in DMSO Solution. Sci. Technol. Adv. Mater. 9:025016.

Jia QX., Mccleskey, TM, Burrell AK, Collis GE, Wang H, Li ADQ, Foltyn SR (2004). Polymer-assisted Deposition of Metal Oxide Films. Nature Materials. 3: www.nature.com/naturematerials.

Mohamed HA, Ali HM (2008). Characterization of ITO/CdO/Glass Thin Films Evaporated by Electron Beam Technique. Sci. Tech. Adv. Mater. 9(025016):9.

Ortega M, Santana G, Morales-Acevedo A (1999). Optoelectronic Properties of CdO-Si Heterojunctions. Superficies y Vacio. 9:294-295.

Rusu RS, Rusu GI (2005). On the Electrical and Optical Characteristics of CdO Thin Films. J. Optoelectronics Adv. Mater. 7(2):823-828.

Vayssieres L (2004). On the Design of Advanced Metal Oxide Nanomaterials. Int. J. Nanotechnol. 1(1/2):6-8.

trans-[PtCl$_2$(NH$_2$C$_6$H$_4$CO$_2$H)$_2$]: a platinum complex forming a two-dimensional hydrogen-bonded network

Ayi A. Ayi[1,2] , Andrew D. Burrows[1], Mary F. Mahon[1] and Anna J. Warren[1]

[1]Department of Chemistry, University of Bath, Claverton Down, Bath BA2 7AY, United Kingdom.
[2]Department of Pure and Applied Chemistry, University of Calabar, P. M. B. 1115, Calabar, Cross River State, Nigeria.

The compound *trans*-[PtCl$_2$(NH$_2$C$_6$H$_4$CO$_2$H)$_2$] has been prepared from the hydrolysis of a platinum diimine complex, and has been characterised by X-ray crystallography. The crystal structure revealed that the compound forms a two-dimensional hydrogen-bonded network in the solid state, with pairs of O–H···O hydrogen bonds between carboxylic acid groups linking the molecules into zigzag tapes, which are cross-linked into sheets by pairs of N–H···Cl hydrogen bonds.

Key words: Metal-organic frameworks, metalloligands, two-dimensional hydrogen-bonded network.

INTRODUCTION

Coordination networks, otherwise known as metal-organic frameworks (MOFs) (Rowsell and Yaghi, 2004; Kitagawa et al., 2004; Férey, 2008; Robson, 2008) are currently attracting considerable interest for a wide range of potential applications (Czaja et al., 2009; Gagnon et al., 2012). Many robust, highly porous MOFs have been prepared that may be particularly useful in gas storage and separations (Rosi et al., 2003; Chae et al., 2004; Bennett et al., 2001). A major challenge in the field is the introduction of complex functionality (e.g., catalysis, luminescence, etc.) into MOF assemblies (Hupp and Poeppelmeier, 2005; Wu et al., 2005, 2004; Kitagawa et al., 2004). Efforts to introduce more complex functionality include modification of the ligand structure and forcing sites of unsaturation or labile solvent coordination at the metal nodes. Among the most attractive strategies for introducing new functionality is the use of "metalloligands"– metal complexes that contain ligands that have uncoordinated donor sites that can themselves act as ligands in MOF synthesis. This strategy has enabled the preparation of bimetallic MOFs, in which different metal centres may have different structural and/or functional roles (Noro et al., 2002; Kitaura et al.,

2004; Vreshch et al., 2004; Chen et al., 2004; Halper et al., 2006; Burrows et al., 2007, 2008; Zhang et al., 2008; Garibay et al., 2009). We were interested in preparing bimetallic MOFs by adopting a "two-step self-assembly" approach, where a metalloligand is synthesized in the first step as a framework linker, and then in the second step, the metalloligand is added to another metal ion, which acts as a node in the framework. By using the platinum diimine complex 1 (Buffin and Kundu, 2003), as a metalloligand in MOF construction, we were able to isolate crystals of *trans*-[PtCl$_2$(NH$_2$C$_6$H$_4$CO$_2$H)2]$_2$, which had formed as a decomposition product of 1. Here we report the characterization of this new platinum(II) complex, which has a two-dimensional hydrogen bonded network in the solid state.

EXPERIMENTAL

Platinum(II) chloride (0.027 g, 0.1 mmol) and 4-aminobenzoic acid (0.027 g, 0.2 mmol) were added to water (2.7 cm^3) and THF (4 cm^3) in a pressure tube, and the mixture was heated at 95°C for 5 days. On cooling to room temperature, crystals of 2 suitable for single

Table 1. Crystal data and structure refinement for *trans*-[PtCl$_2$(NH$_2$C$_6$H$_4$CO$_2$H)$_2$] **2**.

Empirical formula	Cl$_4$ Hl$_4$ Cl$_2$ N$_2$ O$_4$ Pt
Formula weight	540.26
Temperature	173(2) K
Wavelength	0.71073 Å
Crystal system	Triclinic
Space group	*P* -1
Unit cell dimensions	
a =	4.9020(1) Å
b =	5.7840(2) Å
c =	13.5710(4) Å
α =	87.0040(1)°
β =	88.1620(1)°
γ =	89.553(2)°
Volume	384.050(19) Å^3
Z	1
Density (calculated)	2.336 mg/m^3
μ	9.503 mm^{-1}
F(000)	256
Crystal size	0.37 × 0.37 × 0.05 mm^3
Theta range for data collection	3.01 to 27.42°.
Index ranges	-6<=h<=6, -7<=k<=7, -17<=l<=17
Reflections collected	5396
Independent reflections	1747 [R(int) = 0.0765]
Completeness to theta = 27.42°	99.0 %
Refinement method	Full-matrix least-squares on F^2
Data / restraints / parameters	1747 / 0 / 109
Goodness-of-fit on F^2	1.033
Final R indices [l>2σ(l)]	R_1 = 0.0584, w$R2$ = 0.1514
R indices (all data)	R_1 = 0.0583, w$R2$ = 0.1514
Largest diff. peak and hole	5.885 and -3.954 e.Å^{-3}

crystal X-ray analysis were obtained. ^{1}H NMR (DMSO-d$_6$) δ 7.82 (d 4H), 7.28 (d 4H), 3.48 (br 2H), ^{13}C NMR (DMSO-d$_6$) δ 167.82, 153.43, 131.55, 117.28, 112.94. Found: C, 30.2; H, 2.19; N, 5.36%. C$_{14}$H$_{15}$N$_2$O$_{4.5}$PtCl$_2$ (**2**·0.5H$_2$O) requires C, 30.6; H, 2.75; N, 5.10%.

NMR

The ^{1}H NMR and ^{13}C NMR spectra of the complex were recorded with a Bruker Advance 300 instrument using d$_6$-DMSO solutions and TMS as internal standard.

Elemental analysis

The C, N and H content of the compounds were determined using a CE-440 Elemental Analyser by Alan Carver at the University of Bath.

Structural analysis

The structural data were collected on a Nonius KappaCCD diffractometer. Details of the X-ray crystallographic analysis are given in Table 1. Fractional coordinates for the refined atoms and equivalent isotropic thermal parameters are presented in Table 2. The crystal structure was solved using SHELXS-97 and refined using SHELXL-97.

RESULTS AND DISCUSSION

While investigating the use of the platinum diimine complex **1** (Buffin and Kundu, 2003) as a metalloligand in MOF construction, we were able to isolate crystals of *trans*-[PtCl$_2$(NH$_2$C$_6$H$_4$CO$_2$H-4)$_2$] **2**, which had formed as a decomposition product of **1** (Scheme 1). Following identification of **2**, it was found possible to produce this compound *via* a more rational approach by reacting platinum(II) chloride with 4-aminobenzoic acid in a mixture of water and THF. Crystals of **2** were analysed by single crystal X-ray crystallography.

The asymmetric unit of **2** consists of half a platinum

Table 2. Atomic coordinates ($\times 10^4$) and equivalent isotropic displacement parameters for ($\text{Å}^2 \times 10^3$) for *trans*-$[PtCl_2(NH_2C_6H_4CO_2H)_2]$ 2.

Parameter	x	y	z	U(eq)
Pt(1)	10000	10000	0	11(1)
N(3)	11248(19)	7842(16)	1162(9)	19(2)
O(4)	2465(19)	7160(17)	4770(7)	24(2)
O(5)	1603(19)	3909(17)	4003(7)	24(2)
C(6)	6610(20)	8390(20)	3355(9)	20(2)
C(7)	8620(20)	8900(20)	2638(9)	18(2)
C(8)	7770(20)	5271(19)	1896(8)	16(2)
C(9)	9220(20)	7332(18)	1920(8)	14(2)
C(10)	2910(20)	5750(20)	4088(8)	15(2)
C(11)	5710(20)	4780(20)	2597(9)	18(2)
C(12)	5130(20)	6337(19)	3343(8)	15(2)
Cl(1)	13162(5)	12603(4)	460(2)	15(1)

U(eq) is defined as one third of the trace of the orthogonalized U^{ij} tensor

Scheme 1. Formation of 2 from the decomposition of 1.

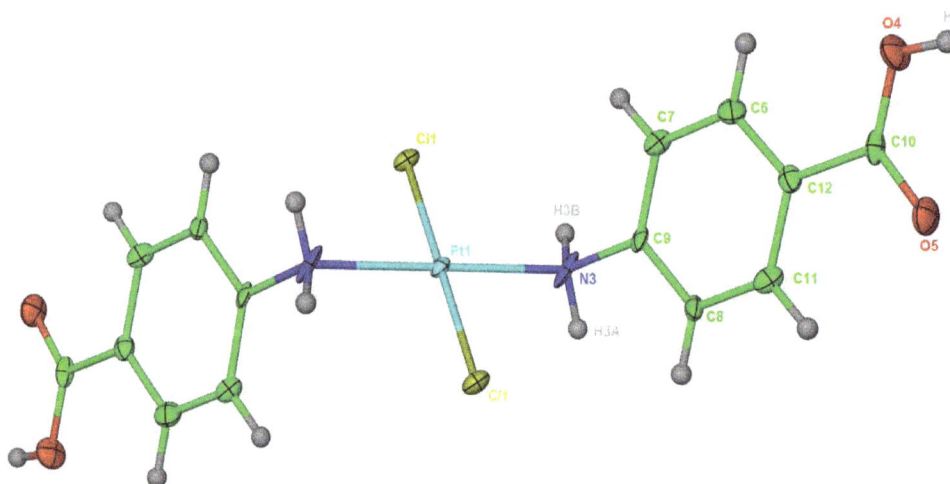

Figure 1. Molecular structure of *trans*-$[PtCl_2(NH_2C_6H_4CO_2H)_2]$ 2.

atom, a coordinated chloride ion and a 4-aminobenzoic acid molecule that is coordinated to the platinum centre through the nitrogen atom. The partial-occupancy platinum atom resides on an inversion centre, which serves to generate the remainder of the molecule. The molecular structure of 2 is shown in Figure 1. The

structural analysis revealed that the platinum centre has the expected distorted square planar geometry (Table 3), with *cis* angles of 87.4(3) and 92.6(3)°. The Pt–N and Pt–Cl bond distances (2.057(10) and 2.293(2) Å respectively) lie within the expected ranges (Orpen et al., 1989). The only previous structurally characterised

Table 3. Selected bond lengths (Å) and angles (°) for 2.

Pt(1)–N(3)	2.057(10)
Pt(1)–Cl(1)	2.293(2)
O(4)–C(10)	1.275(15)
O(5)–C(10)	1.263(15)
N(3)–C(9)	1.435(14)
N(3)–Pt(1)–Cl(1)	87.4(3)
N(3)–Pt(1)–Cl(1)'	92.6(3)
C(9)–N(3)–Pt(1)	115.4(6)

Primed atoms generated by the symmetry operation −x + 2, −y + 2, −z.

Figure 2. Zigzag hydrogen-bonded tapes in the supramolecular structure of *trans*-[PtCl2(NH2C6H4CO2H)2] 2.

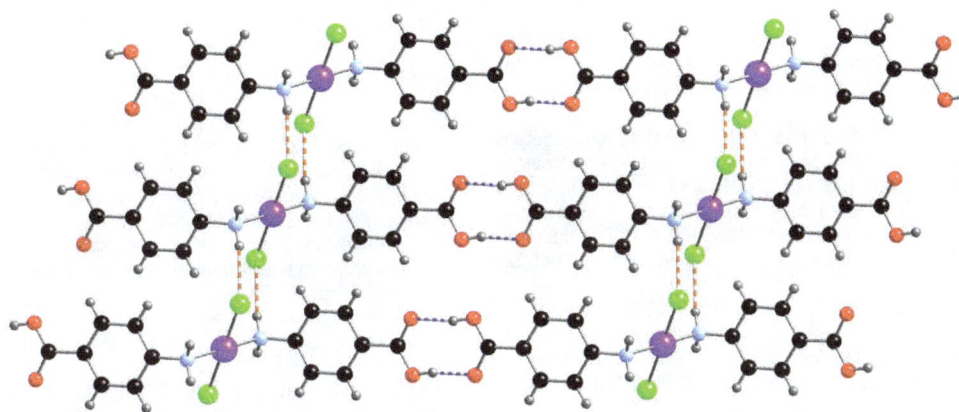

Figure 3. Hydrogen-bonded sheets present in the supramolecular structure of *trans*-[PtCl2(NH2C6H4CO2H)2] 2.

example of a dichlorobis(phenylamine)platinum complex contains 2-(3-aminophenyl)-4,4,5,5-tetramethyl-1,3,2-dioxaborolane (Vogels et al., 1999), and this has a very similar coordination geometry, with identical bond lengths and angles [Pt–Cl 2.290 Å, Pt–N 2.053 Å, N–Pt–Cl 88.1, 91.9°].

Each carboxylic acid group in 2 forms a hydrogen bonded dimer with the equivalent group on a neighbouring molecule with the parameters O(4)···O(5) 2.612 Å, H(4)···O(5) 1.78 Å, O(4)–H(4)···O(5) 170° [O(5) generated by the symmetry operation −x, −y + 1, −z + 1]. The carboxylic acid dimer, described by the graph set $R_2^2(8)$ (Bernstein et al., 1995), is a common motif in crystal engineering, where there is an absence of

competing donors and acceptors (Allen et al., 1999; Ayi et al., 2001; Burrows, 2004), and in the case of 2, these hydrogen bonds link the molecules into zigzag tapes, as shown in Figure 2. The zigzag nature of the tapes arises from the platinum coordination plane and the aminobenzoic acid ligand being non-coplanar, as evidenced by 107° angle between the mean plane containing Pt(1), Cl(1) and N(3), and that containing N(3), C(6)-C(12), O(4) and O(5).

Additional hydrogen bonds between the coordinated amine groups and one of the chloride ligands link the chains into a two-dimensional network, shown in Figure 3. The N–H···Cl hydrogen bonds occur pairwise, generating $R_2^2(8)$ rings with the hydrogen bonding parameters N(3)···Cl(1) 3.339 Å, H(3A)···Cl(1) 2.43 Å, N(3)–H(3A)···Cl(1) 172° [Cl(1) generated by the symmetry operation *x*, *y* − 1, *z*]. These parameters are within the expected range for N–H···Cl interactions when the chloride is coordinated to a metal (Brammer, 2003; Brammer et al., 2001). There are no significant interactions between the sheets, and neither the amino hydrogen atom H(3B) nor the chlorine atom Cl(2) is involved in hydrogen bond formation.

In conclusion, a new platinum(II) complex of aminobenzoic acid, *trans*-[PtCl$_2$(NH$_2$C$_6$H$_4$CO$_2$H-4)$_2$] 2, has been prepared under mild hydrothermal conditions and structurally characterized. A study of 2 as a metalloligand in the formation of mixed-metal MOFs is currently underway.

ACKNOWLEDGEMENT

The Commonwealth Scholarship Commission is thanked for the award of a Commonwealth fellowship to AAA.

REFERENCES

Allen FH, Motherwell WDS, Raithby PR, Shields GP, Taylor R (1999). Systematic analysis of probabilities of formation of bimolecular hydrogen-bonded ring motifs in organic crystal structures. New J. Chem. 23:25-34.

Ayi AA, Choudhury A, Natarajan S, Rao CNR (2001). Cyclic acetate dimmers formed by C-H...O hydrogen bonds in an open-framework zinc phosphate-acetate. New J. Chem. 25:213-215.

Bennett MV, Beauvais LG, Shores MP, Long JR (2001). Expanded Prussian Blue Analogues Incorporating [Re6Se8(CN)6]$^{3-/4-}$ Clusters: Adjusting Porosity via Charge Balance J. Am. Chem. Soc. 123:8022-8032.

Bernstein J, Davis RE, Shimoni L, Chang NL (1995). Patterns in hydrogen bonding: Functionality and graph set analysis in crystals. Angew. Chem. Int. Ed. Engl. 34:1555-1573.

Brammer L (2003). Metals and hydrogen bonds. Dalton Trans. 3145–3157.

Brammer L, Bruton EA, Sherwood P (2001). Understanding the Behavior of Halogens as Hydrogen Bond Acceptors. Cryst. Growth Des. 1:277-290.

Buffin B, Kundu A (2003). Synthesis, characterization, and crystal structure of platinum(II) and palladium(II) chlorides with an acidic α-diimine ligand Inorg. Chem. Commun. 6:680-684.

Burrows AD (2004). Crystal Engineering Using Hydrogen Bonds. Struct. Bonding 108:55-96.

Burrows AD, Cassar K, Mahon MF, Warren JE (2007). The stepwise formation of mixed-metal coordination networks using complexes of 3-cyanoacetylacetonate. Dalton Trans. 2499-2509.

Burrows AD, Mahon MF, Wong CTF (2008). Complexes as metalloligands in network formation: synthesis and characterization of a mixed metal coordination network containing palladium and zinc. Cryst. Eng. Comm. 10:487-489.

Chae HK, Siberio-Pérez DY, Kim J, Go Y, Eddaoudi M, Matzger AJ, O'Keeffe M, Yaghi OM (2004). A route to high surface area, porosity and inclusion of large molecules in crystals Nature. 427:523-527.

Chen B, Fronczek FR, Maverick A (2004). Porous Cu-Cd mixed metal-organic Frameworks constructed from Cu(Pyac)$_2${Bis[3-(4-pyridyl)pentane-2,4-dionato]copper(II)}Inorg. Chem. 43:8209-8211.

Czaja AU, Trukhan N, Müller U (2009). Industrial applications of metal-organic frameworks. Chem. Soc. Rev. 38:1284-1293.

Férey G (2008). Hybrid porous solids: past, present, future. Chem. Soc. Rev. 37:191-214.

Gagnon KJ, Perry HP, Clearfield A (2012). Conventional and Unconventional Metal-Organic Frameworks Based on Phosphonate Ligands: MOFs and UMOFs. Chem. Rev. 112:1034-1054

Garibay SJ, Stork JR, Cohen SM (2009). The Use of Metalloligands in Metal-Organic Frameworks. Prog. Inorg. Chem. 56: 335-378.

Halper SR, Do L, Stork JR, Cohen SM (2006). Topological Control in Heterometallic Metal-Organic Frameworks by Anion Templating and Metalloligand Design. J. Am. Chem. Soc. 128:15255-15268.

Hupp JT, Poeppelmeier KR (2005). Better Living through Nanopore Chemistry. Science 309:2008-2009.

Kitagawa S, Kitaura R, Noro SI (2004). Functional Porous Coordination Polymers. Angew. Chem. Int. Ed. 43:2334-2375.

Kitaura R, Onoyama G, Sakamoto H, Matsuda R, Noro SI, Kitagawa S (2004). Immobilization of a Metallo Schiff Base into a Microporous Coordination Plolymer. Angew. Chem. Int. Ed. 43:2684-2687.

Noro S, Kitagawa S, Yamashita M, Wada T (2002). New microporous coordination polymer affording guest-coordination sites at channel walls. Chem. Commun. pp. 222-223.

Orpen A, Brammer L, Allen F, Kennard O, Watson DG, Taylor R (1989). Supplement. Tables of bond lengths determined by X-ray and neutron diffraction. Part 2. Organometallic compounds and co-ordination complexes of the d- and f-block metals. J. Chem. Soc. Dalton Trans. S1-S83.

Robson R (2008). Design and its limitations in the construction of bi- and poly-nuclear coordination complexes and coordination polymers(aka MOFs): a personal view. Dalton Trans. 5113-5131.

Rosi NL, Eckert J, Eddaoudi M, Vodak DT, Kim J, O'Keeffe M, Yaghi OM (2003). Hydrogen storage in microporous Metal-Organic Frameworks. Science 300:1127-1129.

Rowsell J, Yaghi O (2004). Metal-organic frameworks: a new class of porous materials. Micropor. Mesopor. Mat. 73:3-14.

Vogels CM, Wellwood HL, Biradha K, Zaworotko MJ, Westcott SA (1999). Reactions of aminoboron compounds with palladium and platinum complexes. Can. J. Chem. 77:1196-1207.

Vreshch V, Lysenko A, Chernega A, Howard J, Krautscheid H, Sieler J, Domasevitch K (2004). Extended coordination frameworks incorporating heterometallic squares. Dalton Trans. 2899-2903.

Wu CD, Hu A, Zhang L, Lin WB (2005). A Homochiral Porous Metal–Organic Framework for Highly Enantioselective Heterogeneous Asymmetric Catalysis. J. Am. Chem. Soc. 127:8940-8941.

Wu CD, Ngo HL, Lin WB (2004). Luminescent homochiral silver(I) lamellar coordination networks built from helical chains. Chem. Commun. pp. 1588-1589.

Zhang Y, Chen B, Fronczek FR, Maverick AW (2008). A nanoporous Ag-Fe mixed metal-organic framework exhibiting single-crystal-to-single-crystal transformation upon guest exchange. Inorg. Chem. 47:4433-4435.

7

Simultaneous ultraviolet-visible (UV–VIS) spectrophotometric quantitative determination of Pb, Hg, Cd, As and Ni ions in aqueous solutions using cyanidin as a chromogenic reagent

C. O. B. Okoye, A. M. Chukwuneke, N. R. Ekere* and J. N. Ihedioha

Department of Pure and Industrial Chemistry, University of Nigeria, Nsukka. Nigeria.

The use of cyanidin (3, 3^1 4^1 5, 7 – pentahydroxyflavylium chloride) extracted from a tropical plant *Hibiscus sabradiffa* L. as a chelating reagent for simultaneous spectrophotometric determinations of ions of Pb, Hg, Cd, As and Ni in mixed aqueous solution is reported. The purified extract of the dried calyces of the Roselle plant was characterized and the results compared with literature values. Complexation of the metals ions with cyanidin and scanning through 200-700 nm in a UV –VIS spectrophotometer gave the wavelength of maximum absorption (λ_{max}) of these metal complexes to be 389.6, 360.0, 357.8, 396.8 and 401.0 nm for Pb(II); As(II); Cd(II); Hg(II) and Ni(II) complexes respectively indicating appreciable bathochromic shifts compared to pure cyanidin absorption at 283.2 nm. The effect of pH on the determinations was studied and a pH value of 5 was found to be the optimal. Calibration curve plots of the complexes showed linearity between concentration of 0.1 to 5.0 ppm. This method offer cheap, simple, rapid, sensitive, and eco-friendly technique for simultaneous determination of trace heavy metals in mixed aqueous solutions and have potentials for environmental and biological samples.

Key words: Cyanidin, ultraviolet (UV) spectroscopy, simultaneous, chromogenic, eco friendly, trace metals.

INTRODUCTION

Determination of trace metals is of interest because while some are essential nutrients some others are toxic. Metals like zinc, manganese, copper, chromium, iron and cobalt are essential trace elements for humans, animals and plants; but become toxic if the homeostatic mechanisms maintaining their physiological limit are disrupted. On the other hand, lead, cadmium, nickel, arsenic and mercury etc are toxic even at low levels. The need to estimate the levels of these metals in materials or samples have increased tremendously after reports on different roles they play in human health and disease. Numerous analytical techniques exist in literature for use in trace heavy metals assay in mixed solutions. Among

the methods, atomic absorption spectrophotometry (AAS) is well recognized because of its attractive features such as sensitivity, reliability, versatility, accuracy and precision (Okoye, 2005; Khamms et al., 2009). However, the use of AAS is limited by the need for high technical skills, huge capital and maintenance cost (Strong and Martin, 1990). Furthermore determinations of metals like arsenic and mercury pose health risk to the analyst using AAS.

Some other methods employed in simultaneous determination of metals include polarography, voltammetry, inductively coupled plasma-mass spectroscopy (ICP–MS), inductively coupled plasma-atomic emission spectroscopy (ICP–AES), liquid chromatography (LC) etc (Rouhollahi et al., 2007). Like AAS, these techniques require very costly equipment and reagents that are not easily available in poor nations of the world coupled with very specialized skills needed.

*Corresponding author. E-mail: nwachukwuekere2006@yahoo.com.

Also, most of the reagents are not eco friendly. The use of UV–VIS spectrophotometry in determination of heavy metals in samples is becoming popular in many laboratories because it provides for easy, simple and rapid determination in low to high concentrations at cheap cost (Soomro et al., 2008). Simultaneous determination of metal ions using UV–VIS methods was reported to be difficult without separation due to overlap absorption spectra (Nai-Liang et al., 2005). The problems associated with spectroscopic determination of these trace heavy metals are complicated by the nature of the photochromic or chromogenic ligands which are mainly synthetic and are toxic in the environment. Some of the reported simultaneous determination of metals using UV spectrophotometry were carried out using toxic and expensive reagent (Rouhollahi et al., 2007; Kachbi, 2010; Naguraja et al., 2009). Other works in literature use specific ligands for specific metals at even different pH values for multi-elements determinations.

Cyanidin (3, 3,[1] 4[1] 5, 7 – pentahydroxyflavylium ion) is found in an intensely coloured flowers. It is a water soluble pigment which has been widely used as a food colourant, titrimetric indicator and in disease treatment (Meiers, 2001; Wang and Jiao, 2000; Francis, 1989; Kong et al., 2003; David, 1998; Odigwe et al., 2003; Aoshima et al., 2007). The potentials of cyanidin as chromogenc ligand for metal complexes have been reported by Ukwueze et al. (2009).

This study reports the simultaneous determination of some trace heavy metals by UV-VIS spectrophotometry using cyaniding as a chelating reagent. We relied on the shifts of maximum absorption wavelength (λ_{max}) of cyanidin after complexation with the metals. The objective of the work was to develop a rapid, sensitive, specific and simple method for simultaneous determination of metals in environmental and industrial samples.

MATERIALS AND METHODS

Equipment and reagents

Salts of Pb, Cd, Hg, Ni and oxide of Arsenic were purchased from Rie – del –de Haen, Germany in analytical grade. Jenway (6405-model) Spectrophotometer and Jenway (3015-model) pH meter were used for absorbance and pH measurement respectively. The chemicals used to prepare buffer solutions were also of analytical grade and deionized water was used. All glass wares were first washed with detergents and cupiouslyrinsed with deionized water.

Cyanidin extraction

Cyanidin was extracted from calyces of Roselle plant (*Hibiscus sabdariffa* L.) according to the method of Ukwueze et al. (2009). A 500 g of dry calyces were ground to powder and macerated in 2.5 L methanol: HCl mixture (85:15% v/v) for 72 h and filtered. The filtrate was concentrated to 500 and 100 ml of conc. HCl was added and

content was refluxed for 2 h. The solution was then put in a beaker and cooled in a refrigerator until crystals settled out. The crystals were filtered out under suction and re-crystallized from hot methanol, air dried and weighed.

Preparation of buffer solutions

Buffer solutions of pH 1 to 8 were prepared in accordance with methods described earlier using KCl, HCl, $KHC_8H_4O_4$, KH_2PO_4 in varying concentrations and mixings (Lange, 1973; Meities, 1963).

Determination of λ_{max} of cyanidin

5% cyanidin solution was prepared by dissolving 5 g of the purified crystals in methanol containing 0.01% conc. HCl and made up to 100 cm^3 in a standard flask. 1 cm^3 of this solution was diluted to 10 cm^3 and its λ_{max} was determined by scanning from 200-700 nm using Jenway (6405 model) spectrophotometer in a 1 cm^3 cuvette.

Preparation of stock and working solutions of metals

Stock solutions (1000 ppm) of Pb, Cd, Hg, Ni, and As were prepared from lead nitrate, cadmium, mercury and nickel chlorides and arsenic oxide. The solutions were serially diluted to the required working standards. 2.5 cm^3 of each stock solution was diluted in a 250 cm^3 standard flask to give 10 ppm solution.

Determination of wavelength of maximum absorption (λ_{max}) of metal – cyanidin complexes

5 cm^3 metal solutions (10 ppm) were diluted with 5 cm^3 of the 5% cyanidin solution and the wavelength of absorption was scanned from 200 to 700 nm in a 1 cm^3 cuvette using the spectrophotometer.

Determination of optimum pH (pH $_{opt}$) of metal-cyanidin complexes

Each of the 8 beakers containing 5 cm^3 of 5% cyanidin in methanol were added 5 cm^3 of standard solutions of one of the studied metals. 50 cm^3 of the solution in each beaker was adjusted to a given pH from 1 to 8 using the buffer solutions prepared. The absorbance of each solution was read at the λ_{max} of the analyte metal at the adjusted pH.

Simultaneous determination of various metal – cyanidin complexes at pH$_{(opt)}$

Working standard solutions from 1.0 - 9.0 ppm of Pb(II); Cd(II); Hg(II), As(III) and Ni(II) solutions were prepared from the 10 ppm solution of each metal by diluting appropriately. These were used to plot a calibration graph for each metal-cyanidin complex by determining the absorbance at λ_{max} of 5 cm^3 of standard solution diluted by 5 cm^3 of 5% cyanidin solution in a 1 cm^3 cuvette at pH$_{(opt)}$ determined.

5 cm^3 of 1.0 ppm mixed standard solution was diluted with 5 cm^3 of 5% cyanidin chloride solution adjusted to pH$_{opt}$ and the absorbance read at the λmax of each metal – cyanidin complex. Similar procedures were repeated for 2-9 ppm mixed metal solutions.

Table 1. R_f values (X 100) of TLC and PC of cyaniding in various solvents.

Variable	TLC			PC		
Solvent system	Forestal	2MHCl	BAW	Forestal	2MHCl	BAW
Literature values	48	2.5	52	47	2.5	52
Experimental value	45	5	55	49	3	50

Table 2. Spectral results of Cyanidin extract compared with literature values.

Variable	UV region (λmax)	VIS region (λmax)
Literature values	282.9 nm	530.0 nm
Experimental value	283.2 nm	530.2 nm

Table 3. Effect of metal ions on wavelength of absorption of cyaniding.

Wavelength	Cyanidin (Cy)	Pb(II)-Cy	Hg(II)-Cy	Cd(II)-Cy	As(III)-Cy	Ni(II)-Cy	Pb(II) shift	Hg(II)shift	Cd(II)shift	As(III)shift	Ni(II) shift
UV	283.2	389.6	396.8	357.8	360.0	401	106.4	113.6	74.6	76.8	117.8
VIS	530.2	536.6	538.6	535.6	533.8	555.6	6.4	8.4	5.4	3.6	5.4

RESULTS AND DISCUSSION

Cyanidin extract characterization

The extracted cyanidin was found to conform to the characteristics earlier reported by Ukwueze and others (2009). Table 1 shows the R_f values of thin layer and paper chromatographs (TLC and PC) obtained in various solvents compared to literature values. Both experimental and literature R_f values are very close.

The spectral data obtained from the analysis of the extract in comparison with literature values are shown in Table 2. The R_f values and $\lambda_{max's}$ of the extract showed no significant variation from the authentic literature values (Harborne, 1958).

Effects of the metal ions on the absorption wavelength of cyanidin

The result of the investigation of the effects of the metal ions on the λ_{max} of cyanidin is shown in Table 3. From the table, all the metals caused a marked bathochromic shift in cyanidin absorbance wavelength in the UV region. Shifts in VIS region are very small causing an overlap in absorbances when plotted. The simultaneous determination of these metals can be measured with satisfaction in the UV region. The affinity of metal ions for ligands is controlled by size, charge and electronegativity. These metals are highly polarisable, have lower charge density, large ionic size and their d-orbitals are available

for π- bonding and due to these, they form covalent complexes showing absorption in the UV region.

Selection of pH optimum at λmax of metal ions

The effect of pH on the formation of metal-cyaniding complexes was studied at different pH values in methanol and the result shown in Figure 1.

From the graph, Cd – cyanidin complex has pH = 3 as the optimum pH; Pb – cyanidin has pH = 4; Hg – cyanidin has pH = 5; As – cyanidin has pH = 2 and Ni – cyanidin has pH = 6. At these various pH values, these metal–cyaniding complexes showed reproducible results. However, all the metal complexes showed prominent absorbances at pH = 5. As a midway or compromise between selectivity and sensitivity of the developed method, pH value of 5 was chosen as the optimum pH for the simultaneous determination of these metals complexes with cyanidin in aqueous media.

Simultaneous determination of the metal ions

At the optimal pH value, the UV spectra of the mixed metal ions – cyanidin complexes was studied for possible spectral overlaps or interferences. The results are shown in Table 4.

From the table, only Pb(II), Hg(II) and As(III) were absorbed at their original λ_{max}. It was noted that Cd(II) showed no peaks which could be attributed to the closeness of its λ_{max} value to As(III). The absorption

Table 4. λmax for simultaneous determination of mixed metal – cyanidin complexes.

Λmax (nm)	Pb(II)- Cy	Hg(II)- Cy	Cd(II)- Cy	As(III)- Cy	Ni(II)-Cy
UV	389.6	396.0	—	360.0	496.4
VIS	—	—	—	—	—

Cy = cyaniding.

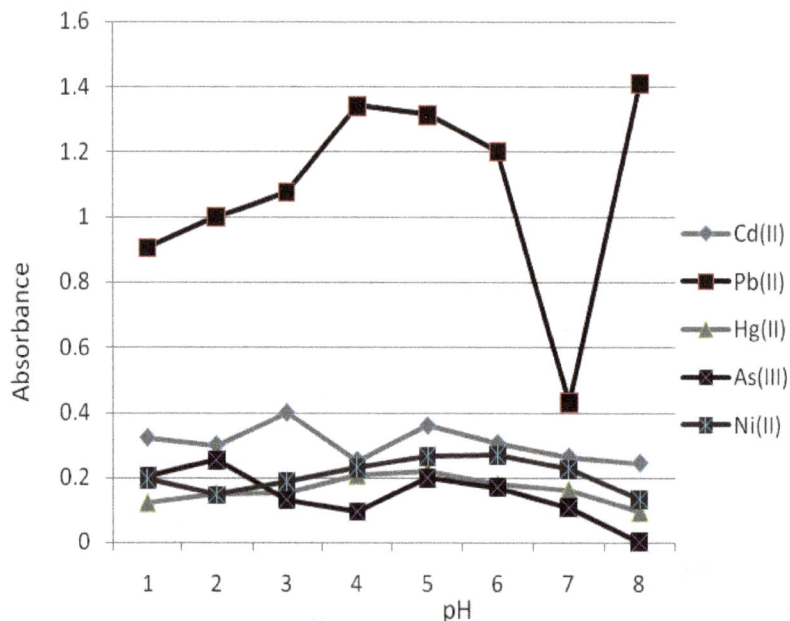

Figure 1. pH variation at λmax of absorption by metal-cyanidin complexes.

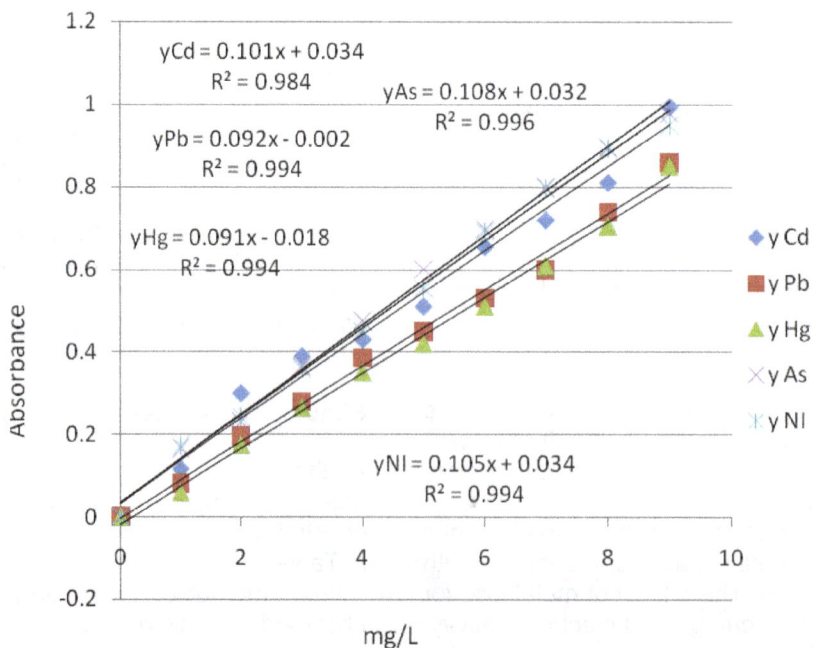

Figure 2. Mixed graph of Abs vs. Conc.Of the metal complexes.

intensity of As(III) was greatly enhanced in the spectra. Ni(II) absorbed at higher wavelength. Such interferences might result from the formation of mixed oxides or from suppression of ionized gaseous metals pressures which occur in the molecular environment adjacent to the chromophore.

Linear detection range of metal – cyanidin complexes

Absorbance of the five complexes at various concentrations at various $\lambda_{max's}$ and optimum pH (pH = 5) were investigated and the result was shown in Figure 2. The calibration curves obtained showed linearity between concentration ranges of 1.0-10.0 ppm. At this range, Beer-Lambart law was obeyed.

Conclusion

Complexation of cyanidin simultaneously with Pb, Hg, Cd, As and Ni ions markedly altered the wavelength of absorption of the cyanidin and this phenomenon can be utilized in spectrophotometric determination of these metal ions. The complexes were stable and yielded reproducible results at pH = 5. It was found that all the metal ions determined show linearity at concentration range of 1.0 to 10.0 ppm indicating that at these concentrations, UV spectroscopy can conveniently be used for their simultaneous determination.

This offers a simple, rapid, sensitive and cheap method that promotes the spirit of green chemistry in chemical quantitative analysis. It has potentials in evaluation of trace heavy metals in environmental, biological and food samples.

REFERENCES

Aoshima H, Hirtu S, Ayile S (2007). Antioxidative and antihydrogen peroxide activities of various herbal tea. Food Chem. 103:617- 622.

David D (1998). Acid – Base indicators, Available at Fill/2A: indicators: htm. P. 2.

Francis F (1989). Food colorants: Anthocyanins. Critical rev. Food Sci. Nutr. 28:273-314.

Harborne JB (1958). Spectral methods of characterizing Antocyanidins. Biochem. J. 70:22-28.

Kachbi A, Benamor M, Aguerssif N (2010). Simultaneous Spectrophotometric Deterimation of Co^{2+}, Ni^{2+} and Cu^{2+} in Industrial Alloys Using Partial Least Squares Method. Curr. Anal. Chem. 6:88-93.

Khamms AA, Al-Ayash AS, Jasin F (2009). Indirect electrothermal atomization AAS Spectrometric determination of drugs, desfeiroxamine in some pharmaceutical preparations using Vanadium (V) as a mediatory element elestial. J. Anal. Chem. 3:257-269.

Kong JM, Chia IS, Goh NK, Chia TF, Brou HR (2003). Analysis of and biological activities of Anthocyanins. Photochemistry 64:922-933.

Lange AA (1973). Handbook of Chemistry. 11th ed. McGraw hill. New York. pp. 70-75.

Meiers S (2001). The Anthocyanidins – cyandin and Delphinidin are potent inhibitors of epidermal growth factor reception. J. Agric. Food Chem. 49:958-962.

Meities L (1963). Handbook of Analytical Chemistry. 1st ed. McGraw Hill, New York. pp. 93-97.

Naguraja P, Al-Tayar SI, Gowda A (2009). Rapid and sensitive spectrophotometric method for the determination of trace amount of thalium(ii) in water and urine samples by new oxidative coupling reaction. J. MexChem Soc. 53(4):201-208.

Nai-Liang H, Hong-wen G, Biao ZI, Guo-Qing Z (2005). Simultaneous Determinations of Cobalt and Nickel in waste water with 2.- Hydroxyl – 5 – benzene azoformoamithiozone by spectral correction Technique. J. Chin. Chem. Soc. 52:1145-1152.

Odigwe IP, Ettarh RR, Adigu A (2003). Chronic administration of aqueous extract of Hibiscus sabradiffa alternatives for Hypertension and reverse Cardiahypertrophy in 2K – IC hypertensive rate. J. Ethnopharmcol. 86:181-185.

Okoye COB (2005). Spectroscopic methods of analysis.Undergraduate Analytical Chemistry.Jolyn Publishers, Nsukka pp. 98-119.

Rouhollahi NS, Ghasem J, Tavakol H, Noroozi M, Hashemi M (2007). Simultaneous spectrophotometric determination of Heavy metal ions using several chemometric methods: Effective of different parameters of Savitzky–Golany and Direct Orthogaonal Signal Correction Filters. Iran J. Chem. Eng. 26(2):41-51.

Soomro R, Jamahiddin MA, Menpou N, Khan H (2009). A simple and selective spectrophotometric method for the determination of trace Gold on real Environmental, Biological, Geological and Soil samples using Bis, (Salicylaldehyde) Orthphenyldiamine. J. Anal. Chem. Insights 3:75-90.

Strong FC, Martin NJ (1990). Rapid determination of zinc and iron in food by flow – injection analysis with flame atomic – absorption spectrophotometry and slurry nebulization. Talanta 7:11-718.

Ukwueze NN, Nwadinigwe AC, Okoye COB, Okoye FBC (2009). Potentials of 3 $3^{1}4^{1}5$ 7 – pentahydroxyflavylium Hibiscus rosa – sinesis L. (malvacea) flowers as ligands in the quantitative determination of Pb, Cd and Cr. Int. J. Phys. 4(2):58-62.

Wang SY, Jiao H (2000). Scavenging Capacity of berry crops on super oxide Radicals, H_2O_2 and OH and Singlet oxygen. Agric. Food Chem. 48:677-684.

8

Mechanical properties of Rowan wood impregnated with various chemical materials

Hakan Keskin[1], Neslihan Süzer Ertürk[2], Mustafa Hilmi Çolakoğlu[3] and Süleyman Korkut[4]*

[1]Department of Woodworking Industry Engineering, Faculty of Technology, Gazi University, 06500 Beşevler, Ankara, Turkey.
[2]Department of Industrial Technology, Faculty of Industrial Arts Education, Gazi University, 06830 Gölbaşi, Ankara, Turkey.
[3]Technology Development Foundation of Turkey (TTGV) 06800 Bilkent, Ankara, Turkey.
[4]Department of Forest Industrial Engineering, Faculty of Forestry, Düzce University, 81620 Düzce, Turkey.

The aim of this research was to determine some of the mechanical properties of Rowan wood impregnated with various impregnation materials. For this purpose, the test samples were prepared from Rowan (*Sorbus aucuparia* Lipsky) wood materials that are of common use in the forest products industry of Turkey according to TS 345. The test samples were treated with Tanalith-E, Vacsol Azure, Imersol-Aqua, and Boron compounds (Borax and Boric acid) by the vacuum impregnation technique in accordance with ASTM-D 1413-76 standards and directives of the manufacturers. After impregnation, each sample was tested for observation of amount of retention, compression strengths, bending strengths, and modulus of elasticity in bending. As a result, compression strength was the highest with Boric acid impregnation, bending strength with Imersol-Aqua impregnation and modulus of elasticity in bending with Tanalith-E impregnation. The lowest value in compression strength was measured with Tanalith-E, bending strength in Borax and modulus of elasticity in bending with Vacsol Azure. So, the impregnations with Tanalith-E have advantages in compression and bending strengths and modulus of elasticity in bending of Rowan wood.

Key words: Rowan wood, chemical materials, compression strength, bending strength, modulus of elasticity in bending.

INTRODUCTION

The durability of wood is affected by several factors. If the wood materials are used without processing by preservative chemicals (with regard to the area of usage), fungal stains, insect infestation, humidity, fire etc damage the wood. As a result of these damages, the woods require to be repaired, maintained or replaced before its economic life ends (Richardson, 1987). For this reason, in most places the wood materials should be impregnated with the proper chemicals. In the case of the wood not been impregnated but only painted and varnished instead, the prevention on the surfaces is limited to a maximum of two years (Evans et al., 1992).

It is reported that, in mines, as a result of the impregnation of the beech and spruce wood with water-soluble salts, the bending, tensile, and impact strength decreased a little whereas compression strength increased (Kollmann, 1959). In another research concerning the impregnation of pine, spruce, fir, beech, and poplar woods with Antrasen, it was found that, the compression strength increased by 6 to 40% and bending strength increased by 10 to 22% (Stabnikov, 1957). In the impregnation of pine and beech wood with tar oil and UA salts, the tar oil increased compression strength by 10% and UA salts increased with a small rate. On the other hand, the tar oil increased the bending strength whereas the UA salts diminished the bending strength (Gillwald, 1961).

Vologdin (1966) declared that, among the materials

---

*Corresponding author. E-mail: suleymankorkut@hotmail.com.

used for the impregnation of pine; sodium pentaclorfenet, copper sulphate, and sodium fluoride increased the compression strength, respectively by 95, 25, and 3% whereas zinc chloride decreased by 9%. Sodium pentaclorfenet also increased the bending strength. In another study, pressure treatment caused a decrease of 8 to 10% in the bending strength of different wood types (Isaacs, 1972). It was assessed that, salty impregnation materials increased the compression strength by 4.6 to 9.6%, whereas decreased the bending strength by 2.9 to 16% (Wazny, 1973). In another study, chromate copper arsenate (CCA) and arsenate copper arsenate (ACA) salts did not caused any significant impact on modulus of elasticity in bending (Bendtsen, 1984).

After the impregnation of Scotch pine wood samples by hot-cold open tank method with eleven preservatives, no significant difference was observed in the bending strength except the decreasing effects of fluotox containing acid florid (Lutomsky, 1976).

Korkut et al. (2009) determined Rowan wood's air dry (801 kg/m^3), oven dry (737 kg/m^3) and basic (635 kg/m^3) densities. Fiber saturation point was calculated to be 23.79%; volumetric shrinkage and swelling were found as 15.048 and 18.465%; bending strength (MOR), modulus of elasticity (MOE), compression strength parallel to grain, impact bending, tensile strength parallel and perpendicular to grain, shear strength, cleavage strength, Janka hardness values (parallel and perpendicular to grain) and surface roughness (Ra) value were determined as 115.571 N/mm^2, 9843.857 N/mm^2, 55.027 N/mm^2, 14.849 J/cm^2, 120.71 N/mm^2, 6.187 N/mm^2, 12.792 N/mm^2, 0.941 N/mm^2, 1.416 and 1.159 kN, 7.239 μm, respectively.

Özçifci et al. (1999) carried out a research on yellow pine, oriental beech, ash-tree, and oak woods bleached with NaOH+H$_2$O, NaOH+Ca(OH)$_2$+H$_2$O$_2$, HClO, and HCl and varnished with synthetic, polyurethane, acid catalytic varnish to research the impact on surface adhesion strength and brightness. These chemicals have no impact on brightness: Wood type and varnish type was important for brightness. The highest brightness of surface was determined with acrylic varnish and the lowest with acid hardened varnish. HCl decreases adhesion strength of varnishes. Uysal et al. (1999) studied static bending strength of yellow pine, oriental beech, ash-tree and oak woods bleached with NaOH+H$_2$O$_2$, NaOH+Ca(OH)$_2$+H$_2$O$_2$, HClO and HCl: They determined that HClO and HCl causes the highest decrease in bending strength.

Aytaşkin (2009) studied the impact of impregnation materials on wood types and determined that the impregnation materials increase the density and heat conduction of woods. Impregnation materials also have impacts on the mechanical properties, decreases the bending strength and modulus of elasticity. Compression strength and impregnation relation depends on the type of impregnation material. Impregnation materials decrease the adhesion strength of woods and also the

weight loss in combustion tests. Kartal (1998) determined that, impregnation with CCA and drying at a temperature 70°C for 72 h have no impact on the mechanical properties of wood, but the differences between the control samples in bending test with fixation at 20 and 70°C was statistically meaningful at level of 5%.

One well-known technique to protect wood is by impregnating it with chemical products. In this study, naturally growing in Turkey, thin in texture, flexible and hard Rowan wood (Sorbus aucuparia Lipsky) was impregnated with Tanalith-E, Imersol Aqua, Vacsol Azure, and Boron compounds and tested in retention, compression strength, bending strength, and modulus of elasticity in bending.

MATERIALS AND METHODS

Wood materials

Test samples were obtained from Kastamonu Forestry Regional Directory, Küre Directory, Kösreli Department number 200. Test samples were cut from the trees in accordance with TS 4176 (1984) standard. Test samples were prepared in accordance with TS 2470 (1976) and TS 53 (1981). Accordingly, non-deficient, knotless, normally growth (without zone line, without reaction wood and without decay and insect mushroom damage) wood materials was selected. Test samples cut to 70 × 70 × 800 mm were air-dried at a temperature of 20 ± 2°C and 65 ± 3% relative humidity conditions up to reaching 12% humidity level.

Impregnation material

Tanalith-E

Tanalith-E is an impregnation material used against the attacks of agent, yeast, insect, and termite. It is a new generation of impregnation material consisting of copper and organic biocide (triazole) and not harmful to plant, animal, and human health. Tanalith-E, light green in color, odorless, pH 7, 1:04 g/cmP^3 density, smooth and completely water-soluble, water-based, non-corrosive to metal parts are available in the form of ready solution. Tanalith-E was applied to woods used in fences, railings, garden furniture, barns, silos, farm buildings, the wood used in children's play areas by vacuum - pressure method (Hickson's Timber Impregnation Co. (GB), 2000).

Vacsol Azure

Vacsol Azure a product of a new technology developed by using active ingredients, used in the process, ground wood materials on the level of fungi, insects (Propiconazole and tebuconazole), and termites (permethrin) to prevent decay by protecting against transparent impregnating agent. This solvent-based material is water-insoluble, pale yellow in color, flammable, density 0.806 g/cm3 at 20 ± 2°C, contains 64% of volatile organic compounds (VOC) (Hickson's Timber Impregnation Co. (GB), 2000).

Imersol-Aqua

Imersol-Agua, used as an impregnation material in this study was supplied from Hemel (Hemel-Hickson Timber Products Limited), Istanbul. Imersol-Aqua is non-flammable, odorless, fluent, water-based and completely soluble in water and non-corrosive material with a pH value of 7 and a density of 1.03 g cm^{-3}. It is available as a ready-made solution. It contains 0.5% w/w tebuconazole, 0.5% w/w propiconazole, 1% w/w 3-Iodo-2-propynl-butyl carbonate, and 0.5% w/w cypermethrin. Before the application of Imersol-Aqua on the wood material, all kinds of drilling, cutting, turning, and milling operations should be completed and the relative humidity should be in equilibrium with the test environment. In the impregnation process, dipping duration should be at least 6 min and the impregnation pool must contain at least 15 liters of impregnation material for 1m^3 of wood. The impregnated wood should be left for drying for at least 24 h at 20 ± 2°C and 65% ± 3 (Hickson's Timber Impregnation Co. (GB), 2000).

Determination of density

The densities of wood materials, used for the preparation of test samples were determined according to TS 2472 (1976). For determining the air-dry density, the test samples with a dimension of 20 × 30 x 30 mm were kept under the conditions of temperature at 20 ± 2°C and 65 ± 3% relative humidity until they reached to a stable weight. The weights were measured with an analytic scale of ± 0.01g sensitivity. Afterwards, the dimensions were measured with a digital compass of ± 0.01mm sensitivity. The air-dried densities (δ_{12}) of the samples were calculated by the formula:

$$\delta_{12} = \frac{W_{12}}{V_{12}} \ g \ cm^{-3} \tag{1}$$

Where, W_{12} is the air-dry weight (g) and V_{12} is the air-dry volume (cm^3).

The samples were kept at a temperature of 103 ± 2°C in the drying oven until they reached to a stable weight for the assessment of full dry density. Afterwards, the full dried samples were cooled in the desiccators containing phosphorus pentoxide (P$_2$O$_5$). Then, they were weighted on a scale of ± 0.01g sensitivity and their dimensions were measured with a digital compass of ± 0.01mm sensitivity. The volumes of the samples were determined by stereo metric method and the densities (δo) were calculated by the formula:

$$\delta o = \frac{Wo}{Vo} \ g \ cm^{-3} \tag{2}$$

Where, Wo is the full dry weight (g) and Vo is the full dry volume (cm^3).

Determination of humidity

The humidity of test samples before and after the impregnation process was determined according to TS 2471 (1976). Thus, the samples with a dimension of 20 × 20 x 20 mm were weighed and then oven dried at 103 ± 2°C till they reach to a constant weight. Then, the samples were cooled in desiccators containing phosphorus pentoxide (P$_2$O$_5$) and weighed with an analytic scale of 0.01 g sensitivity. The humidity of the samples (r) was calculated by the formula;

$$r = \frac{Mr - Mo}{Mo} x100 \tag{3}$$

Where, Mr is the initial weight (g) and Mo is the full-dry weight (g).

Preparation of the test samples

The rough drafts for the preparation test and control samples were cut from the sapwood parts of massive woods and conditioned at a temperature of 20 ± 2°C and 65 ± 3% relative humidity for three months until reaching an equilibrium in humidity distribution. The samples for compression strength test, with a dimension of 20 × 20 × 30 mm were cut from the drafts having an average humidity of 12% according to TS 2595 (1977). The samples for bending strength and modulus of elasticity in bending test, with a dimension of 20 × 20 × 400 mm were cut from the drafts having an average humidity of 12% according to TS EN 408 (2011). The densities and humidity values of all test samples were measured before the impregnation process.

The test samples were impregnated according to ASTM-D 1413-76 (1976), TS 344 (2012) and TS 345 (2012). Accordingly, test samples were subject to 60 cm HgP^{-1P} (vacuum) for 60 min and to diffusion at normal atmospheric pressure. Impregnated test samples were air-dried and then, weighed with 0.001g precision analytical balance and retention amount were calculated. The processes were carried out at 20 ± 2°C. Retention of impregnation material (R) was calculated by the formula:

$$R = \frac{G.C}{V} 10^3 \ kg \ m^{-3} \qquad G = T_2 - T_1 \tag{4}$$

Where, G is the amount of impregnation solution absorbed by the sample (g), T_2 is the sample weight after the impregnation (g), T_1 is the sample weight before the impregnation (g), C is the concentration (%) of the impregnation solution and V is the volume of the samples (cm^3).

Impregnated test samples were kept under a temperature of 20 ± 2°C and 65 ± 3% relative humidity until they reach a stable weight. Five types of impregnation materials (Tanalith-E, Imersol Aqua, Vacsol Azure, Boric acid and Borax), four test types (retention amount, compressive strength, flexural strength and modulus of elasticity) plus one control and ten replications for a total of 250 test samples were prepared.

Compression strength

The tests for compression strength parallel to grains of wood materials were carried out with Universal Testing Machine shown in Figure 1, according to TS 2595 (1977). The capacity of Universal Testing Machine was 400 N. The speed of testing machine was adjusted to 5 mm/min. for crashing to occur in 1 to 2 min. Compression strength was calculated by the formula:

$$\sigma_b = \frac{F_{max}}{ab} \ N.mm^{-2} \tag{5}$$

Where, F_{max} is the breaking load on the scale (N), a is the cross-sectional width of test sample (mm), b is the cross-sectional thickness of the test sample (mm).

Bending strength

The tests for bending strength were carried out with the Universal Testing Machine shown in Figure 2, according to TS 2474 (1976). The capacity of the Universal Testing Machine was 400 N. The speed of the testing machine was adjusted to 5 mm/min for breakage to occur in 1 to 2 min. Bending strength was calculated by the formula:

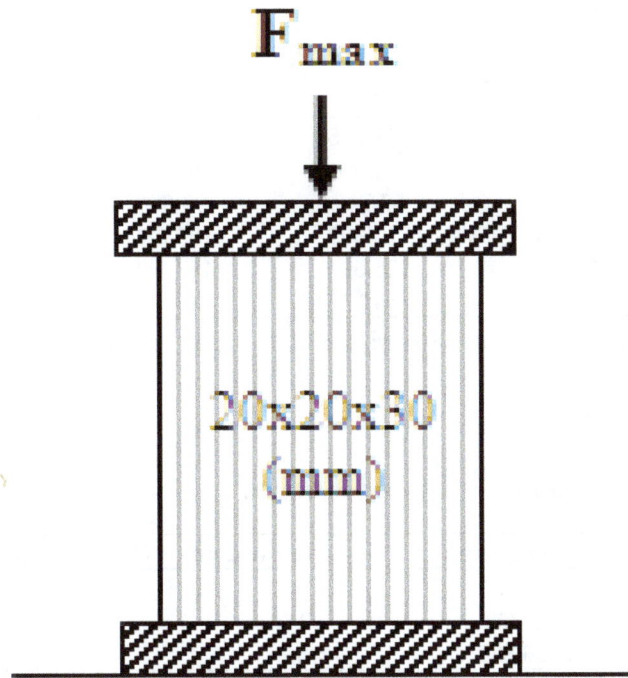

Figure 1. Universal testing machine for compression strength.

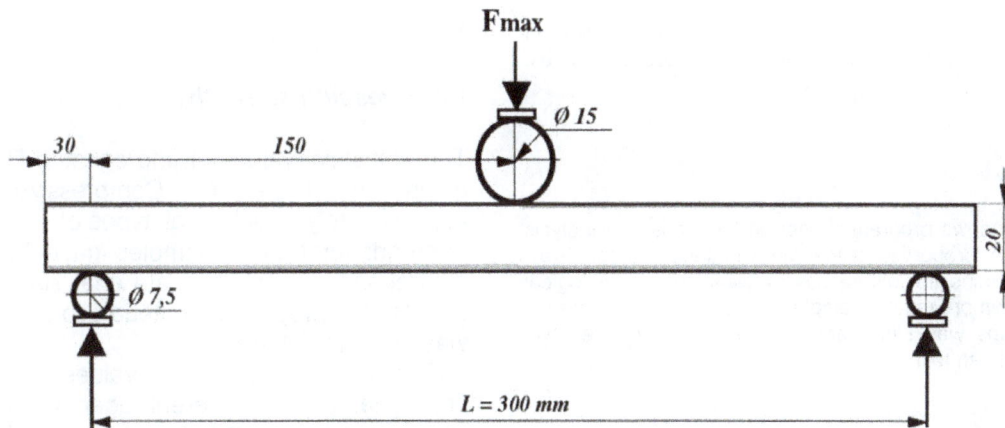

Figure 2. Test equipment for bending strength and MOE (dimensions in mm).

$$\sigma_e = \frac{3F_{max}(L-L_1)}{2bh^2} \ \text{(N mm}^{-2}\text{)} \tag{6}$$

Where, F_{max} is the breaking load on the scale (N), L is the distance between the lower tension rods (mm), L_1 is the distance between two loads (mm), b is the cross-sectional width of test sample (mm), and h is the cross-sectional thickness of the test sample (mm).

Modulus of elasticity (MOE) in bending

The tests for modulus of elasticity in bending were carried out with the Universal Testing Machine shown in Figure 2, according to TS 2478 (1976). The capacity of the Universal Testing Machine was 400 N. Deformations on the test samples were measured in the middle of the sample within a zone of five times the width of the sample by comparator. The deformations by incrementally increasing the forces were assessed with a sensitivity of ± 0.01mm. In the elastic deformation zone, modulus of elasticity was calculated by the formula:

$$MOE = \frac{\Delta F \cdot L^3}{4 \cdot b \cdot h^3 \cdot \Delta f} \ N\,mm^{-2} \tag{7}$$

Where, ΔF is the difference between the arithmetic average of the upper and lower limits of applied force in the elastic deformation zone (N), Δf is the net elastic deflection - difference between the measured elastic deflection in the upper and lower loading

Table 1. Retention amounts of Rowan wood (kg m^{-3}).

Statistical values	Impregnation materials				
	Te	Ia	Va	Ba	Bx
X (kg/m³)	104.083	127.045	151.044	64.887	86.393
Ss (kg/m³)	9.862	7.209	9.440	4.127	6.889
V (s²)	108.078	57.747	99.026	18.933	52.736
min (kg/m³)	89.648	110.608	138.602	58.423	70.734
max (kg/m³)	118.742	135.621	167.095	72.004	93.722
N	10	10	10	10	10

x: Arithmetic mean, v: Variance, Ss: Standard deviation, N: Number of samples, Te: Tanalith-E, Ia: Imersol Aqua, Va: Vacsol Azure, Ba: Boric acid, Bx: Borax.

Table 2. Variance Analysis Results of Amount of Retention (ANOVA).

Source of variance	Sum of squares	Degree of freedom	Mean of squares	F value	Sig.
Between groups	45478.129	4	11369.530	168.926*	0.000
Intra group	3028.703	45	67.304		
Total	48506.833	49			

*P < 0.05.

limits- (mm), L is the span (mm), b is the cross-sectional width of test sample (mm), h is the cross-sectional thickness of the test sample (mm).

Data analysis

SPSS 15.0 for Windows program is used in the statistical analysis of the technological properties of the wood material tested. F-test was used to determine the differences between the technological properties of the impregnated samples. In this case, difference between the groups was significant ($\alpha = 0.05$) confidence level compared with Duncan test.

RESULTS AND DISCUSSION

Retention quantities

The amount of retention by Rowan wood according to the impregnation period is shown in Table 1. Retention amounts varied according to the type of impregnation material. The amount of retention multivariate analysis of variance of the effects of type of impregnation materials are given in Table 2.

According to the F test for the quantities of wood preservatives retention, retention volumes showed statistically significant score differences according to the type of wood preservatives (F (4; 45) = 168 926, P < 0.05). Duncan test results were given to determine the severity of the differences between the two groups in Table 3.

Mechanical properties

Compression strength

The statistical values of impregnated Rowan wood were given in Table 4. Compressive strength was approximately equal for all types of impregnating agents. Impregnation-treated samples gave better results than control samples. Multivariate analysis of variance on the effects of impregnation materials to compression strength was given in Table 5.

Compression strength values of Rowan wood impregnated with different agents showed statistically important differences in F test given in Table 5. (F (5; 54) = 15.142, P < 0.05) Duncan test results to define the degree of importance between the groups and were given in Table 6. Compression strength was the highest with Boric acid followed by Vacsol Azure, Borax, Imersol Aqua, Tanalith-E, and control samples as shown in Figure 3.

Bending strength

Bending strength values of impregnated Rowan wood were given in Table 7. Bending strength values of Rowan wood impregnated with different impregnation materials showed statistically important differences in F test given in Table 8. (F (5; 54) = 10.404, P < 0.05). Duncan test results to define the degree of importance between the groups and were given in Table 9. The highest bending

Table 3. Duncan test results of amount of retention.

Group	N	For sub-groups α = 0.05				
		1	2	3	4	5
Boric acid	10	64.887				
Borax	10		86.393			
Tanalith-E	10			104.083		
Imersol-Aqua	10				127.045	
Vacsol Azure	10					151.044
Significant		1.000	1.000	10.071	1.000	1.000

The amount of retention according to the type impregnation material was highest with Vacsol Azure (151, 044 kg/m^3) followed by Imersol Aqua (127, 045 kg/m^3), Tanalith-E (104, 083 kg/m^3), Borax (86,393 kg/m^3), and Boric acid (64.887 kg/m^3).

Table 4. Statistical value of compression strength test.

Statistical values	Impregnated samples					
	Control	Te	Ia	Va	Ba	Bx
X (N/Mm^2)	53.012	54.622	55.576	57.829	61.902	56.244
Ss (N/Mm^2)	1.649	2.224	2.687	2.157	1.511	3.477
V (S^2)	3.021	5.496	8.024	5.170	2.538	13.433
Min (N/Mm^2)	50.230	50.120	52.340	54.360	60.020	50.230
Max (N/Mm^2)	55.040	58.620	59.190	61.350	64.320	60.320
N	10	10	10	10	10	10

x: Arithmetic mean, v: Variance, Ss: Standard deviation, N: Number of samples, Te: Tanalith-E, Ia: Imersol Aqua, Va: Vacsol Azure, Ba: Boric acid, Bx: Borax.

Table 5. Results of compression strength variance analysis (ANOVA).

Source of variance	Sum of squares	Degree of freedom	Sum of squares	F value	Sig.
Between groups	475.544	5	95.108	15.142*	0.000
Intra group	339.169	54	6.2809		
Total	814.714	59			

*$P < 0.05$.

Table 6. Duncan test results of compression test.

Groups	N	For subgroups α = 0.05			
		1	2	3	4
Control	10	53.012			
Tanalith E	10	54.622	54.622		
Imersol-Aqua	10		55.576	55.576	
Borax	10		56.244	56.244	
Vacsol Azure	10			57.829	
Boric acid	10				61.902
Significant		0.156	0.178	0.0616	1.000

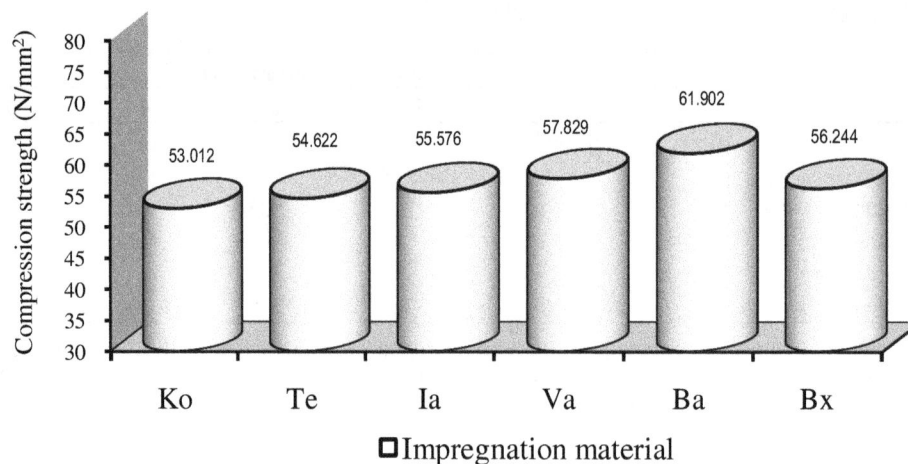

Figure 3. Compression strength for different impregnation materials.

Table 7. Statistical values of bending strength test.

Statistical values	Impregnated samples					
	Co	Te	Ia	Va	Ba	Bx
x (N/Mm^2)	88.331	94.707	98.177	87.941	95.623	85.956
Ss (N/Mm^2)	4.129	2.813	6.320	3.773	5.823	4.059
V (S^2)	18.947	8.797	44.388	15.822	37.680	18.310
Min (N/Mm^2)	83.033	90.102	92.356	80.359	90.234	80.243
Max (N/Mm^2)	96.269	99.625	114.628	94.206	106.335	91.273
N	10	10	10	10	10	10

x: Arithmetic mean, v: Variance, Ss: Standard deviation, N: Number of samples, Te: Tanalith-E, Ia: Imersol Aqua, Va: Vacsol Azure, Ba: Boric acid, Bx: Borax.

Table 8. Variance analysis (ANOVA) results of bending strength test.

Source of variance	Sum of squares	Degree of freedom	Sum of squares	F value	Sig.
Between groups	1248.024	5	249.604	10.404*	0.000
Intra groups	1295.524	54	23.991		
Total	2543.549	59			

*P< 0.05.

Table 9. Duncan test results of bending strength test.

Group	N	For sub-groups α = 0.05	
		1	2
Borax	10	85.956	
Vacsol Azure	10	87.941	
Control	10	88.331	
Tanalith E	10		94.707
Boric acid	10		95.623
Imersol Aqua	10		98.177
Significant		0.312	0.140

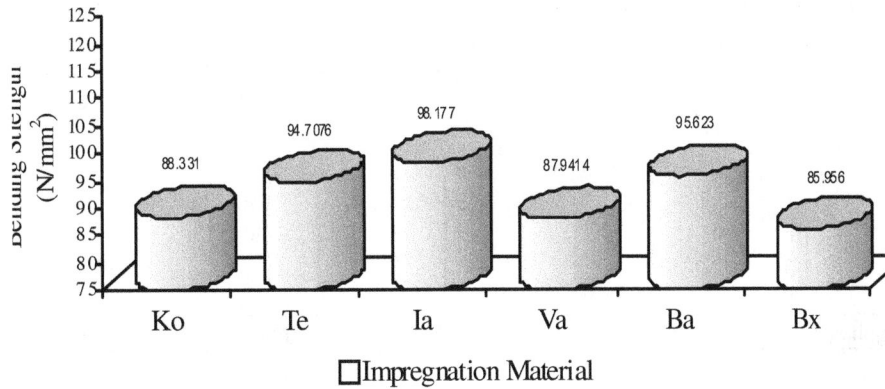

Figure 4. Bending strength values.

Table 10. Statistical values of modulus of elasticity in bending.

Statistical values	Impregnated samples					
	Co	Te	Ia	Va	Ba	Bx
x (N/Mm^2)	9255.810	10224.080	9853.650	8924.700	9456.480	9699.130
Ss (N/Mm^2)	226.604	300.203	340.154	244.804	309.548	320.627
V (S^2)	57055	100135	128560	66588	106466	114224
Min (N/Mm^2)	9028.250	9800.000	9345.900	8500.000	9032.400	9284.340
Max (N/mm^2)	9567.200	10760.900	10442.100	9384.700	9904.310	10255.900

x: Arithmetic mean, v: Variance, Ss: Standard deviation, N: Number of samples, Te: Tanalith-E, Ia: Imersol Aqua, Va: Vacsol Azure, Ba: Boric acid, Bx: Borax.

Table 11. Variance analysis (ANOVA) results of modulus of elasticity in bending.

Source of variance	Sum of squares	Degree of freedom	Sum of squares	F value	Sig
Between groups	10529693.840	5	2105939	22.050*	0.000
Intra group	5157279.104	54	95505.170		
Total	15686972.940	59			

*P< 0.05.

strength was measured in Imersol Aqua, Boric acid, and Tanalith-E' followed by Borax, Vacsol Azure, and control samples. Bending strength values were given in Figure 4.

Modulus of elasticity in bending

Modulus of elasticity in bending values of impregnated Rowan wood was given in Table 10. Modulus of elasticity in bending values of Rowan wood impregnated with different impregnation materials showed statistically important differences in F test F(5; 54) = 22.050, P < 0.05) as shown in Table 11. The highest Duncan test results was found for samples impregnated with Tanalith-E followed by Imersol Aqua, Borax, Boric acid, control, and Vacsol Azure (Table 12). Modulus of elasticity in bending was given in Figure 5.

RESULTS

According to the type of impregnation materials, retention amounts showed a statistically significant difference. This may be due to different concentration of impregnation solutions. The highest value retention was obtained in Vacsol Azure followed by Imersol Aqua, Tanalith-E, Borax, and boric. According to the type of impregnation material, the highest retention amounts were in Vacsol Azure samples (151 044 kg/m^3) followed by Imersol-Aqua (127 045 kg/m^3), Tanalith-E (104 083 kg/m^3), Borax (86 393 kg/m^3), and Boric acid (64 887 kg /m^3). The amount of retention in samples impregnated with Vacsol Azure is higher due to the difference in concentration between the impregnation materials. High amount of retention of samples impregnated with Vacsol Azure was possibly due to the high retention ability of it.

Table 12. Duncan test result of modulus of elasticity in bending.

| Groups | N | For Sub-groups α = 0.05 | | | | |
		1	2	3	4	5
Va	10	8924.701				
Ko	10		9255.807			
Ba	10		9456.488	9456.488		
Bx	10			9699.131	9699.130	
Ia	10				9853.657	
Te	10					10224.080
Significant		1.000	0.152	0.084	0.268	1.000

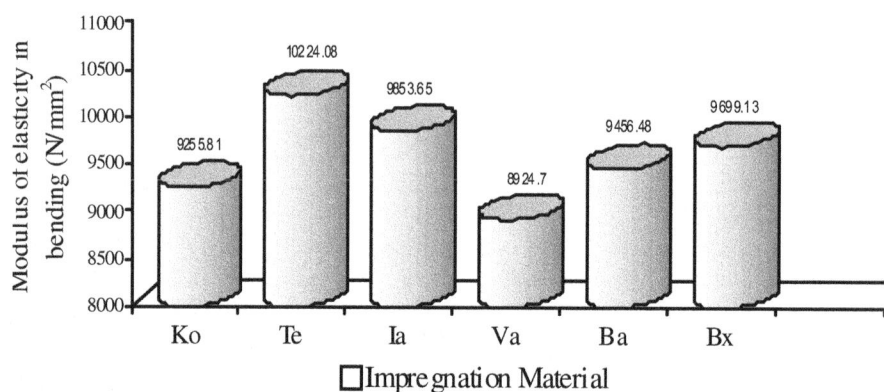

Figure 5. Modulus of elasticity in bending.

Compression strength was highest in Boric acid (61.902 N/mm^2) followed by Vacsol Azure (57.829 N/mm^2), Borax (56.244 N/mm^2), Imersol Aqua (55.576 N/mm^2), Tanalith-E (54.622 N/mm^2) and control sample (53.012 N/mm^2). The difference between Imersol Aqua, Borax, and Vacsol Azure was not found statistically meaningful. Impregnation material did not decrease the compression strength. So, those materials may be used in construction sector.

Bending strength of impregnated materials were found as Imersol Aqua 98.177 N/mm^2, Boric acid 95.623 N/mm^2, Tanalith-E 94.708 N/mm^2, control 88.331 N/mm^2, Vacsol Azure 87.941 116,230 N/mm^2 and Borax 85.926 N/mm^2. According to F test, bending strength showed statistically important differences (F (5; 54) = 10.404, P < 0.05). According to Duncan Test results, the highest bending strength was in Imersol Aqua, Boric acid and Tanalith-E followed by Borax, Vacsol Azure. The reason for lower value in samples impregnated with boron compounds may be due to its acidic property.

Modulus of elasticity in bending was measured as follows; Tanalith-E 102.240 N/mm^2, Imersol Aqua 9853.700 N/mm^2, Borax 9699.130 N/mm^2, Boric acid 9456.480 N/mm^2, control 9255.800 N/mm^2, and Vacsol

Azure 8924.700 N/mm^2. According to F test, modulus of elasticity in bending of impregnated samples was found significant (F (5; 540) = 22, P < 0.05). According to Duncan test, the highest modulus of elasticity was in Tanalith-E followed by Imersol Aqua, Borax, Boric acid, and Vacsol Azure

As a result, the highest values in compression strength of Rowan wood was measured in samples impregnated with Boric acid, in bending strength impregnated with Imersol Aqua and in modulus of elasticity in bending impregnated with Tanalith. The lowest values were with Tanalith-E in compression strength, with Borax in bending strength, with Vacsol Azure in the modulus of elasticity in bending. According to this results Rowan wood may be impregnated with Tanalith-E to be used in the fields where compression strength, bending strength, and modulus of elasticity in bending are critical.

In general the results of this study on the mechanical properties of Rowan wood impregnated with various impregnation materials are compatible with the findings in literature on the effect of various impregnation materials on mechanical properties of different tree species.

Togay (2009) found that except for oak samples, one time impregnation with timbercare aqua due to the increasing MOE and impregnation with timbercare aqua can be useful for the wood material subject to bending stress, which needs high elasticity.

ACKNOWLEDGEMENTS

This study is a part of M.Sc. Thesis prepared by Neslihan SÜZER ERTÜRK, Institute of Science and Technology, Gazi University, Ankara, Turkey.

REFERENCES

ASTM-D 1413-76 (1976). Standard Test Method of Testing Wood Preservatives by Laboratory Soilblock Cultures. Ann. Book ASTM Stand. pp. 452-460.

Aytaşkin A (2009). Some technological properties of wood imptegnated with various chemical substances. M. Sc. Thesis, Karabük University Graduate School of Natural and Applied Sciences Department of Furniture and Decoration Education, Karabük. P. 134.

Bendtsen BA (1984). Mechanical Properties of Longleaf Pine Treated Waterborn Salt Preservatives, USDA Forest Service. USA. P. 434.

Evans PD, Michell AJ, Schmalzl K (1992). Studies of the Degradetion and Protection of Wood Surfaces. Wood Sci. Technol. 26:151-163.

Gillwald W (1961). Der Einfluss Verschiedener Impragnier Mittel Auf Die Physikallischen Und Festigkeitseigen Schaften Des Holzes. Holtechnologie 2:4-16.

Hickson's Timber Impregnation Co.(GB) (2000). Into the 21st. Century, Imersol-Aqua Brochure, Datassheet, 6214, 1-4, Hickson Timber Treatments.

Isaacs CP (1972). The Effect of Two Accerelated Treating Methods on Wood Strength. AWPA, USA. 68:175-182.

Kartal SN (1998). The Biological Resistance, Leachability and Strength Properties of Wood Treated with CCA and CCB Preservatives, Ph. D. Thesis, Istanbul University Instit. Sci. Istanbul. P. 157.

Kollmann F (1959). Die Eigenschaftanderung Von Gruben Holz Nach Schutzsalzimprag-nierung, Forschungsber, Des Landes Nordhrhein, Westfalen-Germany.

Korkut S, Guller B, Aytin A, Kök MS (2009). Turkey's native wood species: Physical and mechanical characterization and surface roughness of Rowan (*Sorbus aucuparia* L.). Wood Res. 54(2):19-30.

Lutomsky K (1976). Effect of Treatment Conditions Using The Hot-Cold Method of Impregnation Pine Wood with Water Solution of Static Bending of Treated Wood, Zest. Probl. Nauk. Rolniczych. P. 178.

Özçifci A, Atar M, Uysal B (1999). The Effects of Wood Bleaching Chemicals on The Surface Gloss and the Adhesion Strength of Varnishes. Turk. J. Agric. For. 23(EK3):763-770.

Richardson BA (1987). Wood Preservation, The Construction Press: Lancaster. England.

Stabnikov VM (1957). Puti Uviliczenia Sroka Sluschby Dreviesinyw Konstrukcjach. Leningrad. pp. 81-185.

Togay A (2009). Effect of Impregnation with Timbercare Aqua on the Properties of Some Woods. J. Appl. Sci. 9(5):956-961.

TS 2470 (1976). Wood - Sampling Methods and General Requirements for Physical and Mechanical Tests. Türk Standartlari Enstitüsü. Ankara.

TS 2471 (1976). Wood, Determination of Moisture Content for Physical and Mechanical Tests, TSE, Ankara.

TS 2472 (1976). Wood - Determination of Density for Physical and Mechanical Tests, TSE, Ankara.

TS 2474 (1976). Wood - Determination of Ultimate Strength in Static Bending. TSE. Ankara.

TS 2478 (1976). Wood-Determination of Modulus of Elasticity In Static Bending. TSE. Ankara.

TS 2595 (1977). Wood-Determination of Ultimate Stress In Compression Parallel to Grain. TSE, Ankara.

TS 344 (2012). Wood preservation-General Rules. TSE. Ankara.

TS 345 (2012). Testing methods for the effects of wood imregnating substances, TSE, Ankara.

TS 4176 (1984). Wood - Sampling Sample Trees and Long for Determination of Physical and Mechanical Properties of Wood in Homogeneous Stands. *T.S.E.* Ankara.

TS 53 (1981). Wood - Sampling and test methods - Determination of physical properties. T.S.E. Ankara.

Uysal B, Atar M, Özçifci A (1999). The Effects of Wood Bleaching Chemicals on the Bending Strength of Wood. Turk. J. Agric. For. 23(6):615-620.

Vologdin AJ (1966). Vlijanije Rozlicznych Antiseptic of Na Fizyko Mechanitchiestkije Svosjstva Dreviesiny Sosny Svotsva Dreviesin, Jejo Zaschita; Novyje Dreviesinyje Materialy. Moskova.

Wazny J (1973). Investigations of The Influence of Wood Preservatives on Strength, Dreviesiny, Sreda. 3:181-185.

9

Mechanical characterization of nickel plated copper heat spreaders with different catalytic activation surface treatment techniques

Victor C. H. Lim[1,2] , Nowshad Amin[1], Foong Chee Seng[3], Ibrahim Ahmad[4] and Azman Jalar[2]

[1]Department of Electrical, Electronic and Systems Engineering, Faculty of Engineering and Built Environment, Universiti Kebangsaan Malaysia, Bangi, Selangor, Malaysia.
[2]Microelectronics Semiconductor Packaging, Institute of Micro Engineering and Nanoelectronics, Universiti Kebangsaan Malaysia, Bangi, Selangor, Malaysia.
[3]Freescale Semiconductor (M) Sdn. Bhd., Petaling Jaya, Selangor, Malaysia.
[4]Department of Electronics and Communication, College of Engineering, University Tenaga Nasional, Kajang, Selangor, Malaysia.

This paper studied the effects of different catalytic activation processes towards intermetallic diffusion and mechanical properties on nickel plated heat spreader after high temperature storage (HTS). Heat spreader performs as medium to dissipate heat from silicon die towards heat-sink and is normally made by copper that is plated with nickel to improve wear resistance and prevent oxidation of copper. Two types of heat spreader that using galvanic initiation and thin nickel-copper electrodeposition surface treatment technique had been studied on their hardness and moduli by using Micro Tester and Nano-indenter. Besides, HTS tests were performed to investigate intermetallic diffusion between the nickel and copper layers. Young's moduli of the heat spreaders which were plated by galvanic initiation and thin nickel-copper strike electroless nickel plating catalytic activation techniques were 45 to 65 GPa and 60 to 80 GPa, respectively. The results found that thin nickel-copper electrodeposition technique gave a higher modulus for the heat spreader and this also increased the mechanical strength of heat spreader. Diffusion also took place with a very slow rate in nickel-copper layer.

Key words: Heat spreader, electroless nickel plating, galvanic initiation, thin nickel-copper electrodeposition, high temperature storage.

INTRODUCTION

Thermal is an important issue for latest semiconductor package. This is due to more energy that have been generated in high power and processing speed packages. Among these packages, flip chip ball grid array (FCBGA) is the packaging type that creates most heat during operation in high frequency application (Samson et al., 2005; Kutz, 2005). The ability for the substrate to transfer heat out from the junction is quite

limited, especially for epoxy and plastic substrates which are made from low conductivity material. Therefore, a highly conductive component like the heat spreader needs to be attached onto the other side of the die to transfer heat of the junction in the opposite direction which was shown in Figure 1 (Samson et al., 2005; Bolanos, 2007; Ohadi and Qi, 2004). Heat spreader is normally made from copper because copper is a material

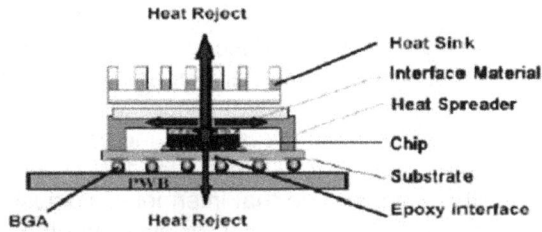

Figure 1. Schematic of thermal packaging architecture.

of high thermal conductivity (Cengel, 2006). However, copper easily oxidises with air, and copper oxide is not a ductile material (Callister, 1999). Copper oxide will affect the performance of the heat spreader. Copper oxide is also an irritant that causes health implications to human body. Therefore, nickel is plated onto the copper heat spreader to prevent copper oxidation from occurring. Although nickel is also a highly reactive element, it reacts very slowly in room temperature and ambient pressure (Callister, 1999). Due to its slow oxidation rate, nickel has been commonly used for the plating of other metals.

Advantages of electroless plated nickel over electroplated nickel include higher corrosion resistance, a very uniform thickness over the most complicated shapes, high after-plated hardness, very high hardness after a heat treatment procedure, high solderability and bondability, and control over magnetic properties (Durkin et al., 1993; Kanungo et al., 2006; Taheri, 2003). Electroless nickel plating is an autocatalytic process that does not use externally applied electric current to deposit a layer of nickel alloy with the reducing agent (Taheri, 2003; Chen et al., 2003). An electroless deposition process uses only one electrode without external source of electrical current. However, the solution for the electroless process involves a reducing agent so that the electrode reaction can take place. The electrons are supplied by this reducing agent, normally hypophosphite that is dissolved in the plating solution. The deposited nickel acts as a catalyst for the continuation of the chemical reaction until the plating process is terminated by taking out the deposited part from the bath (Chen et al., 2003; Kantola, 2006; Van Den Meerakker, 1980; Watanabe and Honma, 1998). Electroless nickel plating process starts spontaneously for some material when immersed into the electroless nickel plating solution, provided that there is a chemically clean surface. For some other materials like plastics, ceramics, silver, copper and copper alloys, catalytic activation is required to initiate the chemical reduction process (Kantola, 2006; Van Den Meerakker, 1980). Prior to the process of electroless nickel plating, copper or copper alloys need to go through several cleaning and activation processes. The purpose of these processes is to clean the oxide layer and other impurities, and provide a suitable surface for chemical reduction process. The success for copper depends upon the activation process used. Some copper

and copper alloys will not catalytically initiate plating in the nickel phosphorus systems without an additional activation process after normal preparation (Durkin et al., 1993; McKinnon, 2003).

Two types of nickel plated copper heat spreader were used in the assembly of flip-chip ball grid array packages. A difference in performance from the package was noticed after the assembly process. After both heat spreaders had been cross sectioned, one of the heat spreader was found to have an additional thin layer between the electroless nickel deposit layer and copper layer. The objective of this paper is to determine the intermetallic diffusion between each layer for different catalyst activation process after high temperature storage. Besides, this paper also studies the moduli and hardness test in determining mechanical properties for heat spreaders.

MATERIALS AND METHODS

Designs of heat spreaders in this project were in accordance to the 33 × 33 mm FCBGA package and the 30 × 30 mm heat sink. The area of the heat spreader was 30 × 30 mm which matched the heat sink. The heat spreaders were made from copper and plated with nickel. The copper heat spreaders were sent to two different metal finishing suppliers which used different catalytic activation processes. One of the suppliers used nickel-copper strike as catalytic activation technique while another supplier used galvanic initiation as their catalytic activation process. The heat spreaders that were plated with nickel-copper strike catalytic activation technique were named as heat spreader A, while the heat spreaders that were plated using the galvanic initiation catalytic activation technique were named as heat spreader B. The intrinsic properties of the heat spreaders were assumed to be the same but difference in catalytic activation technique before electroless plating process.

After the heat spreaders were received from suppliers, most of the heat spreaders were put into the high temperature storage (HTS) oven for the HTS test. HTS tests for heat spreaders were performed at 150°C for 24, 48, 96 and 168 h, respectively. The HTS ovens that were used in this project were HTS oven and profiler oven. Accelerated aging at high temperature was used to promote nickel-copper intermetallic growth. All the samples with same initial catalytic activation techniques were baked under different time ranges to see the growth of catalytic activation layer. After the heat spreaders were taken out from the HTS oven, some heat spreaders were cross-sectioned using the normal metallographic manner. Cross sections of heat spreaders were prepared by using grinding and cross section machine, extra precautions were taken not to smear the thin metal layers. Proper processing steps were required to ensure quality encapsulation of the samples. In addition, the deformation of the thin metal layers due to the polishing process had to be minimized. After the samples had been cross-sectioned, optical images of each sample were taken by high power microscope under x1000 magnification. Finally the thickness of each layer had been taken using camera microscope and was measured by measurement software.

Both types of heat spreaders after HTS test were sent for mechanical properties tests. These tests provided information about the effect of different electroless nickel plating catalytic activation techniques towards mechanical behaviors of the heat spreaders. Young's Modulus of the heat spreaders was determined by the three points bending test using Instron Micro Tester. Specimens

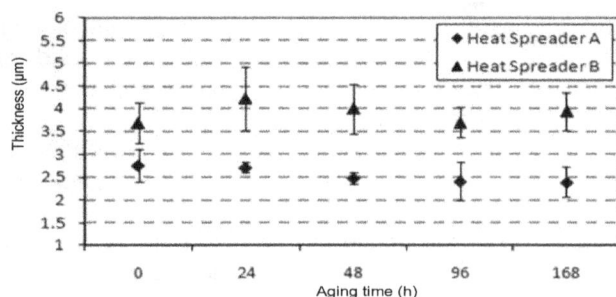

Figure 2. Nickel-phosphorus thicknesses of heat spreader A and B under different time conditions with 95% confidence interval for the mean.

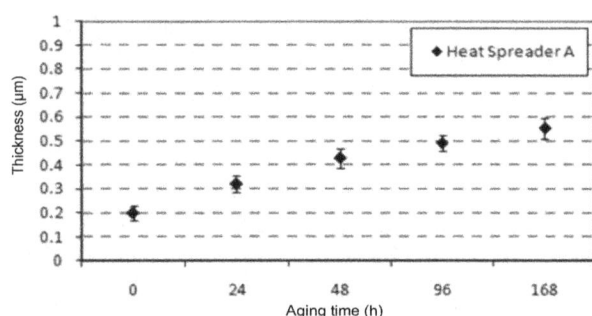

Figure 3. Nickel-copper thicknesses of heat spreader A under different time conditions with 95% confidence interval for the mean.

were supported as a simple beam, with the compressive load being applied at midpoint, and maximum stress and strain were calculated.

Apart from that, the hardness and reduced modulus of heat spreaders were determined by nano indentation test using nano indenter. For indentation, a prescribed load was applied to an indenter in contact with a specimen. The probe was forced into the surface at a selected rate and to a selected maximum force. The depth of penetration was measured when the load was applied. The area of contact at full load was determined by the depth of the impression and the known angle or radius of the indenter. A force-displacement curve had been obtained during indentation provided indications of the sample material's mechanical and physical properties.

RESULTS AND DISCUSSION

Intermetallic diffusion after high temperature storage

The purpose of this study is to evaluate intermetallic diffusions between each layer of heat spreader under HTS thermal aging conditions. Intermetallic diffusions for each heat spreaders layer under HTS conditions were investigated in this section using normal metallographic cross section manner. Figure 2 shows the nickel-phosphorus thicknesses for heat spreader A and B while Figure 3 shows the nickel-copper thicknesses for heat spreader A after HTS thermal aging. Figure 2 illustrates

the electroless nickel plating layer thickness of heat spreader A is slightly reduced from the initial (T0) thickness of 2.8 to 2.4 µm after 168 aging h. Meanwhile, Figure 3 illustrates the nickel-copper electrodeposition layer thickness of heat spreader A increased from the initial thickness of 0.2 to 0.6 µm after 160 h aging time. Both results demonstrated that intermetallic diffusion took place on the nickel-copper layer towards the other layers. The diffusion of nickel-copper electrodeposition layer actually occurred in two directions; that is, in the direction of the nickel-phosphorus and copper layers. The diffusion coefficient, D, is the rate at which atoms diffuse. The equation of the diffusion coefficient is as follows:

$$D = D_0 \exp(-Q_d/RT) \qquad [m^2/s] \qquad (1)$$

Where D_0 is the temperature-independent pre-exponential, Q_d is the activation energy for diffusion, R is the gas constant, 8.31 J/molK and T is the absolute temperature. The nickel-copper metal diffusion rate is slow because of its low temperature-independent pre-exponential ($D_0 = 2.7 \times 10^{-5}$ m^2/s) and high activation energy ($Q_d = 256$ K) compared with other metals (Callister, 1999; Wulff et al., 2004).

Figure 2 illustrates the thickness of nickel-phosphorus layer for heat spreader B are ranging between of 3.6 and 4.3 µm. The results did not indicate any trends in terms of the thickness of the nickel-phosphorus layer. Under all aging conditions, it was found that heat spreader B did not have any diffusion layers forming between the electroless nickel plating layer and the copper layer. The deposited atoms are arranged neatly onto the copper surface during the electroless nickel plating deposition process. There is not enough vacancy for copper atoms to diffuse into the electroless nickel plating layer (Durkin et al., 1993; McKinnon, 2003). Therefore, electroless nickel plating layer provides strong resistance against wear and tear, and is able to act as anti-corrosion material for today's industry.

For the nickel-copper electrodeposition layer in heat spreader A, the nickel and copper atoms created some vacancies during deposition onto the surface of copper. This made the copper and nickel atoms diffuse easily inside the nickel-copper electrodeposition layer. Strike plating method was used in combination with the plating of different metals where the deposits of the strike method become the foundation for subsequent plating processes. Strike method normally uses a high current density and a bath with a low ion concentration (Durkin et al., 1993; Singh et al., 2006; Dini, 1993; Marquis et al., 2006). However, this plating process is extremely slow and takes time. In order to shorten the strike time, manufacturer normally increases the bath ion concentration. Nevertheless, if the deposition rate is too high, poor adhesion and plating quality will occur (McKinnon, 2003; Dini, 1993; Fritz et al., 2001). This is because the deposits do not arrange orderly on the

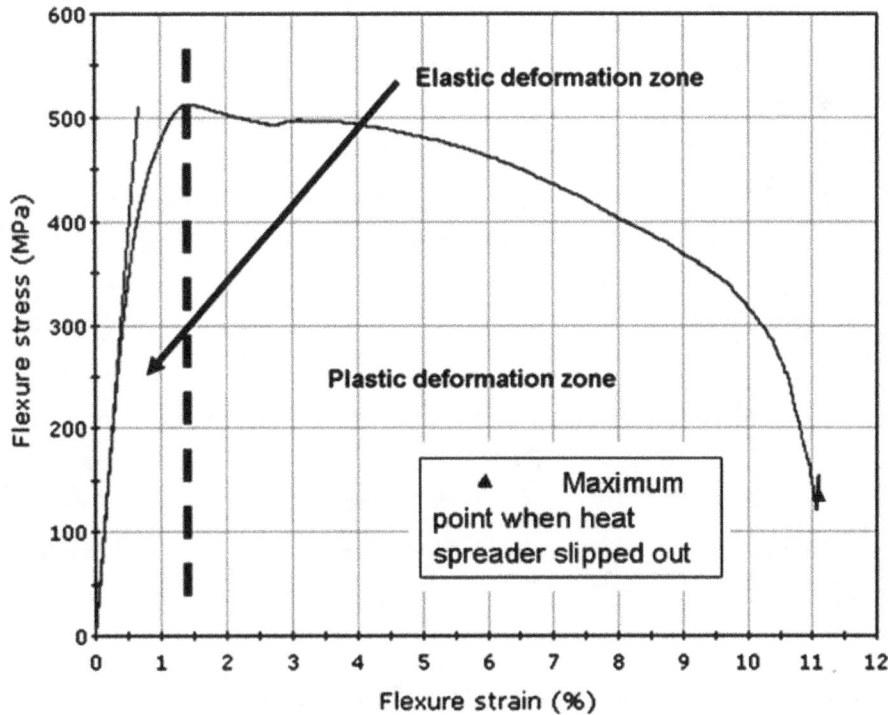

Figure 4. Graph of 3-points bending flexure stress against flexure strain for heat spreader.

substrate and creates vacancies between the deposits. In the case of heat spreader A, increasing the strike rate did not affect the plating quality. Instead, the vacancies between the deposited atoms help to promote intermetallic diffusion of the nickel-copper layers.

Effect of catalytic activation technique towards mechanical properties

Characterization of mechanical properties in this study involved two parts, namely 3-points bending test and nano-indentation test. 3-point bending test provided the Young's Modulus information on heat spreaders while nano-indentation provided hardness and reduced modulus parameters for the heat spreaders. Both of these parameters prove to be useful information in identifying the mechanical properties of heat spreaders which were produced using different catalytic activation processes.

3-point bending test is a flexural stress versus strain test that is able to calculate the modulus of elasticity (Young's modulus). 3-point bending test was chosen as the mechanical tensile test instead of the tension test because the probability of heat spreader being bent in the package during operation is higher than being pulled. This phenomenon happens when the package faces warpage effect due to coefficient of thermal expansion (CTE) mismatch inside the package (Kutz, 2005). The

output of this test is recorded on a strip chart as load versus elongation, which is dependent on the specimen's size. In order to calculate modulus in an easier way, load and elongation are normalized to the flexure stress and flexure strain to minimize the geometrical factors. Flexure stress (σ) is equal to the load force (F) divided by original cross-section area (A_0) which is shows in Equation 2. Flexure strain (ε) is equal to deformation elongation (Δl) divided by original length (l_0) which is shows in Equation 3.

$$\sigma = F / A_0 \qquad\qquad [Pa] \qquad\qquad (2)$$

$$\varepsilon = \Delta l / l_0 \qquad\qquad\qquad (3)$$

Figure 4 shows the graph of flexure stress against flexure strain after normalisation. From the graph, the mechanical behaviour of the heat spreader in the bending test can be divided into two parts which were elastic deformation zone and plastic deformation zone. Young's modulus was calculated by the gradient of the elastic deformation zone, which is flexure stress divided by flexure strain. The mechanical behaviour of the heat spreader was the same as metal and metal alloys by having a high Young's modulus and wide plastic deformation zone (Callister, 1999).

During the test, it was found that the heat spreaders were not broken and the test ended when the heat spreaders started slipping out of the test platform. This is

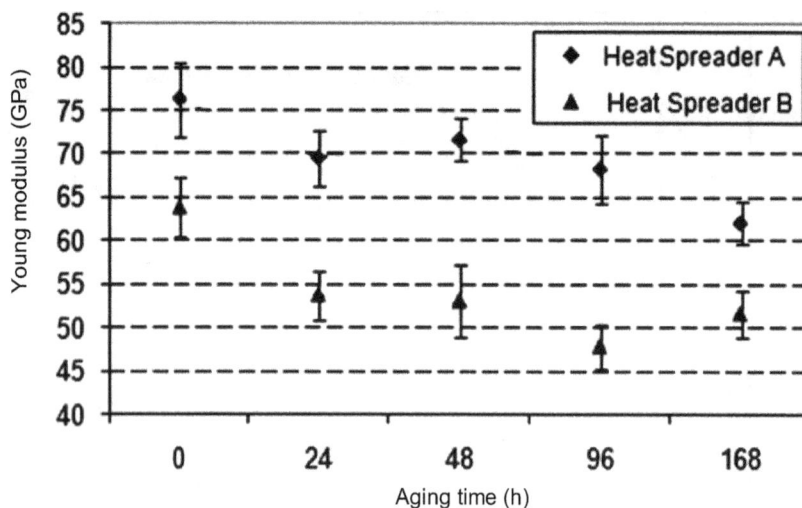

Figure 5. Young's moduli of heat spreader A and B under different time conditions with 95% confidence interval for the mean.

because heat spreader was made by copper, which is a very ductile material that is not easily broken during the bending test. Both heat spreaders' electroless nickel plating layers were checked after the test ended. There were no peelings or cracking lines found on the electroless nickel plating layers. This means that both nickel-phosphorus layers, which were deposited by different catalytic activation processes, still maintained the ductility of the heat spreader.

Young's moduli of the heat spreaders under different HTS conditions were plotted. Figure 5 shows the Young's modulus of thin nickel-copper strike and galvanic initiation heat spreaders (heat spreader A and B). The results show the Young's moduli of the heat spreader A are in the range of 60 and 80 GPa, while Young's moduli of the heat spreader B are in the range of 45 and 65 GPa. The higher Young's modulus of heat spreader A is probably due to the nickel-copper electrodeposition layer, since nickel-copper electrodeposition layer is actually a metal alloy layer made up of a mixture of copper and nickel metals (Callister, 1999; Tench and White, 1984).

Alloys have higher Young's modulus than pure metals because impurity atoms that exist inside the solid solution customarily enforce lattice strains on the surrounding host atoms. Lattice strain field interaction between dislocation and these impurity atoms restrict any dislocative movement. A different size of atom tends to diffuse or segregate the surrounding crystal lattice and this creates a weak bonding effect. The resistance to slip is larger when impurity atoms are present because the overall lattice strain must increase if a dislocation is torn away from them. Moreover, the same lattice strain interactions exist between impurity atoms and dislocations that are in motion during plastic deformation (Callister, 1999; Tench and White, 1984; Hanke, 2001).

Therefore, a higher stress is necessary initially into plastic deformation phase for solid-solution alloys.

Nano-indentation test is a mechanical property test that enables the measurement of hardness and reduced modulus of the material at specific places. This highly localised test is very suitable to characterise the properties of thin coating, and is able to perform a small indentation in nano-scale of depth displacement. The test determines the hardness and reduced modulus of the analytical models from a load-displacement data, where force, displacement and time are recorded throughout the test (Beake and Leggett, 2002; Choi and Suresh, 2003; Dong et al., 2003; Lin et al., 2004; Van Vlient et al., 2004). Figure 6 shows the load-displacement graph plotted from results that were gathered from multiple indentation tests on the surface of the heat spreader. This graph consists of two parts of load-displacement measurement, which were collected during the loading and unloading of the indenter's head.

The load was applied to Berkovich indenter while in contact with the heat spreader. When the load was applied, the maximum measured depth of penetration was used to calculate the hardness (H) of the heat spreader (Nomura et al., 2011; Oyen, 2013; Varughese et al., 2011). In Equation 4, the hardness of the heat spreader coating was derived by dividing the maximum load (P_{max}) by the projected area of contact (A_c). The area of contact at full load was determined by the depth of the impression (h_p) and the known angle or radius of the indenter. For Berkovich indenter, the value of the contact area was calculated in the Equation 5. The reduced modulus measurement of the heat spreader coating was provided by the shape of the unloading curve. The reduced modulus, E_r was calculated according the formula provided by Equation 6, where dP/dh is the

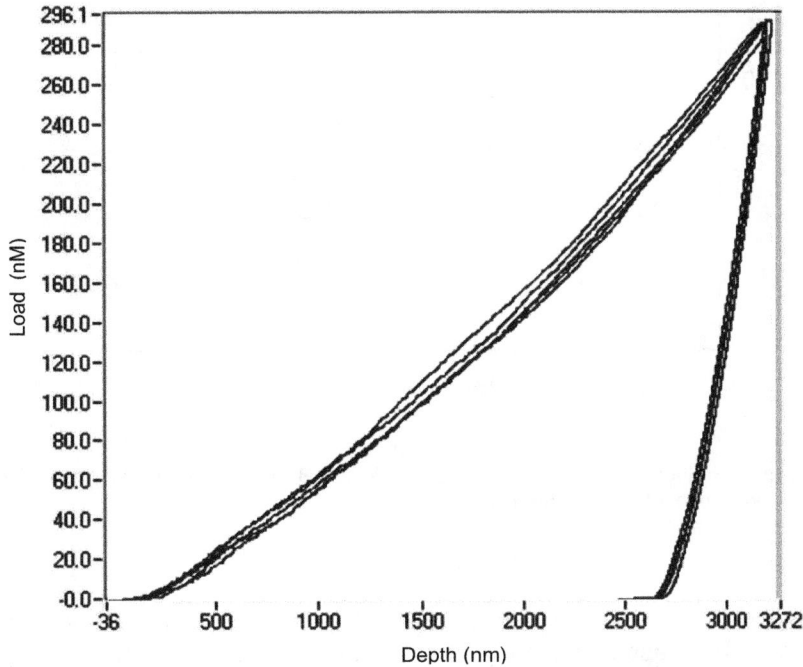

Figure 6. Graph of multiple plotting loads versus depths of heat spreader in nano-indentation test.

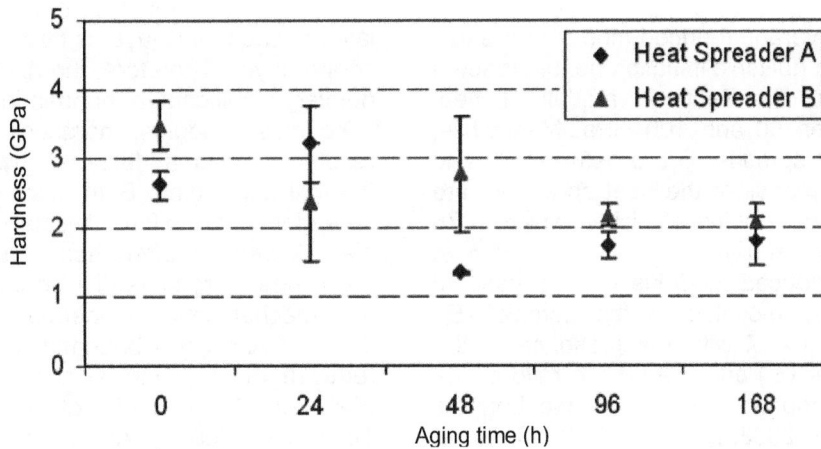

Figure 7. Hardness of heat spreader A and B under different time conditions with 95% confidence interval for the mean.

stiffness of the heat spreader coating, β is the correction index of the Berkovich indenter (Beake and Leggett, 2002; Choi and Suresh, 2003; Dong et al., 2003; Lin et al., 2004; Van Vlient et al., 2004). The following are the related equations for the nano-indentation test to determine hardness and reduced modulus:

$$H = P_{max} / A_c \qquad \text{[Pa]} \qquad (4)$$

$$A_c = 3(\sqrt{3}) \, h_p^2 \tan^2 65.3 = 24.5 \, h_p^2 \, \text{[m}^2\text{]} \qquad (5)$$

$$E_r = (dP / dh) \times ((\sqrt{\pi}) / 2\beta(\sqrt{A_c})) \qquad \text{[Pa]} \qquad (6)$$

After both results were calculated according to the respective formulas, the entire results were restructured and plotted in several graphs. Figure 7 shows the average hardness for thin nickel-copper strike and galvanic initiation heat spreaders (heat spreader A and B) under different HTS time conditions. According to the results shown in Figure 7, hardness of heat spreader A were found to be in the range of 1 and 3.5 GPa, while the

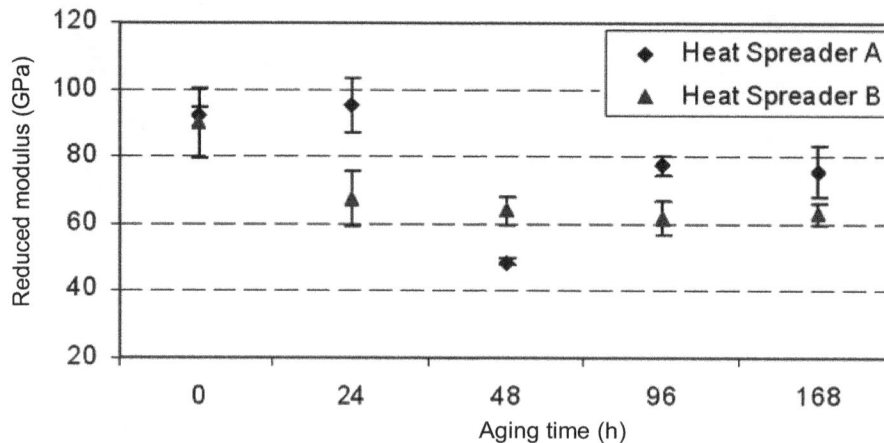

Figure 8. Reduced moduli of heat spreader A and B under different time conditions with 95% confidence interval for the mean.

hardness of heat spreader B were in the range of 2 and 3.5 GPa. Hardness of both heat spreaders are higher than the hardness of nickel, which have Vicker's hardness of 638 MPa and Brinell hardness of 700 MPa (Callister, 1999). This results show that electroless nickel plating provides a very good mechanical strength over heat spreader because electroless nickel plating layer is an alloy.

Figure 8 shows the average reduced moduli of the thin nickel-copper strike and galvanic initiation heat spreaders (heat spreader A and B). Reduced moduli of heat spreader A are between 50 and 100 GPa. Meanwhile, reduced moduli of heat spreader B are between 60 and 90 GPa. The reduced moduli of the heat spreaders are very close with reduced modulus of nickel, which is 76 GPa (Beake and Leggett, 2002).

The results of the reduced modulus can be used to determine the elasticity modulus of the sample (E_s) according to the Equation 7 with the presence of the modulus of the indenter (E_i) and the poison ratio of the sample (v_s) and the indenter (v_i) (Beake and Leggett, 2002; Choi and Suresh, 2003; Dong et al., 2003; Lin et al., 2004; Van Vlient et al., 2004).

$$1 / E_r = ((1-v_s^2) / E_s) + ((1-v_i^2) / E_i) \qquad (7)$$

From the hardness and reduced modulus results that were shown by the related figures, it can be concluded that the hardness and reduced modulus of heat spreader A is slightly higher than heat spreader B. The results show a different trend compared with the Young's modulus results for the bending test because both experiments were performed under different dimensions. The bending test was performed onto the x-y plane of the heat spreaders while the indentation test was just performed in-situ onto one point of the surface of the heat spreaders.

When the heat spreaders were bent in the bending test, the whole nickel-copper deposition layer was bent together with other layers and this created significant differences in terms of results. For indentation test, electroless nickel plating layer was first indented, followed by the nickel-copper layer and lastly, the copper layer. The results were highly dependent on the hardness and the reduced modulus of the electroless nickel plating layer because this layer is thicker compared to the nickel-copper layer. Therefore, the difference of the results was not very significant for nano-indentation test.

Young's modulus, hardness and reduced modulus results showed a decrease with the increase of HTS thermal aging time. Both mechanical properties for heat spreaders with different catalytic activation processes also decreased after thermal stress was applied onto these heat spreaders. These behaviours may be due to the mechanisms of thermal stress-induced vacancy diffusion and grain boundary diffusion. These diffusions reduced the mechanical strength of the heat spreader (Callister, 1999; Wulff et al., 2004; Razeghi, 2002; Tummala, 2001). As a result, both moduli and hardnesses of the heat spreader decreased.

Conclusion

The HTS and mechanical properties of the different catalytic activation techniques were studied and compared. This study shows that under HTS condition, diffusion took place between the Ni-Cu electrodeposition layer and neighbouring layers. Young's moduli of thin nickel-copper strike and galvanic initiation techniques were 60 to 80 GPa and 45 to 65 GPa, respectively. Consequently, thin nickel-copper strike technique provided a higher mechanical strength in bending test. Differences on the heat spreaders' hardness and reduced moduli were not significant for both catalytic activation techniques. Besides, mechanical strengths of the heat spreaders for both catalytic activation techniques decreased when thermal aging time increased.

ACKNOWLEDGMENTS

This study was supported by the Malaysia Government and Universiti Kebangsaan Malaysia, under the IRPA grant of 03-01-01-0088-PR0075/09-08 and Science Fund 03-01-02-SF0495. Special thanks also go to Freescale Semiconductor (M) Sdn. Bhd. for its assistance in providing experimental facilities.

REFERENCES

Beake BD, Leggett GJ (2002). Nanoindentation and nanoscratch testing of uniaxially and biaxially drawn poly(ethylene terephthalate) film. Polymer 43:319-327.

Bolanos MA (2007). Packaging technology trends "applied research opportunities". USA: Texas Instruments. pp. 27-29.

Callister WD Jr (1999). Materials science and engineering: an introduction. 5th ed. USA: John Wiley & Sons, Inc. pp. 92-240.

Cengel YA (2006). Heat and Mass Transfer: A Practical Approach. 3rd ed. USA: MacGraw-Hill. pp. 844-846.

Chen WP, Lu SG, Chan HLW (2003). Influence of electroless nickel plating on I/V characteristics and its implications for reliability in ZnO-based ceramic varistors. Mater. Sci. Eng. B 99:70-73.

Choi Y, Suresh S (2003). Nanoindentation of patterned metal lines on a Si substrate. Scr. Mater. 48:249-254.

Dini JW (1993). Electrodeposition: the material science of coating and substrates. USA: Noyes Publications. pp. 46-89.

Dong S, Beake BD, Parkinson R, Xu B, Hu Z, Bell T (2003). Determination of hardness and Young's modulus of brush plated nano-Al2O3/Ni composite coating by nanoindentation testing. Surf Eng. 19(3):195-199.

Durkin B, Barnstead M, Morcos B (1993). Basic substrate strategies and approaches for electroless nickel. USA: MacDermid, Inc. p. 1.

Fritz T, Mokwa W, Schnakenberg U (2001). Material characterisation of electroplated nickel structures for microsystem technology. Electrochim. Acta 47:55-60.

Hanke LD (2001). Handbook of analytical methods for materials. UK: Mater. Evaluation Eng. Inc. pp. 27-28.

Kantola K (2006). Modelling, estimation and control of Electroless nickel plating process of printed circuit board manufacturing. Helsinki University of Technology, Finland. pp. 3-6.

Kanungo J, Pramanik C, Bandopadhyay S, Gangopadhyay U, Das L, Saha H, Gettens RTT (2006). Improved contacts on a porous silicon layer by electroless nickel plating and copper thickening. Semicond. Sci. Technol. 21:964-970.

Kutz M (2005). Mechanical Engineers' Handbook. 3rd ed. USA: John Wiley & Sons, Inc. pp. 257-299.

Lin JF, Wei PJ, Pan JC, Ai CF (2004). Effect of nitrogen content at coating film and film thickness on nanohardness and Young's modulus of hydrogenated carbon films. Diam. Relat. Mater. 13:42-53.

Marquis EA, Talin AA, Kelly JJ, Goods SH, Michael JR (2006). Effect of current density on the structure of Ni and Ni-Mn electrodeposits. J. Appl. Electrochem. 36:669-676.

McKinnon HW (2003). Nickel Plating: Industry Practices Control Technology and Environmental Management. USA: EPA United States Environmental Protection Agency. pp. 2-11.

Nomura K, Chen YC, Kalia RK, Nakano A, Vashishtaa P (2011). Defect migration and recombination in nanoindentation of silica glass. Appl. Phys. Lett. 99:111906.

Ohadi M, Qi J (2004). Thermal management of harsh-environment electronics. 20th IEEE SEMI-THERM Symposium, pp. 231-240.

Oyen ML (2013). Nanoindentation of Biological and Biomimetic Materials. Exp. Tech. 37:73-87.

Razeghi M (2002). Fundamentals of solid state engineering. USA: Kluwer Academic Publisher. pp. 1-40.

Samson EC, Machiroutu SV, Chang JY, Santos I, Hermerding J, Dani A, Prasher R, Song DW (2005). Interface Material Selection and a Thermal Management Technique in Second-Generation Platforms Built on Intel Centrino Mobile Technology. Intl. Technol. J. 9(1):75-86.

Singh S, Ghosh SK, Basu S, Gupta M, Mishra P, Groverb AK (2006). Structural and Magnetic Study of an Electrodeposited Ni/Cu Thin Film by Neutron Reflectometry. Electrochem. Solid-State Lett. 9(3):J5-8.

Taheri R (2003). Evaluation of Electroless Nickel-Phosphorus (EN) Coatings. PhD dissertation, University of Saskatchewan, Canada. pp. 4-10.

Tench D, White J (1984). Enhanced tensile strength for electrodeposited nickel-copper multilayer composites. Metallurg. Trans. A 15(A):2039-2040.

Tummala RR (2001). Fundamentals of Microsystems packaging. Singapore: McGraw-Hill. pp. 212-263.

Van Den Meerakker JEAM (1980). On the mechanism of electroless plating II: one mechanism for different reductants. J. Appl. Electrochem. 11:395-400.

Van Vlient KJ, Prchlik L, Smith JF (2004). Direct measurement of indentation frame compliance. J. Mater. Res. 19(1):325-31.

Varughese S, Kiran MSRN, Solanko KA, Bond AD, Ramamurty U, Desiraju GR (2011). Interaction anisotropy and shear instability of aspirin polymorphs established by nanoindentation. Chem. Sci. 2:2236-2242.

Watanabe H, Honma H (1998). Direct Electroless Nickel Plating on Copper Circuits Using DMAB as a Second Reducing Agent. IEMT/IMC Proceedings. pp. 149-153.

Wulff F, Breach C, Stephan D, Saraswati, Dittmer K (2004). Characterization of intermetallic growth in copper and gold ball bonds on aluminum metallization. In: Proc. 6th EPTC Singapore. pp.348-353.

10

Flow behaviour of polyvinyl alcohol (PVOH) modified blends of polyvinyl acetate (PVAc)/natural rubber (NR) latexes

Stephen Shaibu Ochigbo

Department of Chemistry, Federal University of Technology, P. M. B. 65, Minna, Niger State, Nigeria.

During compounding processes, polymer latexes are mixed with various colloidal systems (in particular, PVOH) or surface-active agents to modify the flow behaviour in order to suit the manufacturing process. In this study, the flow behaviour of aqueous dispersed PVAc, NR and different mixed dispersions were investigated both in the absence and presence of PVOH as flow modifier, using an Ostwald glass capillary viscometer. The results show an almost linear decrease in the relative viscosity of the NR/PVAc dispersion mixtures in the absence of a modifier, but the presence of different amounts of PVOH gave rise to significant deviations from this behaviour. It was found that these changes in the relative viscosities are governed by a balance between the following four effects: (i) the higher mobility of the NR chains compared to those of PVAc, which generally caused a decrease in the viscosities of the dispersion mixtures; (ii) the physical adsorption of PVOH on the NR chains that slowed down the movement of these chains and caused an increase in relative viscosity; (iii) the chemical interaction between the functional groups on PVAc and PVOH that gave rise to latex thickening and increased the relative viscosity; (iv) a dilution effect which reduced the viscosity, unless counterbalanced by one or more of the other effects.

Key words: Dispersion flow behaviour, viscosity, rubber latex, polyvinyl alcohol, polyvinyl acetate.

INTRODUCTION

Latexes are polymer particles dispersed in water. Their prominence has arisen because of mounting pressure from Environmental Protection Agencies (EPAs) against industrialists and researchers to replace solvent based systems with water based counterparts. Latex-based systems are all water based, without organic content, using only very small amounts of the latter to modify the final film properties, paint flow or rheological properties (Amalvy and Soria, 1996; Nabuurs et al., 1996; Meng et al., 2008). Because water is inexhaustible, cheap and readily available in comparison to organic solvents, latex based systems have corresponding advantages over their solvent based counterparts. In addition, the use of

latex-based systems controls pollution, reduces risks of fire occurrence, and improves aspects of occupational health and safety (Zhaoying et al., 2001; Tambe et al., 2008; Kosto and Schall, 2007; Chen et al., 2005).

Polyvinyl acetate (PVAc) is an example of a synthetic polymer latex, for which the term "emulsion" is usually used in preference to latex, because it is a product of emulsion polymerization of the vinyl acetate monomer. PVAc emulsions have been used in large quantities in several application fields, especially in adhesives of papers and woods (Okaya et al., 1993; Backman and Lindberg, 2004; Kim and Kim, 2006). Because of its structural configuration, PVAc film is resistant to both oils and ultra-violet (UV) radiation. However, the pendant acetate groups restrict free rotation along the C-C axis, thus resulting in a dry film which, besides having poor resistance to water, is too brittle and hence unsuitable in packaging applications unless in the presence of added plasticizer. On the other hand, natural rubber (NR) latex is a natural biosynthetic polymer which is derived from the tree *Hevea brasiliensis*. The latex that is obtained fresh from the tree (field latex) contains about 70% water, part of which should be removed to concentrate the latex for a wide variety of applications. The resulting latex concentrate has excellent physical properties and is used for the manufacture of dipped goods, adhesives/binders, thread, carpets/rugs, and moulded foams. Among these, the dipped goods, which include hand gloves, balloons, condoms, bladders, and catheters/tubes account for up to 60% NR latex usage (Haque et al., 1995; Sanguansap et al., 2005; Yip and Cacioli, 2002). However, the film cast from NR latex, unlike PVAc, is soft and tacky and possesses poor resistance to oil (Sirisinha et al., 2003), as well as poor resistance to both ultra violet (UV) radiation and ozone. Blending of polymer emulsions or dispersions is required in order to balance factors such as ease of application, wettability, drying characteristics, bond strength, clarity, environmental resistance, and especially cost (Meng et al., 2008). NR latex has intrinsic water resistance and toughness and therefore can improve water resistance behaviour of PVAc while the former simultaneously can improve in oil and ozone resistance when the latex and emulsion are blended together.

During the compounding processes, latexes are mixed with various colloid systems, in particular, poly (vinyl alcohol) (PVOH)) or surface-active agents to modify the flow behaviour in order to suit the manufacturing process (Schoeder and Brown, 1951; Brown and Garrett, 1959; Irving, 1990; Geurink et al., 1996; Hellgren et al., 1999. The flow behaviour of latexes is a critical factor, as it provides a guide for the formulation and ease of processing. A latex compound with a low viscosity and some thixotropic nature is good for dipping operations. The count of a latex thread is influenced by its viscosity (Peethambaran et al., 1990; Blackley, 1966; Calvert, 1982). Studies were carried out to examine the effect of PVOH on a PVAc emulsion (Dibbern-Brunelli and Atvars, 1995; Dibbern-Brunelli et al., 1998) and on NR latex (Peethambaran et al., 1990). An emulsion stabilized with PVOH has many advantages over surfactants, including Newtonian fluidity, superior primary wet tackiness, high strength and creep resistant film properties (Nakamae et al., 1999). This study used an Ostwald glass capillary viscometer to examine the flow behaviour of blends of PVAc/NR latexes in the presence of PVOH as a flow modifier. Owing to its simplicity, viscometry is popularly employed for studying compatibility as well as flow behaviour of polymers in solution (Singh and Singh, 1983; Kuleznev et al., 1978; Hourston and Hughes, 1978).

MATERIALS AND METHODS

A PVAc homopolymer emulsion was obtained from Makeean Polymers, South Africa with the trade name ML50 (solids contents = 52.5%, pH = 6.5). Field NR latex was provided by the Rubber Research Institute of Nigeria (RRIN), Iyanomo. It has a solid content of 45%. The PVOH was a Merck grade quality.

A dilute aqueous dispersion of the PVAc emulsion was prepared with a fixed concentration of 10 wt.%, designated PVAc_10. NR latex was prepared in different concentrations of 10 wt.% (NR_10), 5 wt.% (NR_5), and 1 wt.% (NR_1). All the dispersions were prepared relative to the respective dry solids contents of the original latexes. The prepared dispersions were in each case immediately passed through a stainless steel sieve with aperture 150 µm in order to remove any foreign suspended particles that may block the capillary of the Ostwald viscometer. An appropriate amount of PVOH was dissolved in water, under magnetic stirring at 90°C, to obtain a 2 wt.% stock solution from which lower concentrations of 1 wt.% and 0.5 wt.% were in turn prepared by successive dilution using de-ionised water. NR_10/PVAc_10, NR_5/PVAc_10, and NR_1/PVAc_10 were respectively blended into w/w compositions of 0/100, 25/75, 50/50, 75/25 and 100/0. This mixing method whereby the concentration of one component, PVAc (called the host polymer) is held constant while the concentration of the second, NR (called the guest polymer) is varied is known as the polymer-solvent method (Danait and Deshpande, 1995; Papanagopoulos and Dondos, 1996; Haiyang et al., 2000). In this paper, the method was applied to polymer dispersions rather than to conventional solution based polymer systems.

The flow rate (efflux time) of the different samples were directly observed as the time, t, needed for the liquid to pass through the capillary (Billmeyer, 1984). Measurements were made in the absence and presence of PVOH at a temperature of 22 ± 0.5°C. From the efflux times, the relative viscosities were calculated using Equation 1.

$$\eta_{rel} = t/t_o \tag{1}$$

Where, η_{rel} = relative viscosity, t = efflux time of dispersion, t_o = efflux time of pure water.

RESULTS AND DISCUSSION

The variation of relative viscosity with blend composition is represented by NR_10/ PVAc_10 blend, as shown in Figure 1. With increases in NR content, it is observed that the relative viscosity for the untreated sample (0 wt.% PVOH) decreases more rapidly than those for the samples containing different amounts of PVOH. This is attributed to the fact that NR latex is a pseudoplastic fluid, and even at rest the rubber particles are in random motion (Peethambaran et al., 1990). Therefore, as NR content increases, there results a corresponding increase in random motion and hence, an enhanced flow which leads to the observed marked decrease in viscosity of the untreated sample. On the other hand, the dispersions containing the different amounts (0.5, 1 and 2 wt.%) of PVOH show gradual and parallel decreases in viscosity with increasing NR content. It is suggested that the observed gradual and parallel decrease in viscosity of the PVOH-modified samples with increasing NR content, compared to that of the untreated sample, might be due

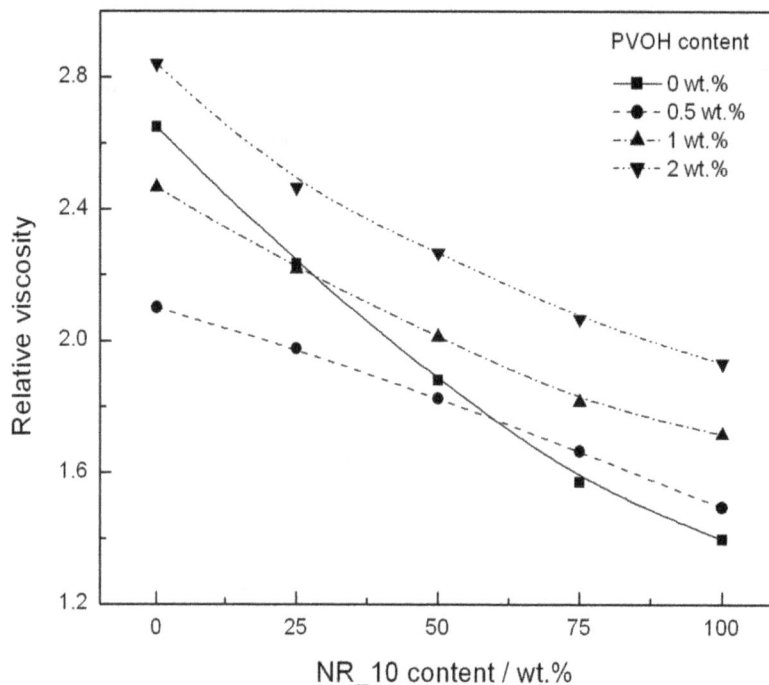

Figure 1. Dependence of relative viscosity on the blends' composition at various contents of PVOH modifier.

to adsorption of PVOH particles onto the NR particles. This adsorption could slow down the speed of the randomly moving NR particles and give rise to the observed trend of decrease in viscosity of the mixtures in the presence of PVOH. This agrees with the results of Peethambaran et al. (1990) in a study carried out to investigate the role of various surfactants on the viscosity of natural rubber latex.

Figure 2 presents the change in relative viscosity of the various unblended dispersions against different PVOH contents. The viscosity of pure PVAc (PVAc_10) is at maximum throughout the concentration range of the surfactant. Primarily, the viscosity of PVAc is intrinsically due to restricted mobility of the chains arising from steric hindrance between the acetate groups on one hand and then intermolecular dipole-dipole attraction between adjacent acetate functional groups. In addition, the high concentration of the PVAC_10 (that is, 10wt.%) which means that there are much chains condensed within a small volume, permitting intermolecular entanglements contributes to observed high viscosity. The viscosity of NR_10 of equivalent concentration (that is, 10 wt.%) by comparison is considerably lower than that of PVAc_10 as observed due to the fact that the particles of natural rubber is in constant random motion (pseudoplastic). Of course, NR_5 and NR_1 have lowered viscosity comparatively due to dilution effect and also pseudoplastic nature of the natural rubber latex as stated earlier.

The addition of the 0.5% PVOH immediately results in a significant drop in the relative viscosity curve for PVAc_10 observed. This concentration of PVOH is very dilute. So the accompanying relatively high water content imparts a plasticizing effect on the PVAc emulsion and thus causing its viscosity to drop sharply as observed. However, with higher concentration of PVOH, the viscosity values of PVAc_10 slightly improve after the depression. This improvement is as a result of the overshadowing of any plasticizing effect of the water content of the modifier by viscosity increase attributed firstly to adsorption of PVOH onto PVAc particles. An additional reason for the increase in viscosity is a consequence of the partial hydrolysis of the acetate functional group into carboxylate ions which in turns allows absorption (dissolution) of PVAc into PVOH (Lloyd, 1989; Hornby and Peach, 1993). These ions are capable of taking part in hydrogen bonded interactions with the hydroxyl groups of the modifier and, therefore, can contribute to the viscosity increase of the mixture. Interactions of this kind between particles of emulsion and aqueous polymeric solutions, leading to viscosity increases, have earlier been reported and the phenomenon is called latex thickening (Brown and Garrett, 1959). Several competitive interactions were postulated to account for the overall rheological behaviour of such mixtures. There were a number of observations providing evidence that the process involved in latex thickening is not merely that of the

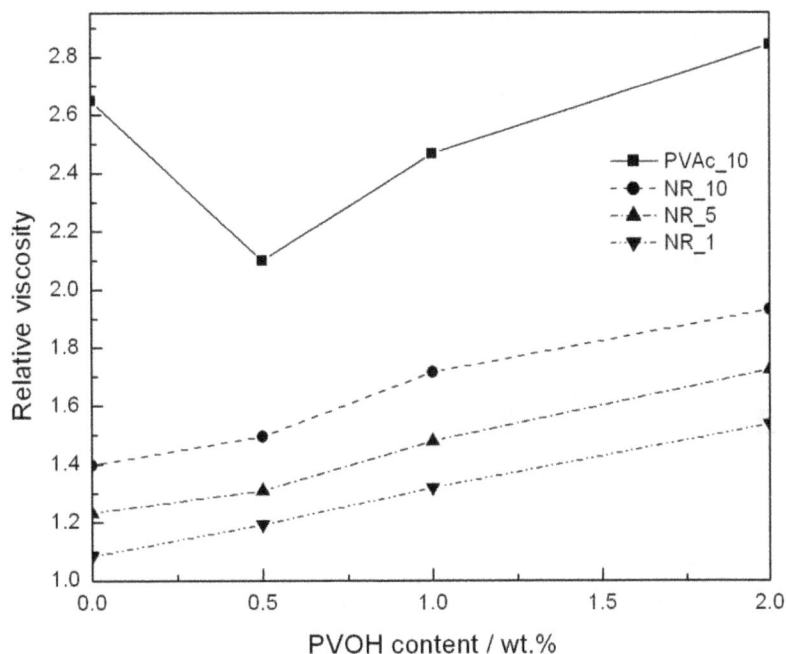

Figure 2. Dependence of relative viscosity on PVOH modifier content for various unblended aqueous dispersions.

enhancement of the viscosity of the continuous (water) phase by the water soluble polymer. There was also very little correlation between the aqueous viscosity of a polymer and its efficiency in thickening latexes. The significant drop in relative viscosity at 0.5% PVOH is attributed to the corresponding drop in concentration of the emulsion (PVAc_10) due to a dilution effect caused by the mixing of the two dispersions, which at this PVOH concentration dominated the latex thickening effect. The viscosity increase observed for the higher concentrated PVOH-modified samples is because of the latex thickening effect becoming more dominant.

The relative viscosities of the NR-based dispersions (with and without PVOH) are generally lower than that of the PVAc dispersion, which is due to the pseudoplasticity of NR, while the differences between the different NR dispersions are attributed to their corresponding differences in concentration. The relative viscosities of all the NR dispersions increase more or less linearly with increasing PVOH content. This behaviour shows that a completely different mechanism decides the interaction between NR and PVOH. As previously reported (Peethambaran et al., 1990), this involves a physical adsorption of PVOH solutes on the surfaces of the NR particles, which reduces the mobility of the NR particles.

The changes in the relative viscosities of the blended dispersions as a function of PVOH content are shown in Figures 3 - 5. These results clearly show the competition between the different influences discussed earlier. For all the blended dispersions, the viscosity generally increases

with increasing PVOH content, but decreases according to the order: NR_10/ PVAc_10> NR_5/PVAc_10 >NR_1/PVAc_10. The increase with increasing PVOH content is the result of a combination of the latex thickening and physical adsorption effects discussed previously, while the decrease with increasing NR_10/5/1:PVAc_10 ratio is the result of the higher mobility of the NR chains. However, a dilution effect comes into play when more dilute NR dispersions are used, and therefore there is a decrease in relative viscosity with decreasing NR content in the NR dispersions that were mixed with the PVAc_10 dispersion at a constant ratio. This dilution effect also manifests itself in the lower than expected viscosities at 0.5% PVOH (as discussed above), but the physical adsorption between PVOH and NR becomes more dominant as the NR_10/5/1:PVAc_10 ratio increases in the dispersion mixtures, and the decrease in viscosity for the 0.5% PVOH containing samples disappears for the 75:25 NR_10/5/1:PVAc_10 dispersion mixture.

Utracki and Kanial (1982) divided viscosity-composition curves into three types on the basis of their deviation from the log-additivity rule ($\ln \eta_b = \Sigma \Phi_i \ln \eta_i$, in which η_b represents the blend viscosity, and Φ_i and η_i are the volume fraction and viscosity of component i in the blend). Based on this rule, Figures 6 - 9 were obtained, showing a comparison of the experimental and theoretical relative viscosities as a function of dispersion composition in the presence of different amounts of PVOH. As seen, all the experimental curves show

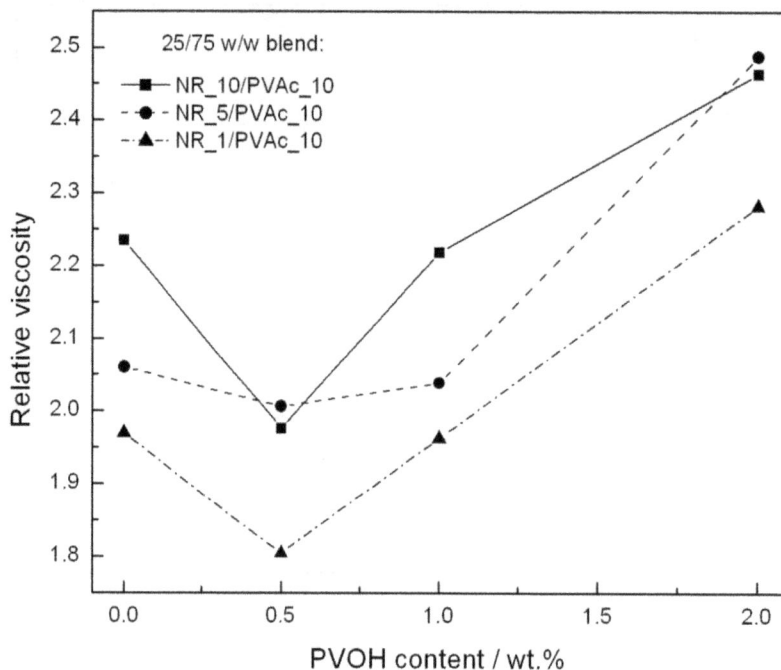

Figure 3. Dependence of relative viscosity on PVOH modifier content for different 25:75 w/w NR/PVAc blends.

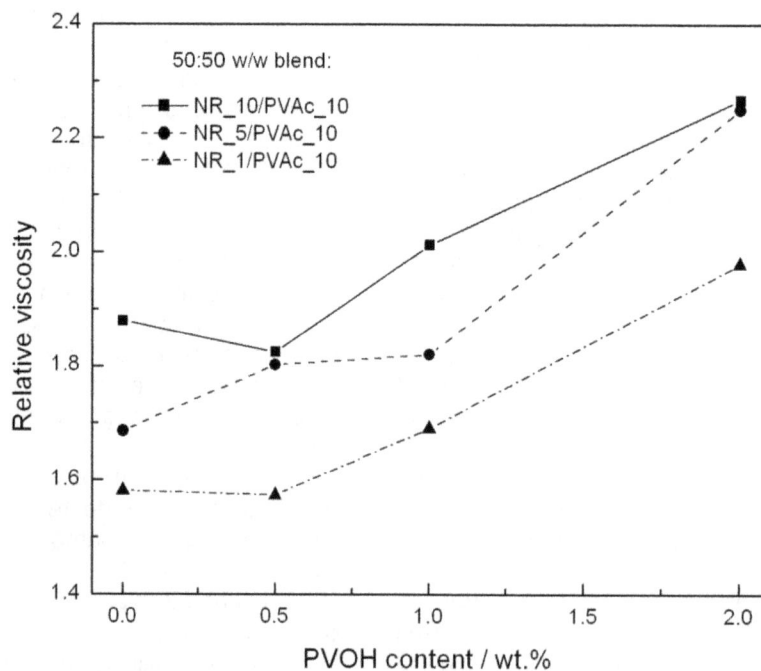

Figure 4. Dependence of relative viscosity on PVOH modifier content for different 50:50 w/w NR/PVAc blends.

deviations from the theoretical values. The linear (theoretical) curves indicate ideal behaviour which is exemplified by a situation in which particles of the mixed components consist of similar sizes in a given medium.

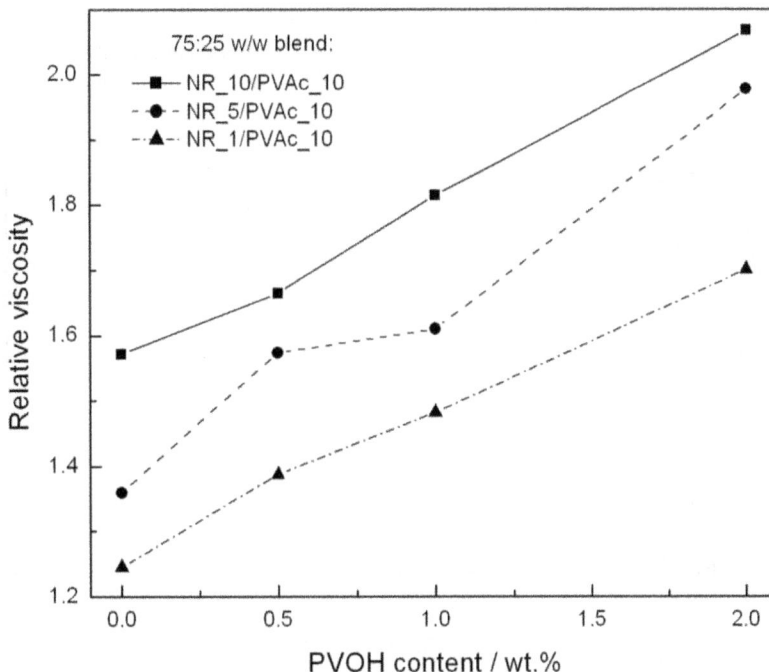

Figure 5. Dependence of relative viscosity on PVOH modifier content for 75:25 w/w NR/PVAc blends.

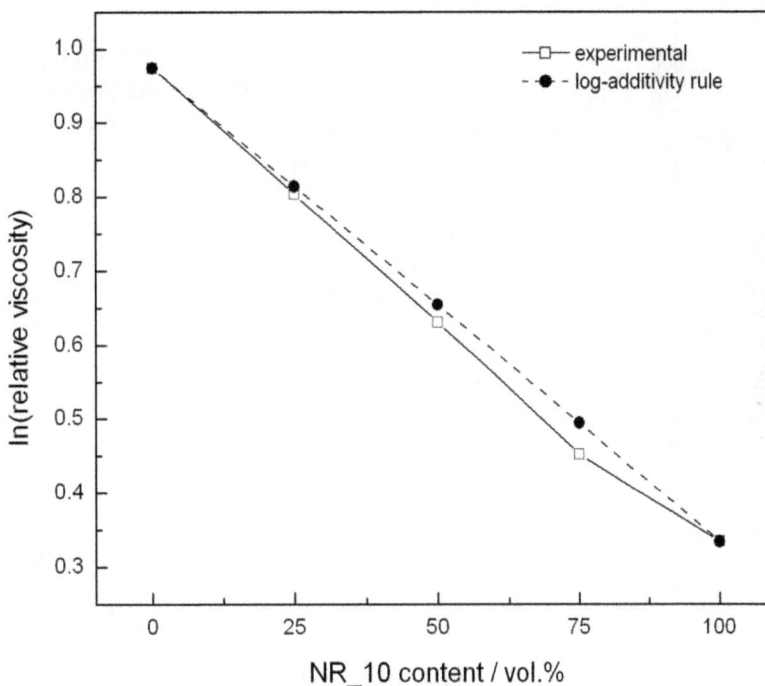

Figure 6. Comparison between experimental and theoretical viscosities for aqueous dispersions of NR_10/PVAc_10 without PVOH.

Such behaviour is in agreement with the additive rule (Wong, 1991). The observed deviations are evidence of interactions taking place between the mixed components, namely NR, PVAc, PVOH and water. Negative deviations

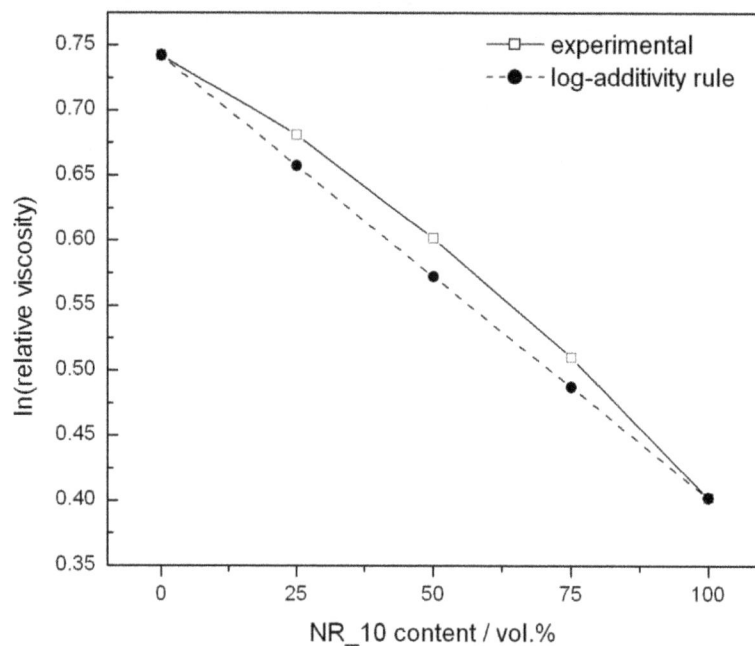

Figure 7. Comparison between experimental and theoretical viscosities for aqueous dispersions of NR_10/PVAc_10 with 0.5 wt.% PVOH.

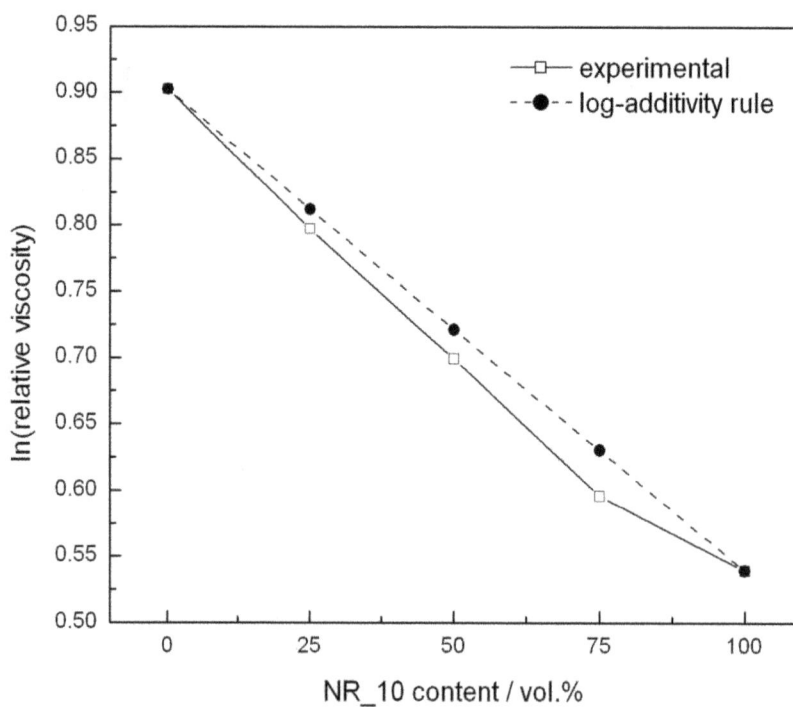

Figure 8. Comparison between experimental and theoretical viscosities for aqueous dispersions of NR_10/PVAc_10 with 1 wt.% PVOH.

are attributed to lowered mobility of particles, and hence reduced interactions between components, whereas positive deviations indicate that there are enhanced interactions between the components due to high mobility

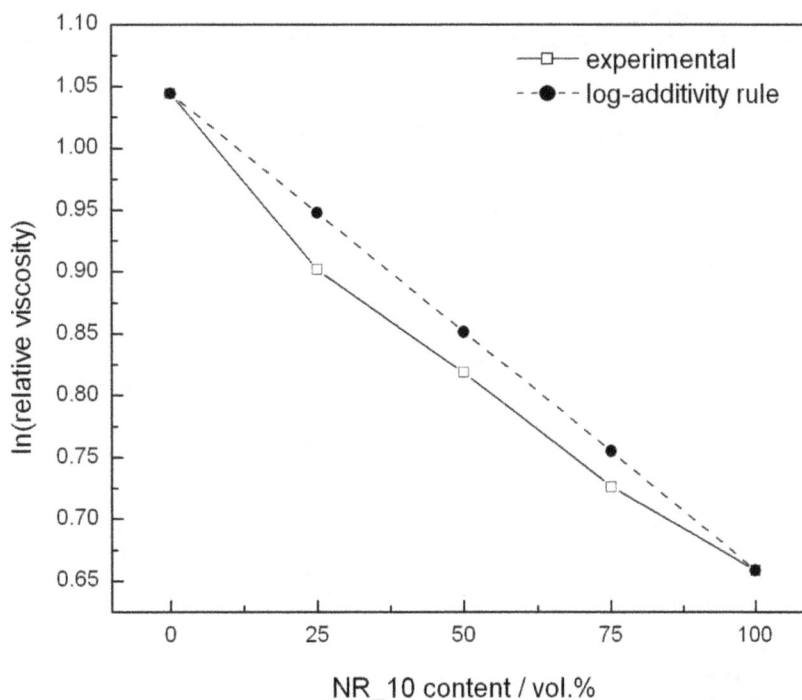

Figure 9. Comparison between experimental and theoretical viscosities for aqueous dispersions of NR_10/PVAc_10 with 2 wt.% PVOH.

of particles. These interactions are not anticipated by theoretical prediction. A number of factors can be adduced for the reduced mobility of the particles. They include repulsive forces between the mixed components because of dissimilarity in their chemical natures, as for example, NR and PVAc, and high viscosity of medium which prevent free mobility, and hence interactions between components. While the first factor might be the reason for the negative deviation observed for the pure blends of NR_10 and PVAc_10 (Figure 6), the case of negative deviations observed with the blends containing 1 and 2% PVOH (Figures 8 and 9, respectively) might be due to high viscosity of the system. In Figure 7, the introduction of 0.5% PVOH enhances freedom of intermolecular interactions between the mixed components. This increased mobility is thus observed as the positive deviation from the theoretical prediction.

Conclusions

The flow behaviours of NR latexes and PVAc dispersions, and mixtures of these dispersions, in aqueous medium, as well as the addition of a PVOH modifier to these dispersions, were investigated. The results clearly showed that changes in the relative viscosities of mixtures of NR, PVAc and PVOH dispersions are governed by a balance between the following four effects:

(i) The higher mobility of the NR chains compared to those of PVAc, which generally caused a decrease in the viscosities of the dispersion mixtures,
(ii) The physical adsorption of PVOH on the NR chains that slowed down the movement of these chains and caused an increase in relative viscosity,
(iii) The chemical interaction between the functional groups on PVAc and PVOH that gave rise to latex thickening and increased the relative viscosity,
(iv) A dilution effect which reduced the viscosity, unless counterbalanced by one or more of the other effects.

ACKNOWLEDGEMENTS

The National Research Foundation of South Africa (GUN 62693) and the University of the Free State are acknowledged for financial support of this research.

REFERENCES

Amalvy JI, Soria DB (1996). Vibrational spectroscopic study of distribution of sodium dodecyl sulphate in latex films. Progr. Org. Coat. 28:279-283.
Backman AC, Lindberg KAH (2004). Interaction between wood and polyvinyl acetate glue studied with Dynamic Mechanical Analysis and Scanning Electron Microscopy. J. Appl. Polym. Sci. 91:3009-3015.
Billmeyer FW Jr. (1984). Textbook of Polymer Science, 3rd ed.; Wiley: New York. P. 149.
Blackley DC (1966). High Polymer Latices, Vol. 2. Applied Science, London.

Brown GL, Garrett BS (1959). Latex thickening: Interactions between aqueous polymeric dispersions and solutions. J. Appl. Polym. Sci. 1:283-295.

Calvert KO (1982). Polymer Latices. Applied Science, London.

Chen J, Spear SK, Huddleston JG, Rogers RD (2005). Polyethylene glycol and solutions of polyethylene glycol as green reaction media. Green Chem. 7:64-82.

Danait A, Deshpande DD (1995). A novel method for determination of polymer-polymer miscibility by viscometry, Eur. Polym. J. 31:1221-1225.

Dibbern-Brunelli D, Atvars TDZ (1995). Study of miscibility of poly (vinyl acetate) and poly (vinyl alcohol) blends by fluorescence spectroscopy. J. Appl. Polym. Sci. 55:889-902.

Dibbern-Brunelli D, Atvars TDZ, Joekes I, Barbosa VC (1998). Mapping phases of poly (vinyl alcohol) and poly (vinyl acetate) blends by FTIR microspectroscopy and optical fluorescence microscopy. J. Appl. Polym. Sci. 69:645-655.

Geurink PJA, van Dalen L, van der Ven LGJ, Lamping RR (1996). Analytical aspects and film properties of two-pack acetoacetate functional latexes, Prog. Org. Coat. 27:73-78.

Haiyang Y, Pingping Z, Feng R, Yuanyuan W, Tiao Z (2000). Viscometric investigations on the intermolecular interactions between poly (methyl methacrylate) and poly (vinyl acetate) in various solvents, Eur. Polym. J. 36:21-26.

Haque ME, Akhtar F, Dafader NC, Al-Siddique FR, Sen AR, Ahmad MU (1995). Characterization of natural rubber latex concentrate from Bangladesh, Macromolecular Reports A32 suppl. 4:435-445.

Hellgren A, Weissenborn P, Holmberg K (1999). Surfactants in waterborne paints. Prog. Org. Coat. 35:79-87.

Hornby M, Peach J (1993). Foundations of Organic Chemistry; Oxford University Press, New York.

Hourston DJ, Hughes ID (1978). Dynamic mechanical and sonic velocity behaviour of polystyrene-poly (vinyl methyl ether) blends. Polymer 19:1181-1185.

Irving S (Ed.) (1990). Handbook of Adhesives Third Ed. Van Nostrand Reinhold, New York.

Kim S, Kim HJ (2006). Thermal stability and viscoelastic properties of MF/PVAc hybrid resins on the adhesion for engineered flooring in under heating system. ONDOL Thermochimica Acta 444:134-140.

Kosto KB, Schall DC (2007). Low-temperature waterborne pavement marking paints: a road assessment of this LOW-VOC option. Presented at the 2007 FutureCoat! Conference, sponsored by Federation of Societies for Coatings Technology, in Toronto, Ont., Canada, October 3-5, 2007.

Kuleznev VN, Melnikova OL, Klykova VD (1978). Dependence of modulus and viscosity upon composition for mixtures of polymers. Effects of phase composition and properties of phases. Eur. Polym. J. 14:455-461.

Lloyd DA (1989). A First Course in Organic Chemistry. Wiley, Chichester.

Meng W, Wu L, Chen D, Zhong A (2008). Ambient self-crosslinkable acrylic microemulsion in the presence of reactive surfactants. Iran. Polym. J. 17(7):555-564.

Nabuurs T, Baijards RA, German AL (1996). Alkyd-acrylic hybrid systems for use as binders in waterborne paints. Progr. Org. Coat. 27:163-172.

Nakamae M, Yuki K, Sato T, Maruyama H (1999). Preparation of polymer emulsions using a poly (vinyl alcohol) as a protective colloid, Colloids and Surfaces A: Physiochem. Eng. Aspects 153:367-372.

Okaya T, Tanaka T, Yuki K (1993). Study on physical properties of poly (vinyl acetate) emulsion films obtained in batchwise and in semicontinuous systems. J. Appl. Polym. Sci. 50:745-751.

Papanagopoulos D, Dondos A (1996). Difference between the dynamic and static behaviour of polymers in dilute solutions. 3. Influence of the host polymer on the dimensions of the guest polymer, Polymer 37:1053-1055.

Peethambaran NR, Kuriakose B, Rajan M, Kuriakose AP (1990). Rheological behaviour of natural rubber latex in the presence of surface-active agents. J. Appl. Polym. Sci. 41:975-983.

Sanguansap K, Suteewong T, Saendee P, Buranabunya U, Tangboriboonrat P (2005). Composite natural rubber based latex particles: A novel approach. Polymer 46:1373-1378.

Schoeder WD, Brown GL (1951). Carboranesiloxane Polymers. Rubber Age 69:433.

Singh YP, Singh RP (1983). Compatibility studies on solutions of polymer blends by viscometric and ultrasonic techniques. Eur. Polym. J. 19:535-541.

Sirisinha C, Limcharoen S, Thunyarittikorn J (2003). Oil resistance controlled by phase morphology in natural rubber/nitrile rubber blends. J. Appl. Polym. Sci. 87:83-89.

Tambe SP, Singh SK, Patri M, Kumar D (2008). Ethylene vinyl acetate and ethylene vinyl alcohol copolymer for thermal spray coating application. Prog. Org. Coat. 62:382-386.

Utracki LA, Kanial MR (1982). Melt rheology of polymer blends. Polym. Eng. Sci. 22:96-114.

Wong ACY (1991). The Study of the Relationships Between Melt Index, Density abd Blend Ratio of Binary Polyethylene Blends. Polym. Eng. Sci. 31(21):549.

Yip E, Cacioli P (2002). The manufacture of gloves from natural rubber latex. J. Allergy Clin. Immunol. 110:S3-S14.

Zhaoying Z, Yuhui H, Bing L, Guangming C (2001). Studies on particle size of waterborne emulsions derived from epoxy resin. Eur. Polym. J. 37:1207-1211.

11

Synthesis, characterization and neuropharmacological activity of novel angular pentacyclic phenothiazine

Odin E. M., Onoja P. K. and Saleh J. F.

Department of Pure and Industrial Chemistry, Kogi State University, Anyigba, Nigeria.

13H-5,14-dihydroquinoxalino[2,3-a]phenothiazine (a new pentacyclic ring system) was synthesized by condensation of diphenylamine and sulphur, which on nitration yielded 1-nitrophenothiazine. Reduction of this compound afforded 1-aminophenothiazine dihydrochloride and subsequent protection of this product with acetic anhydride gave 1-acetylaminophenothiazine. The nitration of this acetylated compound yielded two isomeric compounds: 1-amino-2-nitrophenothiazine and 1-amino-4-nitrophenothiazine. The reduction of the ortho isomer furnished 1,2-diaminophenothiazine which when added to catecol and refluxed with ethanol gave the pentacyclic product 13H-5,14-dihydroquinoxalino[2,3-a]phenothiazine. Structures were established by analytical and spectral data. The results of the neuropharmacological screening data revealed that the novel system possessed neurosedative properties. The ability of this compound to antagonise fortwin – induced climbing behaviour in mice was correlated with neuroleptic potential.

Key words: 13H-5,14-dihydroquinoxalino[2,3-a]phenothiazine, pentacyclic, phenothiazine, neurosedative, fortwin-induced.

INTRODUCTION

The chemistry of phenothiazines has generated intensive scientific interest due to their biological properties (Ujuwala et al., 2012). Great work has been done on the bioactivity of phenothiazine and its derivatives. Phenothiazine itself is found to be a worming agent for livestock. The pesticidal action of phenothiazine results from the fact that they affect the nervous system of insects by inhibiting the breakdown of acetylcholine. The derivatives of phenothiazine have been studied for their antipsychotic properties (Luiza et al., 2007; Whitaker, 2004). They constitute the largest of the five classes of antipsychotic drugs. The antipsychotic activities of phenothiazine have been attributed to the basic nitrogen of the thiazine ring which donates electrons to the biological receptors by a charge transfer mechanism and also the ability of substituting the hydrogen atom to the nitrogen atom by substituent groups which further enhances the pharmacological property (Abdel-Monem and Portoghese, 1972; Martina et al., 2007).

The chemistry of the linear phenothiazine (**1**) is well developed (Abdel-Rahman et al., 2013; Okafor 1971; 1978). The non-linear aza phenothiazine is relatively understudied in spite of their pharmacological importance in medicine, agriculture and industry (Ezema et al., 2012; Okoro et al., 2009).

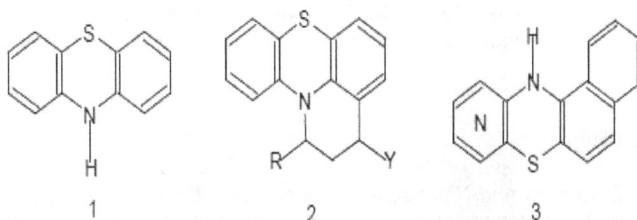

1   2   3

While tetracyclic, N-(2,3-dihydro-1H-pyrido[3,2,1-kl]phenothiazines) of type (**2**) have previously been reported (Alan et al., 1999; Tim et al., 2012), very few of the non linear tetracyclic phenothiazine of type **3** benzo[a]azaphenothiazine and its diaza-analogue type **4** have been reported (Chuan et al., 2012; Okoro et al., 2009).

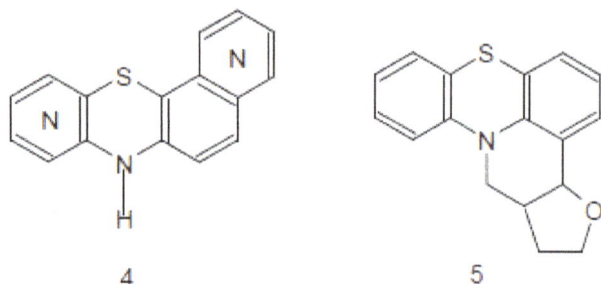

4   5

Still grossly understudied are the non-linear (angular) pentacyclic phenothiazine systems in spite of their known pharmaceutical and industrial applications. Pentacyclic phenothiazines of type **5** (Tetrahydro-3aH-furo[2,3:4,5]pyrido[3,2,1-kl]phenothiazine) have previously been prepared by the treatment of 10-[benzotriazol-1-yl] methyl phenothiazine with cyclic hydrofuran and 3,4-2H-dihydropyran (Abdel-Rahman et al., 2013; Alan et al., 1999). The non-linear pentacyclic aza phenothiazine of type **13** to the best of our knowledge has not been reported. In this paper we report the synthesis of the pentacyclic phenothiazine **13** (a new pentacyclic ring system) and its pharmacological properties.

MATERIALS AND METHODS

All chemicals were obtained from different sources (Lavans, Aldrich, Merck) and were used without further purification. The melting points were determined on a SMP3 melting point apparatus and are reported in degree Celsius uncorrected. Column chromatography was performed in Scharlan silica gel 60 (70 – 230 mesh).^{1}H and ^{13}C –NMR spectra were recorded on a Varian Gemini 2000 spectrophotometer operating 200 and 50 MHz respectively. Chemical shifts were recorded as ∂ values in ppm referenced to the solvent. HPLC separations were performed in a Bulk Scientific 500 apparatus using a reverse phase Lichrospher 100RP-18(5 μm) column at room temperature (eluent: methanol/water-8:2, v/v). The Infrared (IR) spectra were recorded in cm[-1] on a Bulk Scientific 500 Spectrophotometer. The mass spectra were recorded on a Schimadzu GCMS-QP-1000E, mass spectrometer at 70 eV and elementary analysis for C, H, S and N on a Perkin-Elmer analyzer 2400.

Drugs

Diazepam and Nitrazepam were obtained from Roche Nigeria ltd, while pentobarbital and apomorphine were obtained from Sigma Chemical Company, USA. All drugs were freshly prepared. Parallel control experiments were done in each case to correct possible effects caused by the vehicle alone.

Animals

All experiments performed on laboratory animals in this study followed the "Principle of laboratory animal care" (NIH publication No 85-23, revised 1985). Swiss albino mice (20 to 30 g) and wistar rats (180 to 200 g) of either sex were used. All the animals were maintained at the Animal Facility Centre of Kogi State University at standard conditions and temperature (25°C) and fed with standard diet (Ladokum feeds, Ibadan and water *ad libitum*.

Synthesis

The synthetic routes for all compounds are outlined in the scheme 1 and the details are given below:

Phenothiazine (1)

A mixture of the corresponding diphenylamine (13 g, 3.0 mol) and sulphur (1.2 g, 0.834 mol) was heated in a glycerol oil bath to 195°C. After cooling to 100°C, elemental iodine was added while heating continued. The separation of hydrogen sulphide was observed at 170°C and was decomposed by leading in 5% aqueous sodium hydroxide solution. The mixture was heated to 185°C and held at this temperature for 45 min. The bath was cooled to 50°C and was diluted with benzene (100 ml). This was filtered hot in a vacuum pump to remove the excess sulphur. The yellowish filtrate was concentrated using a rotary evaporator. The product was dried and purified by column chromatography. Yield 11.06 g (80.5%), m.p. 184-186°C. IR (Vmax/cm[-1]): 2999 (NH), 2859-2861 (C-H arom.), 717(C-H bending), 1197-1211 (C-H in-plane) and 1300-1411 (C-N arom.). UV: 311 (log Ɛ 3.3281)nm. ^{1}H NMR (200 MHz, DMSO): 7.32 (d, 2H), 7.34 (d, 4H), 7.29 (s, 6H), 7.28 (s, 8H), 7.04 (s, 3H), 7.06 (s, 7H), 6.99-6.92 (m, 1H), 6.98-6.91(m, 9H), 11.48 (s, -NH proton). ^{13}C NMR (50MHz, DMSO): 140.7 (C. Arom. Ring), 120.8 (CNH), 110.2, 113.3, 121.4, 122.6, 125.2, 139.6, 103.2, 118.9, 124.9, 130.6 (CH and C). Anal. Cal. For $C_{12}H_9NS$: C, 72.40; H, 6.30; N, 9.72; S, 22.22%. Found: C.72.29; H, 6.34; N, 9.69; S, 22.14%.

1-Nitrophenothiazine (7)

Concentrated nitric acid (10 ml, 0.5 mol) was placed in a 200 ml round bottom flask, while concentrated sulphuric acid (10 ml, 0.5 mol) was added to it portion wise over 30 min. With efficient stirring at room temperature, compound (**1**)(15 g, mol) was added. The mixture was refluxed in a water bath while the temperature was held at 50°C for 40 min. The product was washed with 500 ml cold water and filtered with suction on a Buchner funnel, dried and purified by column chromatography. Yield 20.40 g (95%), m.p. 160-161°C, IR: 2910 (C-H stretch), 972 (C-H bend), 1611-1462 (Arom. Skeleton), 1580-1550 and 134s4-1332 (aromatic nitro gr. Vibrations),UV: 320 nm. ^{1}H NMR: 7.21 (d,1H), 6.95 (d,2H), 6.91 (s,6H), 6.80 (m, 7H), 6.92 (d,8H), 8.82 (d, 9H), 10.30-9.82 (s, NH protons).^{13}C NMR: 141.8 (C arom. Ring), 121.9 (CNH), 130.8 (NO_2), 111.5, 112.4, 121.5, 123.2, 126.7, 140.1, 115.1, 117.3, 122.8, 123.4 (CH and C). Anal. Cal. For $C_{12}H_8N_2S$: C, 67.89; H, 3.80; N, 13.20; S, 15.11%. Found: C, 67.83; H, 3.78; N, 13.18; S, 15.09%.

1-Aminophenothiazine dihydrochloride (8)

Iron powder (20 g, 0.36 mol) was added portionwise to 1-nitrophenothiazine (17 g, 0.07 mol) suspended in 100 ml warm water containing 5 ml concentrated hydrochloric acid. The mixture was heated to 60°C and held at this temperature for 1½ h. The reaction mixture was filtered hot and the filtrate treated with excess concentrated hydrochloric acid, dried and purified by column chromatography (silica gel, DMSO). Yield: 13.50 g (79.4%); m.p. 151-158°C. IR: 3541 (N-H stretch), 2819-2821 (C-H stretch), 1093 (C-H inplane), 1320 (C-N stretch), 1684-1698 (Arom. skeletal system). UV 312 nm. ^{1}H NMR: 1.15-1.31 (m, 9H), 3.02-3.21 (m, 3H), 4.14-4.50 (m, 2H), 7.00-7.32(m, 8H), 10.30(br. s, 7H), 1.17-1.33 (m, 5H), 3.16-3.39(m, 1H), 3.57(m, 3H), 4.17-4.44(m,10H) 9.76 (s, NH protons), 5.70 (m,NH$_2$ protons). ^{13}C NMR: 141.5 (arom. Ring 5), 118.6 (CNH), 163.9 (CNH$_2$), 114.3, 112.2, 119.5, 122.5, 123.6, 141.5, 115.2, 118.2 121.5,1248 (CH and C). Anal. Cal. For C$_{12}$H$_{10}$N$_2$S: C, 67.25; H, 4.70; N, 13.08; S, 14.97%. Found: C, 67.18; H, 4.50; N, 13.02; S, 14.09%.

1-Acetylaminophenothiazine (9)

In a 100 ml beaker, 3.20 g (0.13 mol) of 1-aminophenothiazine dihydrochloride was added to 30 ml water. The solution was warmed to 50°C and 1.5 ml acetic anhydride added. Aqueous lead acetate prepared from 5 g (0.015 mol) lead acetate in 10 ml water was quickly added to the mixture. The beaker was swirled intermittently and placed in an ice bath for 20 min, filtered and the crystals were washed with cold water, dried and purified by column chromatography (silica gel, DMSO). Yield: 14.80 g (82.31%); m.p. 162-163°C. IR: 3670 (N-H stretch), 2929-2861 (C-H stretch), 979-713 (C-H out of plane), 1462 (C-H in-plane), 1354 (C-N stretch), 1611-1462 (Arom. Skeletal system), 2671 (C=O stretch), 2385 (-CH$_2$ groups). UV: 262.0 nm. ^{1}H NMR: 7.29 (d, 1H), 7.09 (d, 2H), 6.97 (d, 6H), 6.84 (d, 7H), 6.83 (d, 8H), 7.20 (d, 9H), 6.53-8.36 (m, NH protons), 3.98-3.94 (s, OCH$_3$), 2.19-2.26(s, -CH$_3$). ^{13}C NMR: 144.6 (C aromatic ring), 54.6 (CNH), 55.3 (OCH$_3$), 115.6, 113.4, 120.1, 122.6, 124.6, 141.8, 115.10, 118.4, 121.6, 124.9 (CH and C). Anal. Cal. For C$_{14}$H$_{12}$N$_2$OS: C,65.76; H,4.70; N,10.88; O,6.21; S,12.45%. Found: C,65.70; H,4.67; N,10.78; O,6.18; S,12.42%.

1-Amino-2-Nitrophenothiazine (10)

Powdered 1-acetylaminophenothiazine (0.41 g, 0.002 mol) was added to glacial acetic acid (0.4 ml) in a 100 ml beaker. While stirring, concentrated sulphuric acid (0.8 ml) was added to the mixture surrounded by a freezing mixture of ice and salt. At 0°C, a cold mixture of concentrated nitric acid 90.2 ml) and was added dropwise. The mixture was held at room temperature for one hour. After cooling to room temperature, the reaction mixture was poured into 500 ml cold water and allowed to cool for 15 min, then filtered with suction in a Buchner funnel and washed with cold water. The filtrate was heated for 2 h to obtain oily product of two layers which were separated to give two isomeric compounds. Purification was by column chromatographic method (Scheme 1). Yield: 210 ml (96.8%). UV: 540 nm. IR: 3698-3100 (hydrogen bonded N-H), 2912 (Ar,C-H); 1370 (Ar, C-N), 1644, 1473 (Aromatic skeleton). ^{1}H NMR: 7.23 (d, 1H), 6.98 (d, 2H), 6.95 (s,6H), 6.82 (m, 7H), 6.92 (d, 8H), 8.80 (d,9H), 8.30 (m, NH protons), 6.71 (m, NH$_2$ protons). ^{13}C NMR: 142.8 (C aromatic ring), 121.7 (CNH), 167.3 (CNH$_2$), 130.6 (CNO$_2$), 111.6, 112.5, 121.4, 123.2, 126.7, 140.1, 115.2, 117.3, 122.8, 124.1 (CH and C). Anal. Cal. For C$_{12}$H$_9$N$_2$O$_2$S: C, 58.76; H, 3.70; N, 11.42; S,13.07; O, 13.05%. Found: C, 58.72; H, 3.56; N, 11.38; S, 13.02; O, 13.01%.

1-amino-4-nitrophenothiazine (11)

Compound 11 was synthesized by using similar method as in 10 above. Yield: 68 ml (31.3%). UV: 490 nm. IR: 3692-3100 (hydrogen bonded N-H), 2899 (Ar. C-H), 1376 (Ar. C-N), 1642, 1472 (Ar. Skeleton). ^{1}H NMR: 6.80 (d, 1H), 6.20 (d,2H), 7.92 (s,6H), 5,89

(m,7H), 7.32 (d,8H), 7.50 (d,9H), 8.11 (m, NH protons), 6.67 (m ,NH$_2$ protons), ^{13}C NMR: 138.6 (C aromatic ring), 119,2 (CNH), 165.5 (CNH$_2$), 148.4 (CNO$_2$), 112.5, 106.5, 123.4, 132.1, 116.7, 138.6, 116.3, 117.6, 123.9, 120.4 (CH and C). Anal. Cal. for C$_{12}$H$_9$N$_2$O$_2$S: C, 58.76; H, 3.70; N, 11.42; S, 13.07; O, 13.05%. Found: C, 58.70; H, 3.54; N, 11.30; S, 13.00; O, 13.02%.

1,2-diaminophenothiazine trihydrochloride(12)

4 g (0.07 mol) of iron powder was added to a warm suspension of 2-nitro-1-aminophenothiazine (10 ml) in water (40 ml) containing 3 ml concentrated hydrochloric acid. 2 g (0.036 mol) of iron powder was added to the reaction mixture and heated for 50 min in a water bath. The resulting suspension was filtered hot and the filtrate treated with excess concentrated hydrochloric acid. Yield: 290 ml. (94.7%). UV: 312 nm. IR: 3671-3200 (N-H stretch), 809-781 (C-H out of plane), 1051 (C-H in-plane), 1477 (C-N stretch), 1641 and 1477 (Arom. skeletal system). ^{1}H NMR: 7.29 (d, 1H), 7.09 (d, 2H), 6.97 (d, 6H), 6.84 (d, 7H), 6.83 (d, 8H), 7.20 (d, 9H), 7.60 (m, NH protons), 5.70 (m, NH$_2$ protons). ^{13}C NMR: 144.6 (C aromatic ring), 119.5 (CNH), 169.5 (CH$_2$), 115.6, 113.4, 120.1, 122.6, 124.6, 141.8, 115.1, 118.4, 121.6, 124.9 (CH and C). Anal. Cal. for: C$_{12}$H$_{11}$N$_3$S: C, 63.02; H, 4.81; N, 18.24; S, 13.92%. Found: C, 63.01; H, 4.79; N, 18.22; S, 13.90%.

13H-5,14-dihydroquinoxalino[2,3-a]phenothiazine (13)

A mixture of 1,2-diaminophenothiazine hydrochloride (10 ml) and catechol (7.5 g, mol) was refluxed with ethanol (30 ml, 3 times for 1 h) and filtered off. The product was dried and purified by column chromatography. Yield; 15.32 g (98.4%). m.p. 174-175°C. UV: 320 nm. IR: 3773 (N-H stretch), 2910-2819 (Ar, C-H), 713 (C-H out of plane bending), 1098 (C-H in plane bend). ^{1}H NMR: 6.83 (d, 1H), 7.89 (d, 2H), 7.62 (s, 4H), 5.88 (m, NH protons), 7.29 (dd, 8H), 7.08 (d, 9H), 6.98 (dd, 13H), 6.84 (d, 15H), 7.19(d, 16H), 7.93 (s, 3H), 6.85 (d, 14H).^{13}C NMR: 145.7 (C-aromatic ring); 119.7 (CNH), 116.11 (C=C), 126.9, 128.2, 127.3, 127.8, 122.7, 115.8, 116.3, 112.3, 121.1,123.4, 123.9, 142.4, 116.11, 149.3, 122.4, 125.2 (CH and C). Anal. Cal. For C$_{20}$H$_{15}$N$_3$S: C, 72.92; H, 4.59; N, 12.76; S, 9.74%. Found: C, 72.89; H, 4.55; N, 12.57; S, 9.71%.

Acute toxicity studies

The acute toxicity (LD$_{50}$) was determined following the method described by (Amos et al., 2002; Gurad et al., 2011; Lorke, 1983). Animals were divided randomly into six groups of six mice each. The sample was administered intraperitoneally in the range of doses 10, 100, 1000, 1500, 2000, 3500 and 5000 mg/kg. The animals were observed for 72 h. At the end of the experiment, the animals were sacrificed and then autopsied and examined microscopically for any pathological changes.

Studies on exploratory activity in mice

Mice were divided into four groups of six mice each. Groups 1 and 2 were treated with sample at doses 50 and 100 mg/kg i.p respectively, while group 3 received normal saline (10 ml/kg) which served as control. Animals in group 4 were treated with diazepam (a known neurosedative) 2 mg/kg i.p (File and Pellow, 1985; Kota et al., 2010). 30 min after the drugs were administered; the animals were placed individually in an automatic Letica board with 16 evenly spaced holes with a counter (Letica LE3333). The number of head dips by the mice into the holes over a period of 5 min was automatically counted (Koofreh et al., 2012; Perez et al., 1998;

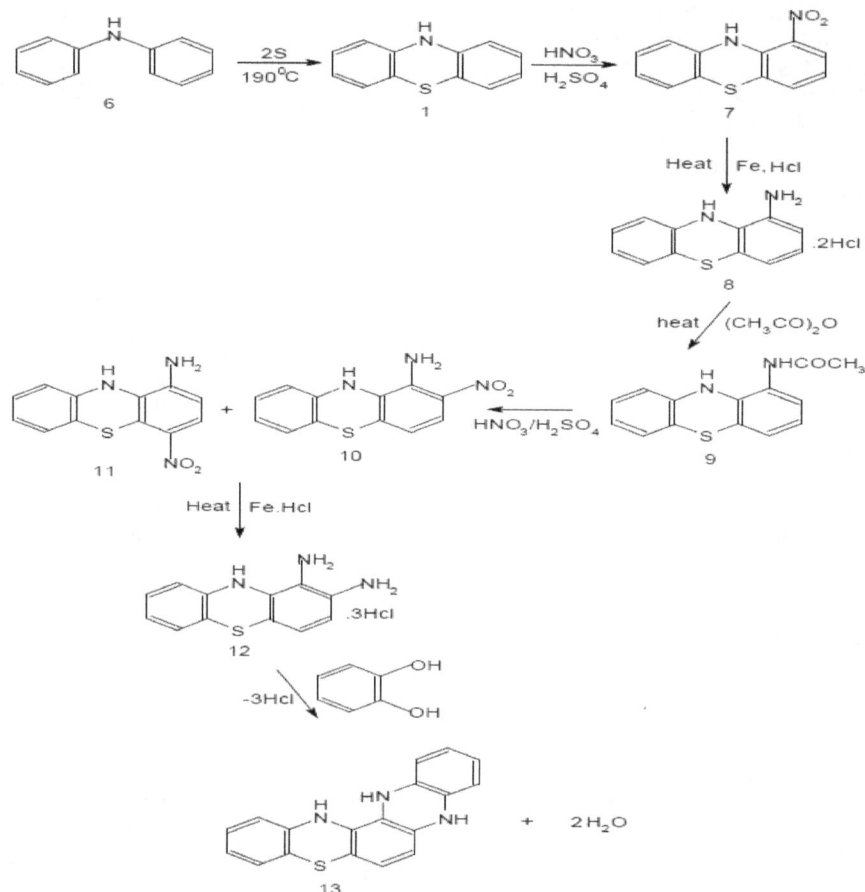

Scheme 1. Synthetic routes for all the compounds.

Wolfman et al., 1994).

Studies on spontaneous motor activity

Adult mice were randomly divided into three groups of 6 mice each. Groups 1 and 2 received the sample at doses 50 and 100 mg/kg i.p, while group 3 received normal saline (10 ml/kg p.o). Motor activity of the mice was recorded using a Letica activity cage floor. The animals were singly placed in the cage and their activity was recorded for 6 min at 30 min intervals for a period of 120 min (Odin et al., 2003). In another experiment, the effect of aminophiline 2 mg/kg i.p was recorded. The effects of the sample on aminophiline – induced hyperactivity were compared to that of chlorpromazine 2 mg/kg i.p.

Studies on pentobarbital induced sleep

Adult rats were divided into 4 groups of 6 rats each. Groups 1 and 2 received 50 and 100 mg/kg i.p of the sample and group 3 was administered normal saline (10 ml/kg p.o) and served as control (Jae-wook et al., 2012; Kaul and Kulerni, 1978; Ngouemo et al., 1994). Diazepam 1 mg/kg i.p was administered to animals in group 4. All the animals were injected with Phenobarbital sodium (phenobarbitone 35 mg/kg), 30 min after the drug treatment. The onset and duration of sleeping time were recorded (Hong et al., 2009; Wambebe, 1985).

Studies on apomorphine (fortwin) – induced climbing in mice

Adult mice were randomly divided into 3 groups of 10 mice each. The first group received normal saline (10 ml/kg p.o) and served as control. Groups 2 and 3 received the sample at doses of 50 and 100 mg/kg i.p. 30 min after treatment, all mice were treated with apomorphine (fortwin) 3 mg/kg (Hong et al., 2009; Protais et al., 1976). Readings were taken at 10, 20, and 30 min after apomorphine administration. The mice were observed for climbing and scored as follows:

0= fore paws on the floor
1= fore feet holding the vertical bars
2= fore feet holding the bars.

RESULTS

Phenothiazine (**1**) was synthesized from the fusion of diphenylamine (**6**) with excess elemental sulphur. The sulphur residue was removed by heating the mixture after benzene was added and was filtered hot. The filtrate is phenothiazine in benzene which was recovered on heating to dryness.The phenothiazine (**1**) was nitrated with a mixed acid at 50°C while avoiding polynitration to give 1- nitrophenothiazine (**7**).

Reduction of (**7**) with iron in dilute hydrochloric acid furnished 1-aminophenothiazine dihydrochloride (**8**), which was subsequently reacted with acetic anhydride to achieve (**9**). This was done to protect the amino group in (**8**) from further nitration. Nitrating compound (**9**) in mixed acid yielded ortho and para nitro phenothiazines, (**10**) and (**11**) respectively. Reduction of (**10**) with iron in dilute hydrochloric acid furnished 1,2-diaminophenothiazine trihydrochloride (**12**), which when added to catecol and refluxed with ethanol gave the novel pentacyclic product:13H-5,14-dihydroquinoxalino[2,3-a]phenothiazine (**13**) (Scheme 1).

The structural assignment of the synthesized compounds is based on the spectral data. In the IR spectrum of Compound **1**, absorption band at 2999 cm^{-1} represents the hydrogen bonded N-H stretching. There were number of peaks at 2859-2861 cm^{-1}, 717 cm^{-1}, 1197-1211 cm^{-1}, 1300-1411 cm^{-1} for aromatic C-H stretching, out of plane C-H bending, in-plane C-H bending and the aromatic C-N stretching respectively. Only two of the aromatic skeletal stretching bands were readily visible at 1300-1473 cm^{-1}. The ^{1}H and ^{13}C NMR studies of Compound 1 confirmed the structure. The Compound **1** reacting with mixed acids (nitration) yielded 1-nitrophenothiazine (**7**). The IR spectrum of compound (**7**) showed absorption for aromatic C-H stretching and in-plane C-H bending at 2910 and 972 cm^{-1} respectively. The aromatic skeletal system is responsible for the bands at 1611 and 1462 cm^{-1}.

Compound (**7**) was reduced with iron to give (**8**)-1-aminophenothiazine dihydrochloride. The N-H stretching and the aromatic C-H stretching appeared at 3541, 2819 and 2821 cm^{-1} respectively. The in-plane C-H band appeared weakly at 1093 cm^{-1}, while the C-N stretching showed at 1320 cm^{-1} which is characteristic of aromatic amines. The bands at 1684 to 1698 cm^{-1} are that of aromatic skeletal system. In ^{1}H NMR spectra data, Compound (**7**) shows a singlet at δ 10.30 to 9.81 due to N-H proton. This was shifted to δ 9.76 in 1-aminophenothiazine hydrochloride (**8**). This shifting towards upfield in compound 8 is ascribed to intramolecular hydrogen bonding as –NH...O=N in Compound 7.

Compound (**8**) was subsequently reacted with acetic anhydride to protect the amino group from further nitration. This yielded 1-acetylaminophenothiazine (**9**). The hydrogen bonded N-H stretching appeared at 3670 cm^{-1}, while the bands at 2929 and 2861 cm^{-1} were for C-H stretching for aromatic systems. The bands at 1354 cm^{-1} is characteristic of aromatic C-N stretching, while bands at 2385 and 2671 cm^{-1} indicated C=O stretching and methylene CH_2 groups. In the ^{1}H NMR spectrum of Compound 9, the multiplet for –NH protons appeared in the region δ 6.53 to 8.36. The –OCH_3 protons and –CH_3 protons in the compound showed a singlet in the region δ 3.98-3.94 and δ 2.19 to 2.26 respectively, indicating a complete acylation of Compound 8.

The nitration of Compound (**9**) yielded two isomers: 2-nitro-1-aminophenothiazine and 4-nitro-1-aminophenoyhiazine (Compounds **10** and **11** respectively). The IR spectra of compounds **10** and **11** showed broad bands at 3698 to 3100 cm^{-1} indicating hydrogen bonded N-H stretching. The absorption band at 2912 cm^{-1} was for aromatic C-H stretching. The band at 1370 cm^{-1} is characteristic of aromatic C-N stretching, while the aromatic skeletal was found at 1644 cm^{-1} and 1473 cm^{-1}. Similarly, in compound 10, N-H_2 proton appeared as multiplet at δ 6.71, while in ^{13}C NMR spectrum, a characteristic signal appeared for (CNH_2) and (CNO_2) in the range of δ 167.3 and δ 130.6 respectively. These were found absent in compound 9 indicating a successful nitration of this compound. The nitro group is responsible for the broad shoulder at 1195 cm^{-1}.

Reduction of (**10**) furnished 1,2-diaminophenothiazine hydrochloride (**12**). The IR spectrum of (**12**) showed a broad band at 3671 to 3200 cm^{-1} for hydrogen bonded N-H stretching. The band at 1477 cm^{-1} indicated C-N stretching, while the aromatic skeletal system was located at 1641 and 14777 cm^{-1}. In the ^{1}H NMR spectrum, Compound 12 showed two signals for NH and NH_2 at δ 7.60 to 5.70 respectively. A characteristic signal appeared for CH_2 in the range of δ 119.5 in the ^{13}C NMR spectrum, while that of CNH_2 was located at δ 169.5.

Refluxing a mixture of Compound (**12**) and catechol with ethanol yielded the pentacyclic product-13H-5,14-dihydroquinoxalino[2,3-a]phenothiazine (**13**). The IR spectrum of Compound (**13**) showed N-H stretching at 3773 cm^{-1}. The bands at 2910 to 2819 cm^{-1} appeared for aromatic C-H stretching, while bands at 713 and 1098 cm^{-1} were for C-H out-of-plane bending and in-plane bending respectively. These clearly support the fact that Compound (**11**) was not used in the synthesis of the novel product (**13**). The absorption at 205 nm in the UV-visible spectrum of Compound (1) resembles that of benzene while the shift in wavelength to 311 nm indicated the presence of auxochrome type –NHR in phenothiazine.

The UV spectrum of (**8**) showed maximum absorption at 312 nm. No appreciable bathochromic shift because the compound is in the form of hydrochloride. Compound (**10**) showed a UV maximum at 540 nm. This powerful bathochromic shift is probably due to the presence of free amino group, while the pentacyclic product (**13**) exhibited a UV maximum at 320 nm characteristic of phenothiazine systems. In the ^{1}H NMR spectrum, compound 13 displayed a signal at δ 5.88 for NH protons, while the multiplet for aromatic protons appeared in the region between δ 6.54-8.36. In ^{13}C NMR spectrum of compound 13, a characteristic signal appeared for (CNH) in the range of δ 119.7. The mass spectrophotometric studies performed on the phenothiazines confirmed the molecular weight values. The results of the pharmacological tests are as presented in Tables 1, 2, 3, 4 and 5.

Table 1. Effect of sample on exploratory activity in mice.

Treatment	Dose mg/kg	Mean score
Normal saline	10 ml/kg	40.8±5.8
Sample	50	21.3±2.4
Sample	100	12.2±2.9
Diazepam	2	15.0±3.5

Table 2. Effect of sample on spontaneous motor activity in mice.

Treatment	Dose mg/kg	Time (minutes)				
		0	30	60	90	120
Normal saline	10 ml/kg	91.2± 3.6	86.2± 2.5	82.8± 1.8	81.8± 2.0	75.5± 2.8
Sample	50	91.8±4.1	35.8±3.1	31.8± 2.3	22.5± 2.8	13.7±2.0
Sample	100	90.7± 1.2	33.0± 2.4	14.7± 1.7	8.2± 1.3	5.8±1.1

Table 3. Effect of sample on aminophilin induced hypermotility in mice.

Treatment	Dose mg/kg	Time (minutes)				
		0	30	60	90	120
Normal saline	10 ml/kg	90.3±2.9	85.2± 2.5	82.5±2.4	81.0±1.5	76.0± 2.9
Aminophillin	2	90.0±1.6	102.1±4.4	126±5.2	118±3.1	104 ±2.9
Aminiphillin + sample	50	91.8±1.6	73.5±2.9	53.5±2.6	33.7±3.1	17.3±2.7
Aminophillin +sample	100	90.5±2.2	62.5±2.2	34.2±2.4	17.3±2.6	14.7±2.5

DISCUSSION

The structural assignment of the synthesized compounds was based on the spectral data. The IR spectrum of the pentacyclic Compound (13) clearly showed that the isomeric Compound (11) was not used in the synthesis of the final product. This was further buttressed by the disappearance of C-O absorption (1200 cm^{-1}) in the spectrum of (13). The pentacyclic product exhibited a UV maximum at 320 nm characteristic of phenothiazine systems. The angular pentacyclic ring system was further identified by the information from the ^{1}H and ^{13}CNMR spectral with the resonances assigned to hydrogen and carbon. Compound **13** (sample) when administered, inhibited the exploratory behaviour in mice dose dependently. The effect was similar to that of diazepam (2 mg/kg), a known neurosedative and significantly different from those of control (Table 1). From Table 2, at 50 and 100 mg/kg i.p, the sample caused a significant time and dose dependent decrease in the spontaneous motor activity in mice.

Similarly, aminophilin induced hypermotility was reduced dose and time dependently (Table 3). When aminophilin alone is administered, at 60 min the mice was very active, 12±5.2, while the combination of 50 mg/kg sample with aminophilin reduced the activities of the mice, 53.5±2.6. The activities of the mice were further reduced, 34.2±2.4 when the dose was increased to 100 mg/kg. Similarly at 120 min, it was noticed that at 50 mg/kg the activities of the mice reduced to 17.3±27, while 100 mg/kg sample plus aminophilin further reduced the activities of mice to 14.7±2.4. Table 4 recorded that the administration of the sample at 50 and 100 mg/kg i.p did not affect the onset of sleep, but significantly prolonged the duration of pentobarbital sleep dose dependently. Similarly, Table 5 showed that 50 and 100 mg/kg administered i.p inhibited fortwin induced climbing dose dependently.

Conclusion

The importance of linear phenothiazine compounds as antipsychotic drugs has long been recognised. Compound **13** is the first angular pentacyclic phenothiazine to possess neurosedative properties. The hole board experiment is a measure of exploratory activity and a decrease in this parameter revealed sedative effects. The procedure has been accepted as a parameter for evaluating anxiety condition. The decrease

Table 4. Effect of sample on pentobarbital induced sleep in mice.

Treatment	Dose{(mg/kg}	Duration of sleep (minutes)
Normal saline	10 ml/kg	52.6±4.2
Sample	50	92.8± 5.6s
Sample	100	126±2.6
Diazepam	1	82.8±3.4

Table 5. Effect of sample on fortwin induced climbing in mice.

Treatment	Time (minutes)		
	1 0	**20**	**30**
normal saline 10ml/kg	0	1	2
Sample 50 mg/kg	1	0	0
Sample 100 ml/kg	0	0	0

The values are expressed as follows: 0= four paws on the floor, 1= fore feet holding the vertical bars, 2= fore feet holding the bars.

in spontaneous motor activity and potentiation of pentobarbital induced sleep strongly suggest central depressant activity. The ability of Compound **13** to antagonise fortwin - induced climbing behaviour in mice has been correlated with neuroleptic potential.

ACKNOWLEDGEMENT

The authors are grateful to Chemistry Laboratory, Kogi State University for the spectroscopic and elemental analysis and Paul Ojodale Samuel for Secretarial assistance.

REFERENCES

Abdel–Monem MM, Portoghese PS (1972). Medical Chemistry, Hampton Press, N. Y., USA. pp.16/208.

Abdel-Rahman A, Kandeel E, Berghot M, Mauwa A (2013). Synthesis and Reactions of Some new Benzo[a]phenothiazine-3,4-dione Derivative. J. Het. Chem. 50:298-303.

Alan RK, Samia A, Baozhen Y, Guotang Q (1999).Synthesis of Tetracyclic and Pentacyclic Phenothiazine via Benzotriazole Methodology. J. Het. Chem. 36:473.

Amos S, Binda A, Vongtan H, Odin EM, Okwute SK (2002). Sedative effect of the methanolic leaf extract of Newbouldia Leavis in mice and rats. Boll. Chim. Farmac 144(6):471– 475.

Chuan D, Xiaofei S, Xingzhao T, Li W, Dan Z (2012). Synthesis of Phenothiazines via Ligand-free CuI-catalyzed cascade C-S and C-N Coupling of aryl ortho-dihalides and ortho-aminobenzenethiols. Chem. Comm. 48:5367-5369.

Ezema B, Okafor C, Ezema C, Onoabedje A (2012). Synthesis of New Diaza Angular and Tetraaza Complex Phenothiazine Rings. Chem. Pro. Eng. Res. 3:107-119.

File S, Pellow S (1985). The effect of Triazolobenodiazepines in TWO Animals of Anxiety on the hole board. Brit. J. Pharm. 86:729–735.

Gurad A, Anshoo G, Pravin K, Abdesh K (2011). Acute Toxicity Studies of Safer and more effective Analogues of N,N-Diethyl-2-Phenylacetamide. J. Med. Entomol. 48(6):1160-1166.

Hong M, Chung-soo K, Yuan M, Ki-wan O (2009). Magnold Enhances Pentobarbital-induced Sleeping Behaviours: Possible involvement of GABA ergic Systems. Phyto. Res. 23(9):1340-1344.

Jae-wook K, Chung-soo K, Zhenzhen H, Ki-wan O (2012). Enhancement of Pentobarbital- induced Sleep by Apigenin through Chloride Ion Channel Activation. Arch. Pharm. Res. 35(2):367-373.

Kaul PN, Kulkarni SK (1978). New Drug Metabolism Inhibitor of Marine Origin. J. Pharm. Sci. 67:1293–1296.

Koofreh D, Christopher E, Justina N, Atim A (2012).: Locomotor and Exploratory Behaviour in Mice with Treated Oral Artemether Suspension. Sci. Acad. Pub. 1(3):17-24.

Kota T, Shozo T, Nobuhiro N (2010). Decreased Exploratory Activity in a Mouse Model of 15 q duplication Syndrome. J. PLOS ONE 5:12.

Lorke D (1983). A new approach to practical acute toxicity. Arch. Toxicol. 54:25–27.

Luiza G, Castelia C, Clavdia M, Ioan A (2007). Microwave Assisted Synthesis of Phenothiazine and Quinoline Derivatives. Int. J. Mol. Sci. 8(2):70-80.

Martina H, Jan S, Anthony J, Kenneth I, Thomas J, Uwe H (2007). Phenthiazine Synthesis and Metallochromic Properties. J. Org. Chem. 72(18): 6714-6725.

N'gouemo P, Nguemby–Bina C, Baldy-Moulinia M (1994). Some Neuropharmacological effect of an Ethanolic Extract of Mapronnea African in Rodents. J. Ethno. 43:161–166.

Odin EM, Okwute SK, Amos S, Gamaliel K (2003). Antimalarial and Neurosedative Properties of Newbouldia Laevis leaf. Int. Wd. J. Sci. Tech. 2(1):18–97.

Okafor CO (1978). A New Synthesis of Three – Branched Diazaphenothiazine Dyes. Dye Pig. 9:427–442.

Okafor CO (1971). The Chemistry of Natural Products. Int. J. Sulph. Chem. 6B:237.

Okoro UC, Onoabedje E, Odin EM (2009). The first Angular Triazaphenothinone and the related diaza – analogue. Int. J. Chem. 19(4):197–221.

Perez GRM, Perez IJA, Gacia D, Sossa MH (1998). Neuropharmacological activity of Solanum Nigrum Fruit. J. Ethno. 62:43.

Protais P, Costertin J, Schwartz JC (1976). Climbing behaviour induced by Apomorphine in Mice. Asimple test for the study of dopamine receptors in the stratum. J. Psycho. 50:1-6.

Tim M, Daniel O, Andrea P, Karl K, Thomas J (2012). Phenothiazinyl Rhodanylidene Merocyanines for Dye-sensitized Solar Cells. J. Org. Chem. 8:300-307.

Ujuwala S, Meghasham N, Mahendra C (2012). Synthesis,

characterization and antimicrobial activity of some 2-(propenone) aryl 3-substituted phenothiazine. Der Pharm. Chem. 4(3):967–971.

Wambebe C (1985). Influence of some agent that Affect 5-HT metabolism and receptorsand nitrazepam induced sleep in mice. Brit. J. Pharm. 84.185–191.

Whitaker R (2004). The case against antipsychotic drugs – A 50year record of doing more harm than good. Med. hypo. 62(1):5 –13.

Wolfman C, Viola H, Paladini AC, Dajas D, Medina J (1994). Possible axxiolytic effects of chrysin, a central benzodiazepine receptor ligand isolated from *Passiflora cocruica*. Pharm. Biochem. Behav. 47: 1.

12

Synthesis and characterization of polyindole with liquid crystalline azobenzene as side chains

Seyed Hossein Hosseini[1] and Maryam Ashjari[2]

[1]Department of Chemistry, Faculty of Science, Islamic Azad University, Islamshahr Branch, Tehran-Iran.
[2]Department of Chemistry, Faculty of Science and Engineering, Islamic Azad University, Saveh Branch, Saveh-Iran.

In this study, a series of azobenzene-functionalized liquid crystalline (LC) polyindole derivatives: poly{2-[N-ethyl-N-[4-[4'-(nitrophenyl)azo]phenyl]amino]ethyl-3-indolyl acetate}, poly(In3AA-RedI), poly{2-[N-ethyl-N-[4-[4'-(nitrophenyl)azo]phenyl]amino]ethyl-3-indolyl acetate-co-indole}, poly(In3AA-RedI-co-In), poly{2-[N-ethyl-N-[4-[4'-(nitrophenyl)azo]phenyl] amino]ethyl-3-indolyl acetate-co-pyrrole}, and Poly(In3AA-RedI-co-py) were synthesized. Novel 3-substituted indole with liquid crystalline side chain (In3AA-RedI) was synthesized by the direct reaction of indole-3-acetic acid with 2-[N-ethyl-N-[4-[4'-(nitrophenyl)azo]phenyl] amino]ethanol (RedI). Chemical polymerization of (In3AA-RedI), and its copolymerization with indole and pyrrole were carried out by using ferric percholorate as oxidizing agent. The composition, structure and thermal property of these LC polyindole derivatives were fully characterized by FTIR, ^{1}H,^{13}C-NMR and UV-Visible spectroscopic methods, and its LC behavior and photoresponsive property were also investigated by polarized optical microscope and differential scanning calorimetry (DSC). The results show that poly(In3AA-RedI) exhibited the smectic A (S_A) and nematic (N) liquid crystalline behavior. Conclusion shifted phase transition temperatures of the poly(In3AA-RedI) in the heating process are as follows: C$\rightarrow$ S_A (161°C), $S_A$$\rightarrow$N (184°C) and N$\rightarrow$I (231°C). Electrical conductivity of polymer [poly(In3AA-RedI)] and two of its copolymers [poly(In3AA-RedI-co-In) and poly(In3AA-RedI-co-Py)], has been studied by four probe methods and produced 8.3×10^{-4}, 6.4×10^{-4} and 4.7×10^{-3} Scm^{-1} conductivities, respectively.

Key words: Conducting polymers, electrical conductivity, liquid crystalline polymer, optical properties, optical materials, polyindole.

INTRODUCTION

Conjugated polymers are well-known for their excellent electrical conductivities in oxidized (doped) state. The recent development of processable conducting polymers has opened the way for large-scale industrial applications. Conjugated polymers have been used widely in many areas such as rechargeable batteries (Heinze, 1991; Roth and Graupher, 1993), condensators (Mohammadi et al., 1986), diodes (Turut and Koleli, 1993; Kolelil et al., 1994), and sensors (Hosseini et al., 2005, 2006). Among these classes of polymers,

polyaniline, polypyrrole, polythiophene, etc. have been studied extensively because of their favorable processability and relative stability (MacDiarmid, 1997; Hosseini and Entezami, 2001, 2003). Heteroaromatic molecules containing nitrogen have very interesting properties. Polyindole is an electro active polymer, which can be obtained from electrochemical oxidation of indole or chemical oxidation using $FeCl_3$ or $CuCl_2$ (Xu et al., 2006). However, only little investigations have been made on chemically synthesized polyindole (John and

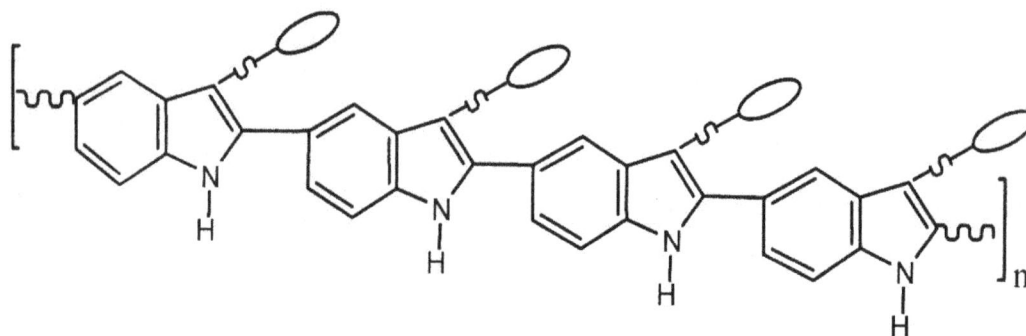

Figure 1. Side chain polyindole.

Palaniappan, 2005). The polymerization efficiency and the conductivity of polyindole are lower than the other known hetero atom containing conducting polymers such as polycarbazole, polyfuran, polyisothianapthene, polybithiophene, etc. Therefore, they did not attract much attention as the other types of conducting materials. There are two main procedures for the preparation of soluble conjugated polymers: one is the incorporation of relatively long and flexible side chains; another is the introduction of large counterions (Zaho and Wang, 2006).

Synthesis of three-substituted azobenzene-functionalized polyindoles has attracted much interest both from synthetic considerations as well as from materials science. Polyindoles with azobenzene groups in three-position will not only have better processability and stability, but also may possess novel electrical, electrochemical and optical properties. Thus, the combination of polyindole backbone with photoactive azobenzene groups can provide a new approach to develop other novel materials with unique electronic and optical properties. Azo chromophore has been demonstrated to be good photoisomerizable units (Matsui et al., 2001) for optical switching, image strage and other electrooptic devices. The introduction of photoresponsive moieties into liquid crystalline polymers is a useful method to provide the liquid crystal (LC) materials with photoresponsive properties. Hu and coworkers (Zhao et al., 2005) synthesized a series of novel LC azobenzene-functionalized polythiophenes with the aims of preparing liquid crystalline thiophene derivatives for photonic applications.

In the previous works, we synthesized a new liquid single crystal (Yousefi et al., 2008) and reported liquid crystalline polymer based N-substituted pyrrole (Hosseini and Mohammadi, 2009). This polymer exhibit liquid crystalline and electrically conductivity properties, as well. In this paper, polyindole was selected as a main chain skeleton. The side chain polyindole contains mesogenic group which shows liquid crystalline property. The 3-substituted of the indole ring was prepared and their liquid crystallinity and thermal properties investigated. Figure 1 shows side chain polyindole. So, a series of

novel LC azobenzene-functionalized polyindoles synthesized with the aims at preparation of liquid crystalline indole derivatives for photonic applications. The synthesis, characterization, and photoresponsive behavior of these chromophore-based LC polyindole derivatives were fully discussed. Molecular structure of the LC polyindole is illustrated in Figure 2, where LC group is introduced into 3-position of the indole unit. The polymer consists of main chain, flexible methylene spacer, linking group and tailing group, as shown in Figure 2.

EXPERIMENTAL PROCEDURE

Physical measurements

^{1}H, ^{13}C-NMR spectra were recorded on a BRUKER 250 NMR spectrometer at 400 MHz in deuterated chloroform-d or dimethylsulfoxide-d_6 with TMS as an internal standard. NMR data are reported in the following order: chemical shift (ppm), spin multiplicity (s=singlet, d=doublet, t=triplet, q=quartet, m=multiplet), and integration. Differential Scanning Calorimetry (DSC) analyses were performed at 5°Cmin^{-1} on a TA instruments using STA 625 DSC. FT-IR spectra were recorded on a 8101-M-Shimadzu and BRUKER-IF-66.5 spectrometer. Vibrational transition frequencies are reported in wave number (cm^{-1}). The UV-Visible spectra were obtained using a UV-Vis recording spectrophotometer (Perkin-Elmer Lambda 15).

Materials

Indole-3-acetic acid (Fluka), indole (Merck) and pyrrole (Fluka, 96%) were distilled prior to use. Tetrahydrofuran (Merck), petroleum ether and ethyl acetate were distilled and dried with molecular sieves (4A°) prior to use. 4-nitroaniline (Merck, 98%), dicyclohexyl carbodiimide (DCC) (Merck, 98%), 4-(dimethylamino)pyridine (DMAP) (Merck, 99%), 2-(N-ethyl aniline)ethanol (Merck, 99%), sodium nitrite (NaNO$_2$) (Merck, 99%), methanol (Fluka) and the other materials used in this work were purchased from Merck chemicals and purified, or prepared according to literature methods.

Preparation of 2-[N-ethyl-N-[4-[(4'-nitrophenyl)azo]phenyl]amino] ethyl (RedI)

7 g (0.05 mol) of 4-nitroaniline was dissolved in a solution of 25 ml

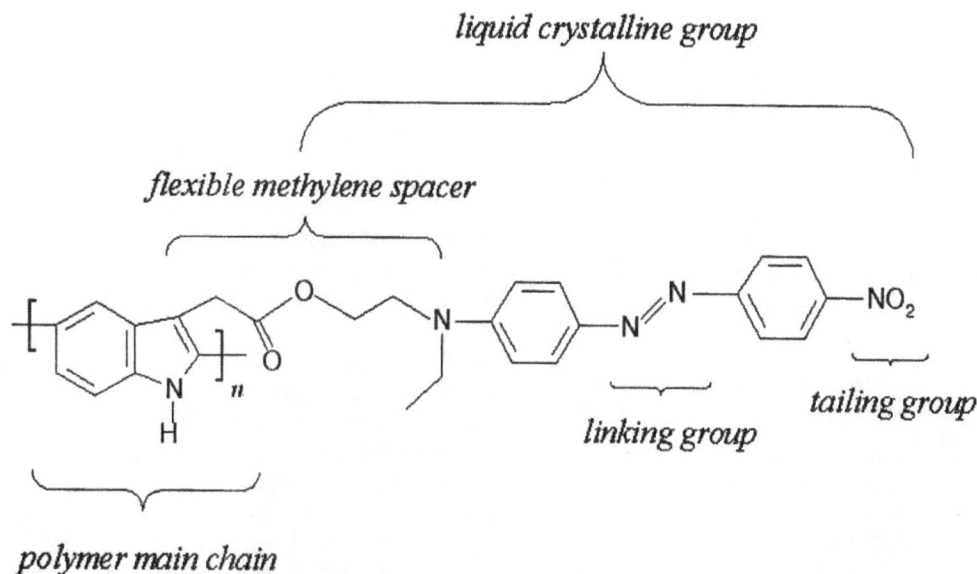

Figure 2. Molecular structure of liquid crystalline polyindole derivative.

of concentrated hydrochloric acid and 150 ml of water. The mixture was cooled to 0°C in an ice-water bath, and then a solution of 3.6 g (0.05 mol) of sodium nitrite in 15 ml of water was added dropwise. The resultant solution of diazonium salt was stirred for 30 min at 0–3°C. Another solution of 5 g (0.03 mol) of N-ethyl-N-hydroxyethyl aniline dissolved in 10 ml of hydrochloric acid (10%) was stirred for 5 min at 0°C. The above solution of diazonium salt was added into this solution within 30 min. The mixture was allowed to stand for 20 min. Then was neutralized with added NaOH solution with slowly and stirring. Reddish crystals were filtered on a Buchner funnel and recrystallised with 2-propanol. Compound 1 was obtained as crimson crystals. Yield 9 g–85%, mp: 170-173°C. UV (THF): λ_{max}=310 nm (0.3 intensity), 470 nm (1.25 intensity). FT-IR (KBr pellets, υ in cm^{-1}), υ: 3434 (υ_{OH}), 3200 (υ_{C-H}, Ar), 2950 (υ_{C-H}, Al), 1599 ($\upsilon_{N=N}$), 1515, 1341 (υ_{NO2}), 1450 ($\upsilon_{C=C}$), 1141 (υ_{C-O}), 800-850 (υ_{C-H}, OOP) cm^{-1}. ^{1}H-NMR (CDCl$_3$): δ 1.26 (3H, t), 1.75 (1H, s), 3.56 (2H, q), 3.62 (2H, t), 3.90 (2H, t), 6.81 (2H, d), 7.88 (2H, d), 7.92 (4H, d), 8.32 (2H, d) ppm. ^{13}C-NMR (CDCl$_3$): δ 12.1, 46.4, 52.8, 60.1, 112.3, 122.5, 124.7, 126.6, 143.8, 147.2, 151.6, 157.2 ppm.

Preparation of 2-[N-ethyl-N-[4-[(4'-nitrophenyl)azo]-phenyl]amino]ethyl-3-indolyl acetate, (In3AA-RedI)

A total 2.275 g (0.013 mol) of indole-3-acetic acid and 3.454 g (0.011 mol) of RedI (crystallized in isopropyl alcohol) were dissolved in 50 ml of dry THF. Then 2.269 g (0.011 mol) of N, N-dicyclohexyl carbodiimide (DCC) and 0.0916 g (0.75 mmol) of 4-(dimethylamino) pyridine (DMAP) were added to the vigorously stirred solution. The stirring continued for 5 h. The mixture was then filtered, and the solvent was removed by rotary evaporator under vacuum. The product was purified by column chromatography (silica gel, petroleum ether: ethyl acetate=1:4, v/v), followed by recrystallization from petroleum ether/ethyl acetate to yield red crystals (compound 2 was obtained). Yield: 50%, mp: 138-140°C. UV (THF); λ_{max} = 285 nm (0.70 intensity), 335 nm (0.4 intensity) and 535 nm (0.65 intensity). FT-IR (KBr pellets, υ in cm^{-1}), υ: 3384 (υ_{N-H}, In), 3100 (υ_{C-H}, Ar), 2930 (υ_{C-H}, Al), 1726 ($\upsilon_{C=O}$), 1601 ($\upsilon_{N=N}$), 1515, 1334 (υ_{NO2}), 1627, 1458 ($\upsilon_{C=C}$), 1139 (υ_{C-O}), 828 (υ_{C-H}, OOP) cm^{-1}. ^{1}H-NMR (d^6- DMSO): δ 1.06 (3H, t), 2.50 (2H, d), 3.43 (2H, q), 3.74

(2H, t), 4.28 (2H, t), 6.93 (2H, d), 7.81 (2H, d), 7.94 (2H, d), 8.36 (2H, d), 7-7.5 (5H indole, m) 10.85 (1H, s) ppm. ^{13}C-NMR (d^6-DMSO): δ 12.4, 31.5, 45.4, 48.8, 62.2, 106.6, 111.9, 112.1, 118.4, 119, 121.6, 123, 124.6, 125.4, 126.5, 127, 137, 147.6, 149.5, 152.7, 157.5, 172 ppm.

Preparation of Poly{2-[N-ethyl-N-[4-[(4'-nitrophenyl)azo]-phenyl]amino] ethyl-3-indolyl acetate}, Poly(In3AA-RedI)

Polymerization was carried out as follows: 0.5 g (1.06 mmol) of monomer (In3AA-RedI) in 25 ml THF was added dropwise to a suspension 1.126 g, (3.18 mmol) of Fe(ClO$_4$)$_3$ in 20 ml of THF under nitrogen atmosphere. The mixture was stirred at 50°C temperature for 24 h. The polymerization mixture was added dropwise into 100 ml of methanol. The precipitates were filtered, dissolved in 5 ml of THF, and reprecipitated into 100 ml of methanol. This procedure was repeated three times until the unreacted monomer was completely removed. Finally, the polymer was dried under vacuum at 25°C to constant weight (Compound 3 was obtained). UV (THF): λ_{max}=280 nm (1 intensity), 330 nm (0.3 intensity), 475 nm (0.2 intensity) and 680 nm (0.4 intensity). FT-IR (KBr pellets, υ in cm^{-1}), υ: 3383 (υ_{N-H}, indole), 3200 (υ_{C-H}, aromatic), 2917 (υ_{C-H}, aliphatic), 1750 ($\upsilon_{C=O}$), 1627 ($\upsilon_{N=N}$), 1514, 1335 (υ_{NO2}), 1100 (υ_{C-O}) cm^{-1}. ^{1}H-NMR (d^6- DMSO): δ 1.09 (3H, t), 1.24 (2H, d), 3.63 (2H, q), 3.72 (2H, t), 4.27 (2H, t), 6.88-7.36 (6H, m), 7.82 (2H, d), 7.95 (2H, d), 8.38 (2H, d), 10.91 (1H, s) ppm. ^{13}C-NMR (d^6-DMSO): δ 11.8, 30.6, 44.8, 48.2, 61.6, 106.6, 111.3, 111.6, 118.3, 118.4, 121.0, 122.4, 124.1, 124.8, 126.0, 127.0, 136.0, 142.8, 146.8, 151.5, 155.1, 171.5 ppm.

Preparation of Poly{2-[N-ethyl-N-[4-[(4'-nitrophenyl)azo]-phenyl]amino] ethyl-3-indolyl acetate-co-indole}, Poly(In3AA-RedI-co-In)

The typical synthesis procedures utilized can be described as follows: 0.1 g (0.2123 mmol) of monomer (In3AA-RedI) and 0.0248 g (0.2123 mmol) of monomer indole in 10 ml of THF was added dropwise to a suspension 0.1503 g (0.4246 mmol) of Fe(ClO$_4$)$_3$ in

10 ml of THF under nitrogen atmosphere. The mixture was stirred at 50°C temperature for 24 h. The copolymer in solution was precipitated by addition of excess methanol. The precipitate was extracted using boiling absolute ethanol. The precipitate was dried under vacuum. FT-IR (KBr pellets, υ in cm^{-1}), υ: 3429 (υ_{N-H}, In), 3200 (υ_{C-H}, Ar), 2927 (υ_{C-H}, Al), 1738 ($\upsilon_{C=O}$), 1627 ($\upsilon_{N=N}$), 1515, 1335 (υ_{NO2}), 1150 (υ_{C-O}) cm^{-1}. ^{1}H-NMR (d^6- DMSO): δ 1.17-1.3 (broad), 1.3-1.9 (broad), 6.5-9 (broad) ppm.

Preparation of Poly{2-[N-ethyl-N-[4-[(4'-nitrophenyl)azo]-phenyl]amino] ethyl-3-indolyl acetate-co-pyrrole}, Poly(In3AA-RedI-co-Py)

Poly(In3AA-RedI-co-Py) was synthesized using the same synthetic procedures as for Poly(In3AA-RedI-co-In). The (In3AA-RedI) monomer (0.1 g, 0.2123 mmol) and pyrrole (0.0142 g, 0.2123 mmol) in anhydrous tetrahydrofuran (10 mL) was added dropwise to a suspension of Fe(ClO$_4$)$_3$ (0.1503 g, 0.4246 mmol) in tetrahydrofuran (10 ml) under nitrogen. The mixture was stirred at 50°C temperature for 24 h. The polymer in solution was precipitated by addition of excess methanol. The precipitate was extracted using boiling absolute ethanol. The precipitate was dried under vacuum. FT-IR (KBr pellets, υ in cm^{-1}), υ: 3250 (υ_{N-H}, indole), 3100 (υ_{C-H}, aromatic), 2989 (υ_{C-H}, aliphatic), 1745 ($\upsilon_{C=O}$), 1603 ($\upsilon_{N=N}$), 1516, 1414 (υ_{NO2}), 1103 (υ_{C-O}), 828 (υ_{C-H}, oop) cm^{-1}.

RESULTS AND DISCUSSION

All synthetic routes of RedI, In3AA-RedI and poly(In3AA-RedI) shown are in Scheme 1. Therefore, schematic copolymerization of In3AA-RedI with indole and pyrrole has been shown in Scheme 2.

Structural characterization

In order to obtain polymers with higher molecular weights, the polymerization and copolymerization were carried out at 50°C temperature with dropwise addition of monomers. This polymer was found to be soluble in chloroform, THF and methylene chloride, but two copolymers synthesized by chemical oxidative copolymerization using Fe(ClO$_4$)$_3$ are not soluble in common organic solvents. Figure 3 illustrates the FT-IR spectra of poly(In3AA-RedI). The peaks of about 3100-3600, 3020 and 2917 cm^{-1} was related to N-H, C-H (aromatic) and C-H (aliphatic) stretching vibrations, respectively. The peaks at around 1750 and 1100 cm^{-1} are due to the carbonyl (C=O) stretching vibration and C-O-C stretching vibration respectively.

Figure 4 shows the ^{1}H-NMR spectrum of poly(In3AA-RedI). On the basis of comparison with the spectrum of monomer, we can assign the following peaks for poly(In3AA-RedI): 1.09 (-CH$_3$), 1.24 (-CH$_2$-COO-), 2.52 (DMSO), 3.31 (H$_2$O),3.63 (-CH$_2$-CH$_3$), 3.72 (-CH$_2$-N), 4.27 (-O-CH$_2$-), 7.37, 7.82, 7.95 and 8.38 ppm (protons of phenylene groups). The lines refer to the aromatic protons of the indole ring that are located at 6.88-7.36 and 10.85 (H-N) ppm. The oxidative polymerization of ß-

substituted indole monomer with Fe(ClO$_4$)$_3$ always leads to two different types of couplings: head-to-tail and head-to-head. Thus, the α-methylene protons directly attached to the indole ring (ß-position) and can be incorporated into a polymer chain with the above two diads. The two peaks located at 1.2-1.5 ppm arise from the methylene protons between the ester group and the indole ring, showing that poly(In3AA-RedI) has a stereo random chain structure with almost equal distribution of head-to-tail and head-to-head linkages along the polymer chain.

The ^{13}C-NMR spectrum of the poly(In3AA-RedI) is shown in Figure 5. Using INEPT- pulse sequence, we were able to distinguish the proton bonded carbons from all other carbons present in the molecule. On the basis of a comparison with the spectra of poly(In3AA-RedI), the following assignments are proposed: The line at the lowest field (171.5 ppm) corresponds to the carbon of the ester group. The lines at 155.1, 151.5, 146.8, 142.8, 125.0, 124.8, 122.4 and 111.6 ppm are assigned to the aromatic carbons in the RedI moiety. The remaining eight lines at 136.0, 127.0, 124.1, 121, 118.4, 118.3, 111.3 and 106.6 ppm originate from the carbons of the indole ring. In the aliphatic part, the lines of low intensity at 11.2, 30.6, 44.8, 48.2 and 61.6 ppm correspond to the RedI substituent. The peaks are located at 38-40 ppm arise from the d^6-DMSO solvent.

Figure 6 shows ^{1}H-NMR spectrum of the polymer poly(In3AA-RedI-co-In) in (d^6-DMSO) solvent. In this spectrum, peaks of aliphatic protons (-CH$_3$), (-CH$_2$-COO-) and (-CH$_2$-CH$_3$) observed in 1.1 to 2.2 ppm. Peak of protons (-CH$_2$-N) and (-O-CH$_2$-) are not observed in spectra, because of overlap with peak of DMSO and water around 2.49 and 3.3 ppm, respectively. Protons of aromatic ring and indole are characterized between 6.5 to 8 ppm. Of course peaks of aromatic ring are not clear precisely because they are blocked copolymer, but they appeared in the area related to aromatic compounds. Copolymer has a low solubility, so ^{1}H-NMR is not clear. The great adherence of peak is a result of the greater involvement of indole monomers than In3AA-RedI monomer in polymeric chain. The peaks in aromatic region confirm performance of copolymerization.

Figure 7(a) shows UV-Visible spectrum of In3AA-RedI in THF as a solvent. According to Figure 7a, the In3AA-RedI as a monomer has three peaks in 285, 335 and 535 nm. The first absorption band is related to an n-π* transition and the second one is associated with a π-π* transition. Upon UV irradiation, a trans-cis isomerization is induced, leading to two absorption bands centered near 335 and 535 nm, respectively. The photochemical properties of the poly(In3AA-RedI) were examined also in THF as a solvent (Figure 7b). As can be seen in Figure 7b, the poly(In3AA-RedI) exhibits an absorption band centered at 280 nm which is related to the π-π* transition of trans configuration of the azobenzene moieties and a broad absorption bands around 550-750 nm which can be related to the π-π* transition of the highly conjugated

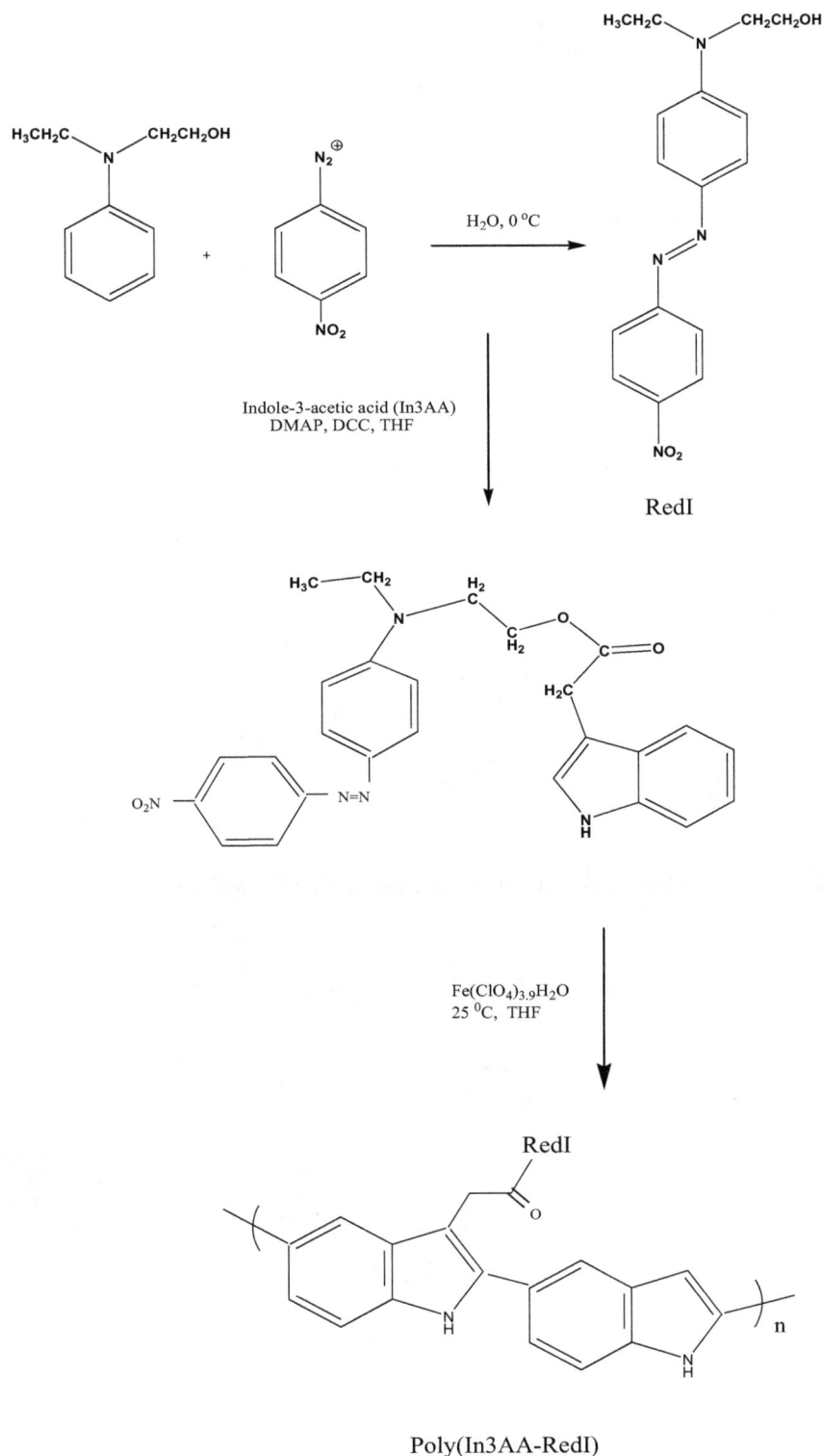

Scheme 1. Schematic reactions for route synthesis of Poly(In3AA-RedI).

polyindole units. Because of low solubility of poly(In3AA-RedI-co-In) and poly(In3AA-RedI-co-Py), we were not able to investigate UV-Visible or ^{13}C-NMR spectra of them.

Liquid crystalline and thermal properties

Liquid crystallinity and thermal analysis of the polymer were conducted by optical absorption, optical polarizing

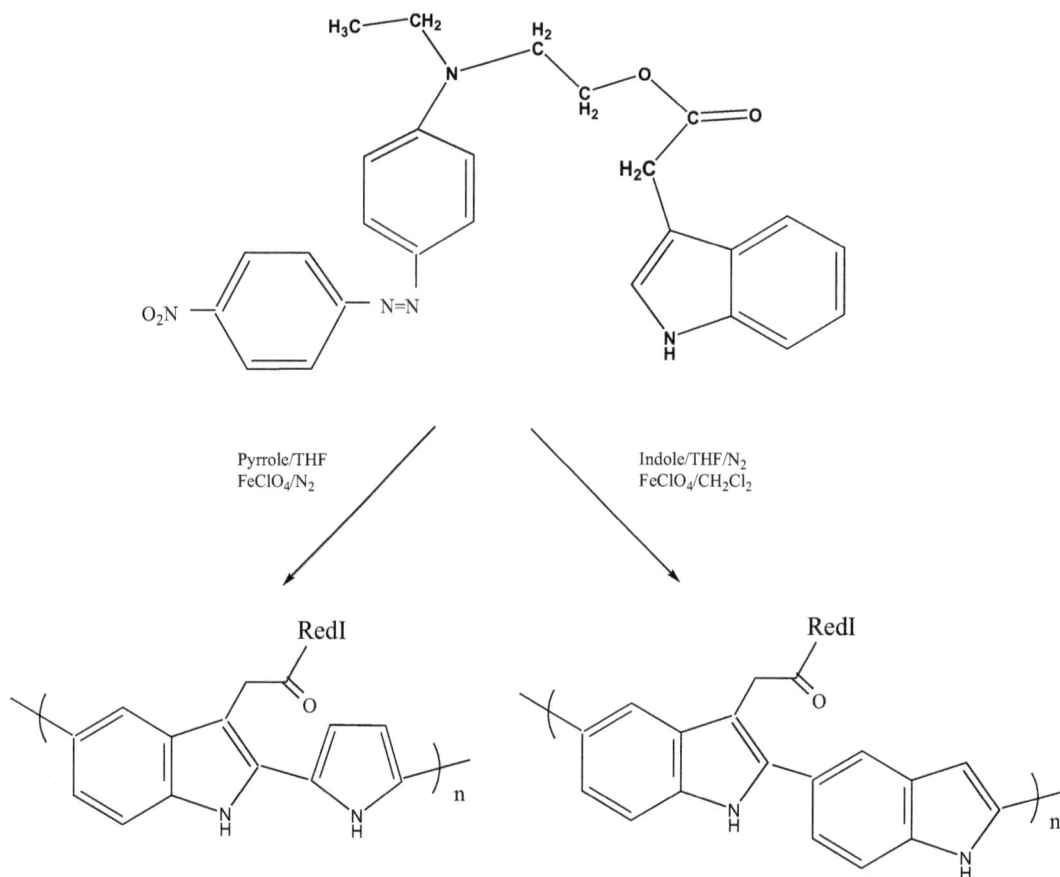

Scheme 2. Schematic reactions for copolymerization of poly(In3AA-RedI) with pyrrole and indole.

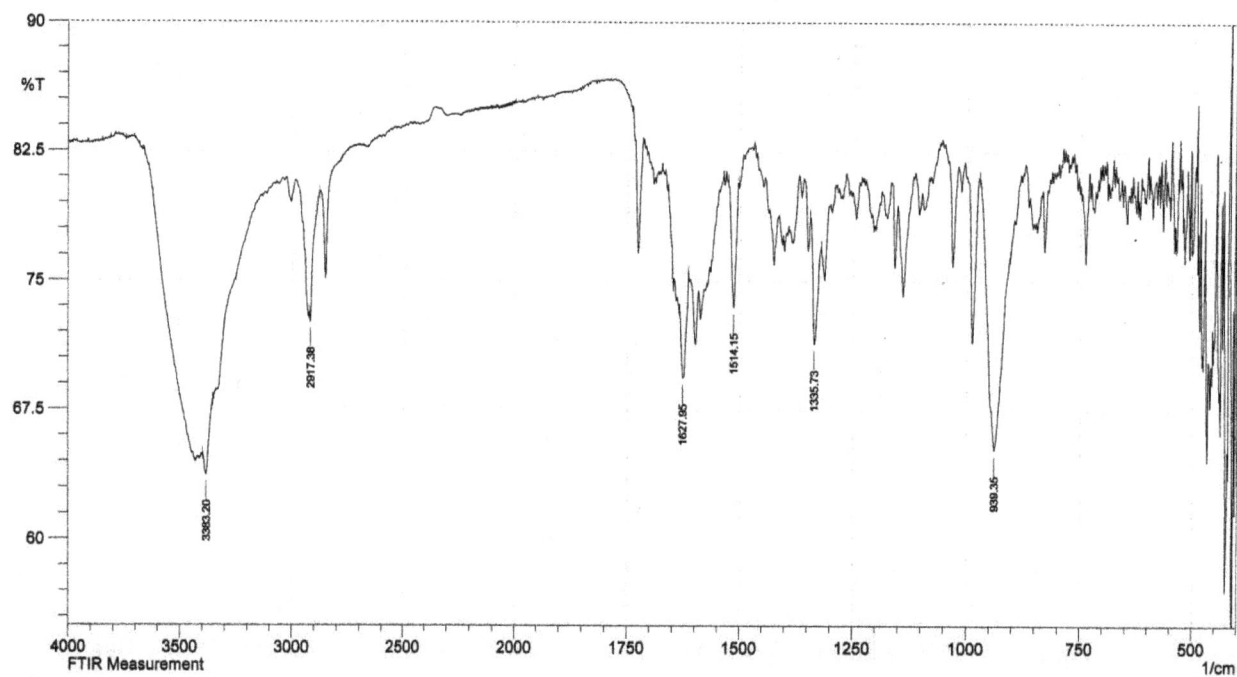

Figure 3. FT-IR Spectrum of poly(In3AA-RedI).

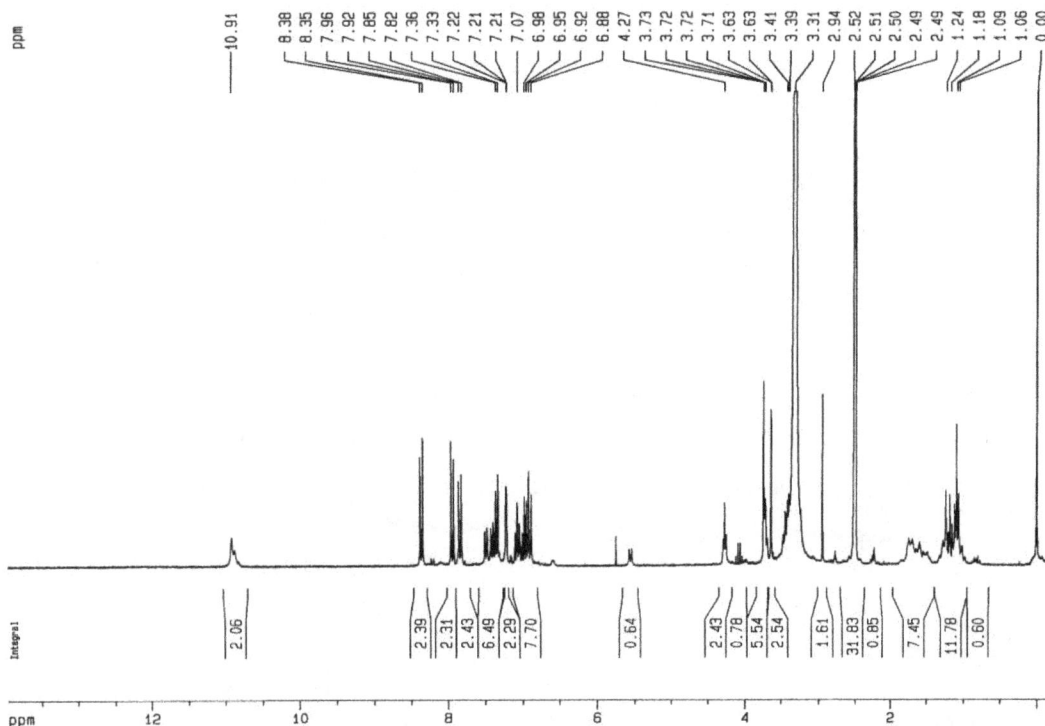

Figure 4. FT^{1}H-NMR Spectra of poly(In3AA-Redl).

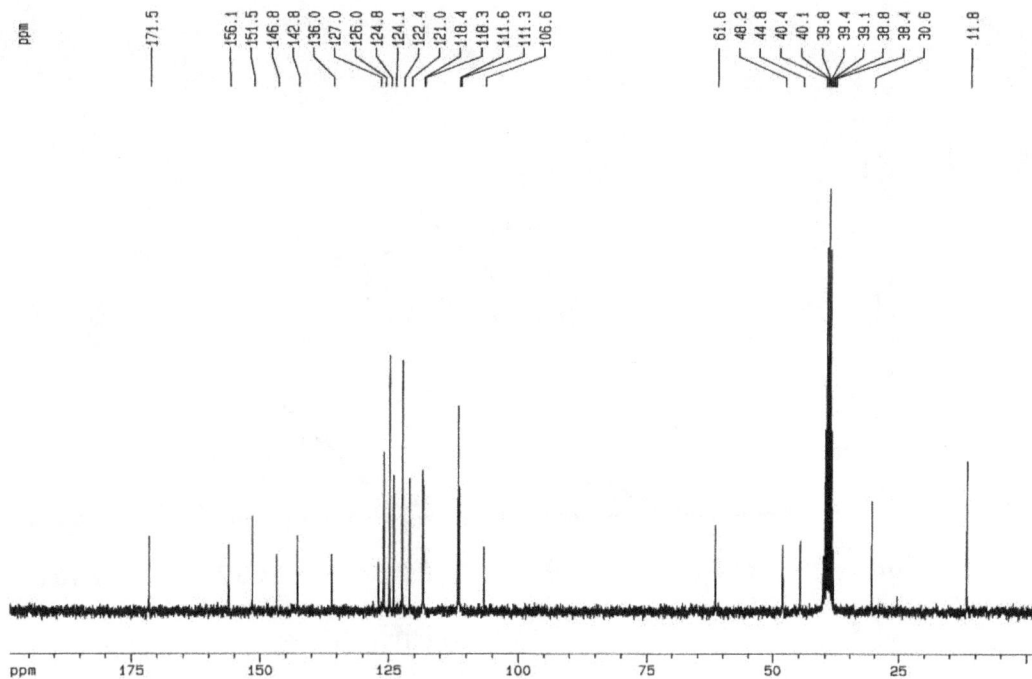

Figure 5. FT^{13}C-NMR Spectrum of poly(In3AA-Redl).

microscope, and differential scanning calorimetry (DSC). Phase transition temperatures were determined by DSC measurement. All DSC runs in this research were made under nitrogen atmosphere with a heating rate of

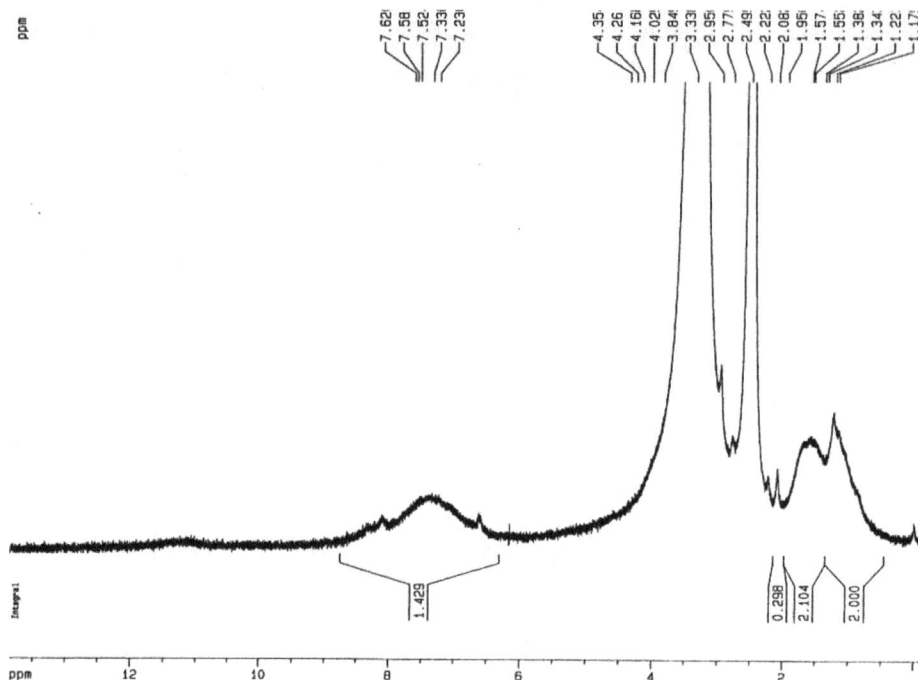

Figure 6. FT¹H-NMR Spectra of poly(In3AA-RedI-co-In).

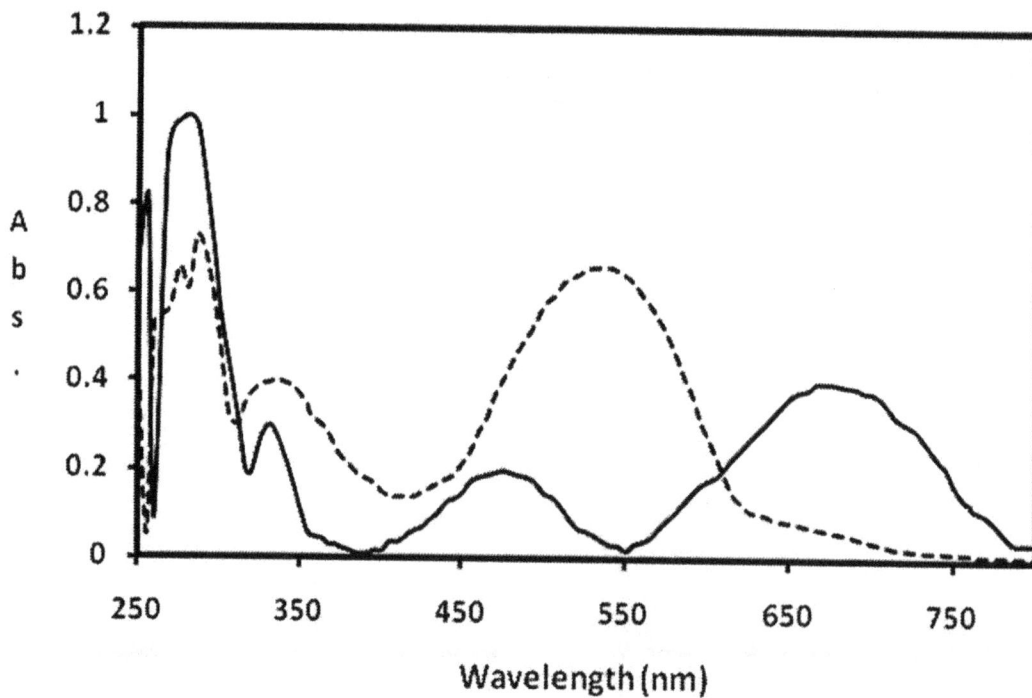

Figure 7. UV-Visible spectra of a) In3AA-RedI (...) and b) poly(In3AA-RedI) (-).

5°C/min. DSC thermograms of RedI in heating and cooling process are shown in Figures 8(a, b). Figure 8a shows thermogram and endothermic peak at 174.71°C which is related to the melting process of RedI and also a transformation of crystalline phase to isotropic state. Crystalline phase of this compound is changed directly to

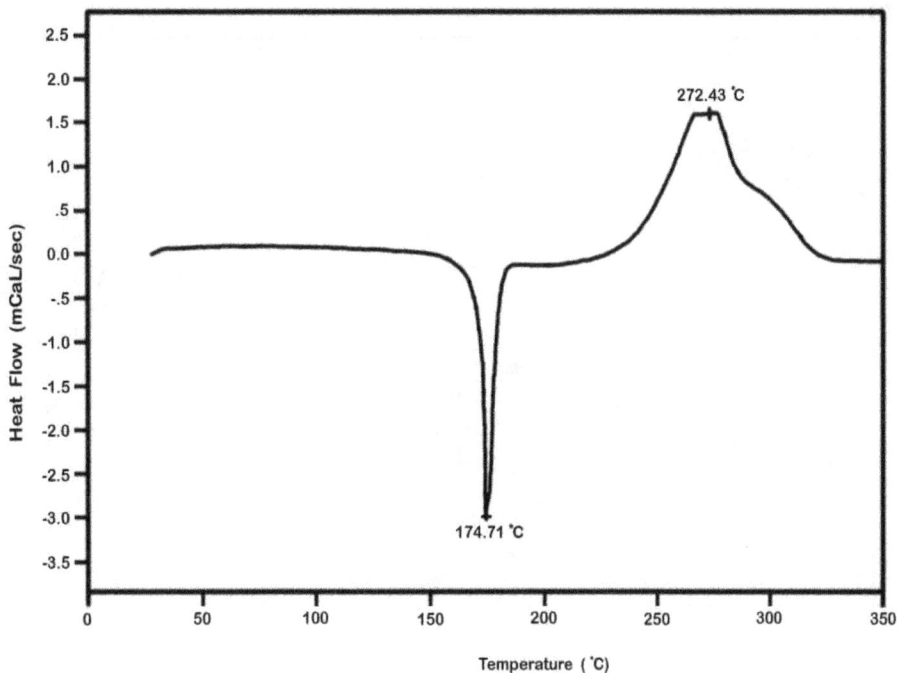

(a)

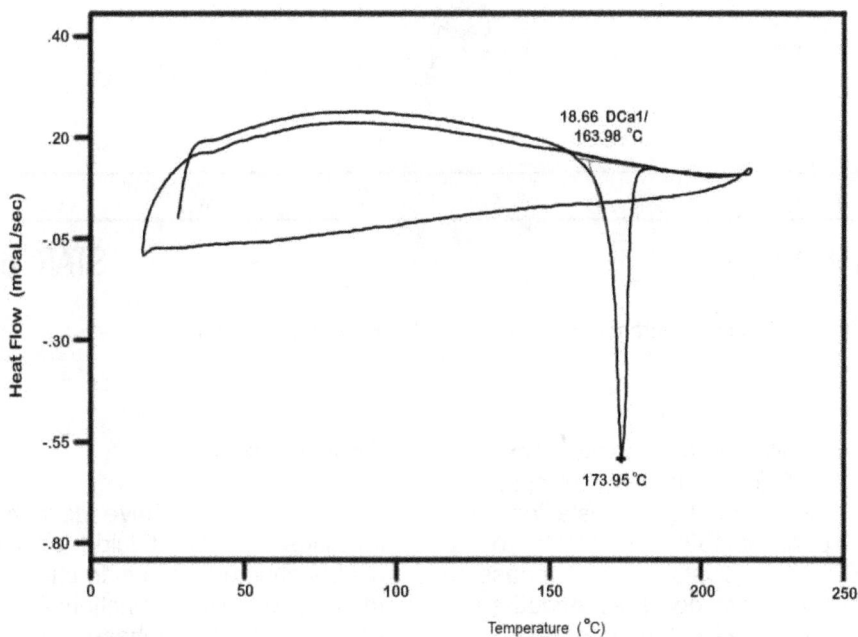

(b)

Figure 8. DSC Thermmogram of RedI a) heating and b) heating and cooling with rate of 5°C/min.

isotropic liquid as a result of temperature effect. Therefore, it shows exothermic peak at 272.43°C which is related to the degrading process of RedI. In cooling process no peak is observed probably because the sample decomposition occurs at higher temperature or it does not have clear LC behaviour. By heating the sample again no peak is observed, this confirms probable sample decomposition (Figure 8b).

Transition temperatures for poly(In3AA-RedI) were obtained using DSC shown as in Figure 9. DSC Thermogram of poly(In3AA-RedI) showed 5 glass transition temperature about 59, 91, 151, 181 and 231°C.

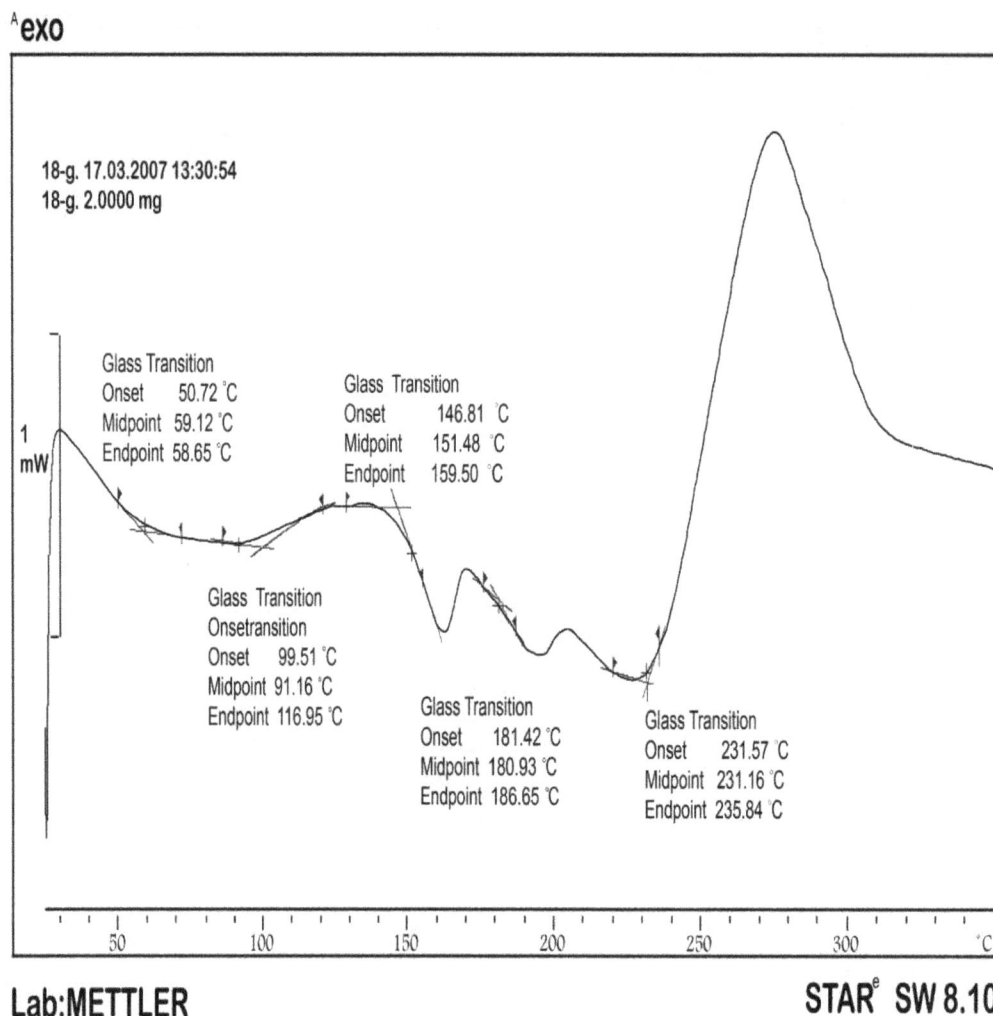

Figure 9 DSC Thermmogram of poly(In3AA-RedI) with rate of 5°C/min.

Therefore, it has three endothermic peaks. The first endothermic peak at 162°C is an indication of changing of crystalline phase to smectic A liquid crystalline state (C→S$_A$). The second peak at 193°C is related to the changing of smectic A liquid crystalline mesophase to nematic state (S$_A$→N). The last observed endothermic peak in this curve is about 225°C that is related to transition phase of nematic liquid crystalline to isotropic liquid (N→I). In the first and the second transitions, the stability of liquid crystalline mesophases are about 31°C and 32°C, respectively. The images of polarizing optical microscope of poly(In3AA-RedI) demonstrate the smectic A, nematic and isotropic liquid crystalline state in the ranges 150-168°C and 184-201°C and 205-236°C temperatures respectively. This indicates that poly(In3AA-RedI) is a monotropic or isotropic compound. Typical crystalline, smectic A, nematic textures and isotropic state of poly(In3AA-RedI) are shown in Figure 10(a-f).

Conclusion

In this paper, we have described the preparation and investigation of liquid crystalline polyindole with azobenzene group in side chain. The results indicate that these azobenzene-functionalized polyindole derivatives have wide mesophase temperature ranges. DSC thermogram of poly(In3AA-RedI) showed 5 semi glass transition and three endothermic peaks. The endothermic peaks are indicative of changing of C→S$_A$, S$_A$→N and N→I. Images of polarizing optical microscope of poly(In3AA-RedI) demonstrated the smectic A, nematic and isotropic liquid crystalline states. Chemical, structural, thermal and morphological studies indicated that high quality polymer film can be obtained. As-formed polymer film was thoroughly soluble in polar solvent such as DMSO. According to IR and ^{1}H-NMR spectra, the existence of N–H bond implies that coupling between the monomer units occurred at the C2 and C5 positions.

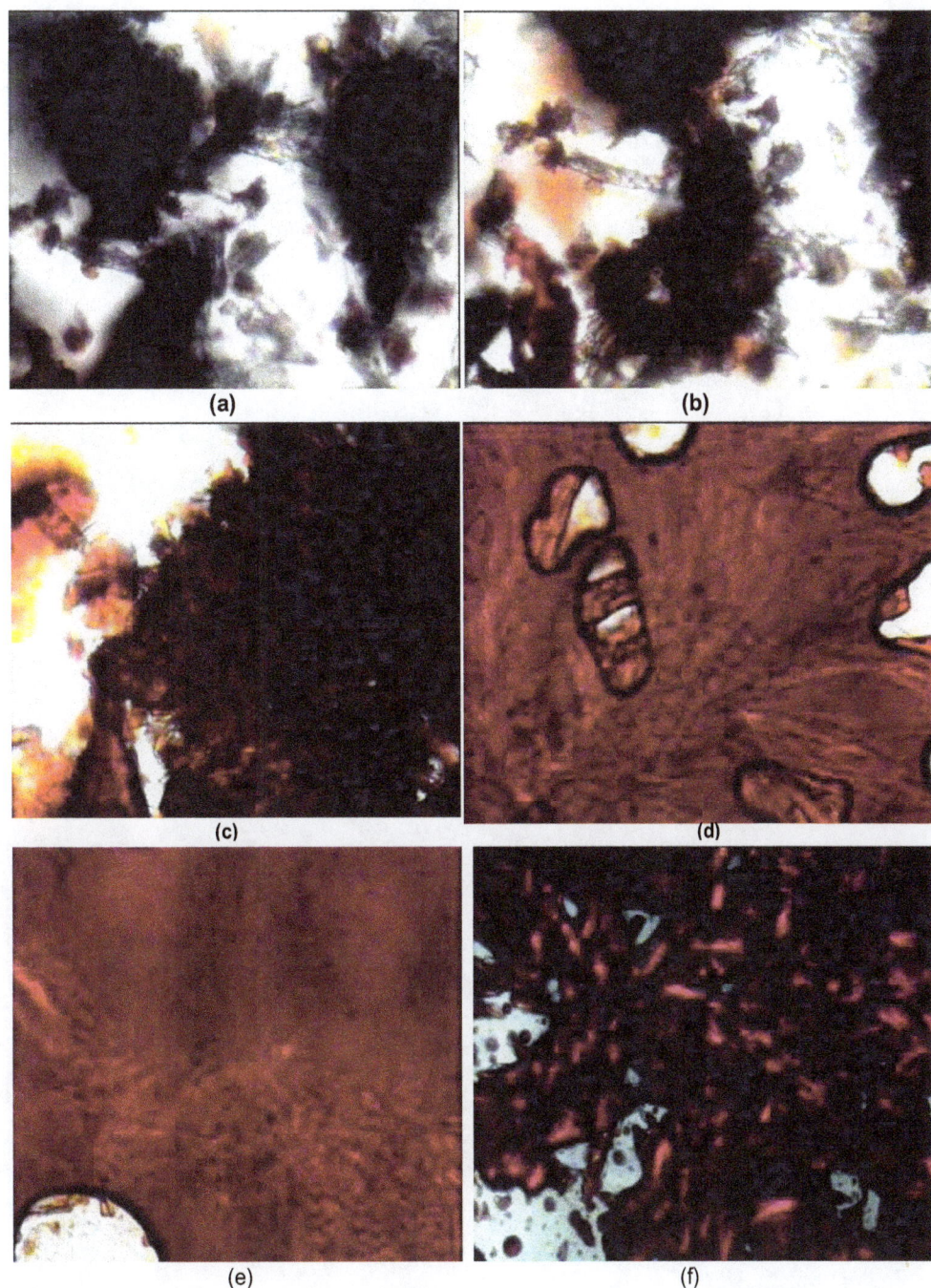

Figure 10. Cross-polarized optical micrograph of the polymer sample, smectic C phase a) 152°C, b) 160°C, c) 170°C, d) 185 °C, e) 195°C and f) 205°C.

REFRENCES

Heinze J (1991). Electrochemistry of conducting polymers. Synth. Met. 43(1-2):2805-2823.

Hosseini SH (2006). Investigation of sensing effects polystyrene graft polyaniline for cyanide compounds. J. Appl. Polym. Sci. 101(6):3920-3926.

Hosseini SH, Abdi Oskooe SH, Entezami AA (2005). Toxic Gas and Vapour Detection with Polyaniline Gas Sensors. Iranian Polym. J. 14(4):333-343.

Hosseini SH, Entezami AA (2001). Chemical and electrochemical synthesis of polymer and copolymers of 3-methoxyethoxythiophene with aniline. Polym. Adv. Technol. 12:524-534.

Hosseini SH, Entezami AA (2003). Chemical and electrochemical synthesis of conducting polyDi-heteroaromatics from pyrrole, indole, carbazole and their mixed containing hydroxamic acid groups and studies of its metal complexes. J. Appl. Polym. Sci. 90:63-71.

Hosseini SH, Mohammadi M (2009). Preparation and characterization of new polypyrrole having side chain liquid crystalline moieties. Mater. Sci. Eng. C. 29:1503-150.

John A, Palaniappan S (2005). Synthesis and characterization of soluble poly (N-heptylindole). Polymer 49:12037-12039.

Kolelil F, Saglam M, Turut A (1994). Polyindole Based Schottky. Turk J Chem 18(1):22-28.

MacDiarmid AG (1997). Poly aniline and polypyrrole: Where are we headed?. Synth. Met. 84:27-34.

Matsui T, Nagata T, Ozaki M, Fujii A, Onoda M, Teraguchi M (2001). Novel properties of conducting polymers containing azobenzene moieties in side chain. Synth. Met. 119:599-600.

Mohammadi A, Hassan MA, Liedberg B, Hungstorm L, Saleneck R (1986). Chemical vapour deposition (cvd) of conducting polymers: polypyrrole. Synth. Met. 14:189-197.

Roth S, Graupher W (1993). Conductive polymers: Evaluation of industrial applications. Synth. Met. 57:3623-3631.

Turut A, Koleli F (1993). Metallic polythiophene/inorganic semiconductor Schottky diodes. Phys. B. 192:279-283.

Xu J, Hou J, Zhang S, Zhang R, Nie G, Pu S (2006). Electrosynthesis of high quality poly(5-methylindole) films in mixed electrolytes of boron trifluoride diethyl etherate and diethyl ether. Eur. Polym. J. 42:1384-1395.

Yousefi M, Hosseini SH, Amani V, Khavasi HR (2008). (E)-2-{N-ethyl-4-[(4-nitrophenyl)diazenyl]aniline}ethyl acrylate. Acta. Cryst. E64:o789.

Zaho X, Wang M (2006). Synthesis and photoresponsive behavior of azobenzene-functionalized polythiophene film. Eur. Polym. J. 42:247-253.

Zhao X, Hu X, Zheng PJ, Gan LH, Lee CKP (2005). Synthesis and characterization azobenzene as side chain. Thin Solid Films. 477:88-94.

13

Modification of some Albanian wood properties through chemical treatment

Ajdinaj D., Lato E., Quku D. and Cota H.

Department of Wood Industry, Faculty of Forest Sciences, Agricultural University of Tirana, Kodër-Kamëz 1029, Tirana, Albania.

A study was carried out to evaluate the influence of acetylation on physical properties and gluing ability of some Albanian woods. Three woods were studied: beech (*Fagus sylvatica L.*), poplar (*Populus alba L.*) and fir (*Abies alba Mill.*). For acetylation, pyridine and acetic anhydride were used. After treatment, all the samples were conditioned to 100% relative humidity of air and were measured according to radial and tangential directions. After that, the samples were ovendried and were weighed and measured again, calculating the shrinkage and wood percent gain. Again, the samples were reconditioned to 100% relative humidity of air, calculating the swelling. The volumetric shrinkages of acetylated poplar and beech resulted 4.05 and 8.5%, meanwhile for non acetylated samples 9.8 and 16.5%. Tangential shrinkage of acetylated fir was 2.2% and for non acetylated 6.5%. The volumetric swellings of acetylated poplar and beech resulted 4.2 and 9.3%, meanwhile for the non acetylated 10.8 and 19.8%. Tangential swelling of acetylated fir wood was 1.7 and 5.8% for non acetylated. It seems that acetylation reduced more than 50% the moisture-related dimensional changes of wood. Regarding gluing the shear strength of acetylated beech resulted 2.4 times less comparing with non acetylated one.

Key words: Acetylation, gluing, shrinkage, strength, swelling, wood.

INTRODUCTION

Wood is a hygroscopic material, sensible to atmospheric humidity changes and it can be submitted to considerable dimensional changes according to its grain directions. Its swelling and shrinkage values are different according to the sectional cut. Swelling and shrinkage values in tangential direction are almost twice bigger than those of radial direction, while in axial direction these values are inconsiderable (Giordano, 1981). These dimensional changes can cause splits or deformations.

Improvement of dimensional stability of wood has always been a primary interest, because its swelling and shrinkage's tendency, caused by variations of humidity, is considered its most negative property. Dimensional

changes of wood caused by moisture content changes are considerable. Wood's cell wall swells about 45% from the 0% moisture content to saturation fibres point (Skaar, 1988). In tangential directions, the shrinkage from green to 6% moisture content varies from 4 to 9%, while in radial direction, for the same conditions, varies from 1.8 to 6% (Kollmann and Côté, 1968). This can cause deformations and splits to wood boards, during their drying and their use. With the aim to improve its dimensional stability and to reduce its volumetric changes, the wood has been modified by means of different methods.

One of the main methods that reduce the dimensional

changes of wood is acetylating (Rowell, 2006). This method consists in chemical treatment and has been industrially applied since 1961 in boards used in buildings (Rowell, 1985). According to this technique, one hydroxyl group (–OH) of wood is substituted by a radical CH_3COO of an anhydride acetic molecule $(CH_3CO)_2O$. This substitution reduces the number of free positions –OH that can react with water molecules, reducing the water sensibility of wood. The reaction of wood's acetylation can be presented below:

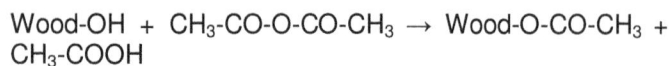

Wood-OH + CH_3-CO-O-CO-CH_3 → Wood-O-CO-CH_3 + CH_3-COOH

Chemically, hydroxyl group (–OH) can be found to the functional groups of phenol, benzene and carbohydrate alcohols. The alcoholic hydroxyls can be primary or secondary, while the phenols can be connected with aromatic chains which contain different types of substitutes. As result, each of these groups can present a different reaction regarding anhydrite.

The reaction's product between anhydrite acetic and wood is acetic acid (HOac). It is known that a concentration of acetic acid lower than 10%, accelerates the reaction and the opposite happens for higher concentration (Navi and Heger, 2005). The size of radical CH_3-CO-O is bigger than of hydroxyl group (–OH), causing so the swelling of wood at the beginning of the treatment. According to results regarding to pine wood, is noted that acetyl treatment increases more than 55% the dimensional stability of wood, but reduces about 50% its tensile modulus (Ramsden et al., 1997). It is found that anti shrinkage efficiency (ASE) of acetyl treated pine, spruce, birch and beech arrive from 45 to 50%, and the module of elasticity and rupture are reduced about 15% (Epmeier et al., 2001; Epmeir et al., 2003). Other studies show that Brinell hardness of acetylated pine is increased up to 20%, but Janka hardness is not significantly affected (Larsson and Simonson, 1994; Papadopoulos and Tountziarakis, 2011).

This treatment improves wood's resistance against biodegradation and colour changes (Imamura, 1993; Sakuragawa, 1996; Nasheri et al., 2005; Papadopoulos et al., 2010; Mohebby and Militz, 2010). Tests have proven a good resistance against termite attacks and marine organisms (Suttie et al., 1999; Westin et al., 2004). Besides this, wood's darkening does not happen, which characterise resins treatments.

Regarding to the method mentioned above, a study is carried out to evaluate the influence of acetylating on physical properties (specifically shrinkage and swelling) and gluing ability of some of the most important Albanian woods. It aims to give information about opportunities of application of this treatment, especially for outdoor applications, as well as for a better selection of wood material, referring to Albanian native woods. It is motivated by increasing demand for high quality and long-term performance wood products. The study is focused respectively:

(i) For physical properties; on beech wood (*Fagus sylvatica* L.), poplar (*Populus alba* L.) and fir (*Abies alba* Mill.). These species are selected taking into account that they are the most important species used in Albania. The research is focused on both groups of softwoods and hardwoods. Also, a comparison is made between species from the same group, but noticeably with different density (poplar with beech).
(ii) For gluing; on beech wood (*F. sylvatica* L.). This selection is conditioned because beech is the main wood used for production of solid wood based panels not only in Albania, but in many other countries of the region, and also, gluing is one of the most important chains of their production.

MATERIALS AND METHODS

The study was based on comparative laboratory method, cause-consequence (Creswell, 2003). The method consisted of quantity evaluation of a specific phenomenon caused by a provocative factor and after that, the evaluation of the same phenomenon in the situation of the factor's absence. In our case, the phenomenon was the dimensional variability (shrinkage and swelling), and the provocative factor was the acetylation of the sample. Laboratory tests consisted in preparation of samples, treatment (acetylation), conditioning of samples in different conditions of relative air humidity and measurements of their dimensional changes as well as tests of gluing shear strength. The study was carried out at the Faculty of Forestry Sciences of Tirana, during the period February-June 2009.

Preparation of samples

Regarding to physical properties, the samples were selected from pieces of kiln dried boards without deformations or structure defects. For each wood, there was selected one single board's piece. From each piece, one strip with dimensions of section 2 × 1 cm, and length the same as the piece of board, was sawn. After that, the strips were sawn in pieces with dimensions 2 × 1 × 1 cm and then the pieces were cut in samples by a single use microtome knife. The samples presented clear radial, tangential and cross-sections, with dimensions in longitudinal and tangential directions 1 cm, whereas their dimension in radial direction varied from 0.5 to 1 cm (Figure 1). All samples were marked with numbers, taking into account their positions in pieces. Samples with odd numbers were acetylated, respectively 48 for beech, 44 for fir and 38 for poplar. With regard to gluing, the samples were prepared from beech boards without deformations or structure defects, which could have influence on gluing resistance. Pieces with dimensions 5 × 2 × 1 cm, and with grain direction parallel with the longer edge were produced. The pieces were weighed and selected in couples, in manner that pieces with approximate density could be glued together. Their faces had to present flat and smooth surfaces. Thickness tolerance higher than 0.1 mm was not permitted, making possible a good pressure during the hardening of glue. Based on the standard EN 205 (Determination of tensile shear strength of lap-joints), the pieces were selected with angles from 30° to 90° between annual rings and bonding surface (EN205, 1991). Almost half of them were acetylated.

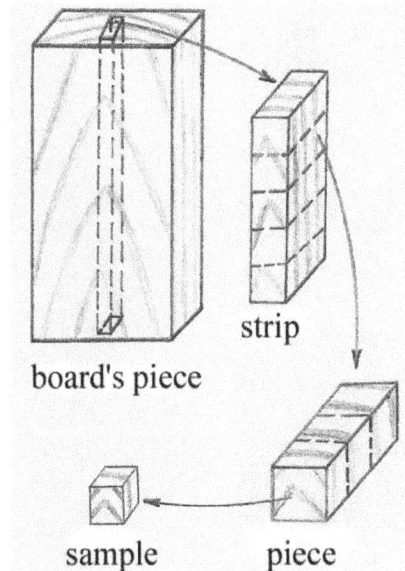

Figure 1. Preparation of samples.

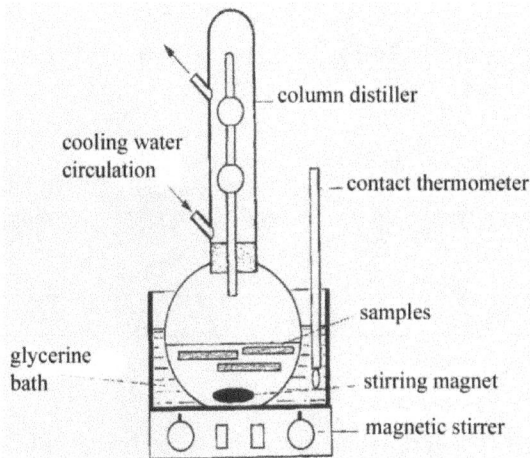

Figure 2. The acetylating system.

Acetylation and tests

Acetylation was performed according to the laboratory procedure, adequate for small dimension samples with a total weight of 150 g (Navi and Heger, 2005). Oven dried samples were weighed and after that they were soaked in a mixture of acetic anhydride and pyridine. There was 500 ml acetic anhydride used (95%, MERCK, Germany) for acetylation and 900 ml pyridine (99%, MERCK, Germany) as catalyst. The wood-liquid system was kept in 90 °C for 4 h. The boiling point of pyridine was between 114 and 116 °C, requiring the utilisation of a system for cooling and steams circulation. The system used for acetylation is showed in Figure 2. After 4 h, the samples were taken off from the balloon and were cooled down. After that, the measurement points on tangential and radial faces were marked on all acetylated or not acetylated small samples, selected for physical properties study, and after were conditioned. Conditioning consisted in three steps:

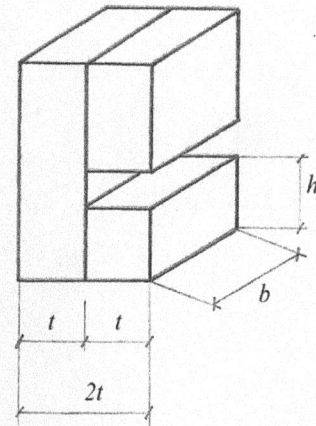

Figure 3. Finished samples. h – height of test's surface, b – width of test's surface, $2t$ – width of test piece.

(i) 100% relative humidity of the air: Dessicator was used. To create an environment with 100% humidity of the air, the bottom of the dessicator was filled with distilled water and after was applied the vacuum −0.9 atm. The samples stayed in these conditions until they reached equilibrium state, which corresponds to the fibers saturation point, and by means of an indicator they were measured according to tangential and radial directions. It must be noted that fir's samples were not measured in radial direction, because their dimensions were higher than the measurement's step of indicator.
(ii) Oven drying in temperature 103 ± 1 °C, until they reached equilibrium state, corresponding to 0% moisture content. After that, the samples were weighed and measured again. Weight percent gain (WPG) and shrinkage was calculated.
(iii) 100% relative humidity of the air and after respective measurements, the swelling was calculated. Volumetric swelling (αv) was calculated below:

$$\alpha_v = \frac{\beta_v}{1 - \dfrac{\beta_v}{100}}$$

β_v – volumetric shrinkage for each sample.
　With regard to gluing tests, the procedure continued thus:

(1) After treated pieces were cooled down and the smell was removed, they were oven dried, weighed and equilibrated to environment room conditions. WPG of acetylated pieces resulted 13.5%,
(2) Application of adhesive. PVA glue (NEON, Albania) was used. The quantity of glue was referred to industrial application, from 170 to 240 g/m², resulting in 0, 2 g per piece. It was verified by weighing the pieces before and after application of glue,
(3) Pressure. It was applied with hand grip vice for 24 h,
(4) Preparation of samples and tests.

There were prepared 32 acetylated samples (test pieces) and 37 non-acetylated. After gluing, canals were sawed on samples, as shown in Figure 3. The samples were tested with a mechanical test

Figure 4. Testing photo.

Table 1. Results of shrinkage.

Wood species	Treatment	Weight gain (%)	Stand. dev.	Radial shrink.	Stand. dev.	Tang. shrink.	Stand. dev.	Volum. shrink.	Stand. dev.
Fir	Untreated	0	-	-	-	6.5	2.99	-	-
	Acetylated	21	1.07	-	-	2.2	0.59	-	-
Poplar	Untreated	0	-	3.3	0.45	6.4	0.41	9.8	0.64
	Acetylated	20.6	0.4	1.3	0.25	2.7	0.4	4.05	0.48
Beech	Untreated	0	-	4.9	0.54	11.6	0.64	16.5	1.02
	Acetylated	13.6	0.92	2.6	0.16	5.9	0.64	8.5	0.65

machine (CONTROLAB, France) to rupture (Figure 4). The shear resistance of gluing for each test was calculated:

$$S = F/A \quad [N/mm^2]$$

F- rupture load [N], A- gluing surface [mm^2].

RESULTS

Table 1 summarised results of shrinkage for three studied species present, while Table 2 shows results of swelling. Table 3 summarised results of gluing tests present.

DISCUSSION

The volumetric shrinkages of acetylated poplar and beech resulted 4.05 and 8.5%, meanwhile for non acetylated samples the shrinkages were 9.8 and 16.5%. According to these results, anti shrinkage efficiency

(ASE) resulted, respectively 59% for poplar and 49% for beech. The tangential shrinkage of acetylated fir was 2.2% and for non acetylated 6.5%. About volumetric swellings, acetylated poplar and beech resulted 4.2 and 9.3%, meanwhile the non acetylated resulted 10.8 and 19.8%. Anti swelling efficiency resulted respectively 61% for poplar and 53% for beech. Tangential swelling of acetylated fir wood was 1.7 and 5.8% for non acetylated.

It was noted that acetylation reduced moisture-related dimensional changes of poplar and beech wood more than 50%. The treatment's efficiency for poplar was almost 10% higher than beech. Taking into consideration results of other studies, we can say that the efficiency of acetylation depends on wood density. Much greater to be the density, smaller is the effect of acetylation (Rowell et al., 1986; Beckers and Militz, 1994). This means that in woods with high density, the acetyl molecules penetrate with more difficulty, making possible the blockage of a smaller number of free hydroxyl (–OH) groups comparing to low density woods. Also, it must be noted

Table 2. Results of swelling.

Wood species	Treatment	Weight gain (%)	Stand. dev.	Radial swell.	Stand. dev.	Tang. swell.	Stand. dev.	Volum. swell.	Stand. dev.
Fir	Untreated	0	-	-	-	5.8	0.85	-	-
	Acetylated	21	0.77	-	-	1.7	0.37	-	-
Poplar	Untreated	0	-	3.2	0.44	6.1	0.52	10.8	0.79
	Acetylated	20.6	0.4	1.5	0.21	2.5	0.43	4.2	0.52
Beech	Untreated	0	-	5.2	0.24	12.5	0.89	19.8	1.45
	Acetylated	13.6	0.92	2.5	0.13	5.8	0.73	9.3	0.78

Table 3. Results of gluing tests.

Treatment	Weight gain (%)	Standard deviation	Mean value of shear strength (N/mm^2)	Standard deviation
Untreated	0	-	6.72	0.95
Acetylated	13.5	0.73	2.79	0.67

that concentration of impregnation solution is another factor which play a significant role on the treatment degree (Keskin et al., 2013).

Regarding gluing tests, the shear strength of acetylated wood was reduced by 59%. The destruction of samples happened to the adhesive layer, typically for weak gluing. In case of a good adhesion, wood near the adhesive layer will be destroyed, but not the adhesive layer. A good adhesion gives a higher resistance than glued wood. From results we can say that adhesion of acetylated wood with PVA glues can be classified as not qualitative. A recent study shows that acetylated finger jointed beech wood produced with PVA glue, present lower values of bending strength comparing with those of non acetylated (Papadopoulos, 2008). However, cross-linking PVA glues develop bonds of high shear strength of acetylated wood, in tests of dry strength (Rowell, 2005).

In general, the results of the study are compatible with those findings in literature regarding to the effect of acetylation on different wood species. As expected wood's acetylating is a chemical modification with clear positive effects on dimensional stability (Rowell, 2006). It improves notably the physical properties of wood reducing the tendency of interchanges with air humidity. From all other modifications developed to improve the dimensional stability of the wood, this method offers the highest advantages, because of small negative impact on others desirable properties of wood (Epmeir et al., 2003, Korkut, 2008).

It seems that Albanian woods react very well regarding to this modification, opening so a positive perspective for industrial application for windows, stairs and outside doors. The efficiency of acetylation related to dimensional changes is more than 50%. Considering the demand of timber as raw material for constructions in Albania, the acetylated wood appears as a product which presents many advantages. By the other side, acetylation reduces some of wood's technological properties like ability of gluing. Application of PVA glues with acetylated beech is not advisable, because in this case the reduction of gluing resistance is more than 50%, excluding this combination for industrial structural applications.

REFERENCES

Beckers EPJ, Militz H (1994). Acetylation of solid wood. Initial trials on lab and semi industrial scale. In: Second Pacific Rim Bio-Based Composites Symposium. Vancouver, Canada. pp. 125-135.

Creswell WJ (2003). Research Design-Qualitative, Quantitative and Mixed Methods Approaches, Second Edition. SAGE Publications Thousand Oaks, London, New Delhi.

EN205 (1991). Test methods for wood adhesives for non-structural applications - Determination of tensile shear strength of lap-joints. European Committee for Standardization.

Epmeier H, Bengtsson C, Westin M (2001). Effect of Acetylating and Heat Treatment on Dimensional Stability and MOE of Spruce Timber. In Navi P (eds): Proceedings of 1st International Conference of the European Society for Wood Mechanics. Lausanne, Switzerland, pp. 205-214.

Epmeir H, Westin M, Rap AO, Nilson T (2003). Comparison of Properties of Wood Modified by 8 Different Methods – Durability, Mechanical and Physical Properties. In: Van Acker J, Hill C (eds): Proceedings of the First European Conference on Wood Modification. Ghent University (RUG), Belgium. pp. 121-142.

Giordano G (1981). Tecnologia del legno. Unione Tipografico-Editrice Torinese. pp. 767-794.

Imamura Y (1993). Morphological changes in acetylated wood exposed to weathering. Wood Res. Kyoto. 79:54-61.

Keskin H, Süzer EN, Çolakoğlu MH, Korkut S (2013). Mechanical properties of Rowan wood impregnated with various chemical materials. Int. J. Phys. Sci. 8(2):73-82.

Kollmann FFP, Côté WR (1968). Principles of Wood Science and Technology – I – Solid Wood. Springer-Verlag Berlin, Heidelberg, New-York.

Korkut S (2008). The effects of heat treatment on some technological properties in Uludağ fir (*Abies bornmuellerinana* Mattf.) wood. Build. Environ. 43(4):422-428.

Larsson P, Simonson R (1994). A study of strength, hardness and deformation of acetylated Scandinavian softwoods. Holz als Roh und Werkstoff. 52:83-86.

Mohebby B, Militz H (2010). Microbial attack of acetylated wood in field soil trials. Int. Biodeterior. Biodegr. 64:41-50.

Nasheri K, Durbin G, Singh A, O'Callahan D (2005). Stability and decay resistance of acetylated wood. ENSIS edition, Wood Processing. 36:15-17.

Navi P, Heger F (2005). Comportement thermo-hydromécanique du bois – Applications technologiques et dans les structures. Presses polytechniques et universitaires romandes.

Papadopoulos AN (2008). The effect of acetylation on bending strength of finger jointed beechwood (*Fagus sylvatica* L.). Holz Roh Werkst. 66:309-310.

Papadopoulos AN, Militz H, Pfeffer A (2010). The biological behaviors of *pine wood* modified with linear chain carboxylic acid anhydrides against soft rot fungi. Int. Biodeterior. Biodegradation 64:409-412.

Papadopoulos AN, Tountziarakis P (2011). The effect of acetylation on the Janka hardness of pine wood. Eur. J. Wood Prod. 69(3):499-500.

Ramsden MJ, Blake FSR, Fey NJ (1997). The effect of acetylation on the mechanical properties, hydrophobicity and dimensional stability of *Pinus Sylvestris*. Wood Sci. Technol. 31:97-104.

Rowell R (1985). The Chemistry of Solid Wood, Advanced in Chemistry Series. American Chemical Society, Washington D.C.

Rowell R, Tillman A, Simonson R (1986). A simplified procedure for the acetylation of hardwood and softwood flakes for flakeboard production. J. Wood Chem. Tech. 6(3):427-448.

Rowell R (2005). Handbook of wood chemistry and wood composites. CRC Press, Taylor & Francis Group London, New-York, Singapore.

Rowell R (2006). Chemical modification of wood: A short review. Wood Mater. Sci. Eng. 1:29-33.

Sakuragawa S (1996). Dyeing of wood and protection of discoloration. Mokuzai Kogyo. 51(3):102-106.

Skaar C (1988). Wood-Water Relations. Springer-Verlag Berlin, Heidelberg New-York, London, Paris, Tokyo.

Suttie ED, Hill CAS, Jones D, Orsler RJ (1999). Chemically modified solid wod. Resistance to *Hylotropes bajulus* attack. Material und Organismen 33(2):81-90.

Westin M, Rapp AO, Nilsson T (2004). Durability of pine modified by 9 different methods. International Research Group on Wood preservation, Doc, No. IRG/WP 04-40288.

14

Utilization of paper sludge in clay bricks industry to obtain lightweight material: Evidence from partial replacement of feldspar by paper sludge

Bachir Chemani[1] and Halima Chemani[2]

[1]Department of Process Engineering, Faculty of Engineering Sciences, University M'Hamed BOUGARA of Boumerdes – 35000 – Algeria.
[2]Department of Materials Engineering, Faculty of Engineering Sciences, University M'Hamed BOUGARA of Boumerdes – 35000 – Algeria.

In recent years, scientific issues related to environmental preservation have acquired great importance and the major challenge to be met is the recycling of materials discarded, causing various damages to the environment by technological development through a disposal of waste. This paper offers a possibility of recovery of waste and byproducts of paper manufacturing (paper sludge) and their use as an addition for obtaining a new formulation of ceramic tiles. The raw materials, paper sludge and clay were mixed together in different proportions. The ceramic samples were characterized with respect to water absorption, the porosity, the linear shrinkage and rupture strength. A new formulation was carried using four mixtures (M1, M2, M3 and M4) where feldspar (FT) was substituted by waste sludge (SP). Substituting FT per paper wastes, the rate of FT is reduced by 50%, thereby improving the rheological properties and physic-mechanical ceramic tiles. The influence of the paper sludge on physical-mechanical properties of final fired product has been studied. Paper waste has proven to be a good raw material and the corresponding formulations were shown to be viable and acceptable in the manufacture of ceramic tiles.

Key words: Recycling, sludge, waste materials, tile, ceramic formulation.

INTRODUCTION

The use of waste sludge in ceramic industry is a field with high potential in the years to come. The majority of waste sludge comes principally from the manufacturing industries and mineral processing. There are classified and selected sludge as from the crushing process of gneiss, sludge from the cutting and polishing process of varvite, sludge from the process of filtration-clarification of potable water and clay as waste, from waste tannery, from glass cullet, etc. The meaningful utilization of industrial waste materials has been acclaimed world over not only as an economic opportunity (Mir, 1982), but also as a step towards solving problems of environmental pollution. For the interest of using these wastes in the ceramic industry, many attempts were made to incorporate sludges into raw materials mixtures in order to produce different ceramic products.

Waste sludge has been considered as low cost resource material for alumino-silicate and many authors

(Marcis et al., 2005) have tried to use waste mill sludge as a partial substitution of clay in the development of ceramic tiles. The fabrication of products from waste is an advantage that may give to the manufacturer a highly competitive position in the market due to economic issues involved and the opportunity of marketing this principle particularly with regard to the ecological aspect (Modesto et al., 2003).

Sustainable development requires a reduction in consumption of natural raw materials that are not renewable. The use of wastes from the beneficiation of coal fly ash and paper mill sludge has been investigated, usually by the addition of up to 70% mass of these wastes into clayish products (Kumar et al., 2001; Olgun et al., 2005). The wastes may be used to replace conventional flux materials, with the advantage of controlling the plasticity and shrinkage of the ceramic body without producing any negative effect on the product properties, and allowing sintering at low temperatures, thus resulting in energy conservation. The paper presents the results of a feasibility study on the use of paper sludge by integrating them into a ceramic product.

The purpose of this study was to evaluate the use of primary sludge from waste paper in ceramic tile manufacturing. For this purpose, it is necessary to control parameters such as the rate of incorporation of sludge in different mixtures and the substitution effect of feldspar (FT) by waste paper sludge as an addition. The rheological and physico-mechanical properties of each mixture were studied and compared to reference formula industry. Incorporation of sludge waste paper into mixture of ceramic tiles mass production has resulted in a new formulation with significantly improved properties.

Finally, this work aims to develop new ceramic tiles from waste paper sludge allowing the withdrawal of these residues from the environment and give them a nobler destination.

MATERIALS AND METHODS

The selected wastes are sludge from the Company Stationery and cardboard Saida "PAPCAS" (Western Algerian region), whose physical properties are:

(i) The natural water content of paper sludge is estimated between 26 and 30%;
(ii) Specific Gravity is approximately 2.36 g/cm^3.

The mixtures were prepared with the raw materials, according to predetermined amounts. The study is carried out using combinations of various types clay (yellow clay (YC), gray clay (GC), ordinary clay (OC) hereafter called YC, GC and OC respectively), which are added the FT and the sand (SB) from TIMEZRIT (Eastern Algerian region) and BOU SAÂDA (Southern Algerian region) respectively.

The study is carried on the use of the following raw materials: clays YC, GC and OC, FT, sand. The elementary chemical analyses were carried out with X-ray fluorescence spectrometer,

PW2540 type, Vrc, Dy-1189. The mineralogical analyses were carried out in a diffractometer type X'Pert Pro Detector X'celerator, Dy-2233-0525.

Particle size analysis is performed by pipette "Robinson". The method used is a wet process. Raw materials (plastic, not plastic) were ground for 6 h in a moist environment with humidity of 41% in pots to jars "Gerhards" type TPR, using bodies grinding in the form of alumina beads. Particle size analysis is performed by pipette "Robinson." The process used is a process by wet. Raw materials (plastic, not plastic) were ground wet with humidity of 41% in mills jars "Gerhards," type TPR with grinding body into alumina beads with a milling time of 6 h. The research protocol is based on 04 types of mixtures (M1, M2, M3 and M4). The FT was substituted by the paper mill sludge of waste. The content of paper sludges varied in 5, 10, and 15 to 20%. The viscosity is determined using a Ford Cup 4 mm in diameter, by measuring the flow time for 100 ml of the suspension. The slips are dried in an oven for 24 h at 110°C, then ground into powder. The powder obtained was wetted with moisture content in 5 to 6%. The pressing is done using a hydraulic press semi automatic type Gabrielli 262367. The drying of tiles is performed, at a temperature of 110°C, in an oven of laboratory Memmert type. The cooking of ceramic tiles is carried out in a muffle oven model: N200A at quick-cooking, type NR: 55269, according to three cooking temperatures: 105, 1100 and 1150°C, with a temperature landing of 45 min. The mechanical strength is determined by a device force - Model: 424 CRAB Type Gabrielli.

RESULTS AND DISCUSSION

Chemical analysis

The chemical analysis of the raw materials was determined by X-ray fluorescence analysis, as shown in Table 1. It can be seen from analysis of the clays body YC, GC and OC, high levels of silica, the absence practically of calcium oxide and the low levels of magnesium oxide. It is observed that the predominant elements in the two clays have a content almost identical among which we mention: SiO_2, which must come from different mineral origins (FTs, muscovite and tourmaline) or of free silica. A high concentration of this oxide means that is present in the raw material under these two forms. It can be observed also that Al_2O_3 is present in clays under a low content (<14.1%). The three types of clays are very few refractory. This oxide was correlated to the presence of low clay fractions offering low plasticity for the material. In view of high levels of Fe_2O_3, these clays are of ferruginous type. Iron plays the role of flow element during firing, forming a eutectic melting at low temperature. It communicates with products ranging in color from red to dark brown. The presence of relatively high CaO in two clays indicates that the mineral contains plagioclase. CaO, derived from carbonates, plays the role as melting element and combines with silicates during firing. Alkali oxides Na_2O and K_2O derived primarily from feldspars, illites, micas and smectites. They favor the formation of vitreous phase in low melting point during cooking, playing the role of energy fluxes. Their associations with iron oxide Fe_2O_3 during cooking,

Table 1. Chemical analysis of materials used.

Component	SiO_2	Al_2O_3	Fe_2O_3	CaO	MgO	SO_3	K_2O	Na_2O	TiO_2	Loss ignition
Clay YC	47.17	11.88	5.51	14.19	1.80	0.64	1.63	0.67	1.33	15.69
Clay GC	46.91	11.99	5.63	14.63	1.58	0.37	1.55	0.62	-	15.90
Clay OC	53.90	14.04	5.57	8.15	1.29	-	1.98	1.37	0.16	13.11
Feldspar (FT)	72.90	14.11	1.93	0.72	0.51	-	4.98	2.16	0.73	2.00
Sand (SB)	97.10	0.83	1.19	0.09	0.05	-	0.09	0.11	0.11	0.50

Table 2. Mineralogical analysis of the three types of clay.

Clays (YC + GC)		Clay OC	
Mineral	Presence	Minerals	Presence
Montmorillonite	++	Montmorillonite	++
Illite / muscovite	++	Illite / muscovite	++
Chlorite	++	Chlorite	++
Quartz	+	Quartz Quartz	+
Calcite / dolomite	++	Calcite	+
Iron oxide	+	Iron oxide	+
Soluble salts	+	Soluble salts	+

Table 3. Mineralogical analysis of feldspar and sand.

Feldspar (FT)			Sand (SB)	
Mineral	Presence	Contents (%)	Minerals	Presence
Quartz	+++	41.0	Quartz	+++
Potash feldspar	++	29.5		
Sodium feldspar	++	18.5		
Biotite	+	11		
Chlorite	+	11		
Iron oxide	+	11	Iron oxide	+

Legend: (+) Low percentage ; (+ +) Mean percentage; (+ + +) High percentage.

causes sinterizing reactions which give the products their definitive qualities.

It is noteworthy that the high loss on ignition in two clays (YC and GC) is due specifically to the release of CO_2 from the decomposition of carbonates and liberation of SO_3 gas present in clays. It is observed that in the OC clay, the SiO_2 and Al_2O_3 content are more important. Also we notice that the CaO content is lower with no sulfur element which explains a lower loss on ignition. The alkalis content was higher allowing us to deduce that this clay has higher "grésification" (cementation process of grains performed by precipitation and crystallization of the salt dissolved in pore water).

The FT has a SiO_2 and alkalis content relatively high compared to other elements. The ignition loss is almost negligible and is correlated with low content CaO and the absence of SO_3 gas.

Mineralogical analysis

The results of mineralogical analyses (Tables 2 and 3) are in accordance with chemical analyses. The results show a predominance of a mineral inter laminate consisting of montmorillonite and illite, the clay fraction consisting. A considerable rate of montmorillonite may be the cause of a high drying shrinkage along with certain sensitivity to drying but could be corrected by adding degreaser. We also detected the presence of muscovite and chlorite iron. The clay fraction is composed of calcite in a high amount and traces of dolomite.

We also note every contents low in soluble salts that are difficultly identifiable without any detrimental effects. Clay OC is without dolomite but has the same minerals as YC and GC. The X-ray diffraction patters of the waste paper sample material under investigation are given in

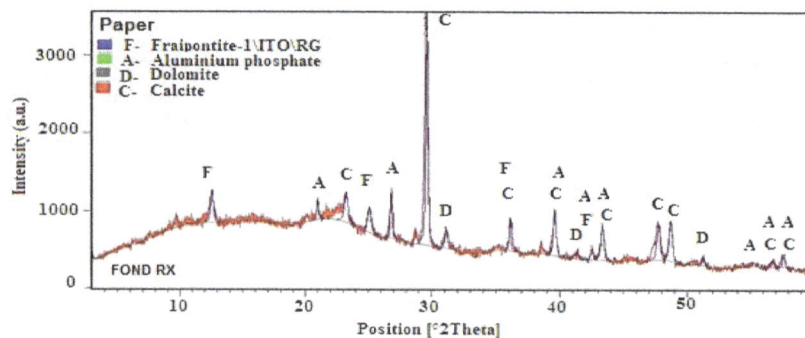

Figure 1. X-ray diffraction pattern of the waste paper sample.

Figure 1. In the paper waste the main constituent is Calcite *(CaCo₃),* followed by Aluminum Phosphate ($4AlPO_4$), Fraipontite [$(Zn,Al)_3(Si,Al)_2O_5(OH)_4$] and Dolomite [$CaMg(CO_3)_2$]. These mineralogical data confirm that paper waste is mostly adequate for the mixing with the ceramic body.

Morphology of raw materials

Figure 2 shows the micrographs of the YC, GC, OC and FT. The micrograph to FT shows a predominance of large aggregates belonging to alumina. The FT is surrounded by spaces inter granular which consist mainly of aggregates belonging to silica which are added to his neighborhood of small particles which includes specific components of FT, namely the potassium and sodium. The empty spaces account for dark areas of the picture. Given the density of large aggregates, the FT requires a grinding that is intense enough during its treatment.

Technological characteristics

Preparation of suspensions

In order to determine rheological properties, a mixture of 04 preparations M1, M2, M3 and M4 were developed. The grinding is performed in a ball mill for 06 h with a moisture content of 41% suspensions. A residue on the sieve 63 μm determined. The weight of the composition is determined in Table 4. The values of rheological properties are shown in Table 5.

Preparation of powders

A traditional preparation of the powders was applied. This was carried out in two stages:

(i) Drying of slurries in the drying oven,
(ii) Crushing and screening the resulting cake.

The homogenization of the powder obtained was carried

out on the choice of a mixture of 04 fractions: fractions (1 - 0.5) = 20%, fractions (0.5 - 0.25) = 45%, fractions (0.25 - 0.15) = 20% and fractions < 0.15 = 15%. A mixture of different particle size classes of particles provides the advantage of reducing the quantity of second phase necessary to the flow and leading to a more compact pile, which reduces the sintering shrinkage (Boch, 2001). After mixing the different fractions, the obtained powder is humidified with 5% humidity and allowed to stand during 24 h before pressing.

Confections of tiles

The pressing is carried out using a semi-automatic hydraulic press, type-Nassetti VIS801. Two pressures were applied on the tiles: One of 9 MPa for desecrating the powder and the second of 22 MPa for compacting. The drying process was carried out into a dryer rolls type NASSITI WHO. The firing is carried out in a roller furnace at temperatures of 1050, 1100 and 1150°C.

Characterization of physical-mechanical properties of dry and fired tiles

The strength values on raw, dried and fired material are shown in Table 6. Considering these values, we notice that the M3 mixture composed of 15% waste paper gives the best resistance to the raw, dried and fired at a temperature of 1100°C (Figure 3). Increasing the temperature to 1150°C, overcooking occurs resulting in a decrease in resistance accompanied by a partial melting of the constituents of the ceramic tile.

Furthermore, considering the mineralogical analysis of waste paper (Figure 1), these contain a rate of dolomite and calcite that are not insignificant. Increasing the concentration of this addition in the mixture led to an appearance of excess liquid phase that in turn led to a reopening of the pores and decreasing in strength.

The values of flexural strength of dry products show a marked increase in the mixture M2. Besides this

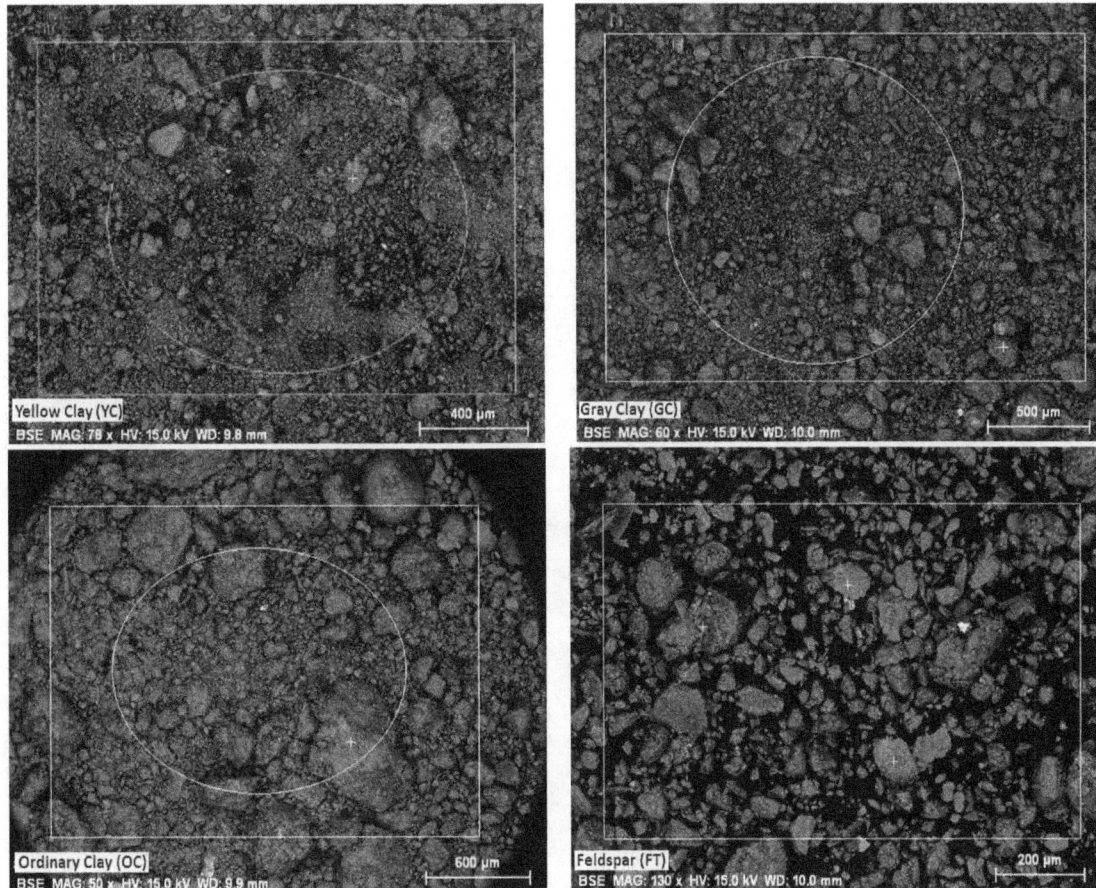

Figure 2. SEM micrographs of the surfaces of the yellow, green clays and feldspar.

Table 4. Compositions of mass.

Content (%)	Mixtures				Natural moisture (%)
	M1	M2	M3	M4	
Clay　(YC+GC)	15	15	15	15	13.85
Clay　　OC	35	35	35	35	9.68
Sand　(SB)	15	15	15	15	1.48
Feldspar (FT)	30	25	20	15	5.6
STPP	0.25	0.25	0.25	0.25	-
Waste paper (SP)	5	10	15	20	-

Table 5. Variation of rheological properties depending on the addition of waste paper.

Mixtures		M1	M2	M3	M4
Refusal (%)		8.82	6.83	5.88	3.20
Density (g/cm^3)		1.65	1.594	1.594	1.584
Viscosity (cp)		20.65	15.79	18.07	19.92
Thixotropy	15 min	21.78	17.20	18.29	21.02
	30 min	22.46	18.42	18.58	22.60
	45 min	23.10	19.12	19.03	23.05

Table 6. Variation of the flexural strength as a function of the addition of waste paper and temperature.

Mixture		M1	M2	M3	M4
Bending strength on raw (MPa)		0.8	0.82	0.97	0.93
Bending strength on dry (MPa)		1.93	3.23	2.84	1.66
Bending strength on fired (MPa)	1050°C	11.75	13.62	16.18	12.52
	1100°C	12.81	14.55	16.27	14.04
	1150°C	12.01	13.87	13.08	12.24

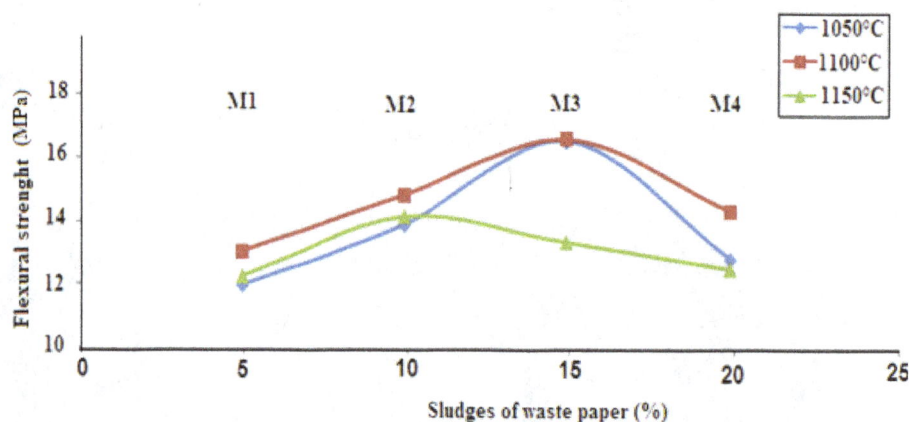

Figure 3. Flexural strength of the various mixtures versus in wastes paper sludge content and the firing temperature.

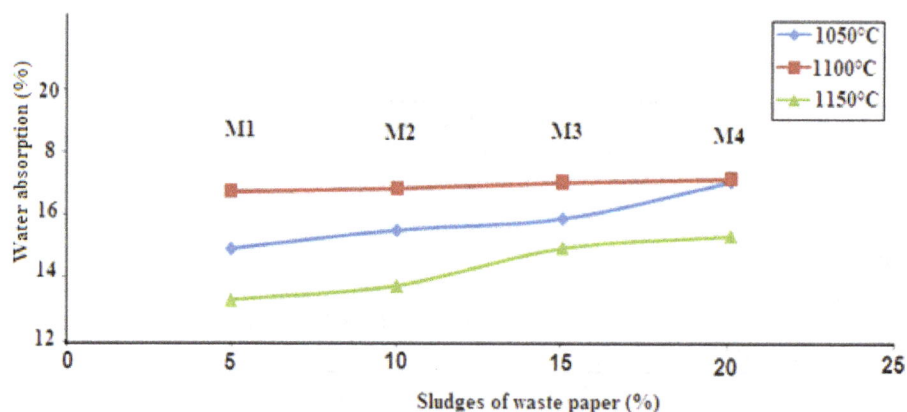

Figure 4. Water absorption of the various mixtures versus in wastes paper sludge content and the firing temperature.

concentration the resistance decreases due to the high evaporation of the physically bound water of the water in the clay suspension (mixture) and one contained in the waste paper. For fired products the mechanical properties increase in the range temperature 1050 - 1100°C. Over 1100°C, these properties are strongly dependent on the addition of waste which provides an excess of liquid phase which decreases mechanical properties. The diagrams of "Grésification" of the various compositions were prepared in terms of water absorption and density, porosity open, closed and total. The dependence of these parameters as a function of firing temperature is shown in Figures 3 to 6.

Wang et al. (2000) have shown that if sintering duration

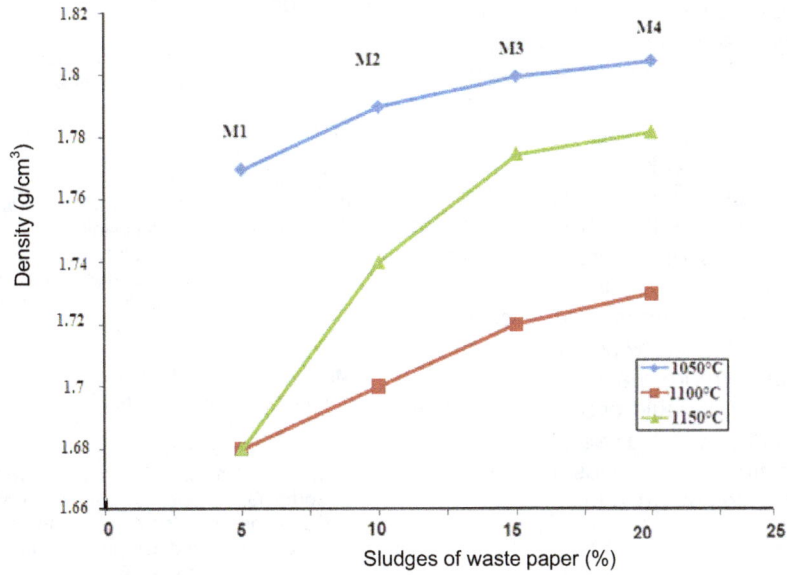

Figure 5. Density of the various mixtures versus in wastes paper sludge content and the firing temperature.

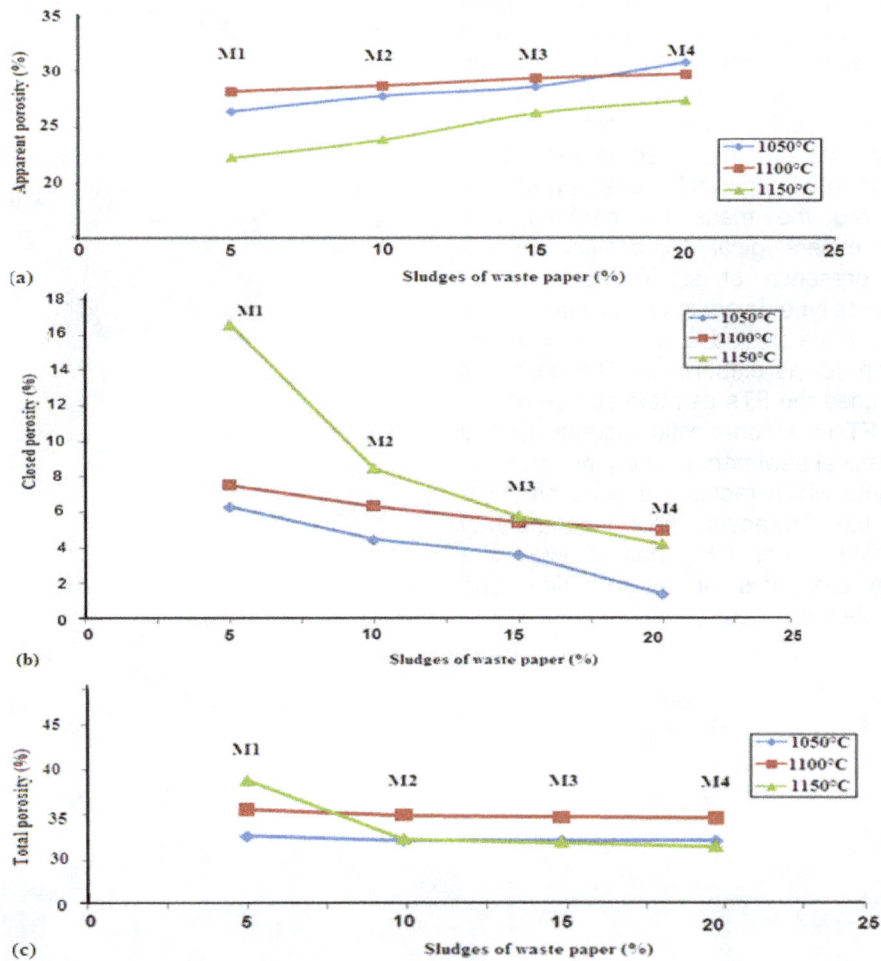

Figure 6. (a) Apparent porosity, (b) Closed porosity, (c) Total porosity of the various mixtures versus in wastes paper sludge content and the firing temperature.

increases, the density final product increases. In the case where the sintering stage decreases the opposite case occurs (Won et al., 2002). Developments specific structural characteristics and mechanical properties of clay products are attributed to their chemical and mineralogical composition (El Yakoubi et al., 2006). Waste paper contains a significant concentration of calcite and dolomite involved in the formation of the glassy phase. From the values of absorption and open porosity of mixtures M2 and M3, we can say that in the mixture M2, the tiles are less absorbent and less porous and their mechanical strength is lower. As mentioned above, the mixture contains more waste M3 acting as flows that produce a quantity of liquid phase due to their higher content (15%). This is what gives M3 mixture a greater mechanical strength, and be chosen as optimal formulation for the ceramic tiles manufacture.

Conclusion

The use of industrial wastes and by-products is broadly recognized as major privileged options for achieving sustainable development. Environment policies in effect require the recovery of waste instead abandoning them to the landfill. In view of the considerable rate of waste thrown in the nature by paper industry, it becomes important to recycle them and use them in other industry sectors. In this study we are interested in recovering sludge from the treatment of paper mill effluents and their use as additions into the mass for ceramic tiles manufacturing. The mineralogical composition of the sludge shows the presence of significant levels of alkaline - earth elements type dolomite and calcite. These chemical compounds have played a role of flux in the masses for ceramic products elaboration. The ceramics industry has always used the FTs as main source of flux elements. Because FT is a non-plastic material that is very tough, its mechanical treatment is often the cause of damage of equipments which requires a good crushing and grinding of the flux. Moreover, the replacement of 50% of FT by waste paper (PS) has allowed the improvement of the properties of ceramic tiles and obtaining a new formulation.

REFERENCES

Boch P (2001). Frittage et microstructure des céramiques. In : BOCH P. Ed. Matériaux et Processus Céramiques, Paris. Hermès Sci. Pub. pp. 75-112.

El Yakoubi N, Aberkan M, Ouadia M (2006). Potentialité d'utilisation d'argiles marocaines de Jbel Kharrou dans l'industrie céramique. C R Geosci. 338(10):693-702.

Kumar S, Singh KK, Ramachandrarao P (2001). Effects of fly ash additions on the mechanical and other properties of porcelanaised stoneware tiles. J. Mater. Sci. 36:5917-5922.

Marcis C, Minichelli D, Bruckner S, Bachiorrini A, Maschio S (2005). Production of monolithic ceramics from incinerated municipal sewage sludge, paper mill sludge an steelworks slag. Ind. Ceram. 25(2):89-95.

Mir AA (1982). Utilization of industrial wastes - An economic opportunity. Ind. Waste. Proc. 14th Mid-Atlantic Conf. Eds. James E. Alleman and Joseph T. Kavanagh, Am. Arbor Science Publishers, Anu Arbor, Michigan. 2:8-12.

Modesto C, Bristot V, Menegali G, De Brida M, Mazzucco M, Mazona A, Borba G, Virtuoso J, Gastaldon M, Novaes de Oliveira A (2003). Collection and Characterization of Ceramic Materials from Industrial Solid Wastes. Ind. Ceram. 8(4):14-18.

Olgun A, Erdogan Y, Ayhan Y, Zeybek B (2005). Development of ceramic tiles from coal fly ash and tincalore waste. Ceram. Int. 31:153-158.

Wang SW, Chen LD, Hirai T (2000). Densification of Al_2O_3 powder using spark plasma sintering. J. Mater. Res. 15(4):982-987.

Won JH, Kim KH, Chae JH, Shim KB (2002). Sintering of attrition-milled TiN powders using a spark plasma sintering technique. J. Ceram. Process. Res. 3(3):166-170.

15

Effect of milling conditions on the formation of $ZnFe_2O_4$ nanocrystalline

O. M. Lemine

Department of Physics, College of Sciences, Al imam Mohammad Bin Saud Islamic University (IMSIU), P. O. Box 90950 11623 Riyadh, Saudi Arabia.

The effects of milling parameters on the formation of $ZnFe_2O_4$ nanocrystalline are studied. Powder mixtures of ZnO and Fe_2O_3 were milled in high energy vibrant ball milling for different balls to powders mass ratio and milling times. X-ray diffraction and scanning electron microscopy are used to characterize the powders. A crystalline size and lattice parameter are obtained from Scherrer formula and the Nelson–Riley function respectively. It was found that the balls to powder mass ratio plays a key role on the formation of $ZnFe_2O_4$ nanocrystalline more than other parameters (type of mill and rotation speed). A spinel phase appears from 6 h of milling duration with balls to powder ratio of 10:1 but nevertheless, the Fe_2O_3 and ZnO disappear if the milling time was further increased to 24 h. For balls to powders mass ratio of 20:1, the obtained zinc ferrite phase seems to be pure after 12 h milling. The effect of the different parameters is discussed in detail.

Key words: Zinc-Ferrite, nanocrystalline, ball milling, scanning electron microscope (SEM), X-ray diffraction.

INTRODUCTION

Zinc ferrite ($ZnFe_2O_4$) is of interest not only to basic research in magnetism, but also has great potential in technological application, such as magnetic materials, gas sensors, catalysts, photocatalysts and absorbent materials (Xu et al., 2011). The structure of the spinel oxides AB_2O_4 consists of the close-packed face cubic centred (FCC) arrangement of oxygen atoms, with two non-equivalent crystallographic tetrahedral A and octahedral B sites. In zinc ferrite, the Zn^{2+} and Fe^{3+} ions can be distributed over the A and B sites, and therefore the formula is sometimes represented by $(Zn_{1-\delta}Fe_\delta)[Zn_\delta Fe_{2-\delta}]O_4$, where the part between the round brackets represent, are the atoms at the A sites, the part between square brackets are the atoms at the B sites and δ is a measure of the fraction of Fe^{3+} on the A sites and is called the inversion parameter. δ is 1 for a perfectly inverse spinel, 2/3 for the random arrangement and 0 for a perfectly normal spinel. Several methods have been used for the preparation of $ZnFe_2O_4$ nanocrystalline such as co-precipitation, aerogel and hydrothermal method (Jiang et al., 1999; Shenoy et al., 2004; Yu et al., 2003). Among all this techniques mechanical alloying (MA) turns to be the mostly used to prepare nanocrystalline due to its simplicity, low cost and its ability to produce large volumes. Mechanical alloying influences phase ratio, particle morphology, particle size distribution, and microstructure of the final product. Milling parameters of importance include type of mill, milling container, speed, time, grinding medium, ball-to-powder weight ratio, milling atmosphere, process control agent and milling temperature. However, the most important parameters are the milling time, speed and ball-to-powder ratio (Suryanarayama, 2000). Several groups have been interested by the production of $ZnFe_2O_4$ nanocrystalline by mechanical alloying using a start materials ZnO and Fe_2O_3 powders (ZHAO et al., 2010; Verdier et al., 2005; Jean and Nachbaur, 2008; Nachbaur et al., 2009). For example Nachbaur et al. (2009) and Verdier et al. (2005) studied the effect of milling times with a fixed balls to powders mass ratio (20:1) and speed rotation. Jean and Nachbaur (2008) were interested by the influence of disc and vials velocity on the end products. ZHAO et al. (2010) obtained zinc ferrite after 40 h milling with balls to powder ratio of 20:1.

A common point between all the works that most of them used high energy planetary milling to prepare the sample. The aim of this work is to give an additional contribution to understand the influence of the type of mill (vibrant mill), the balls to powders mass ratio and the milling times on the formation of $ZnFe_2O_4$ nanocrystalline.

EXPERIMENTAL

Commercially powders of hematite (α-Fe_2O_3) and zinc oxide (ZnO) are used with equal molar (1:1) and were introduced into a stainless steel vials with stainless steel balls (12 and 6 mm in diameter) in a high energy mill (SPEX 8000 mixer mill). Different milling times were considered (6, 12 and 24) and two values of the balls to powders mass ratio were used (10:1 and 20:1). X-ray powder diffraction (XRD) measurements were performed using Shimadzu diffractometer (θ-2θ) equipped with Cu-Kα radiation (λ=1.5418 Å). The crystallite was calculated from the Scherer formula:

$$D = \frac{K\lambda}{B\cos\theta}$$

(1)

where λ=1.5418 Å, K is a constant whose value is approximately 0.9 and B (rad) is the width of peak. B (in rad) was determined as full width at half-maximum (FWHM). Particles morphology of our samples was investigated using Nova 200 NanoLab field emission scanning electron microscope (FE-SEM).

RESULTS AND DISCUSSION

Figure 1 shows the XRD patterns for Zinc oxides, hematite and the initial hematite and zinc oxide mixture, before ball milling. The patterns of un-milled ZnO, α-Fe_2O_3 and (ZnO + α-Fe_2O_3) mixture powders show a series of strong and narrow peaks characteristic for high quality ZnO and α-Fe_2O_3 crystals. The crystallite size for the started powders ZnO and α-Fe_2O_3 are 150-300 nm and 200-300 nm, respectively.

Figure 2 shows the XRD patterns for the sample milled for different times (6-24 h) and balls to powder ratio of 10:1. XRD pattern of the un-milled mixture is added to see the line positions of the initial oxides. Table 1 shows the phases obtained with different milling parameters. With increasing the milling time, the diffraction peaks became broader and their relative intensity decreases, as shown in Figure 2b-d. This effect is a typical behaviour of

materials after milling and attributed usually the presence of crystallite with small size and internal strain induced by mechanical deformation. Similar results were obtained for milling of ZnO and α-Fe_2O_3 powders separately (Lemine, 2009). Regarding the decrease of the relative intensity of the peaks, it can be seen that the decrease of the ZnO peaks was more obvious, compared with that of α-Fe_2O_3 for longer milling times. That can be explained by the amorphization of ZnO or may be by the diffusion of ZnO into α-Fe_2O_3. But the second reason seemed more reasonable due to all milling times, the ZnO peaks remained. In addition, the equal molar (1:1) of ZnO and α-Fe_2O_3 used in the preparation of the samples explained that ZnO diffused into Fe_2O_3 crystal lattice during mechanical alloying process.

After 6 h milling, it can be seen that the zinc ferrite phase, $ZnFe_2O_4$ appears and coexists with the remaining hematite and Zinc oxides phases. The presence of zinc ferrite is revealed by the lines around 17.5° (spinel phase) and can be explained by the solid state reaction of ZnO and α-Fe_2O_3. The same result was obtained for longer milling times (12 and 24 h) and α-Fe_2O_3 and ZnO are still present. For the same milling times, several research groups obtained a pure zinc ferrite (ZHAO et al., 2010; Verdier et al., 2005; Jean and Nachbaur, 2008; Nachbaur et al., 2009). For example Nachbaur et al. (2009) observed that after 12 h of milling, the main phase is zinc ferrite. After 24 of milling the phase obtained is a pure zinc ferrite. Recently ZHAO et al. (2010) obtained zinc ferrite phase after 30 h. The difference with the current results can be explained by the value of balls to powders mass ratio used by all the groups (20:1, in this study 10:1 was used). Nevertheless, the Fe_2O_3 and ZnO disappear if the milling time was further increased to 24 h. We can conclude that the balls to powder mass ratio plays a key role of the formation of $ZnFe_2O_4$ nanocrystalline more than other parameters (type of mill and rotation speed).

In order to compare the effect of the type of mill, we have prepared one sample with the same ball to powders mass ratio used by the other groups (20:1). Considering that the rotation speed of high energy vibrant mill (SPEX 8000D) is higher than that used for planetary milling (500 rpm), we used only one milling time of 12 h and the XRD patterns are shown in Figure 3. It is clear from the figure that (h k l) values of each peak corresponding to the zinc ferrite. XRD pattern of the sample exhibits the formation of single-phase cubic spinel structure. The characteristic peaks for the spinel Zn-ferrite appear as the main crystalline phases. The average crystallite size (10 nm) was estimated from the most intense peak, corresponding to (311) reflection by using the Debye-Scherer formula. The interplanar spacing, dhkl (h, k, l are Milles indices), were calculated from the Bragg equation (Cullity, 1979).

$$2d_{hkl}\sin\theta = n\lambda$$

(2)

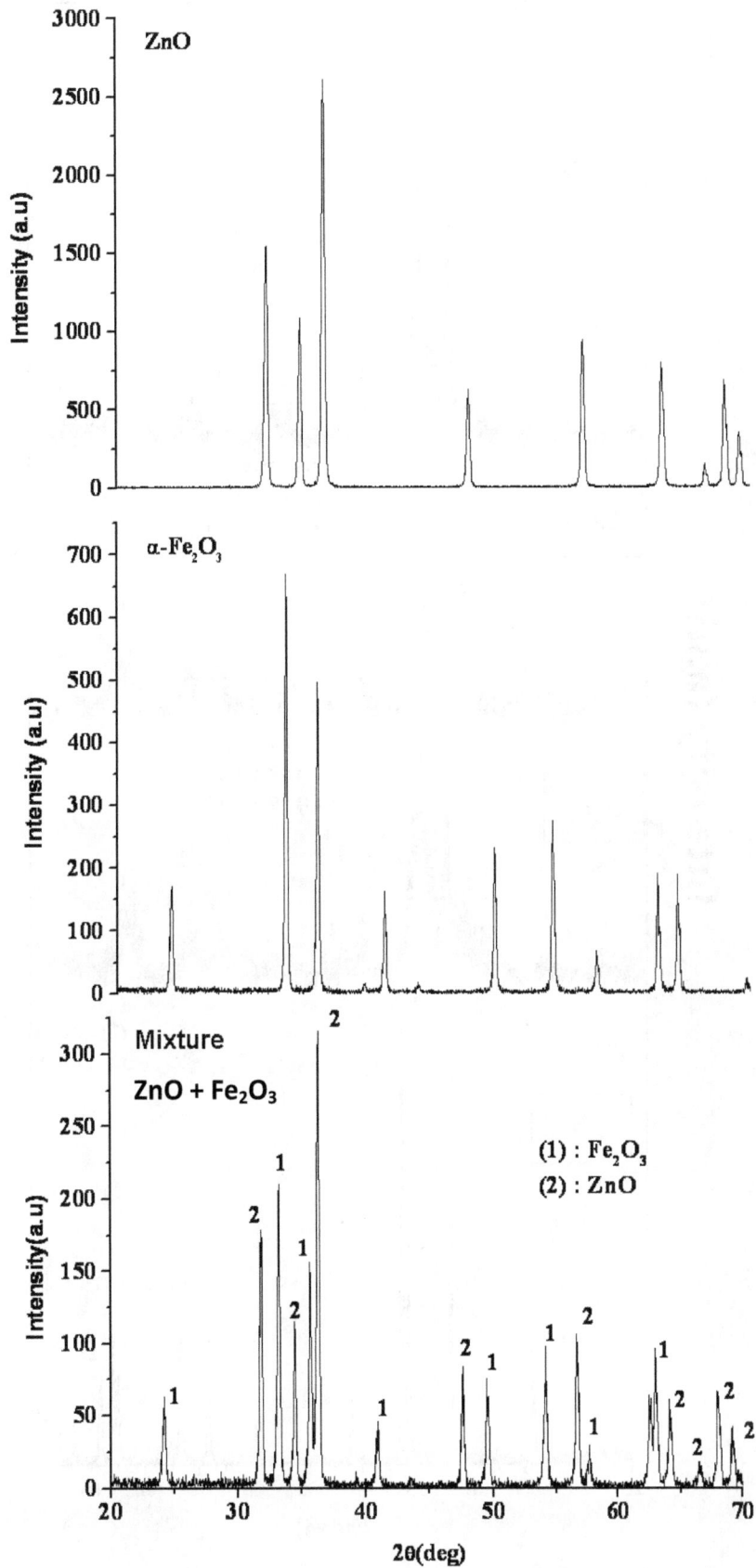

Figure 1. XRD patterns of ZnO, α-Fe$_2$O$_3$ and ZnO- α -Fe$_2$O$_3$ mixture.

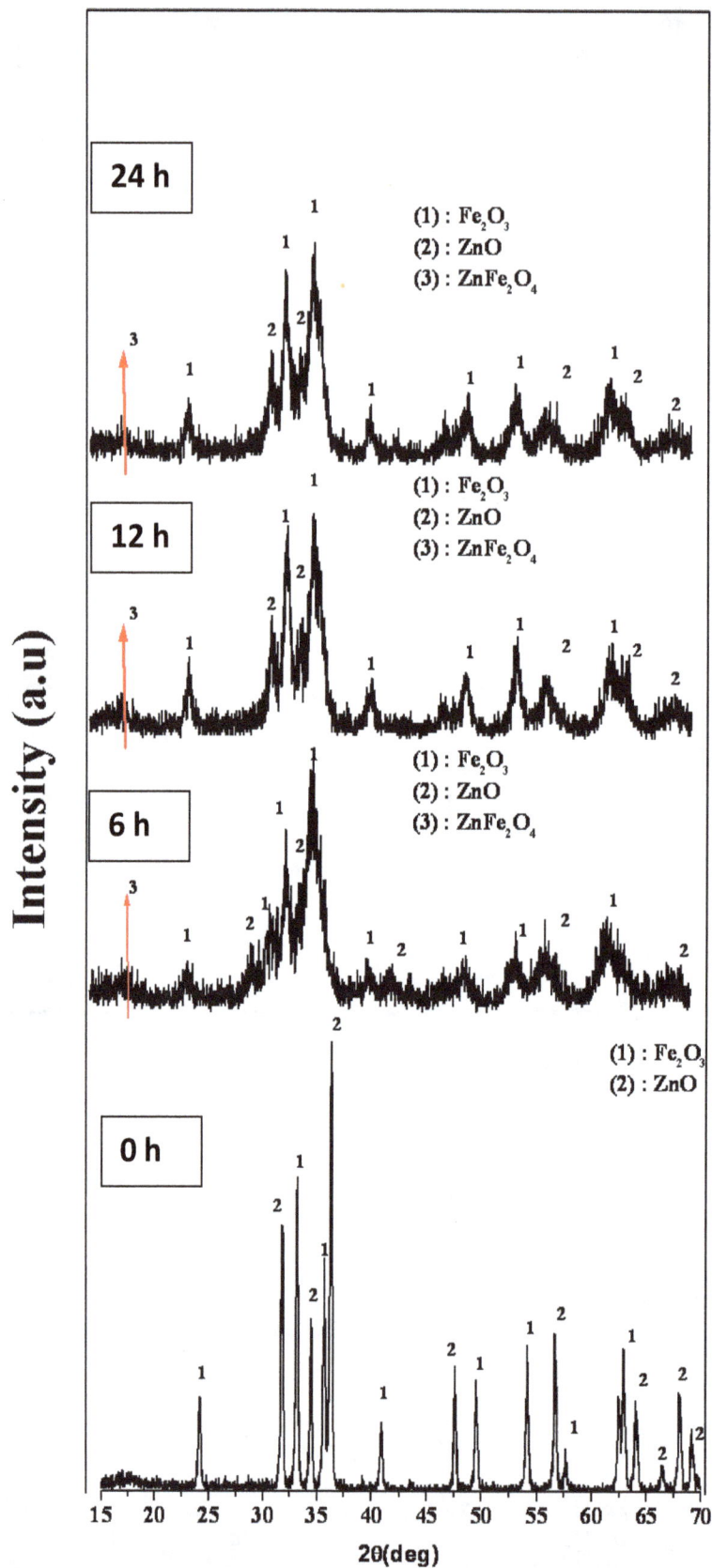

Figure 2. X-ray diffraction patterns of ball-milled ZnO–Fe₂O₃ mixture. For different milling times and balls to powder mass ratio of 10:1.

Table 1. The phase obtained with different milling parameters.

Variable	Milling times (h)	Phase
Balls to powders mass ratio 10:1	6	$ZnFe_2O_4$, ZnO and Fe_2O_3
	12	$ZnFe_2O_4$, ZnO and Fe_2O_3
	24	$ZnFe_2O_4$, ZnO and Fe_2O_3
Balls to powders mass ratio 20:1	12	$ZnFe_2O_4$

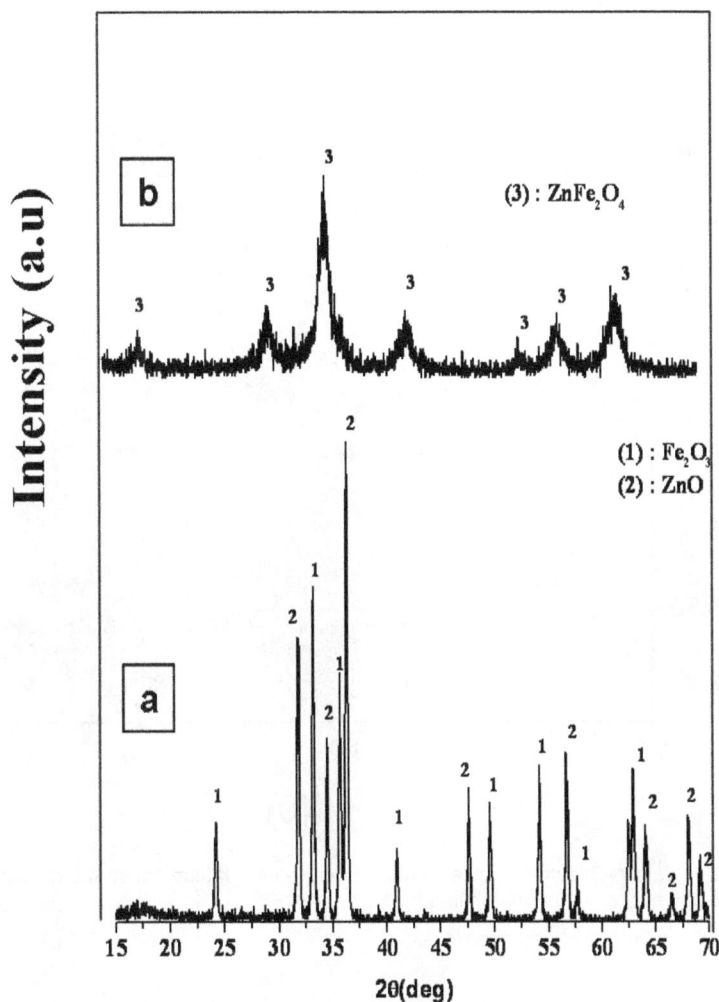

Figure 3. X-ray diffraction patterns of ball-milled $ZnO–Fe_2O_3$ with balls to powder mass ratio of 20:1 : (a) before milling and (b) 12 h.

The lattice parameter, a, was determined by the relationship:

$$a = d_{hkl}(h^2 + k^2 + l^2)^{1/2}$$

(3)

The values of the inter-planar spacing and zinc-ferrite lattice are reported in Table 2. The corrected values of

lattice parameter are estimated from the Nelson–Riley extrapolation method (Cullity, 1979). The values of the lattice parameter obtained from each reflected plane are plotted against the Nelson–Riley function f(θ):

$$f(\theta) = \frac{1}{2}(\frac{\cos^2\theta}{\theta} + \frac{\cos^2\theta}{\sin\theta})$$

(4)

Table 2. Structural characteristics parameters.

Variable	Sample milled for 12 h			
(hkl)	2θ(deg)	d_{hkl}(Å)	a_{hkl}(Å)	a (Å): Nelson-Riley function
(111)	18.26	4.85	8.400	
(220)	30.12	3	8.485	
(311)	35.46	2.53	8.391	
(400)	43.02	2.1	8.400	8.418
(422)	53.36	1.71	8.377	
(511)	57	1.61	8.365	
(440)	62.06	1.5	8.485	

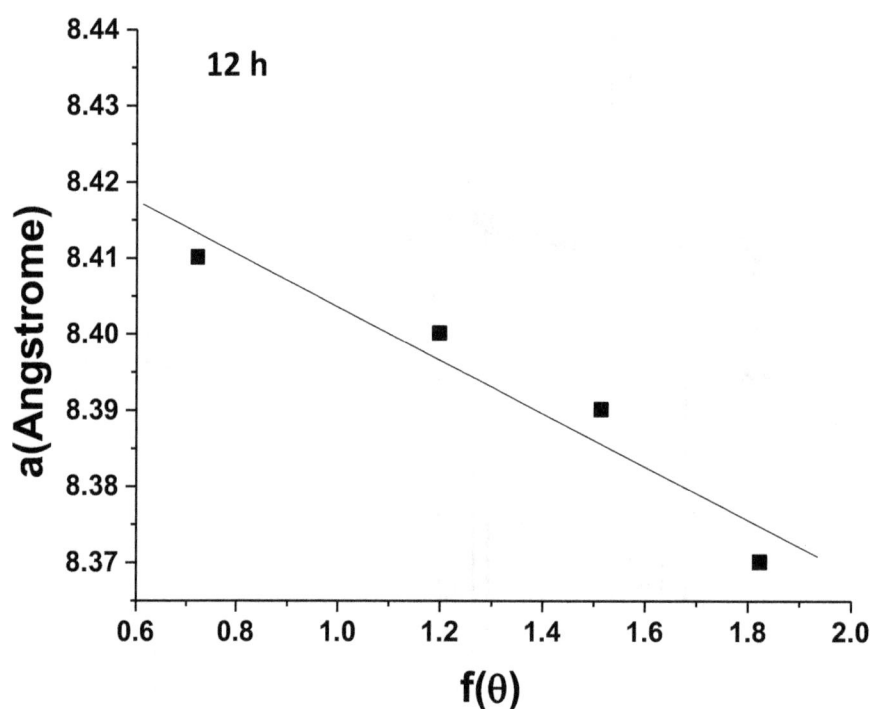

Figure 4. Nelson-Riley plots for accurate measurement lattice constants of sample with balls to powder ratio 20:1 and milled for 12 h.

Where Θ is Bragg's angle and as straight line is obtained. The value of the lattice parameter was estimated from the extrapolation of the straight line to $f(\Theta) = 0$ (Figure 4). The obtained value was a = 8.418 Å and it is slightly smaller than the 8.441 Å value reported in Joint Committee on Powder Diffraction Standards (JCPDS) file (JCPDS 22-1012) but in agreement with value of a = 8.437 Å reported recently by Nachbaur et al. (2009). Other groups obtained a values slightly larger than the value reported for bulk zinc ferrite, a = 8.445 Å (Hakim et al., 2011) and a=8.449 Å (Ehrhardt et al., 2002). The contraction of the lattice parameter observed in our case and in the works published by Nachbaur et al., 2009) and

(Verdier et al., 2005) can be explained by the interpretation proposed by O'Neill (1992). For the compounds obtained by ceramic route with inversion parameters λ lower than 0.2, the cell parameter decreases with λ according to the following equation: a (nm) = 0.84423-0.001247 λ. Sepelak et al. (1997) reported that this inversion leads to a contraction of the crystal lattice from 8.4432 (ceramic sample) to 8.4136 Å (same sample milled for 24 min) but in our case it is difficult to conclude about an inversion phenomenon from our XRD results or to have a value for the inversion parameter. But it is interesting to note that in contrast to the bulk compound, the nanocrystalline $ZnFe_2O_4$ system

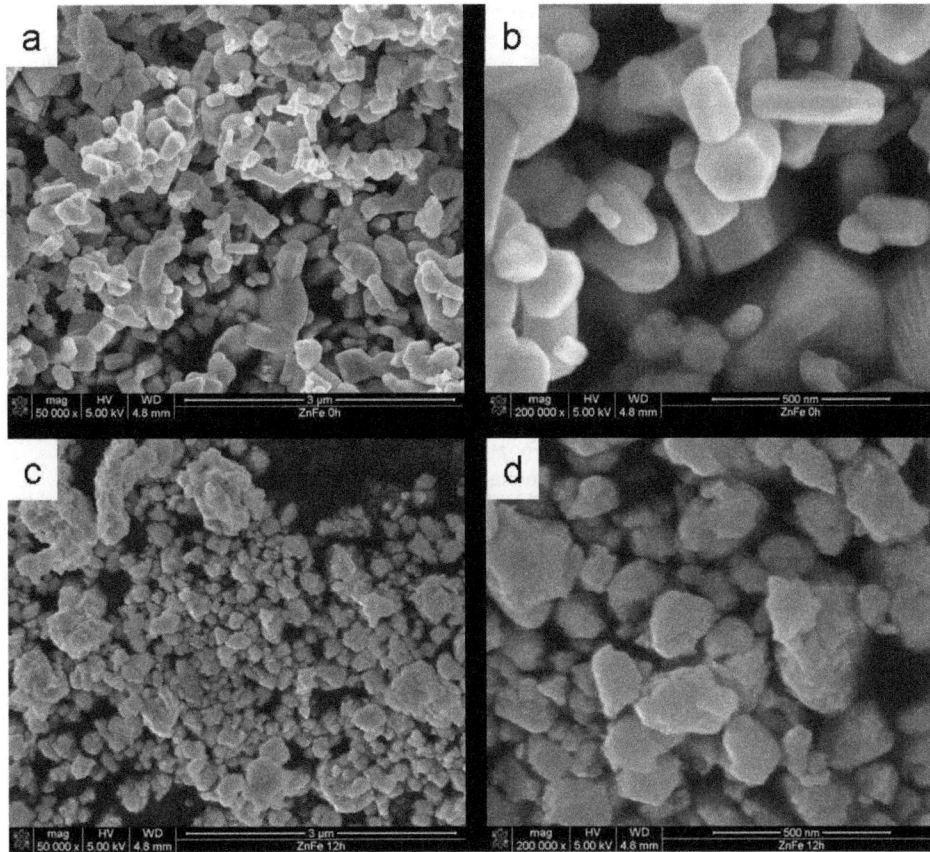

Figure 5. (a) SEM micrographs of mixtures (zinc oxides +hematite) powders as received (b) as received at high magnification (c) milled for 12h (c) milled for 12 h high magnification.

always shows up as a mixed spinel, of which the value of the inversion parameter λ is largely dependent on the synthesis procedure.

In general the process of synthesis of nanocrystalline zinc ferrite by ball milling of mixtures ($ZnO–α-Fe_2O_3$) could be described as:

$$ZnO + α\text{-}Fe_2O_3 \rightarrow ZnFe_2O_4$$

The difference in the results (milling times producing zinc ferrite or crystallites size) from one study to other depend on the type of milling used and the milling parameters such as balls to powder ratio, milling times and molar ratio between ZnO and Fe_2O_3. Our results are similar with those obtained by Nachbaur et al. (2009) but they used planetary milling type with rotation speed of 500 rpm (in our case the vibrant milling with rotation speed of 1200 rpm) and that can explained the difference of milling times values.

SEM micrographs of the samples before and after milling are shown in Figure 5. It is clear that unmilled powder shows a different chap of powders due to zinc oxides and hematite powders (Figure 5a and b). After milling, a reduction of the crystallite size can be observed

(Figure 5c). High magnification images (Figure 5d) reveal clearly that large particles are in fact agglomerates of much smaller particles.

Conclusion

The synthesis process of $ZnFe_2O_4$ nanocrystalline by high energy ball milling the mixture of ZnO and Fe_2O_3 is affected by the milling parameters such as balls to powders mass ratio and milling time. It was found that the balls to powder mass ratio plays a key role of the formation of $ZnFe_2O_4$ nanocrystalline more than other parameters (type of mill and rotation speed). For balls to powder ratio of 10:1 and from 6 h milling, $ZnFe_2O_4$ appears and coexists with the remaining hematite and Zinc oxides phases. For balls to powder ratio of 20:1 the spinel Zn-ferrite appear as the main crystalline phases after 12 h of milling duration.

REFERENCES

Cullity BD (1979). Elements of X-ray Diffraction, Addison-Wesley, Reading, Massachusetts P. 356.

Ehrhardt H, Campbell SJ, Hofmann M (2002). Structural evolution of ball-milled $ZnFe_2O_4$. J. Alloys Compd. 339(2002):255-260.

Hakim MA, Haque M, Huq M, Nordblad P (2011). Spin-glass-like ordering in the spinel $ZnFe_2O_4$ ferrite. 406(1):48-51.

Jean M, Nachbaur V (2008). Determination of milling parameters to obtain mechanosynthesized $ZnFe_2O_4$. J. Alloys Compd. 454(1-2):432-436.

Jiang JZ, Wynn P, Morup S, Okada T, Berry FJ (1999). Magnetic structure evolution in mechanically milled nanostructured $ZnFe_2O_4$ particles. Nanostruct. Mater. 12:737-740.

Lemine OM (2009). Microstructural characterisation of nanoparticles using, XRD line profiles analysis, FE-SEM and FT-IR. Superlattices Microst. 45:576-582.

Nachbaur V, Tauvel G, Verdier T, Jean M (2009). Mecanosynthesis of partially inverted zinc ferrite. 473(1-2):303-307.

O'Neill HSC (1992). Temperature dependence of the cation distri bution in zinc ferrite ($ZnFe_2O_4$) from powder XRD structural refinements. Eur. J. Miner. 4:571.

Sepelak V, Tkacova K, Boldyrev VV, Wibmann S, Becker KD (1997). Mechanically induced cation redistribution in $ZnFe_2O_4$ and its thermal stability Physica B 234-236 (1997) 617.

Shenoy SD, Joy PA, Anantharaman MR (2004). Effect of mechanical milling on coprecipitaded ultrafine zinc ferrites. J. Magn. Magn. Mater. 269:217-226.

Suryanarayama C (2000). Mechanical alloying. Progr. Mater. Sci. 46:1-184.

Verdier T, Nachbaur V, Jean M (2005). Mechanosynthesis of zinc ferrite in hardened steel vials: Influence of ZnO on the appearance of Fe(II). J. Solid State Chem. 178:3243-3250.

Xu Y, Liang Y, Jiang L, Wu H, Zhao H, Xue D (2011). Preparation and Magnetic Properties of $ZnFe_2O_4$ Nanotubes, J. Nanomaterials. 2011.

Yu SH, Fujino T, Yoshimura M (2003). Hydrothermal synthesis of $ZnFe_2O_4$ ultrafine particles with high magnetization, J. Magn. Magn. Mater. 256(2003):420-424.

Zhao ZW, Ouyang K, Wang M (2010). Structural macrokinetics of synthesizing ZnFe 2O 4 by mechanical ball milling, Trans. Nonferrous Met. Soc. China 201131-1135.

16

Nanocarbonfibril in rice flour charcoal

Sumrit Mopoung*, Anchalee Sirikulkajorn, Duennapa Dummun and Pichyawee Luethanom

Department of Chemistry, Faculty of Science, Naresuan University, Phitsanulok, Thailand.

This paper presents the results of a study of nanocarbonfibrils in charcoals, which were prepared from rice flour with diameters of 90, 75 and 38 µm, respectively. The rice flours were carbonized for 30 min at a temperature of 600 to 800°C. The carbonized products were characterized by scanning electron microscope (SEM), transmission electron microscope (TEM), Fourier transform infrared spectrometer (FTIR), energy dispersive X-ray spectrometer (EDS) and X-ray diffractometer (XRD). Nanocarbonfibrils with diameters of 59.42 to 68.38 nm were found in all carbonized products. The other nanocarbons exhibited by the products are carbon granules with diameters ranging from 0.9 to 1.1 µm. The most functional groups on surface charcoals are OH-stretching, C=C stretching, P-O-C stretching and Si-O-C stretching. The charcoal products are composed of amorphous carbon together with SiO_2, K_2CO_3, K_2O, KO_2, Mg and P.

Key words: Rice flour charcoal, nanocarbonfibril, nanostructure carbon, carbonization.

INTRODUCTION

Starch granules are composed of two types of *alpha*-glucan, amylose and amylopectin. Cereal starches contain integral lipids in the form of lysophospholipids and free fatty acids. Purified starches contain <0.6% protein and small quantities (<0.4%) of such minerals (as calcium, magnesium, phosphorus, potassium and sodium) (Tester et al., 2004). Starch structures consist of granule (2 to 100 µm), growth ring (120 to 500 nm), blocklet (20 to 500 nm), amorphous and crystalline lamellae (9 nm) and amylopectin and amylase chain levels (0.1 to 1.0 nm), according to Vandeputte and Delcour (2004). Carbon spheres have been synthesized from corn starch by a two stage process (oxidation and carbonization) after which the products kept the original shape of corn starch perfectly and the diameters of carbon spheres ranged from 5 to 25 µm (Zhao et al., 2008). The activated carbon spheres, that were prepared from potato starch by carbonization followed by activation with KOH, were hollow and retained the original morphology of potato starch with a decrease in size (Zhoa et al., 2009). Xue et al. (2004) prepared a carbon atom wire by pyrolysis of starch. The carbon

atom wire is composed of winding lines with a diameter of around 20 nm. Moreover, activated carbons prepared from cationic starch using KOH, $ZnCl_2$ and $ZnCl_2/CO_2$ activation were used for the electrodes of a supercapacitor. The supercapacitor exhibited excellent capacitance characteristics in 30 wt% KOH aqueous electrolytes and showed a high specific capacitance of 238 F/g at 370 mA/g, which was nearly twice that of the commercial activated carbon (Wang et al., 2008). Additionally, the reinforcing potentials of cellulose nanofibers obtained from a starch-based thermoplastic polymer have been found to show diameters in the range of 10 to 80 nm and lengths of several thousand nanometers (Alemdar and Sain, 2008). Carbon fibers are sought for many applications, e.g. supercapacitor electrodes (Wang et al., 2008), reinforcement (Alemdar and Sain, 2008), ultraviolet (UV) absorbents (Ma et al., 2009) and hydrogen storage (Shindo et al., 2004).

In this research, the nanocarbonfibrils in rice flour charcoal were investigated by scanning electron microscope (SEM), transmission electron microscope (TEM), Fourier transform infrared spectrometer (FTIR), energy dispersive X-ray spectrometer (EDS) and X-ray diffractometer (XRD). The effects of sizes of rice flour and carbonization temperature to nanocarbon in charcoals were studied.

---

*Corresponding author. E-mail: sumritm@nu.ac.th.

Table 1. Percent yield of rice flour charcoals from different sizes of fresh rice flour and carbonized temperature.

Charcoals from	Percent yield of rice flour charcoals carbonized at		
	600 °C	700 °C	800 °C
90 μm rice flour	22.27	21.78	20.12
75 μm rice flour	18.92	18.50	17.03
38 μm rice flour	17.01	16.90	16.17

MATERIALS AND METHODS

Dried Zhao flour starch powders (obtain from Phitsanulok market, Thailand) were sieved to sizes of 170 mesh (90 μm), 200 mesh (75 μm) and 400 mesh (38 μm). The starch powders of each size were carbonized at 600, 700 and 800 °C in electric furnace (Fisher Scientific Isotemp® Muffle Furnace). The temperature was increased with a rate of 15 °C/min up to the desired temperature and was kept constant for 30 min. At last, the powders were cooled to room temperature. The carbonized products were analyzed for percent yield and were characterized by a Fourier transform infrared spectrometer (FTIR, Spectrum GX, Perkin Elmer), scanning electron microscope and energy dispersive X-ray spectrometer (SEM/EDS, LEO 1455 VP), transmission electron microscope (TEM, Phillips, Tecnai12) and X-ray diffractometer (XRD, PW 3040/60, X' Pert Pro MPD).

RESULTS AND DISCUSSION

Percent yield of charcoals

The results of percent yield of rice flour charcoals carbonized at 600 to 800 °C are shown in Table 1. They reveal that the percent yield of charcoal decreases as the carbonized temperature increases and the size of rice flour decreases. The increase of the carbonization temperature leads to a higher thermal degradation, resulting in a decreasing percent yield of charcoal (Trompowsky et al., 2005). The percent yield of rice flour charcoals is size-dependent, e.g. gradually decreasing from 22.27% in the 90 μm fraction to 17.01% in the 38 μm one for carbonization at 600 °C. An explanation for this behavior is that the smaller sizes of rice flours have a higher surface area which progressively oxidizes and that they probably exhibit a more branched structure created by the small grains (Francioso et al., 2011).

FTIR spectroscopy

The FTIR spectra of fresh rice flour and rice flour charcoals with carbonization at 600 °C are shown in Figure 1. This figure shows the effect of starch powder size on carbonization. The broad band of fresh rice flour (sample) at 3400 cm^{-1} represents the –OH stretching vibration, possibly including H_2O, alcoholic OH, phenolic OH and/or carboxylic OH (Ascough et al., 2011), and the peaks at 2931 cm^{-1} are due to stretching vibrations of aliphatic CH (Guo and Bustin, 1998). The peak of OH is narrower after carbonization (Figure 1A to C), which suggests an increase in the number of oscillation modes that could be attributed to the presence of new hydrogen bonding interactions (Liu et al., 2011). The peak at 2931 cm^{-1} almost disappeared in the carbonized products, indicating that the hydrogen element was removed to a large extent (Wang et al., 2011). The peak at about 1700 cm^{-1} (C=O stretching band) in fresh powder starch is attributed to either the acetyl and uronic ester groups of the hemicelluloses or the ester linkage of carboxylic group of the ferulic and p-coumeric acids of hemicelluloses (Alemdar and Sain, 2008). This peak disappears completely in the starch powder charcoals, suggesting a destruction of these groups during the carbonization process. The peak between 1600 and 1500 cm^{-1} can be attributed to aromatic C=C ring stretching (Guo and Bustin, 1998). The absorption at 1300 to 900 cm^{-1} could be tentatively ascribed to the phosphorus species. These peaks may be assigned to ionized linkage P–O in acid phosphate esters (Puziy et al., 2002) or in form of phosphate ester groups (Pérez et al., 2005). The absorption in this region is also characteristic for silicon compounds and silicates, which is ascribed to the Si–C or Si–O fundamental stretching vibration (Qian et al., 2004). These peaks are still present in charcoals, suggesting the contribution of phosphorus compounds and silicon compounds in rice flour charcoals. The bands region located from 1200 to 1000 cm^{-1}, which is often considered as the vibration modes of C–C and C–O stretching and the bending mode of C–H bonds or the stretching vibration of C-O bond in C-O-H and C-O-C group, were attributed to saccharide (Liu et al., 2011) or the anhydrous glucose ring (Chen et al., 2008). The band due to aliphatic C–O–C and alcohol–OH in the range of 1060 to 1030 cm^{-1} in the spectrum of fresh rice flour, which represents oxygenated functional groups of cellulose (Guo and Bustin, 1998), is completely eliminated after carbonization. The FTIR spectra of all charcoals from rice flour powder with different sizes are the same (Figure 1A to C), thus indicating that the size of rice flour does not affect the behavior degradation of surface groups on starch powder.

Figure 2 shows the effect of carbonization temperature (600 to 800 °C) on the FTIR spectra of rice flour charcoals. The FTIR spectrum of rice flour charcoal formed at 600 °C shows similar characteristics to that

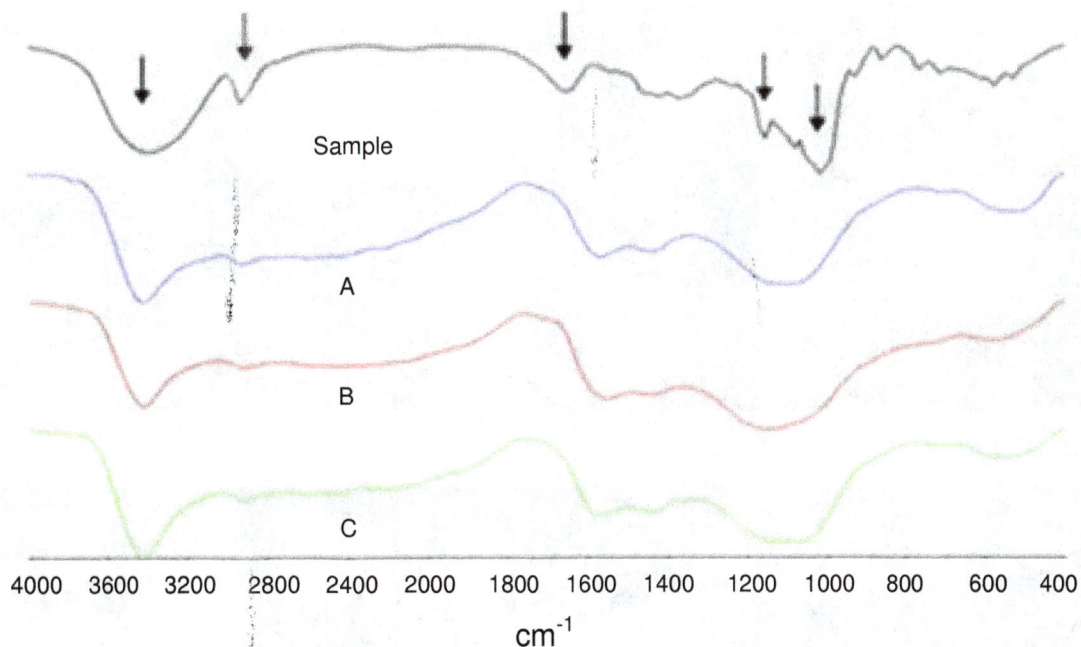

Figure 1 FTIR transmission spectra of charcoals produced from rice flour powder by carbonization at 600 °C: sample = fresh rice flour, A = 90 μm rice flour charcoal, B = 75 μm rice flour charcoal, C = 38 μm rice flour charcoal.

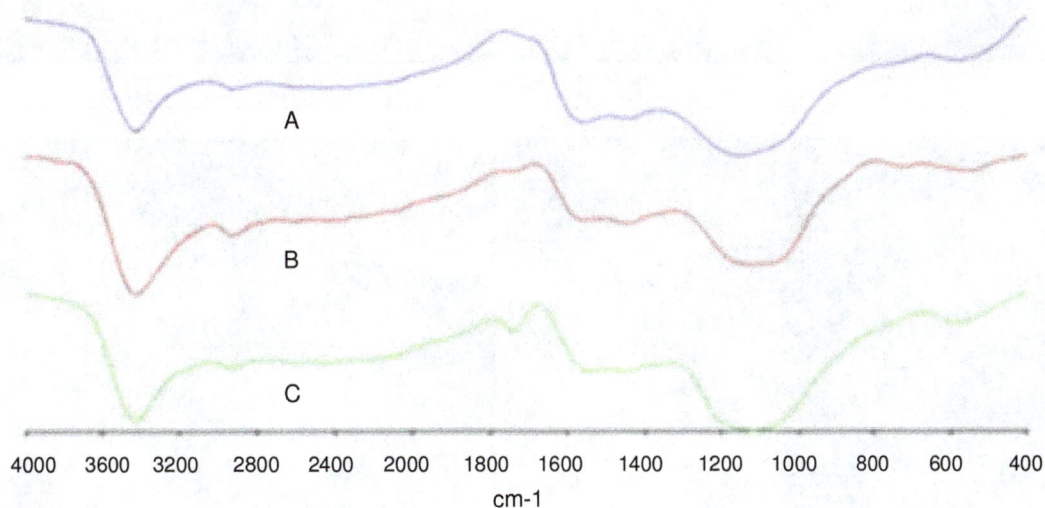

Figure 2. FTIR transmission spectrum of 75 μm charcoals produced from rice flour by carbonization at 600 to 800 °C: A = rice flour charcoal at 800 °C, B = rice flour charcoal at 700 °C and C = rice flour charcoal at 600 °C.

formed at 700 and 800 °C, but those of acid C=O groups (at about 1750 cm^{-1}) of rice flour charcoal formed at 600 °C recede at higher carbonization temperatures. This phenomenon probably originates from rearrangement and cyclization reactions of organic matter at high temperatures, where it vanishes completely after carbonization above 700 °C.

SEM micrographs

The SEM micrograph of fresh rice flour is as shown in Figure 3a. The starch granules were polygonal with sharp angles and edges. The granule sizes are 3 to 6 μm in diameter. A selection of SEM micrographs of the charcoal surfaces is as shown in Figure 3b to f. It clearly shows

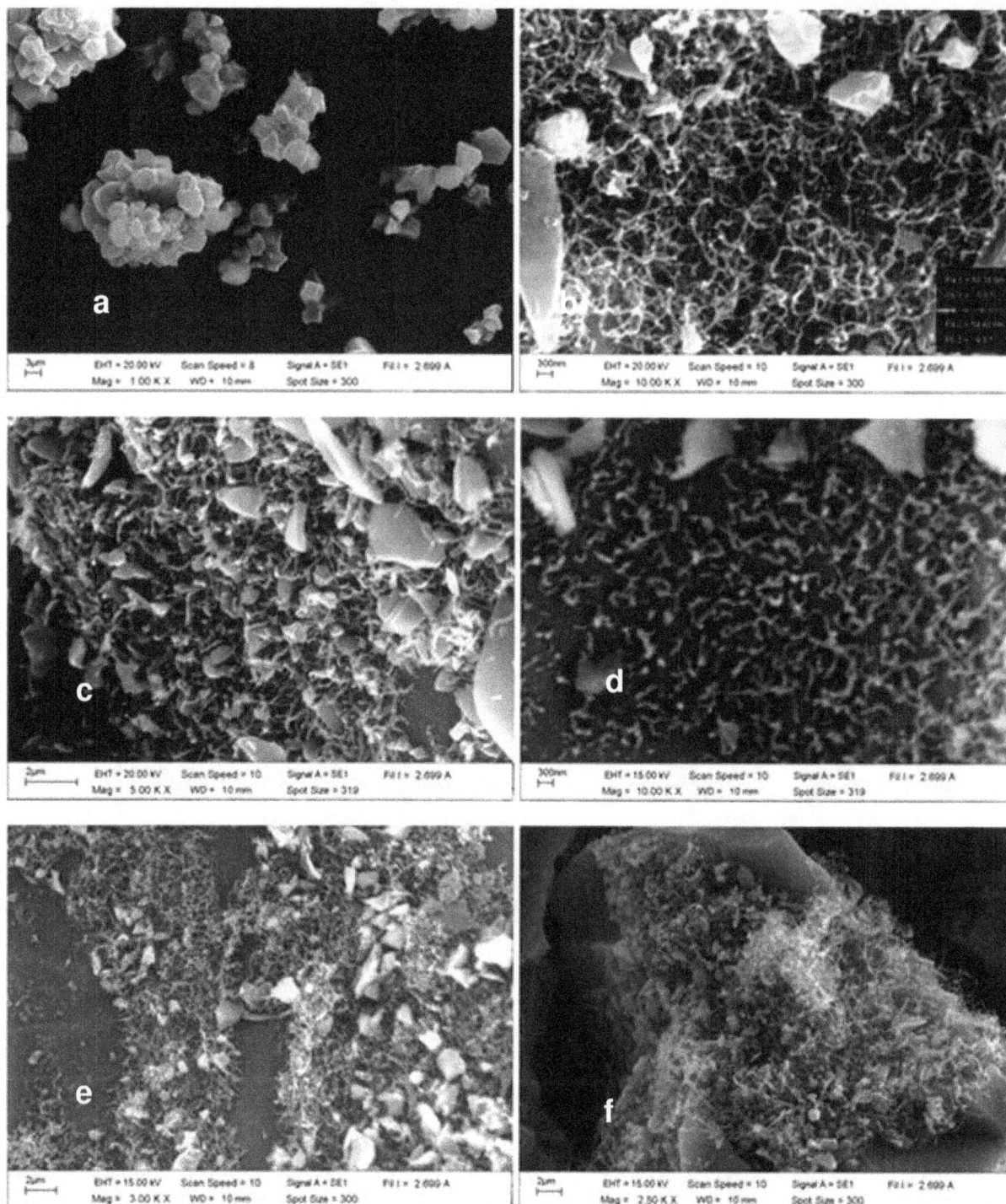

Figure 3. SEM micrographs of (a) fresh rice flour, (b) 90 µm rice flour charcoal carbonized at 600 °C, (c) 38 µm rice flour charcoal carbonized at 600 °C, (d) 38 µm rice flour charcoal carbonized at 700 °C, (e) 90 µm rice flour charcoal carbonized at 800 °C, (f) 38 µm rice flour charcoal carbonized at 800 °C.

the shape and size distribution of the agglomerate nanocarbonfibrils and some carbon granules existed outside of the charcoal surfaces, even at different carbonized temperatures (Figure 3c, d and f) and starch powder sizes (Figure 3b, c, e and f). The SEM images of charcoals show that the surface texture of the sample consisted of uniformly distributed nanocarbonfibrils. The nanocarbonfibrils morphology looks like microfibrillated cellulose (Ray and Bousmina, 2005) with diameters of 59. 42 to 68.38 nm and high length (micrometer scale).

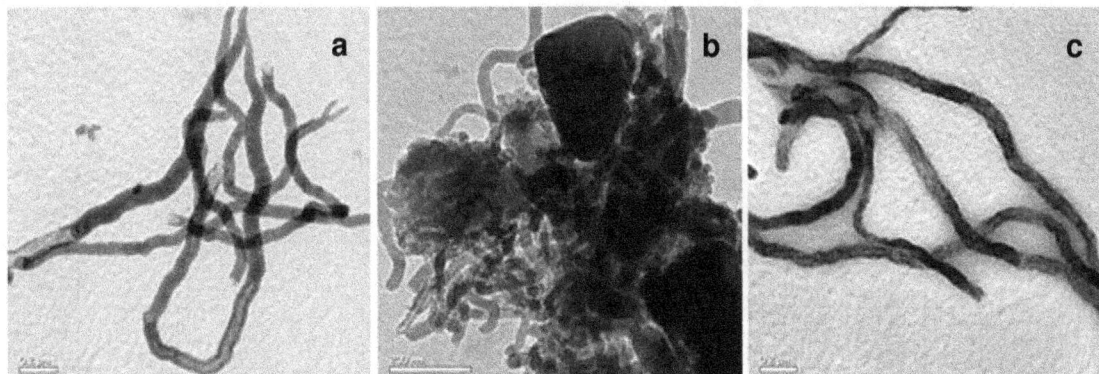

Figure 4. TEM photographs of (a and b) 90 µm rice flour charcoal carbonized at 600°C, (c) 90 µm rice flour charcoal carbonized at 700°C.

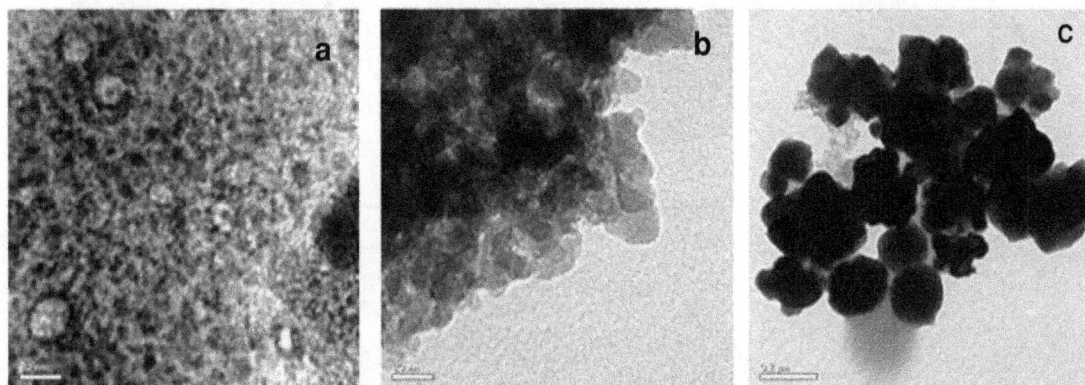

Figure 5. TEM photographs of (a and b) 75 µm rice flour charcoal carbonized at 600°C, (c) 75 µm rice flour charcoal carbonized at 700°C.

The nanocarbonfibrils were formed from the cellulose nanofibrils content of original starch. The size of carbon granules are 0.9 to 1.1 µm with smaller size than that of fresh granules.

TEM micrographs

The particular structures of the nanocarbonfibrils observed by TEM are shown in Figure 4. It can be seen that the nanocarbonfibrils are shaped like a snake and warp with nanosizes. Diameter and length of the nanocarbonfibrils were found to be in the range of 75 to 120 nm and 1.5 to 3.5 µm, respectively. A tendency to agglomeration was also observed from the TEM data (Figure 4b) in association with SEM data.

The other structures of charcoal particles are quasispherical and polygonal in shape with a diameter of 0.05 to 2 µm (Figure 5c). Further TEM examination at high magnification (Figure 5a and b) revealed that the carbon has a typical cryo-fractured surface.

EDS spectrum

Figure 6 shows the EDS spectrum from the rice flour charcoal carbonized at 600°C. From the spectrum, C is seen to be the main element (77.06 wt%), whereas O (16.81 wt%), Si (1.05 wt%), P (3.77 wt%) and K (1.30 wt%) are of secondary importance. The charcoal shows slightly high carbon content, indicating an increase in the amount of polycondensed aromatic structures which reflects the progressive aromatization of the original materials. Zhang et al. (2002) explained that above 400°C, the starch structure is destroyed and the product appears structurally similar to thermally crosslinked/decomposed phenolic/furfuryl alcohol resins. The carbon structures were generated at 600°C. The low content of O is due to intermolecular rearrangements caused by C–C bond cleavage (Trompowsky et al., 2005). The P, Si and K contents are due to incorporation of lysophospholipids and minerals from original rice flour (Tester et al., 2004). These findings match well the FTIR spectrum data.

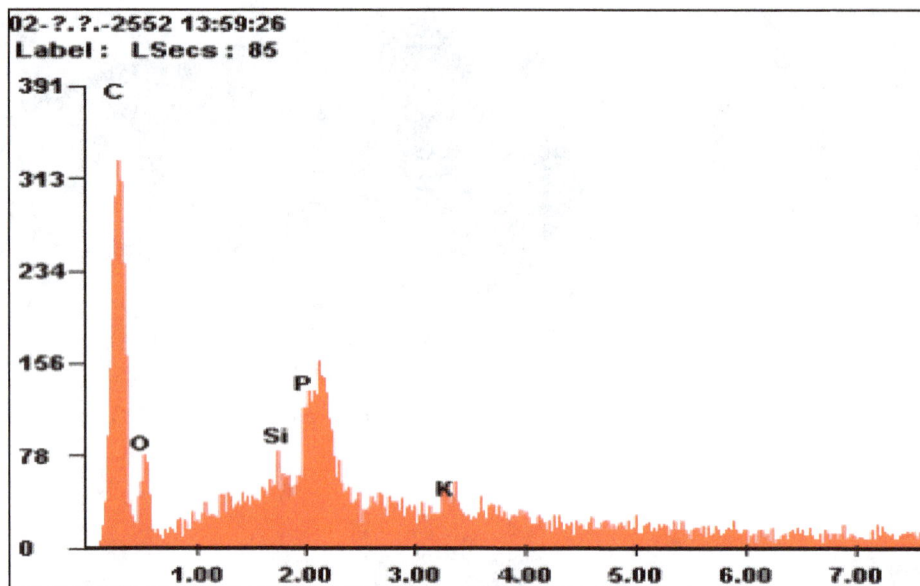

Figure 6. EDS spectrum of 90 μm rice flour charcoal carbonized at 600°C.

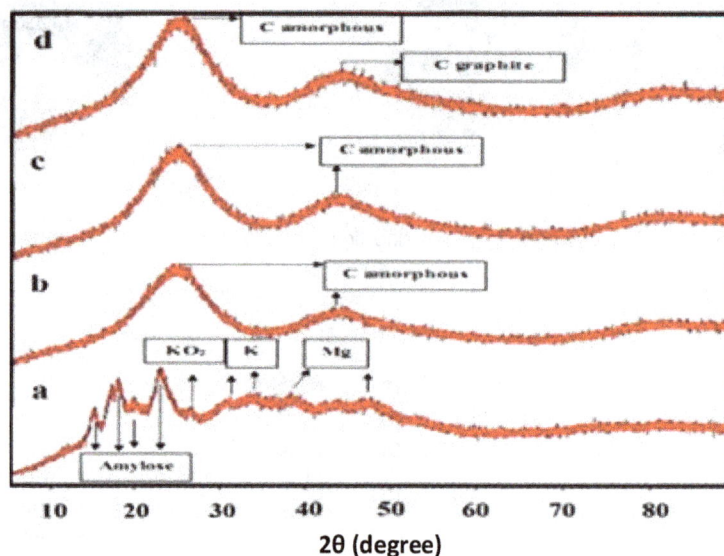

Figure 7. X-ray powder diffraction spectra of (a) fresh rice flour, (b) 90 μm rice flour charcoal with carbonization at 600°C, (c) 75 μm rice flour charcoal with carbonization at 600°C and (d) 38 μm rice flour charcoal with carbonization at 600°C.

XRD analysis

Figure 7 shows the effects of starch powder size on the XRD patterns. The XRD patterns show ordered structures of the crystalline on the fresh rice flour. The diffractogram (Figure 7a) shows the content of amylase and amylopectin (cellulose) in fresh rice flour by peaks at $2\theta = 15, 18, 20.3$ and $23.5°$ (Zabar et al., 2009). These peaks almost disappear after carbonization (Figure 7b to

d). This shows that the crystalline of fresh rice flour have been destructed and degraded to be amorphous carbon during the carbonization. It can be seen that two broad peaks centered around 26 and 44° occur, which correspond to the 002 and 10 reflections of disordered micrographite stacking (Zhang et al., 2008). Furthermore, this reveals the same graphite diffraction peak (a small sharp peak at 44°), which indicates that some amorphous carbons transform to graphite crystalline structures

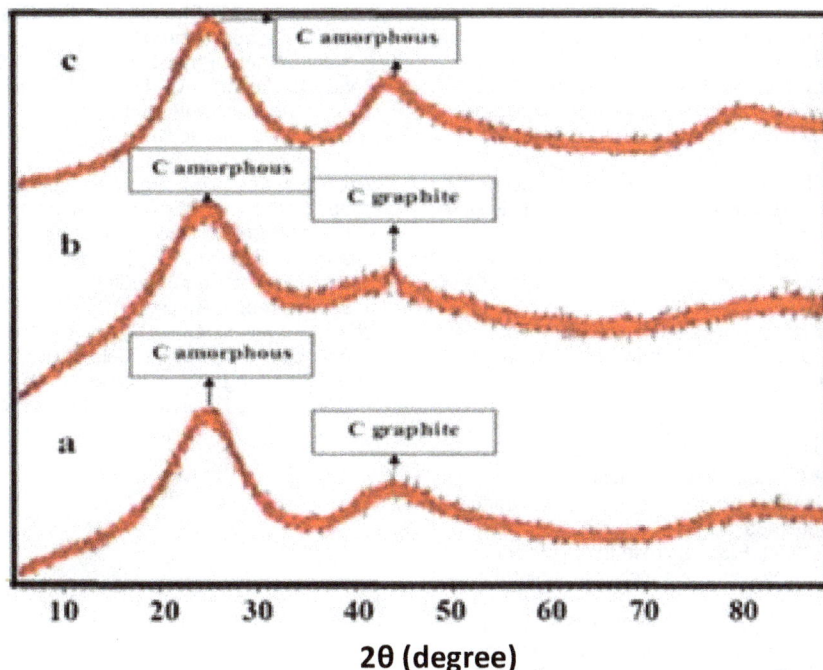

Figure 8. X-ray powder diffraction spectra of (a) 38 μm rice flour charcoal with carbonization at 600 °C, (b) 38 μm rice flour charcoal with carbonization at 700 °C and (c) 38 μm rice flour charcoal with carbonization at 800 °C.

(Xingzhong et al., 2010). The very weak peak at about 32° is the pattern of the principal K_2CO_3 compound which occurs from KO_2 and K after carbonization (Díaz-Terán et al., 2003). The peak of Mg occurs at 38 and 47° of fresh rice flour.

The XRD patterns for all charcoals with different size (Figure 7) and carbonization temperatures (Figure 8) are very similar. This indicates that the 600 to 800 °C carbonization temperatures and 38 to 90 μm starch powder sizes do not affect the structure of charcoals.

Conclusion

Starch charcoals with different sizes of powder were prepared by carbonization of starch at 600, 700 and 800 °C. The results of this research show that the resulting carbonized products are amorphous carbon. The agglomerate nanocarbonfibrils and some carbon granules exist outside the surfaces of all charcoals. The nanocarbonfibrils are shaped like a snake and warp with diameters and lengths in the range of 75 to 120 nm and 1.5 to 3.5 μm, respectively. Other structures of charcoal particles are quasispherical and polygonal in shape with diameters in the range of 0.05 to 2 μm. There is a possibility that the natural nanocarbonfibriles are promising candidates in such applications as reinforced materials, adsorbents, supercapacitors and hydrogen storage.

ACKNOWLEDGEMENT

Funding for this research was supported by Naresuan University.

REFERENCES

Alemdar A, Sain M (2008). Biocomposites from wheat straw nanofibers: Morphology, thermal and mechanical properties. Compos. Sci. Technol., 68: 557-565.

Ascough PL, Bird MI, Francis SM, Lebl T (2011). Alkali extraction of archaeological and geological charcoal: evidence for diagenetic degradation and formation of humic acids. J. Archaeol. Sci., 38(1): 69-78.

Chen Y, Cao X, Chang PR, Huneault MA (2008). Comparative study on the films of poly(vinyl alcohol)/pea starch nanocrystals and poly(vinyl alcohol)/native pea starch. Carbohyd. Polym., 73(1): 8-17.

Díaz-Terán J, Nevskaia DM, Fierro JLG, López-Peinado AJ, Jerez A (2003). Study of chemical activation process of a lignocellulosic material with KOH by XPS and XRD. Micropor. Mesopor. Mater., 60(1-3): 173-181.

Francioso O, Sanchez-Cortes S, Bonora S, Roldán ML, Certini G (2011). Structural characterization of charcoal size-fractions from a burnt *Pinus pinea* forest by FT-IR, Raman and surface-enhanced Raman spectroscopies. J. Mol. Struct., 994(1-3): 155-162.

Guo Y, Bustin RM (1998). FTIR spectroscopy and reflectance of modern charcoals and fungal decayed woods: implications for studies of inertinite in coals. Int. J. Coal Geol., 37(1-2): 29-53.

Liu H, Chaudhary D, Yusa S, Tadé MO (2011). Glycerol/starch/Na+-montmorillonite nanocomposites: A XRD, FTIR, DSC and ^{1}H NMR study. Carbohyd. Polym., 83(4): 1591-1597.

Ma X, Chang PR, Yang J, Yu J (2009). Preparation and properties of glycerol plasticized-pea starch/zinc oxide-starch bionanocomposites. Carbohyd. Polym., 75: 472-478.

Pérez E, Schultz FS, Delahaye EP de (2005). Characterization of some properties of starches isolated from Xanthosoma sagittifolium (tannia) and Colocassia esculenta (taro). Carbohyd. Polym., 60(2): 139-145.

Puziy AM, Poddubnaya OI, Martínez-Alonso A, Suárez-García F, Tascón JMD (2002). Synthetic carbons activated with phosphoric acid: I. Surface chemistry and ion binding properties. Carbon, 40(9): 1493-1505.

Qian J, Wang J, Jin Z (2004). Preparation of biomorphic SiC ceramic by carbothermal reduction of oak wood charcoal. Mat. Sci. Eng. A-Struct., 371(1-2): 229-235.

Ray SS, Bousmina M (2005). Biodegradable polymers and their layered silicate nanocomposites: In greening the 21st century materialsworld. Prog. Mater. Sci., 50(8): 962-1079.

Shindo K, Kondo T, Sakurai Y (2004). Dependence of hydrogen storage characteristics of mechanically milled carbon materials on their host structures. J. Alloy. Compd., 372(1-2): 201-207.

Tester RF, Karkalas J, Qi X (2004). Starch – composition, fine structure and architecture. J. Cereal. Sci., 39: 151-165.

Trompowsky PM, Benites V de M, Madari BE, Pimenta AS, Hockaday WC, Patrick G, Hatcher PG (2005). Characterization of humic like substances obtained by chemical oxidation of eucalyptus charcoal. Org. Geochem., 36(11): 1480-1489.

Vandeputte GE, Delcour JA (2004). From sucrose to starch granule to starch physical behaviour a focus on rice starch. Carbohyd. Polym., 58: 245-266.

Wang H, Zhong Y, Li Q, Yang J, Dai Q (2008). Cationic starch as a precursor to prepare porous activation carbon for application in supercapacitor electrodes. J. Phys. Chem. Solids, 69: 2420-2425.

Wang X, Liang X, Wang Y, Wang X, Liu M, Yin D, Xia S, Zhao J, Zhang Y (2011). Adsorption of Copper (II) onto activated carbons from sewage sludge by microwave-induced phosphoric acid and zinc chloride activation. Desalination, 278(1-3): 231-237.

Xingzhong G, Lingjie Z, Liqing Y, Hui Y, Lin Z (2010). Preparation of silicon carbide using bamboo charcoal as carbon source. Mater. Lett., 64(3): 331-333.

Xue KH, Tao FF, Shen W, He CJ, Chen QL, Wu LJ, Zhu YM (2004). Linear carbon allotrope-carbon atom wires prepared by pyrolysis of starch. Chem. Phys. Lett., 385: 477-480.

Zabar S, Lesmes U, Katz I, Shimoni E, Bianco-Peled H (2009). Studying different dimensions of amylose-long chain fatty acid complexes: Molecular, nano and micro level characteristics. Food Hydrocolloil., 23: 1918-1925.

Zhang SJ, Feng HM, Wang JP, Yu HQ (2008). Structure evolution and optimization in the fabrication of PVA-based activated carbon fibers. J. Colloid. Interf. Sci., 321(1): 96-102.

Zhang X, Golding J, Burgar I (2002). Thermal decomposition chemistry of starch studied by ^{13}C high-resolution solid-state NMR spectroscopy. Polymer, 43: 5791-5796.

Zhao S, Wang C, Chen M, Shi Z (2008). Preparation of carbon sphere from corn starch by a simple method. Mater. Lett., 62: 3322-3324.

Zhoa S, Wang CY, Chen MM, Wang J, Shi ZQ (2009). Potato starch-based activated carbon spheres as electrode material for electrochemical capacitor. J. Phys. Chem. Solids, 70: 1256-1260.

17

Synthesis and characterization of nano-hydroxyapatite (n-HAP) using the wet chemical technique

Arunseshan Chandrasekar, Suresh Sagadevan and Arivuoli Dakshnamoorthy

Crystal Growth Centre, Anna University, Chennai-600 025, India.

Hydroxyapatite nanoparticles were synthesized, using a wet chemical technique with diammonium hydrogen phosphate and calcium nitrate tetrahydrate precursors, respectively. The pH of the system was maintained at 10.8 throughout the stirring process, by using 0.1 M sodium hydroxide. The mixture was allowed to remain stirred overnight and a white precipitate was formed. The precipitate was vacuum dried and cleaned with distilled water and ethanol simultaneously three or four times. The prepared powder was used for further characterization. The prepared nano-hydroxyapatite powder was characterized for phase composition, using X-ray diffractometry; elemental dispersive X-ray and Fourier transform infrared spectroscopy. The elemental compositions of the nano-hydroxyapatite were analyzed and confirmed by elemental dispersive X-ray (EDX). The particle size and morphology were studied using the scanning electron microscope (SEM) and transmission electron microscopy (TEM). The particle size of the nano-hydroxyapatite was also analyzed, using the dynamic light scattering (DLS) experiment.

Key words: Hydroxyapatite, wet chemical, elemental dispersive X-ray (EDX), Fourier transform infrared spectroscopy (FTIR), scanning electron microscope (SEM), transmission electron microscopy (TEM).

INTRODUCTION

Tissue engineering is a multidisciplinary science, encompassing diverse fields like materials engineering and molecular biology, in the effort to develop biological substitutes for failing tissues and organs. Tissue engineering thus seeks to replace diseased and damaged tissues of the body. An important factor for the success of tissue engineering is the ability of developing materials, which can interface with tissues structurally, mechanically, and bio functionally (Yang et al., 2001). Many biomaterials lack the desired functional properties to interface with biological systems, in spite of numerous uses of materials in tissue engineering. Thus, developing new materials to reach these issues is necessary. Composites of hydrophilic polymers and inorganic minerals like hydroxyapatite can be good materials for biomedical applications. Bone apposition and differentiation of mesenchymal cells to osteoblasts can be promoted by the attachment of hydroxyapatite nanoparticles to a polymer surface (Sinha and Guha, 2008). Hydroxyapatite is a significant biomaterial in the health care industry. Its chemical and mineral phases are analogous to those of natural bone and hence, its usage in the field of dentistry and orthopedics has been explored (Zhao and Ma, 2005; Chen et al., 2002; Hornez et al., 2007). Properties like osteoconductivity and osteoinductivity enhance bone regeneration and make hydroxyapatite an important material in tissue engineering (Burg et al., 2000), and its biocompatibility leads to its use as bioactive coating over implants (Wang et al., 2006; Ye et al., 2007). Bone is considered as a nanocomposite of minerals and proteins, and recently nano-level hydroxyapatite has been investigated and demonstrated as having a good impact on cell biomaterial interaction (Elliot, 1994; Webster et al., 2000).

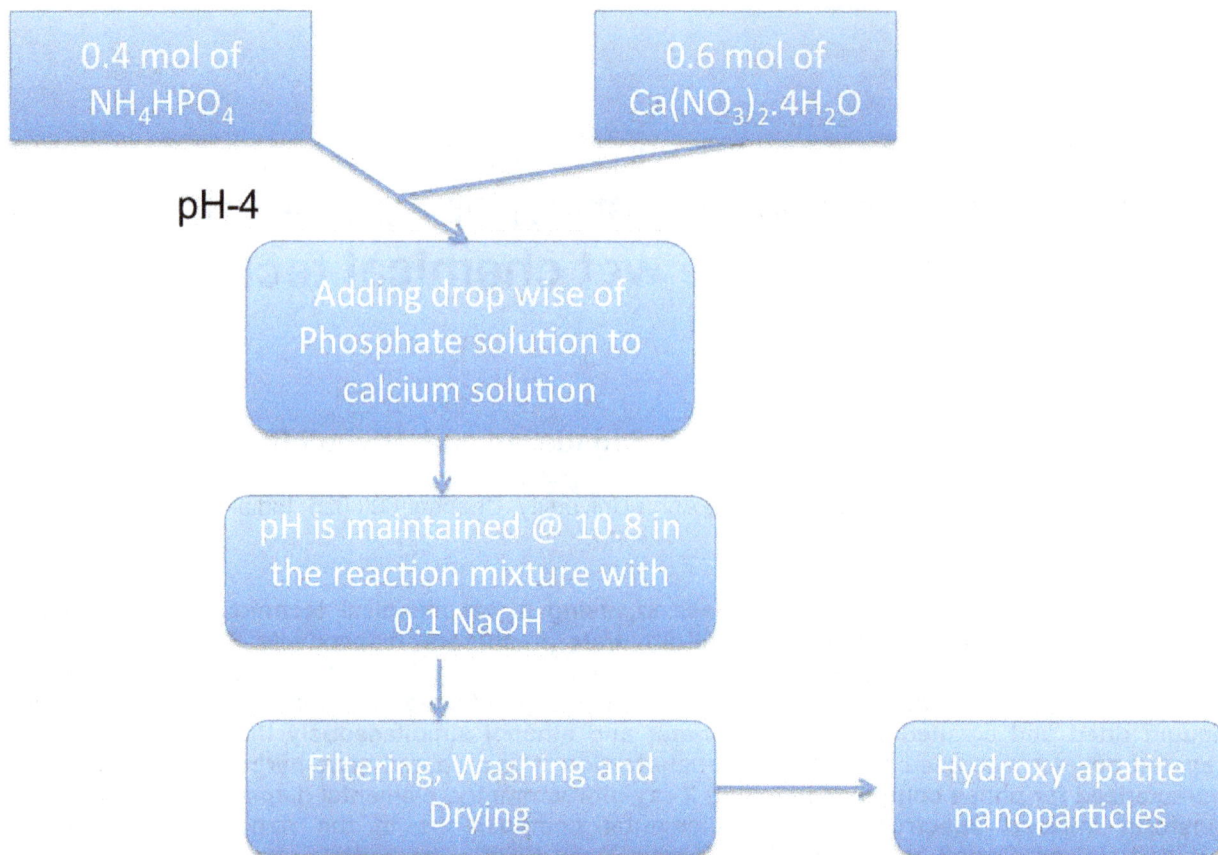

Figure 1. Flow chart of hydroxyapatite nanoparticles preparation by wet chemical method.

However, the migration of the nano- hydroxyapatite particles from the implanted site into the surrounding tissues might cause damage to healthy tissue (Miyamato et al., 1998). To find a solution, composites of nano-hydroxyapatite and polymers were researched to find a material that retained the good properties of nano-hydroxyapatite and prevented the nano- hydroxyapatite particles from migrating. In the present investigation, we report the synthesis and characterization of nano-hydroxyapatite by a wet chemical synthesis technique. The as prepared hydroxyapatite nanoparticles are characterized by X- ray diffraction, Fourier transform infrared spectroscopy (FTIR), elemental dispersive X-ray (EDX) analysis, scanning electron microscopy (SEM), transmission electron microscopy (TEM), and dynamic light scattering (DLS) analysis.

EXPERIMENTAL PROCEDURES

Synthesis of hydroxyapatite nanoparticle

Hydroxyapatite nanoparticles were synthesized by the wet chemical method. 500 ml of 0.4 mol of diammonium hydrogen phosphate with pH-4.0 was vigorously stirred in 2 L beaker at room temperature and 500 mL of 0.6 mol calcium nitrate tetrahydrate with

pH=7.4 was added drop-wise over 4 h. The pH of the system was maintained at 10.8 throughout the stirring process, by using 0.1 M sodium hydroxide. The mixture was allowed to remain stirred overnight. A white precipitate was formed. The precipitate was vacuum dried and cleaned with distilled water and ethanol simultaneously for three or four times. The prepared powder was used for further characterization. The schematic presentation of the procedure is given in Figure 1. This precipitation reaction for synthesis of hydroxyapatite nanoparticles was first proposed by Yagai and Aoki, as indicated by Bouyer et al. (2000).

RESULTS AND DISCUSSION

XRD analysis

The structural analysis of sample was done by the powder X-ray diffraction. The XRD patterns of the synthesized nano hydroxyapatite are shown in Figure 2. The XRD pattern of nano hydroxyapatite shows sharper peaks which indicate better crystallinity. The peak positions are in good agreement with the JCPDS (896438). As can be seen, hydroxyapatite XRD patterns, with the diffraction peaks, obtained with d-spacing values of 2.82Å, 2.79Å and 2.72Å and the other d-spacing values match exactly with the hexagonal system with

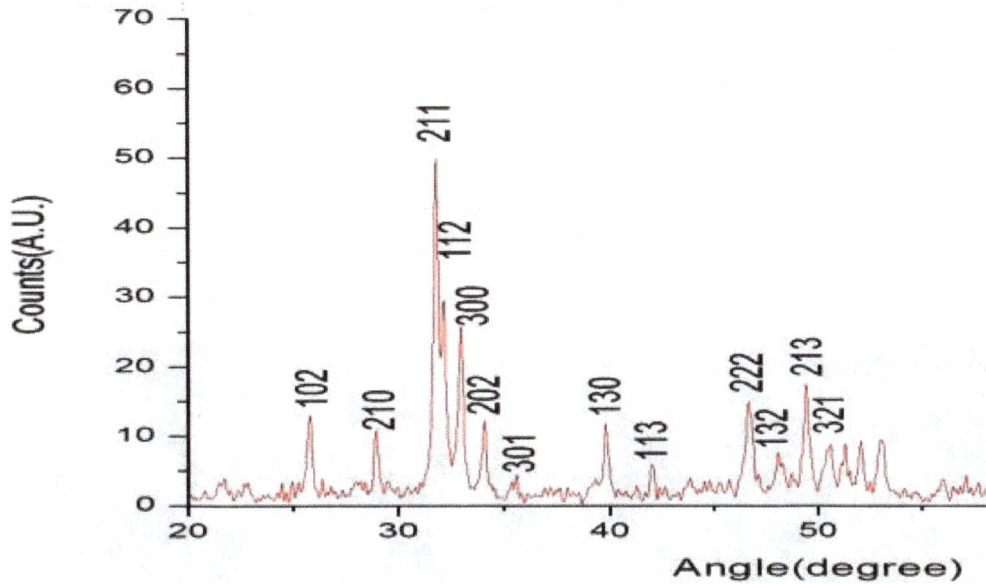

Figure 2. XRD spectra of hydroxyapatite nanoparticles.

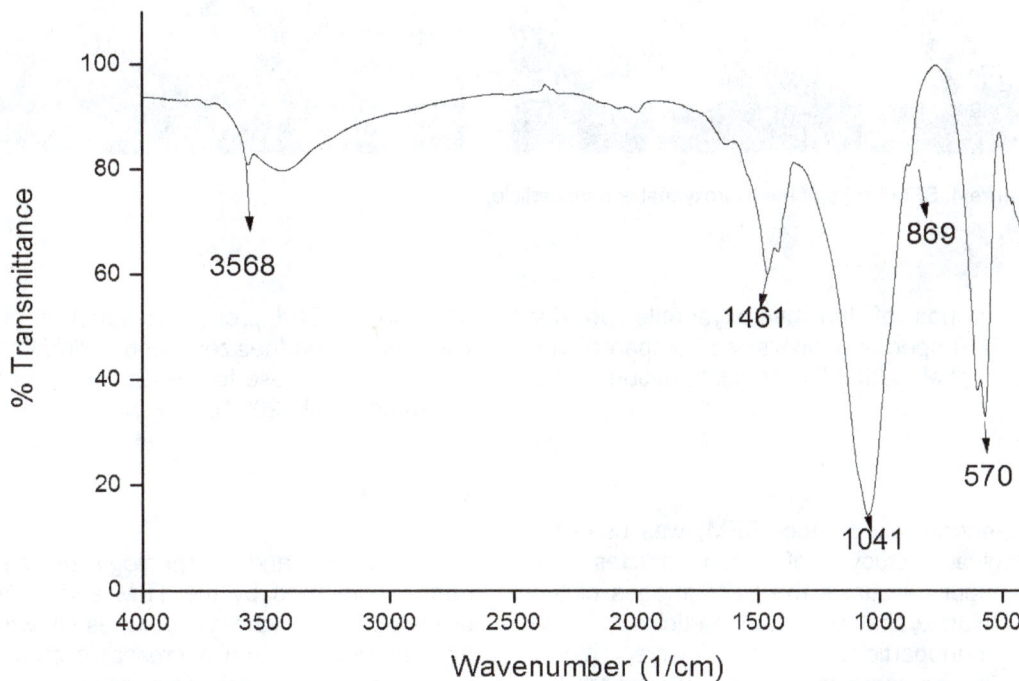

Figure 3. FTIR spectrum of nano- hydroxyapatite.

primitive lattice. The results of XRD analysis obtained in the present investigation are in good agreement with the reported results (Bouyer et al., 2000).

FTIR analysis

Functional groups associated with hydroxyapatite were identified by FTIR spectroscopy. The FTIR spectra of the prepared samples are given in Figure 3. The ion stretching vibration around 3568 cm^{-1} confirms the presence of a hydroxyl group. Likewise, the other stretching vibrations for carbonyl and phosphate groups were also observed as reported earlier (Cengiz et al., 2008). The observed functional groups and their corresponding assignments are presented in Table 1.

Table 1. Some important functional group assignments of hydroxyapatite nanoparticles.

Wavenumber cm^{-1}	Stretching mode	Functional group
3568	Ion Stretching	OH^-
1461	Asymmetric stretching	CO_3^{2-}
1041	Asymmetric stretching	PO_4^{3-}
869	Out of plane bending mode	CO_3^{2-}
570	Asymmetric bending vibration	PO_4^{3-}

Figure 4. SEM Image of the hydroxyapatite nanoparticle.

The functional groups of the hydroxyapatite powder predicted from FTIR spectra analysis are compared with the results of Choi et al. (2004) for the confirmation.

SEM analysis

The scanning electron microscope (SEM) was used for the morphological study of nanoparticles of hydroxyapatite. Figure 4 shows the SEM images of the as-prepared hydroxyapatite nanoparticles. The hydroxyapatite nanoparticles formed were highly agglomerated. The agglomeration of the nanoparticles might be because of Ostwald ripening. The spherical shaped particles with clumped distributions are visible from the SEM analysis. The SEM images show the spherical shaped particles as confirmed by Ferraz et al. (2004) for reported results of hydroxyapatite nanoparticles.

EDX analysis

In Figure 5a and b, and Table 2, the standard EDX spectra recorded on the examined nano-hydroxyapitate

are shown. The presented spectrum shows the Ca/P value of the synthesized nano-hydroxyapatite to be 1.68, which is quite close to the Ca/P ratio of the human bone (Trommer et al., 2007).

TEM analysis

The structure and morphology of the samples were further confirmed by the TEM and TEM images of the prepared nano-hydroxyapatite, as shown in Figure 6. The transmission electron microscopic analysis confirms the presence of the spherical shape morphology of the prepared hydroxyapatite nanoparticle with the particle size of around 60 to 70 nm (Figure 7). The particle size is also found to be in agreement with the report results of Ferraz et al. (2004).

DLS studies

Dynamic light scattering (DLS) is an important tool for characterizing the size of nanoparticles in a solution. The DLS measures the light scattered from a laser that passes through a colloidal solution and by analyzing the

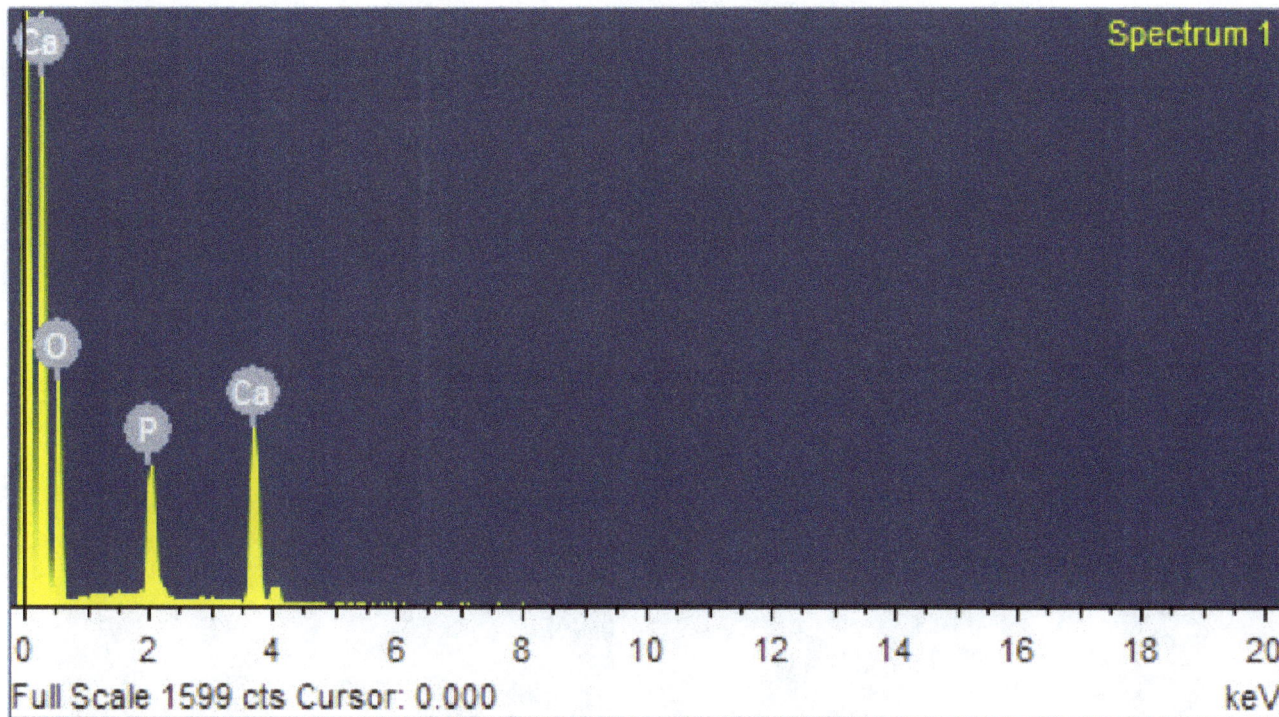

Figure 5a. EDX spectrum of nano-hydroxyapitate.

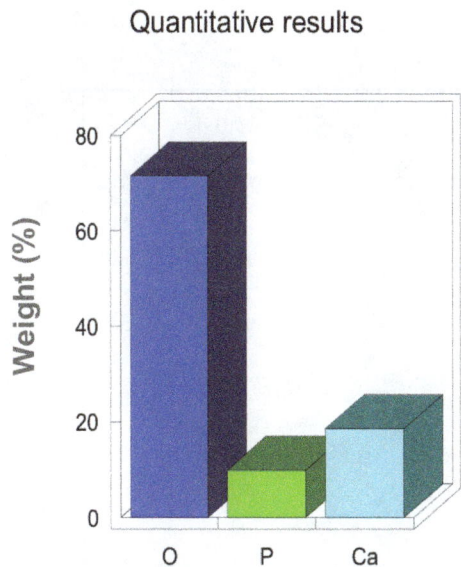

Figure 5b. Chart showing the weight % and quantitative results of EDX spectrum of nano-hydroxyapitate.

modulation of the scattered light intensity as a function of time, the hydrodynamic size of the particles and particle agglomerates can be determined. Larger particles will diffuse slower than smaller particles, and the DLS instrument measures the time dependence of the scattered light, to generate a correlation function that can be mathematically linked to the particle size. The DLS is a valuable tool for determining and measuring the agglomeration state of the nanoparticles as a function of time or suspending solution. When the DLS sizing data is compared to the transmission electron microscopy images, the aggregation state of the particles can be determined. In an unagglomerated suspension, the DLS measured diameter will be similar or slightly larger than the TEM size. If the particles are agglomerated, the DLS measurement is often much larger than the TEM size, and can have a high polydispersity index (large variability in the particle size). The dynamic light scattering experiment shows that the particle size distribution is in the range of 50 to 70 nm, which is well supported by the TEM analysis. Dynamic light scattering is used to monitor the size of the precipitating particles and to provide information about their concentration (De Bruyn et al., 2013).

Conclusion

Nano-hydroxyapatite has been successfully synthesized using the wet chemical technique. The formation of hydroxyapatite nano particles was confirmed by X-ray diffraction (XRD) and Fourier transform infrared spectroscopy (FT-IR).The elemental compositions were examined using the EDX analysis. The size and morphology of the samples were characterized using

Table 2. Elemental composition of nano-hydroxyapitate.

Element	App Conc.	Intensity Conc.	Weight (%)	Weight (%) Sigma	Atomic (%)
OK	2.26	0.7087	73.52	0.96	85.09
PK	0.56	1.2858	9.88	0.55	6.07
Ca K	0.82	0.9935	16.60	0.70	8.84
Total			100.00		

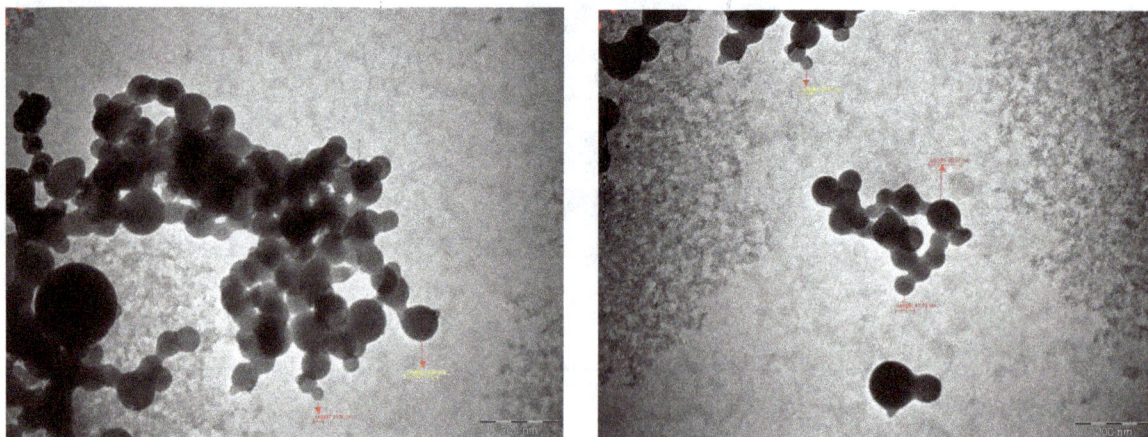

Figure 6. TEM images of hydroxyapatite nanoparticles.

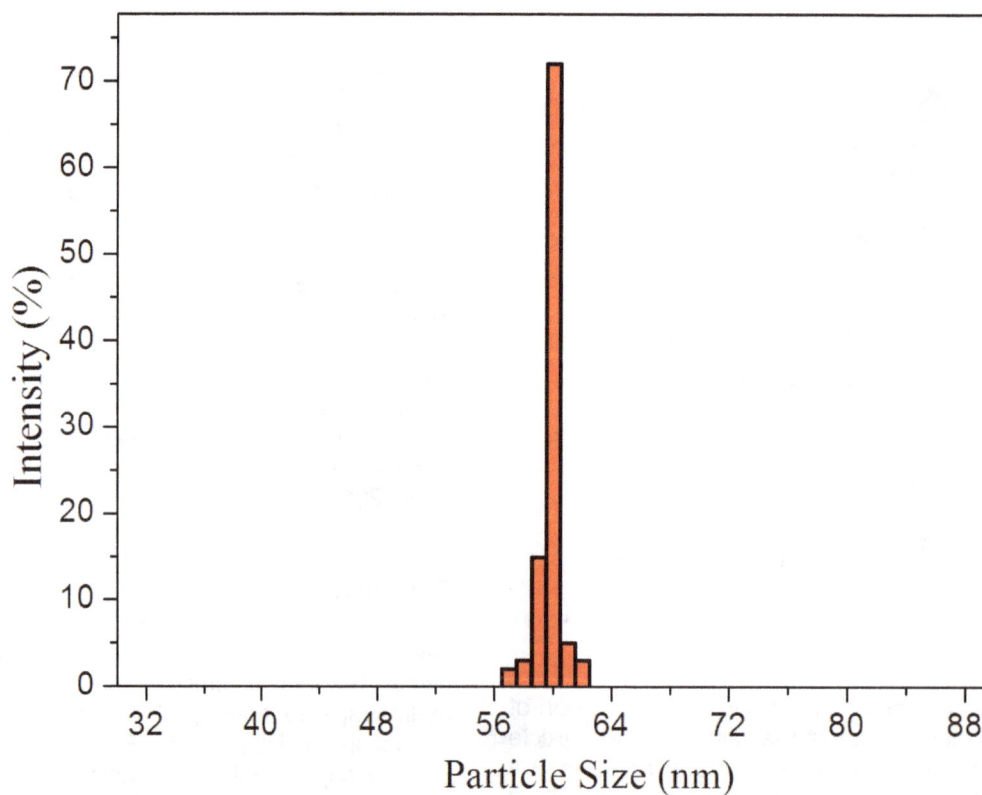

Figure 7. Particle size of hydroxyapatite nanoparticles.

scanning and transmission electron microscopy (SEM and TEM). The spherical shaped particles were confirmed through the SEM analysis. The transmission electron microscopic analysis confirms the prepared hydroxyapatite nanoparticles with the particle size of around 60 to 70 nm. The particle size of the nano-hydroxyapatite range of 50 to 70 nm was determined using the dynamic light scattering (DLS) experiment which is good agreement with the TEM analysis.

ACKNOWLEDGEMENTS

One of the authors (Arunseshan Chandrasekar) acknowledges with thanks to the support rendered by the Council of Scientific and Industrial Research, India for providing him the Senior Research Fellowship (CSIR-SRF).

REFERENCES

Bouyer E, Gitzhofer F, Boulos MI (2000). Morphological study of hydroxyapatite nanocrystal suspension. J. Mater. Sci. Mater Med. 11: 523-531.

Burg KJL, Porter S, Kellam JF (2000). Biomaterial developments in bone tissue engineering. Biomaterials. 21:2347–2359.

Cengiz B, Gokce Y, Yildiz N, Aktas Z, Calimli A (2008). Synthesis and characterization of hydroxyapatite nanoparticles. Colloids and Surfaces A: Physicochemical and Engineering Aspects, 322(1-3):29–33.

Chen F, Wang ZC, Lin CJ (2002). Preparation and characterization of nano-sized hydroxyapatite particles and hydroxyapatite/chitosan nano-composite for use in biomedical materials. Mater. Lett. 57(4):858–861.

Choi D, Marra K, Kumta PN (2004). Chemical synthesis of hydroxyapatite/poly (caprolactone) composite. Materials Research Bulletin. 39:417-432.

De Bruyn JR, Goiko M, Mozaffari M, Bator D, Dauphinee RL (2013). Dynamic light scattering study of inhibition of nucleation and growth of hydroxyapitate crystals by osteopontin. Plus One. 8(2):e 56764.

Elliot JC (1994). Amsterdam: Structure and Chemistry of the Apatites and Other Calcium Orthophosphates. Elsevier Science.111.

Ferraz MP, Monteiro FJ, Manuel CM (2004). Hydroxyapatite nanoparticles: A review of preparation methodologies. J. Appl. Biomater. Biomech. 2:74-80.

Hornez JC, Chai F, Monchau F, Blanchemain N, Descamps M, Hildebrand HF (2007). Biological and physico-chemical assessment of hydroxyapatite (HA) with different porosity. Biomol. Eng. 24:505–509.

Miyamato Y, Ishikawa KI, Takechi M, Toh T, Yuasa T, Nagayama M, Suzuki K (1998). Basic properties of calcium phosphate cement containing atelocollagen in its liquid or powder phases.Biomaterials.19:707–715.

Sinha A, Guha A (2008). Biomimetic patterning of polymer hydrogels with hydroxyapatite nanopartcles. Mater. Sci. Eng. C. 29:1330-1333.

Trommer RM, Santos LA, Bergmann CP (2007). Alternative technique for hydroxyapatite coatings. Surface Coatings Technol. 201(24):9587–9593.

Wang YJ, Chen JD, Wei K, Zhang SH, Wang XD (2006). Surfactant-assisted synthesis of hydroxyapatite particles. Mater. Lett. 60:3227–3231.

Webster TJ, Ergun C, Doremus RH, Siegel RW, Bizios R (2000). Enhanced functions of osteoblasts on nanophase ceramics. Biomaterials. 21:1803–1810.

Yang S, Leong KF, Du Z, Chua CK (2001). The design of scaffolds for use in tissue engineering. Part I. Traditional factors.Tissue Engineering. 7:679-689.

Ye W, Wang XX (2007). Ribbon-like and rod-like hydroxyapatite crystals deposited on titanium surface with electrochemical method. Mater. Lett. 61:4062–4065.

Zhao YF, Ma J (2005). Triblock co-polymer templating synthesis of mesostructured hydroxyapatite. Micro. Meso. Mater. 87:110–117.

18

Second-order kinetic model for the adsorption of divalent metal ions on *Sida acuta* leaves

Oboh I. O.[1] , Aluyor E. O.[2] and Audu T. O. K.[2]

[1]Department of Chemical and Petroleum Engineering, University of Uyo, Uyo, Nigeria.
[2]Department of Chemical Engineering, University of Benin, Benin city, Nigeria.

The removal of metal ions from effluents is of importance to many countries of the world both environmentally and for water re-use. A comparison was made of the linear least-squares method of the widely used pseudo-second order kinetic model for the sorption of some divalent metal ions onto *Sida acuta* leaves. Four pseudo-second order kinetic linear equations were used for this study. The results obtained from the experimental data for *S. acuta* leaves of 1.0 g with particle size range of 0.25 to 0.5 mm and at 300 rpm with initial concentrations of 5.0, 4.0, 20 and 2.5 mg/dm^3 for Cu^{2+}, Ni^{2+}, Pb^{2+} and Zn^{2+} ions, respectively showed that when the four pseudo-second order kinetic linear equations were compared, the Type 2 model with regression values for Cu^{2+}, Ni^{2+}, Pb^{2+} and Zn^{2+} ions as 0.979, 0.995, 0.991 and 0.965 respectively showed best fit for the sorption onto *S. acuta* leaves. This provided the evidence that the divalent metal ions sorption unto *S. acuta* followed the pseudo-second – order kinetic expression.

Key words: *Sida acuta*, sorption, divalent metal ions, kinetic model, effluents.

INTRODUCTION

Elements in every group of the periodic table have been found to be stimulatory to animals. Most metals in the fourth period are carcinogenic. It can be assumed that the carcinogenicity is related to the electronic structure of transition and inner transitional metals (Luckey and Venugopal, 1977). Sorption of pollutants from wastewater has long been studied. The rate at which sorption takes place is of the utmost importance when designing batch sorption systems, consequently it is important to establish the time dependence of such systems under various process conditions (Ho, 2006).

The sorption of metal ions from aqueous solution plays an important role in water pollution control, and in recent years there has been considerable interest in the use of low-cost sorbents such as peat for this purpose.

However, although the sorption kinetics of individual metal ions onto this type of material has been examined, the processes which occur are not completely understood, for instance, the rate limiting step and the bonding mechanism (Ho and McKay, 2000).

The application of low-cost sorbents including carbonaceous materials, agricultural products and waste by-products has been investigated. In recent years, agricultural by-products have been widely studied for metal removal from water. These include peat, wood, pine bark, banana pith, rice bran, soybean and cottonseed hulls, peanut shells, hazelnut shell, rice husk, sawdust, wool, orange peel and compost and leaves. Most of this work has shown that natural products can be good sorbents for heavy metals (Ho et al., 2002).

Table 1. A comparison of the second order models.

References	Linear form	Plot
Sobkowsk and Czerwinski (1974)	$\dfrac{\theta}{1-\theta} = k_2 t$	$\dfrac{\theta}{1-\theta}$ vs. t
Ritchie (1977)	$\dfrac{q_\infty}{q_\infty - q} = \alpha t + 1$	$\dfrac{q_\infty}{q_\infty - q}$ vs. t
Blanchard et al. (1984)	$\dfrac{1}{n_0 - n} - \alpha_\beta = kt$	$\dfrac{1}{n_0 - n}$ vs. t
Ho (1995)	$\dfrac{t}{q_t} = \dfrac{1}{kq_e^2} + \dfrac{1}{q_e} t$	t/q_t vs. t

Sida acuta is a malvaceous weed that frequently dominates improved pastures, waste and disturbed places roadsides (Mann et al., 2003). The plant is native to Mexico and Central America but has spread throughout the tropics and subtropics (Holm et al., 1977). A comparison of the second order model is shown in Table 1. Sobkowsk and Czerwinski (1974) used a second-order rate equation based on the sorption capacity of a solid for a higher concentration of the solid for the rate of reaction of carbon dioxide sorption on a platinum electrode. Ritchie (1977) used a second-order empirical equation to test the sorption of gases onto a solid. The Ritchie equation has also been applied in solution/solid sorption systems. Blanchard et al. (1984) reported a similar rate equation for the overall exchange reaction of NH_4^+ ions fixed in zeolite by divalent metallic ions in solution. The second-order expression of Blanchard et al. (1984) was used to describe the kinetics of exchange processes between the sodium ions from zeolite A and cadmium, copper, and nickel ions from solutions. In recent years, Ho (2006) described sorption, which included chemisorption and provided a different idea to the second-order equation called a pseudo-second-order rate expression.

The purpose for this study is to know if the sorption of some divalent metal ions onto *S. acuta* leaves used as sorbent follows the pseudo-second –order kinetic expression.

MATERIALS AND METHODS

Preparation of *S. acuta* leaves

The leaves were dried at room temperature for a period of three days. The sorbent was screened to obtain a geometrical size of 0.25 - 0.5 mm. This was to allow for shorter diffusion path, thus allowing the sorbent to penetrate deeper into the effluent more quickly, resulting in a higher rate of adsorption (Adeyinka et al., 2007).

Phytochemical screening

Phytochemical tests were carried out on the powdered plant material employing standard phytochemical procedures to establish the presence or otherwise of secondary metabolites such as alkaloids, steroids, flavonoids, tannins and saponin glycosides (Sofowora, 1982; Evans, 1989).

Preparation of synthetic wastewater

Stock solutions of Nickel, Lead, Copper, Zinc and Aluminium were prepared with distilled water and Nickel (II) Sulphate, Lead (II) Nitrate, Zinc (II) Sulphate, and Copper (II) Sulphate respectively. All working solutions were obtained by diluting the stock solution with distilled water. The pH of the effluent was adjusted to a pH of 5 to prevent hydrolysis by the use of relevant acids and bases. The concentration of metal ions in effluent was analyzed by Atomic Absorption Spectrophotometer.

For quality control purpose, the diluted water were digested and analyzed with every sample group to track any possible contamination source. A duplicate was analyzed for every sample to track experimental error and show capability of reproducing results (Marshall and Champagne, 1995).

Adsorption experiment

The experiments were carried out in the batch mode for the measurement of adsorption capabilities. The bottles with 500 ml capacity were filled with 50 ml of the synthetic wastewater, and 1 g of *S. acuta* dried leaves (ground). The bottles were shaken for a predetermined period at room temperature in a reciprocating shaker for 2 h at 300 rpm.

The separation of the adsorbents and solutions was carried out by filtration with Whatman filter paper No. 42 and the filtrate stored in sample cans in a refrigerator prior to analysis. The residual metallic ion concentrations were also determined using an Atomic Absorption Spectrophotometer (AAS).

RESULTS AND DISCUSSION

Table 2 shows that *S. acuta* contained bioactive constituents such as alkaloids, flavonoids, cardiac and

Table 2. Results of phytochemical tests.

Test	Observation	Inference
Molisch's test for carbohydrates	Deep violet ring was observed at the interface	Carbohydrates present
Fehling's solution test for reducing sugar	A brick-red precipitate was observed	Glycoside present
Frothing test for saponin glycosides	Persistent frothing was observed	Saponin glycoside present
Blood haemolysis test for saponin glycoside	Clear zones of haemolysis was observed	Saponin glycoside confirmed
Borntrager's test for anthraquinone glycosides	No pink colouration was observed	Anthraquinone glycoside absent
Test for cyanogenetic glycosides	Yellow colour of sodium picrate paper retained.	Cyanogenetic glycoside absent
Keller-killiani test for deoxy-sugar	A brown ring was observed at the interface	Deoxy-sugar present in cardiac glycosides
Kedde's test for lactone ring	A violet colour that faded gradually with the deposition of whitish crystalline solid was observed	Lactone ring present in cardiac glycosides
Lieberman's test for steroidal ring	A colour change from violet to blue to green was observed	Steroidal ring present in cardiac glycosides
Salkowski's test for steroidal ring	A reddish-brown colour was observed at the interface	Steroidal ring present in cardiac glycosides
Tests for flavonoids	A yellow colour which turned to colourless was observed	Flavonoids present
Aqueous ferric chloride test for Tannins	No blue black, green or blue green precipitate or colouration observed	Tannins absent
Test for Phlobatannins	No red precipitate was observed	Phlobatanins absent
Test for alkaloids using water, methanol and chloroform as extracting solvents	Wagner's, Hager's and Dragendorff's reagent gave characteristic precipitates with methanol and chloroform extracts	Alkaloidal base present

saponin glycosides. An expression of the pseudo-second-order rate based on the solid capacity has been presented for the kinetics of sorption of divalent metal ions onto peat (Ho, 2006):

$$q_t = \frac{q_e^2 kt}{1 + q_e kt},$$

(1)

where k is the pseudo-second-order rate constant (g/mg min), q_e is the amount of cadmium ion sorbed at equilibrium (mg/g), and q_t is amount of cadmium ion on the surface of the tree fern at any time, t,(mg/g). Equation 1 can be rearranged to obtain

$$q_t = \frac{t}{1/kq_e^2 + t/q_e},$$

(2)

This has a linear form of

$$\frac{t}{q_t} = \frac{1}{kq_e^2} + \frac{1}{q_e}t.$$

(3)

If the initial sorption rate, as $h = q_t/t$ when t approaches 0, h (mg/g min), is

$$h = kq_e^2.$$

(4)

Table 3. Pseudo-second order kinetic model linear forms.

Type	Linear form	Plot	Parameter
Type 1	$\dfrac{t}{q_t} = \dfrac{1}{kq_e^2} + \dfrac{1}{q_e}t$	t/q_t vs. t	$q_e = 1/\text{slope}$ $k = \text{slope}^2/\text{intercept}$ $h = 1/\text{intercept}$
Type 2	$\dfrac{1}{q_t} = \left(\dfrac{1}{kq_e^2}\right)\dfrac{1}{t} + \dfrac{1}{q_e}$	$1/q_t$ vs $1/t$	$q_e = 1/\text{intercept}$ $k = \text{intercept}^2/\text{slope}$ $h = 1/\text{slope}$
Type 3	$q_t = q_e - \left(\dfrac{1}{kq_e^2}\right)\dfrac{q_t}{t}$	q_t vs q_t/t	$q_e = \text{intercept}$ $k = -1/\text{intercept} \times \text{slope}$ $h = -\text{intercept}/\text{slope}$
Type 4	$\dfrac{q_t}{t} = kq_e^2 - kq_e^2 q_t$	q_t/t vs q_t	$q_e = -\text{intercept}/\text{slope}$ $k = \text{slope}^2/\text{intercept}$ $h = \text{intercept}$

Equation 2 can be rearranged to obtain

$$q_t = \frac{t}{1/h + t/q_e},$$
(5)

and

$$\frac{t}{q_t} = \frac{1}{h} + \frac{1}{q_e}t.$$
(6)

In order to distinguish the kinetics equation based on the concentration of a solution from the sorption capacity of solids, this second-order rate equation has been called a pseudo-second-order one since it was represented.

The pseudo-second-order rate constants can be determined experimentally by plotting t/q_t against t. Although there are many factors which influence the sorption capacity, including the initial sorbate concentration, the reaction temperature, the solution pH value, the sorbent particle size and dose, and the nature of the solute, a kinetic model is concerned only with the effect of observable parameters on the overall rate (Ho, 2006). In this study, regression, r^2, was used to test the best-fitting of the kinetic model to the experimental data.

The least squares method is used for finding the parameters for kinetic models. The pseudo-second–order kinetic model has been linearized into four different types which were shown in Table 3 and a simple linear regression could result in different parameter estimates (Kinniburgh, 1986; Longhinotti, 1998; Ho, 2004). The most popular linear used is Type 1.

Figures 1 to 4 show experimental data with linear equations of the four pseudo-second–order kinetic models obtained by using the linear method for the sorption of the divalent metal ions under study onto S. acuta. Values of the pseudo-second – order kinetic model constant, k, the amount of the divalent metal ions under study sorbed at equilibrium, q_e, and the initial sorption rate, h, are listed in Table 4. The regression (r^2) values obtained for Type 3 indicated that there was a strong positive evidence that the divalent metal ions sorption unto S. acuta followed the pseudo-second –order kinetic expression. It is clear that transformations of non-linear pseudo-second –order kinetic models to linear forms implicitly alter their error structure and may also violate the error variance and normality assumptions of standard least-squares method (Longhinotti et al., 1998; Ho, 2004). In a linear analysis, different linear forms of the same model would significantly affect calculations of the parameters.

Conclusion

The following conclusion can be drawn as follows:

(1) The results show sorption of divalent metal ions used

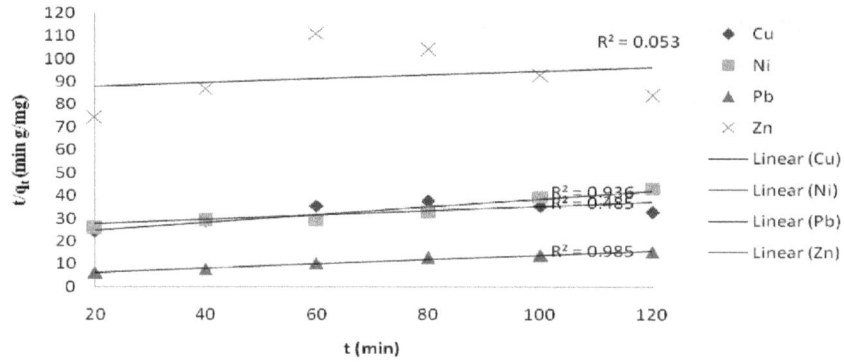

Figure 1. Type 1 pseudo-second- linear equations obtained by using the linear equations obtained from the linear method for the sorption of divalent metal ions onto sida acuta leaves.

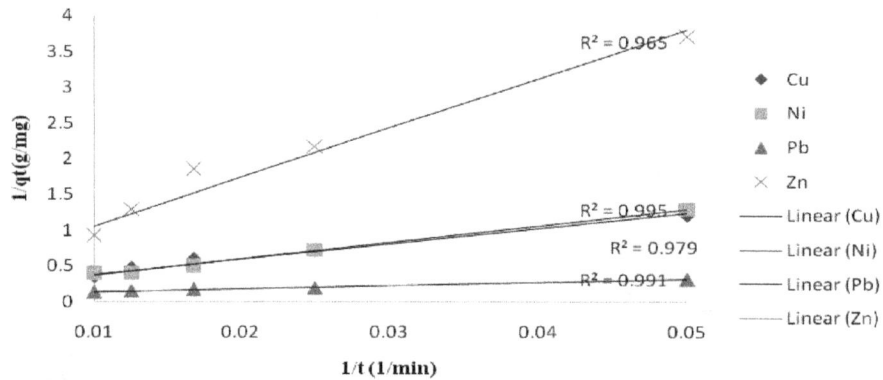

Figure 2. Type 2 pseudo-second- linear equations obtained by using the linear equations obtained from the linear method for the sorption of divalent metal ions onto sida acuta leaves.

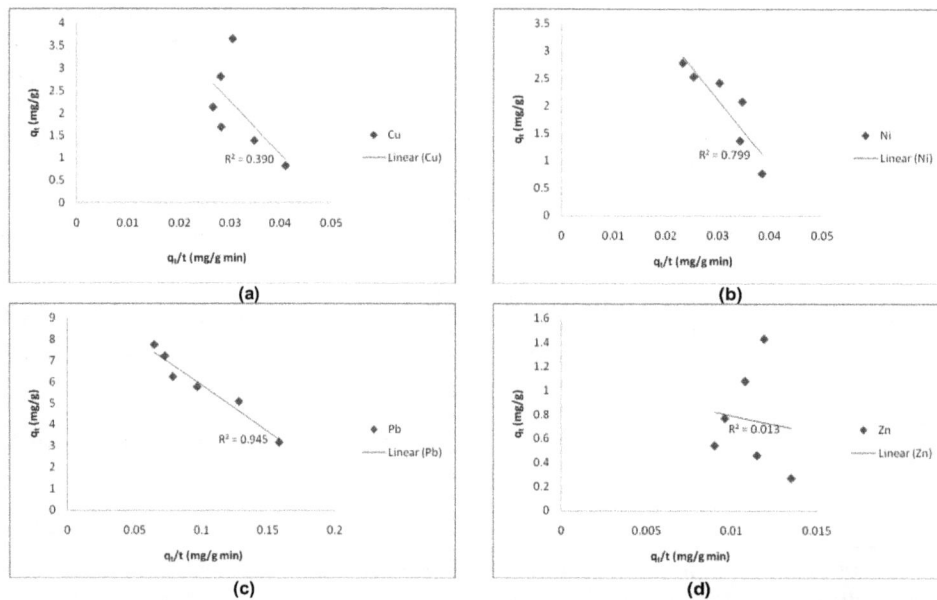

Figure 3. Type-3 pseudo-second- linear equations obtained by using the linear equations obtained from the linear method for the sorption of (a) Cu^{2+} ions, (b) Ni^{2+} ions, (c) Pb^{2+} ions, and (d) Zn^{2+} ions onto sida acuta leaves.

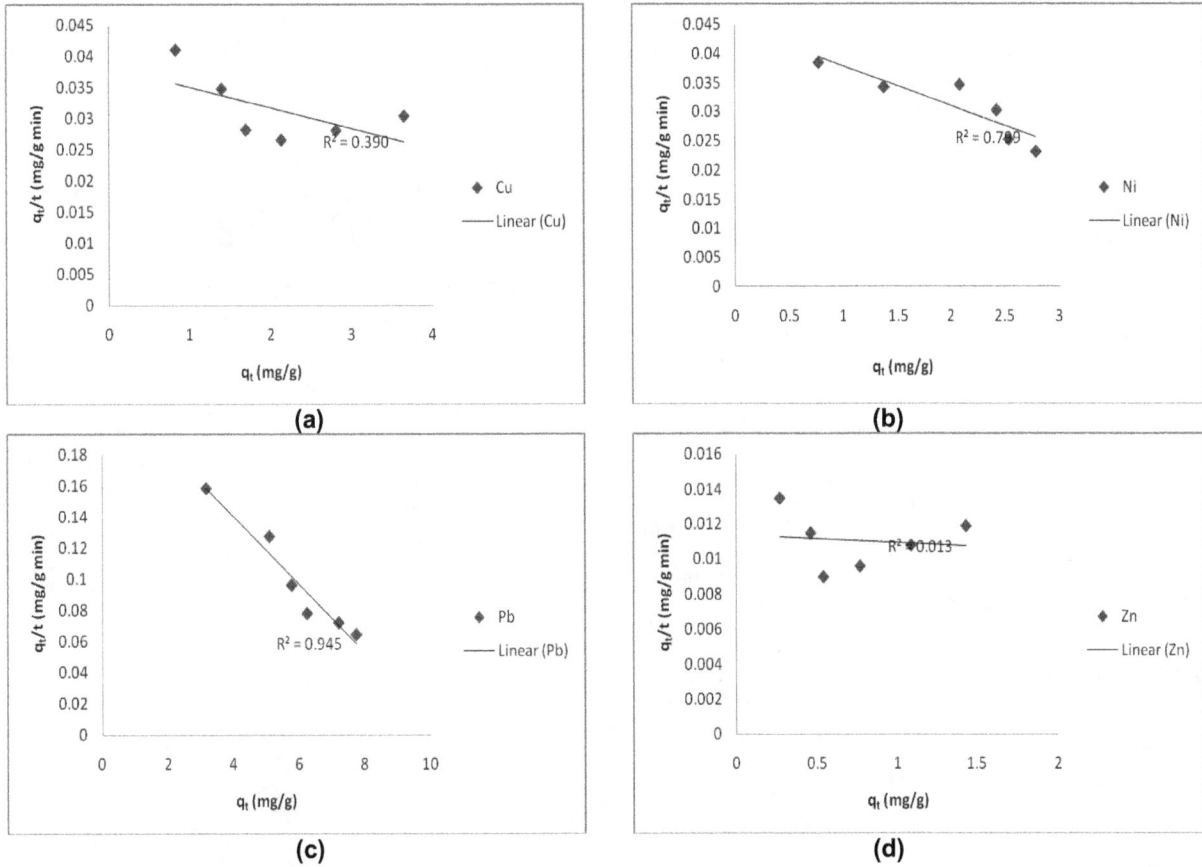

Figure 4. Type-4 pseudo-second- linear equations obtained by using the linear equations obtained from the linear method for the sorption of (a) Cu^{2+} ions, (b) Ni^{2+} ions, (c) Pb^{2+} ions, and (d) Zn^{2+} ions onto sida acuta leaves.

Table 4. Pseudo-second-order kinetic parameters obtained by using the linear methods for divalent metal ions.

Type	Parameter	Cu^{2+}	Ni^{2+}	Pb^{2+}	Zn^{2+}
Type 1	q_e (mg/g)	10.783	5.805	10.556	11.884
	k (g/mg min)	0.00033	0.00139	0.00200	0.00008
	h (mg/g min)	0.0385	0.0469	0.2232	0.0116
	r^2	0.485	0.985	0.936	0.053
Type 2	q_e (mg/g)	6.040	6.8630	10.315	2.668
	k (g/mg min)	0.00127	0.00093	0.00217	0.00205
	h (mg/g min)	0.0465	0.0436	0.2310	0.0146
	r^2	0.979	0.995	0.991	0.965
Type 3	q_e (mg/g)	5.770	5.620	10.206	1.099
	k (g/mg min)	0.00148	0.00152	0.00225	0.02955
	h (mg/g min)	0.0493	0.0481	0.2347	0.0357
	r^2	0.390	0.799	0.945	0.013
Type 4	q_e (mg/g)	11.522	6.529	10.456	25.850
	k (g/mg min)	0.00029	0.00105	0.00208	0.00002
	h (mg/g min)	0.0385	0.0447	0.2273	0.0114
	r^2	0.390	0.799	0.945	0.013

for the study on to *S. acuta* leaves by agitation; a pseudo-second order model can describe the sorption process. (2) The sorption equilibrium capacity, the sorption rate constant and the initial sorption rate can be a function of the metal ion present.

REFERENCES

Adeyinka A, Liang H, Tina G (2007). Removal of Metal Ion form Waste Water with Natural Waste, Sch. Eng. Technol. 33(4):1-8.

Blanchard G, Maunaye M, Martin G (1984). Removal of heavy metals from waters by means of natural zeolites, Water Res. 18(12):1501–1507.

Evans WC (1989). Trease and Evan's Pharmacognosy. English Language Book Society Bailliere Tindale, London. 13th edn., pp.39, 54-55, 131-132, 387, 546, 800-803.

Ho YS (1995). Adsorption of heavy metals from waste streams by peat. Ph.D. thesis, University of Birmingham, Birmingham, UK.

Ho YS, Mckay G (2000). The Kinetics of Sorption of Divalent Metal Ions onto Sphagnum Moss Peat, *Wat. Res.* 34(3):735-742.

Ho YS, Huang CT, Huang HW (2002). Equilibrium sorption isotherm for metal ions on tree fern. Process Biochem. 37:1421-1430.

Ho YS (2004). Selection of optimum sorption isotherm. Carbon 42(10): 2115-2116.

Ho YS (2006). Second-order kinetic model for the sorption of cadmium onto tree fern: A comparison of linear and non-linear Methods. Water Res. 40:119-125.

Holm LG, Plucknett DL, Pancho JV, Herberger JP (1977). The Worlds Worst Weeds: distribution and biology (University Press of Hawaii, Honolulu, USA).

Kinniburgh DG (1986). General purpose adsorption isotherms, Environ. Sci. Technol. 20(9):895-904.

Longhinotti E, Pozza F, Furlan L, Sanchez MDND, Klug M, Laranjeira MCM, Favere VT (1998). Adsorption of anionic dyes on the biopolymer chitin. J. Brazilian Chem. Soc. 9(5):435-440.

Luckey TD, Venugopal B (1977). Metal Toxicity in Mammals, Physiologic and Chemical Basis for Metal Toxicity, New York: Plenum Press, Vol. 1.

Mann A, Gbate M, Umar AN (2003). *Sida acuta* subspecie *acuta*. *Medicinal and economic plant of Nupeland*. Jube Evans Books and Publication, P. 241.

Marshall WE, Champagne TE (1995). Agricultural Byproducts as Adsorbents for Metal Ions in Laboratory Prepared Solutions and in Manufacturing Wastewater, *Journal of Environmental Science and Health, Part A:* Environ. Sci. Eng. 30(2):241-261.

Ritchie AG (1977). Alternative to the Elovich equation for the kinetics of adsorption of gases on solids. J. Chem. Soc. Faraday Trans. I 73(10):1650-1653

Sobkowsk J, Czerwinski A (1974). Kinetics of carbon dioxide adsorption on a platinum electrode, J. Electroanal. Chem. 55(3):391-397.

Sofowora A (1982). Medicinal Plants and Traditional Medicine in Africa. John Wiley and Sons, Chichester. pp. 142-145.

19

Determination of trace elements in nutrition materials in Kingdom of Saudi Arabia

Badriah Saad Al-Farhan

Department of Chemistry, Faculty of Girls for Science, King Khalid University, Abha, KSA, Saudi Arabia.

Four types of edible tubercular roots cultivated in Saudi Arabia are analysed through sequential determination of certain essential and toxic trace elements by inductively coupled plasma atomic emission spectrometry (ICP-AES). Comparable runs carried out using both flame and graphite atomic absorption spectroscopy (AAS). Radish proved to contain the highest concentration level of iron (>21 µg/g), onion contains high concentration of zinc and strontium (~ 6 and 9 µg/g) and potato was found to contain the highest concentration level of copper (~2 µg/g). Samples of carrots, radish and potato collected from different locations are also analysed to study the effect of cultivation area on the concentration levels of trace elements in edible tubercular roots. Variation in the concentration levels of iron, zinc, copper, cobalt, strontium, cadmium and lead in each type of test samples seem not to be significant. Detailed studies seem necessary to throw further light on the effect of different of sample location on the concentration levels of both essential and toxic trace elements in different vegetable materials; especially those cultivated in areas neighboring various industrial and other human activities in Saudi Arabia.

Key words: Tubercular roots, trace elements, inductively coupled plasma atomic emission spectrometry (ICP-AES), atomic absorption spectroscopy (AAS) analysis.

INTRODUCTION

Human beings are encouraged to consume more food materials such as vegetables and fruits, which are good sources of vitamins and minerals beneficial to human health (Mohamed et al., 2005). Heavy metal contamination in agricultural environments can come from atmospheric fall-out, pesticide formulations, contamination by chemical fertilizer and irrigation with water of poor quality (Marcovecchio et al., 2007).

Trace elements in nutrition materials play significant role in human health. Trace concentration of iron, zinc, copper, manganese, nickel, cobalt, molybdenum, selenium, iodine, and fluorine are considered essential for human life (Clemente et al., 1977; Roberts, 1981). The absence or deficiency of one of these elements in certain body organs leads to physiological abnormalities in a number of biological processes which can be remedied by addition of limited quantities from the deficient element (Cotzias, 1970). Few other elements such as lead, cadmium, tin and mercury are highly toxic for both animal and human lives and may lead to death when ingested with high doses. The presence of different concentration levels of several trace elements (including those with toxic effects) in individual food articles and integrated human diets is mainly due to the uncontrolled release of various types of toxic pollutants in the different environmental compartment from increased industrial and other human activities (Underwood, 1971). It is therefore necessary to monitor the concentration level of toxic and essential elements in common food items for daily intake (Qureshi et al., 1990; Noel et al., 2011). In the present work, four edible tubercular roots including carrots, onion, potato, radish that are mostly consumed by a wide spectrum

Table 1. Analysed vegetable tubercular roots.

Common name	Family	Botanic name
Carrots	Umbelliferae	*Daucus carota*
Potato	Solanacea	*Solanum tuberosum*
Onion	Amarylilaceae	*Allium cepa*
Radish	Cruciferae	*Raphanus sativum*

of the King of Saudi Arabia population have been analysed to comment on their suitability for human intake.

EXPERIMENTAL

Sampling and sample preparation

The test samples were collected from a number of agricultural areas. A list of test species is given in Table 1, with their botanical names and respective families. To investigate the effect of sample location on the concentration of trace elements in test items; carrots, radish, potato were collected from three areas. From the cultivated part of the southern region in Saudi Arabia, from the northern region, and from eastern region as shown in Figure 1. For sample preparation, collected samples were thoroughly washed and air dried at room temperature. After recording the wet weight, each species was oven- dried at 60°C for 72 h) (Zaidi et al., 1990) and the corresponding dry weight and moisture content determined. Representative dried samples were powdered by using a teflon ball mill, sieved to ≅200 mesh and finally stored in pre-cleaned polyethylene capped bottles. Nitric acid – hydrogen peroxide – perchloric acid mixture was used to digest different test samples.
For (2 to 10 g) of dried matter, the mixture used includes 20 ml of 14.4 mol l^{-1} nitric acid, 10 ml of 30% hydrogen peroxide and 10 ml of 9.9 mol l^{-1} perchloric acid. In addition, 18.0 mol l^{-1} sulphuric acid (for 10 g of dried matter, 2.0 ml of acid was added to prevent losses of metal halides by volatilization (Feinberg and Ducauze, 1980; Erwin and Ivo, 1992). Digestion normally took place in all glass containers under reflux at 170°C until a clear digest was obtained after approximately 3 h (Yaman and Gucer, 1995). The digest was centrifuged to separate the clear solution and the residue washed with bidistilled water and re-centrifuged to prevent any elemental losses. The first washing was added to the original solution before being diluted to known volume.

Instrumentation

(i) Inductively coupled plasma atomic emission spectrometry (ICP-AES) measurements were done with a compact tuned – oscillator coupled with high resolution Echelle grating spectrometer, minicomputer control services, peristaltic pump and an automated sample changer. The system includes a plasma spectrometer, type Leeman from USA, 2.5 KW generator, a three -turn copper load coil and a Hidebrand Grid nebulizer. The spectrometric system is of a fixed optics model with a PMT for sequential operation (type f18 Echelle), with a single pass prism / lens used for stray light reduction to cover a wavelength range from 190 to 800 nm.
(ii) The atomic absorption spectroscopy (AAS) measurements were carried out with AA spectrometer, model Z -8100 polarized Zeeman, manufactured by Hitachi, Ltd., from Japan Hitachi single – element hollow cathode lamps were used with air- acetylene flow rate ranging from 0.5 to 4.0 L/min with an auxiliary oxidant gas pressure ranging from 140 to 120 kpa. The instrument is provided with

Figure 1. Samples location

temperature regulation device and automated sampling by a built in auto sampler, type SSC -200. Selection of wavelength ranged from 190 to 900 nm.

Spectroscopic measurements

(i) ICP- measurements were done in sequential multi- element mode. An analytical programme was established both for calibration and routine analysis. The selected analytical wavelengths represent the characteristic lines which are almost free from spectral interference to eliminate any correction at the concentration levels of interest, these are:

(a) Iron - 259.94 (nm)
(b) Copper -   324.75
(c) Zinc - 213.86
(d) Cadmium - 214.44

Measurements were done in triplicates according to the following operating conditions:

(a) Forward r.f. power   1.00 KW (0.5A)
(b) Argon flow rate   12 L/min
(c) Nebulizer gas   0.3-0.5 L/min
(d) Sample uptake rate 1 L/min

(ii) AAS measurements were carried out under a constant air flow rate or (15.0 L/min), according to the following operational condition for each element as in Table 2.

Chemicals and reagents

All chemicals used were of A.R or extra pure grades. A set of standards were prepared from readily made standard solutions provided from Merck, AG, Darmstadt, Germany by dissolution in, or adequate dilution with dilute nitric acid solution. Bidistilled water in all glass apparatus was used for preparation of different solutions, used standards and for final glass ware washing. In the digestion procedure, concentrated nitric acid (65%, 14.4 mol l^{-1}), sulphuric acid (98%, 18 mol l^{-1}), hydrogen peroxide (30%) and perchloric acid (65%, 9.9 mol l^{-1}) were used.

Table 2. Operational conditions for AAS measurements.

Condition	Fe	Zn	Cu	Co	Sr	Cd	Pb
Wavelength, nm	248.3	213.9	324.8	240.7	460.7	228.8	283.3
Lamp current ,mA	15	7.5	7.5	15	10	7.5	10
Slit width	0.2	1.3	1.3	0.2	0.5	1.3	1.3
Acetylene flow rate,Lmin^{-1}	1.5	1.5	1.7	1.7	1.7	-----	-----
Heating program drying temp., °C	-----	-----	-----	-----	-----	80-120	80-120
Time/sec	-----	-----	-----	-----	-----	30	30
Ashing temp., °C	-----	-----	-----	-----	-----	300	400
Time/sec	-----	-----	-----	-----	-----	30	30
Atomization temp., °C	-----	-----	-----	-----	-----	1700	2100
Time/sec	-----	-----	-----	-----	-----	7	7
Cleaning temp., °C	-----	-----	-----	-----	-----	2600	3000
Time/sec	-----	-----	-----	-----	-----	30	30

Table 3. Concentration of trace elements in edible tubercular roots*).

Element	Carrots	Onion	Radish	Potato	Intake levels **)
a)Assessment of trace elements by ICP-AES (in µg/g wet weight)					
Iron	4.68±0.1	16.25±0.1	20.84±1.2	6.63±0.1	25-75 mg
Zinc	2.21±0.1	5.74±0.0	2.82±0.2	3.43±0.0	10-20 mg
Copper	1.51±0.2	1.42±0.2	0.31±0.02	1.66±0.05	--------
Cobalt	0.32±0.0	0.54±0.01	0.36±0.0	0.56±0.02	150-580 µg
Cadmium	0.17±0.01	0.34±0.02	0.25±0.01	0.41±0.01	--------
b)Assessment of trace elements by FAAS (in µg/g wet weight)					
Iron	4.79±1.2	16.18±0.4	21.26±0.5	5.82±0.7	25-75 mg
Zinc	2.68±0.2	5.63±0.1	2.67±0.1	3.45±0.2	10-20 mg
Copper	1.39±0.1	1.44±0.3	0.38±0.0	1.92±0.1	--------
Cobalt	0.30±0.0	0.50±0.0	0.38±0.1	0.57±0.02	150-580 µg
Strontium	3.37±0.0	9.19±0.1	4.94±0.05	3.66±0.05	42-1240 µg
c)Assessment of trace elements by GAAs (in µg/g wet weight)					
Cadmium	0.19±0.01	0.34±0.02	0.22±0.01	0.44±0.01	--------
Lead	0.55±0.0	0.68±0.01	0.31±0.04	0.67±0.01	54-500 µg

*) the results are mean of at least triplicate measurements; based on determination of each trace element in aliquot portions of sample solution containing known amounts of respective dried tuber, final concentration levels and the results based on wet weight are calculated from respective dry weight results.**) acceptable levels of daily intake concentration.

RESULTS AND DISCUSSION

Trace elements in tubers

The results in Table 3, show that iron, zinc, copper, cobalt, strontium, cadmium and lead proved to be present in different concentration levels in various types of tubercular roots. The examined species including carrots, onion, radish and potato are among the common vegetables for human nutrition in Saudi Arabia. The choice of these species aims to define the role of different soils, fertilizers and mode of irrigation as possible pathways for trace elements to man through the food chain (Husain et al., 1995; Ozores-Hampton et al., 1997; Millour et al., 2011). To get reliable and comparable results, the assessment of trace elemental concentrations in different samples is based on atomic spectroscopy using ICP-AES, flame and graphite AAS techniques. ICP-AES has the advantage of being rapid in providing analytical data for several elements in a single run. All used techniques proved to give comparable and reliable results. This is clearly illustrated by the results of iron,

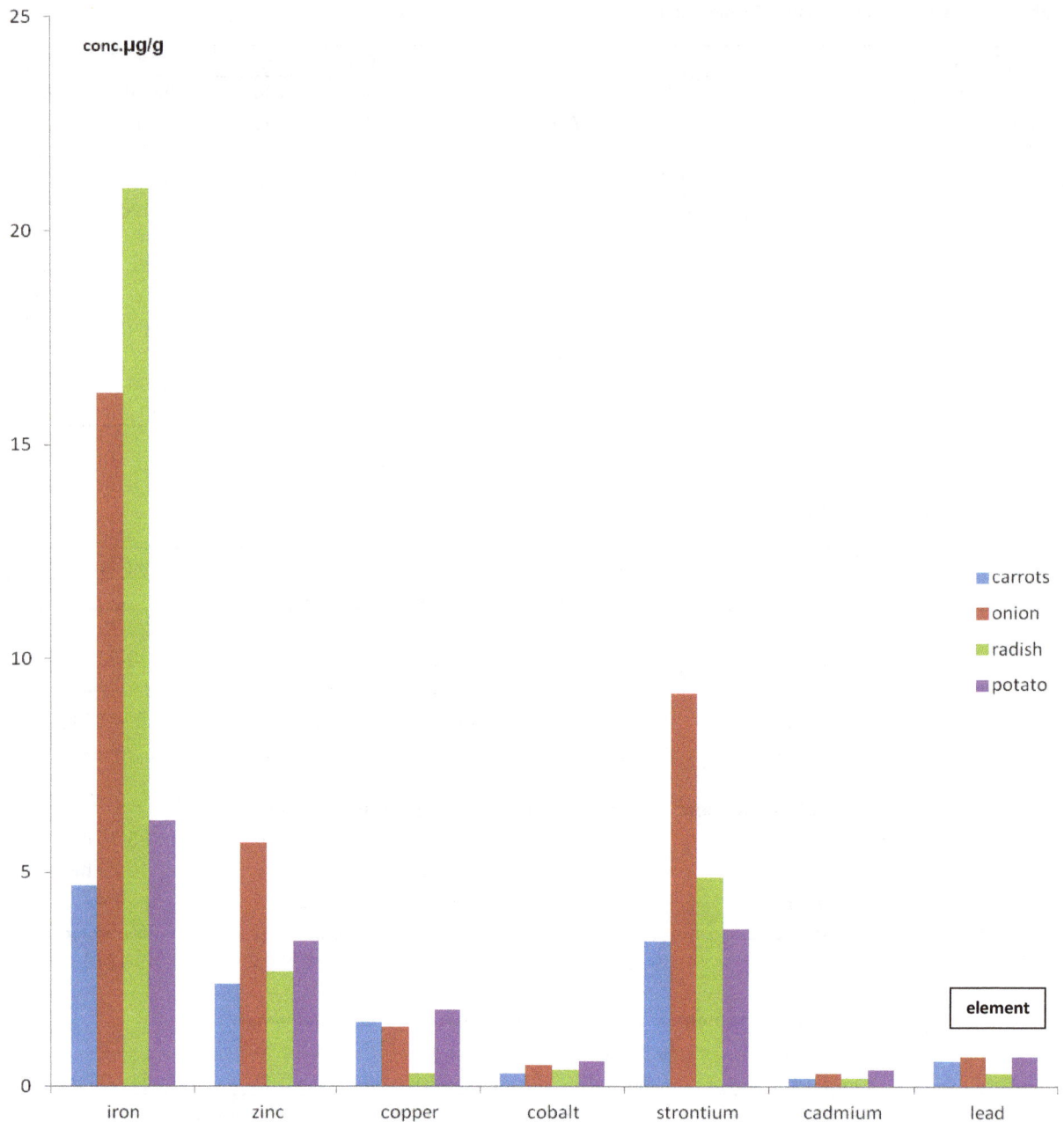

Figure 2. Concentration levels of Trace elements in test samples

zinc, copper, cobalt and cadmium in all types of tested species which proved to be subject to almost the same mean standard deviation for each analyte.

The results (expressed in terms of µg/g of wet weight) showed that radish contains the highest concentration levels of iron (21.26 µg/g), onion proved to contain the highest concentration of zinc (5.74 µg/g) and strontium (9.19 µg/g). Potato contains the highest levels of copper (1.92 µg/g). For toxic elements, it was found that potato contains the highest concentration levels of both cadmium and lead (0.44 and 0.67 µg/g), (Figure 2).

The concentration levels of iron, zinc and copper are almost of the same order or even less than those previously reported by several workers (Finch and Monsen, 1974; Thomas et al., 1952). While the concentration of cobalt and strontium is several orders of magnitude higher than the values reported by other workers (Schroeder et al., 1967; Wikelsk et al., 1993).

In general, one assumes that changes in the concentration levels of trace elements in the examined species can be mainly attributed to changes in the chemical composition of water used for irrigation, the type

Table 4. Trace element concentrations in different carrot, radish and potato samples[*].

Element	Sarat Ebeda	Alehssa	Algoff	Mean δ_n
a) Carrots				
Iron	4.6±0.2	5.03±0.1	4.76±0.5	0.2
Zinc	2.33±0.1	3.13±0.1	2.6±0.5	0.2
Copper	0.78±0.1	1.03±0.1	0.806±0.1	0.0
Cobalt	0.28±0.2	0.33±0.1	0.30±0.2	0.05
Strontium	3.16±1.0	3.57±1.0	3.37±1.2	0.1
Cadmium	0.14±0.01	0.21±0.01	0.21±0.01	0.0
Lead	0.50±0.01	0.59±0.01	0.55±0.01	0.0
b) Radish				
Iron	20.46±1.0	21.8±1.2	21.53±1.0	0.1
Zinc	2.26±0.1	2.90±0.2	2.86±0.1	0.05
Copper	0.33±0.1	0.33±0.05	0.47±0.1	0.02
Cobalt	0.35±0.1	0.42±0.2	0.38±0.1	0.05
Strontium	4.56±1.0	5.50±1.0	4.76±0.5	0.2
Cadmium	0.21±0.01	0.24±0.01	0.22±0.01	0.0
Lead	0.30±0.01	0.32±0.01	0.32±0.01	0.0
c) Potato				
Iron	5.03±0.2	6.43±0.1	6.00±0.1	0.05
Zinc	2.93±0.1	4.26±0.1	3.16±0.2	0.05
Copper	1.56±0.5	2.36±0.1	1.83±0.2	0.2
Cobalt	0.51±0.1	0.63±0.1	0.58±0.2	0.05
Strontium	3.03±0.5	4.17±1.0	3.77±0.5	0.2
Cadmium	0.40±0.02	0.49±0.02	0.44±0.02	0.0
Lead	0.67±0.01	0.71±0.01	0.67±0.01	0.0

[*] concentrations in µg/g, on wet weight basis.

of soil in various agricultural areas, and seasonal changes in ambient temperature.

Under comparable irrigation conditions, however, trace elemental concentrations may be affected by the sorptive capacity of different roots, the physical characteristics of the edible body in different species and the chemical composition of organic compounds in each type that might form different complexes with various metallic species. Thus, the increased concentration levels of cobalt, strontium than the mean values so far reported may be attributed to cultivation in areas rich with different minerals, especially when using water contaminated with industrial waste effluents, including trace concentrations of either or more of these elements. Never the less, the high concentration levels determined in all test samples, do not exceed the acceptable levels for daily intake. These are almost about 150 to 580 µg for cobalt, and 42 to 1240 µg for strontium. For iron, zinc and lead on the other hand, the concentration levels determined are far below those reported for daily intake lying within 25 to 75 mg for iron, 10 to 20 mg for zinc and 54 to 500 µg for lead

(Dabeka et al., 1987; Galal-Gorchev, 1991).

Effect of sample location

To study the effect of sample location on the concentration levels of both essential and toxic trace elements in tubercular roots, carrots, radish and potato were collected as test samples from three different areas. These include an agricultural area near the industrial zone of the eastern region of Saudi Arabia (Alehssa), an agricultural area free from any industrial activities at the southern region of Saudi Arabia (Sarat Ebeda), and from the northern region of Saudi Arabia (Algoff). The essential trace elements investigated include iron, zinc, copper, cobalt and strontium, and the toxic elements are represented by cadmium and lead. This was specifically verified by successive triplicate analysis using not only ICP-AES but also, flame and graphite AAS measurements.

In Table 4, the concentration levels of iron, zinc, copper, cobalt, strontium, cadmium and lead in different samples of carrots, radish and potatoes collected from the three above mentioned areas are presented. It is observed that iron has a mean concentration value of 4.78 µg/g in carrots with a highest concentration level of 5.03 µg/g in samples collected from Alehssa and a lowest concentration level of 4.6 µg/g in those from Sarat Ebeda. For radish and potato, iron proved to have a mean value of 21.26 and 5.82 µg/g respectively. For zinc, the same trend is followed, showing mean concentration levels 2.68, 2.67 and 3.45 µg/g, the highest values of 3.13, 2.90 and 4.26 µg/g for carrots, radish and potato, respectively. On the other hand, copper, cobalt, strontium, cadmium and lead also follow the same trend.

It was found that the concentration levels of the trace elements determined in carrots, radish and potato samples collected from the three different areas are almost of the same order. Samples collected from areas neighboring several industrial activities in (Alehssa) proved to contain slightly higher concentration levels of all the tested elements than those collected from (SaratEbada) and from (Algoff).

Conclusion

The results obtained show that radish contains the highest concentration level of iron, onion contains high concentrations and strontium and potato was found to contain the highest concentration levels of copper. The standard deviations in the results obtained for almost all determined elements are relatively low ranging from 1.5 for strontium determined by FAAS in onion to 0.0 in case of determination of several elements by the different used techniques indicating the reliability of ICP, flame and graphite AAS techniques for the determination of the concerned elements.

In the light of the obtained data, one may conclude that samples cultivated in areas far from industrial and other human activities tend to contain lower concentration levels of both essential and toxic elements than others. This can be noted in particular by considering the change in the concentration levels of the different elements determined in test samples collected from the agricultural area of Sarat Ebeda which seem to be lower than those collected from areas neighboring the industrial zone of Alehssa.

The difference in the concentration levels of all analysed trace elements collected from the different areas give a mean standard deviation not exceeding 0.2 with a percentage difference ≤10%, which lies within the experimental error in optical measurements especially when dealing with low concentration levels of different analytes. It is recommended from the present study that the work needs further investigation referring to soil types and irrigation water that affect on sample location results.

REFERENCES

Clemente GF, Cigna Rossi L, Santaroni GP (1977). Trace element intake and excretion in the italian population. J. Radioanal. Chem. 37:549-558.

Cotzias GC (1970). Trace subst. Environ. Health-Proc. University Mo. Annual Conference, 1 st ed, P. 5.

Dabeka RW, McKenzie AD, Lacroix GM.A (1987). Dietary intakes of lead, cadmium, arsenic and fluoride by Canadian adults: a 24-hour duplicate diet study. Food additives contaminants 4:89-101.

Erwin JM, Ivo N (1992). Determination of Lead in tissues: A pitfall due to wet digestion procedures in the presence of sulphuric acid. Analyst 23:117.

Feinberg M, Ducauze C (1980). High temperature dry ashing of foods. J. Anal. Chem. 52:207.

Finch CA, Monsen ER (1974). Iron nutrition and the fortification of food with iron. J. American Medical Association 219:1462.

Galal-Gorchev H (1991). Dietary intake of pesticide residues, cadmium, mercury and lead. Food additives and contaminants, 6:793-806.

Husain A, Baroon Z, Al-Ati T, Sawaya W (1995). Heavy metals in fruits and vegetables grown in Kuwait during the oil well fires. Arab Gulf J. Sci. Res. 3:535-542.

Marcovecchio JE, Botté SE, Freije RH (2007). Heavy metals, major metals, trace elements. In L.M. L. Nollet (Ed.), Handbook of water analysis (2 nd ed, pp. 270-311). Boca Raton: CRC Press.

Millour S, Noel L, Kadar Ali, Chekri R, Vastel C, Sirot V, Guerin T (2011). Pb, Hg, Cd, As, Sb and Al levels in foodstuffs from the 2nd French total diet study. J. food chem. 126:1787-1799.

Mohamed AE, Rashed MN, Mofty A (2005). Assessment of essential and toxic elements in some kinds of vegetables. J. Ecotoxi. Environ. Safety 55:251-260.

Noel L, Millour S, Kadar Ali, Chekri R, Vastel C, Guerin T (2011). Simultaneous of 21 elements in foodstuffs by ICP-MS after closed-vessel microwave digestion:Method validation. J. food chem. 24: 111-120.

Ozores-Hampton M, Hanlon E, Bryan H, Schaffer B (1997). Cadmium, copper, lead, nickel and zinc concentrations in tomato and squash grown in MSW compost-amended calcareous soil. Compost Sci. Utilization, 4:40-45.

Qureshi IH, Mannan A, Zaidi JH, Arif M, Khalid N (1990). A Comparative Determination of Heavy Metals in Moss Tissue by Atomic. Int. J. Anal. Chem., 38:565-577.

Roberts HR (1981). Food safety. Wiley, New York. Chapter 3, pp. 77.

Schroeder HA, Nason AP, Tipton IH (1967). Essential Trace Metals in Man: Cobalt. J. Chronic Diseases, 20:869.

Thomas B, Thompson A, Oyenuga VA, Armstrong RH (1952). dardization of Analytical Methodology for Feeds: Proc. Int. The ash constituents of some herbage plants at different stages of Workshop, Ottawa, ON. 12–14 Mar. 1979. Rep. IRDC-134e. Int. maturity. Emp. J. Exp. Agric. 20:10–22.

Underwood EJ (1971). Trace elements in human and animal nutrition. Academic Press, New York, USA. P. 545.

Wikelsk M, Gall B, Trillmich F (1993). Ontogenetic changes in food intake and digestion rate of the herbivorous marine iguana. J. Oecologia, 94:373-379.

Yaman M, Gucer S (1995). Determination of cadmium and lead in vegetables after activated carbon enrichment by atomic absorption spectrometry. Analyst 120:101-105.

Zaidi JH, Qureshi IH, Arif M, Fatima I (1990). Critical evaluation of a Trace elements. Intr. J. Anal Chem. 43:25.

20

Studies on the potentials of *Balanites aegyptiaca* seed oil as raw material for the production of liquid cleansing agents

Manji A. J., Sarah E. E. and Modibbo U. U.

Department of Industrial Chemistry, Modibbo Adama University of Technology, P. M. B. 2076, Yola, Adamawa State, Nigeria.

The seed oil of *Balanites aegyptiaca* was extracted and its chemical and physical properties were evaluated. The chemical parameters investigated include: saponification value (SV), iodine value (IV), peroxide value (PV) and free fatty acid (%FFA). These were found to be 168.6 mgKOH/g, 78.7 gI$_2$/100 g, 6.0 mEq/kg and 0.18 mgKOH/g, respectively. The physical parameters evaluated include: percentage yield (49.9%), specific gravity (0.927), refractive index (1.4784) and moisture content (0.27%). Saponin was also screened for and was found to be present. Thus the results have shown that the oil is non-drying and contain saponin; this then suggests that the oil is a good raw material for the formulation of liquid cleansing agents. Liquid soap and shampoo were then formulated using the oil and the properties of the products were evaluated. From the results, it was found that the products compared favourably to similar products sold in the market in terms of pH, colour, percentage alkali and solubility in water.

Key words: Seed oil, liquid soap, saponin, Nigeria.

INTRODUCTION

Balanites aegyptiaca is a tree, which belongs to the Balanitaceae family of plants. Its English name is Desert date, Høba call it Baha, Hausa call it Adu'a and the Fulanis call it Tanni. The tree is wide spread in the drier regions of Africa from Mauritania to Nigeria, Eastward to Ethiopia, Somalia and East Africa. It is also found in Israel, Arabia, India and Pakistan (Sulaiman and Jackson, 1959). There are about 25 known species of the plant widely distributed from tropical Africa to Burma (Usher, 1984). It is a savannah tree, which attains a height of more than 6 m; it has a spherical crown and tangled mass of long thorny branches (Sulaiman and Jackson, 1959). The leaves are subsessile or shortly petiolate, grey green in colour, orbicular, rhomboid or obviate in shape, often measuring 3.6 - 6 by 2 - 5 cm, the apex is acute or rarely obtuse. Spines are simple or very rarely bifurcate, up to 5 cm long and alternate in the axils (Brown, 1979). The flowers are yellow-green in colour and up to 4 cm long and 2.5 cm in diameter. The fruit has thin brittle epicarp, a fleshy mesocarp and a woody endocarp containing the oil seed or kernel.

The term oil is used in generic sense to describe all substances that are greasy or oily fluids at room temperature. They are non-volatile and are insoluble in water but are soluble in organic solvents. Oils from seeds or kernels or nuts along with proteins and carbohydrates, constitute the majority of foodstuffs. They are also found in wide industrial applications, like formulation of soap,

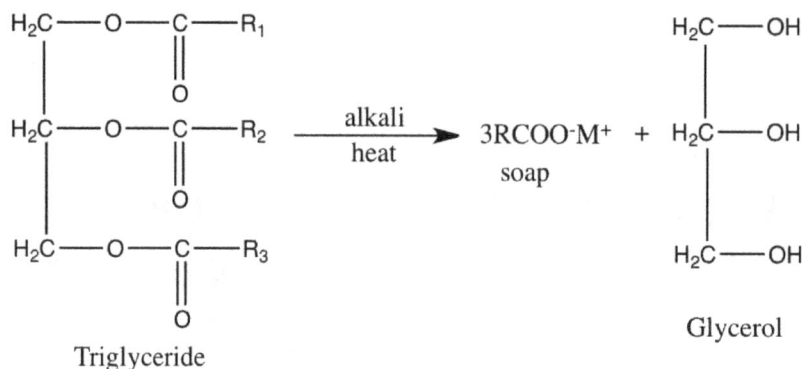

Scheme 1. Saponification process. Triglyceride (oleic or stearic or palmitic or lauric acid), alkali (NaOH-hard soap or KOH-soft/liquid soap), M (Na or K).

toiletries, paints, varnishes, bio-diesels and lubricant.

The criteria for the selection of oil for industrial use are: presence of natural characteristic aroma, clarity, good natural colour, very low moisture content, freedom from solid particles and freedom from flat and rancid (unpleasant) odour (Okoye, 1999). Studies have shown that *Balanites roxburghi,* a native of India and Pakistan, locally referred to as 'Higin ingudi' yielded an edible seed oil (Zachun oil), and the fruits were found to contain saponin; thus it is locally used to wash silk (Usher, 1984). *B. aegyptiaca* specie, which belongs to the same family with *B. roxburghi* is available in the northeastern part of Nigeria and so the investigation of its oil as source of raw material for soap making is worth trying.

Soap is a surfactant used in conjunction with water for washing and cleaning. It usually comes in solid moulded form, termed bars due to its historic and most typical shape. The use of thick liquid soap has also become widespread, especially from dispensers in public washrooms. When applied to a soiled surface, soapy water effectively holds particles in suspension so the whole of it can be rinsed off with clean water. Many soaps are mixtures of sodium (soda-solid soap) or potassium (potash-liquid soap) salts of fatty acids which can be derived from oils or fats by reacting them with an alkali (such as sodium or potassium hydroxides) at 80 - 100°C in a process known as saponification. The fats are hydrolyzed by the base, yielding glycerol and crude soap (Scheme 1).

Historically, the alkali used was potassium made from the deliberate burning of vegetation such as bracken, or from wood ashes. The use of the word "soap" has become such a household name that even cleaning solutions for the body that do not have soap in the ingredients are referred to as soap (Wikipedia, 2007). The need for soap as a cleansing agent has been felt ever since man became aware of the necessity for cleanliness (Donkor, 1986). Soap therefore acquired the status of a basic necessity of modern civilised world.

Soap when used with water, decreases surface tension loosening unwanted particles, emulsify grease and absorb dirt and grime into foam. Its use has increased over the years until its manufacture has become an industry essential to the comfort and health of civilised societies. It is consumed in large quantities on daily basis for laundry, hair dressing, personal and hygiene, in homes and commercial cleansing operations. Textile mills also consume considerable quantity of soap in boiling cotton, scouring wool and silk agitation to remove impurities prior to finished operation and to assist the level of application of softening agents used to improve fabric feed. Thus, because of increased demand for cleansing agents, there is need to evaluate and develop raw materials that have good properties for making varieties of such agents.

Studies have shown that some plants have been used as substitutes for soap. These are plants that contain saponin in sufficient quantities to produce leather (when mashed plant parts are beaten in water) and can be used in either soap or shampoos. The soap plant group (amole root, soap plant root, soap root bulb), guaiac leaves, papaya leaves, Quillaia bark, Red campion root and leaves, Atripex root, Sapindus fruit, soap pod fruit, Mojave yucca root, soapwort root, Our Lord's Candle root, wild gourd fruits are good examples (Wikipedia, 2007). The annual yield of the fruits of *B. aegyptiaca* is very high, thus the availability of the oil is assured. But little is known about the physico-chemical properties of the oil, thus its industrial potentialities are not harnessed.

The present work is therefore aimed at investigating the properties of the seed oil of this plant and to screen for the presence of saponin in the oil since saponin was found in the mesocarp of *B. roxburghi*.

MATERIALS AND METHODS

The fruits were purchased at Girei market in Girei local government area of Adamawa Sate in Nigeria. The chemicals used for the

Table 1. Chemical properties of *Balanites a.* seed oil.

Sapon. value (mgKOH/g)	Iodine value (gI₂/100 g)	%Free fatty acid (mgKOH/kg)	Peroxide value (mgEq/kg)	Saponin
168.6	78.7	0.18	6.0	Present.

Table 2. Physical properties of *Balanites* a. seed oil.

Oil yield (%)	Moisture content (%)	Specific gravity	Refractive index	Class of oil
49.9	0.27	0.927	1.4784	Non-drying

analyses are of Analar grade. The oil was extracted using traditional method as described by Abu-Al-Futuh (1989): seed – crushing to obtain kernel – drying – toasting – grinding – addition of water – extraction of oil – drying the oil. The oil obtained was kept for subsequent investigation. The traditional method is adopted because it gives good oil yield and it checks possible contamination by solvents through solvent extraction. This method is also cost effective, hence even the less privileged can have access to the oil for their small-scale soap production and other uses.

Chemical parameters

Saponification value

This was carried out according to American Oil Chemists Society (AOCS) method (1987). The crude oil (2.0 g) was weighed into a 250 ml conical flask and 10% alcoholic KOH (25 ml) was added. A reflux condenser was attached to the flask and refluxed over steam for 1 h with occasional swirling. Phenolphthalein indicator (1 ml) was added at the end of the refluxing time and the solution was titrated with 0.5M HCl. The result is recorded in Table 1.

Iodine value (IV)

This was carried out as described by Nielsen (1994). The crude oil (2.0 g) was measured into a 100 ml conical flask and Dams iodine (5 ml) was added to it, the flask was corked and placed in a dark cupboard for 5 min. 10% KI (5 ml) was added followed by distilled water (20 ml). The solution was titrated with 6.6% sodium thiosulphate in the presence of 1% starch indicator (1 ml) until the blue colour turned colourless. The value is recorded in Table 1.

Free fatty acid (FFA)

This is the percentage by weight of specified fatty acid (for example, % oleic acid). The method used is as described by Harold et al. (1990). A well-mixed portion of the crude oil (2.0 g) was weighed into a conical flask, a neutralised 1:1 mixture of 95% ethanol and diethyl ether (10 ml) was added and mixed thoroughly. The solution was titrated using 0.1M NaOH and phenolphthalein indicator, shaking constantly until a pink colour, which persisted for 15 s, was observed. The value is recorded in Table 1.

Peroxide value (PV)

This parameter was determined using the method described by Pomeranz and Meloan (1987). The crude oil (1 g) was weighed into

a clean dry 100 ml conical flask; to this powdered KI (1 g) was added followed by a mixture of 2:1 glacial acetic acid and chloroform (20 ml). The flask was placed on a steam bath for 30 s and the content was quickly poured into another flask containing 5% KI solution (20 ml). The solution was then titrated using 0.02M sodium thiosulphate solution and starch indicator. The value is recorded in Table 1.

Iodine number-saponification factor (INS)

This factor is expressed as the difference between the iodine number and saponification value of an oil or fat. It is used to predict the quality of soap obtain from oils or fats. The factor ranges from 15 - 250 for soap making oils and fats. The oils with high-unsaturated fatty acids have low factors while those with low molecular weight saturated acids have high factors. The lathering solubility properties of liquid soap are found to be dependent on the INS factor for the oil or oil blend. Oils and fats with an INS of 130 - 160 are individually unsuitable for liquid soap making on the account of lathering.
INS = saponification value – iodine value (NASCO, 1994). Result is recorded in Table 3.

Physical parameters

Moisture content (M.C.)

The moisture content of oil is expressed as the percentage weight loss when the oil is dried to a constant weight at 110°C. A dry crucible was weighed and the dried oil (5 g) was poured into it. The crucible and content were dried in an oven at 110°C and cooled in a dessicator and weighed. This process was repeated until constant weight was attained. The result is recorded in Table 2.

Refractive index (RI)

The refractive index of oil is a function of molecular structure and impurity. RI provides a quick and easy method to identify oil and determine its purity (Bailey, 1951; Apple White and Bailey, 1985). Abbey refractometer was used and the refractive index determined as explained by Rossel (1971) and the result is recorded in Table 2.

Qualitative determination of saponin

This was done as described by Usher (1984). 1 g of the oil was weighed into a conical flask and 10 ml of distilled water was added and boiled for 10 min. The solution was filtered and 2.5 ml of the

Table 3. Chemical/physical properties of the soap and shampoo.

Property	Liquid soap	Shampoo
Appearance.	Creamy	Creamy
Colour	Pale yellow	Light yellow
Solubility	Very soluble	Very soluble
INS	69.97	69.97
Odour	Pleasant	Pleasant
pH	8.92	9.0
% Free alkali	0.45	0.60

Table 4. Comparative study of the liquid soap and similar products sold in the market.

Property	B. aegyptiaca seed oil liquid soap.	Morning fresh	Ultra liquid soap
Appearance	Liquid	Liquid	Liquid
Colour	Light yellow	Green	Light blue
Odour	Pleasant	Pleasant	Pleasant
Solubility in water	Soluble	Soluble	Soluble
% Free alkali	0.45	0.40	0.50

filtrate was added to 1 ml-distilled water in a test tube. The tube was shaken vigorously for 30 s; a honeycomb - like froth was formed which persisted for 30 min.

Formulation of liquid soap

The semi-boiled method was adopted for the formulation as described by Moore (1970). The crude oil (30.0 ml) was poured into a 500 ml round bottom flask to which a solution of KOH (8.2 g) in distilled water (54.5 ml) was added. Ethanol (3.2 ml) was added to the flask and was swirled carefully. The flask was connected to a reflux condenser and refluxed for 40 min over water bath at 85°C. Perfume (2.5 ml) was added at the end of the saponification.

Formulation of shampoo

The shampoo was prepared using the semi-boiled process as in the case of the soap as follows. The crude oil (30.0 ml) was poured into a 500 ml round flask and KOH (7.0 g) dissolved in distilled water (63 ml) was added to the flask. Formaldehyde (0.1 ml) was added and the flask was connected to a reflux condenser and refluxed for 40 min over a steam bath at 85°C. Perfume (2.5 ml) was added at the end of the saponification.

Characterisation of the cleansing agents

The cleansing agents were characterised based on the following parameters

pH

The liquid soap was poured into a clean dry beaker and a standardised pH meter was used to determine its pH as explained

by Donkor (1986). The same procedure was adopted for the shampoo. The results are as presented in Table 3.

Alkalinity

This was determined as described by NASCO (1994). The cleansing agents were separately titrated against 0.1 M standardised HCl acid using methyl orange indicator. The result was expressed as percentage free alkali.

$$\%Free\ alkali\ =\ \frac{V \times M \times 4.7}{SW}$$

Where V = volume of acid, M = molarity of acid, SW = sample weight, 4.7 = conversion factor. Results are shown in Table 3.

RESULTS AND DISCUSSION

Tables 1 - 6 present the chemical and physical properties of the crude oil and the formulated liquid soap and shampoo and similar products sold in the market. The chemical properties of B. aegyptiaca seed oil are shown in Table 1. The saponification value of the oil is 168.67 mgKOH/g which is comparable to the values of certain vegetable oils like; sesame, neem, groundnut, palm kernel, castor oils, etc (Table 6). The iodine value (IV) is relatively low, thus the oil is non-drying; a property that make the oil good raw material for the formulation of liquid soap. The peroxide value is also very low, indicating that the oil would be stable (to a large extent), to oxidative degradation. Rancidity begins to be

Table 5. Comparative study of the shampoo and similar products sold in the market.

Property	*B. aegyptiaca* seed oil	Dop	Petals
Appearance	Creamy	Creamy	Creamy
Odour	Pleasant	Pleasant	Pleasant
Solubility in water	Soluble	Soluble	Soluble
pH	9.0	9.2	8.9
% Free alkali	0.60	0.40	0.60
Colour	Light yellow	Yellow	White

Table 6. Analytical properties of some crude vegetable oils and fats used for the manufacture of liquid soap and shampoo.

S/N	Oils/Fats	Refractive index	Saponification value	Iodine value
1	Sesame seed	1.465 – 1.469	167 – 195	104 – 120
2	Neem	1.465	194.5	71
3	Groundnut	1.460 – 1.465	187 – 196	80 – 106
4	Palm kernel	1.460 – 1.472	230 – 254	14.5 – 19
5	Castor bean	1.466 – 1.473	176 – 187	81 – 91
6	Mustard seed	1.461 – 1.469	170 – 184	92 – 125
7	Bone tallow	1.456 – 1.457	189 – 200	31 – 38
8	Coconut	1.448 – 1.450	248 – 265	6.11
9	Olive	1.467 – 1.471	184 – 196	–
10	Soya bean	1.467 – 1.470	188 – 195	120 – 143
11	Palm fruit	1.449 – 1.455	190 – 209	50 – 55
12	Babassis kernel	1.448 – 1.457	245 – 256	10 – 18
13	Balanites kernel	1.4784	168.6	78.7

Source: Rossel, 1971.

noticeable when the peroxide value reaches 20 - 40 meq/kg (Charles and Guy, 1991). The low %FFA reduces the tendency of the oil to undergo hydrolytic activities. In most oils, the level of free fatty acid which causes deterioration is noticed when the %FFA calculated as oleic acid falls within the range of 0.5 - 1.5% (Rossel, 1971). Saponin was found to be present suggesting that the oil can produce lathering products like soap and hair shampoos. The physical properties of the oil are shown in Table 2.

The oil yield was found to be 49.9% indicating that the oil content is high, a factor that is favourable for industrial application of the oil. The moisture content and specific gravity of the oil are very low, therefore its stability is guaranteed. The oil has appealing appearance-golden yellow colour, and it remained liquid at room temperature, thus adding to the good qualities required of industrial oil raw material. The refractive index of the oil also has fallen within the same range with other vegetable oils used to formulate liquid soap (Table 6).

The chemical and physical properties of the crude cleansing agents are as presented in Table 3. The pH of the soap and shampoo are 8.92 and 9.0 respectively, these have fallen within tolerable pH range (6.5 - 9.4) (Poucher, 1984). The percentage free alkali is also very low for the two products. The INS value is very low as was observed, thus the high solubility of the cleansing agents in water and good lathering property. The products are light yellow and they are found to be creamy, this also suggests that the oil can be used for solid soap production.

Tables 4 and 5 present the results of a comparative study of the formulated products and similar products sold in Nigerian markets. As is seen in these tables, the products compare favourably well with products like Dop and Petals shampoo, and morning fresh and ultra liquid soaps sold in Jemita market in Adamawa State of Nigeria in terms of nature, colour, odour, solubility and %free alkali.

Similarly, the properties of some crude vegetable oils used for the formulation of liquid soaps and shampoos are presented in Table 6. From the properties presented it can be seen that *B. aegyptiaca* seed oil competes favourably with the oils already used to formulate liquid cleansing agents.

Conclusion

The seed of this plant has high oil content as was revealed by the % yield (almost 50%). Thus it can be a good source of raw material for many oil based products (soap, shampoo, bio-diesel, lubricants, etc). The oil is non-drying; therefore it may not be a good raw material for paints and related products. The presence of saponin suspected at the beginning of the research was confirmed and its presence has added to the good lathering property of the formulated liquid soap.

ACKNOWLEDGEMENT

The authors are grateful to their respective families and friends.

REFERENCES

Abu-Al-Futuh (1989). Study on the processing of *Balanites* aegyptiaca fruits for food, drug and feed industry. Chapman and Hall, London. pp. 272-278.

AOCS Official method (1987). Sampling and analysis of commercial fats and oils. *JAOCS* Chicago pp.801-805.

Apple-White TH, Bailing AE (1985). Bailey Industrial Oil and Fats Products. John Wiley Inter Sci. Publ. pp.340-349.

Bailey AE (1951). Industrial Oil and Fats Products, Science Publishers incorporated, pp. 220-235.

Brown GO (1979). *Balanites aegytiaca*. Draft final report No. F/79/42, UNIDO Vienna.

Charles A, Guy L (1991). Food Biochemistry. Llis Howard Ltd., Market Cross-House, Cooper Street Chichester, West Sussex Po/9/EB, England, pp. 89-92.

Donkor P (1986). Small Scale Soap Making. University of Science and Technology Kumasi, China, IIT Publication pp. 56-70.

Harold E, Ronald JK, Sawyer R (1990). Pearson's Chemical Analysis of Food (18th ed). Longman Sci. Technol. pp. 513-514.

Moore E (1970). Detergents: vol. 1 Information Division Unilever Ltd. London, pp. 10-15.

Nasco (1994). Analytical Method for Raw Material and Finished Products, Soap and   Detergents- a review, pp. 22-30.

Nielsen SS (1994). Introduction to Chemical Analysis of Food. Jones and Bartlet Publishers, Boston, pp. 181-186.

Okoye WE, Okobi AO, Okonkwo EN (1999). Extraction and Characterization of Oils from Thine Lesser-known Nigeria Oil Seeds. In Processing of 23rd Annual NIFST Conference, 25th –29th October, pp. 231-233.

Pomeranz Y, Medean CE (1987). Food Analysis theory and Practice. (2nd Ed), Reinhold, New York, pp. 431-440.

Poucher WA (1984). Modern Perfumes, Cosmetics and Soap. Revised by GM, Howard. Chapman and Hall Ltd. 3:218-280.

Rossel JB (1971). Vegetable Oils and Fats. Blackie and Sons Ltd., Glasgow. pp. 263-270.

Sulaiman AE, Jackson JK (1959). The Tree *Balanites aegytiaca*. Sudan Sihawa No.9, vol. 1 Leaflet No. 6.

Usher G (1984). A Dictionary of Plant Used by Man. CBS Publishers and Distributors, New Delhi. pp. 74-80.

Wikipedia (2007). Soap - From Wikipedia, the free encyclopaedia (Redirected from Liquid Soap). http://en.wikipedia.org/wiki/liquid_soap.

21

Primordial radionuclides in potable water from former tin-mining areas with elevated activity

O. I. Adekoya[1,2]

[1]Physics Department, University of Ibadan, Oyo, Nigeria.
[2]Physical Science Department, Yaba College of Technology, Lagos, Nigeria.

The activity concentrations of the primordial radionuclides in potable water from 2 former mining areas (Bisichi and Bukuru) in Jos, Plateau state in Nigeria have been studied. The activities were determined by a non-destructive analysis using a computerized gamma ray spectrometry system with high purity germanium (HPGe). The results show the average activity concentrations for ^{226}Ra, ^{232}Th and ^{40}K for Bukuru and Bisichi to be respectively 1.20 ±0.02, 1.93 ± 0.01, 4.75 ± 0.14 and 2.03 ± 0.14, 2.20 ± 0.13 and 3.26 ± 0.06 Bq/l. The corresponding annual effective doses for both locations are respectively 0.59 and 0.80 mSv/year which are much higher than the reference level of a dose of 0.1mSv/year from the intake of drinking water.

Key words: Activity concentration, radionuclides, drinking water.

INTRODUCTION

Environmental radiation originates from a number of naturally occurring and man-made sources. The United Nations Scientific Committee on the Effects of Atomic Radiation (UNSCEAR) has estimated that exposure to natural sources contributes more than 98% of the radiation dose to the population (excluding medical exposure) (UNSCEAR, 1998)

The global average human exposure from natural sources is 2.4 mSv/year (UNSCEAR 1993). There are however, large local variations in this exposure depending on a number of factors, such as height above sea level, the amount and type of radionuclides in the soil, and the amount taken into the body in air, food and water.

Research reports on environmental radioactivity studies in the Jos Plateau have indicated high gamma radiation dose rates several orders of magnitude higher than world average value (Oresegun and Babalola, 1990, 1993; Farai and Jibiri, 2000). Majority of these reports attributed these high levels primarily on the influence of tin and its mining activities in the area (Farai and Jibiri, 2000; Jibiri et al., 2007a, b; Ademola, 2008).

The negative impact of tin mining activities such as occurred in Bukuru and Bisichi areas in Jos, Plateau, Nigeria, on the environment is mainly due to the excavation of large amounts of sand and the eventual accumulation of a large volume of tailings (Banat et al., 2005; Remon et al., 2005; Akinlua et al., 2006; Birkefeld et al., 2006; Nyarko et al., 2006), which significantly alter the natural constituents of radionuclides in the soil and

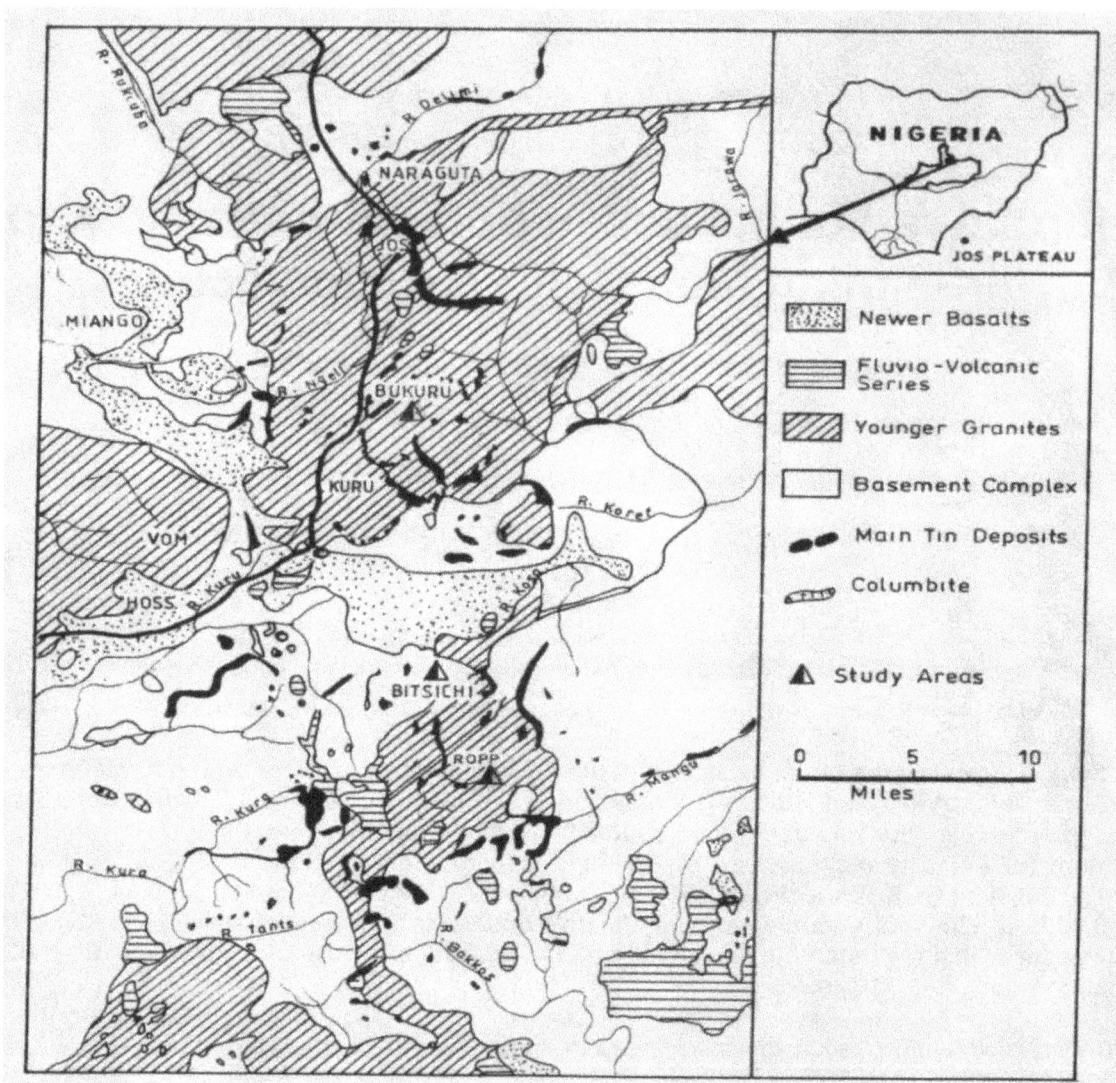

Figure 1. The map of Jos Plateau showing the study locations (Jibiri et al., 2011).

thus affect the terrestrial ecosystem.

Indiscriminate and improper deposition of tailings, especially on steep slopes, increases their mobility and hence the risk of being transported to large inhabited areas (Henriques and Fernandes, 1991). Due to leaching and re-suspension processes, ^{238}U and ^{232}Th from abandoned dumping sites find their way in surface and ground water (Ragnarsdottir and Charlet, 2000). Consequently, this makes mine tailings a source of pollution to the ground and surface water, and to the soil in their vicinities (Hector et al., 2006).

The process of leaching as well as washing away of tailings due to erosion activities into surface and ground water is what goes on at Bukuru and Bisichi areas of Jos, Plateau in Nigeria. The tailings may accumulate to extents that could be detrimental to human lives, where this water is either drunk directly or used in processing foods.

The present research focuses on the assessment of radioactivity in potable water from dams and wells in Bukuru and Bisichi areas of Jos, Plateau, as well as the effective doses to the dwellers in these areas.

MATERIALS AND METHODS

Sample collection and preparation

Drinking water samples were collected from dams and wells from two former tin-mining locations –Bukuru and Bisichi, in Jos, Plateau state, Nigeria. Other water bodies in these areas were discarded once it was established that they were not sources of drinking water. The map of the locations in question is shown in Figure 1.

The bottles were filled to the brim without any head space to prevent trapping of radon gas. For activity concentration measurement, the water samples were also transported to the laboratory and prepared into 1 L Marinelli beakers. The samples were filtered prior to preparation and measurements. The beakers

Table 1. Activity concentration of water samples.

S/No	Description	^{226}Ra(Bq/l)	^{232}Th(Bq/l)	^{40}K(Bq/l)
1	BS-Dam	1.66 ± 0.23	2.00 ± 0.30	0.25 ± 0.01
2	BS-Well water 1	2.11 ± 0.08	3.14 ± 0.05	9.33 ± 0.16
3	BS-Well water 2	2.33 ± 0.10	1.46 ± 0.04	0.20 ± 0.01
4	BK Dam1	<MDL	0.17 ± 0.01	0.16 ± 0.01
5	BK Dam 2	1.62 ± 0.04	3.08 ± 0.02	5.38 ± 0.22
6	BK Dam 3	3.17 ± 0.02	2.77 ± 0.01	13.24 ± 0.32
7	BK Dam 4	<MDL	1.70 ± 0.01	0.20 ± 0.01

BS, Bitsichi; BK, Bukuru, MDL, minimum detection limit.

were thick enough to prevent the permeation of radon. The beakers were closed by screw caps and plastic tape was wrapped over the caps and then stored for measurement. This step was necessary to ensure that radon gas is confined within the volume and that the daughters will also remain in the sample. The samples were sealed for thirty days in order to allow for Radon and its short-lived progenies to reach secular radioactive equilibrium prior to gamma spectroscopy.

Sample measurement and analysis of spectra

All measurements were carried out at the Ghana Atomic Energy Commission, Accra. The activity concentrations of the water samples were determined by a non-destructive analysis using a computerized gamma ray spectrometry system with high purity germanium (HPGe). The relative efficiency of the detector system was 25%, and resolution of 1.8 keV at 1.33 MeV of ^{60}Co. The gamma spectrometer is coupled to conventional electronics connected to a multichannel analyzer card (MCA) installed in a desk top computer. A software program called MAESTRO- 32 was used to accumulate and analyze the data manually using spread sheet (Microsoft Excel) to calculate the natural radioactivity concentrations in the samples. The detector is located inside a cylindrical lead shield of 5 cm thickness with internal diameter of 24 cm and height of 60 cm. The lead shield is lined with various layers of copper, cadmium and Plexiglas, each 3 mm thick.

A counting time of 36,000 s (10 h) was used to acquire spectral data for each sample. The activity concentrations of the uranium-series were determined using γ-ray emissions of ^{214}Pb at 351.9 keV (35.8%) and ^{214}Bi at 609.3 keV (44.8%) for ^{226}Ra, and for the ^{232}Th-series, the emissions of ^{228}Ac at 911 keV (26.6%), ^{212}Pb at 238.6 keV (43.3%) and ^{208}Tl at 583 keV (30.1%) were used. The ^{40}K activity concentration was determined directly from its emission line at 1460.8 keV (10.7%).

Calibration of gamma spectrometry system

Prior to the measurements, the detector and measuring assembly were calibrated for energy and efficiency to enable both qualitative and quantitative analysis of the samples to be performed. The energy and efficiency calibrations were performed using mixed radionuclide calibration standard homogenously distributed in the form of solid water, serial number NW 146 with approximate volume 1000 ml and density 1.0 g cm^{-3} in a 1.0 L Marinelli beaker. The standard was supplied by DeutscherKalibrierdienst (DKD-3), QSA Global GmBH, Germany and contains radionuclides with known energies (^{241}Am (59.54 keV), ^{109}Cd (88.03 keV), ^{57}Co (122.06 keV), ^{139}Ce (165.86 keV), ^{203}Hg (279.20 keV), ^{113}Sn (391.69 keV), ^{85}Sr

(514.01 keV), ^{137}Cs (661.66 keV), ^{60}Co (1173.2 keV and 1332.5 keV) and ^{88}Y (898.04 keV and 1836.1 keV) and activities in a 1000 ml Marinelli beaker was used.

Calculation of activity concentration

The specific activity concentrations (A$_{sp}$) of ^{226}Ra, ^{232}Th, ^{40}K in Bq l^{-1} for the water were determined using the following expression (Beck et al., 1972):

$$A_{sp} = \frac{N_{sam}}{P_E . \varepsilon . T_c . M} \qquad (1)$$

Where; N_{sam} = net counts of the radionuclide in the sample; P_E = gamma ray emission probability (gamma yield); ε = total counting efficiency of the detector system; T_c = sample counting time; M = mass of sample (kg) or volume (L)

Minimum detectable activity

The minimum detectable activity (MDA) of the γ-ray measurements were calculated according to the formula:

$$MDA = \frac{\sigma \sqrt{B}}{\varepsilon P T W} \text{ (Bqkg}^{-1}) \qquad (2)$$

Where σ is the statistical coverage factor equal to 1.645 confidence level 95%, B is the background counts for the region of interest of a certain radionuclide, T is the counting time in seconds, P is the gamma yield for any particular element, W is the weight of the empty Marinelli beaker and ε is the efficiency of the detector.

The minimum detectable activity (MDA) derived from background measurements was approximately 0.11 Bq kg^{-1} for ^{226}Ra, 0.10 Bq kg^{-1} for ^{232}Th and 0.15 Bq kg^{-1} for ^{40}K. Concentration values below these detection limits have been taken in this work to be below the minimum detection limit (MDL).

RESULTS AND DISCUSSION

Primordial radionuclide activity in water

The results for the primordial radionuclide activity in the drinking water samples are shown in Table 1 and Figures 2 to 4.

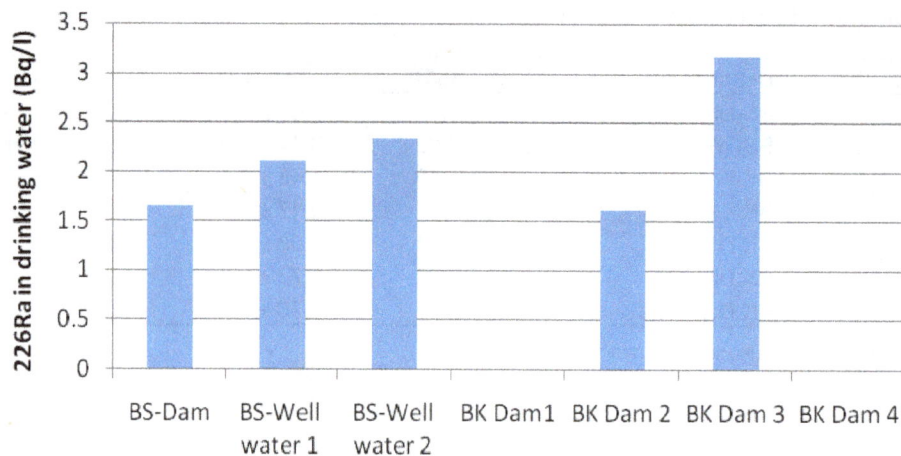

Figure 2. Activity concentration of ^{226}Ra in water samples.

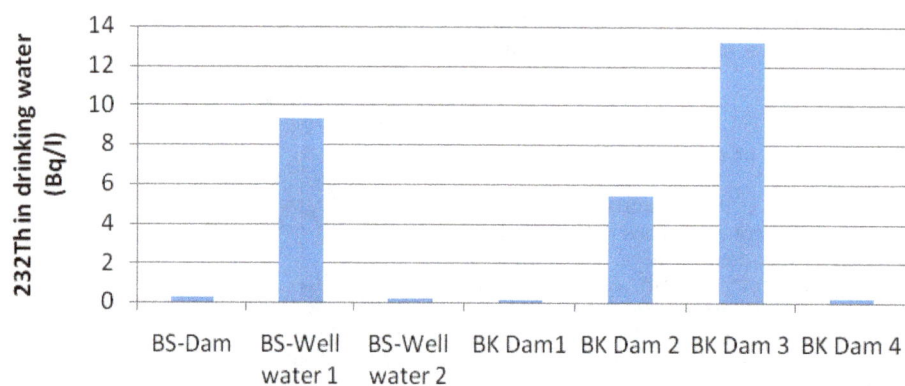

Figure 3. Activity concentration of ^{232}Th in water samples.

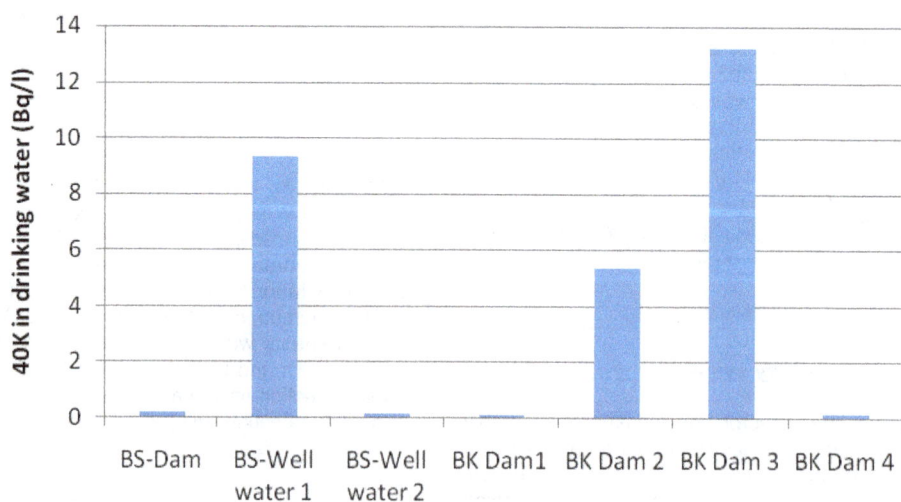

Figure 4. Activity concentration of ^{40}K in water samples.

The highest activities for the radionuclides of interest were noticed in the water sample from the dam 3 in Bukuru. This could only suggest that mining activities was very much pronounced in that vicinity. The water sample

Table 2. Annual effective dose from ingested water in Bukuru and Bisichi.

Location	Average activity concentration (Bq/l)			Annual effective dose (µSv/year)			
	^{226}Ra	^{232}Th	^{40}K	^{226}Ra	^{232}Th	^{40}K	Total
Bukuru	1.20	1.93	4.75	245.3	324	21.5	590.8
Bitsichi	2.03	2.20	3.26	415	369	14.8	798.8

Table 3. Effective doses from ingested water in different European counties (Vesterbacka, 2005).

Country	Effective dose (mSv)	Radionuclides included in the dose estimation
Finland	0.39	^{222}Rn,^{226}Ra,^{238}U, ^{234}U, ^{210}Po,^{210}Pb
Sweden	0.51	^{222}Rn,^{226}Ra,^{228}Ra,^{210}Po,^{238}U
Ukraine	0.22	^{222}Rn,^{226}Ra,^{238}U
Denmark	0.16	^{222}Rn, ^{226}Ra
Switzerland	0.03	^{238}U,^{226}Ra, ^{228}Ra
Scotland	0.05	^{222}Rn
Austria	0.12	^{226}Ra
Spain	4.2	Long lived radionuclides
Hungary	<0.1	^{222}Rn,^{226}Ra
Greece	<0.05	^{222}Rn
Bukuru (this work)	0.59	^{226}Ra, ^{232}Th ,^{40}K
Bisichi (this work)	0.80	^{226}Ra, ^{232}Th ,^{40}K

from well 1 in Bitsichi showed a similar trend except that Ra-226 activity in the sample was lower compared to that in well water 2. The least activities were observed in water samples from dam 1 in Bukuru, where Ra-226 was below minimum detection limit. The average activity concentrations for ^{226}Ra, ^{232}Th and ^{40}K for Bukuru and Bisichi are respectively 1.20 ±0.02, 1.93 ± 0.01, 4.75 ± 0.14Bq/l and 2.03 ± 0.14, 2.20 ± 0.13, 3.26 ±0.06 Bq/l.

It follows from the results that the activities of ^{226}Ra and ^{232}Th on the average are higher in Bitsichi than in Bukuru, unlike ^{40}K which shows the reverse. Furthermore, the concentration of ^{226}Ra on the average is more in well water compared to the dams. The reason for the latter is due to the fact that ground water (well water) flows through fractured rock carrying radioactive materials and other elements from the solid to the liquid phase.

The annual effective dose from radionuclide in drinking water was computed using the following equation, assuming a daily water intake of 2 L/day (EPA, 2000-2005):

$$Reference\ Concentration\,(\text{Bq/l}) = \frac{Dose\,(Sv/yr)}{730\,(litre/yr)*dose\ conversion\ factor\,(Sv/Bq)} \quad (3)$$

The dose conversion factors of 2.8×10^{-7}(Sv/Bq), 2.3×10^{-7}(Sv/Bq) and 6.2×10^{-9}(Sv/Bq) were respectively used for ^{226}Ra, ^{232}Th and ^{40}K (ICRP, 1996) (Table 2).

A comparison of the results with those for different European countries (Table 3) shows that the values obtained in this work are higher. The results obtained in this work are approximately 0.59 and 0.80 mSv/year for Bukuru and Bitsichi respectively which is much higher than the reference level of a dose of 0.1 mSv/year from the intake of potable water.

Conclusion

The activity concentrations of the primordial radionuclides in potable water from Bitsichi and Bukuru, former tin-mining areas in Jos, Plateau were investigated. The results show that internationally recommended minimum acceptable values were exceeded; the contributory factor being likely as a result of the tin- mining activities that had been carried out in the area in the past. There is need for proper water treatment in the areas in order to reduce health risks due to ingested radionuclides.

Conflict of Interest

The authors have not declared any conflict of interest.

ACKNOWLEDGMENTS

The author acknowledges the help of Pastor Dickson of Deeper Life Bible Church, Jos who assisted with the transportation to the former tin mining areas as well as the collection of samples from the dams in Bitsichi and Bukuru areas of Jos, Plateau state. The efforts of David Okoh and Nicholas Sackitey of the Radiation Protection Institute, Ghana Atomic Energy Commission, Accra in the measurement of the samples is also appreciated.

REFERENCES

Akinlua A, Ajayi TR, Adeleke BB (2006). Preliminary assessment of rare earth element contents of Niger-Delta oils. J. Appl. Sci. 6:11–14. http://dx.doi.org/10.3923/jas.2006.11.14

Banat KM, Howari FM, Al-Hamad AA (2005). Heavy metals in urban soils of central Jordan: should we worry about their environmental risks? Environ. Res. Section A 97:258–273. http://dx.doi.org/10.1016/j.envres.2004.07.002

Beck HL, Decompo J, Gologak J (1972). *In situ* Ge (ii) and NaI(Tl) gamma ray spectrometry, In: Otoo*et al*, 2011, Assessment of Natural Radioactive Materials used along the Coast of Central region of Ghana. Res. J. Environ. Health Sci. 3(3):261-268. http://dx.doi.org/10.2172/4599415

Birkefeld A, Schulin R, Newack B (2006). In situ investigation of dissolution of heavy metal containing mineral particles in an acidic forest soil. Geochimicaet. Cosmo. chimica. Acta. 70:2726–2736.

Farai IP, Jibiri NN (2000). Baseline studies of terrestrial outdoor gamma dose rate levels in Nigeria. Radiation Protect. Dosimetry 88:247-254. http://dx.doi.org/10.1093/oxfordjournals.rpd.a033042

Hector MC, Angel F, Raquel A (2006). Heavy metal accumulation and tolerance in plants from mine tailings of the semiarid Cartagena-La Union mining district (SE Spain). Sci. Total Environ. 366:1–11. http://dx.doi.org/10.1016/j.scitotenv.2005.12.008

Henriques FS, Fernandes JC (1991). Metal uptake and distribution in rush (Juncusconglomeratus L.) plants growing in pyrites mine tailings at Lousal, Portugal. Sci. Total Environ.102:253–260. http://dx.doi.org/10.1016/0048-9697(91)90319-A

ICRP (1996). Age-dependent Doses to the Members of the Public from Intake of Radionuclides Part5, Compilation of Ingestion and Inhalation Coefficients. International Commission of Radiological Protection Publication 72. Pergamon Press, Oxford.

Jibiri NN, Farai IP, Alausa SK (2007a). Estimation of annual effective dose due to natural radioactive elements in ingestions of foodstuffs in tin mining area of Jos-Plateau, Nigeria. J. Environ. Radioact. 94:31-40. http://dx.doi.org/10.1016/j.jenvrad.2006.12.011

Jibiri NN, Farai IP, Alausa SK (2007b). Activity concentrations of 226Ra, 228Th, and 40K, in different food crops from a high background radiation area in Bitsichi, Jos Plateau, Nigeria. Radiation Environ. Biophy. 46:53–59. http://dx.doi.org/10.1007/s00411-006-0085-9

Jibiri NN, Alausa SK, Owofolaju AE, Adeniran AA (2011). Terrestrial gamma dose rates and physical-chemical properties of farm soils from ex- tin mining locations in Jos-Plateau, Nigeria. Afr. J. Environ. Sci. Technol. 5(12):1039-1049. December 2011. http://dx.doi.org/10.5897/AJEST11.245

Nyarko BJB, Adomako D, Serfor-Armah Y, Dampare SB, Adotey D Akaho EHK (2006). Biomonitoring of atmospheric trace element deposition. Radiat. Phys. Chem. 75:954-958. http://dx.doi.org/10.1016/j.radphyschem.2005.08.021

Oresegun MO, Babalola IA (1990). Occupational radiation exposure associated with milling of Th-U rich Sn in Nigeria. Health Phy. 58:213-215.

Oresegun MO, Babalola IA (1993). The environmental gamma radiation level of Jos, Nigeria. Nig. J. Sci. 27:263–268.

Remon E, Bouchardon JL, Cornier B, Guy B, Leclerc JC, Faure O (2005). Soil characteristics, heavy metal availability and vegetation recovery at a formal metallurgical landfill: implication in risk assessment and site restoration. Environ. Pollut. 137:316–323. http://dx.doi.org/10.1016/j.envpol.2005.01.012

Ragnarsdottir KV, Charlet L (2000). Uranium behaviour in natural environments. In: Environmental Mineralogy: Microbial Interactions, Anthropogenic Influences, Contaminate Land and Waste Management. The Mineralogical Society of Great Britain & Ireland.

U.S Environmental Protection Agency (EPA). Dallas, Tx (2000 – 2005) 'Chapter 3: Exposure Scenarion Selection' (http://www.epa.gov/earth1r6/6pd/rcra_c/pd-o/chap3 pdf

UNSCEAR (1993). Sources, Effects and Risks of Ionizing Radiation. United Nations Scientific Committee on the Effects of Atomic Radiation, New York. Report to the General Assembly on the Effects of Atomic Radiation, United Nations.

UNSCEAR (1998). Sources, Effects and Risks of Ionizing Radiation. United Nations Scientific Committee on the Effects of Atomic Radiation, New York. Report to the General Assembly on the Effects of Atomic Radiation, United Nations.

Vesterbacka P (2005). U-238 series radionuclides in Finnish groundwater-based drinking water and effective doses. Academic dissertation.

22

Physicochemical characterization and inhibitive performance evaluation of *Commiphora kestingii* gum exudate in acidic medium

Paul Ocheje Ameh

Physical Chemistry Unit, Department of Chemistry, Nigeria Police Academy, Wudil, P. M. B. 3474, Kano State, Nigeria.

Gas chromatographic-mass spectrometry (GCMS), fourier transformed infra-red spectroscopy (FTIR) and physicochemical analysis of *Commiphora kestingii* gum have been carried out. The corrosion inhibition characteristics of *C. kestingii* (CK) gum on aluminium in sulphuric acid media were also investigated at 303 and 333K using gravimetric and scanning electron microscopic studies (SEM) method. The study revealed that the gum contains significant amount of Sucrose, Octadecanoic acid, Alpha camphorenal, Nerolidolisobutyrate, Diisopropenyl-1-methyl-1-vinyl cyclohexane, Abetic acid, Oleic acid, Verbenol, 2,6-dimethylhepta-1,5-diene, Naphthalene, Limonene, 7-hexadecenal and 10-methyl-8-tetradecen-1-ol acetate. Corrosion inhibition tests suggest that GS gum is a good inhibitor for the corrosion of Al in solution of H_2SO_4. The inhibition efficiency of this inhibitor increased with increasing temperature suggesting chemisorption mechanism. Also, the inhibition efficiency of these extract was found to increase in the presence of CK. Inhibition of Aluminium by *C. kestingii* gum occurred through synergistic adsorption of the various components of the gums hence the formation of multiple adsorption layer is proposed.

Key words: *Commiphora kestingii* gum, corrosion inhibitors, adsorption, aluminium.

INTRODUCTION

Aluminium has wider application ranging from all sorts packaging, vital in powerlines, the building and construction industry and common household objects (Hatch, 1984). The key features that lend aluminium to these uses are low density, ductility, electrical conductivity and strength in alloys. Aluminium has a natural corrosion protection from its oxide layer. However, the following factors may affect the stability of the aluminium oxide and thereby cause corrosion:

i. The oxide is not stable in acidic (pH < 4) or alkaline (pH

> 9) environments (Shimizu et al., 1991),

ii. Aggressive ions (chlorides, fluorides) may attack the oxide locally,

iii. Certain elements (Ga, Tl, In, Sn, Pb) may become incorporated in the oxide and destabilize it (Nisancioglu, 1992).

Several methods have been employed to improve and enhance the metal's life span, with the use of corrosion inhibitor being the most practical (Migahed and Nassar, 2008; Fouda and Ellithy, 2009). Corrosion inhibitors are compounds or chemicals that react with metal surface and corrosion medium to protect the metal against corrosion. Most inhibitors are organic compounds synthesized from cheap raw materials. These non-toxic, benign, inexpensive, renewable and readily available alternative corrosion inhibitors have been found in different parts of plant extracts (Okafor et al., 2010; Ameh et al., 2012a, b; Eddy et al., 2012a, b).

Commiphora kestingii commonly called, Fula-fulfulde (fulani), Mbiji (Igbo), Nupe (Esha), Yoruba (Origbo) and Hausa (árár(r)ábií) belongs to the Burseraceas family. The tree usually approaches height of 2 to 5 m and is found mostly in rocks within the low altitudes dry woodland and bush of the savanna from Togo to Nigeria, and on to Ubangi-Shari. Notably, in most developing nations, the washed bark mixed with salt is applied to snakebites, and the fruit provides remedy for stomach ailments (Meer, 1980). The tree exudes a pale yellow liquid, which soon hardens to form yellowish red or reddish brown tears or masses that are then collected when the bark is incised (Meer, 1980).

The present study is aimed at elucidating the chemical structure of *C. kestingii* gum and investigating the corrosion inhibition and adsorption potentials of *C. kestingii* gum for the corrosion of Al in 0.1 M H_2SO_4 using gravimetric, SEM and fourier transformed infra-red spectroscopy (FTIR) techniques.

Collection of sample

Crude gum of *C. kestingii* (CK) was obtained as dried exudates from the parent trees grown at Samaru, Zaria in Sabon Gari LGA of Kaduna State. The outer bark of the tree was broken using a small axe. The cut was extended upward and downward to a significant depth and the gum formed was collected. The rate of formation of gum was found to depend on weather conditions. Gum formation was favoured by dry and retarded by cold or wet weather. The first set of gum was collected six weeks after tapping, while other samples were collected at an interval of two to three weeks. Gum droplets collected, were 2.0 to 7.5 cm in diameter. They dried and hardened on exposure to atmosphere.

Purification of gum

The crude sample of the gum consisted of a mixture of large and small modules and other impurities. These were hand sorted to remove fragments of bark and other visible impurities and then were spread out in the sun to dry for one to two weeks. The gum was then dissolved in cold distilled water and the solution strained through muslin, and centrifuged to obtain a small quantity of dense gel. The straw coloured supernatant liquor obtained was separated and acidified to a pH of 2 with dilute hydrochloric acid. Ethyl alcohol was added until it

was 80%. The gum precipitated out was removed by centrifugation at a rate of 2000 revolution per minute, washed with alcohol, ether and then dried in a desiccator.

Determination of physiochemical properties

Preliminary tests were performed to confirm the nature of the gum obtained. The test that were conducted are pH and solubility test.

Determination of pH

The pH of 2% w/v of the gum mucilage was determined using a Jenway pH meter (Model; 3505). The pH meter electrode was dipped into a buffer at a room temperature of 28°C after which it was removed, shaken a little to remove droplets of the buffer before immersion into the gum mucilage. The reading on the meter was recorded in triplicate measurements were made.

Determination of solubility in various solvents

The solubility of the gum was determined in cold and hot distilled water, acetone and chloroform. 10 mg sample of the studied gum exudate was added to 10 ml each of the solvents and left overnight. 5 ml of the clear supernatant was taken in small pre-weighed evaporating dishes and heated to dryness over a digital thermostatic water bath (Model. HHS, McDonald Scientific International). The weight of the dried residue with reference to the volume of the solutions was determined using a digital analytical balance (Model. XP-300, Denver instrument, USA) and expressed as the percentage solubility of the gum in the solvents (Carter, 2005).

MATERIALS AND METHODS

Corrosion studies

Aluminum alloy sheet of composition (wt. %, as determined by quantiometric method) Mn (1.28), Pb (0.064), Zn (0.006), Ti (0.029), Cu (0.81) and Si (0.381), Fe (0.57), and Al (96.65%) was used. The sheets were mechanically pressed cut into different coupons, each of dimension, 5 x 4 x 0.11 cm. Each coupon was degreased by washing with ethanol, cleaned with acetone and allowed to dry in the air before preservation in a desiccator. All reagents used for the study were analar grade and double distilled water was used for their preparation.

Gravimetric studies

The clean and dried previously weighed aluminum alloy coupon was completely immersed in 250 ml of the test solution in an open beaker. After every 24 h the corrosion product was withdrawn from the electrolyte, kept for 3 to 4 min in 70% nitric acid, washed thoroughly with distilled water and then dried and weighed. The

Table 1. Physiochemical properties of *Commiphora kestingii* gum.

Parameter	
Colour	Pale yellow
Odour	Odourless
Taste	Bland
pH (28°C)	5.2
Percentage yield (%w/w)	63.5
Solubility (%w/v)	
Cold water	14.00
Hot water	11.00
Acetone	2.00
Chloroform	0.00
Ethanol	2.00

experiment was repeated at 333K. In each case, the difference in weight for a period of 168 h was taken as the total weight loss. From the average weight loss (mean of three replicate analysis) results, the inhibition efficiency (%I) of the inhibitor, the degree of surface coverage (θ) and the corrosion rate of aluminum (CR) were calculated using 3.1 to 3.3, respectively (Ameh et al., 2012a).

$$\%I = \left(1 - \frac{W_1}{W_2}\right) \times 100 \tag{1}$$

$$\theta = \left(1 - \frac{W_1}{W_2}\right) \tag{2}$$

$$CR = \frac{\Delta W}{At} \tag{3}$$

where W_1 and W_2 are the weight losses (g) for aluminum in the presence and absence of the inhibitor, θ is the degree of surface coverage of the inhibitor, $\Delta W = W_2 - W_1$, A is the area of the aluminum coupon (in cm^2), t is the period of immersion (in hours) and ΔW is the weight loss of aluminum after time, t.

Scanning electron microscopic studies (SEM)

The aluminium samples immersed in sulphuric acid and in the inhibitor solution for a period of one day were removed, rinsed with double distilled water, dried and observed under a scanning electron microscope to examine the surface morphology. The surface morphology measurements of carbon steel were examined using JSM-5600 LV scanning electron microscope (SEM) of JEOL, Tokyo, Japan. The sample was mounted on a metal stub and sputtered with gold in order to make the sample conductive, and the images were taken at an accelerating voltage of 10 kV.

Chemical analysis of samples

FTIR analysis

FTIR analyses of the gum and that of the corrosion products (in the absence and presence of gum) were carried out using Scimadzu FTIR-8400S Fourier transform infra-red spectrophotometer. The sample was prepared in KBr and the analysis was done by

scanning the sample through a wave number range of 400 to 4000 cm^{-1}.

Gas chromatography–mass spectrometry (GC-MS) analysis

GC-MS analysis was carried out on a GC Clarus 500 Perkin Elmer system. Interpretation on mass spectrum GC-MS was conducted using the database of National Institute Standard and Technology (NIST) Abuja, having more than 62,000 patterns. The spectrum of the unknown component was compared with the spectrum of the known components stored in the NIST library. The name, molecular weight and structure of the components of the test materials were ascertained. Concentrations of the identified compounds were determined through area and height normalization.

RESULTS AND DISCUSSION

Physiochemical parameters of *CK* gum

Table 1 presents physical (pH, colour, taste and percentage yield) and chemical properties (solubility in ethanol, acetone, chloroform, cold and hot water) of CK gum. The gum is odourless, pale yellow in colour (Figure 1) and has a characteristic bland taste. The gum has a pH of 5.2 indicating that it is mild acidic. It is soluble in both cold and hot water. However, its solubility tends to decrease with increase in temperature indicating that the gum is ionic. It has been found that the solubility of some ionic solutes tends to decrease due to the change of properties and structure of liquid water; the lower dielectric constant results in a less polar solvent. Also, an increase in temperature will increase the motion of dissolved particles by weakening the forces acting between the solute and solvent. This allows the dissolved gum to re-coagulate without dissolution hence the decrease in solubility (Ebbing and Gammon, 2005). CK gum is also found to be slightly soluble in ethanol and acetone. The solubility of the gum in ethanol may be due to the presence of polar and non-polar ends in ethanol,

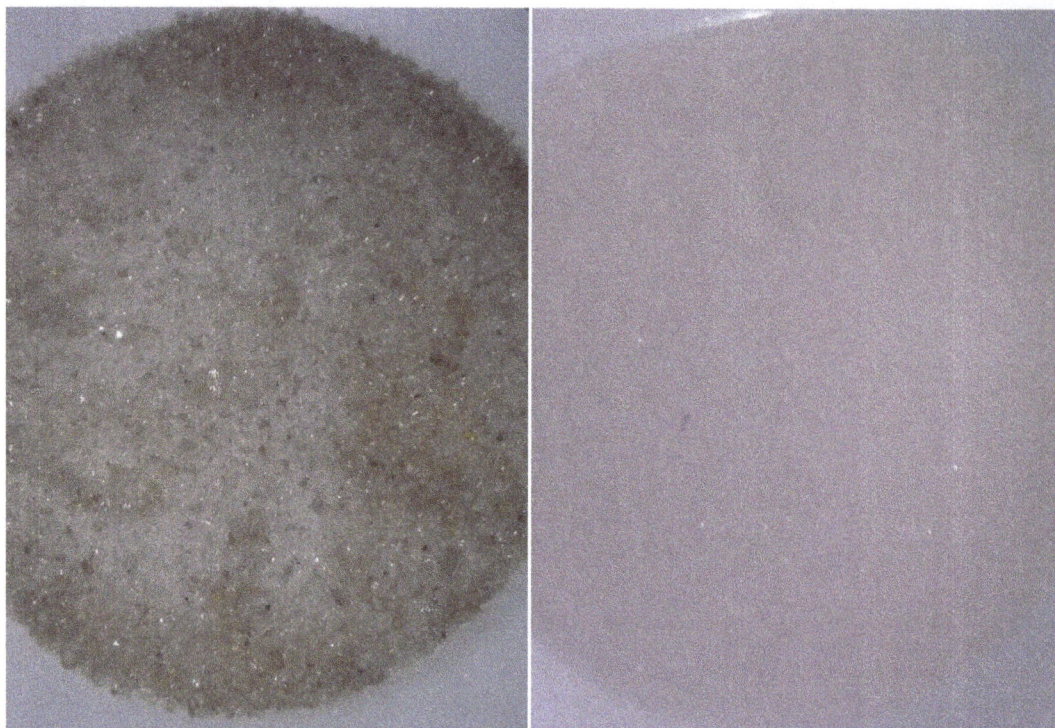

Figure 1. Crushed crude dry and purified sample of *Commiphora kestingii* gum.

which makes it to dissolve some polar and non-polar compounds. On the other hand, aprotic solvents such as acetone tend to have large dipole moment (separation of partial positive and partial negative charges within the same molecule) and solvate positively charged species via their negative dipole. This probably explains why CK gum is soluble in acetone to some extent.

GC-MS study on CK gum

Figure 2 shows the GC-MS spectrum of CK gum. In Table 2, IUPAC names as well as concentrations of compounds identified from various lines in the spectrum are presented while their chemical structures are presented in Figure 3. The gas chromatographic-mass spectrometry (GCMS) spectrum of CK displayed 14 significant peaks. Since area under the chromatogram is proportional to concentration, area normalization was carried out and percentage concentrations of the respective chemical constituents were evaluated. The results obtained shows that the most abundant component of CK gum is sucrose (37.11%), followed by Octadecanoic acid (20.44%). The least abundant constituent was found to be 0.82% alpha camphorenal. Others constituents of the gum included nerolidolisobutyrate (1.72%), diisopropenyl-1-methyl-1-vinyl cyclohexane (13.10%), abetic acid (4.10%), oleic acid (2.14%), verbenol (1%), 2,6-dimethylhepta-1,5-diene

(6.11%), Naphthalene (1.40%), Limonene (2.34%), 7-hexadecenal (1.71%) and 10-methyl-8-tetradecen-1-ol acetate (6.20%). The molar mass and chemical formulas of these compounds are also presented in Table 2. Resolutions of each line in the spectrum of CK gum also revealed the occurrence of several fragmentation peaks for each fraction.

FTIR

Figure 4 presents the FTIR spectrum of CK gum. The spectrum displayed peaks due to C-H bend in phenyl ring (722.37 cm^{-1}), C-H stretch due to alkene at 981.8 cm^{-1}, C-O stretch vibrations due to carboxylic acid, alcohol, ether or esters at 1164.08 and 1239.31 cm^{-1}, C-H scissoring and bending vibrations at 1378.18 and 1465.95 cm^{-1}, C = O stretch due to aldehyde, ketone, carboxylic acid or esters at 1745.64 cm^{-1}, OH stretch due to carboxylic acid at 2853.78 and 2927.08 cm^{-1} and OH stretch due to alcohol or phenol at 3474.88 cm^{-1}.

Corrosion study

Effect of CK gum on the corrosion of Al in 0.1 M H$_2$SO$_4$

Figures 5 and 6 show the variation of weight loss with time for the corrosion of Al in 0.1 M H$_2$SO$_4$ containing

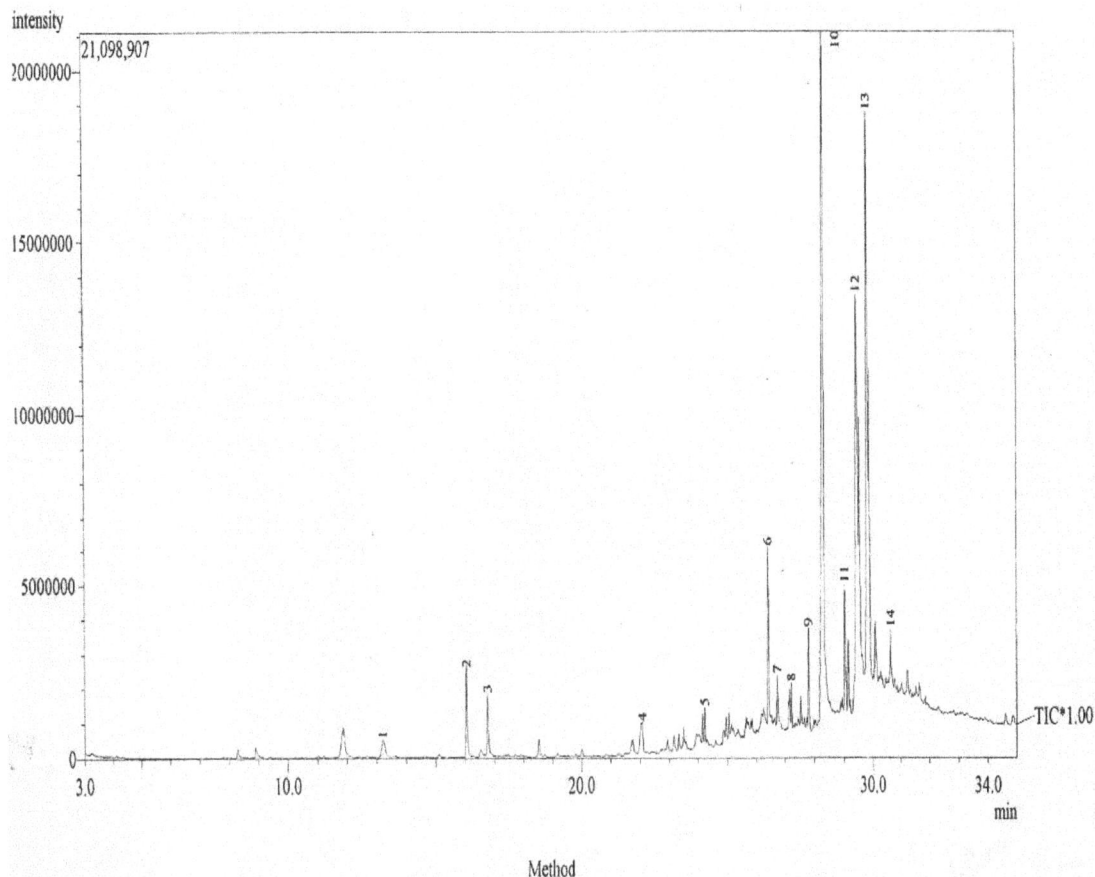

Figure 2. Chemical structures of compounds identified in the GC-MS spectrum of *Commiphora kestingii gum*.

various concentrations of CK gum at 303 and 333K. The figures revealed that the rate of corrosion of Al in solutions of H_2SO_4 increases with the period of contact but decreases with increase in the concentration of CK gum, indicating that CK gum is an adsorption inhibitor for the corrosion of aluminium. It has been found that for an adsorption inhibitor, the corrosion rate decreases with increase in the concentration of the inhibitor (Emregul and Hayvali, 2006).

Calculated values of corrosion rate of Al, inhibition efficiency of CK gum and degree of surface coverage calculated from Equations 1 to 3 are presented in Table 4. The results confirmed that the corrosion rate decreases with increase in temperature, while the inhibition efficiency increases with increase in temperature indicating chemical adsorption. For chemical adsorption mechanism, inhibition efficiency increases with temperature while for physical adsorption mechanism, inhibition efficiency of an inhibitor decreases with temperature (Ebenso et al., 2008). The inhibition efficiency of CK gum was found to range from 65.40 to 83.20% and from 74.90 to 83.22%, indicating that CK gum is a good inhibitor compare to values obtained for other inhibitors (Ameh et al., 2012b).

Kinetic study

Most corrosion reactions have been found to be first order and are consistent with the following equation (Ameh et al., 2012a)

$$-\log (\text{weight loss}) = k_1 t / 2.303 \tag{4}$$

where k_1 is the first order rate constant which is related to the half-life ($t_{1/2}$) through Equation 5:

$$t_{1/2} = 0.693 / k_1 \tag{5}$$

Figures 7 and 8 depict plots for the variation of $-\log$ (weight loss) versus time for the corrosion of Al in solutions of H_2SO_4 containing various concentrations of CK gum at 303 and 333K, respectively. Kinetic parameters deduced from Equations 4 and 5 are presented in Table 5. The results obtained indicated that the corrosion of Al in solutions of H_2SO_4 obeys Equations 4 and 5. The half-lives in the presence of the CK gum are higher than those for the blanks at all temperature and concentrations of CK gum indicating that CK gum has the

Table 2. Characteristics of suggested compounds identified from GC-MS of *Commiphora kestingii* gum.

Line no	C (%)	Compound	MF	MW	RT	Fragmentation peaks
1.	6.20	10-methyl-8-tetradecen-1-ol acetate	$C_{15}H_{28}O_2$	240	8.2	27(15%), 41(15%), 43(15%), 65(10%), 77(50%), 93(100%), 105(2%), 121(2%), 136(10 %)
2	1.81	n-Hexadecanoic acid	$C_{16}H_{32}O_2$	256	10.1	27(15%), 41(20%), 53(10%), 67(100%), 77(35%), 93(100%), 105(10%), 121(20%), 136(10%)
3	1.71	7-hexadecenal	$C_{16}H_{30}O$	238	35.2	41(80%), 55(90%), 71(78%), 85(50%), 98(40%), 121(40%), 135(20%), 141(10%)
4	2.34	Limonene	$C_{10}H_{16}$	136	9.8	27(40%), 39(60%), 53(45%), 68(100%), 79(40%), 93(65%), 107(20%), 121(20%), 136(25 %).
5	1.40	Naphthalene	$C_{20}H_{32}$	272	10.1	27(10%), 41(20%), 67(10%), 81(30%), 105(40%), 121(20%), 136(50%), 143(20%), 157(20%), 171(10%), 185(20%), 213(30%), 241(40%), 259 (50%)
6	6.11	2,6-dimethylhepta-1,5-diene	$C_{10}H_{16}$	136	10.5	27(20%), 41(100%), 53(20%), 69(80%), 77(25%), 93(100%), 107(8%), 121(10%), 136(5 %).
7	1.00	Verbenol	$C_{10}H_{16}O$	152	11.6	27(35%), 41(65%), 43(40%), 59(55%), 79(60%), 94(100%), 109(90%), 119(30%), 137(20%)
8	0.82	alpha-campholenal	$C_{10}H_{16}O$	152	12.0	27(15%), 39(20%), 55(20%), 67(25%), 81(20%), 93(55%), 108(100%), 119(5%), 137(2%), 152(2%)
9	2.14	Oleic acid	$C_{18}H_{34}O_2$	282	12.8	25(35%), 39(60%), 43(40%), 55(50%), 78(60%), 94(100%), 108(100%), 119(30%), 135(20 %)
10	37.11	Sucrose	$C_{12}H_{22}O_{11}$	152	14.3	25(20%), 39(40%), 43(20%), 67(20%), 78(100%), 94(50%), 108(40%), 119(30%), 134(20 %), 151(10%)
11	4.10	Abetic acid	$C_{20}H_{30}O_2$	302	14.8	25(10%), 39(32%), 55(35%), 69(20%), 81(30%), 83(55%), 108(100%), 119(20%), 134(10 %), 152(10%)
12	13.10	diisopropenyl-1-methyl-1vinylcyclohexane	$C_{15}H_{24}$	204	31.8	27(32%), 41(100%), 53(60%), 68(100%), 81(100%), 93(80%), 107(40%), 121(35%), 133(15 %), 147(20%), 161(15%), 189(15%)
13	20.44	Octadecanoic acid	$C_{18}H_{36}O_2$	284	33.2	27(12%), 39(18%), 43(38%), 59(100%), 81(50%), 93(70%), 107(40%), 121(30%), 135(22 %), 147(10%), 161(40%), 189(20%), 204(10%).
14	1.72	nerolidoliisobutyrate	$C_{19}H_{32}O_2$	292	35.3	41(42%), 43 (100%), 69(25%), 71(50%), 93(30%), 107(10%), 121(40%), 127(5 %), 143(2%), 161(2%)

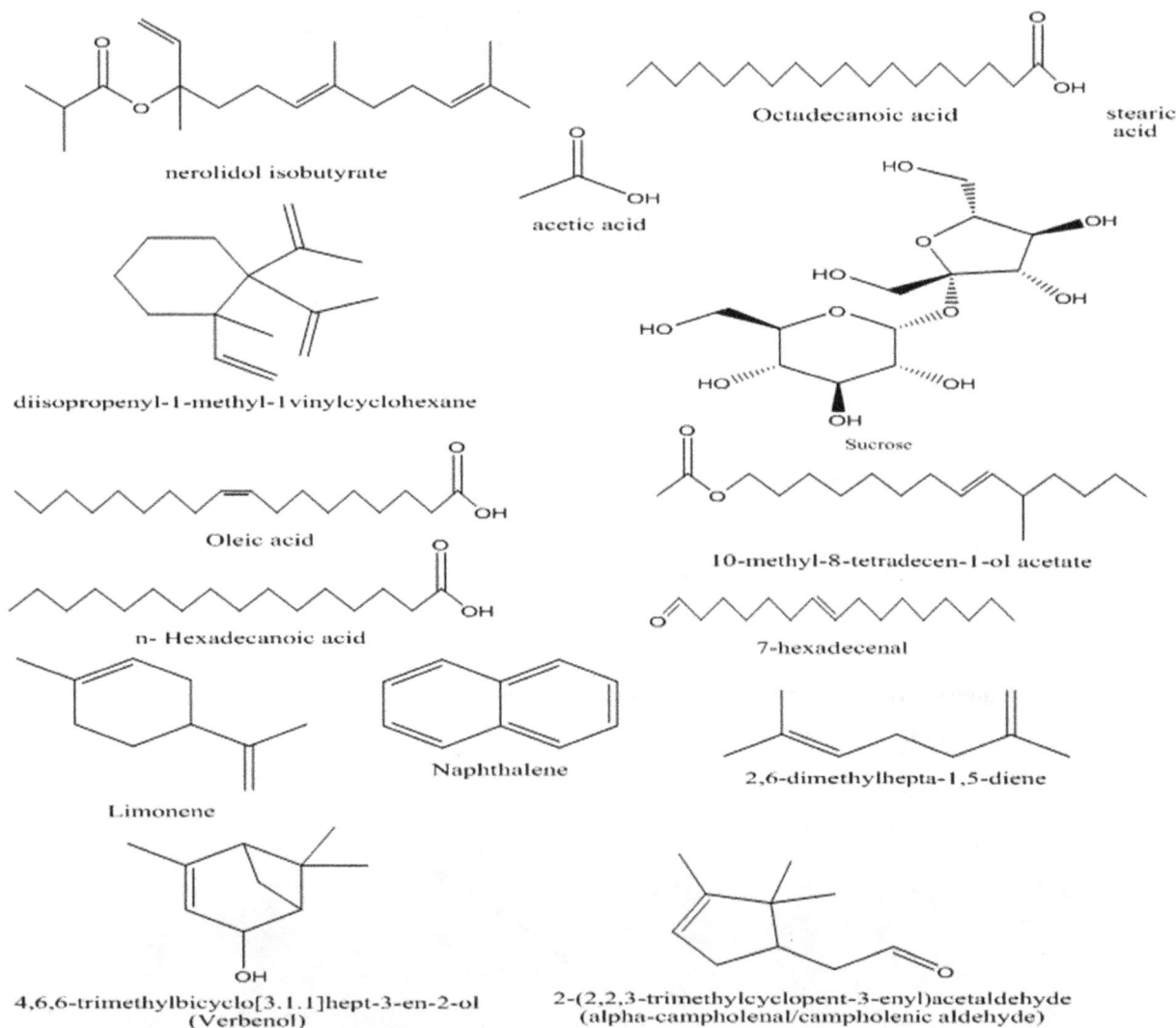

Figure 3. Chemical structures of compounds identified in GC-MS spectrum of CK gum.

tendency of extending the half-life of Al in solution of H_2SO_4, hence it is a good inhibitor.

The effect of temperature on the rate of corrosion of Al in solutions of H_2SO_4 containing various concentrations of CK gum was investigated using the Arrhenius equation, which can be written as follows (Ebenso, 2003a, b):

$$CR = A\exp(-E_a/RT)$$
(6)

where A is the pre-exponential factor, E_a is the activation energy, R is the gas constant and T is the temperature. By taking logarithm of Equations 6 and 7 was obtained:

$$\log(CR) = \log A - E_a/R$$
(7)

Between the temperature range of 303 (T_1) and 333 K (T_2) and the corresponding corrosion rates of CR_1 and

CR_2, Equation 7 can be simplified to Equation 8 as follows:

$$\log\frac{CR_2}{CR_1} = \frac{E_a}{2.303R}\left(\frac{1}{T_1} - \frac{1}{T_2}\right)$$
(8)

In Table 6, calculated values of activation energy are recorded. The activation energies are relatively low when compared to the threshold value (80 kJ/mol) required for the mechanism of chemical adsorption. According to Oguzie (2006), lowered value of E_a in inhibited system compared to the blank as observed in the study (Table 6) is indicative of chemisorption possibly because some of the energy is used up in the chemical reaction. E_a tends to decrease in the presence of an inhibitor by decreasing the available reaction area.

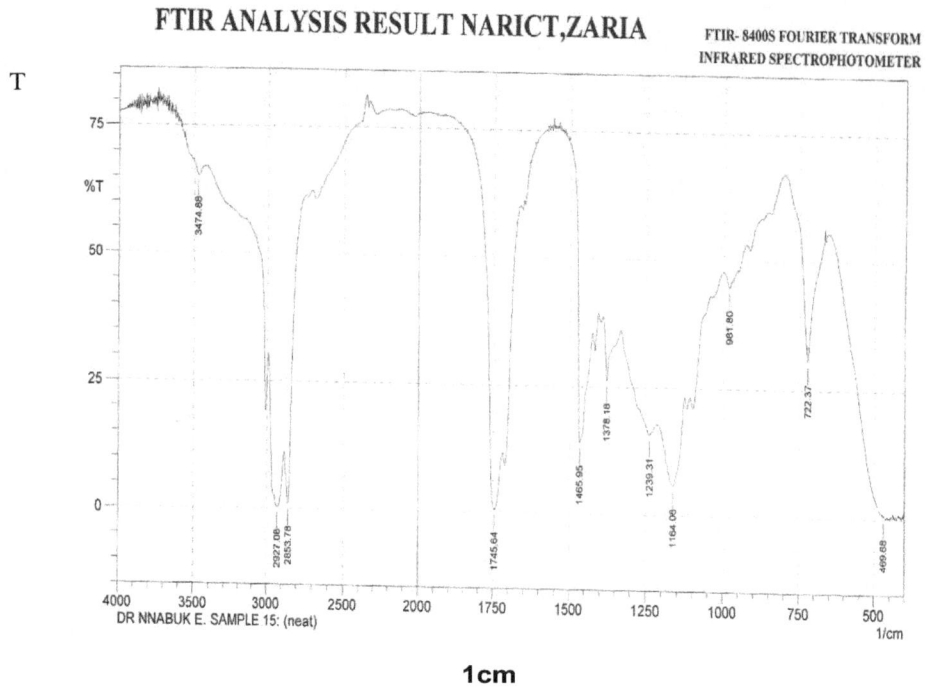

Figure 4. FTIR spectrum of *Commiphora kestingii* gum.

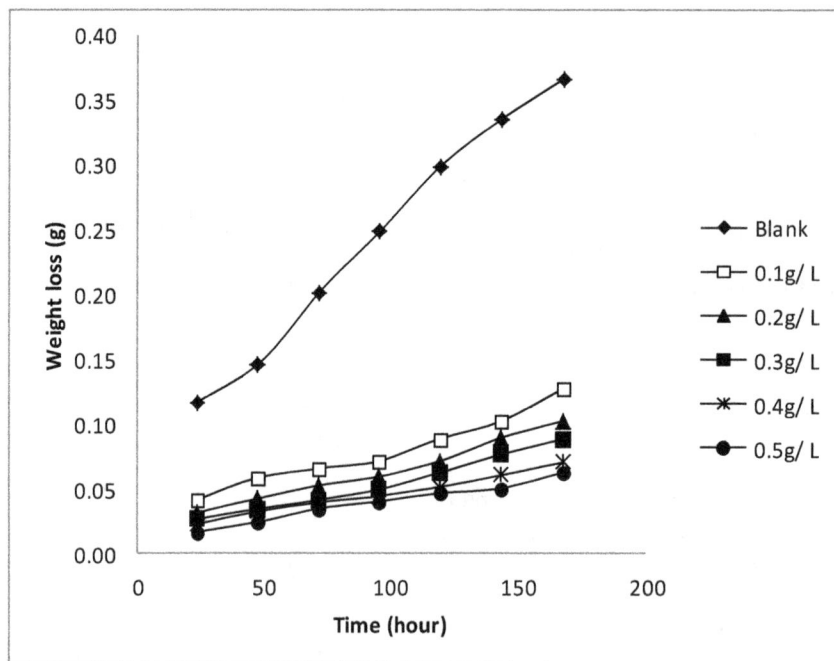

Figure 5. Variation of weight loss with time for the corrosion of Al in HCl containing various concentrations of *CK* gum at 303K.

It is also significant to note that E_a values tend to increase with increase in the concentration of CK gum indicating better adsorption strength with increasing concentration.

Thermodynamic/adsorption study

The heat accompanying the adsorption of CK gum on

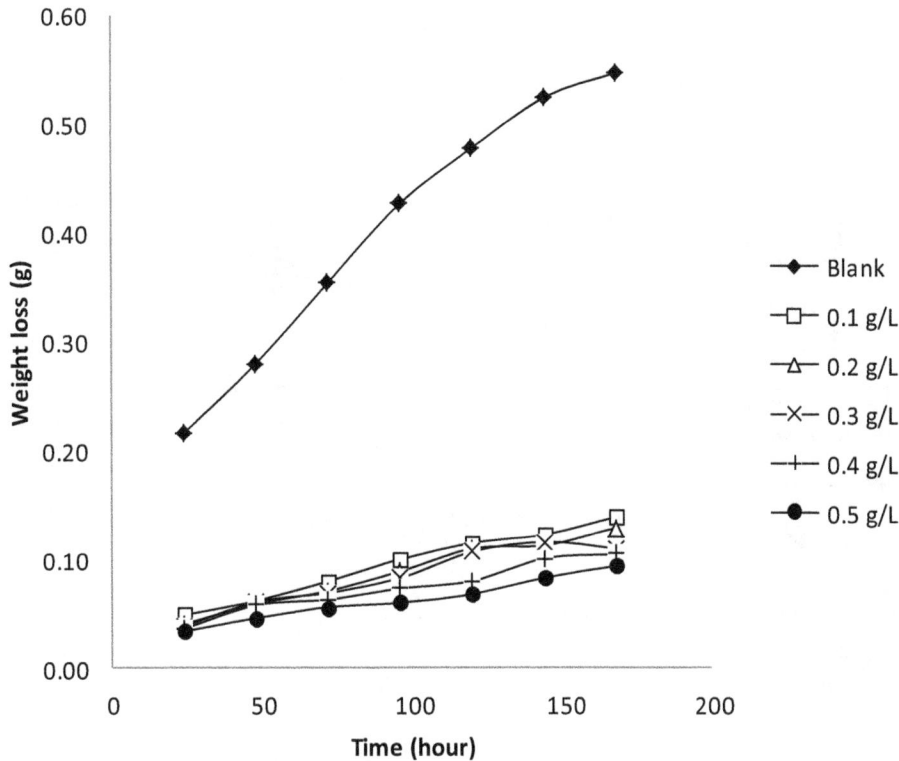

Figure 6. Variation of weight loss with time for the corrosion of Al in HCl containing various concentrations of *CK* gum at 333 K.

Table 3. Peaks and intensity of adsorption of FTIR by *CK* gum.

Peak	Intensity	Area	Assignments/functional groups
469.68	0.619	10.211	
722.37	30.649	39.968	C-H bend due to phenyl ring substitution
981.8	44.661	16.031	C-H stretch due to alkene
1164.08	5.805	84.851	C-O stretch due to carboxylic acid, alcohol, ether, and esters
1239.31	15.641	76.575	C-O stretch due to carboxylic acid, alcohol, ether, and esters
1378.18	26.289	18.176	C-H scissoring and bending due to alkane
1465.95	13.963	15.642	C-H scissoring and bending due to alkane
1745.64	0.593	42.869	C=O stretch due to aldehyde, ketones, carboxylic acid and esters
2853.78	0.957	86.542	OH stretch due to carboxylic acid
2927.08	0.321	47.798	OH stretch due to carboxylic acid
3474.88	65.27	5.111	OH stretch due to alcohol or phenol

Table 4. Corrosion rates for Al, inhibition efficiency and degree of surface coverage of *CK* gum in 0.1M H_2SO_4.

C (g/L)	CR ($g/cm^3/h$) at 303K	%I (303K)	θ (303K)	CR ($g/cm^3/h$) at 333 K	%I (333K)	θ (333K)
Blank	1.09×10^{-4}	-	-	1.63×10^{-4}	-	-
0.1	3.77×10^{-5}	65.40	0.654	4.10×10^{-5}	74.90	0.749
0.2	303×10^{-5}	72.30	0.723	3.80×10^{-5}	76.70	0.767
0.3	2.61×10^{-5}	76.10	0.761	3.24×10^{-5}	80.20	0.802
0.4	2.08×10^{-5}	81.00	0.810	3.09×10^{-5}	81.10	0.811
0.5	1.84×10^{-5}	83.20	0.832	2.76×10^{-5}	83.22	0.831

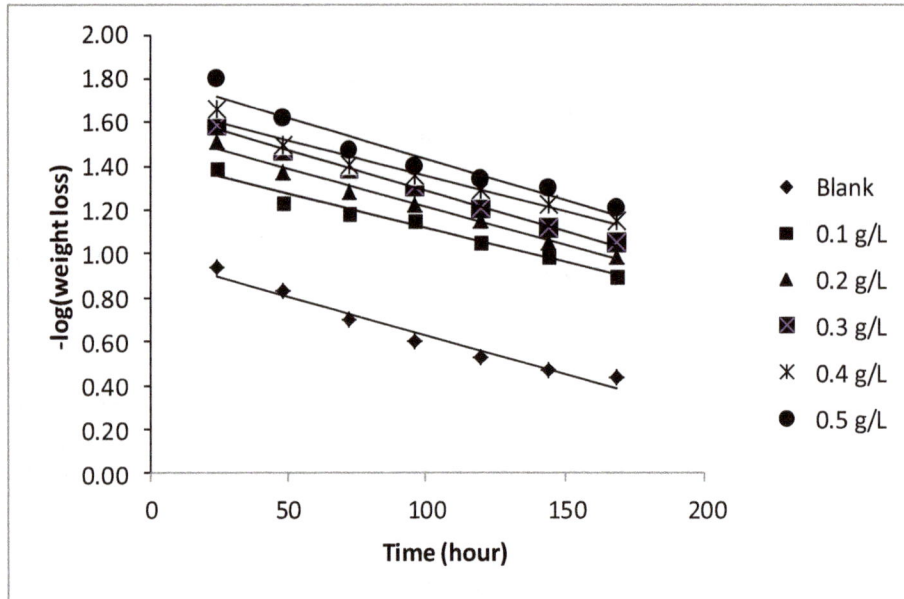

Figure 7. Variation of –log(weight loss) with time for the corrosion of Al in 0.1 M HCl containing various concentrations of CK gum at 303K.

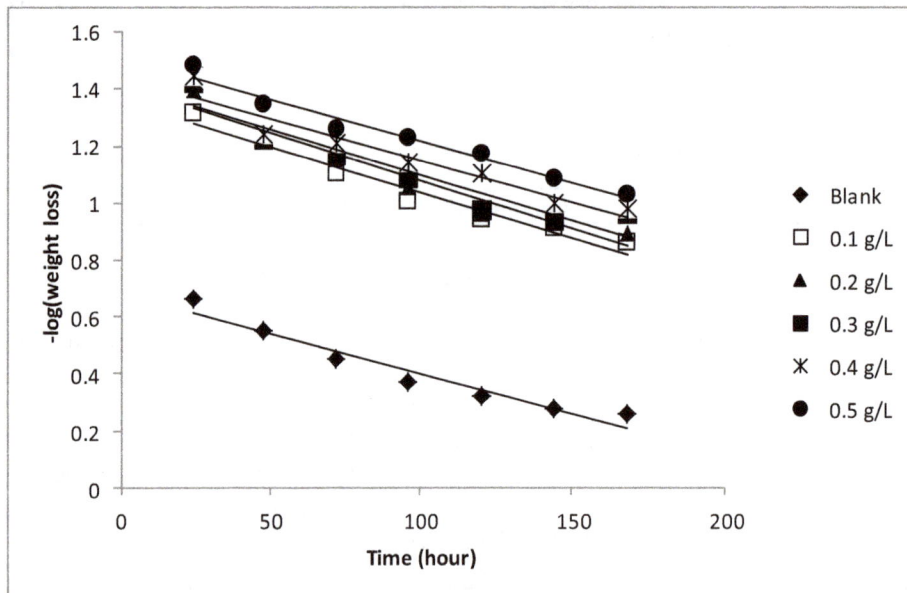

Figure 8. Variation of –log(weight loss) with time for the corrosion of Al in 0.1 M HCl containing various concentrations of CK gum at 333K.

aluminum surface was calculated using the following equation (Eddy et al., 2012b):

$$Q_{ads} = 2.303R\left(\frac{\theta_2}{1-\theta_2} - \frac{\theta_1}{1-\theta_1}\right) \times \left(\frac{T_1 \times T_2}{T_2 - T_1}\right) \quad (9)$$

Where Q_{ads} is the heat of adsorption, R is the universal

gas constant, θ_2 and θ_1 are the degree of surface coverage at the temperatures T_1 (303K) and T_2 (333K), respectively. Calculated values of Q_{ads} are also recorded in Table 6. The results obtained reflect an endothermic process. Adsorption characteristics of a corrosion inhibitor can be simplified, using adsorption isotherms. In order to achieve this, the degrees of surface coverage at

Table 5. Kinetic parameters for the corrosion of Al in 0.1 M H_2SO_4 containing various concentrations of CK gum.

T (K)	C (g/L)	Slope	Intercept	k_1	R^2	$t_{1/2}$ (day)
303 K	Blank	-0.0036	0.986	0.008291	0.9634	1
	0.1	-0.0031	1.4282	0.007139	0.9764	2
	0.2	-0.0035	1.5612	0.008061	0.9883	2
	0.3	-0.0037	1.6571	0.008521	0.996	2
	0.4	-0.0032	1.6797	0.00737	0.9658	2
	0.5	-0.0038	1.8088	0.008751	0.9353	2
333 K	Blank	0.0028	0.6843	0.006448	0.9377	1
	0.1	0.0032	1.3563	0.00737	0.9583	2
	0.2	0.0034	1.4186	0.00783	0.94559	2
	0.3	0.0032	1.4147	0.00737	0.8920	2
	0.4	0.0029	1.4427	0.006679	0.9270	2
	0.5	0.0029	1.5094	0.006679	0.9700	2

Table 6. Activation energy and heat of adsorption of various concentrations of CK gum on Al surface.

C (g/L)	E_a (kJ/mol)	Q_{ads} (kJ/mol)
Blank	11.27	
0.1	2.35	52.83
0.2	6.34	32.93
0.3	6.35	41.85
0.4	11.08	1.34
0.5	11.15	3.46

various concentrations of the inhibitor were used to test for the fitness of Langmuir, Freudlich, Flory-Huggins, El awardy et al., Temkin and Frumkin adsorption isotherms. The tests indicated that Freundlich, Temkin and Flory-Huggins adsorption isotherms are applicable to the adsorption of CK gum on aluminum surface.
Expression of Flory-Huggins isotherm is given by Equation 10:

$$log\left(\frac{\theta}{c}\right) = logb + xlog(1 - \theta) \qquad (10)$$

where x is the number of inhibitor molecules occupying one site (or the number of water molecules replaced by one molecule of the inhibitor). Other parameters are as defined earlier. From Equation 10, a plot of $log(\theta/C)$ versus $log(1-\theta)$ should produce a straight line if the adsorption of the inhibitor follows Flory-Huggins isotherm. Flory-Huggins isotherm for the adsorption of CK gum is presented in Figure 9. Adsorption parameters deduced from the plots are presented in Table 7. From the results, it can be seen that R^2 values are very close to unity confirming the application of the Flory-Huggins adsorption parameters.

Temkin adsorption isotherm operates on the assumptions that relate the concentration of the inhibitor to the degree of surface coverage according to Equation 11:

$$exp(-2a\theta) = bC \qquad (11)$$

where a is molecular interaction parameter; θ is degree of surface coverage, C is inhibitor concentration and b is equilibrium constant of the adsorption process. From the logarithm of both sides of Equation 11 and 12 is obtained

$$\theta = -\frac{lnK}{2a} - \frac{lnC}{2a} \qquad (12)$$

From Equation 12, a plot of θ versus logC should give a straight line with slope equal to 2.303 x a/2 and intercept equal to 2.303 a logK/2 provided assumptions of Temkin isotherm are valid (Emregul et al., 2003, Emregul and Hayvali, 2006). Figure 10 presents Temkin isotherm for the adsorption of CK gum on aluminum surface while Temkin adsorption parameters are recorded in Table 7. The results indicated that the interaction parameters, 'a' are positive indicating the attractive behaviour of the inhibitor. The parameter tend to increase with increasing

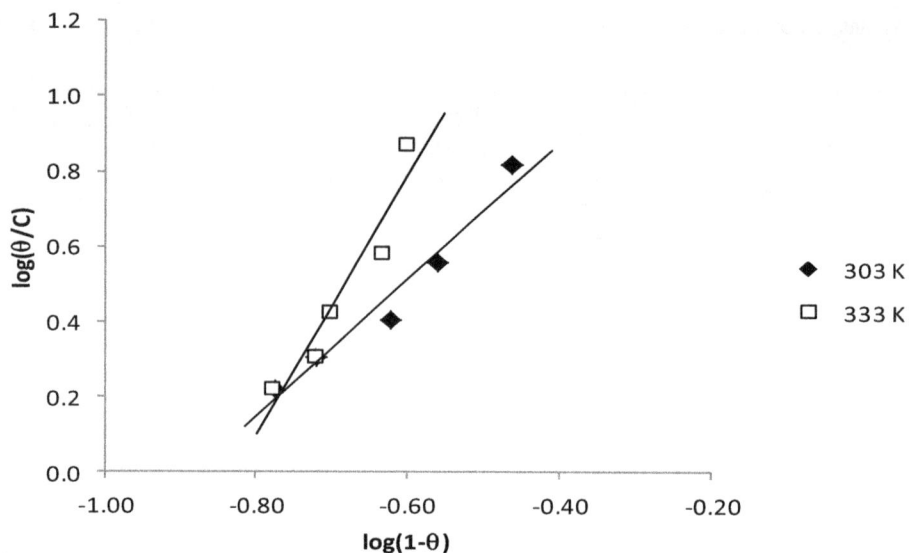

Figure 9. Flory-Huggins isotherm for the adsorption of CK gum on aluminum surface at 303 and 333K.

Table 7. Adsorption parameters for Flory-Huggins, Temkin and Freundlich adsorption isotherms.

Isotherm	T (K)	Slope	Intercept	a or x	ΔG_{ads} (kJ/mol)	R^2
Florry-Huggins	303	0.2563	0.9057	0.26	-30.62	0.9901
	333	0.1208	0.8631	0.12	-51.57	0.9459
Temkin	303	1.8266	1.6068	4.49	-19.44	0.9584
	333	3.4301	2.8426	9.53	-26.61	0.9169
Freundlich	303	0.1508	1.965	6.63	-21.52	0.9946
	333	0.065	1.9365	15.38	-21.35	0.9552

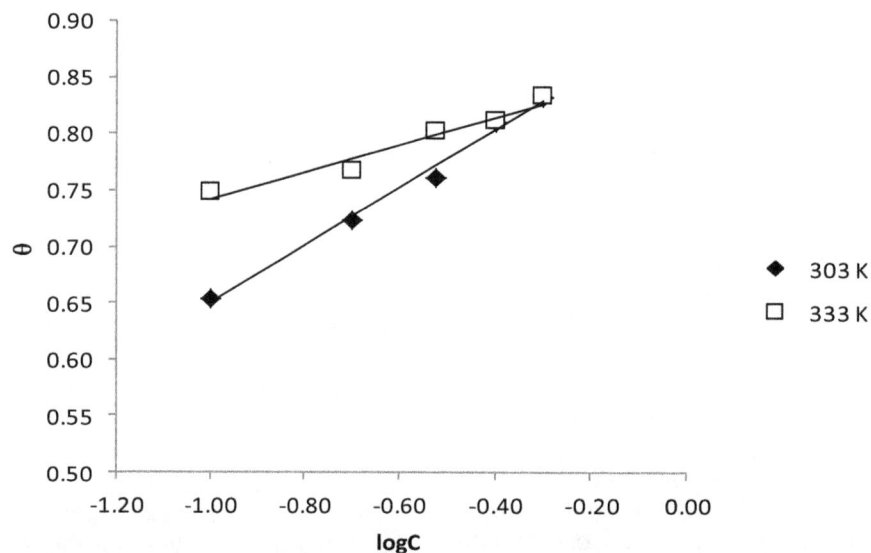

Figure 10. Temkin isotherm for the adsorption of CK gum on aluminum surface at 303 and 333K.

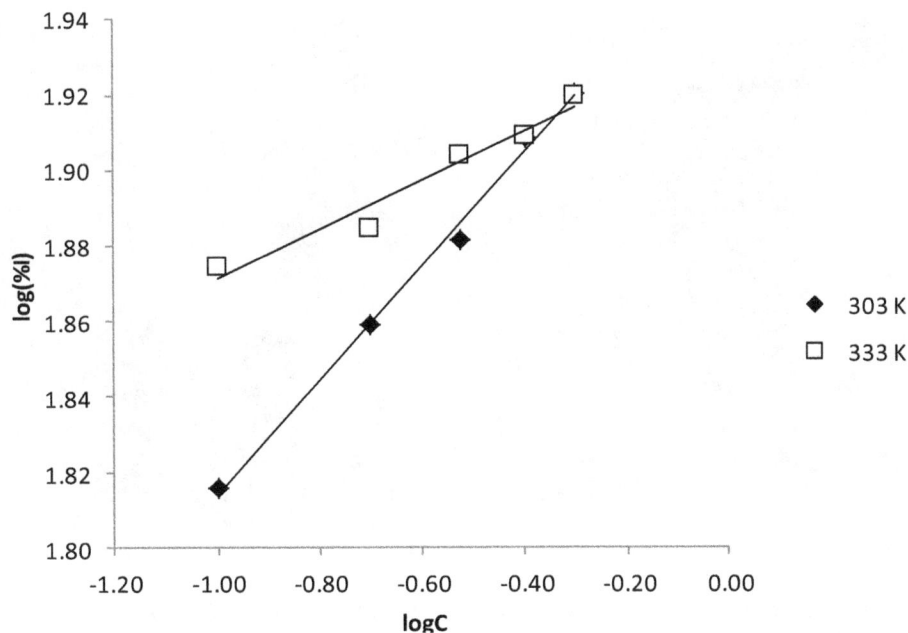

Figure 11. Freundlich isotherm for the adsorption of CK gum on aluminum surface at 303 and 333K.

temperature, which also indicate that the strength of adsorption increases with temperature and points toward chemisorption, which is expected to have chemical bond between the inhibitor and the metal surface.

Freudlich adsorption isotherm is an established isotherm for adsorption through physiosorption. The assumptions establishing the isotherm given in Equation 13 which simplifies to 14:

$$\frac{x}{m} = kC^{\frac{1}{n}} \tag{13}$$

$$log\left(\frac{x}{m}\right) = logk + \frac{1}{n}logC \tag{14}$$

The fraction x/m in Equation 14 is approximately the inhibition efficiency of a given inhibitor. Therefore from Equation 14, a plot of log[inhibition efficiency] versus logC should give a straight line if Freundlich isotherm is obeyed. Figure 11 shows Freundlich plots for the adsorption of CK gum on Al surface. Adsorption parameters deduced from the plots are also presented in Table 7.

Adsorption of organic adsorbate on the aluminum electrode surface is regarded as substitutional adsorption process between the organic molecule in the aqueous phase (Org_{ads}) and the water molecules adsorbed on the Al surface ($H_2O)_{ads}$:

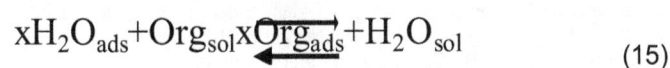

$$xH_2O_{ads} + Org_{sol} x \overleftarrow{Org_{ads}} + H_2O_{sol} \tag{15}$$

where x is the size ratio, that is, the number of water molecules replaced by one organic molecule.

From the Flory-Huggins adsorption isotherm, the value of x was computed to be 0.26 and 0.12 at 303 and 333K and the corresponding number of available adsorption site (n) computed from Freundlich adsorption isotherms were 6.63 and 15.38, respectively. These results show that the fraction of CK molecule adsorbed is temperature dependent, which point towards chemisorption, a mechanism whose extent of adsorption is expected to increase with increasing temperature.

Scanning electron microscopy (SEM) study

The scanning electron micrograph of the polished aluminium, uninhibited and inhibited aluminium in the presence of acidare shown in Figure 12a to c, respectively. It can be observed that there are distinct differences between the three SEM microphotographs. The SEM micrograph of aluminium surface (control) shows the smoothness of the metal surface implying the absence of any corrosion product formed on the metal surface.

The micrograph of the bared aluminum (Figure 12b) compared to the inhibited one (Figure 12c) shows that the former had suffered more severe corrosion than inhibitor protected one. The surface of the aluminum in the absence of the inhibitor was completely damaged after the exposure time while the CK–modified aluminium sheet was not so severely damaged. This means the

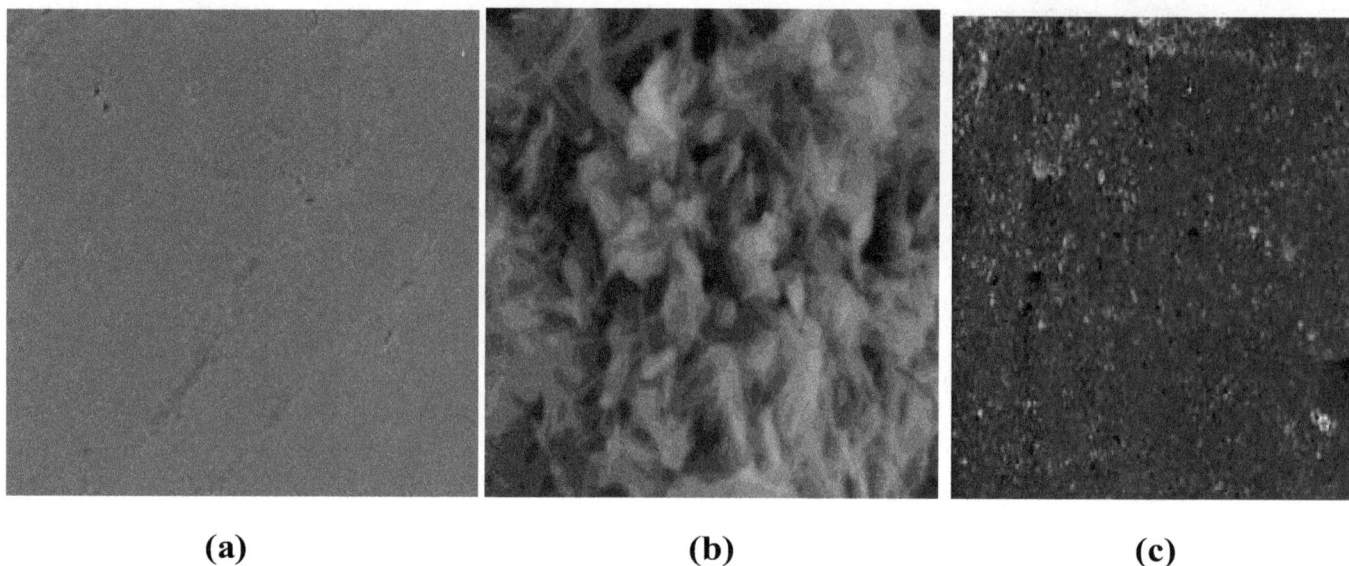

(a)   (b)   (c)

Figure 12. (a) Scanning electron micrograph of polished Al; (b) Scanning electron micrograph of Al in 0.1 M H_2SO_4 (without inhibitor); (c) Scanning electron micrograph of Al in 0.1 M H_2SO_4+ CK gum.

presence of CK gum can partially protect aluminum from corrosion.

Conclusion

This study shows that CK gum contains significant amount of Sucrose, Octadecanoic acid, Alpha camphorenal, Nerolidolisobutyrate, Diisopropenyl-1-methyl-1-vinyl cyclohexane, Abetic acid, Oleic acid, Verbenol, 2,6-dimethylhepta-1,5-diene, Naphthalene, Limonene, 7-hexadecenal and 10-methyl-8-tetradecen-1-ol acetate. Corrosion inhibition tests suggest that CK gum is a good inhibitor for the corrosion of Al in solution of H_2SO_4. The inhibition efficiency of this inhibitor increased with increasing temperature suggesting chemisorption mechanism. Also, the inhibition efficiency of these extract was found to increase in the presence of CK. Inhibition of aluminium by *C. kestingii* gum occurred through synergistic adsorption of the various components of the gums hence the formation of multiple adsorption layer is proposed.

Conflict of Interests

The author(s) have not declared any conflict of interests.

REFERENCES

Ameh PO, Magaji L, Salihu T (2012a). "Corrosion inhibition and adsorption behaviour for mild steel by *Ficus glumosa* gum in H_2SO_4 solution" Afr. J. Pure Appl. Chem. 6(7):100-106.

Ameh PO, Eddy NO, Gimba CE (2012b). Physiochemical and rheological studies on some natural polymers and their potentials as corrosion inhibitors. Published by Lambert Academic Publishing. UK.

Carter SJ (2005). Tutorial Pharmacy: Solution. Great Britain: Pitman Press: 1-8.

Ebenso EE (2003a). "Effect of halide on the corrosion inhibition of steel in H2SO4 using methyl red-Part1". Bull. Electrochem. 19(5):209-216.

Ebenso EE (2003b). "Synergistic effect of halides ions on the corrosion inhibition of aluminium in H_2SO_4 using 2-acetylphenothiazine". Mater. Chem. Phys. 79:58-70. http://dx.doi.org/10.1016/S0254-0584(02)00446-7

Ebenso EE, Eddy NO, Odiongenyi AO (2008). "Corrosion inhibitive properties and adsorption behaviour of ethanol extract of Piper guinensis as a green corrosion inhibitor for mild steel in H_2SO_4". Afr. J. Pure Appl. Chem. 4(11):107-115.

Eddy NO, Ameh PO, Gimba EC, Ebenso EE (2012a). "Chemical information from GCMS of Ficusplatyphylla gum and its corrosion inhibition potential for mild steel in 0.1 M HCl". Int. J. Electrochem. Sci. 7:5677–5691.

Eddy NO, Odiongenyi AO, Ameh PO Ebenso EE (2012b). "Corrosion inhibition potential of Daniella Oliverri gum exudate for mild steel in acidic medium". Int. J. Electrochem. Sci. 7:7425-7439.

Emregul KC, Hayvali M (2006). "Studies on the effect of a newly synthesized Schiff base compound from phenazone and vanillin on the corrosion of steel in 2M HCl". Corros. Sci. 48:797-812. http://dx.doi.org/10.1016/j.corsci.2005.03.001

Emregul KC, Kurtaran CR, Atakol O (2003). An investigation of chloride-substituted Schiff bases as corrosion inhibitors form steel. J. Corros. Sci. 45:2803-2817. http://dx.doi.org/10.1016/S0010-938X(03)00103-3

Fouda AS, Ellithy AS (2009). "Inhibition effect of 4-phenylthiazole derivatives on corrosion of 304L stainless steel in HCl solution". Corros. Sci. 51:868–875. http://dx.doi.org/10.1016/j.corsci.2009.01.011

Hatch JE (1984). Aluminium - Properties and physical metallurgy, ASM, Ohio: 242-264.

Meer W (1980). In hand book of water soluble Gums and Resins, Ed. R. L. Davidson, Mc Graw-Hill, New York. 8(1-8):24.

Migahed MA, Nassar IF (2008). "Corrosion inhibition of tubing steel during acidization of oil and gas wells". Electrochim. Acta. 53:2877-2882. http://dx.doi.org/10.1016/j.electacta.2007.10.070

Nisancioglu K (1992). Corrosion of aluminium alloys. Proceedings of

ICAA3, Trondheim, NTH and SINTEF. 3:239-259.

Oguzie EE (2006). "Adsorption and corrosion inhibitive properties of Azadirachta indica in acid solutions". Pigment Resin Technol. 35(6):334-340. http://dx.doi.org/10.1108/03699420610711335

Okafor PC, Ebenso EE, Ekpe UJ (2010). "*Azadirachta indica* extracts as corrosion inhibitor for mild steel in acid medium". Int. J. Electrochem. Sci. 5:973–998.

Shimizu K, Furneaux RC, Thompson GE, Wood GC, Gotoh A, Kobayashi K (1991). On the nature of "easy paths" for the diffusion of oxygen in thermal oxide films on aluminium, Oxidation of aluminium, 35(5/6):427-439.

23

Silver nanoparticles biogenic synthesized using an orange peel extract and their use as an anti-bacterial agent

Manal A. Awad[1], Awatif A. Hendi[2], Khalid M. O. Ortashi[3], Dalia F. A. Elradi[4], Nada E. Eisa[5,6], Lamia. A. Al-lahieb[7], Shorog. M. Al-Otiby[8], Nada M. Merghani[9] and Abdelelah A. G. Awad[10]

[1]King Abdullah Institute for Nanotechnology, King Saud University, Saudi Arabia.
[2]Department of Physics, Faculty of Science, King Saud University, Saudi Arabia.
[3]Department of Chemical Engineering, King Saud University, Saudi Arabia.
[4]Department of Microbiology, Faculty of Medicine, Princess Nora Bint Abdulrahman University. Saudi Arabia.
[5]Department of Physics, Al -Dammam University Kingdom of Saudi Arabia.
[6]Department of Physics, Omdurman Ahlia University, Sudan.
[7]Department of Chemistry Organic, Faculty of Science, Nature Product, Al Qassim University, Saudi Arabia.
[8]Department of Chemistry Organic, Faculty of Science, Nature Product, Princess Nora Bint Abdulrahman University, Saudi Arabia.
[9]Central Laboratory, College of Science, King Saud University, Saudi Arabia.
[10]Faculty of Animal Production, University of Khartoum, Sudan.

Synthesis of nanoparticles by green methods with antibacterial properties is of great researchers' concern in the explored of new pharmaceutical and biomedical products. In this study, we synthesized a new product of nanosized particles of silver, non-toxic economy, clean, and conservator for energy. An environmentally friendly route is used for synthesizing silver nanoparticles (AgNPs) using an orange peel extract as both reducing and stabilizing agent at room temperature. The synthesized NPs were characterized using ultraviolet (UV)-Vis spectrophotometer, Zitasizer which measures the average size of the particles at about 91 nm, Fourier transform infrared (FT-IR) spectroscopy, scanning electron microscopy (SEM) equipped with the energy dispersive spectroscopy (EDS) and characterization using Transmission electron microscopy (TEM). The results confirmed that the orange peel extract is a very good bioreductant for the synthesis of Ag NPs and we investigated the synthesized nanoparticles as an antibacterial which showed that the biogenic synthesized AgNPs exhibit inhibition, and had significant antibacterial against both gram-positive and gram-negative bacterial strains.

Key words: Silver nanoparticles, biogenic synthesis, orange peel, anti bacterial, gram-positive, gram-negative bacteria streamers.

INTRODUCTION

Nanotechnology mainly deals with the fabrication of nanoparticles having various shapes, sizes and managing their chemical and physical parameters for further use in human benefits with their growing applications in various fields (Bhyan et al., 2007). Preparation of metal nano-sized, usually ranging in size from 1 to 100 nanometers (nm), is amongst the most emerging areas in the field of nanotechnology. Currently, the application of nano materials is becoming increasingly important in order to solve the problems associated with material sciences, including solar energy conversion, photonics (Calvo et al., 2006; Cao et al., 2010), catalysis (Chandan et al., 2011), microelectronics (Dastjerdi et al., 2010), antimicrobial functionalities (Du et al., 2009), and water treatment (Huang et al., 2007).

Nanoparticles usually have better or different properties than the bulk material of the same elements. The antibacterial effect of silver nanoparticles (AgNPs) is greatly enhanced because of tiny size. Nanoparticles have immense surface area relative to volume. Therefore, minuscule amounts of AgNPs can lend antimicrobial effects to hundreds of square meters of its host material. Nanomaterials are the leading requirement of the rapidly developing field of nanomedicine, and bionanotechnology. Nanoparticles are being utilized as therapeutic materials tools in infections against microbes thus, the properties of nanoparticles and their effect on microbes are essential to clinical applications. Among noble metal nanoparticles, AgNPs have received considerable attention owing to their attractive physicochemical properties (Ip et al., 2006).

The AgNPs have various and important applications. Historically, silver has been known having a disinfecting effect and has been found in applications ranging from traditional medicines to culinary items. It has been reported that AgNPs are non-toxic to human and most effective against bacteria, virus and other eukaryotic micro-organism at low concentrations and without any side effects (Jeong et al., 2005; Kamyar et al., 2012). Moreover, several salts of silver and their derivatives are commercially manufactured as antimicrobial agents (Khandelwal et al., 2010). A small concentration of silver is safe for human cells, but lethal for micro organisms (Krutyakov et al., 2008). Antimicrobial capability of AgNPs allows them to be suitably employed in numerous household applications such as textiles disinfection in water treatment, food storage containers, home appliances and in medical devices (Marambio-Jones and Hoek, 2010). The most important application of silver and AgNPs is in medical industry such as tropical ointments to prevent infection against burn and open wounds (Muhammad et al., 2012).

Biological synthesis of nanoparticles by plant extracts is at present under exploitation as some researchers worked on it (Palanivel et al., 2013; Savage and Diallo, 2005) and testing for antimicrobial activities (Savithramma et al., 2011; Saxena et al., 2010; Setua et al., 2007). For the last two decades, extensive work has been done to develop new drugs from natural products because of the resistance of micro-organisms to the existing drugs. Nature has been an important source of a products currently being used in medical practice (Sharma et al., 2009).

A number of synthetic methods have been employed for the synthesis of silver-based nanoparticles involving physical, chemical (Singh et al., 2010) and biochemical techniques (Sinha et al., 2009). Chemical reduction method is widely used to synthesize AgNPs because of its readiness to generate AgNPs under gentle conditions and its ability to synthesize AgNPs on a large scale (Thirumurgan et al., 2010). However, these chemical synthesis methods employ toxic chemicals in the synthesis route which may have adverse effect in the medical applications and hazard to environment. Therefore, preparation of AgNPs by green synthesis approach has advantages over physical and chemical approaches as it is environmental friendly, cost effective and the most significant advantage is that the conditions of high temperature, pressure, energy and no toxic chemicals are required in this synthesis protocol (Thirumurguan et al., 2009; Yugang et al., 2003).

In this present work, we report the biogenic synthesis of AgNPs by using waste biomaterial orange peel extract, which was used as green reducing agent and stabilizer. The efficacy of the synthesized AgNPs as antibacterial agent was studied.

EXPERIMENTAL

Chemicals materials and bio extract

For green synthesis of AgNPs, the reagent in this work is of analytical grade and is used as received without further purification. Silver nitrate ($AgNO_3$) from Techno Pharmchem, India is used. Orange peel was washed and cut into small pieces, then boiled with deionized water for 3min then filtered.

Synthesis of silver nanoparticles

Green AgNPs were synthesized by bio reduction of Ag^+ by using fresh suspension of (5 ml) orange peel extract (greenish in color). The emulsion color was turned to dark brown after adding to 1 mM $AgNO_3$ and stirring at room temperature.

Microorganisms and antibacterial activity method

Pure culture of *Escherichia coli, Pseudomonas aeruginosa, Klebsiella pneumoneae* and *Salmonella* are types of bacteria. The antibacterial activities of biosynthesized AgNPs were carried out by disc diffusion method. Nutrient agar medium plates were prepared, sterilized and solidified. After solidification, bacterial cultures were swabbed on these plates. The sterile discs were dipped in AgNPs solution (5 mg/ml) and placed in the nutrient agar plate and kept for incubation at 37°C for 24 h. Zones of inhibition for control, were measured. The experiments were repeated 3 times and mean values of zone diameter were determined (Jeong et al., 2005).

Characterization of biogenic silver nanoparticles

Biogenic AgNPs were characterized spectrophotometrically using ultraviolet (UV)-Vis spectroscopy analyses as function of time at room temperature using Perkin Elmer UV-Vis spectrometer, Nicolet 6700, Fourier transform infrared (FT-IR) spectrophotometer was recorded, the size of synthesized AgNPs was analyzed through

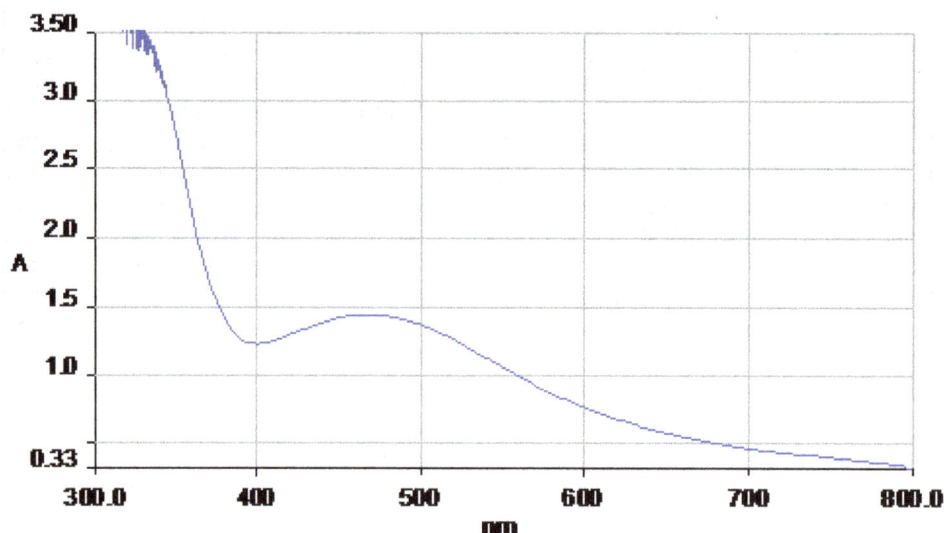

Figure 1. UV-vis spectra of reduced Ag ions to AgNPs with orange peel extract.

Zetasizer, Nano series, HT Laser, ZEN3600 from Molvern Instrument, UK, Scanning electron microscopy (SEM) has been employed to characterize the shape and morphologies of formed biogenic synthesized of AgNPs, JEOL-FE-SEM, and Energy dispersive spectrometer (EDS) analysis for the confirmation of elemental silver was carried out for the detection of elemental silver. The samples were dried at room temperature and then analyzed for samples composition of the synthesized nanoparticles. Elemental analysis on single particles was carried out using Oxford Instrument, Incax-act, equipped with SEM. Transmission electron microscopy (TEM) has been employed to characterize the size, shape and morphologies of formed biogenic synthesized of AgNPs, which was prepared by drop of AgNPs solution on carbon coated copper grid and the film on grid was dried. The TEM was operated and the measurements were performed at accelerating voltage of 100 KV.

RESULTS AND DISCUSSION

The biogenic synthesis of AgNPs by an orange peel extract was carried out. Silver nitrate used has distinctive properties such as good conductivity, catalytic and chemical stability. The formation of AgNPs was found to be successful as suggested by initial changes in color. It is well known that AgNPs exhibit brown color in aqueous solution due to excitation of surface plasmon vibrations in AgNPs.

The synthesis of green AgNPs had been confirmed by measuring the UV-Vis spectrum of colloidal solution which has absorbance peak at 466 nm; and the expanding of peak indicated that the particles are mono-dispersed as shown in Figure 1.

The FT-IR measurements were provided to describe and confirm the possible formation of bio reduction and efficient stabilization of green synthesized AgNPs by using an orange peel extract. The reduction compounds of the extract were confirmed by FT-IR spectra. FT-IR bands of orange peel were inferred at 3270.82, and 1634.24cm^{-1} in blue color (Figure 2) and FT-IR spectrum of the AgNPs shows peaks at 3260.70, 1634.62, 1376.62 and 1243.76 cm^{-1} in red color. Intense absorption is observed at 1634.24 cm^{-1} and is characteristic of the C=C stretching aromatic ring and this result agree with the result of the Thin layer chromatography (TLC) test, which refers to the active ingredient in the orange peel that causes the reduction of Ag^{+} ions, we found that the effective group is Flavonoids which led to the bio reduction of aqueous silver ions (Ag^{+}).

As shown in Figure 3, the average size of the formed biogenic AgNPs was measured by Zitasizer and it was 91.89 nm with monodispersity.

Figure 4a and b illustrate TEM images recorded at high magnification. Morphology of the AgNPs synthesized by using an orange peel extract indicates that the nanoparticles are spherical in shape with a smooth surface morphology.

SEM is shown in Figure 5a was employed to analyze the structure and morphology of the nanoparticles to give further insight into the features of the AgNPs obtained from the proposed biogenic synthesis method, the image showed relatively spherical shape of the formed nanoparticles. The EDS microanalysis is shown in Figure 5b and confirms the presence of AgNPs which is known to provide information on the chemical analysis of the elements or the composition at specific locations. The spectrum analysis reveals signal in the silver region and then confirms the formation of AgNPs. Metallic silver nanocrystals generally show a typical optical absorption peak at approximately 3 keV due to the surface plasmon resonance (Ip et al., 2006; Bar et al. 2009; Magudapathy et al., 2001). This result confirmed that the produced nano-structures are pure silver as shown in Table 1.

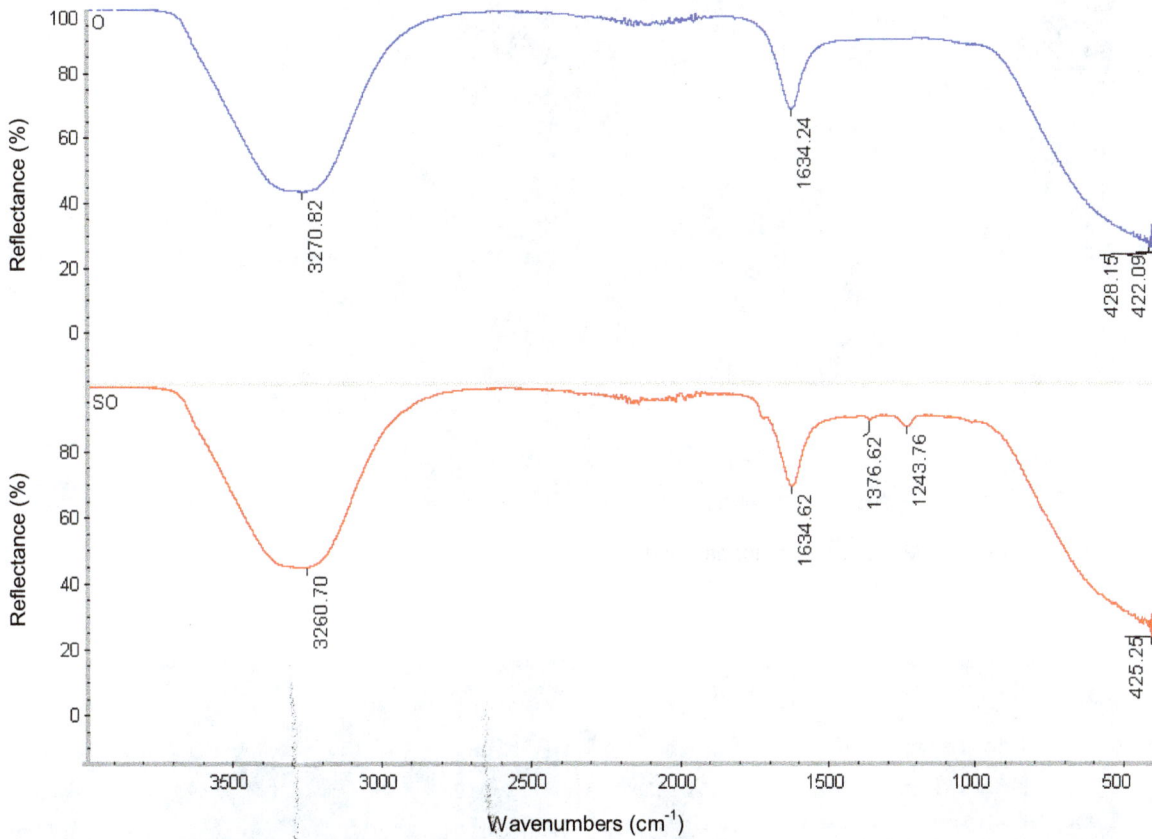

Figure 2. FTIR adsorption spectra of AgNPs prepared by orange peel.

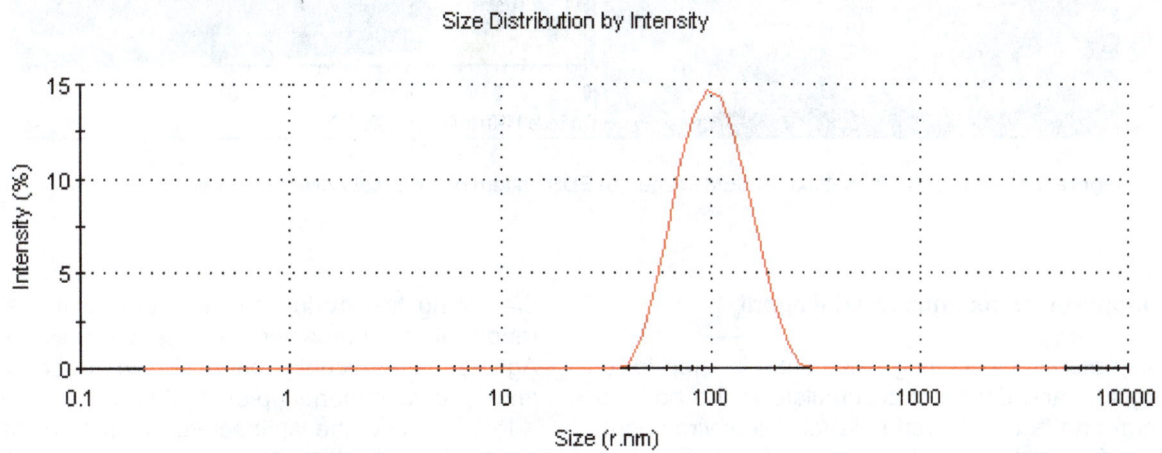

	Diam. (nm)	% Intensity	Width (nm)
Z-Average (r.nm): 91.89	**Peak 1:** 107.4	100.0	39.86
PdI: 0.147	**Peak 2:** 0.000	0.0	0.000
Intercept: 0.869	**Peak 3:** 0.000	0.0	0.000
Result quality : Good			

Figure 3. Zitasizer of the formed AgNPs.

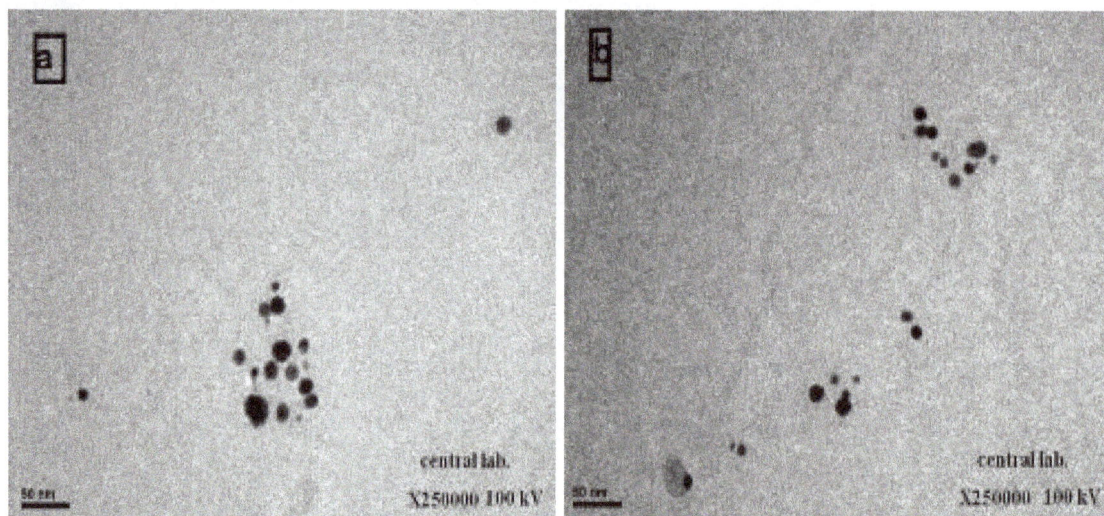

Figure 4a and b. TEM images of the formed biogenic AgNPs.

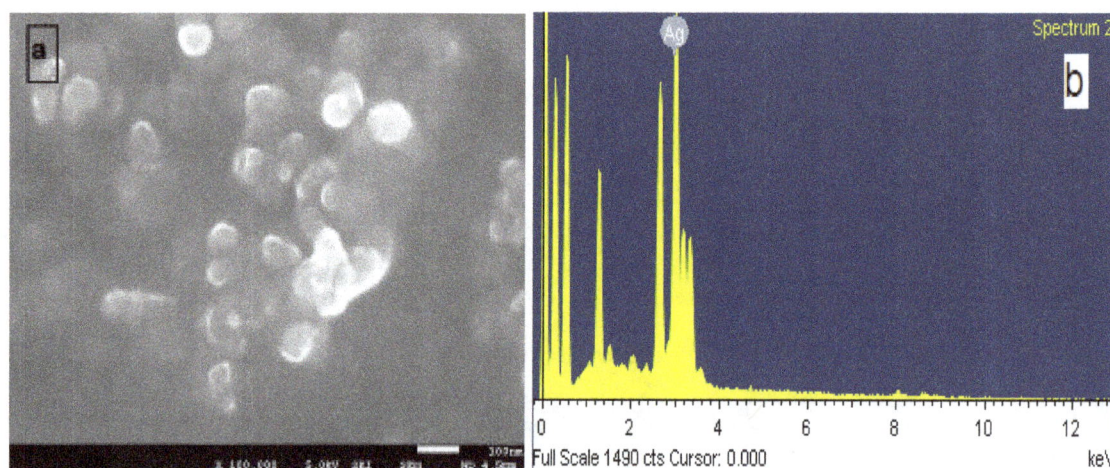

Figure 5. Green synthesis method (a) SEM image, (b) EDS pattern of spherical AgNPs prepared.

Silver nanoparticles as antibacterial agent

Silver, a naturally occurring element, is non-toxic, hypoallergenic, and does not accumulate in the body to cause harm and is considered safe for the environment. Many manufactured goods like washing machines, air conditioners and refrigerators are using linings of AgNPs for their antimicrobial qualities. Sportswear, toys and baby articles, food storage containers, HEPA filters, laundry detergent etc. are made with AgNPs. The products also with AgNPs were used, such as heart valves and other implants, medical face masks, wound dressings and bandages.

Nanomaterials are the leaders in the field of nanomedicine, bio-nanotechnology and have a great importance in nano toxicology research. Silver exhibits the strong toxicity in various chemical forms to a wide range of microorganism that is very well known and AgNPs have recently been shown to be a promising antimicrobial material (Ip et al., 2006).

In this work, the antibacterial activity of the biogenic synthesized AgNPs. The analysis results showed that nanoparticles exhibited low toxicity against *Klebsiella* which the zone of inhibition around AgNP saturated disc for bacterial culture, and the numerical value of diameter of inhibition zone was presented in Table 2, also the results showed maximum sensitivity against *E. coli*, *Pseudomonas*, and *Salmonella*.

The results in Table 2 confirmed that the successfully biogenic synthesized AgNPs showed antibacterial activity on both gram-positive and gram-negative bacteria and the analysis of bacterial growth showed that the toxicity of

Table 1. EDS elemental micro-analysis of the AgNPs.

Element	Weight (%)	Atomic (%)
Ag L	100.00	100.00
Total	100.00	

Table 2. Zone of inhibition (mm) of nanoparticles against different bacterial strains

Reagent	*E. coli* interpretation zone diameters (mm)			*Klebsiella* interpretation zone diameters (mm)			*Pseudomonas* interpretation zone diameters (mm)			*Salmonella* interpretation zone diameters (mm)		
AgNPs	6	6	6	3	2	3	6	6	6	6	6	6

AgNPs spherical shape are higher than that of gold nanoparticles spherical shape.

Hence, AgNPs synthesized by this method should be prospect further for antimicrobial applications for examples in wastewater treatment, food and water storage and manufacturing of medical supplies such as wound dressings or beds, bandages. The biological method used here in preparation is recognized by saving huge amount of energy, eco-friendly, economic, clean, and has no any toxic chemicals for the synthesis.

Conclusion

Present work demonstrated the rapid extracellular biogenic synthesis of green AgNPs using an orange peel extract and their use as an antibacterial agent. The used biogenic method here is non-toxic, environmentally friendly, simple, low cost and has no toxic chemicals. The results confirmed that orange peel plays an important role in the reduction and stabilization of silver. The formation of AgNPs was determined by UV-Vis spectroscopy where surface plasmon absorption maxima can be observed at 466 nm from the UV-Vis spectrum. Zitasizer shows the average size of the produced nanoparticles to be 91 nm. The bio produced AgNPs were characterized using FT-IR spectroscopic, TEM, SEM and EDS techniques.

For technical view, the successfully biogenic synthesized AgNPs showed antibacterial activity on both gram-positive and gram-negative bacteria and this may be useful in a wide variety of applications in pharmaceutical, biomedical fields, industrial appliances like bandage, food and water storage and wastewater treatment in a low price.

ACKNOWLEDGEMENTS

The authors extend their appreciation to the Deanship of Scientific Research at King Saud University for funding this work through research group no RGP- VPP-278.

REFERENCES

Bar H, Bhui DH, Sahoo PG, Sarkar P, De PS, Misra A (2009a). Green synthesis of silver nanoparticles using latex of Jatrapha curcas. Colloids Surf. A Physicochem. Eng. Asp. 339:134–139.

Bhyan SB, Alam MM, Ali MS (2007). Effect of plant extracts on Okra mosaic virus incidence and yield related parameters of Okra. J. Agric. Res. 1:112-118.

Calvo MA, Angulo E, Costa-Batllori P, Shiva C, Adelantado C, Vicente A (2006). Natural plant extracts and organic acids: synergism and implication on piglet's intestinal microbiota. Biotechnol. 5: 137-142.

Cao XL, Cheng C, Ma YL, Zhao CS(2010). Preparation of silver nanoparticles with antimicrobial activities and the researches of their biocompatibilities. J. Mater. Sci. Mater M 21:2861–2868.

Chandan Singh, Vineet Sharama, Pradeep KR Naik, Vikas KHandelwal, Harvinder Singh(2011). A green biogenic approach for synthesis of gold and silver nanoparticles using Zingiber officinale. Digest J. Nanomaterials Biostructures 6(2):335-542.

Dastjerdi R, Montazer M, Shahsavan S (2010). Size-controlled preparation of silver nanoparticles by a modified polyol method, Colloids Surf. A Physicochem. Eng. Aspects 366:197–202.

Du WL, Niu SS, Xu YL; Xu ZR, Fan CL (2009). Antibacterial activity of chitosan tripolyphosphate nanoparticles loaded with various metal ions. Carbohydr. Polym. 75:385–389.

Huang J, Li Q, Sun D, Lu Y, Su Y, Yang X, Wang H, Wang Y, Shao W, He N, Hong J, Chen C (2007). Biosynthesis of silver and gold nanoparticles by novel sundried Cinnamomum camphora leaf. Nanotechnology, 18:105–106.

Ip M, Lui SL, Poon VKM, Lung I, Burd A (2006). Antimicrobial activities of silver dressings: an *in vitro* comparison. J. Medical Microbial. 55:59-63.

Jeong SH, Yeo SY, Yi SC (2005). The effect of filler particle size on the antibacterial properties of compounded polymer/silver fibers. J. Mat. Sci. 40:5407-5411.

Kamyar S, Mansor BA, Seyed DJ, Parvaneh S, Parvanh S, Hossein J, Yadollah GS (2012). Investigation of antibacterial properties silver nanoparticles prepared via green method. Chemistry central J. 6:73.

Khandelwal N, Singh A, Jain D, Upadhyay M.K., Verma HN (2010). Green synthesis of silver nanoparticles using Argimone mexicana leaf extract and Evaluation of their antimicrobial activities. J. Nanomater. Biostruct. 5:483-489.

Krutyakov YA, Kudrynskiy A, Olenin AY, Lisichkin GV (2008). Extracellur biosynthesis and antimicrobial activity of silver nanoparticles. Russ. Chem. Rev.77:233-236.

Magudapathy P, Gangopadhyay P, Panigrahi BK, Nair KGM, Dhara S (2001). Electrical transport studies of Ag nanoclusters embedded in glass matrix. Physica B, 299(1-2):(142–146).

Marambio-Jones C, Hoek EMV (2010). A review of the antibacterial effects of silver nanomaterials and potential implications for human health and the environment. J. Nanopart. Res, 12:1531-1551.

Muhammad A, Farooq A, Muhammad Ramzan SAJ, Muhammad AI,

Umer R (2012). Green Synthesis of Silver Nanoparticles through Reduction with *Solanum xanthocarpum* L. Berry Extract: Characterization, Antimicrobial and Urease Inhibitory Activities against *Helicobacter pylori*. Int. J. Mol. Sci. 13:9923-994.

Palanivel V, Sang-Myung L, Mahudunan L, Kui-Jae L, Byung-Taek O (2013). Pine cone-mediated green synthesis of silver nanoparticles and their antibacterial activity against agricultural pathogens. Appl. Microbiol. Biotechnol. 97:361–368.

Savage N, Diallo MS (2005). J. Nanomaterials and water purification. opportunities and challenges. Nanopart. Res. 7:331–342.

Savithramma N, Linga RM, Rukmini K, Suvarnalatha PD (2011). Antimicrobial activity of silver nanoparticles synthesized by using medicinal plants. Int. J.Chem. Technol. Res. 3(3):1394-1402.

Saxena A, Tripathi RM, Singh RP (2010). Biological Synthesis of silver nanoparticles by using Onion (Allium cepa) extract and their antibacterial activity. J. Nanomater. Biostruct. 5:427-432.

Setua P, Chakraborty A, Seth D, Bhatta MU, Satyam PV, Sarkar N (2007). Synthesis, optical properties, and surface enhanced Raman scattering of silver nanoparticles in nonaqueous methanol reverse micelles. N. J. Phys. Chem. C. 111:3901–3907.

Sharma VK, Yngard RA, Lin Y (2009). Silver nanoparticles: Green synthesis and their antimicrobial activities. Adv. Coll. Int. Sci. 145:83-96.

Singh A, Jain D, Upadhyay MK, Khandelwal N, Verma HN, Verma HN (2010). Green synthesis of silver nanoparticles using Argemone mexicana leaf extract and evaluation of their antimicrobial activities. Dig J Nanomater Bios, 5:483–489.

Sinha S, Pan I, Chanda P, Sen SK (2009). Nanoparticles fabrication using ambient biological resources. J. Appl. Biosci. 19:1113–1130.

Thirumurgan A, Tomy NA, Jai GR, Gobikrishnan S (2010). Biological reduction of silver nanoparticles using plant leaf extracts and its effect an increased antimicrobial activity against clinically isolated organism. Phar. Chem. 2:279-284.

Thirumurguan G, Shaheedha SM, Dhanaraju MD (2009). *In vitro* evaluation of antibacterial activity of silver nanoparticles synthesized by using phytothora infestans. Int. J. Chem Tech Res. 1:714-716.

Yugang S, Mayers B, Xia YP (2003). Synthesis of uniform silver nanowires: A plausible growth mechanism and the supporting evidence. Nano Lett. 3:955–960.

24

Effects of adhesive thicknesses on the stress distribution in prismatic plug-in joints

Sinan Aydin[1], Serdar Mercan[2], Murat Yavuz Solmaz[3] and Aydin Turgut[4]

[1]Faculty of Technology, Cumhuriyet University, Sivas/Turkey.
[2]Faculty of Technical Education, Cumhuriyet University, Sivas/Turkey.
[3]Faculty of Engineering, Firat University, Elaziğ/Turkey.
[4]Faculty of Engineering, Bingöl University, Bingöl/Turkey.

The aim of this study is to research the effects of adhesive thickness, which is one of the factors that affect the stress distribution in the prismatic plug-in joints. The effect of adhesive thickness was theoretically researched in the study. Epoxy-based and acrylic-based adhesives, which are widely used in combining metal joints, were applied. After testing the mechanical characteristics of the adhesives, the models of the prismatic plug-in joints combined with adhesive were designed in the Pro-engineer program. The mechanical analysis of the models was performed in Ansys Workbench and the results are presented as a comparison.

Key words: Adhesive, mechanic properties, bulk specimens, acrylic, epoxy.

INTRODUCTION

Adhesives are generally mixtures composed by chemically mixing materials such as epoxy, phenol, polyamide, polyimide and silicone, which produce the desired design features when at least two different materials are combined by bonding (Morrisey and Johnson, 1985).

Before being compounded, the adhesives may be in various forms such as film, putty, liquid and powder. The adhesives cured by chemical reaction are called structural adhesives. The structural adhesives are load-bearing, flexible, heat-resistant adhesives with high shear strength. The structural adhesives are frequently used in many industries such as aerospace, automotive, shipbuilding and so on. The chemical adhesives cured by chemical reaction and in use today are anaerobics, cyanoacrylates, acrylics, silicones, polyurethanes, epoxies and phenolics (Ciba-Geigy, 1993).

Adhesives that are most widely-used for metal bonding

are epoxies. Epoxy adhesives are composed of resin and hardener, and these adhesives provide extremely strong bonding. They are available in three different forms: single component, double component and film. Their deep hardening rate is very good and they can be used for bonding various materials (Solmaz, 2008).

Acrylic adhesives are cured with an activator in an anaerobic environment. Because the adhesive is cured only when it comes in contact with the activator, it is not required to keep open waiting period short or use the mixed materials immediately. The width of the bonding surface must be at least 5 mm in order to break off when in contact with oxygen. Depending on the adhesive type, the adhesive and activator can be applied on the bonding surfaces separately and used for bonding various materials (Loctite, 1998).

When determining the mechanical properties of an adhesive, samples are used in either bulk or joint forms.

Although, using samples in joint form represents the original loading type in the application area, the adhesion level of the adhesive is tested rather than any mechanical properties. These kinds of disadvantages may be eliminated by using bulk samples (Aydin et al., 2004).

Factors such as overlap distance, adhesive thickness, surface roughness etc. have an effect on the strength of the joints when combined with adhesives. Goland and Reissner (1944) called the shear stress that is parallel to the adhesive layer and normal stress that is vertical to the plane of shear stress as tearing stresses. This study set a reference for many researches relating to adhesive-adherend overlap joints.

The effect of adhesive thickness was firstly examined by Bascom at al. (1977); later, many researchers examined fracture mechanics and traditional stress analysis.

Mall and Ramamurth (1989) double cantilever beam specimens fracture and crack progression by applying cyclic loads of different thicknesses examined and thick adhesive layers for the crack growth rate is expressed at high level.

Turgut and Sancaktar (1991) examined the effects of curing and loading conditions relating to fiber-matrix adhesion on composite materials.

Tamblin et al. (2001) presented the shear results between the adhesive thicknesses of 0.4 to 3 mm via thick adherend shear test (TAST). They explicitly stated that shear strength decreases as the thickness of the adherend pieces increases.

Jarry and Shenoi (2006) 0.1 to 10 mm butt strap adhesive bonding studies used a methacrylate adhesive and increase in the thickness of the adhesive with the significantly reduced load failure.

Grant et al. (2009) accepted that shear strength decreases linearly when the adhesive thickness is increased from 0.1 to 0.3 mm and the reason for this is the flexural stress occurring on the thick adhesive layer.

Davies et al. (2009) carried out research relating to the effect of adhesive thickness and he examined the properties of aluminum joints by using epoxy adhesives via various test techniques. As a result of mechanical analysis, they found that the tensile strength decreases as the adhesive thickness increases as seen in the previous researches and the ideal thickness should be 0.8 mm and below.

Nemeş and Lachaud (2010) researched the effect of the thickness of the adhesive double-lap connection. Adhesive thickness increases, the maximum stress is low, with the exception of shear and peel stresses endpoints expressed uniformly distributed on all the overlap distance.

Fracture mechanics analysis by Daghiyani et al. (1995), Abou-Hamda et al. (1998), and Kawashita et al. (2008) increased the strength of the connections with the increase of the thickness of the adhesive, while, Bascom et al. (1975), and Schmueserand Johnson (1990) stated that the thickness of the adhesive is not important and Chai (1988), Kahramana et al. (2008), and Da Silva et al. (2009) expressed strength reduced (Azari et al., 2011).

Adin (2012) examined the mechanical behavior of the scarf lap joints bonded with adhesive under a tensile loads.

Kimiyoshi et al. (2012) examined the effect of adhesive thickness on tensile and shear strength of a polyimide adhesive. The tensile strength of the butt joints decreased with increasing adhesive thickness. In contrast, adhesive thickness did not seem to affect the shear strength of single lap joints.

In this study, the effects of adhesive thickness and different adhesive types to the stress distribution on prismatic plug-in joints were tried and determined theoretically.

MATERIALS AND METHODS

In this study, the mechanical analysis of prismatic plug-in metal joints combined by using epoxy and acrylic based adhesives were examined. In the study, two adhesives with different properties (1 unit of epoxy-based (Akfix-E300) and 1 unit of acrylic-based (Erde GTR) adhesive were used. Both adhesives include two components and their mixing ratio is 1:1.

Primarily, bulk samples were produced from adhesives in sizes suitable for ISO 527-2 (1993) standards, then tension testing procedures defined on ASTM D1002 (1983) were applied on these samples. The mechanical properties of adhesives and combined pieces obtained as a result of these tests are given in Table 1 and their stress-strain graphics are given in Figure 1.

As shown in Figure 1, Erde GTR displays a nonlinear material behaviour. Thus, elasto-plastic analysis will be made for this adhesive in ANSYS (Academic Teaching Advanced, Version 12.0) software. The stress-strain data required for elasto-plastic analysis was selected from Figure 1b and these values are given in Table 2.

In the study, 3 different overlap distances, 3 different sample widths and 3 different adhesive thicknesses were used. The analysis parameters used are summarized in Table 3.

Models in the sizes given in Figure 2 (Aydin et al., 2012) were prepared in accordance with these parameters. Elastic and elasto-plastic stress analyses of the models were made by using ANSYS Workbench 12.0, the finite element program. 3-Dimensional solid models of prismatic plug-in joints were prepared in the Pro-Engineering program and these models were transferred to ANSYS Workbench for finite element analysis. The solid models of prismatic joints prepared in 3 different overlap distances are shown in Figure 3.

The mesh structure, boundary conditions and loading condition of the models prepared by ANSYS are given in Figure 4. In the analysis carried out, different elements were used for metal and adhesive materials: Solid 187 for metal materials, Surf154 for adhesive materials, Conta174 and Targe 170 for the contact surfaces of metal and adhesive materials. A more sensitive mesh process was carried out by comparing the area where the binding process that is critical for stress distributions was carried out (Figure 5).

The von-Mises flow criteria given in Equation 1 were used for calculating the equivalent stress values occurring on the adhesive layer and the materials adherend.

$$\sigma_{sqv} = \sqrt{\sigma_x^2 + \sigma_y^2 - \sigma_x \cdot \sigma_y + 3\tau_{xy}^2}$$

(1)

Table 1. Mechanical properties of adhesives and combined pieces.

Material	Testing method			
	ISO 527	ISO 527	ISO 527	ISO 178
	Yield strength (MPa)	Tensile strength (MPa)	Poisson ratio	Elasticity module (Mpa)
Akfix E300	-	34.1	0.32	758
Erde GTR	2.01	7.9	0.36	93
St 60	-	600	0.35	210000

Table 2. ERGE GTR stress-strain values.

Node	ε (%)	σ(MPa)
1	0.003	0.732
2	0.007	1.787
3	0.011	2.389
4	0.013	2.942
5	0.017	3.569
6	0.019	3.927
7	0.022	4.459
8	0.025	4.871
9	0.029	5.324
10	0.035	5.891
11	0.041	6.404
12	0.048	6.817
11	0.516	7.010
12	0.059	7.388
13	0.073	7.909
14	0.077	7.759
15	0.082	7.544

Table 3. Parameters used in the study.

Male specimen overlap (plug-in) distance: a (mm)	Female specimen width: b (mm)	Adhesive thickness: t (mm)
10	10.2	0.1
20	10.6	0.3
30	11	0.5

While determining the damage load in the finite elements analysis the tensile strengths given below were taken into consideration and shown as σ^* in the graphics.

$\sigma_{Akfix} = 34.1 Mpa$ $\qquad$ $\sigma_{EGTR} = 7.9 Mpa$

The critical areas on the prismatic plug-in joints combined with an adhesive are the contact surface of the adhesive and interfaces of the material adherend. Thus, the A-C line shown in Figure 6 was taken into consideration for the stress analysis of the adhesive layer. In order to compare the stress distributions occurring on the adhesive layer, the stress distributions obtained from the adhesive layer on the A-C line were normalized by dividing this value with the tensile strength value obtained by tensiling the bulk sample of each adhesive uniaxially (σ_{Akfix}, σ_{EGTR}). In the same way, in order to compare the stress distributions occurring on different overlap distances, the horizontal coordinate value of the point (x), the stress distribution of which was calculated, was normalized by being divided with its own overlap distance (a). As a result of the stress analysis, the stress distribution occurring on the A-C line is given in Figure 7.

The graphics of the stresses occurring were prepared by using

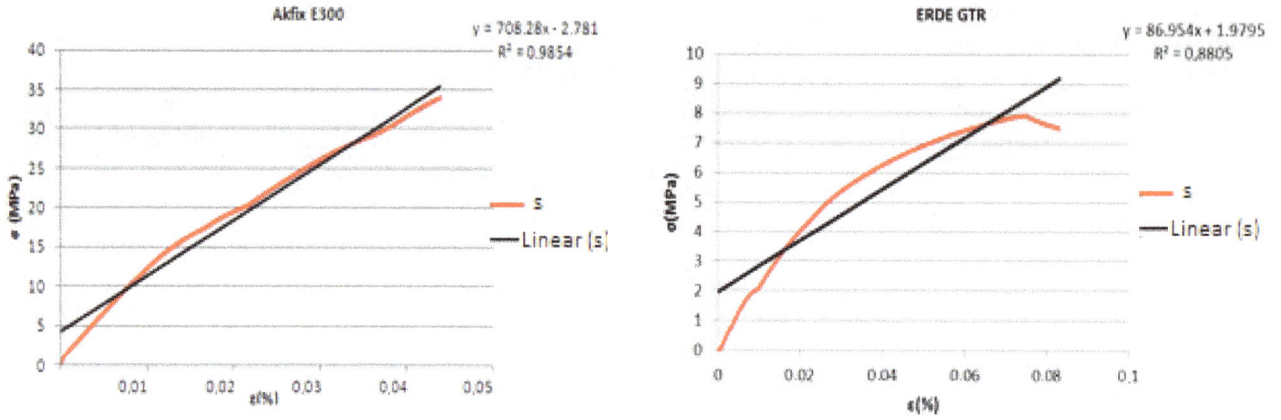

Figure 1. Stress-strain behaviors of adhesives: (a), Akfix E300; (b), Erde GTR.

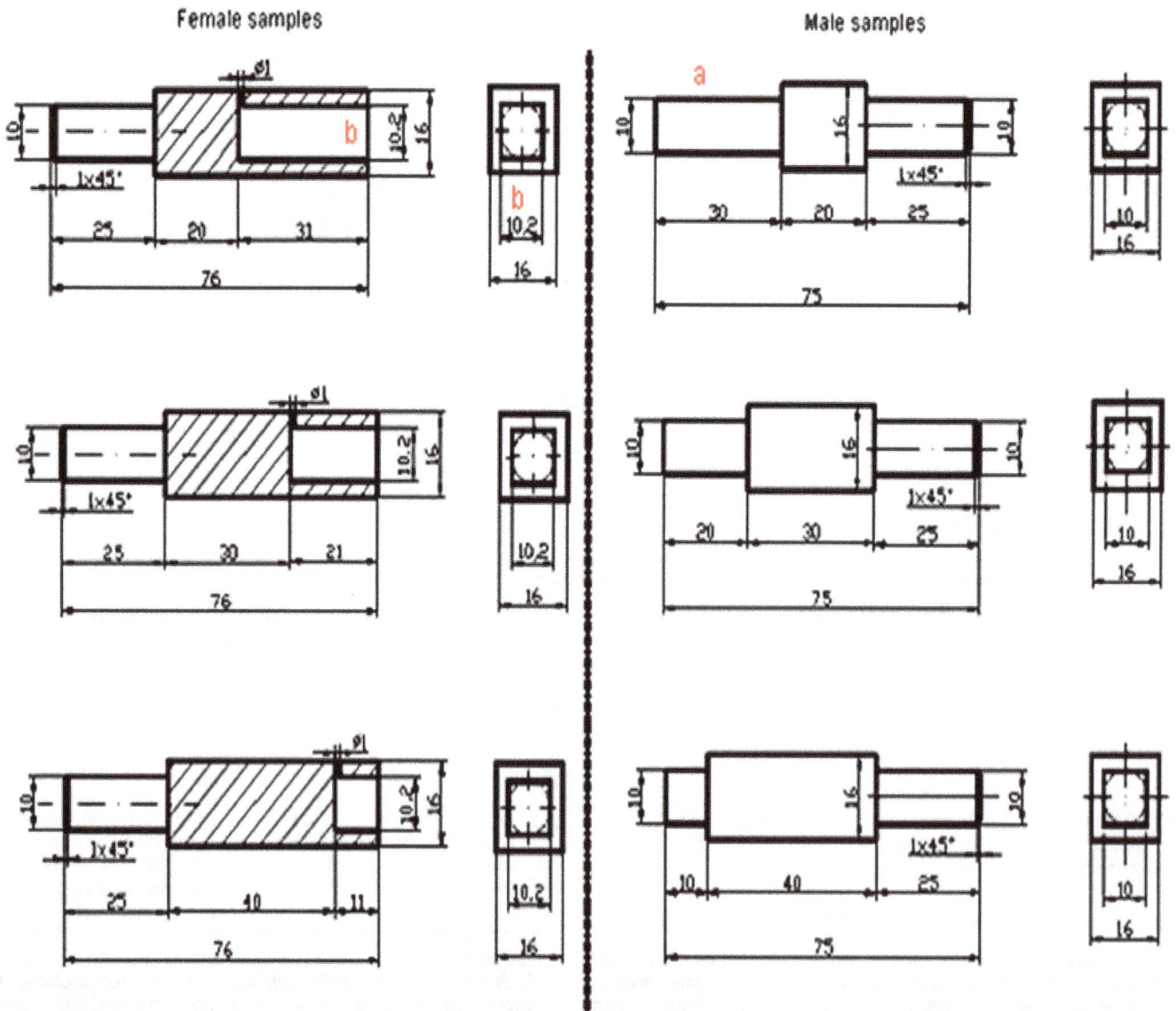

Figure 2. Dimensions of analyses models.

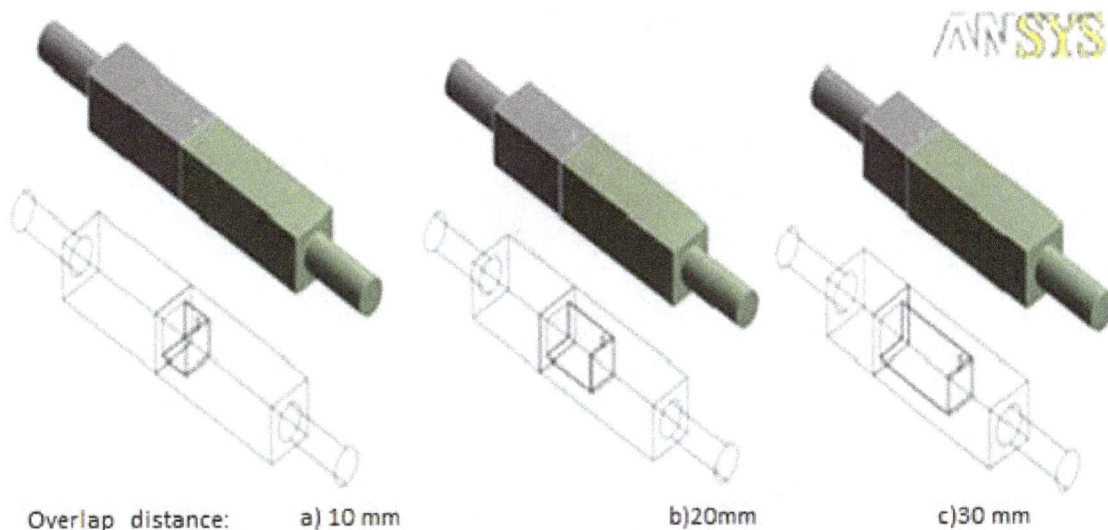

Overlap distance: a) 10 mm   b)20mm   c)30 mm

Figure 3. Perspective view of the models.

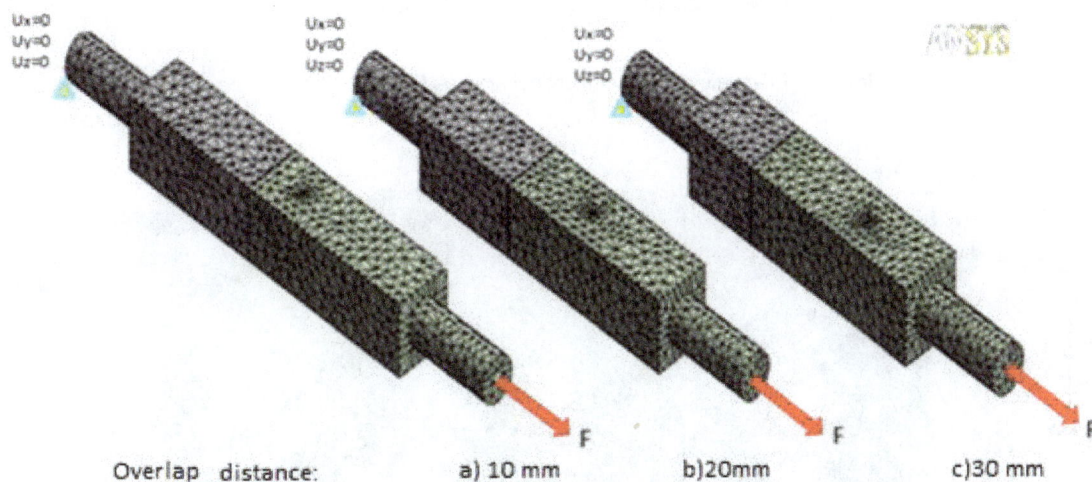

Overlap distance: a) 10 mm   b)20mm   c)30 mm

Figure 4. Mesh structure and boundary conditions of the models.

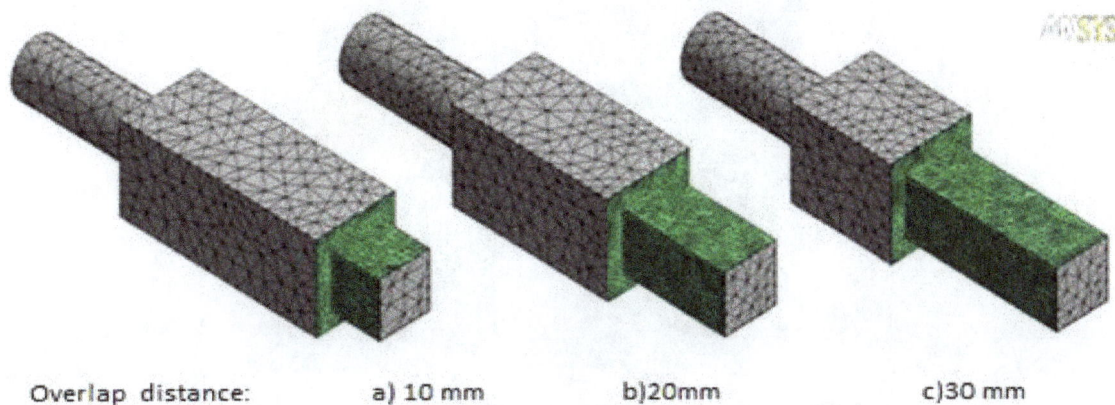

Overlap distance: a) 10 mm   b)20mm   c)30 mm

Figure 5. Adhesion areas and meshed male models.

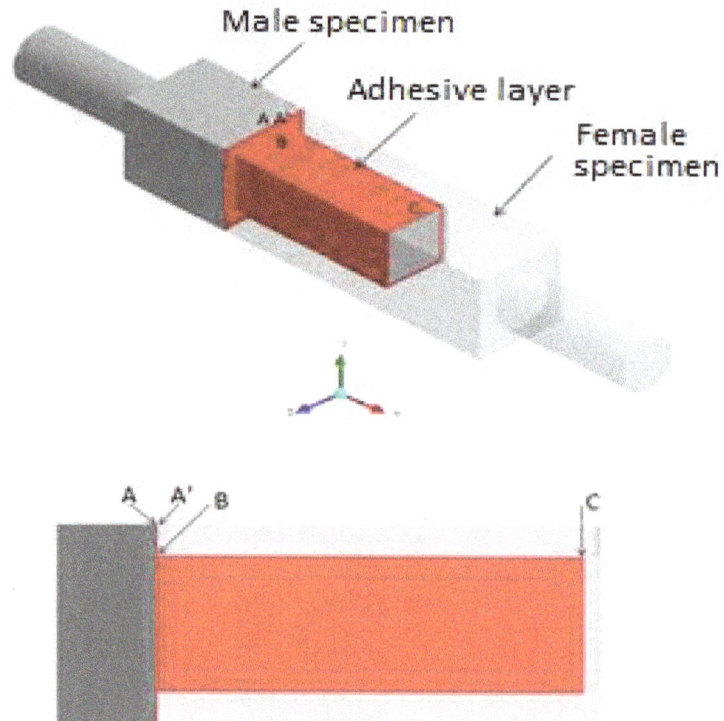

Figure 6. Prismatic plug-in joint critical area (A-C line).

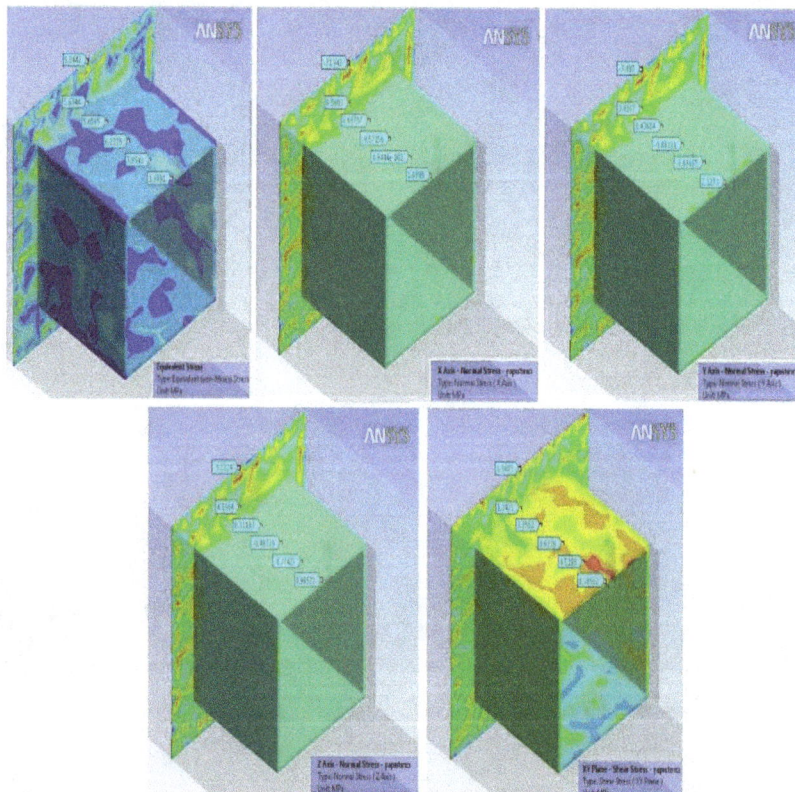

Figure 7. The stress values occurring on adhesive layers as a result of ANSYS analysis (σ_{eqv}, σ_x, σ_y, σ_z, τ_{xy}).

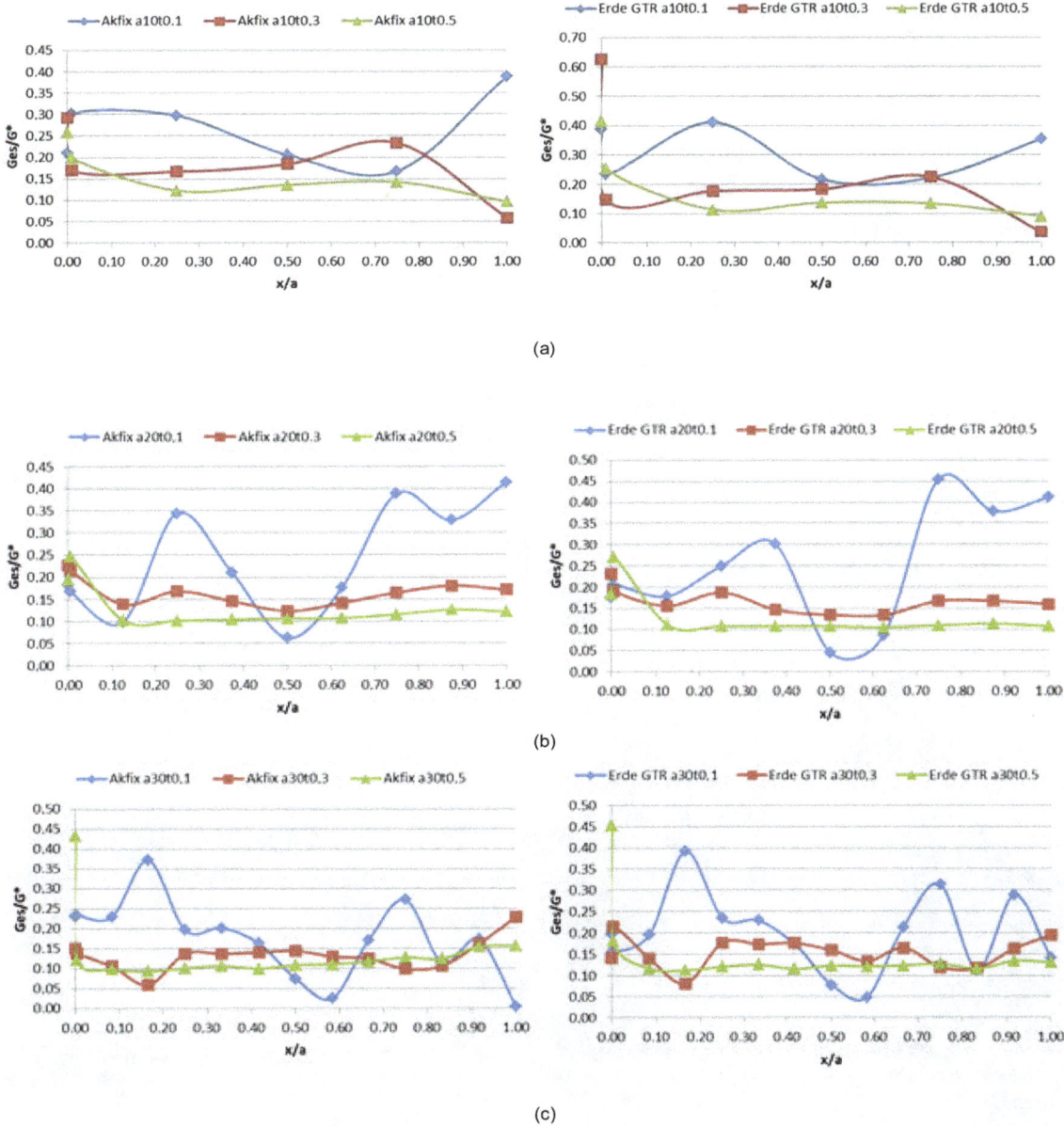

(a)

(b)

(c)

Figure 8. Equivalent stress distribution occurring throughout the A-C line depending on the overlap distance and adhesive type. a) 10 mm overlap distance, b) 20 mm overlap distance, c) 30 mm overlap distance.

the stress values occurring on the A-C line.

RESULTS AND DISCUSSION

Akfix E300 was combined with Erde GTR, the equivalent stress distribution ratios occurring on the adhesive surface throughout the A-C line for the prismatic plug-in joints with a = 10, 20, 30 overlap distance and t = 0.1, 0.3, 0.5 mm adhesive thickness, normal stress distribution ratios, peeling-stress distribution ratios and shear stress distribution ratios are given in Figures 8, 9, 10, and 11.

As shown in Figure 8, for the 10 mm overlap distance, while the equivalent stress ratio gets the maximum value in both adhesives for x/a = 0 and minimum value for x/a = 1 when the adhesive thickness is t = 0.3 mm or t = 0.5 mm, it gets the minimum value for x/a = 0 and the maximum value for x/a = 1 when t = 0.1 mm. The equivalent stress ratio decreases as the adhesive thickness increases throughout the A-C line.

For 20 mm overlap distance, while both adhesives are in a corrugated form throughout the A-C line for the adhesive thickness of t = 0.1, it is seen that the equivalent stress is distributed regularly for the adhesive thicknesses of t = 0.3 and t = 0.5. The equivalent stress

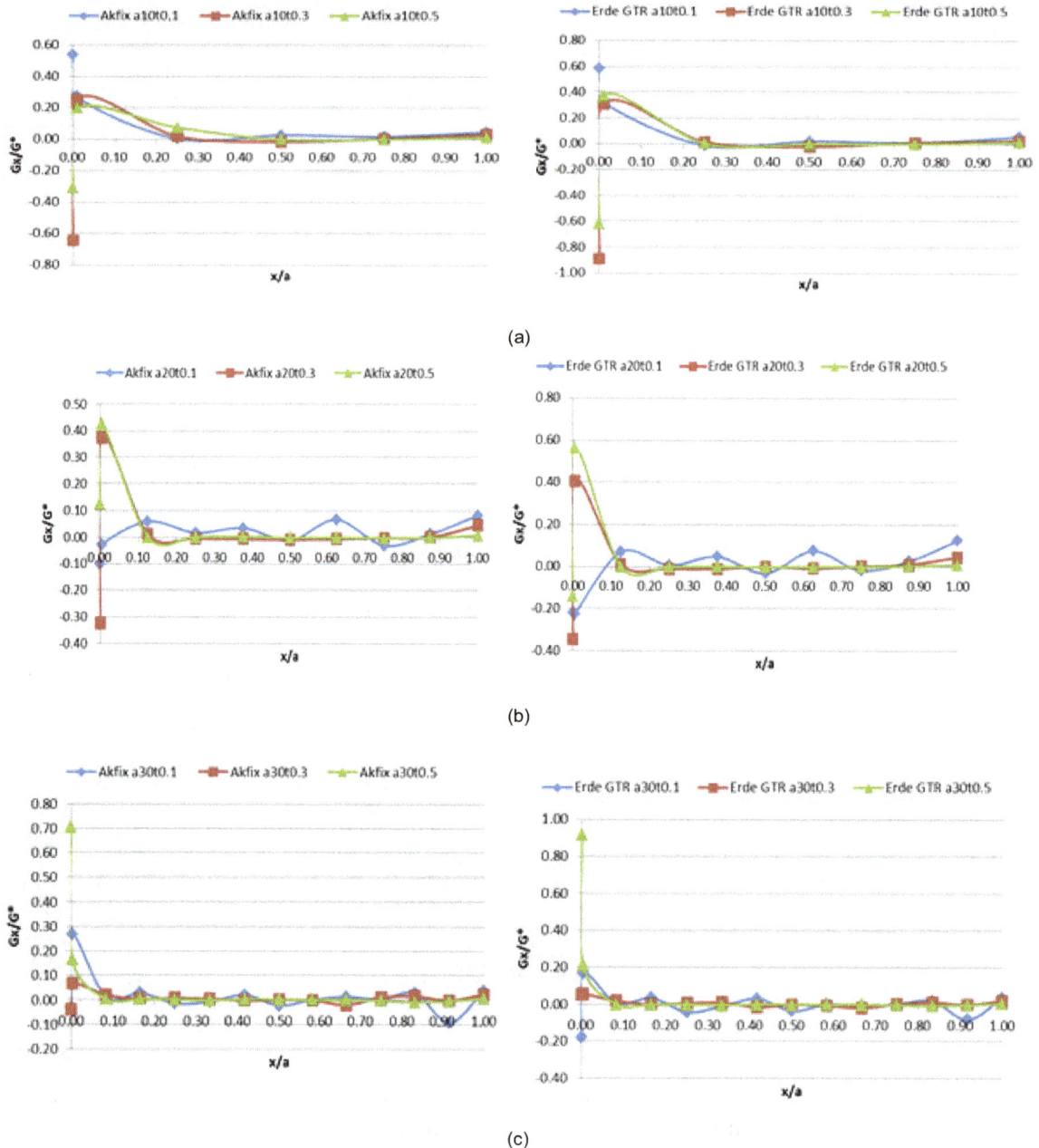

(a)

(b)

(c)

Figure 9. Normal stress distribution occurring throughout the A-C line depending on the overlap distance and adhesive type. a) 10 mm overlap distance, b) 20 mm overlap distance, c) 30 mm overlap distance.

ratio decreases as the adhesive thickness increases.

For 30 mm overlap distance, while both adhesives are in a corrugated form throughout the A-C line for the adhesive thickness of t = 0.1, it is seen that the equivalent stress is distributed more regularly for the adhesive thicknesses of t = 0.3 and t = 0.5.

In Figure 9, for the 10 mm overlap distance, normal stress ratio gets the maximum values for all thickness values in both adhesives for the ratio of x/a = 0. In the ratio of x/a = 0, while normal stress becomes effective as

much as tensile stress for t = 0.1, the effect is as much as for the compressive stress for t = 0.3 and t = 0.5. It was seen that the normal stress ratio decreases as the adhesive thickness increases.

For the 20 mm overlap distance, normal stress ratio gets the maximum value in both adhesives in the ratio of x/a = 0 for t = 0.1 and the small corrugated distribution continues until the end of the joint. It was seen that the normal stress ratio decreases as the adhesive thickness increases.

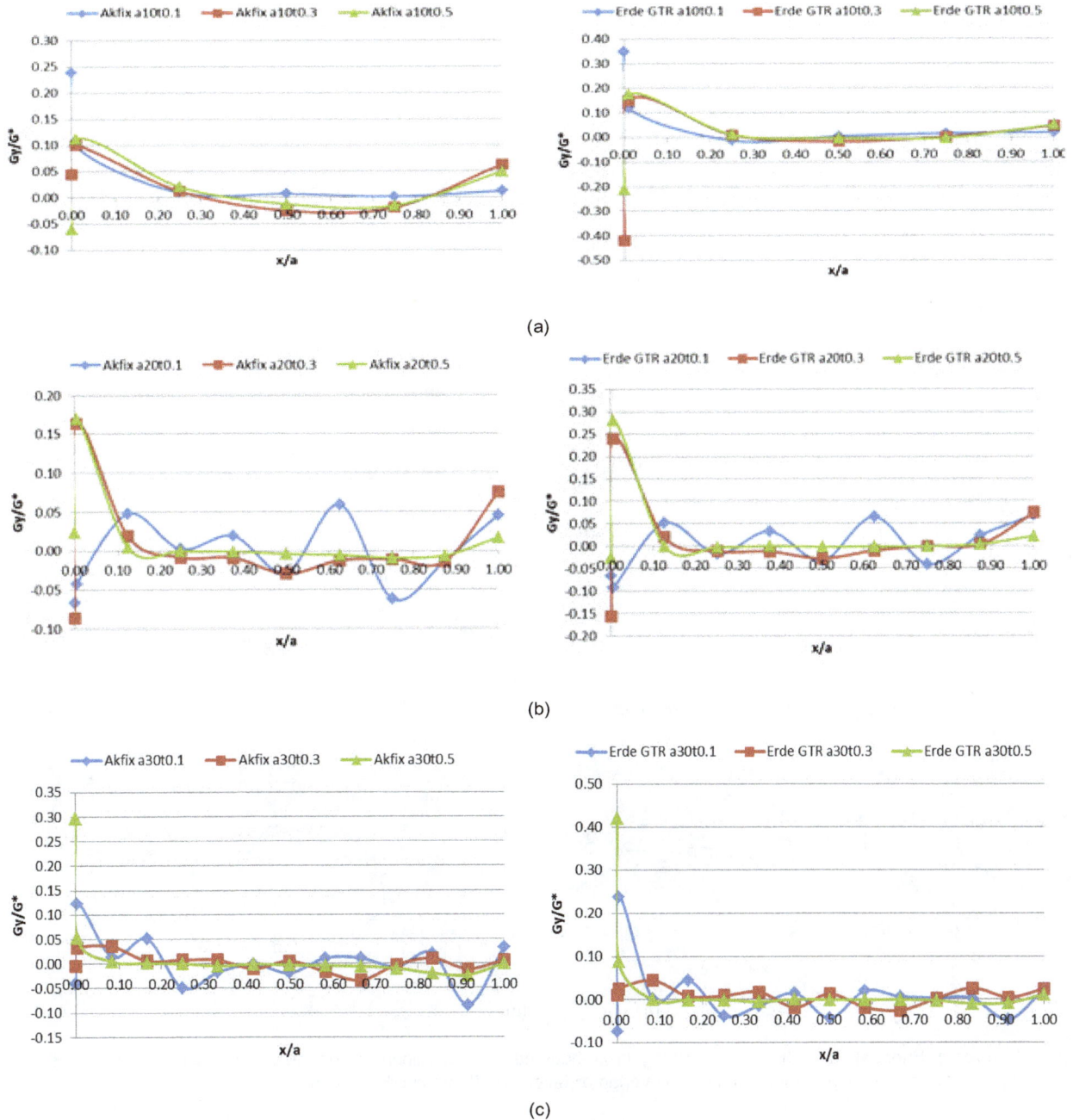

(a)

(b)

(c)

Figure 10. Peeling stress distribution occurring throughout the A-C line depending on the overlap distance and adhesive type. a) 10 mm overlap distance, b) 20 mm overlap distance, c) 30 mm overlap distance.

For the 30 mm overlap distance, the highest stress ratio for both adhesives was obtained for the adhesive thickness of t = 0.1. It was seen that the normal stress ratio decreases as the adhesive thickness increases.

In Figure 10, for the 10 mm overlap distance, peeling stress ratio gets the maximum values for all thickness values in both adhesives for the ratio of x/a = 0, while the peeling stress effects as tensile stress or t = 0.1 and t = 0.3, it effects as compressive stress for t = 0.5.

For the 20 mm overlap distance, the situation of both adhesives is similar to Figure 9b in terms of tensile stress.

For the 30 mm overlap distance, the situation of both adhesives is similar to Figure 9c in terms of tensile stress.

In Figure 11, for the 10 mm overlap distance, while the stress values obtained in the ratio of x/a = 0 for all adhesive thickness values are close to each other,

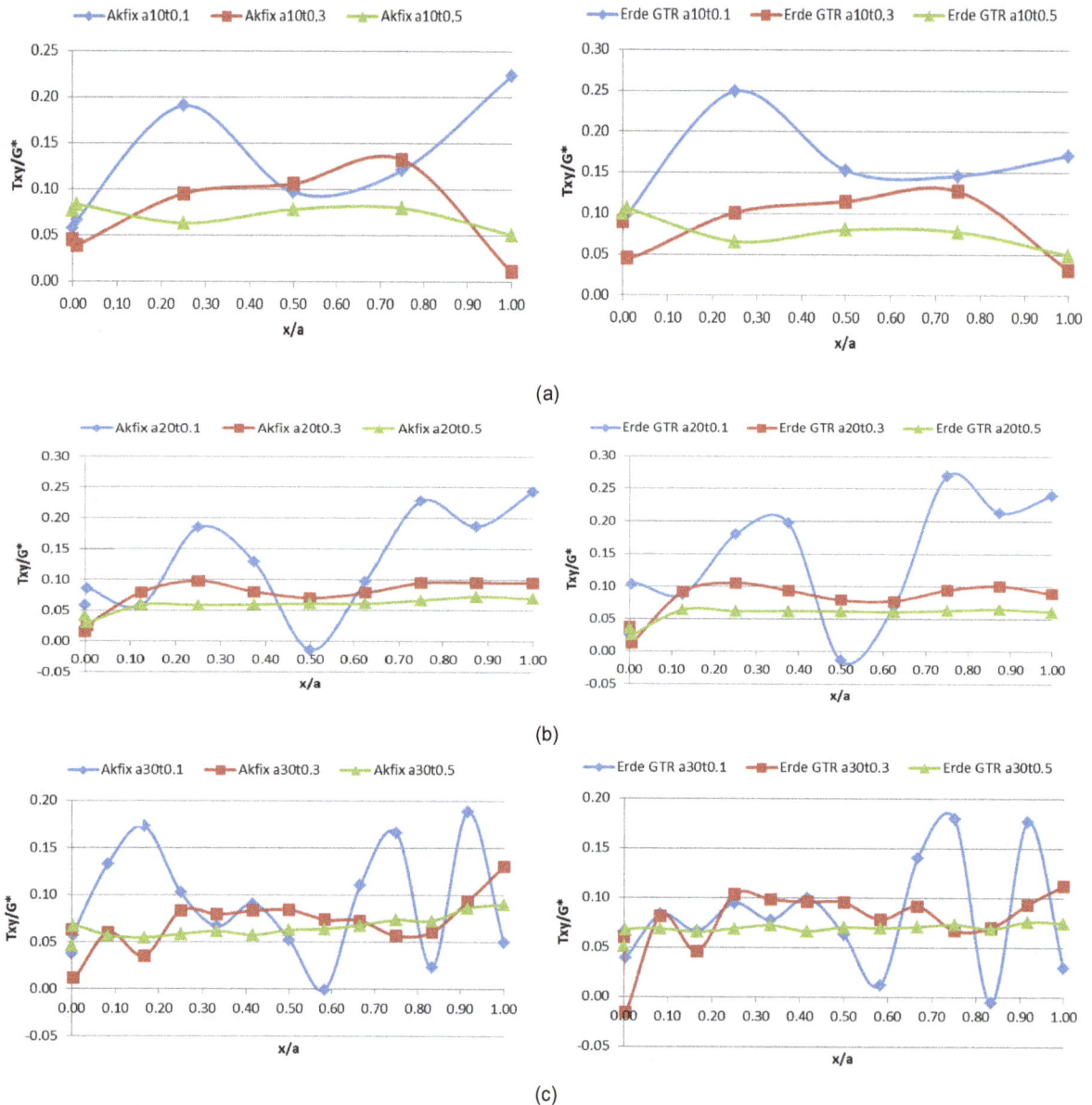

Figure 11. Shear stress distribution occurring throughout the A-C line depending on the overlap distance and adhesive type. a) 10 mm overlap distance, b) 20 mm overlap distance, c) 30 mm overlap distance.

different values are obtained at the end of the joint.

For the 20 mm overlap distance, shear stress spreads throughout the A-C line by being distributed regularly in both adhesives. This situation becomes more apparent as the adhesive thickness increases. The thickness values of t = 0.3 and t = 0.5 provide a more regular formation in terms of shear stress.

In the 30 mm overlap distance, it is seen that shear stress is disrupted in both adhesives and the stress distribution becomes more regular as the adhesive thickness increases. For the adhesive thickness of t = 0.1, the stress is not distributed regularly on the adhesive layer and gets the value of 0 at the centre of the joint by

spreading in a corrugated form. It is seen that shear stress distribution is more regular for the adhesive thickness of t = 0.5.

Conclusions

When the distribution of equivalent stress (σ_{eqv}) over the adherend materials are examined according to the results of the ANSYS finite element analysis, it was seen that adhesive thickness is highly effective on the distribution of the stress as a result of the analysis made for the different adhesive thickness values with the same overlap distance. Although, all stress values occurring on the adherend materials and interfaces of the adhesive materials decrease as the adhesive thickness increases, stress distribution forms a regular structure throughout the adhesion line. While the equivalent stress distribution is in a corrugated form and has the maximum value for the adhesive thickness of t = 0.1, the stress distribution is highly regular and has the minimum value for the adhesive thickness of t = 0.5. The maximum equivalent stress values get the maximum values in Erde GTR and minimum values in Akfix E300. Even in the joints on which Erde GTR shows its lowest performance, equal stress values are obtained with Akfix E300. And this means that Erde GTR which is more flexible and shows nonlinear characteristics can carry more loads.

When normal stresses (σ_x) are considered, it is seen that normal stress values increase as the adhesive thickness increases in the same overlap distance. In different overlap distances, the normal stress value also increases as overlap distance increases. Maximum values for normal stress are mostly obtained with Erde GTR.

The structure seen in the distribution of peeling stress (σ_y) is similar to normal stress (σ_x). When compared with normal stress, the difference of peeling stress is the fact that it shows a small decrease with all adhesives ($\sigma_x/\sigma^* > \sigma_y/\sigma^*$). When peeling stress values are examined, it is seen that Erde GTR gets the maximum values. This situation is due to the fact that even though the tensile strength of the other adhesive is higher than Erde GTR, Erde GTR, which is a more flexible adhesive, takes extensive deformations over by distributing more evenly the peeling strengths occurring at the edge of joints.

When shear stress values (τ_{xy}) are examined, it is seen that the adhesive thickness is highly effective on the values and distribution of shear stress. The shear stress values also decrease as the adhesive thickness increases. Increasing overlap distance has an effect on the distribution and change of the maximum and minimum points of the shear stress rather than its intensity. It is seen that the adhesive thickness of t = 0.5 mm is suitable in terms of the distribution of shear stress. When shear stress values are examined, it is seen that Erde GTR gets maximum values. Although shear stress values show small differences depending on the overlap distances and adhesive thickness, these values are obtained as Erde GTR > Akfix E300.

REFERENCES

Abou-Hamda MM, Megahed MM, Hammouda MMI (1998). Fatigue crack growth in double cantilever beam specimen with an adhesive layer. Eng. Fract. Mech. 60:605-614.

Adin H (2012). The Effect of Angle On The Strain of Scarf Lap Joints Subjected to Tensile Loads. Appl. Math. Model. 36(7):2858-2867.

ANSYS ® (V12.0). The general purpose finite element software. Swanson Analysis Systems. Houston, TX.

ASTM D1002 (1983). Standard test method for strength properties of adhesives in shear by tension loading (metal-to-metal).

Aydin MD, Temiz Ş, Özel A (2004). It determined Experimental Methods in Mechanical Properties of Structural Adhesives. Eng. Mach. 45(536):18-24.

Aydin S, Solmaz Y, Turgut A (2012). The effects of adhesive thickness, surface roughness and overlap distance on joint strength in prismatic plug-in joints attached with adhesive. Int. J. Phys. Sci. 7(17):2580-2586. DOI: 10.5897/IJPS12.208.

Azari S, Papini M, Spelt JK (2011). Effect of adhesive thickness on fatigue and fracture of toughened epoxy joints - Part I:Experiments. Eng. Fract. Mech. 78:153-162.

Bascom WD, Cottington RL, Jones RL, Peyser P (1975). The fracture of epoxy and elastomer-modified epoxy polymers in bulk and as adhesives. J. Appl. Polym. Sci. 19:2545-2562.

Bascom WD, Cottington RL, Timmons CO (1977). Fracture reliability of structural adhesives. J. Appl. Polym. Sci. Appl. Polym. Symp. 32:165-88.

Chai H (1988). Fracture work of thin bondline adhesive joints. J. Mater. Sci. Lett. 7:399-401.

Ciba-Geigy (1993). Ciba composites:Redux Bonding Technology, Duxford-Chambridge, Pub. No. RGU 201A.

Daghiyani HR, Ye L, Mai Y-W (1995). Mode-I fracture behaviour of adhesive joints Part I relationship between fracture energy and bond thickness. J. Adhes. 53:149-162.

Da Silva LFM, Carbas RJC, Critchlow GW, Figueiredo MAV, Brown K (2009). Effect of material surface treatment and environment on the shear strength of single lap joints. Int. J. Adhes. Adhes. 29:621-632.

Davies P, Sohier L, Cognard Y, Bourmaud A, Choqueuse D, Rinnert E, Créac'hcadec R (2009). Influence of adhesive bondline thickness on joint strength. Int. J. Adh. Adh. 29:724-736.

Goland M, Reissner E (1944). The Stresses in Cemented Joints. J. App. Mec. 11:59-45.

Grant LDR, Adams RD, DaSilva LFM (2009). Experimental and numerical analysis of single lap joints for the automotive industry. Int. J. Adh. Adh. 29:405-413.

ISO 527-2 (1993). Plastics-Determination of tensile properties-Part 2: Test conditions for Mouldingand extrusion plastics. International Standard ISO 527-2:1993 (E).

Jarry E, Shenoi RA (2006). Performance of butt strap joints for marine applications. Int. J. Adhes. Adhes. 26:162-176.

Kahramana R, Sunarb M, Yilbas B (2008). Influence of adhesive thickness and filler content on the mechanical performance of aluminum single-lap joints bonded with aluminum powder filled epoxy adhesive. J. Mater. Process Technol. 205:183-189.

Kawashita LF, Kinloch AJ, Moore DR, Williams JG (2008). The influence of bond line thickness and peel arm thickness on adhesive fracture toughness of rubber toughened epoxy-aluminium alloy laminates. Int. J. Adhes. Adhes. 28:199-210.

Kimiyoshi N, Mutsumi O, Yasuo K (2012). The effect of adhesive thickness on tensile and shear strength of polyimide adhesive. Int. J. Adhes. Adhes. 36:77-85.

Loctite (1998). Worldwide Design Handbook. 2nd Edition on CD.

Mall S, Ramamurthy G (1989). Effect of bond thickness on fracture and fatigue strength of adhesively bonded composite joints. Int. J. Adhes. 9(1):33-37.

Morrisey MA, Johnson WR (1985). Douglas Aircraft Company Design Handbook. Adhesive and Cements, California.

Nemeş O, Lachaud F (2010). Double-lap adhesive bonded-joints assemblies modeling. Int. J. Adhes. Adhes. 30:288-297.

Schmueser DW, Johnson NL (1990). Effect of bondline thickness on mixed-mode debonding of adhesive joints to electroprimed steel surfaces. J. Adhes. 32:171-191.

Solmaz MY (2008). Mechanical Analysis And Design Of Adhesive Bonded Joints. PhD Thesis Firat University Graduate School of Natural and Applied Sciences.

Tamblin JS, Yang C, Harter P (2001). Investigation of thick bondline adhesive joints. DOT/FAA/AR- June.01:33.

Turgut A, Sancaktar E (1991). The Effects of Cure and Loading Conditions on Fiber matrix Adhesion. Adhes. Soc. 41:24-26.

25

Theoretical studies of CO converted to CH$_3$OH by ZINDO/1-DFT methods

Leila Mahdavian

Department of Chemistry, Doroud Branch, Islamic Azad University, P.O. Box: 133. Doroud Iran.

The aim of this study is to simulate co-catalysts of titanium nitride nanotubes (TiN-NTs) with Cu nano-particles for converting carbon monoxide (CO) and water into methanol. This method may provide a new way to reduce carbon-monoxide levels in the atmosphere rising due to our planet's heavy use of fossil fuels as well as to produce alternative fuels. Using TiN-NTs has shown the efficiency to remove CO from the atmosphere. There are four positions for CO and TiN-NTs (4, 4) that we have investigated, passing of CO endohedrally and between the nanotubes. We study the structural, total energy, thermodynamic and conductive properties of converted CO on Cu nano-particles in TiN-NTs to methanol. The electronic and geometric structures and thermodynamic properties are quantum mechanically calculated for these situations by DFT methods at the semi-empirical ZINDO/1 level, B3LYP/6-31G basis sets. The thermodynamic properties show which interactions are endothermic. The heat of exhaust can be used for CO conversion to CH$_3$OH or other products.

Key words: Carbon monoxide (CO), Methanol, Titanium nitride nanotubes (TiN-NTs), ZINDO/1, environment.

INTRODUCTION

Carbon monoxide (CO) is a colorless, odorless, and tasteless gas that is somewhat less dense than air. It is dangerous to humans and animals when withstood in larger levels, though it is also produced in standard pet metabolism in minimal amounts, and is considered to have some standard organic functions. In the air, it is spatially variable and temporary, having a role in the formation of ground-level ozone (Riduan et al., 2009).

Solid state gas sensors are most important in semiconductor processing, medical diagnosis, environmental sensing, and personal safety etc. (Kolmakov et al., 2003). The high surface to volume ratio of nano-forms makes them natural contenders as new sensors and suitable surface to remove and converted pollutants.

TiN-nanotubes (TiN-NTs) have attracted great interest due to their unique electronic properties and nanometer size. Because of these unique properties, they are great potential candidates in many important applications such as nanoscale electronic devices, chemical sensors and field emitters. The effect of gas adsorption on the electrical resistance of a TiN-NT has received great attraction because of fast response, good sensitivity of chemical environment gases and low operating temperature (Paulose et al., 2006).

The products of the carbon monoxide (CO) and water

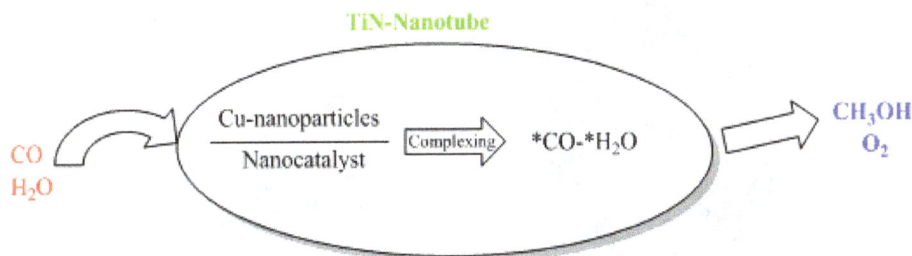

Figure 1. The conversion of CO into methanol.

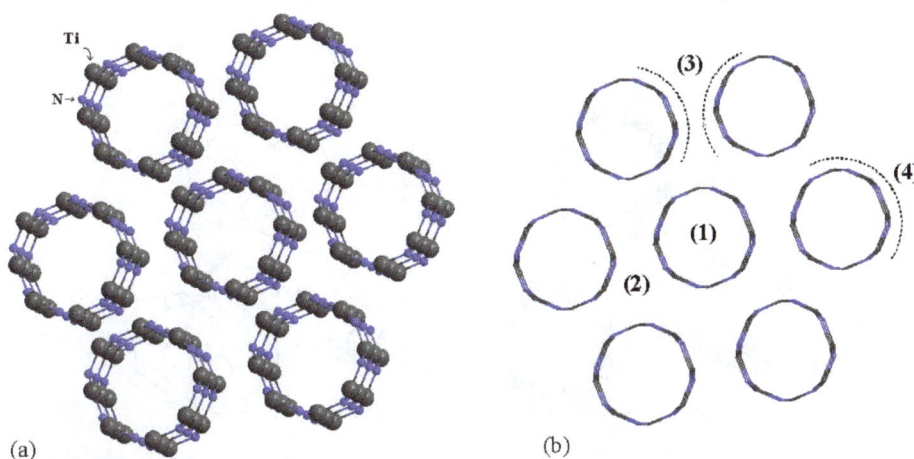

Figure 2. There are four situations for interaction CO and H_2O in TiN-nanotube, (a) Ball-and-stick and (b) stick models Configuration for them.

(H_2O) reaction depend on metal oxide catalysis, temperature and pressures [Li et al., 2007]. The most important challenge is that of the catalyst. The current commercial catalysts are co-catalysts of titanium nitride nanotubes with Cu- nanoparticles that help converts carbon monoxide and water into methanol using sunlight as the power source for sun gas conversion (Figure 1). There are some methods such as options for carbon monoxide and carbon dioxide removal from the atmosphere include afforestation and chemical approaches like direct air capture of CO from the atmosphere or reactions of CO with minerals to form carbonates [Varghese et al., 2009; Farha et al., 2010; Yang et al., 2012; Aresta and Dibenedetto, 2004; Hohenberg and Kohn, 1964]. Methane and methanol are major products of the chemical industry and also a feedstock for many chemicals. However CO conversion to methanol and other product is challenging.

As shown in Figure 2, there are four situations in which CO and H_2O can pass between TiNNTs-Cu. In this work, the 1st and 2nd situations are investigated for them. In Figures 3 and 4, TiN- nanotubes with Cu-nanoparticle simulated by ball-and-stick models, CO and H_2O converted to CH_3OH and O_2.

Interaction between CO and H_2O on Cu nanoparticle in TiN-nanotubes is investigated using ZINDO/1 method by semi empirical. We study the structural, total energy, thermodynamic properties of them in room temperature. All the geometry optimization structures were carried out using Gaussian program package. Density Functional Theory (DFT) optimized their intermediates and transient states. The results show a sensitivity enhancement in resistance and capacitance when CO and H_2O are converted to CH_3OH and O_2.

THE COMPUTATIONAL METHODS

The computational approach consists of three stages: First, all the geometry optimization structures in present work were performed by employing DFT, using Becke's three parameter hybrid with the Lee-Yang-Parr correlation functional, B3LYP (Becke, 1993; Yang and Parr, 1988) method with the 6-31G basis sets. Then the thermodynamic properties of them were calculated for different distances. We can use the information obtained from semi-empirical calculations to investigate many thermodynamic aspects of chemical processes. The heat of formation is calculated by subtracting atomic heats of formation from the binding energy. ZINDO/1 has been used widely to calculate heats of formation, molecular geometries, dipole moments, ionization energies,

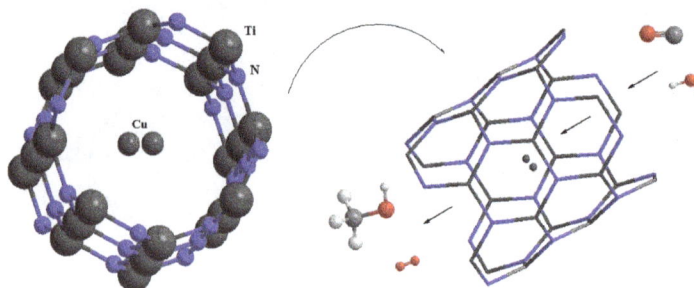

Figure 3. Ball-and-stick models configuration: Top-view of first situation passing CO and H₂O endohedral of TiN-nanotube with Cu-nanoparticle.

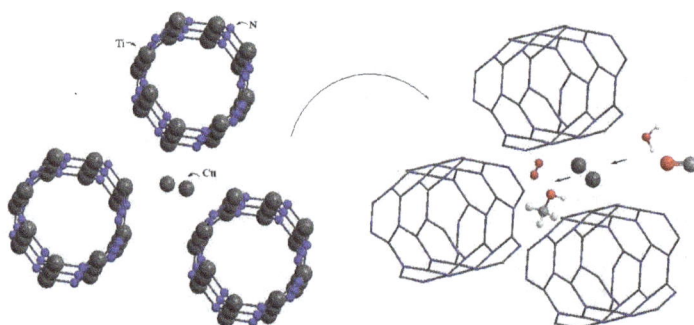

Figure 4. Ball-and-stick models configuration: Top-view of second situation passing CO and H₂O between TiN-nanotube with Cu-nanoparticle.

electron affinities, and other properties (Ridley and Zerner, 1973; Zerner, 1991). At the end, the reaction pathway is analyzed in order to confirm the structure of the transition state obtained.

In this study, entry carbon monoxide and water on Cu nanoparticles between titanium nitride nanotubes have been simulated in three steps (Figures 3 and 4), and then in steps 4 and 5 are transition state conversion of them to CH_3OH and O_2. The conversion was completed in 6th step, methanol and oxygen are produced, and then it is excretion from TiN-NTs in 7th or 8th step. The electronic structure and the thermodynamic properties are calculated for all steps by ZINDO/1.

RESULTS AND DISCUSSION

The electronic structures and properties of CO converted to CH_3OH have been investigated using hybrid density functional theories with the basis set B3LYP/6–31G use correct logical order. In this work, we use TiN-NTs from kind of armchair TiN nanotubes (4, 4) that show in Figures 3 and 4. When CO gas from the exhaust exits to the filter of TiN-NTs with Cu nanoparticles (into or between of TiN-NTs), electron exchange between them occurs and CO converts to CH_3OH or other products. Table 1 show thermodynamice parameters of monoxide carbon and water endohedral passing in TiN-NT by

seven steps and their passing between TiN-nanotubes (second state) shown in Table 2. The thermodynamic properties for all steps are without the effect of the catalyst in the reaction for example:

$$\Delta E_{total} = E_{(TiN\text{-}Cu + CO + water)} - E_{(TiN\text{-}Cu)} \qquad (1)$$

The study includes conformational searches (and further refinement by DFT) and semi- empirical by ZINDO/1 methods. The most significant property is the ZINDO/1, which is, finding a good correlation between the ZINDO/1 and the substitution pattern on this conversation. ZINDO/1 has been used widely to calculate heats of formation, molecular geometries, dipole moments, ionization energies, electron affinities, and other properties (Reslan et al., 2012; Dondela et al., 2005).

In Tables 1 and 2, the dipole moment (D) measurement gives an idea about the degree of polarity for approached of CO and H_2O to Cu nanoparticles into and between TiN-NT. When CO passing in nanotube, dipole moment (D) decrease, so in 4th and 5th steps increase polar because in these steps occur exchange electron between them and is formed CH_3OH and O_2 for two situations. The root mean square (RMS) gradient (kcal/mol. Å) is different for this interaction at room temperature.

Table 1. The thermodynamic properties of interaction CO and H_2O into Cu-TiN nanotube (4, 4) to CH_3OH at 298K (ZNDO/1).

	The CO and H_2O endohedral passing in Cu-TiN nanotube						
	E_{total} (kJ/mol)	E_{nuc} (kJ/mol)	Dipol Moment (D)	RMS kcal/mol.$^{\circ}$A	E_{bin} (kJ/mol)	H (kJ/mol)	G_{elec} (kJ/mol)
TiN-Cu	1845450	8280830	3012	1.63×10^4	2984390	3007790	3064840
Steps	ΔE_{total} (kJ/mol)	ΔE_{nuc} (kJ/mol)	Δ(Dipol Moment) (D)	Δ(RMS) kcal/mol.$^{\circ}$A	ΔE_{bin} (kJ/mol)	ΔH (kJ/mol)	ΔG_{elec} (kJ/mol)
1	234770	1021660	25	-0.06×10^4	34010	341740	374420
2	462610	1163420	10	-0.08×10^4	56794	569580	333890
3	330140	1351290	4	-0.07×10^4	43547	437100	486360
4	51650	1604210	247	-0.12×10^4	15698	158620	739190
5	582100	2020920	337	-0.13×10^4	73393	736260	685150
6	588500	1910460	22	-0.09×10^4	74033	742660	629180
7	404040	1632390	69	-0.09×10^4	55584	558170	584790

Table 2. The thermodynamic properties of interaction CO and H_2O between Cu-TiN nanotube (4,4) to CH_3OH at 298K (ZNDO/1).

	The CO and H_2O passing between Cu-TiN nanotube						
	E_{total} (kJ/mol)	E_{nuc} (kJ/mol)	Dipol Moment (D)	RMS kcal/mol.$^{\circ}$A	E_{bin} (kJ/mol)	H (kJ/mol)	G_{elec} (kJ/mol)
3TiN-Cu	18844550	39060150	1741	8335	21706320	2177513	962877
Steps	ΔE_{total} (kJ/mol)	ΔE_{nuc} (kJ/mol)	Δ(Dipol Moment) (D)	Δ(RMS) kcal/mol.$^{\circ}$A	ΔE_{bin} (kJ/mol)	ΔH (kJ/mol)	ΔG_{elec} (kJ/mol)
1	-4033420	1210270	328	2125	-3928100	-3926440	2497420
2	-3017480	1560550	214	1461	-5879250	-2910500	2180900
3	2203560	1654000	24	122	2308870	2310530	-262480
4	-4737300	1778780	435	2345	-4631990	-4630340	3103440
5	-3521920	1960840	422	2565	-3416610	-3414950	2611290
6	2488480	2776170	20	-94	2640300	2642640	137030
7	3152810	2448490	58	181	3304630	3306970	-335820
8	1828020	1608430	66	-149	1979840	1982190	-106150

The total energy (E_{total}) is minimum amount for 4th step, is 51650 (endohedral) and -4737300 kJ/mol (between tube), after its conversion, E_{total} increase to 582100 kJ/mol in 5th step for endohedral passing and -3521920 kJ/mol for between passing (Figures 5 and 6).

The ΔE_{nuc}, ΔE_{bin}, ΔH and ΔG_{elec} increase in 5th step that methanol molecule is formed in this interaction for tow situations. Thermodynamic equilibrium constants, K, for these interactions were calculated by the related standard Gibbs free energy difference (ΔG_{elec}):

$$K = \exp\ (-\Delta G_{elec}/RT) \qquad (2)$$

Where, T is the transition temperature and R is gas constant. The entropy difference (ΔS) at the phase transitions are given by:

$$\Delta S = \frac{\Delta H}{T} \qquad (3)$$

Where, ΔH is the electronic enthalpy difference. In Table 3 is shown the other thermodynamic properties for these interactions.

Thermodynamic parameters (ΔG_{ele}, ΔH_{ele} and ΔS_{ele}) for CO and H_2O converted to CH_3OH and O_2 were calculated and the results suggest that the nature of adsorption is endothermic.

This method needs energy that can be provided from heat of exhaust gas or solar energy. The calculation of thermodynamics is evaluated due to the comparison with experimental values. The calculated data are in good agreement with the experimental spectra (Varghese et al., 2009).

Conversion of CO by TiN-nanotube is caused changing in the electric structure of nanotube, because CO and

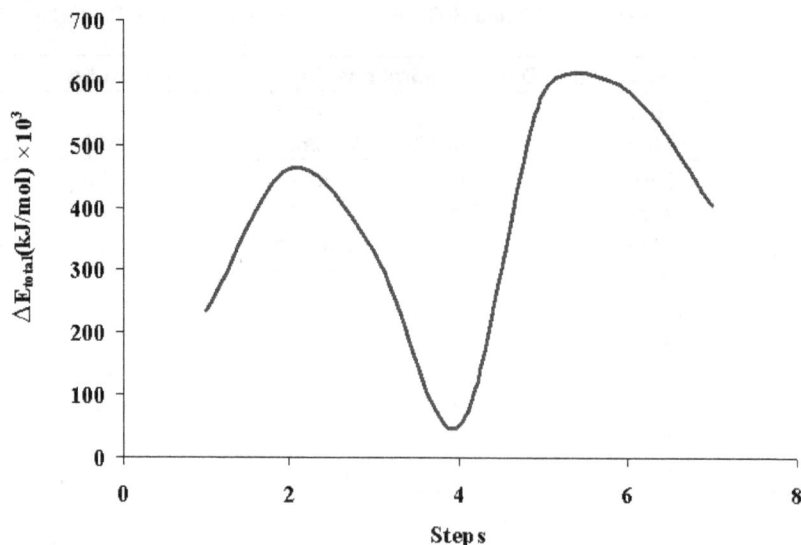

Figure 5. The total energy (MJ/mol) of converted CO and H_2O to CH_3OH into Cu-TiN nanotube.

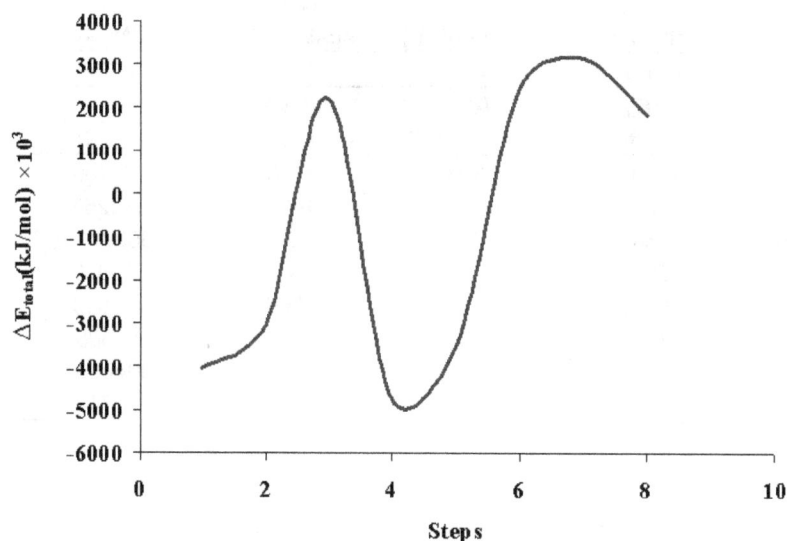

Figure 6. The total energy (kJ/mol) of converted CO and H_2O to CH_3OH between Cu-TiN nanotubes.

Table 3. The thermodynamic properties of CO and H_2O passing Cu-TiN and convert to CH_3OH and O_2.

	ΔG_{ele} (kJ/mol)	ΔH_{ele} (kJ/mol)	ΔS_{ele} (kJ/mol)	K
CO and H_2O passing into Cu-TiN	113740	-765660	-2950	-45907.93
CO and H_2O passing between Cu-TiN	1033320	1215390	610	-417068.65

H_2O are adsorbed on Cu in TiN-NT by weak bond (Figures 3 and 4). The electric resistance for them is following as:

$$E_{elec} = RI \qquad (4)$$

Where, E_{elec} is electric energy (V), R (Ω) is electric resistance and I (A) is electric intensity that is $I = \frac{q}{t}$ and q(C) is electric charge and t is time interaction, in experimental data, it is so:

Time (s)

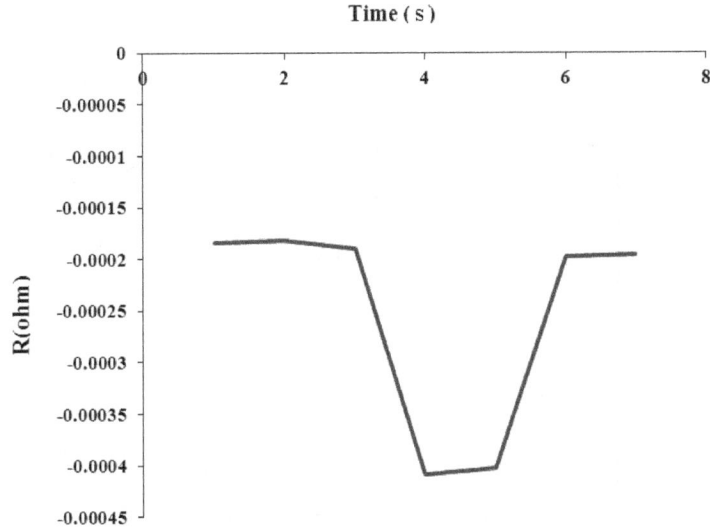

Figure 7. The resistance (Ω), interaction CO and H_2O into Cu-TiN nanotube at 298K.

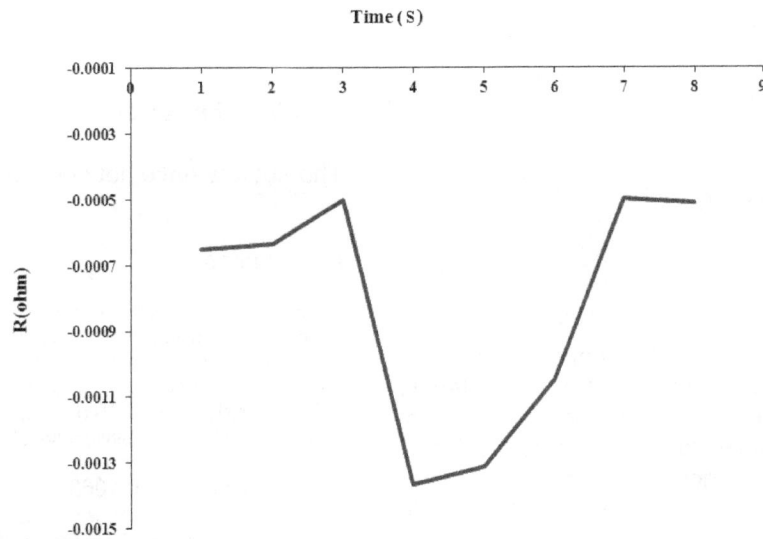

Time (s)

Figure 8. The resistance (Ω), interaction CO and H_2O between Cu-TiN nanotubes at 298K.

$$R = \frac{E_{elec}t}{nF} \qquad (5)$$

Where, n, F and t are electron number of conversion, faraday constant and time (h) respectively (Lee, 2005). After adsorption of CO and H_2O on surface of Cu nanoparticles in TiN-NT and transition electron between them, the electric resistance to time (s) decreased that showed in Figures 7 and 8, when methanol is formed, the electric resistance is increasing.

TiN-nanotube is used as a photocatalyst in photocatalysis process, because is a semiconductor (Varghese et al., 2009). The absorption of a photon (hv) with ultra-band energy from UV irradiation source is caused TiN activation (Figure 9). The transmission of an electron (e^-) from the valence band to the conduction band is caused highly reactive positive holes (h^+) in the valence band, therefore this transfer is been to adsorption and conversion to pollutants on the Cu nanoparticles into low-risk products in environment:

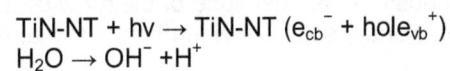

$$TiN\text{-}NT + hv \rightarrow TiN\text{-}NT\ (e_{cb}^- + hole_{vb}^+)$$
$$H_2O \rightarrow OH^- + H^+$$

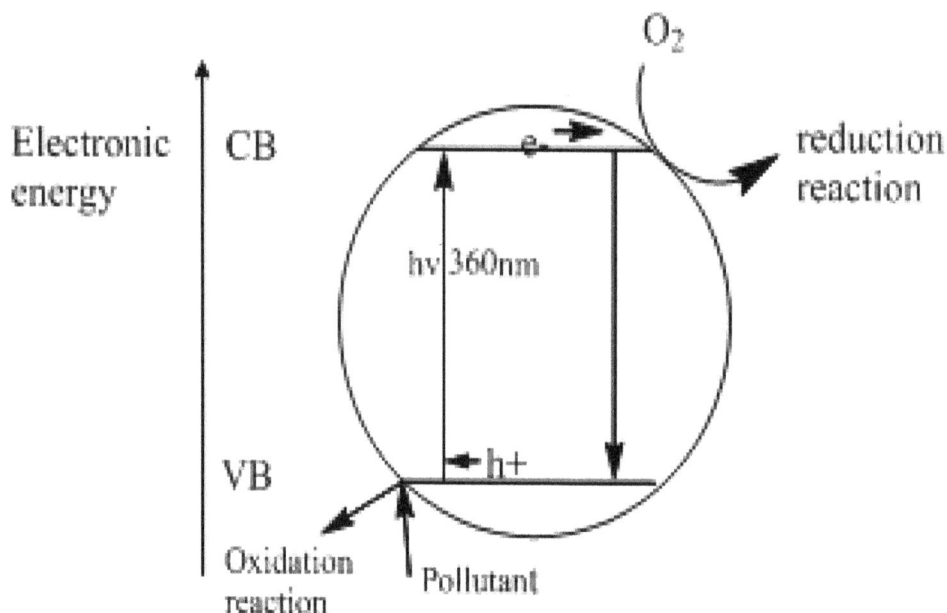

Figure 9. The electron transition from valance band (VB) to conduction band (CB) in semiconductor Cu-TiN-nanotube.

Oxidative reaction:

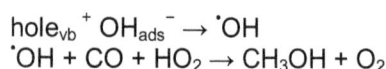

$$hole_{vb}^{+} \; OH_{ads}^{-} \rightarrow {}^{\bullet}OH$$
$${}^{\bullet}OH + CO + HO_2 \rightarrow CH_3OH + O_2$$

Conclusion

After adsorption of CO onto the nanosurface and transference of electrons between them, the electric resistance decreased. The TiN-NTs are semiconductors, photo-active catalysts, able to utilize near-UV light, biologically, chemically inert, photo-stable and inexpensive, therefore TiN-NT is appropriate for this conversion in environment.

The co-catalysts of Cu and TiN with a nano structure significantly enhance the photocatalytic reduction of CO with H_2O to CH_3OH and O_2. In this manuscript this conversion to CH_3OH is calculated, the local minimum geometries of them were determined at B3LYP/6-31G level. The ZINDO/1 method is used for calculation of thermodynamic parameters these interactions. The calculation which shows heat reaction formation (ΔH) is endothermic for these reactions. These reactions need sun, photo active or other energy in presence of visible light. We propose, TiN nanotubes with Cu nanoparticles to form filter to be installed in exhaust of automobile and heat of exit gases provides energy to the converter. These techniques have been highly considered by researchers and industrial men because of their low cost, high adsorption efficiency, selective operation, easy and hazardless application.

Conflict of Interest

The authors have not declared any conflict of interest.

REFERENCES

Aresta M, Dibenedetto A (2004). The contribution of the utilization option to reducing the CO_2 atmospheric loading: research needed to overcome existing barriers for a full exploitation of the potential of the CO2 use. Catal. Today. 98:455–462. http://dx.doi.org/10.1016/j.cattod.2004.09.001

Becke D (1993). Densityfunctional thermochemistry. III. The role of exact exchange. J.Chem.Phys. 98:5648-5652. http://dx.doi.org/10.1063/1.464913

Dondela B, Peszke J, Śliwa W (2005). Semiempirical ZINDO/S, AM1 and ab initio HF/STO-3G and HF/6-31G study of quaternary salts of diazaphenanthrenes with ethyl bromide and 1,3-dibromopropane. J. Mol. Structure. 753(1–3):154–164. http://dx.doi.org/10.1016/j.molstruc.2005.06.016

Farha OK, Yazaydin O, Eryazici I, Malliakas C, Hauser B, Kanatzidis MG, Nguyen ST, Snurr RQ, Hupp JT (2010). De novo synthesis of a metal-organic framework material featuring ultrahigh surface area and gas storage capacities. Nat. Chem. 2:944–948. http://dx.doi.org/10.1038/nchem.834

Hohenberg P, Kohn W (1964). Inhomogeneous electron gas. Phys. Rev.136:B864-B871. http://dx.doi.org/10.1103/PhysRev.136.B864

Kolmakov A, Zhang Y, Cheng G, Moskovits M (2003). Detection of CO and O2 using Tin Oxide Nanowire Sensors. Adv. Mater. 15(12):997-1000. http://dx.doi.org/10.1002/adma.200304889

Lee S (2005). Encyclopedia of chemical processing and design. Taylor & Francis, 2005. ISBN: 0824755634, 9780824755638.

Li J, Zhao Z, Kazakov A, Chaos M, Dryer FL, Jr JJS (2007). A comprehensive kinetic mechanism for CO, CH2O, and CH3OH combustion. Int. J. Chemical Kinetics. 39(3):109–136. http://dx.doi.org/10.1002/kin.20218

Paulose M, Shankar K, Yoriya S, Prakasam HE, Varghese OK, Mor GK,

Latempa TA, Fitzgerald A, Grimes CA (2006). Anodic Growth of Highly Ordered TiO_2 Nanotube Arrays to 134 μm in Length. J. Phys. Chem. B 110(33):16179-16184. http://dx.doi.org/10.1021/jp064020k

Reslan R, Lopata K, Arntsen C, Govind N, Neuhauser D (2012). Electron transfer beyond the static picture: A TDDFT/TD-ZINDO study of a pentacene dimer. HE. J. Chem. Phys. 137:22A502-6. http://dx.doi.org/10.1063/1.4729047

Ridley J, Zerner M (1973). An intermediate neglect of differential overlap technique for spectroscopy: Pyrrole and the azines. Theoretica chimica acta. 32(2):111-134. http://dx.doi.org/10.1007/BF00528484

Riduan SN, Zhang Y, Ying JY (2009). Conversion of carbon dioxide into methanol with silanes over N-Heterocyclic Carbene Catalysts. Angew. Chem. 121:3372–3375. http://dx.doi.org/10.1002/ange.200806058

Varghese OK, Paulose M, LaTempa TJ, Grimes AG (2009). High-rate solar photocatalytic conversion of CO_2 and water vapor to hydrocarbon fuels. Nano. Lett. 9(2):731-737. http://dx.doi.org/10.1021/nl803258p

Yang DA, Cho HY, Kim J, Yang ST, Ahn WS (2012). CO_2 capture and conversion using Mg-MOF-74 prepared by a sonochemical method. Energy Environ. Sci. 5:6465–6473. http://dx.doi.org/10.1039/c1ee02234b

Yang LW, Parr RG (1988). Development of the colle-salvetti correlation-energy formula into a functional of the electron density Phys. Rev. B: Condens. Matter. 37:785-789. http://dx.doi.org/10.1103/PhysRevB.37.785

Zerner M (1991). Reviews in computational chemistry, Volume 2, Eds. K. B. Lipkowitz and D. B. Boyd, VCH, New York, P. 313. http://dx.doi.org/10.1002/9780470125793.ch8

26

Wet chemical synthesis of Tin Sulfide nanoparticles and its characterization

S. Suresh

Crystal Growth Centre, Anna University, Chennai-25, India.

Nanostructured materials have attracted much attention in various fields of science and technology. Tin sulfide (SnS) nanoparticles were successfully synthesized by wet chemical method. The as-prepared tin sulfide nanoparticles were characterized by X-ray diffraction, scanning electron microscopy (SEM), transmission electron microscopy (TEM) and dielectric studies. The SEM measurements show the aggregates of small nanoparticles. TEM images showed the presence of spherical tin sulfide nanoparticles of size in the range of 15 nm. The optical properties were obtained from UV-VIS absorption spectrum and the optical band gap was calculated. The dielectric constant calculated by varying the frequencies at different temperatures.

Key words: Preparation, SnS nanoparticles, scanning electron microscopy (SEM), transmission electron microscopy (TEM), optical property and dielectric constant.

INTRODUCTION

The consumption of fossil fuels has increased immensely in the recent years which made the role of renewable energy sources relevant. The other forms of energy such as coal, oil and gas are at the stage of extinction because of the extensive usage. Among the alternative energy sources, photovoltaics is known to be an almost maintenance free clean energy technology. In recent years, semiconductor nanostructures such as nanoparticles, nanorods, nanotubes, nanowires and nanobelts have attracted intensive interest due to their novel physiochemical properties that differ greatly from their bulk counterparts (Alivisatos et al., 1991). IV-VI semiconductor compounds such as PbSe, SnSe and SnS have been important role in different areas of materials science for several decades (Unger et al., 1986).

Nanomaterials were widely studied due to their unique physical and chemical properties and also its potential applications in different areas (Simon et al., 1998). These properties and potential applications have stimulated the search for new synthetic techniques for these materials. In recent years, great resources were devoted to the preparation of nanocrystals using a wide variety of methods including electrodeposition (Natter et al., 1998), solvothermal route (Zhang et al., 2003), thermal decomposition (Nayral et al., 1999) and chemical reduction (Yang et al., 2000). These efforts have led to the successful synthesis of many nanocrystals including metals (Shafi et al., 1998) oxides (Liu et al., 2001; Haubold et al., 2001), as well as sulfides (Price et al., 2000) which have already been used as optoelectronic

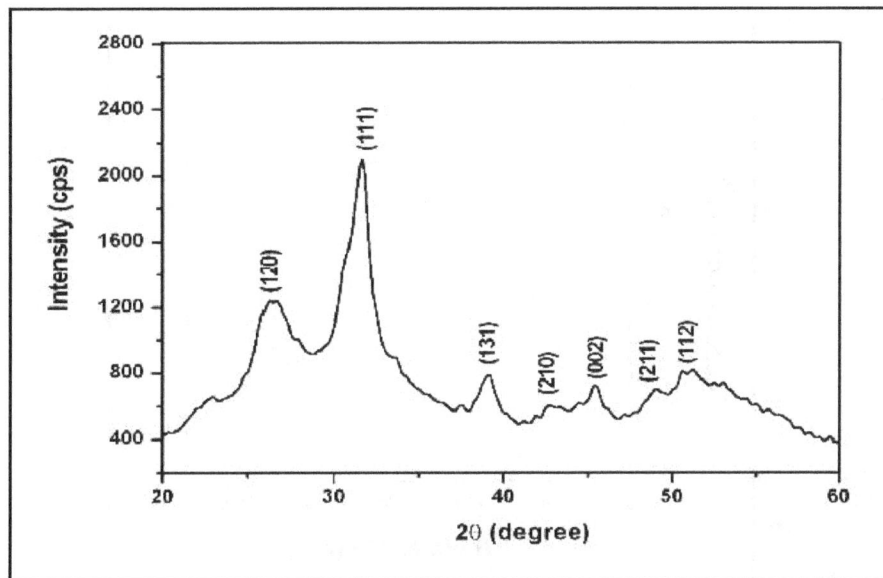

Figure 1. XRD pattern of tin sulfide nanoparticles.

materials in sensors, laser materials, solar cells and other devices. In this paper, we report a wet chemical method to prepare tin sulfide nanoparticles. Powder X-ray diffraction (XRD), scanning electron microscopy (SEM) and transmission electron microscopy (TEM) show the formation of tin sulfide nanoparticles possessing orthorhombic structure. The as-prepared tin sulfide nanoparticles display blue-UV emission, promising for applications in optical devices. Dielectric constant studies are carried out at different temperatures.

SYNTHESES OF TIN SULFIDE NANOPARTICLES

The tin sulfide nanoparticles were synthesized through wet chemical method. Tin (II) chloride ($SnCl_2$. $2H_2O$) and sodium sulfide (Na_2S) were taken as tin and sulfur sources respectively and deionized water was used as solvent. 1.2 g of tin (II) chloride and 1.72 g of sodium sulfide were dissolved in deionized water. Sodium sulfide solution was added drop wise into the solution. The colorless tin (II) chloride solution turns dark brown color with the addition of sodium sulfide solution. This indicates the formation of SnS nanoparticles. This reaction was carried out at room temperature for two hours. The precipitates were centrifuged and washed with deionized water and ethanol for several times and dried at room temperature.

RESULTS AND DISCUSSION

Structural studies

XRD is a non-destructive analytical method which identifies and determines various crystalline forms of materials. According to studies, the solution of nanoparticles obtained was purified by repeated

centrifugation at 10, 000 rpm followed by re-dispersion of the pellet of nanoparticles into distilled water. After freeze drying of the purified particles, the structure and composition of nanoparticles were analyzed by XRD. As waves interact with a regular structure the diffraction occurs. Figure 1 shows the XRD pattern of tin sulfide nanoparticles. All the diffraction peaks are indexed to pure orthorhombic phase of tin sulfide. This is due to agglomeration of the particles in the powdered sample and hence, XRD was used for phase identification only. The strong and sharp diffraction peaks indicate that the product is well crystallized. Phase purity is confirmed from powder X-ray diffraction. No other impurity peaks are observed. The broadening of the peaks indicates the nanocrystalline nature of SnS. By Knowing the wavelength (λ) full width at half maximum (FWHM) of the peaks β and the diffracting angle θ, particle size (D) was calculated by using the Scherrer formula,

$$D = \frac{0.9\lambda}{\beta \cos \theta} \qquad (1)$$

The average grain size of SnS is determined using Scherrer relation and it was found to be around 15 nm.

Optical studies

Optical absorption measurement was carried out on tin sulfide nanoparticles. Figure 2 shows the variation of the optical absorbance with the wavelength of the as-prepared tin sulfide nanoparticles. The optical absorption coefficient has been calculated in the wavelength range

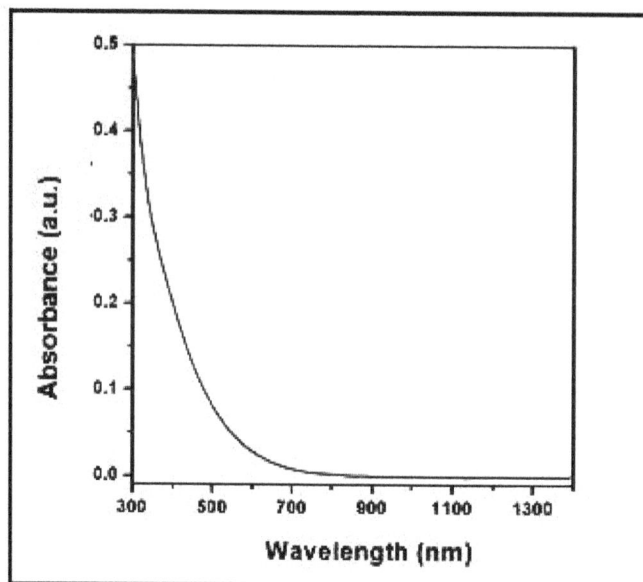

Figure 2. Optical absorption spectrum of SnS nanoparticles.

of 300 to 900 nm. It is clearly observed that the nanoparticles have a wide absorption range from the NIR to the UV, which means it is good for absorption of the sunlight. The absorption edge has been obtained at a shorter wavelength.   The band gap energy gap was estimated from the Equation (1)

$$E_g=1.243\times10^3/\lambda_{max} \qquad (2)$$

The band gap of tin sulfide nanoparticles was found to be 1.8 eV. In semiconductor nanoparticles, the band gap increases as the particle size decreases.

Scanning electron microscopy (SEM) analysis

External morphology, chemical composition, crystalline structure and orientation of materials making up the sample are revealed by SEM. Figure 3 shows the SEM image of tin sulfide nanoparticles. This SEM image reveals that the particles are in aggregation state due to their extremely small dimensions and high surface energy. It can be seen that the particles adopt irregular morphology with different sized particle. From the image it is clear that the particles were highly agglomerated in nature. This might be due to the fact that the agglomeration may be induced during the crystal growth itself because of the small size regime which is evident from the XRD analysis.

Transmission electron microscopy (TEM) analysis

TEM is commonly used for imaging and analytical

characterization of the nanoparticles to assess the shape, size, and morphology. The particle size distribution was also measured from the bright-field TEM image shown in Figure 4. The detection and measurement of the nanoparticles (segmentation) on this type of samples is difficult because thickness changes locally and diffraction from different crystal orientations introduce large contrast variations. Size of tin sulfide nanoparticles measured from TEM image is 15 nm.

Dielectric constant studies

Dielectric studies show the effects of temperature and frequency on the conduction phenomenon in nanostructured materials. Dielectric behavior can effectively be used to study the electrical properties of the grain boundaries. The dielectric properties of materials are mainly due to contributions from the electronic, ionic, dipolar and space charge polarizations. Among these, the most important contribution to the polycrystalline materials in bulk form is from the electronic polarization, present in the optical range of frequencies. The next contribution is from ionic polarization, which arises due to the relative displacement of the positive and negative ions. Dipolar or orientation polarization arises from molecules having a permanent electric dipole moment that can change its orientation when an electric field is applied. Space charge polarization arises from molecules having a permanent electric dipole moment that can change its orientation when an electric field is applied. The dielectric parameters, like the dielectric constant (ε_r) is the basic electrical properties of the tin sulfide nanoparticles. The measurement of the dielectric

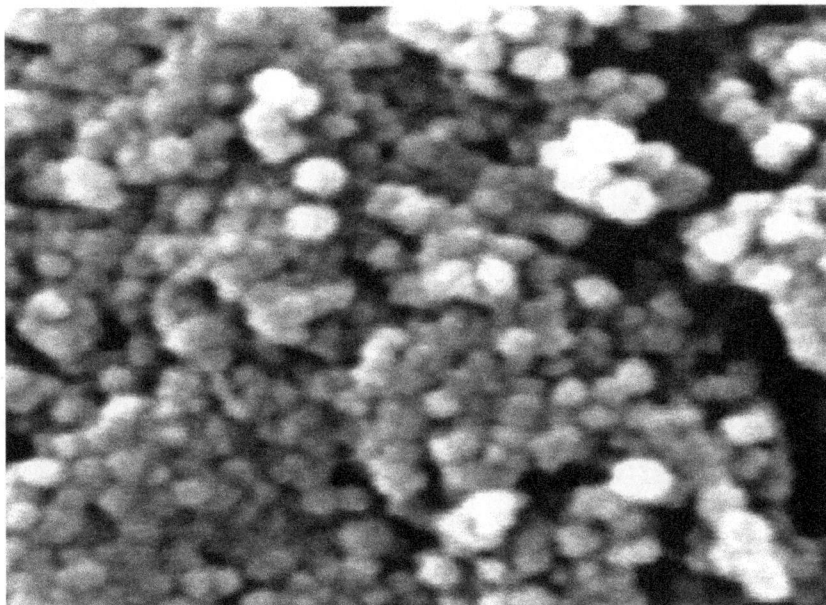

Figure 3. SEM image of tin sulfide nanoparticles.

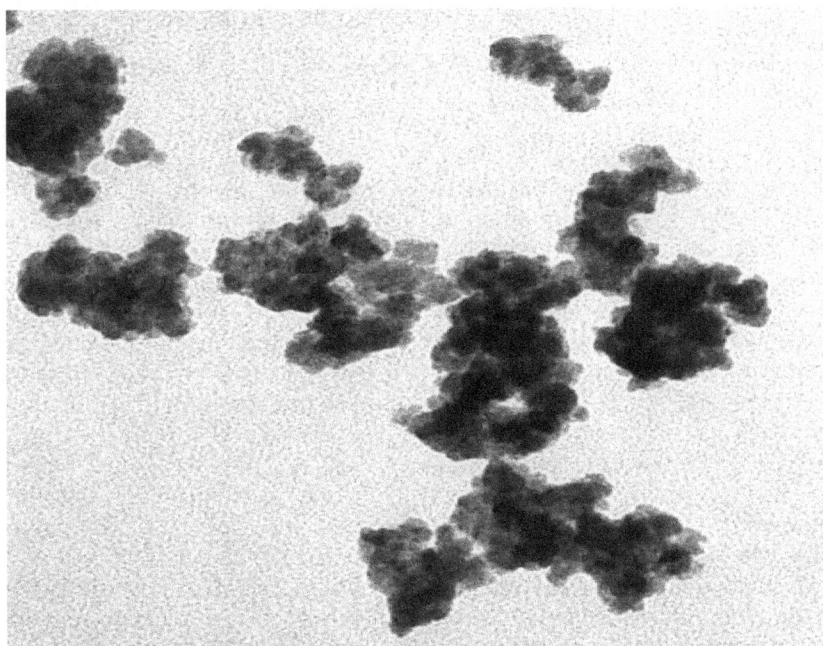

Figure 4. Typical TEM image of tin sulfide nanoparticles.

constant as a function of frequency and different temperatures reveals the electrical processes that take place in tin sulfide nanoparticles and these parameters have been measured. The variations of the dielectric constant of the tin sulfide nanoparticles at frequencies of 50Hz to 5 MHz and at different temperatures of 40, 50, 100 and120°C are displayed in Figure 5. The dielectric constant is evaluated using the relation:

$$\varepsilon_r = \frac{Cd}{\varepsilon_0 A} \tag{3}$$

Where d is the thickness of the sample and A, is the area of the sample. The results suggest that the dielectric constant strongly depend on the frequency of the a.c. signal and the different temperatures of the tin sulfide

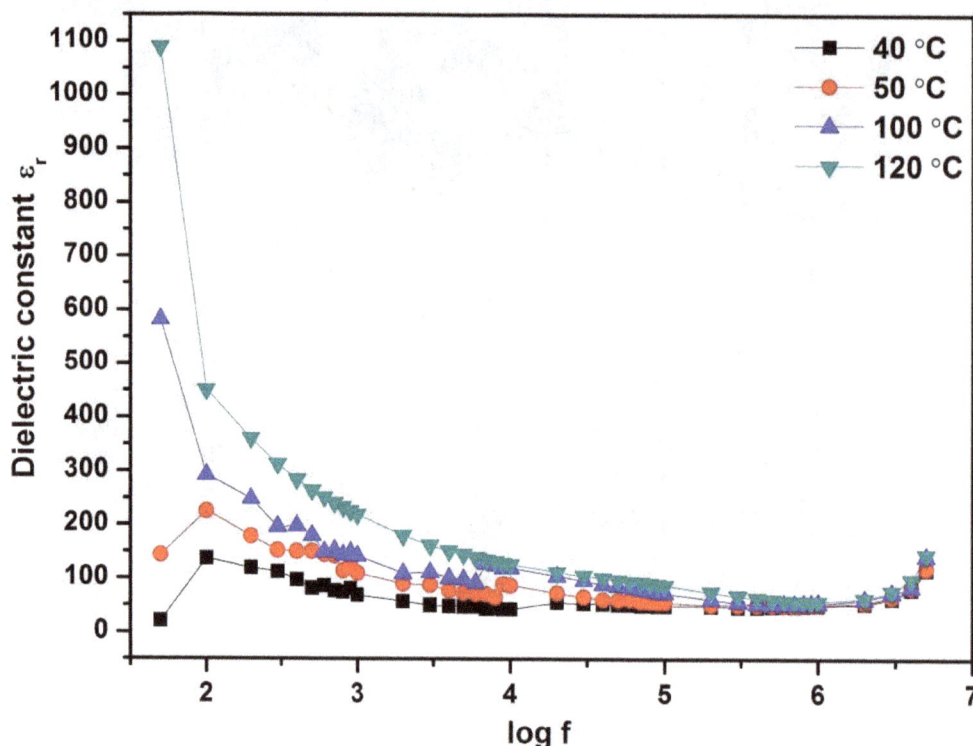

Figure 5. Variation of dielectric constant with log frequency.

nanoparticles. The dielectric constant has higher values in the lower-frequency (50 Hz) and then it decreases up to the high frequency (5 MHz). The dielectric constant of is high at lower frequencies due to the contribution of the electronic, ionic, dipolar and space charge polarizations, which depend on the frequencies (Xue et al., 2002). Space charge polarization is generally active at lower frequencies and indicates the purity and perfection of the nanoparticles. Its influence is strong at higher temperature and is noticeable in the low frequency range (Smyth et al., 1956).

The Figure 5 shows the variation of dielectric constant as function of frequency and temperatures. It is clear from Figure 5 that dielectric constant increases with the increase in temperature. This increase in dielectric constant as a result of increase in temperature can be explained on the basis of phenomenon that as the temperature increases, the dipoles relatively become free and they respond to the applied electric filed. Consequently the polarization increased and hence dielectric constant also increases with the increase in temperature. The variation of dielectric constant with frequency may be explained on the basis of space-charge polarization phenomenon (Rezlescu et al., 1974). According to this, dielectric material has well conducting grains separated by highly resistive grain boundaries. On the application of electric field, space charge accumulates at the grain boundaries and voltage drops

mainly at grain boundaries (Gul et al., 2010).

Most of the atoms in the nanocrystalline materials reside in the grain boundaries, which become electrically active as a result of charge trapping. The dipole moment can easily follow the changes in the electric field, especially at low frequencies. Hence, the contributions to the dielectric constant increase through space charge and rotation polarizations, which occur mainly in the interfaces. Therefore, the dielectric constant of nanostructured materials should be larger than that of the conventional materials. One of the reasons for the large dielectric constant of nanocrystalline materials at sufficiently high temperature is the increased space charge polarization due to the structure of their grain boundary interfaces. As the temperature increases, the space charge and ion jump polarization decrease, resulting in a decrease in the dielectric constant.

CONCLUSION

The tin sulfide nanoparticles were synthesized through wet chemical method. The XRD pattern revealed the orthorhombic structure of tin sulfide nanoparticles. SEM micrograph shows the aggregation state of tin sulfide nanoparticles. TEM images showed the presence of spherical tin sulfide nanoparticles of size in the range of 15 nm. In UV-VIS absorption spectrum shows that the nanoparticles have a wide absorption range and 1.8 eV direct allowed transition energy gap. The as-prepared tin sulfide nanoparticles have good crystalline and show strong blue-UV emission, promising for applications in optical devices. The dielectric property studied at different temperatures.

Conflict of Interest

The authors have not declared any conflict of interest.

REFERENCES

Alivisatos AP (1996). Semiconductor clusters, nanocrystals and quantum dots. Science 271: 933. http://dx.doi.org/10.1126/science.271.5251.933

Gul H, Maqsood A, Naeem M, Naeem AM (2010). Optical, magnetic and electrical investigation of cobalt ferrite nanoparticles synthesized by co-precipitation route. J. Alloys. Compd. pp. 507,201. http://dx.doi.org/10.1016/j.jallcom.2010.07.155

Haubold S, Haase M, Kornowski A, Weller H (2001). Strongly Luminescent InP/ZnS Core-Shell Nanoparticles. Chem. Phys. Chem. 2:331 http://dx.doi.org/10.1002/1439-7641(20010518)2:5<331::AID-CPHC331>3.0.CO;2-0

Unger K, Verbindungshalbletier A, Verlagsanstalt L (1986). AkademischeVerlagsanstalt, Leipzing, 339.PMCid: PMC460295.

Liu Y, Zheng C, Wang W, Yin C, Wang G (2001). Synthesis and Characterization of Rutile SnO_2 Nanorods. Adv. Mater. 13:1883 http://dx.doi.org/10.1002/1521-4095(200112)13:24<1883::AID-ADMA1883>3.0.CO;2-Q

Natter H, Schmelzer M, Hempelmann R (1998). Nanocrystalline nickel and nickel-copper alloys: Synthesis, characterization, and thermal stability. J. Mater. Res. 13:1186 http://dx.doi.org/10.1557/JMR.1998.0169

Nayral C, Ould-Ely T, Maisonnat A, Chaudret B, Fau P, Lescouzeres L, Peyre-Lavigne A (1999). A Novel Mechanism for the Synthesis of Tin /Tin Oxide Nanoparticles of Low Size Dispersion and of Nanostructured SnO2 for the Sensitive Layers of Gas Sensors. Adv. Mater. 11:61. http://dx.doi.org/10.1002/(SICI)1521-4095(199901)11:1<61::AID-ADMA61>3.0.CO;2-U

Price LS, Parkin IP, Field MN, Hardy AME, Clark RJH, Hibbert TG, Molloy KC (2000). Atmospheric pressure chemical vapour deposition of tin (II) sulfide films on glass substrates from $Bu^n_3SnO_2CCF_3$ with hydrogen sulfide. J. Mater. Chem. 10:527. http://dx.doi.org/10.1039/a907939d

Rezlescu N, Rezlescu E (1974). Dielectric properties of copper containing ferrites. Physics Status Solid A, 23:575 http://dx.doi.org/10.1002/pssa.2210230229

Shafi KYPM, Gedanken A, Prozorov R (1998). Sonochemical preparation and characterization of nanosized amorphous Co–Ni alloy powders. J. Mater. Chem. 8:769 http://dx.doi.org/10.1039/a706871i

Smyth CP (1956). Dielectric behavior and structure. Acta Cryst 9:838-839. http://dx.doi.org/10.1107/S0365110X56002382

Xue D, Kitamura K (2002). Dielectric characterization of the defect concentration in lithium niobate single crystals. Solid State Commun 122:537–541.http://dx.doi.org/10.1016/S0038-1098(02)00180-1

Yang CS, Liu Q, Kauzlarich SM (2000). Synthesis and characterization of Sn/R, Sn/Si-R, and Sn/SiO2 core/shell nanoparticles. Chem. Mater. 12:983 http://dx.doi.org/10.1021/cm990529z

Zhang P, Gao L (2003). Synthesis and Characterization of CdS Nanorods via Hydrothermal Microemulsion. Langmuir 19:208 http://dx.doi.org/10.1021/la0206458

27

Evaluation of properties of coal slurry from different solvents

Ugwu K. E. and Ofomatah A. C.

National Center for Energy Research and Development, University of Nigeria, Nsukka, Enugu State, Nigeria.

Coal slurries were separately prepared with benzene, ethanol, hexane and water, and the fuel properties of the slurries were examined. The ability of the solvents to dissolve the coal was also evaluated using spectrophotometer in the ultra-violet and visible regions. Some of the slurry properties indicated that the liquid coal have comparable properties to petroleum-based conventional diesel oil. Benzene and hexane were highly flammable and therefore hazardous. The liquid coal can be easily transported in pipelines based on the choice of suitable solvents as evaluated. The coal was extracted with the solvents in decreasing order of ethanol>benzene>water>hexane.

Key words: Coal, properties, slurries, solvents.

INTRODUCTION

Coal is a chemically and physically extremely complex and heterogeneous material (Schobert et al., 1991), consisting of two-phase systems, the organic and the inorganic constituents (Dolinska et al., 2011). Coal is a primary fuel utilized in several countries for generation of electricity, heating and for metallurgical purposes. It is also a feedstock in the making of some chemicals. Coal is found in many countries including United States of America, Nigeria, South Africa, China and Russia. The world's proved international recoverable coal reserves at the end of 2002 from each continent is over 100,000 million tons (World Energy Council, 2004). In Nigeria, as with several countries, coal is grossly under-utilized as crude oil is largely preferred to coal in modern utility. With the increasing crises surrounding crude oil availability and pricing, researches are ongoing for direct replacement of heavy petroleum oil with coal (IChemE, 1983). The conversion of solid coal to liquid form will improve utilization of coal including its transportation in pipelines and usage in engines. The methods of production of liquid fuels from coal are through direct liquefaction and the Fischer-Tropsch (F-T) processes.

Direct liquefaction approach entails hydrogenation while F-T process requires the gasification of the coal and catalytic conversion of the gases generated to liquid products. The conversion of coal to liquids using suitable solvents is attractive as various experiments have given positive prospects for their use in utility boilers, industrial boilers and in blast furnace (Nunez et al., 2010). In this study, coal slurries were separately prepared with benzene, ethanol, hexane and water. The nature of the slurries was examined. These solvents were assessed for suitability in preparation of slurries that can be applied as potential substitutes for conventional petroleum-based diesel oil. This investigation was carried out at the National Center for Energy Research and Development Laboratory, University of Nigeria, Nsukka, between April, 2012 and July, 2012.

MATERIALS AND METHODS

The coal used for this study was from Ogwashi-uku mine, Nigeria and provided by the Nigerian coal corporation. This received coal was ground to a fine powder of 250 microns (60-mesh).

Table 1. Proximate analysis for Ogwashi-Uku coal, Nigeria.

Parameter	%
Moisture content	35.0
Ash content	6.85
Volatile matter	38.1
Fixed carbon	20.05

Table 2. Slurry properties of the coal with different solvents versus the typical No. 2 diesel.

Properties	Benzene	Ethanol	Hexane	Water	No. 2 Diesel +
Density	0.895	0.927	0.849	1.062	0.952*
Kinematic					
Viscosity (cSt)	1.36	2.46	1.19	1.63	2.6-4.1
Flash point (°C)	<0	18	<0	No flash	74
Boiling point (°C)	25	30	20	86	188 to 343

+ http://www.afdc.energy.gov/afdc/fuels/propertierties.html; *Density of Nigerian commercial diesel oil.

Proximate analysis

Moisture, ash, and volatile matter, fixed carbon: This was carried out on the sample following standard methods (Speight, 2005).

Slurry preparation

Two non-polar solvents (benzene and hexane) and two polar solvents (ethanol and water) were separately used to prepare the coal slurries. The slurries were prepared by measuring 14 g of the coal sample into a beaker. 28 g of benzene was poured into the beaker of coal at 1:2 and the mixture was mechanically agitated with a magnetic stirrer set at high speed for 20 min. This procedure was repeated separately for ethanol, hexane and water. The following tests were carried out on the slurries:

(a) **Viscosity:** This was investigated by using Oswald portable capillary viscometer. This investigation was conducted at room temperature (3°C),
(b) **Density:** The density was measured using a density bottle. The density of the slurry was calculated from the simple relationship between the measured volumes of a fluid and the mass of the fluid. Density = mass/volume.
(c) **Flash point:** This was measured with Pemsky Martin semi-automatic multi flash closed cup flash point tester (Made in Japan),
(d) **Boiling point:** Boiling point was determined with 0-350°C thermometer and
(e) **Solvent extraction test:** This test was done to determine the ability of the solvents to dissolve the coal. The slurries prepared as described above were dried in an oven to evaporate the solvent and left in an incubator. 0.8 g of each of the dried coal was dissolved in 15 g ethanol. The absorbance of each of the extracts was measured with a UV/Visible spectrophotometer (Jenway, model 6305) in which ethanol was used as a blank at 285 nm.

RESULTS AND DISCUSSION

Proximate analysis

From Table 1, the moisture content of the coal is 35%.

This value is appropriate for coal slurry as it is recommended that the moisture content for coal liquid fuel could be as high as 30 to 50%. The ash content at 6.85% is also suitable for coal liquid mixture that can be used for boilers (Vickers and Ivatt, 1983). It could be as high as 10%. In essence, this coal meets the recommended moisture and ash contents for coal slurry fuel. The volatile matter (38.1%) and the fixed carbon (20.05%) contents further shows that the coal is an appropriate fuel whether in solid or in liquid form.

Slurry fuel properties

The densities of the slurries as given in Table 2 showed that the values were within the level for typical No. 2 diesel oil such as the measured commercial diesel (0.952 g/cm^3) which was purchased in a gas station at Nsukka. These values gave an indication of good ignition property for the slurries. Viscosity is a measure of fluid's resistance to flow. Viscosity affects the stability and the atomization of a fuel upon injection into the combustion chamber. The viscosity of the slurries which was measured in centistokes (cSt) showed that the slurries could be pumped and this would ease transportation of the coal in pipelines. The values of 1.36 cSt (Coal-benzene), 1.19 cSt (Coal-hexane) and 1.63 cSt (Coal-water) are lower than the recommended limit for typical No. 2 diesel. But the viscosity of the coal-ethanol slurry was within the limit. Low viscosity of a fuel will not give sufficient lubrication to the engine. The viscosity values of the slurries were low and may not cause any corrosive effect on the internal parts of any compression engine when used as fuels (Knothe and Dunn, 2001).

The flash point is used to assess the overall flammability hazard of a fuel. It also gives indication of

Table 3. Absorbance of the coal slurries in ethanol at 285 nm.

Slurries	Absorbance
Coal-benzene	1.834
Coal-ethanol	1.999
Coal-hexane	1.804
Coal-water	1.813

how easy a chemical may burn. The flash point of coal-benzene mixture and coal-hexane mixture was below 0 °C and that of ethanol at 18 °C was low. These low flash point values indicate that the slurries are likely to ignite accidentally and within the range to pose extreme fire hazard (Abayeh and Shekarau, 2010). But the coal-organic solvents slurries are easily ignited, while the coal-water slurry did not flash. The boiling point values as given in Table 2, showed that coal water slurry has the highest boiling point of the slurries at 86°C and in decreasing order, ethanol>benzene> hexane.

These values were low compared to standard diesel boiling point at between 188 and 343°C but with blending; it may be possible to raise the boiling point. The boiling point values implied that at ambient temperature, the benzene and hexane slurries will volatilize. The examined slurries except coal-water slurry volatilized at a fast rate. In considering the choice of solvents based on volatility for transporting coal in pipelines so as to subsequently use the coal as solid fuel, then in decreasing order, hexane>benzene>ethanol>water. In this way, the option may be based on the solvent that easily evaporates.

Solvent extraction test

The ability of the various solvents to dissolve the coal, thereby forming a liquid coal was evaluated by the simple solvent extraction test. The solvents dissolved the coal in the decreasing order of ethanol>benzene>water>hexane. This is indicated by the absorbance values as given in Table 3. Solvents with molecules containing unpaired electrons show higher ability to disrupt the hydrogen bonds in coal (Shin and Shen, 2007). This might have accounted for higher dissolution of coal with ethanol.

Conclusion

To improve coal utilization, the liquid form of coal was prepared. The slurries were of low viscosity which is important characteristic for pumping. The flash point and boiling point values indicated that the slurries are volatile and prone to easy ignition thereby posing a fire hazard. From the solvent extraction test, ethanol extracted more of the coal than the other solvents. A high extraction power indicates a high compatibility with coal. This study

demonstrated that the coal-solvent slurry is a potential substitute for the conventional petroleum-based diesel oil. More researches are however desirable to improve on some of the characteristics of the slurries.

ACKNOWLEDGEMENT

The support of the National Center for Energy Research and Development, University of Nigeria, Nsukka, Nigeria, is hereby acknowledged.

REFERENCES

Abayeh OJ, Shekarau JI (2010). Transesterified water melon seed oil as a Bio diesel fuel, J. Chem. Soc. Niger. 35(1):38-42.

Dolinska S, Lovas M, Znamenackova I, Hredzak S, Jakabsky S, Matik M (2011). 11th International Multidisciplinary Scientific GeoConference and published in the SGEM2011 Conference Proceedings. 1:1083-1090

http://www.afdc.energy.gov/afdc/fuels/propertierties.html.

IChemE (The Institution of Chemical Engineers) (1983). Coal Liquid Mixtures, IChemE Symposium Series No. 83, EFCE Publication Series No. 34, Oxford: Pergamon Press, preface.

Knothe G, Dunn RO (2001). In: Gunstone FD, Hamilton RJ (eds.), Oleochemical manufacture and applications, Sheffield: Academic press. P. 106.

Nunez GA, Briceno MI, Joseph DDA (2010). Colloidal coal in water suspensions. Energy Environ. Sci. 3:629.

Schobert HH, Bartle KD, Lynch LJ (1991). Coal Science, Symposium Series 461: American Chemical Society, Washington, DC. Chapter 7.

Speight JG (2005). Handbook of Coal Analysis, New Jersey: John Wiley $ Sons. New pp. 41-64.

Shin Y, Shen Y (2007). Preparation of coal slurry with organic solvents. Chemosphere 68:389-393.

Vickers F, Ivatt S (1983). The Preparation of low sulphur coal In: *Coal Liquid Mixtures*, IChemE Symposium Series No. 83, EFCE Publication Series No. 34, Pergamon Press, Oxford, England. pp. 55-74.

World Energy Council (2004). 2004 Survey of Energy Resources, 20th ed., Amsterdam: Elsevier.

Permissions

All chapters in this book were first published in IJPS, by Academic Journals; hereby published with permission under the Creative Commons Attribution License or equivalent. Every chapter published in this book has been scrutinized by our experts. Their significance has been extensively debated. The topics covered herein carry significant findings which will fuel the growth of the discipline. They may even be implemented as practical applications or may be referred to as a beginning point for another development.

The contributors of this book come from diverse backgrounds, making this book a truly international effort. This book will bring forth new frontiers with its revolutionizing research information and detailed analysis of the nascent developments around the world.

We would like to thank all the contributing authors for lending their expertise to make the book truly unique. They have played a crucial role in the development of this book. Without their invaluable contributions this book wouldn't have been possible. They have made vital efforts to compile up to date information on the varied aspects of this subject to make this book a valuable addition to the collection of many professionals and students.

This book was conceptualized with the vision of imparting up-to-date information and advanced data in this field. To ensure the same, a matchless editorial board was set up. Every individual on the board went through rigorous rounds of assessment to prove their worth. After which they invested a large part of their time researching and compiling the most relevant data for our readers.

The editorial board has been involved in producing this book since its inception. They have spent rigorous hours researching and exploring the diverse topics which have resulted in the successful publishing of this book. They have passed on their knowledge of decades through this book. To expedite this challenging task, the publisher supported the team at every step. A small team of assistant editors was also appointed to further simplify the editing procedure and attain best results for the readers.

Apart from the editorial board, the designing team has also invested a significant amount of their time in understanding the subject and creating the most relevant covers. They scrutinized every image to scout for the most suitable representation of the subject and create an appropriate cover for the book.

The publishing team has been an ardent support to the editorial, designing and production team. Their endless efforts to recruit the best for this project, has resulted in the accomplishment of this book. They are a veteran in the field of academics and their pool of knowledge is as vast as their experience in printing. Their expertise and guidance has proved useful at every step. Their uncompromising quality standards have made this book an exceptional effort. Their encouragement from time to time has been an inspiration for everyone.

The publisher and the editorial board hope that this book will prove to be a valuable piece of knowledge for researchers, students, practitioners and scholars across the globe.

List of Contributors

Nuha S. Mashaan
Center for Transportation Research, Faculty of Engineering, University of Malaya, Kuala Lumpur 50603, Malaysia

Asim Hassan Ali
Center for Transportation Research, Faculty of Engineering, University of Malaya, Kuala Lumpur 50603, Malaysia

Mohamed Rehan Karim
Center for Transportation Research, Faculty of Engineering, University of Malaya, Kuala Lumpur 50603, Malaysia

Mahrez Abdelaziz
Center for Transportation Research, Faculty of Engineering, University of Malaya, Kuala Lumpur 50603, Malaysia

Samo Lubej
Faculty of Civil Engineering, University of Maribor, Smetanova 17, 2000 Maribor, Slovenia

Milan Radosavljevic
School of Construction Management and Engineering, University of Reading, Whiteknights, P. O. Box 219, Reading, RG6 & AW, United Kingdom

Semire Banjo
Department of Pure and Applied Chemistry, Faculty of Pure and Applied Sciences, Ladoke Akintola University of Technology, P. M. B. 4000, Ogbomoso, Oyo State, Nigeria

Adeoye Idowu Olatunbosun
Department of Pure and Applied Chemistry, Faculty of Pure and Applied Sciences, Ladoke Akintola University of Technology, P. M. B. 4000, Ogbomoso, Oyo State, Nigeria

Zulfiqar Ali
College of Earth and Environmental Sciences, University of the Punjab, Lahore, Pakistan
Nano Science and Catalysis Division, National Centre for Physics, Islamabad, Pakistan

Syed Tajammul Hussain
Nano Science and Catalysis Division, National Centre for Physics, Islamabad, Pakistan

Muhammad Nawaz Chaudhry
College of Earth and Environmental Sciences, University of the Punjab, Lahore, Pakistan

Syeda Adila Batool
Department of Space Science Punjab University Lahore, Pakistan

Tariq Mahmood
Nano Science and Catalysis Division, National Centre for Physics, Islamabad, Pakistan

B. A. Ezekoye
Crystal Growth and Characterization Laboratory, Department of Physics and Astronomy, University of Nigeria, Nsukka, Enugu State, Nigeria

V. A. Ezekoye
Crystal Growth and Characterization Laboratory, Department of Physics and Astronomy, University of Nigeria, Nsukka, Enugu State, Nigeria

P. O. Offor
Department of Metallurgical and Materials Engineering, University of Nigeria, Nsukka, Enugu State, Nigeria

S. C. Utazi
Crystal Growth and Characterization Laboratory, Department of Physics and Astronomy, University of Nigeria, Nsukka, Enugu State, Nigeria

Ayi A. Ayi
Department of Chemistry, University of Bath, Claverton Down, Bath BA2 7AY, United Kingdom
Department of Pure and Applied Chemistry, University of Calabar, P. M. B. 1115, Calabar, Cross River State, Nigeria

Andrew D. Burrows
Department of Chemistry, University of Bath, Claverton Down, Bath BA2 7AY, United Kingdom

Mary F. Mahon
Department of Chemistry, University of Bath, Claverton Down, Bath BA2 7AY, United Kingdom

Anna J. Warren
Department of Chemistry, University of Bath, Claverton Down, Bath BA2 7AY, United Kingdom

C. O. B. Okoye
Department of Pure and Industrial Chemistry, University of Nigeria, Nsukka. Nigeria

A. M. Chukwuneke
Department of Pure and Industrial Chemistry, University of Nigeria, Nsukka. Nigeria

N. R. Ekere
Department of Pure and Industrial Chemistry, University of Nigeria, Nsukka. Nigeria

J. N. Ihedioha
Department of Pure and Industrial Chemistry, University of Nigeria, Nsukka. Nigeria

Hakan Keskin
Department of Woodworking Industry Engineering, Faculty of Technology, Gazi University, 06500 Besevler, Ankara, Turkey

Neslihan Süzer Ertürk
Department of Industrial Technology, Faculty of Industrial Arts Education, Gazi University, 06830 Gölbasi, Ankara, Turkey

Mustafa Hilmi Çolakoğlu
Technology Development Foundation of Turkey (TTGV) 06800 Bilkent, Ankara, Turkey

Süleyman Korkut
Department of Forest Industrial Engineering, Faculty of Forestry, Düzce University, 81620 Düzce, Turkey

Victor C. H. Lim
Department of Electrical, Electronic and Systems Engineering, Faculty of Engineering and Built Environment, Universiti Kebangsaan Malaysia, Bangi, Selangor, Malaysia.
Microelectronics Semiconductor Packaging, Institute of Micro Engineering and Nanoelectronics, Universiti Kebangsaan Malaysia, Bangi, Selangor, Malaysia

Nowshad Amin
Department of Electrical, Electronic and Systems Engineering, Faculty of Engineering and Built Environment, Universiti Kebangsaan Malaysia, Bangi, Selangor, Malaysia

Foong Chee Seng
Freescale Semiconductor (M) Sdn. Bhd., Petaling Jaya, Selangor, Malaysia

Ibrahim Ahmad
Microelectronics Semiconductor Packaging, Institute of Micro Engineering and Nanoelectronics, Universiti Kebangsaan Malaysia, Bangi, Selangor, Malaysia

Azman Jalar
Department of Electronics and Communication, College of Engineering, University Tenaga Nasional, Kajang, Selangor, Malaysia

Stephen Shaibu Ochigbo
Department of Chemistry, Federal University of Technology, P. M. B. 65, Minna, Niger State, Nigeria

E. M Odin
Department of Pure and Industrial Chemistry, Kogi State University, Anyigba, Nigeria

P. K. Onoja
Department of Pure and Industrial Chemistry, Kogi State University, Anyigba, Nigeria

J. F. Saleh
Department of Pure and Industrial Chemistry, Kogi State University, Anyigba, Nigeria

Seyed Hossein Hosseini
Department of Chemistry, Faculty of Science, Islamic Azad University, Islamshahr Branch, Tehran-Iran

Maryam Ashjari
Department of Chemistry, Faculty of Science and Engineering, Islamic Azad University, Saveh Branch, Saveh-Iran

D. Ajdinaj
Department of Wood Industry, Faculty of Forest Sciences, Agricultural University of Tirana, Kodër-Kamëz 1029, Tirana, Albania

E. Lato
Department of Wood Industry, Faculty of Forest Sciences, Agricultural University of Tirana, Kodër-Kamëz 1029, Tirana, Albania

D. Quku
Department of Wood Industry, Faculty of Forest Sciences, Agricultural University of Tirana, Kodër-Kamëz 1029, Tirana, Albania

H. Cota
Department of Wood Industry, Faculty of Forest Sciences, Agricultural University of Tirana, Kodër-Kamëz 1029, Tirana, Albania

Bachir Chemani
Department of Process Engineering, Faculty of Engineering Sciences, University M'Hamed BOUGARA of Boumerdes – 35000 – Algeria

Halima Chemani
Department of Materials Engineering, Faculty of Engineering Sciences, University M'Hamed BOUGARA of Boumerdes – 35000 – Algeria

O. M. Lemine
Department of Physics, College of Sciences, Al imam Mohammad Bin Saud Islamic University (IMSIU), P. O. Box 90950 11623 Riyadh, Saudi Arabia

Sumrit Mopoung
Department of Chemistry, Faculty of Science, Naresuan University, Phitsanulok, Thailand

Anchalee Sirikulkajorn
Department of Chemistry, Faculty of Science, Naresuan University, Phitsanulok, Thailand

Duennapa Dummun
Department of Chemistry, Faculty of Science, Naresuan University, Phitsanulok, Thailand

Pichyawee Luethanom
Department of Chemistry, Faculty of Science, Naresuan University, Phitsanulok, Thailand

Arunseshan Chandrasekar
Crystal Growth Centre, Anna University, Chennai-600 025, India

Suresh Sagadevan
Crystal Growth Centre, Anna University, Chennai-600 025, India

Arivuoli Dakshnamoorthy
Crystal Growth Centre, Anna University, Chennai-600 025, India

I. O. Oboh
Department of Chemical and Petroleum Engineering, University of Uyo, Uyo, Nigeria

E. O. Aluyor
Department of Chemical Engineering, University of Benin, Benin city, Nigeria

T. O. K. Audu
Department of Chemical Engineering, University of Benin, Benin city, Nigeria

Badriah Saad Al-Farhan
Department of Chemistry, Faculty of Girls for Science, King Khalid University, Abha, KSA, Saudi Arabia

A. J. Manji
Department of Industrial Chemistry, Modibbo Adama University of Technology, P. M. B. 2076, Yola, Adamawa State, Nigeria

E. E. Sarah
Department of Industrial Chemistry, Modibbo Adama University of Technology, P. M. B. 2076, Yola, Adamawa State, Nigeria

U. U. Modibbo
Department of Industrial Chemistry, Modibbo Adama University of Technology, P. M. B. 2076, Yola, Adamawa State, Nigeria

O. I. Adekoya
Physics Department, University of Ibadan, Oyo, Nigeria
Physical Science Department, Yaba College of Technology, Lagos, Nigeria

Paul Ocheje Ameh
Physical Chemistry Unit, Department of Chemistry, Nigeria Police Academy, Wudil, P. M. B. 3474, Kano State, Nigeria

Manal A. Awad
King Abdullah Institute for Nanotechnology, King Saud University, Saudi Arabia

Awatif A. Hendi
Department of Physics, Faculty of Science, King Saud University, Saudi Arabia

Khalid M. O. Ortashi
Department of Chemical Engineering, King Saud University, Saudi Arabia

Dalia F. A. Elradi
Department of Microbiology, Faculty of Medicine, Princess Nora Bint Abdulrahman University Saudi Arabia

Nada E. Eisa
Department of Physics, Al -Dammam University Kingdom of Saudi Arabia
Department of Physics, Omdurman Ahlia University, Sudan

Lamia. A. Al-lahieb
Department of Chemistry Organic, Faculty of Science, Nature Product, Al Qassim University, Saudi Arabia

Shorog. M. Al-Otiby
Department of Chemistry Organic, Faculty of Science, Nature Product, Princess Nora Bint Abdulrahman University, Saudi Arabia

Nada M. Merghani
Central Laboratory, College of Science, King Saud University, Saudi Arabia

Abdelelah A. G. Awad
Faculty of Animal Production, University of Khartoum, Sudan

Sinan Aydin
Faculty of Technology, Cumhuriyet University, Sivas/ Turkey

Serdar Mercan

Faculty of Technical Education, Cumhuriyet University, Sivas/Turkey

Murat Yavuz Solmaz
Faculty of Engineering, Firat University, Elaziğ/Turkey

Aydin Turgut
Faculty of Engineering, Bingöl University, Bingöl/Turkey

Leila Mahdavian
Department of Chemistry, Doroud Branch, Islamic Azad University, P.O. Box: 133. Doroud Iran

S. Suresh
Crystal Growth Centre, Anna University, Chennai-25, India

K. E. Ugwu
National Center for Energy Research and Development, University of Nigeria, Nsukka, Enugu State, Nigeria

A. C. Ofomatah
National Center for Energy Research and Development, University of Nigeria, Nsukka, Enugu State, Nigeria

A Researcher's Guide to Aerospace Engineering

A Researcher's Guide to Aerospace Engineering

Edited by Natalie Spagner

CLANRYE
INTERNATIONAL
www.clanryeinternational.com

Clanrye International,
750 Third Avenue, 9ᵗʰ Floor,
New York, NY 10017, USA

ISBN: 978-1-63240-776-4

Cataloging-in-Publication Data

A researcher's guide to aerospace engineering / edited by Natalie Spagner.
    p. cm.
Includes bibliographical references and index.
ISBN 978-1-63240-776-4
1. Aerospace engineering. 2. Aeronautics. I. Spagner, Natalie.
TL546 .R47 2019
629.1--dc23

For information on all Clanrye International publications
visit our website at www.clanryeinternational.com

CLANRYE
INTERNATIONAL

Contents

Preface

Aerospace engineering is concerned with the construction of aircrafts and spacecrafts. The two main branches of aerospace engineering are aeronautical and astronautical engineering. Aerospace engineering is a multidisciplinary field that encompasses theories and principles of prominent disciplines such as aerodynamics, mathematics, control engineering, materials science, etc. While understanding the long-term perspectives of the topics, the book makes an effort in highlighting their impact as a modern tool for the growth of the discipline. The chapters included herein are a valuable compilation of topics, ranging from the basic to the most complex advancements in the field of aerospace engineering. It contains some path-breaking studies related to this field and unfolds the innovative aspects of aerospace engineering which will be crucial for the progress of this discipline in the future. This book is appropriate for students seeking detailed information in this area as well as for experts.

After months of intensive research and writing, this book is the end result of all who devoted their time and efforts in the initiation and progress of this book. It will surely be a source of reference in enhancing the required knowledge of the new developments in the area. During the course of developing this book, certain measures such as accuracy, authenticity and research focused analytical studies were given preference in order to produce a comprehensive book in the area of study.

This book would not have been possible without the efforts of the authors and the publisher. I extend my sincere thanks to them. Secondly, I express my gratitude to my family and well-wishers. And most importantly, I thank my students for constantly expressing their willingness and curiosity in enhancing their knowledge in the field, which encourages me to take up further research projects for the advancement of the area.

Editor

1

Calculation of the Flight Characteristics of the Aircraft, AN-225

Ahmed Soliman M.Sherif*

Novosibirsk State Technical University, Russia

Abstract

Flight dynamics - The science of the laws of motion of aircraft under the influence of wind, gravity, and reaction forces. It is a combination of mainly three classic disciplines: solid mechanics, fluid dynamics, and mathematics. Among the wide range of problems in the dynamics of flight of great practical importance are the problems connected with the study of the steady rectilinear motion of the aircraft. The solution allows them to determine the flight characteristics of the aircraft, characterized by the range of possible speeds and heights, rate of climb, range, flight time, and so on.

Keywords: Calculation aerodynamic characteristics of the aircraft, AN-225; Thrust required and thrust available; Practical ceiling of aircraft; Building a polar flight; Flight dynamics

Introduction

Building a polar flight, making level flight at various speeds (Mach number 0.4 to 0.9) and on the same altitude, the aircraft as it passes from one polar to another, it is the flight of the aircraft polar.

From the equilibrium conditions of the lift Y_a gravity (weight) G (G = mg) in a horizontal flight:

$$C_{Y_a} = \frac{2G}{\rho S V^2} = \frac{2G}{\rho S a^2 M^2} = \frac{A}{M^2}$$

Where, $A = \dfrac{2G}{\rho S V^2}$ on the height and a constant weight of the aircraft, the value is constant.

All calculations are carried out in SI.

From this formula, it follows that, in a steady horizontal flight, each Mach number M complies to a specific lift coefficient C_{Y_a}.

For the aircraft, AN-225 with turbojet engines must use the curves of required and available thrust. Calculation and construction of required thrust P_{Req} by the formula:

$$P_{Req.} = \frac{G}{K}, K = \frac{C_Y}{C_X}$$, the aerodynamic quality of the aircraft.

When determining the flight characteristics of the aircraft used by the equation of power in the projection on the axis of the trajectory of the coordinate system, considering at the same plane as the material point of variable mass. And when the aircraft stability and controllability of the calculations it is regarded as solid [1-4].

Initial data for the implementation of the research is of course work in aerodynamics,

"Calculation of aerodynamic characteristics of the aircraft AN-225", its geometrical parameters, the aerodynamic characteristics and polar cruising.

Research includes calculations, graphics and drawings, explanation and justification of the calculation of performance, the characteristics of longitudinal stability and controllability of the aircraft.

To calculate the flight characteristics of the aircraft AN-225, used the method of N.E. Zhukovsky,

A method based on the construction of curves thrust required and thrust available, which is determined by the parameters of steady flight modes.

Initial Data

Characteristics of the standard atmosphere are as follows:

Calculate thrust required and thrust available

Calculation of the algorithm:

Specifies flight height, H, m.; H=0

Specifies the number of flight Mach; M=0.3

Determine the relative density of the air; Δ (Table 1); Δ=1

Determine the density of the air; ρH (Table 1); $\rho H = 1.225 \dfrac{kg}{m^3}$.

Determine the ratio; K_a (Table 1); $K_a = 1$

Determine the speed of sound at a given height with a given number M, (m/s).

aH= [aH=0 *Ka]=340.28×1=340.28 m/s.

Where, $a_{H=0}$=340.28 m/s - the speed of sound at sea level H=0

Determine the flight speed (m/s); V= [M*$a_{H=0}$] =0.3×340.28=102.08 m/s.

Determine the dynamic pressure (N/m²); $q = \dfrac{\rho H \times V^2}{2} = \dfrac{1.225 \times 102.08^2}{2} = 6382 \dfrac{N}{m^2}$

Determine the average gross weight of the aircraft, (N).

$$G_{Average} = \left(m_0 - \frac{m_f}{2}\right) \times 10^3 \times 9.81 = \left(640 - \frac{128}{2}\right) \times 10^3 \times 9.81 = 5650560 N$$

***Corresponding author:** Ahmed Soliman M.Sherif, Researcher, Novosibirsk State Technical University, pr. Karla Marksa, 20, Novosibirsk, Novosibirskaya-oblast', Russia, E-mail: engsherifsoliman78@gmail.com

Height, H, (km)	Relative density, Δ	Relative speed of sound, K_a	Density, $\rho H, \frac{kg}{m^3}$	$\rho * a^2 \frac{kg}{m^2 * S}$
0	1	1	1.225	$141.8*10^3$
2	0.822	0.977	1.0067	$111.3*10^3$
4	0.699	0.954	0.8194	$86.3*10^3$
6	0.538	0.930	0.6602	$66, 08*10^3$
8	0.429	0.905	0.5259	$50*10^3$
10	0.337	0.880	0.4136	$37.06*10^3$
11	0.298	0.867	0.3648	$31.75*10^3$
12	0.2536	0.867	0.3156	$27.48*10^3$
14	0.185	0.867	0.2306	$20.08*10^3$
16	0.135	0.867	0.1654	$14.4*10^3$
18	0.0983	0.867	0.1207	$10.51*10^3$
20	0.0718	0.867	0.0889	$7.74*10^3$

Table 1: Characteristics of the standard atmosphere.

Changing the ratio of the polar blade, $K_a(M)$	Mach number (M)	Change parasitic drag coefficient, $K_{CX0}(M)$	Changing the maximum lift coefficient, $K_{cymax}(M)$
1	0	1	1
1	0.2	1	1
1	0.4	1	1
1.09	0.6	1.03	0.94
1.16	0.7	1.2	0.89
1.27	0.8	1.5	0.81
1.4	0.9	1.9	0.73
1.6	1.0	2.0	0.65

Table 2: Odd changes in the number of M.

Take-off weight, m_0	Fuel weight, m_f	Wing area, S, m²	Take-off thrust, P_0, k N	Parasitic drag coefficient, C_{x0}	Wingspan, L_{wing}, m	Specific fuel consumption, $C_{spec}, \frac{kg}{(N*hr)}$
640	128	905	6*234	0.021	88.4	0.057

Table 3: Personal data on the aircraft AN-225.

$C_{y\ max\ cruiser}$	M_{max}	$Q_{qmaxmax}, \frac{kN}{m^2}$	$C_{max,\ T.off}$	$K_{T.off}$	$C_{y\ max\ Const}$	K_{const}
1.7	0.88	22	2.5	8.5	3.1	5.0

Table 4: Personal data on the aircraft AN-225.

where, W_0= 640 ton-takeoff weight of the aircraft;

W_f =128 ton-mass of the fuel

The coefficient of aerodynamic lift in a horizontal flight.

$$C_{y.h.f.} = \frac{G_{Av.}}{s*q} = \frac{5650560}{905 \times 6382} = 0.98 \leq 0.9 * C_{y\ max\ cruiser} = 0.9*1.7 = 1.53$$

where,

$C_{ymax\ cruiser}$ is the maximum lift coefficient of the wing when stowed mechanization (Tables 2-4).

Determine the effective extension of the wing:

$$\lambda ef = \left[\frac{L_{wing}^2}{S} \times (1-\delta) \right] = \left[\frac{88.4^2}{905} \times (1-0.03) \right] = 8.37$$

Where, δ =0.02....0.04;

S=905 m²- wing area;

L_{wing} = 88.4 - wingspan

Determine the ratio of the blade of the polar;

$$A_0 = \frac{1}{\pi * \lambda_{ef.}} = \frac{1}{\pi * 8.37} = 0.038$$

Determine the rate of change of blade polar depending on the number of M, K_A (Table 2); K_A=1

Determine the rate of change of parasitic drag coefficient as a function of the number of M, K_{CX0} (Table 2).

$$K_{CX0} = 1$$

Determine the drag coefficient in horizontal flight;

$$C_{xah.f.} = K_{CX0} \times C_{X0} + A_0 \times K_a \times C_{yh.f.}^2 = 1 \times 0.021 + 0.038 \times 0.98^2 = 0.0574$$

where, $C_{x0}=0.021$ - parasitic drag coefficient (Table 3).

Define flight aerodynamic efficiency;

$$K = \frac{C_{y\,h.f.}}{C_{xa\,h.f.}} = \frac{0.98}{0.0574} = 17.04$$

Identify thrust required for level flight, N.

$$P_{Req.} = \frac{G_{Av}}{K} = \frac{5650560}{17.04} = 331510N$$

Determine the ratio of thrust change the number of M.

$$\xi = 1 - 0.32 * M + 0.4 * M^2 - 0.01 * M^3 =$$
$$=> 1 - (0.32 * 0.3) + (0.4 * 0.3^2) - (0.01 * 0.3^3) = 0.94$$

Determine the takeoff thrust engines (N) (Table 3).

$$P_0 = 6 * 234000 = 1404000N$$

Identify the thrust available in horizontal flight.

When,

$H \leq 11km$ -- $\rightarrow P_{Avail} = \xi \times \Delta^{0.85} \times P_0 = 0.94 \times 1^{0.85} \times 1404000 = 1319760N$.

$H > 11km$ -- $\rightarrow P_{Avail} = \xi \times \Delta \times 1.2 \times P_0 = 0.94 \times 1 \times 1.2 \times 1404000 = 1583712N$.

By algorithm, using a program in Excel.

The calculation results are shown in Tables 5 - 13 and in Figures 1 - 13.

"Area velocity values, at which horizontal flight is possible at a fixed weight of the aircraft and the altitude, it called horizontal flight speed range.

At this altitude:

$$V_{min} = V_{most\ advantageous} = V_{max}$$

At this altitude, the curve of Thrust available (or power) not intersects the curve of thrust required, but only touches it."

Determination of the Flight Range of the Aircraft H = f (v)

At the intersection point of the curve Thrust Required and Thrust Available, we define the boundaries of the possible limits of the aircraft (Table 14).

Drawing a schedule of possible aircraft flight boundaries under the conditions of thrust required and thrust available (Figures 11-13).

On the left side, should be restrictions on the minimum flight speed of the conditions for safe values of the coefficient of lift of the aircraft.

$$C_{y\ safe\ values} = 0.9 * C_{ymaxcruiser} = 0.9 \times 1.7 = 1.53$$

The significance of this factor determines the minimum speed of horizontal flight of the conditions of a possible lift of an aircraft wing.

Flight altitude, км	0									
The number of flight Mach	0.1	0.2	0.3	0.4	0.6	0.7	0.8	0.85	0.9	0.95
Relative density of the air, Δ	1	1	1	1	1	1	1	1	1	1
The air density, ρH, kg/m³	1.225	1.225	1.225	1.225	1.225	1.225	1.225	1.225	1.225	1.225
Coefficient K_a	1	1	1	1	1	1	1	1	1	1
The speed of sound a, m/S	340.28	340.28	340.28	340.28	340.28	340.28	340.28	340.28	340.28	340.28
Flight speed , V, m/s	34.03	68.06	102.08	136.11	204.17	238.20	272.22	289.24	306.25	323.27
Dynamic pressure, N/m²	709	2837	6383	11347	25532	34752	45390	51241	57447	64007
The average gross weight of the aircraft, Gav. N	5650560	5650560	5650560	5650560	5650560	5650560	5650560	5650560	5650560	5650560
The coefficient of aerodynamic lift in horizontal flight, $C_{yh.f.}$	8.80	2.20	0.98	0.55	0.24	0.18	0.14	0.12	0.11	0.10
Effective of wing extension , λ	8, 38	8.38	8, 38	8, 38	8.38	8.38	8.38	8.38	8.38	8.38
Factor Blade of the polar, A_0	0.038	0.038	0.038	0.038	0.038	0.038	0.038	0.038	0.038	0.038
Coefficient K_A	1	1	1	1	1.09	1.16	1.27	1.33	1.4	1.5
Coefficient K_{cx0}	1	1	1	1	1, 03	1.2	1.5	1.7	1.9	1.95
The coefficient of aerodynamic drag in horizontal flight, C_{xa} h.f.	2.9664	0.2051	0.0574	0.0325	0.0241	0.0266	0.0324	0.0365	0.0405	0.0415
Flight aerodynamic quality, K	2.97	10.73	17.05	16.93	10.14	6.75	4.24	3.34	2.68	2.35
Thrust required $P_{required}$ N	1903984	526541	331364	333815	557029	837300	1331466	1690318	2107042	2403497
The Rate of changes in the thrust of the number of flight Mach , ξ	0.972	0.952	0.940	0, 935	0, 950	0.969	0.995	1.011	1.029	1.048
Take-off thrust engines, P_0 N,	1404000	1404000	1404000	1404000	1404000	1404000	1404000	1404000	1404000	1404000
Thrust available, P_{avail} , N	1364674	1336496	1319381	1313245	1333575	1359872	1396812	1419246	1444309	1471990
Vertical speed, V_y m/s	-3.25	9.76	17.85	23.59	28.06	22.03	3.15	-13.88	-35.92	-53.29

Table 5: The calculation of thrust required and thrust available, H = 0 km.

Flight altitude, км	2									
The number of flight Mach	0.1	0.2	0.3	0.4	0.6	0.7	0.8	0.85	0.9	0.95
Relative density of the air, Δ	0.822	0.822	0.822	0.822	0, 822	0.822	0.822	0.822	0.822	0.822
The air density, $\rho H, \dfrac{kg}{m^3}$	1.0067	1.0067	1.0067	1.0067	1.0067	1.0067	1.0067	1.0067	1.0067	1.0067
Coefficient K_a	0.977	0.977	0.977	0.977	0.977	0.977	0.977	0.977	0.977	0.977
The speed of sound a, m/S	332.454	332.454	332.454	332.454	332.454	332.454	332.454	332.454	332.454	332.454
Flight speed , V, m/s	33.25	66.49	99.74	132.98	199.47	232.72	265.96	282.59	299.21	315.83
Dynamic pressure, N/m²	556	2225	5007	8901	20028	27260	35605	40195	45063	50209
The average gross weight of the aircraft, G_{Av}, N	5650560	5650560	5650560	5650560	5650560	5650560	5650560	5650560	5650560	5650560
The coefficient of aerodynamic lift in horizontal flight, $C_{yh.f.}$	11.22	2.81	1.25	0.70	0.31	0.23	0.18	0.16	0.14	0.12
Effective of wing extension , λ	8.38	8.38	8.38	8.38	8.38	8.38	8.38	8.38	8.38	8.38
Factor Blade of the polar, A_0	0.038	0.038	0.038	0.038	0.038	0, 038	0.038	0.038	0, 038	0.038
Coefficient K_A	1	1	1	1	1.09	1.16	1.27	1.33	1.4	1.5
Coefficient K_{cx0}	1	1	1	1	1.03	1.2	1.5	1.7	1.9	1, 95
The coefficient of aerodynamic drag in horizontal flight, $C_{xa h.f.}$	4.8078	0.3202	0.0801	0, 0397	0.0257	0.0275	0.0330	0.0369	0.0409	0.0418
Flight aerodynamic quality, K	2.33	8.76	15.57	17.67	12.15	8.32	5.32	4.21	3.39	2.97
Thrust required $P_{Required}$, N	2420616	644803	362940	319796	465019	678749	1062836	1342999	1668846	1900779
The Rate of changes in the thrust of the number of flight Mach , ξ	0.972	0.952	0, 940	0.935	0.950	0.969	0.995	1.011	1.029	1.048
Take-off thrust engines, P_0, N	1404000	1404000	1404000	1404000	1404000	1404000	1404000	1404000	1404000	1404000
Thrust available, P_{avail} , N	1155234	1131380	1116892	1111698	1128908	1151169	1182439	1201430	1222647	1246080
Vertical speed, V_y, m/s	-7.44	5.73	13.31	18.64	23.44	19.46	5.63	-7.08	-23.63	-36.59

Table 6: The calculation of thrust required and thrust available, $H = 2$ km.

Flight altitude, км	4									
The number of flight Mach	0.1	0.2	0.3	0.4	0.6	0.7	0.8	0.85	0.9	0.95
Relative density of the air, Δ	0.699	0.699	0.699	0.699	0.699	0.699	0.699	0.699	0.699	0.699
The air density, $\rho H, \dfrac{kg}{m^3}$	0.8194	0.8194	0.8194	0.8194	0.8194	0.8194	0.8194	0.8194	0.8194	0.8194
Coefficient K_a	0.954	0.954	0.954	0.954	0.954	0.954	0.954	0.954	0.954	0.954
The speed of sound a, m/S	324.627	324.627	324.627	324.627	324.627	324.627	324.627	324.627	324.627	324.627
Flight speed , V, m/s	32.46	64.93	97.39	129.85	194.78	227.24	259.70	275.93	292.16	308.40
Dynamic pressure, N/m²	432	1727	3886	6908	15543	21156	27632	31194	34972	38966
The average gross weight of the aircraft, G_{Av}, N	5650560	5650560	5650560	5650560	5650560	5650560	5650560	5650560	5650560	5650560
The coefficient of aerodynamic lift in horizontal flight, $C_{yh.f.}$	14.46	3.62	1.61	0.90	0.40	0.30	0.23	0.20	0.18	0.16
Effective of wing extension , λ	8.38	8.38	8.38	8.38	8.38	8.38	8.38	8.38	8.38	8.38
Factor Blade of the polar, A_0	0.038	0.038	0.038	0.038	0.038	0.038	0.038	0.038	0.038	0.038

Coefficient K_A	1	1	1	1	1.09	1.16	1.27	1.33	1.4	1.5
Coefficient K_{cx0}	1	1	1	1	1.03	1.2	1.5	1.7	1.9	1.95
The coefficient of aerodynamic drag in horizontal flight, $C_{xa.h.f.}$	7.9686	0.5177	0.1191	0.0520	0.0283	0.0290	0.0340	0.0377	0.0416	0.0424
Flight aerodynamic quality, K	1.81	6.98	13.49	17.37	14.19	10.16	6.65	5.31	4.29	3.78
Thrust required $P_{Required,}$ N	3113632	809179	418897	325377	398284	555998	849348	1065002	1316496	1495674
The Rate of changes in the thrust of the number of flight Mach , ξ	0.972	0.952	0.940	0.935	0.950	0.969	0.995	1.011	1.029	1.048
Take-off thrust engines P_0, N	1404000	1404000	1404000	1404000	1404000	1404000	1404000	1404000	1404000	1404000
Thrust available, $P_{avail,}$ N	1006548	985764	973141	968616	983611	1003006	1030252	1046799	1065285	1085702
Vertical speed, Vy, m/s	-12.11	2.03	9.55	14.78	20.18	17.98	8.31	-0.89	-12.99	-22.38

Table 7: The calculation of thrust required and thrust available, $H = 4$ km.

Flight altitude, км	6									
The number of flight Mach	0.1	0.2	0.3	0.4	0.6	0.7	0.8	0.85	0.9	0.95
Relative density of the air, Δ	0.538	0.538	0.538	0.538	0.538	0.538	0.538	0.538	0.538	0.538
The air density, $\rho H, \frac{kg}{m^3}$	0.6602	0.6602	0.6602	0.6602	0.6602	0.6602	0.6602	0.6602	0.6602	0.6602
Coefficient K_a	0.93	0.93	0.93	0.93	0.93	0.93	0.93	0.93	0.93	0.93
The speed of sound a, m/S	316.46	316.46	316.46	316.46	316.46	316.46	316.46	316.46	316.46	316.46
Flight speed , V, m/s	31.65	63.29	94.94	126.58	189.88	221.52	253.17	268.99	284.81	300.64
Dynamic pressure, N/m²	331	1322	2975	5289	11901	16199	21157	23885	26777	29835
The average gross weight of the aircraft, $G_{Av,}$ N	5650560	5650560	5650560	5650560	5650560	5650560	5650560	5650560	5650560	5650560
The coefficient of aerodynamic lift in horizontal flight, $C_{yh.f.}$	18.89	4.72	2.10	1.18	0.52	0.39	0.30	0.26	0.23	0.21
Effective of wing extension , λ	8.38	8.38	8.38	8.38	8.38	8.38	8.38	8.38	8.38	8.38
Factor Blade of the polar, A_0	0.038	0.038	0.038	0.038	0.038	0.038	0.038	0.038	0.038	0.038
Coefficient K_A	1	1	1	1	1.09	1.16	1.27	1.33	1.4	1.5
Coefficient K_{cx0}	1	1	1	1	1.03	1.2	1.5	1.7	1.9	1.95
The coefficient of aerodynamic drag in horizontal flight, $C_{xah.f.}$	13.5773	0.8683	0.1884	0.0740	0.0330	0.0317	0.0357	0.0392	0.0428	0.0434
Flight aerodynamic quality, K	1.39	5.44	11.14	15.96	15.88	12.14	8.27	6.68	5.45	4.82
Thrust required $P_{Required,}$ N	4062046	1039072	507185	354010	355765	465442	683629	846343	1037020	1173100
The Rate of changes in the thrust of the number of flight Mach , ξ	0.972	0.952	0.940	0.935	0.950	0.969	0.995	1.011	1.029	1.048
Take-off thrust engines, P_0, N	1404000	1404000	1404000	1404000	1404000	1404000	1404000	1404000	1404000	1404000
Thrust available, $P_{avail,}$ N	805738	789101	778996	775373	787377	802903	824713	837959	852756	869100
Vertical speed, Vy, m/s	-18.24	-2.80	4.57	9.44	14.50	13.23	6.32	-0.40	-9.29	-16.17

Table 8: The calculation of thrust required and thrust available, H = 6 km.

Flight altitude, км	8									
The number of flight Mach	0.1	0.2	0.3	0.4	0.6	0.7	0.8	0.85	0.9	0.95
Relative density of the air, Δ	0.429	0.429	0.429	0.429	0.429	0.429	0.429	0.429	0.429	0.429
The air density, ρH, kg/m³	0.5259	0.5259	0.5259	0.5259	0.5259	0.5259	0.5259	0.5259	0.5259	0.5259
Coefficient K_a	0.905	0.905	0.905	0.905	0.905	0.905	0.905	0.905	0.905	0.905
The speed of sound a, m/S	307.953	307.953	307.953	307.953	307.953	307.953	307.953	307.953	307.953	307.953
Flight speed , V, m/s	30.80	61.59	92.39	123.18	184.77	215.57	246.36	261.76	277.16	292.56
Dynamic pressure, N/m²	249	997	2244	3990	8977	12219	15960	18017	20199	22506
The average gross weight of the aircraft, G_{AV}, N	5650560	5650560	5650560	5650560	5650560	5650560	5650560	5650560	5650560	5650560
The coefficient of aerodynamic lift in horizontal flight, $C_{Y\,h.f.}$	25.04	6.26	2.78	1.56	0.70	0.51	0.39	0.35	0, .1	0.28
Effective of wing extension , λ	8.38	8.38	8.38	8.38	8.38	8.38	8.38	8.38	8.38	8.38
Factor Blade of the polar, A_0	0.038	0.038	0.038	0.038	0.038	0.038	0.038	0.038	0.038	0.038
Coefficient K_A	1	1	1	1	1.09	1.16	1.27	1.33	1.4	1.5
Coefficient K_{CXO}	1	1	1	1	1.03	1.2	1.5	1.7	1.9	1.95
The coefficient of aerodynamic drag in horizontal flight, $C_{xa.h.f}$	23.8454	1.5100	0.3151	0.1141	0.0417	0.0367	0.0389	0.0418	0.0450	0.0453
Flight aerodynamic quality, K	1.05	4.15	8.83	13.72	16.69	13.92	10.06	8.30	6.87	6.12
Thrust required $P_{Required, N}$	5381413	1363125	640062	411870	338526	405953	561663	681076	822303	923415
The Rate of changes in the thrust of the number of flight Mach , ξ	0.972	0.952	0.940	0.935	0.950	0.969	0.995	1.011	1.029	1.048
Take-off thrust engines, P_0, N	1404000	1404000	1404000	1404000	1404000	1404000	1404000	1404000	1404000	1404000
Thrust available, P_{avail} , N	664688	650963	642627	639639	649541	662349	680341	691268	703475	716958
Vertical speed, V_y, m/S	-25.71	-7.76	0.04	4.97	10.17	9.78	5.17	0.47	-5.83	-10.69

Table 9: The calculation of thrust required and thrust available, H = 8 km.

Flight altitude, км	10									
The number of flight Mach	0.1	0.2	0.3	0.4	0.6	0.7	0.8	0.85	0.9	0.95
Relative density of the air, Δ	0.337	0.337	0.337	0.337	0.337	0.337	0.337	0.337	0.337	0.337
The air density, $\rho H, \dfrac{kg}{m^3}$	0.4136	0.4136	0, 4136	0.4136	0.4136	0.4136	0.4136	0.4136	0.4136	0.4136
Coefficient K_a	0.88	0.88	0.88	0.88	0.88	0.88	0.88	0.88	0.88	0.88
The speed of sound a, m/S	299.446	299.446	299.446	299.446	299.446	299.446	299.446	299.446	299.446	299.446
Flight speed , V, m/s	29.94	59.89	89.83	119.78	179.67	209.61	239.56	254.53	269.50	284.47
Dynamic pressure, N/m²	185	742	1669	2967	6676	9086	11868	13398	15020	16735
The average gross weight of the aircraft, $G_{Av.}$ N	5650560	5650560	5650560	5650560	5650560	5650560	5650560	5650560	5650560	5650560
The coefficient of aerodynamic lift in horizontal flight, $C_{yh.f.}$	33.67	8.42	3.74	2.10	0.94	0.69	0.53	0.47	0.42	0.37
Effective of wing extension , λ	8.38	8.38	8.38	8.38	8.38	8.38	8.38	8.38	8.38	8.38
Factor Blade of the polar, A_0	0.038	0.038	0.038	0.038	0.038	0.038	0.038	0.038	0.038	0.038
Coefficient K_A	1	1	1	1	1, 09	1, 16	1, 27	1, 33	1, 4	1, 5
Coefficient K_{CXO}	1	1	1	1	1.03	1.2	1.5	1.7	1.9	1.95

The coefficient of aerodynamic drag in horizontal flight, $C_{xa\,h.f.}$	43.1065	2.7138	0.5529	0.1893	0.0579	0.0460	0.0449	0.0467	0.0491	0.0489
Flight aerodynamic quality, K	0.78	3.10	6.77	11.12	16.16	14.93	11.73	9.98	8.47	7.63
Thrust required $P_{Required}$, N	7234021	1821721	835106	508293	349600	378392	481800	565957	667341	740384
The Rate of changes in the thrust of the number of flight Mach , ξ	0.972	0.952	0.940	0.935	0.950	0.969	0.995	1.011	1.029	1.048
Take-off thrust engines, P_0, N	1404000	1404000	1404000	1404000	1404000	1404000	1404000	1404000	1404000	1404000
Thrust available, P_{avail}, N	541395	530216	523426	520992	529058	539490	554145	563045	572988	583970
Vertical speed, V_y, m/S	-35.47	-13.69	-4.96	0.27	5.71	5.98	3.07	-0.13	-4.50	-7.87

Table 10: The calculation of thrust required and thrust available, H = 10 km.

Flight altitude, км	11									
The number of flight Mach	0.1	0.2	0.3	0.4	0.6	0.7	0.8	0.85	0.9	0.95
Relative density of the air, Δ	0.298	0.298	0.298	0.298	0.298	0.298	0.298	0.298	0.298	0.298
The air density, $\rho H, \frac{kg}{m^3}$	0.3648	0.3648	0.3648	0.3648	0.3648	0.3648	0.3648	0.3648	0.3648	0.3648
Coefficient K_a	0.867	0.867	0.867	0.867	0.867	0.867	0.867	0.867	0.867	0.867
The speed of sound a, m/S	295.023	295.023	295.023	295.023	295.023	295.023	295.023	295.023	295.023	295.023
Flight speed , V, m/s	29.50	59.00	88.51	118.01	177.01	206.52	236.02	250.77	265.52	280.27
Dynamic pressure, N/m²	159	635	1429	2540	5715	7779	10161	11470	12859	14328
The average gross weight of the aircraft, G_{Av}, N	5650560	5650560	5650560	5650560	5650560	5650560	5650560	5650560	5650560	5650560
The coefficient of aerodynamic lift in horizontal flight, $C_{yh.f.}$	39.33	9.83	4.37	2.46	1.09	0.80	0.61	0.54	0.49	0.44
Effective of wing extension , λ	8.38	8.38	8.38	8.38	8.38	8.38	8.38	8.38	8.38	8.38
Factor Blade of the polar, A_0	0.038	0.038	0.038	0.038	0.038	0.038	0.038	0.038	0.038	0.038
Coefficient K_A	1	1	1	1	1.09	1.16	1.27	1.33	1.4	1.5
Coefficient K_{cx0}	1	1	1	1	1.03	1.2	1.5	1.7	1.9	1.95
The coefficient of aerodynamic drag in horizontal flight, $C_{xah.f.}$	58.8019	3.6948	0.7467	0.2506	0.0711	0.0536	0.0497	0.0507	0.0524	0.0518
Flight aerodynamic quality, K	0.67	2.66	5.85	9.81	15.37	14.97	12.36	10.74	9.26	8.42
Thrust required $P_{Required}$, N	8448432	2123422	965534	576114	367586	377343	457240	526053	610317	671356
The Rate of changes in the thrust of the number of flight Mach , ξ	0.972	0.952	0.940	0.935	0.950	0.969	0.995	1.011	1.029	1.048
Take-off thrust engines, P_0, N	1404000	1404000	1404000	1404000	1404000	1404000	1404000	1404000	1404000	1404000
Thrust available, P_{avail}, N	487655	477586	471470	469277	476542	485939	499139	507156	516112	526004
Vertical speed, V_y, m/S	-41.56	-17.19	-7.74	-2.23	3.41	3.97	1.75	-0.84	-4.43	-7.21

Table 11: The calculation of thrust required and thrust available, H = 11 km.

Flight altitude, км	12									
The number of flight Mach	0.1	0.2	0.3	0.4	0.6	0.7	0.8	0.85	0.9	0.95
Relative density of the air, Δ	0.2536	0.2536	0.2536	0.2536	0.2536	0.2536	0.2536	0.2536	0.2536	0.2536
The air density, $\rho H, \dfrac{kg}{m^3}$	0.3156	0.3156	0.3156	0.3156	0.3156	0.3156	0.3156	0.3156	0.3156	0.3156
Coefficient K_a	0.867	0.867	0.867	0.867	0.867	0.867	0.867	0.867	0.867	0.867
The speed of sound a, m/S	295.023	295.023	295.023	295.023	295.023	295.023	295.023	295.023	295.023	295.023
Flight speed , V, m/s	29.50	59.00	88.51	118.01	177.01	206.52	236.02	250.77	265.52	280.27
Dynamic pressure, N/m²	137	549	1236	2198	4944	6730	8790	9923	11125	12396
The average gross weight of the aircraft, $G_{Av,}$ N	5650560	5650560	5650560	5650560	5650560	5650560	5650560	5650560	5650560	5650560
The coefficient of aerodynamic lift in horizontal flight, $C_{yh.f.}$	45.46	11.36	5.05	2.84	1.26	0.93	0.71	0.63	0.56	0.50
Effective of wing extension , λ	8.38	8.38	8.38	8.38	8.38	8.38	8.38	8.38	8.38	8.38
Factor Blade of the polar, A_0	0.038	0.038	0.038	0.038	0.038	0.038	0.038	0.038	0.038	0.038
Coefficient K_A	1	1	1	1	1.09	1.16	1.27	1.33	1.4	1.5
Coefficient K_{CX0}	1	1	1	1	1.03	1.2	1.5	1.7	1.9	1.95
The coefficient of aerodynamic drag in horizontal flight, $C_{xa\,h.f.}$	78.5576	4.9295	0.9906	0.3278	0.0877	0.0631	0.0559	0.0557	0.0567	0.0554
Flight aerodynamic quality, K	0.58	2.31	5.10	8, 67	14.40	14.69	12.72	11.29	9.91	9.09
Thrust required $P_{Required,}$ N	9764610	2450941	1108159	651889	392361	384584	444301	500309	570447	621625
The Rate of changes in the thrust of the number of flight Mach , ξ	0.972	0.952	0.940	0.935	0.950	0.969	0.995	1.011	1.029	1.048
Take-off thrust engines, P_0 , N	1404000	1404000	1404000	1404000	1404000	1404000	1404000	1404000	1404000	1404000
Thrust available, $P_{avail,}$ N	425163	416384	411052	409141	415475	423667	435176	442165	449973	458598
Vertical speed, Vy, m/S	-48.76	-21.25	-10.92	-5.07	0.72	1.43	-0.38	-2.58	-5.66	-8.09

Table 12: The calculation of thrust required and thrust available, H = 12 km.

Flight altitude, км	12.4									
The number of flight Mach	0.1	0.2	0.3	0.4	0.6	0.7	0.8	0.85	0.9	0.95
Relative density of the air, Δ	0.2399	0.2399	0.2399	0.2399	0.2399	0.2399	0.2399	0.2399	0.2399	0.2399
The air density, $\rho H, \dfrac{kg}{m^3}$	0.2986	0.2986	0.2986	0.2986	0.2986	0.2986	0.2986	0.2986	0.2986	0.2986
Coefficient K_a	0.867	0.867	0.867	0.867	0.867	0.867	0.867	0.867	0.867	0.867
The speed of sound a, m/S	295.023	295.023	295.023	295.023	295.023	295.023	295.023	295.023	295.023	295.023
Flight speed , V, m/s	29.50	59.00	88.51	118.01	177.01	206.52	236.02	250.77	265.52	280.27
Dynamic pressure, N/m²	130	520	1170	2079	4678	6367	8317	9389	10526	11728
The average gross weight of the aircraft, $G_{Av,}$ N	5650560	5650560	5650560	5650560	5650560	5650560	5650560	5650560	5650560	5650560
The coefficient of aerodynamic lift in horizontal flight, $C_{yh.f.}$	48.05	12.01	5.34	3.00	1.33	0.98	0.75	0.67	0.59	0.53
Effective of wing extension , λ	8.38	8.38	8.38	8.38	8.38	8.38	8.38	8.38	8.38	8.38
Factor Blade of the polar, A_0	0.038	0.038	0.038	0.038	0.038	0.038	0.038	0.038	0.038	0.038
Coefficient K_A	1	1	1	1	1.09	1.16	1.27	1.33	1.4	1.5
Coefficient K_{CX0}	1	1	1	1	1.03	1.2	1.5	1.7	1.9	1.95

The coefficient of aerodynamic drag in horizontal flight, $C_{xah.f.}$	87.7547	5.5044	1.1041	0.3637	0.0954	0.0676	0.0587	0.0581	0.0586	0.0571
Flight aerodynamic quality, K	0.55	2.18	4.84	8.26	13.99	14.51	12.79	11.46	10.12	9.32
Thrust required $P_{Required,}$ N	1E+07	2589322	1168646	684376	403975	389474	441832	493270	558414	606117
The Rate of changes in the thrust of the number of flight Mach , ξ	0.972	0.952	0.940	0.935	0.950	0.969	0.995	1.011	1.029	1.048
Take-off thrust engines, P_0, N	1404000	1404000	1404000	1404000	1404000	1404000	1404000	1404000	1404000	1404000
Thrust available, P_{avail} , N	405560	397185	392099	390276	396318	404133	415110	421777	429226	437452
Vertica' speed, V_y m/S	-51.77	-22.89	-12.16	-6.14	-0.24	0.54	-1.12	-3.17	-6.07	-8.37

Table 13: The calculation of thrust required and thrust available, H = 12.4 km.

Height Flight, H, km	0	2	4	6	8	10	11	12	12.4
Minimum speed, $Vmin$, m/s (left)	46	55	60	70	94	118	140	170	200
Maximum(Full) speed, $Vmax$, m/s (right)	277	275	275	267	264	255	248	230	200

Table 14: Calculation of a possible aircraft flight boundaries

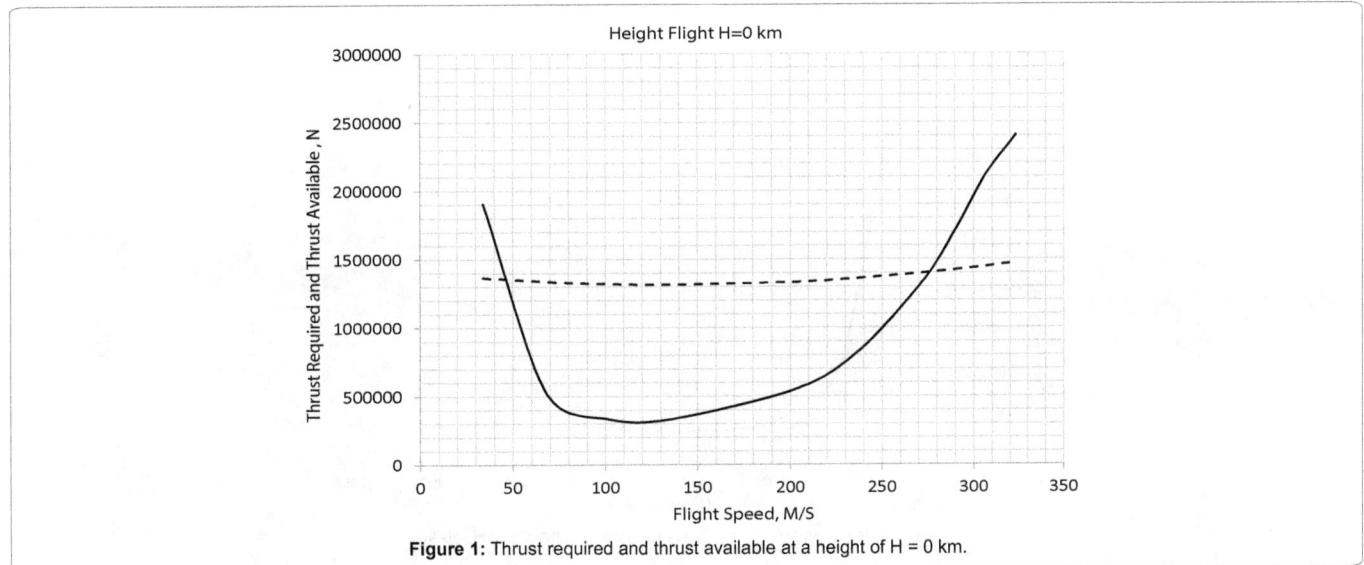

Figure 1: Thrust required and thrust available at a height of H = 0 km.

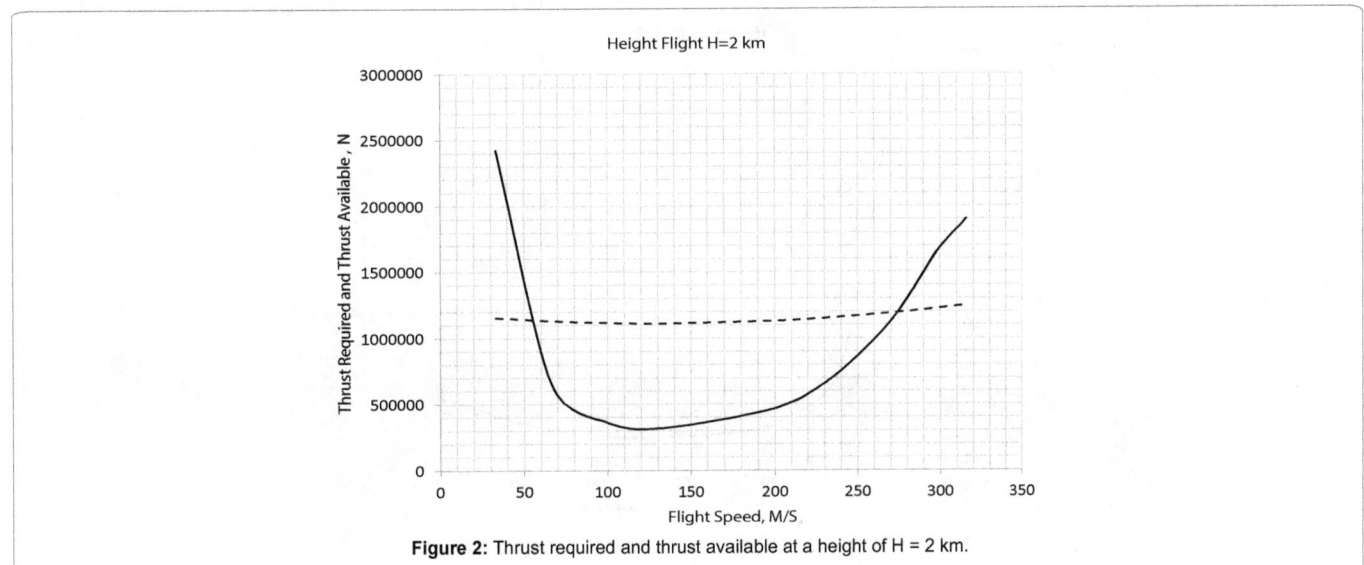

Figure 2: Thrust required and thrust available at a height of H = 2 km.

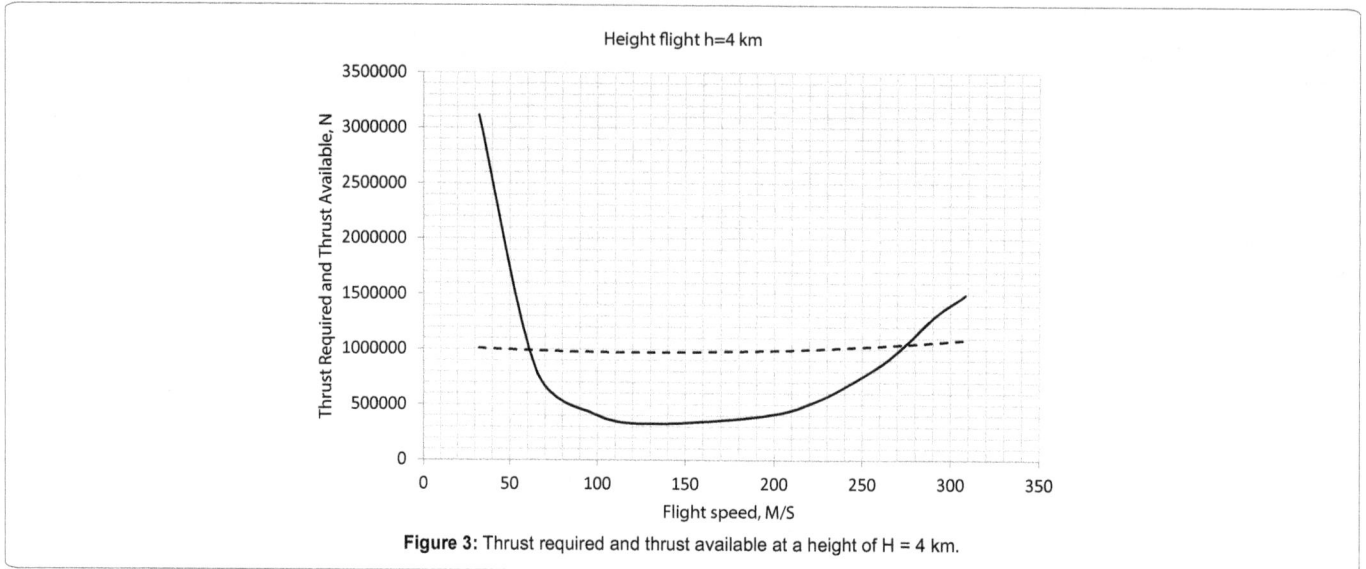

Figure 3: Thrust required and thrust available at a height of H = 4 km.

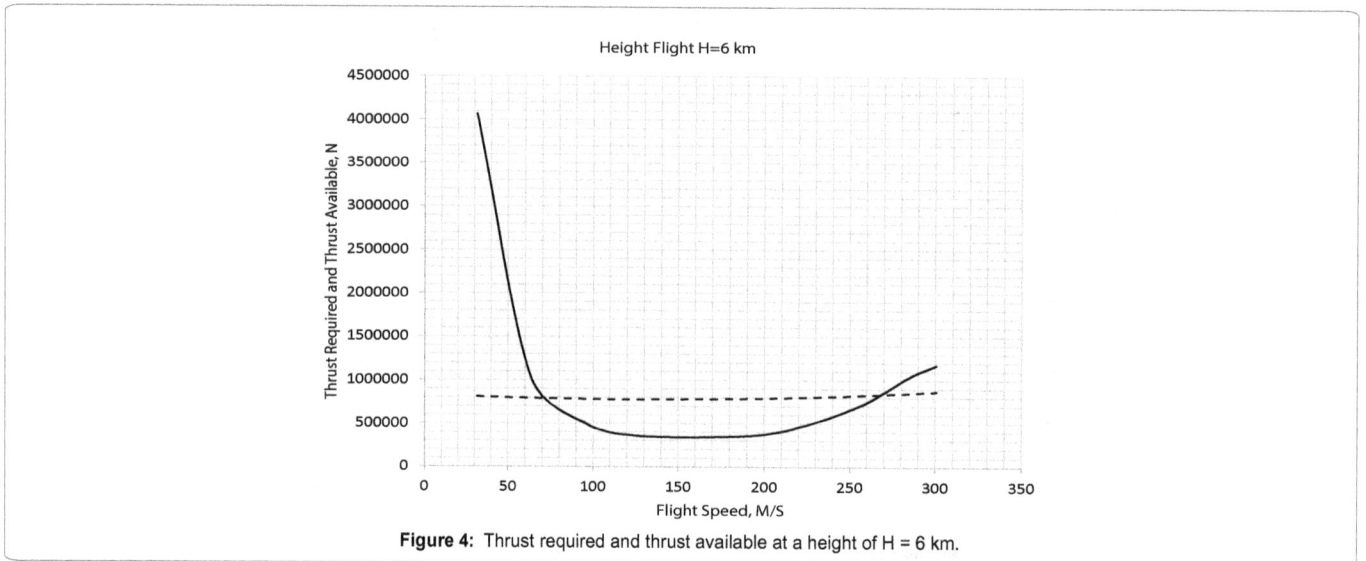

Figure 4: Thrust required and thrust available at a height of H = 6 km.

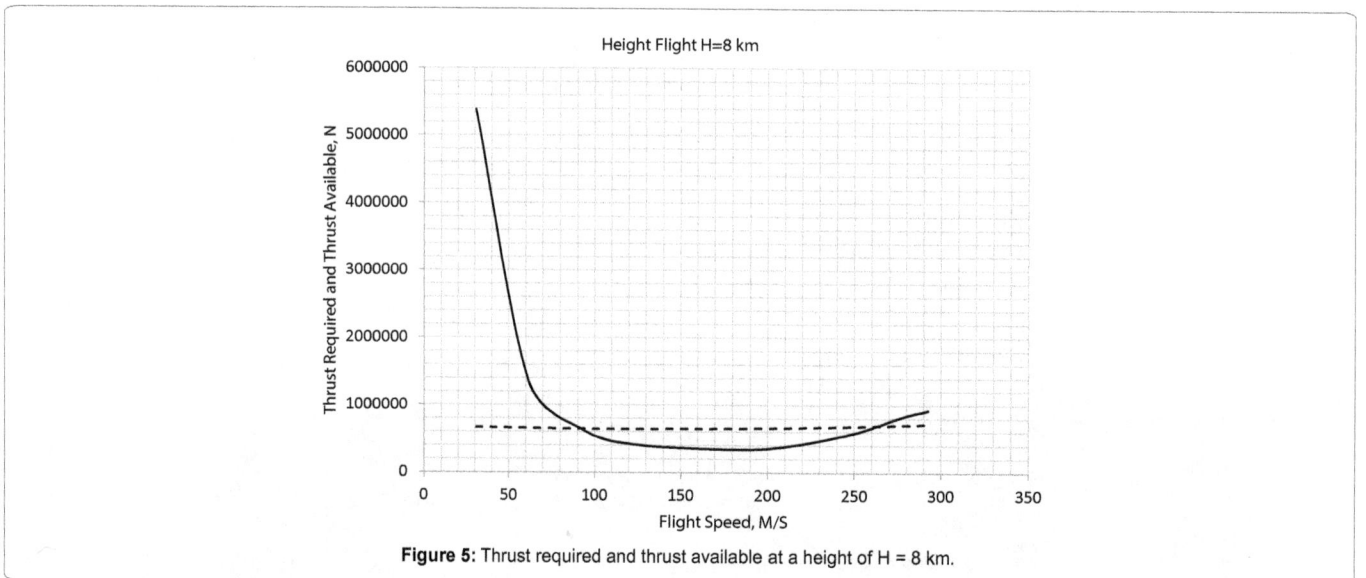

Figure 5: Thrust required and thrust available at a height of H = 8 km.

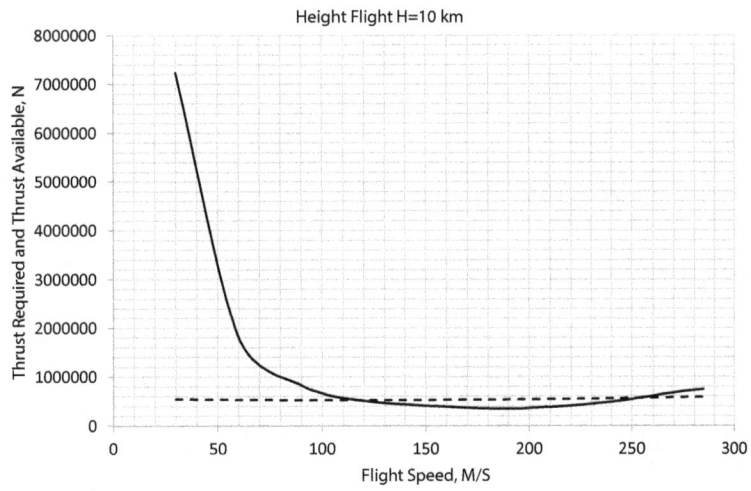

Figure 6: Thrust required and thrust available at a height of H = 10 km.

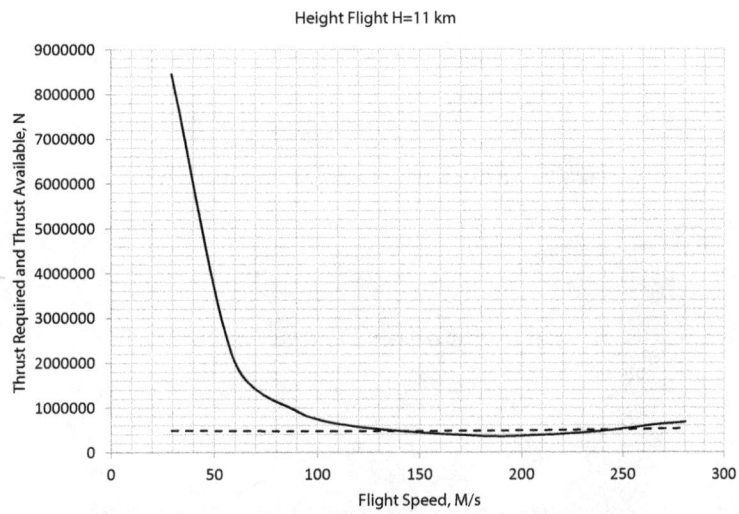

Figure 7: Thrust required and thrust available at a height of H = 11 km.

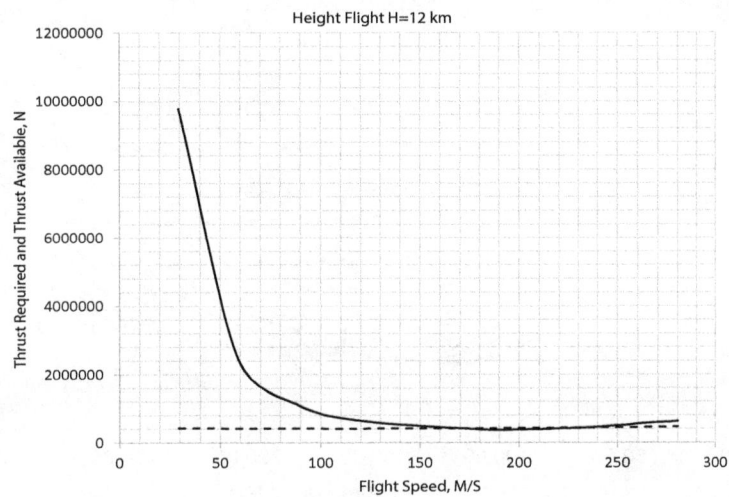

Figure 8: Thrust required and thrust available at a height of H = 12 km.

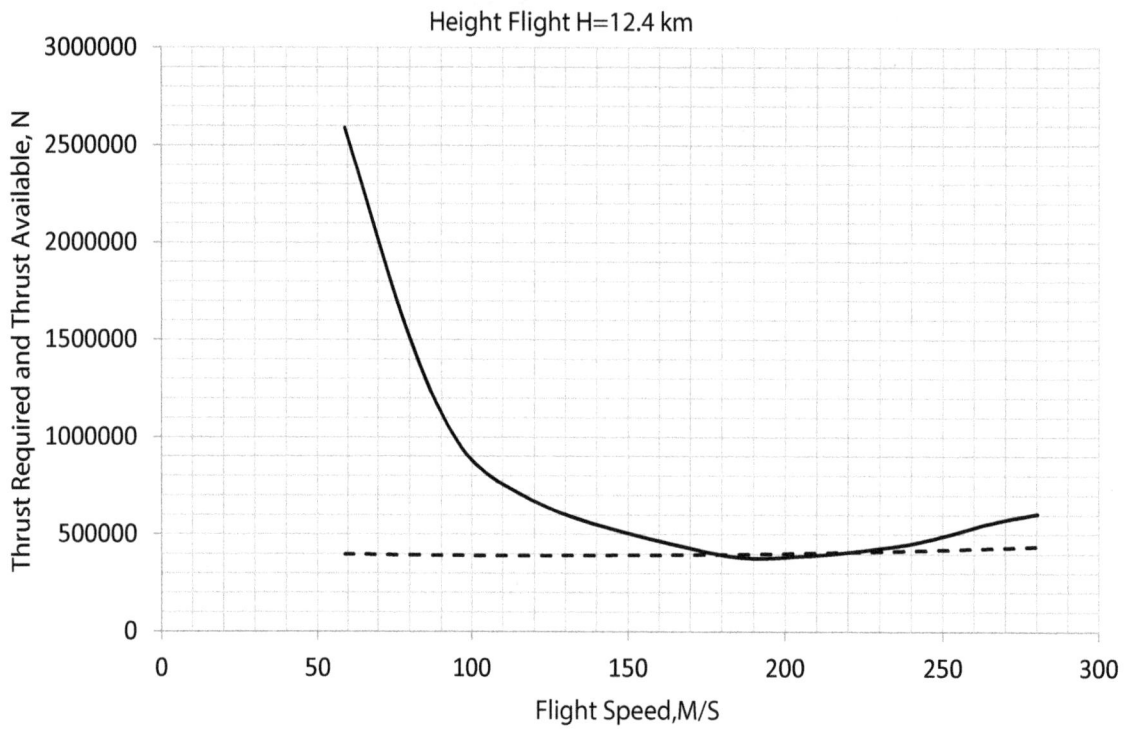

Figure 9: Thrust required and thrust available at a height of H = 12.4 km.

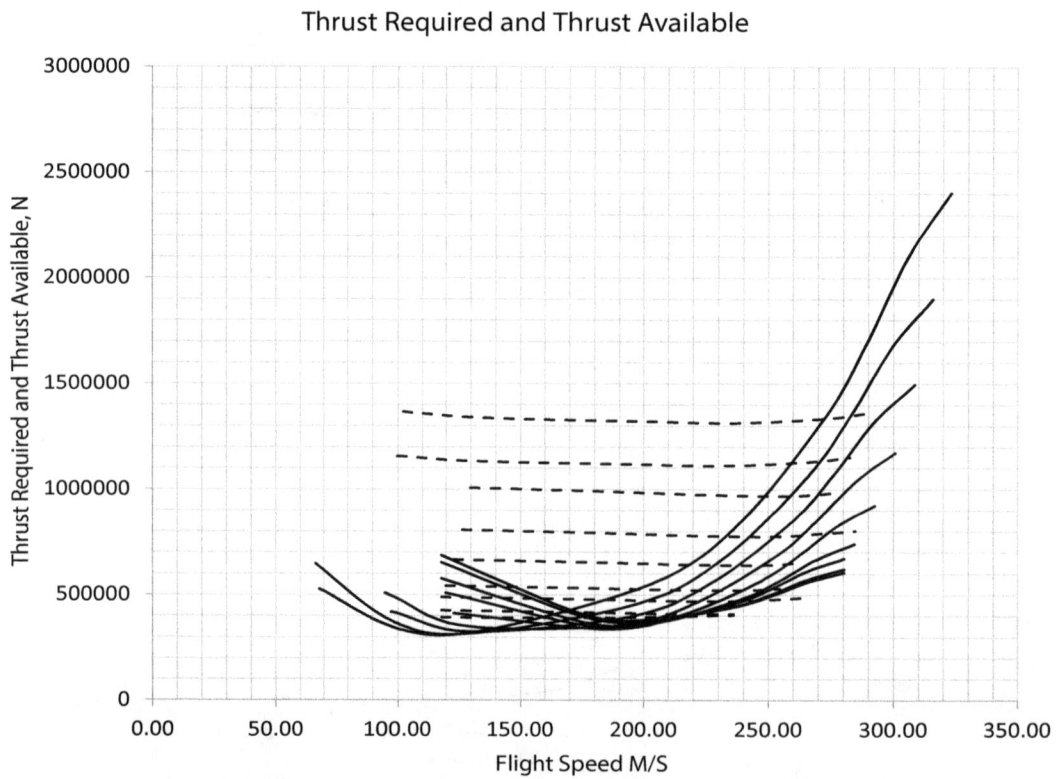

Figure 10: Thrust required and thrust available at a height of H = 0…. 12.4 km.

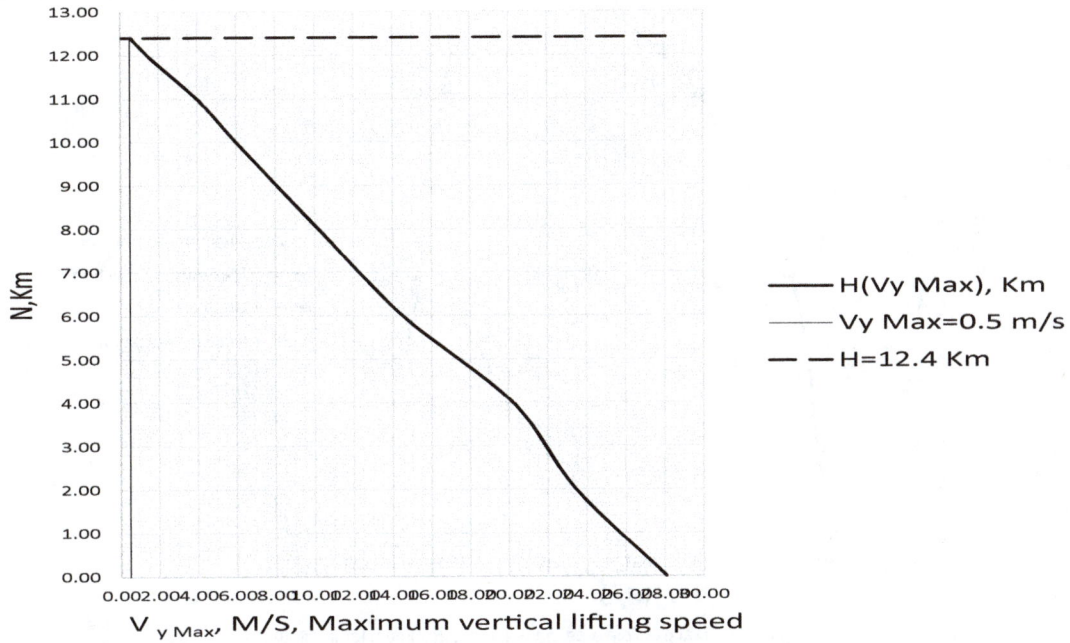

Figure 11: Finding a practical ceiling of the aircraft.

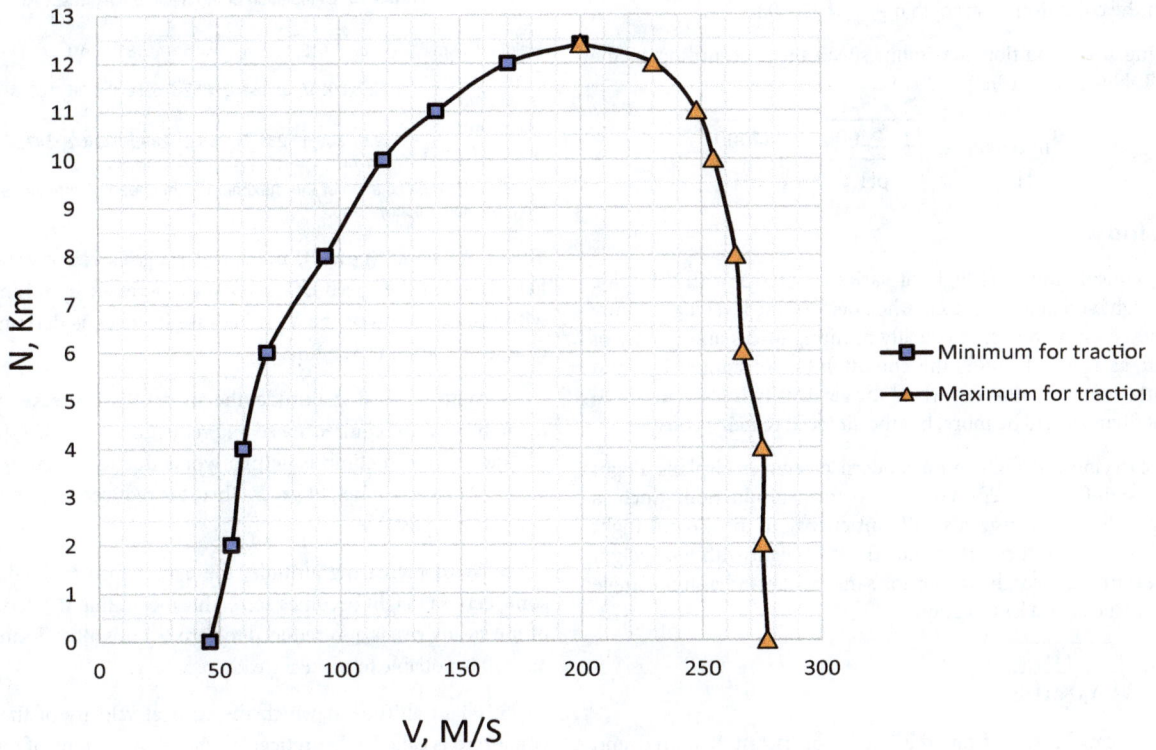

Figure 12: Real possible area of the aircraft (maximum for traction).

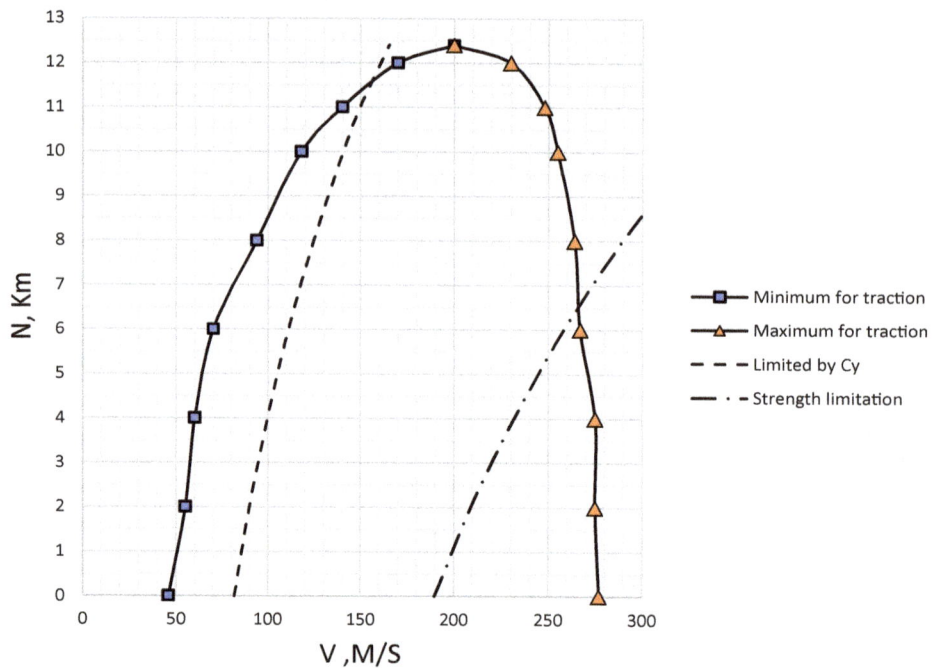

Figure 13: Real possible area of the aircraft (minimum for traction).

The minimum flight speed by the formula:

$$V_{min} = \sqrt{\frac{2 \times G_{Av.}}{C_y \times \rho H \times S}} = \sqrt{\frac{2 \times 5650560}{1.53 \times \rho H \times 905}} = \sqrt{\frac{8161.7}{\rho H}}$$

On the right side, should be limits on the N/m² maximum velocity head from the condition of strength $q_{max,max} = 22000$.

Defining this condition maximum speed flight according to the formula (Tables 15 and 16):

$$Vmax = \sqrt{\frac{2 \times q_{maxmax}}{\rho H}} = \sqrt{\frac{2 \times 22000}{\rho H}} = \sqrt{\frac{44000}{\rho H}}$$

Conclusion

1. By, considering level flight at various altitudes with the same weight of flight and angle of attack, when performing level flight at any altitude is necessary to ensure equality of lifting forces and gravity of the aircraft, as $Y_a = G$. To fulfill this condition for constant weight and angle of attack at high altitudes where the air density is less true speed horizontal flight should be more, but the airspeed remains constant.

2. In carrying out flight on a modern passenger airplane Flight weight is significantly reduced due to fuel production. Such a change of flight mass causes a significant change in the aircraft flight characteristics. To perform horizontal flight of flight with less weight requires less lifting force, hence for the same attack and altitude angle requires less speed and less traction.

$$V_{h.f.} = \sqrt{\frac{2 \times G}{C_{Y_a} \times \rho H \times S}}$$

3. As can be seen from the graphs of Thrust Required and

Height, H, km	0	2	4	6	8	10	11	12	12.4
Density of air, ρH, kg/m3	1.225	1.0067	0.8194	0.6602	0.5259	0.4136	0.3648	0.3156	0.2986
Minimum speed, V_{min},m/s	81.6	90	99.8	111.2	124.6	140.5	149.6	160.8	165.3

Table 15: Calculation of the minimum flight speed.

Height, H, km	0	2	4	6	8	10	11	12	12.4
Density of air, ρH, kg/m3	1.225	1.0067	0.8194	0.6602	0.5259	0.4136	0.3648	0.3156	0.2986
Minimum speed, V_{min},m/s	189.5	209.1	231.7	258.2	289.3	326.2	347.3	373.4	383.9

Table 16: The calculation of the maximum flight speed of conditions limitations on the maximun velocity head.

Thrust Available (power), the speed range is reduced by raising at height, So all the speed characteristics increases by raising at height, with the exception of the V_{max}, because its value is determined by the characteristics of the engine.

4. With increasing altitude, the air density decreases, which leads to an increase in required speed and reduction of vertical speed(climb). Characteristics of climb is getting worse due to the fall of the engine thrust. At a certain height excess thrust is reduced to zero, so a further climb is not possible.

5. With increasing altitude, the excess thrust is reduced and at some certain height becomes zero. This means that the vertical velocity of the steady rise is also reduced to zero. At this altitude and above the aircraft is not able to make a steady recovery.

6. Flight altitude at which the vertical velocity of the steady rise equal zero is called a theoretical (or static), the ceiling of the aircraft.

7. There's not an excess thrust On a theoretical ceiling therefore the only possible is horizontal flight, and only the most advantageous angle of attack (and only in the most advantageous rate) at which the lowest Required thrust power. Speed range at this moment equal zero.

8. With the steady rise of the plane, almost cannot reach the theoretical ceiling, because as you get closer to it excess thrust becomes so small, that in order to set the height of the rest needs to spend too much time and fuel. Due to the lack of excess flying thrust on a theoretical ceiling is almost impossible, because any violation of the flight mode cannot be eliminated without excessive traction. For example, when randomly formed even small roll plane loses a considerable height (falls). Therefore, in addition to theoretical concepts (static) Ceiling introduced the concept of practical ceiling.

References

1. Mkhitaryan AM (1978) Flight dynamics. (2ndedn) Mechanical Engineering Press, Moscow.

2. Salenko SD, Obuhovsky AD (2014) Flight dynamics. NSTU Press, Novosibirsk.

3. Ostoslavsky IV, Strazheva IV (1969) Flight dynamics. The trajectories of the summer-enforcement apparatus. Mechanical Engineering Press, Moscow.

4. Bochkarev AF (1985) Aeromechanics Aircraft. Mechanical Engineering press, Moscow.

Design and Realization of Payload Operation and Application System of China's Space Station

Wang H*, Guo L and Wu P

Key Laboratory of Space Utilization, Technology and Engineering Center for Space Utilization, Chinese Academy of Sciences, PR China

Abstract

China's Space Station will be launched in the year 2018; this space station is China's largest space science experiment and application platform until now. This paper mainly introduces the function composition of payload operation and application system of China's space station, system architecture, hardware architecture and the new technology we use to implement the payload operation and application ground system of China's space station.

Keywords: Human spaceflight; China's space station; Ground operations system; Telescience

Introduction

China's space station will be launched at year 2018; this space station is China's largest space science experiment and application platform until now. It will stay in space orbit more than a decade and have many complex scientific space experimental tasks to do step by step in the future. This would mean that we are facing many challenges for our payload operation and application ground system, including complex mission planning, high-speed mass data processing, health management, payload status monitoring, remote support for tele-Science and the ground system dynamically updates with the scientific mission changes all the time. This paper mainly introduces the function composition of payload operation and application system of China's space station, system architecture, hardware architecture and the new technology we use to implement the payload operation and application ground system of China's space station [1-4].

System Overall Design

Components and design of the system

Payload operation and application system is a very important sub system of the ground system of the China's space station. It responsible for establishing a communication link with the ground station, spacecraft control center and science center, receiving telemetry data and various types of application data, real-time processing payload data and telemetry data to monitor the health status of the payload, planning and scheduling payload work timeline According to various scientific experiments request and spacecraft resources, providing space station payload telescience support system for providing scientific experiments remote operation, remote scene and remote analysis; providing technical support for science education and international cooperation. Overall, it is a necessary infrastructure of the space station project [5,6]. Payload operation and application ground system is a typical information system, the function including front-end data communications, integrated monitoring and real-time data processing, planning and scheduling payload work plan, data simulation, health management and fault diagnosis for payload, supporting remote scientific experiments.

System architecture design

The software architecture of the payload operation and application ground system is SOA (Service oriented architecture). All system functions are designed as service components to achieve software functional reuse in the system. All system service components exchange information and data between various functional components through standardized interfaces by data distribution service bus. The feature of this system is reuse, scalability and flexibility. The payload operation and application ground system's architecture showed as below, this architecture including four layers. They are UI layer, integration layer, application layer and resource layer (Figure 1).

System hardware architecture design

Payload operation and application ground system is a typical information system, composed mainly by the computing systems, network devices, storage system, its high performance computing and mass storage devices are integrated together by virtual software technology for sharing of resources and easy to manage (Figure 2).

Seamless upgrade mechanism

China space station construction process is gradually finish by building space capsule one by one for a long time, during this time, all payload scientific experiments are done step by step, which means that payload scientific experiments on the China space station are different and change all the time.

Payload operation and application ground system needs to support new scientific experiments tasks by dynamically upgrade while the system working. So this system needs flexible system architecture to deal with this situation. System architecture design principles follow the norms modularity, encapsulation, loose coupling, and separation of concerns, reuse and composition (Figure 3).

Payload operation and application ground system's soft architecture is SOA (service-oriented architecture) , This service-oriented architecture (SOA) is an architectural pattern in computer software design in which application components provide services to other components via a communications protocol, typically over an

*Corresponding author: Wang H, Key Laboratory of Space Utilization, Technology and Engineering Center for Space Utilization, Chinese Academy of Sciences, PR China, E-mail: beyondwhfei@163.com

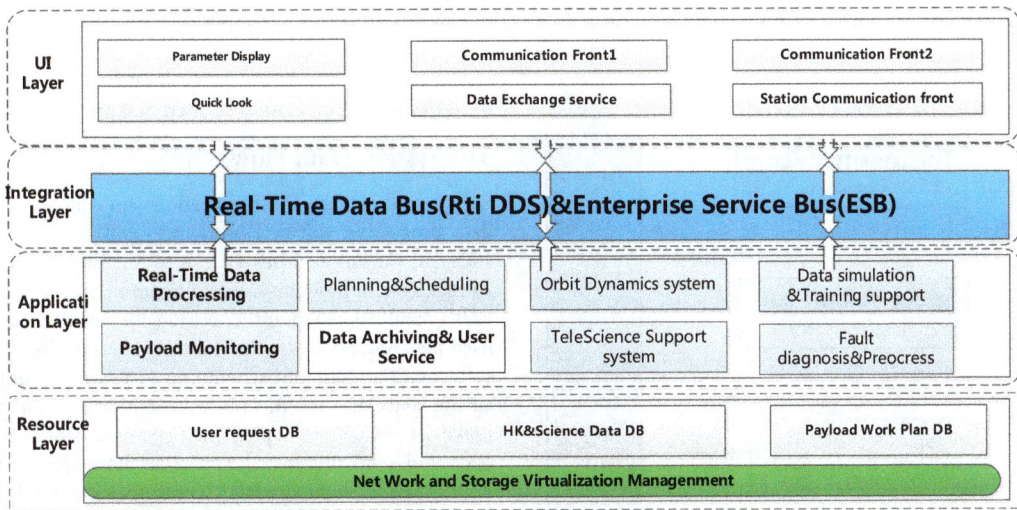

Figure 1: Architecture of the system.

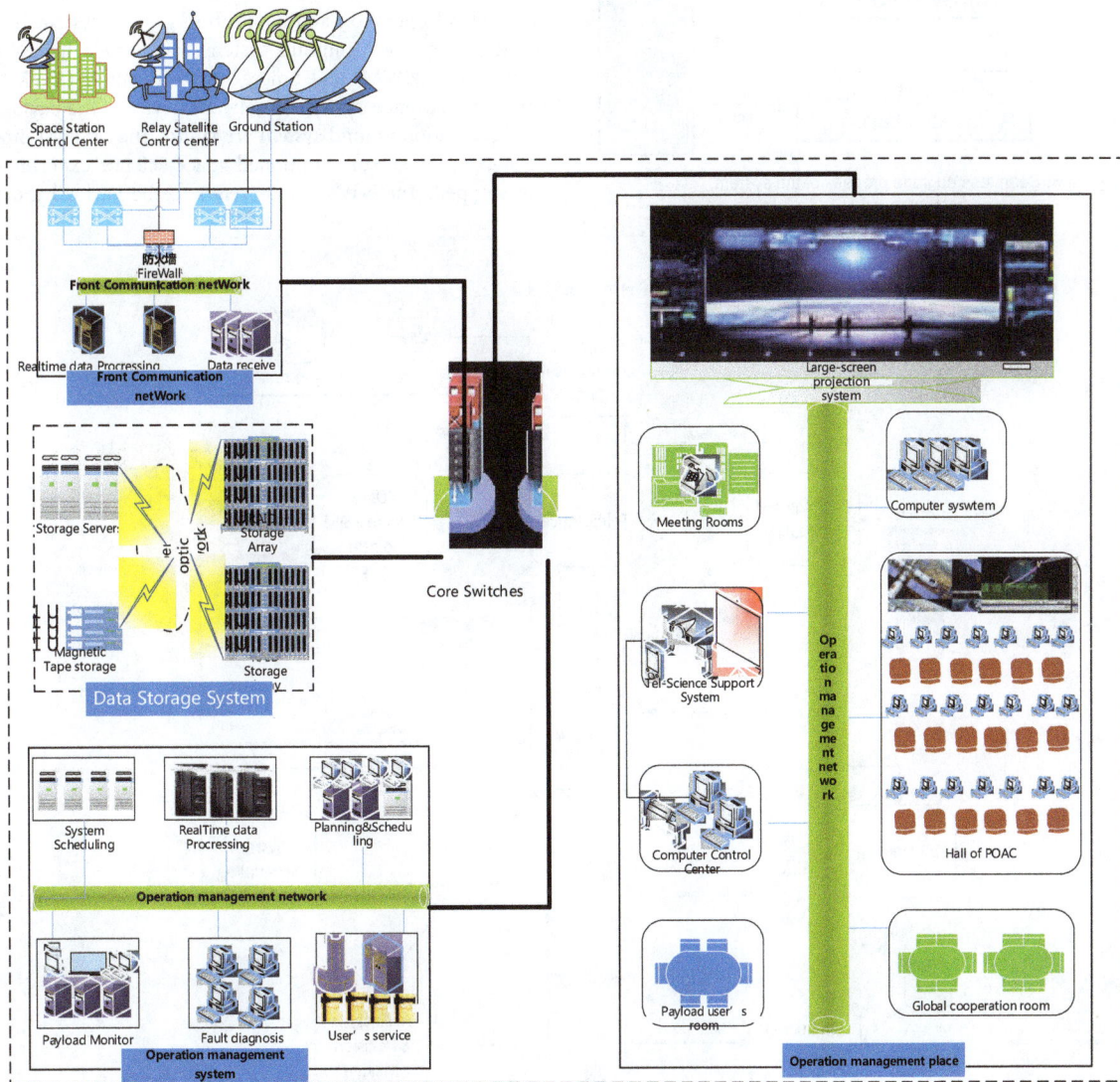

Figure 2: Design of the system hardware architecture.

enterprise service bus (ESB). The principles of service-orientation are independent of any vendor, product or technology. A service is a self-contained unit of functionality. SOA makes it easier for software

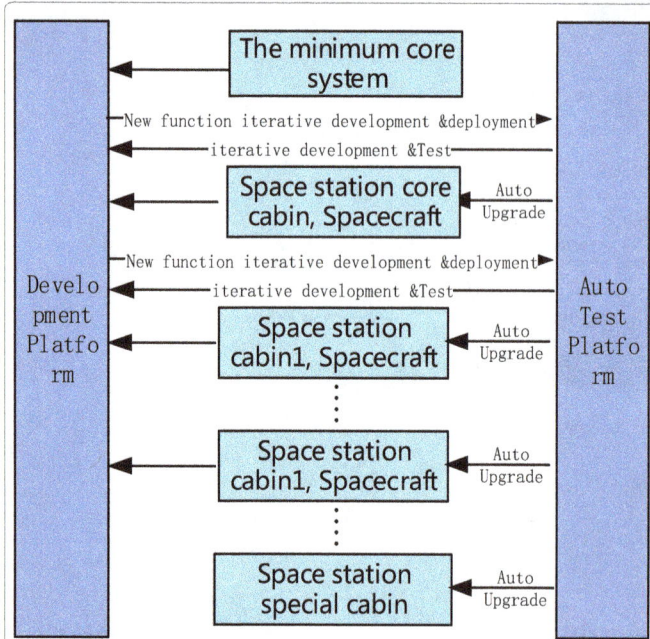

Figure 3: Seamless upgrade process of the system.

components on computers connected over an ESB to cooperate. Every computer can run any number of services, and each service is built in a way that ensures that the service can exchange information with any other service in the ESB without human interaction and without the need to make changes to the underlying program itself.

The System Data Flow

Front-end data communication software send data to real-time data processing software and integrated comprehensive monitoring software via the data bus DDS and then archive the raw data to the storage system to archive. Planning and scheduling software receiving user's application request, checking the legality and then archive the user' plan request files to the storage system via the File and data archive software. Planning and scheduling software planning and scheduling all the space and ground resources to support the payload work and make the payload work plan, then encoding work plan into instruction code and send this instruction code file to payload on orbit via the Beijing Space Flight Control Center uplink ground control stations or relay communication satellites uplink channel (Figure 4).

System Operating Mode

Payload operation and application ground system has four work modes. They are Minimum system work mode (MSWM), Normal work mode (NWM) and Emergency work mode (EWM). When the software, computer system or network device of the payload operation and application ground system are failure, the MSWM mode required to start to make sure continued space science experiments mission not stopped. The NWM mode refers to the payload monitoring and

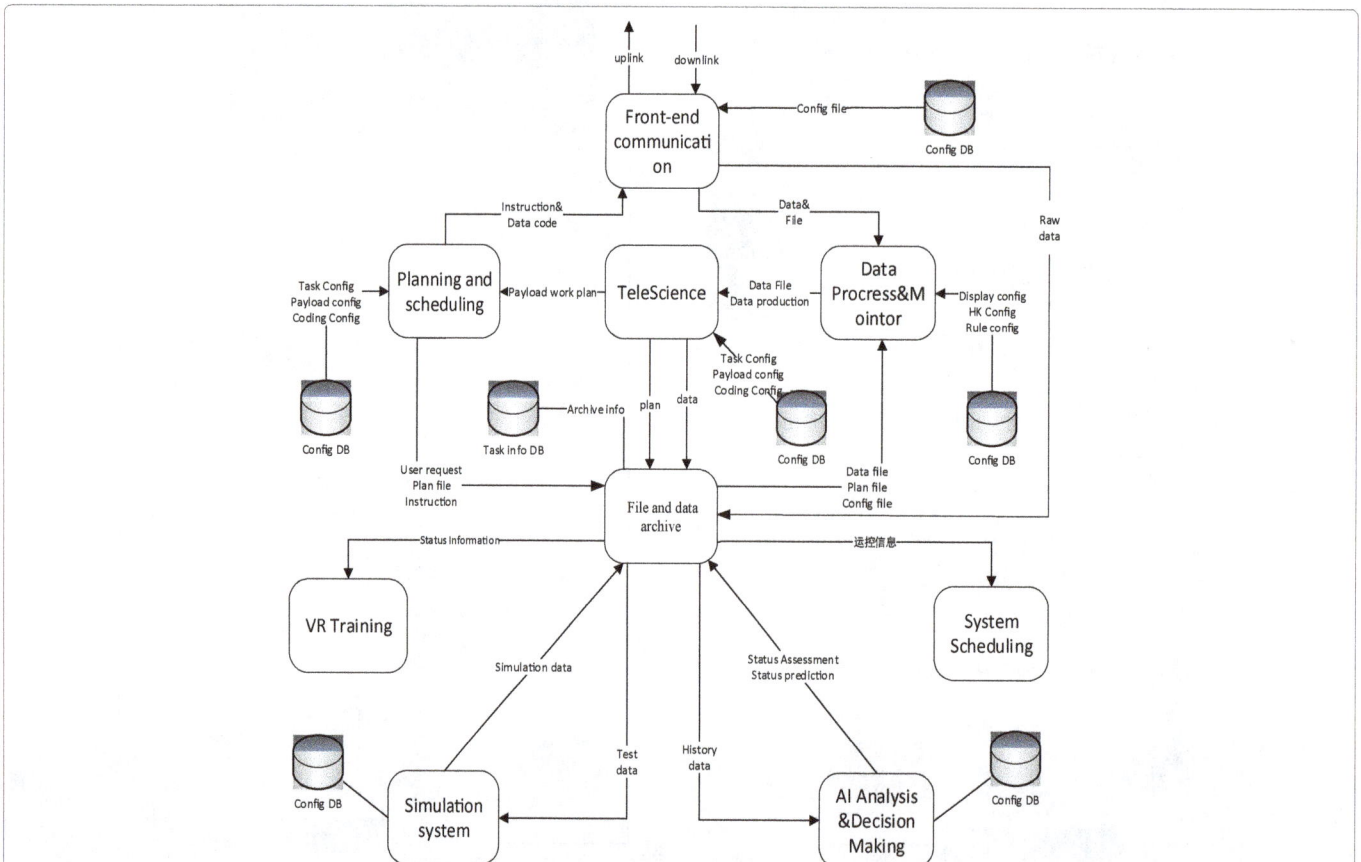

Figure 4: System data flow.

management in accordance with the normal operation control flow and data processing flow to work. It is the most conventional and most comprehensive work mode of our ground system during the space station on orbit. The EWM mode is a special work mode of our system. It will start to work when the payload of the Space station will be broken or has broken. In this mode, we will deal with temporarily interrupted the normal operation control process (Including data processing and task Planning) by fault plan manual requirement to rescue the payload equipment.

Conclusion

China's space station will be launched at year 2018; this space station is China's largest space science experiment and application platform until now. It will stay in space orbit more than a decade and have many complex scientific space experimental tasks to do step by step in the future.

The system uses a lot of new information technology and has all complete function of the payload ground support system. We hope this payload operation and application ground system of China's space station will support the payload application system to obtain significant scientific and technological achievements and improve the effectiveness of the payload work in the future.

References

1. CCSDS (2005) Report concerning mission operations services concept CCSDS 520.0-G-1 Green book.

2. Data Distribution Service (2015) Real-time System V1.2. OMG Released Versions of DDS

3. Suzuki H (1977) A statistical model for urban radio propagation. IEEE Transactions on Communications 25:673-680.

4. NDDS International. NDDS [EB/OL].

5. The Real-time publish-subscribe wire protocol, DDS Interoperability Wire Protocol Specification (DDS-RTPS).

6. Walton J (1993) Models for the management of satellite based sensors. Massachusetts Institute of Technology Massachusetts.

Paradoxical Variation of the Solar Day Related to Kepler/Newton System

Luiz Sampaio Athayde Junior*

Professor at School of Accounting Sciences, Federal University of Bahia, Brazil

Abstract

According to the first law of Kepler, the planets orbit the sun in elliptical path. This ellipse causes a slowdown in the world when it goes from the nearest point of the sun to the farthest point and also causes acceleration when the opposite occurs. This variation of the velocity of the planet combined with the inclination of its imaginary axis creates the analemma chart, which can be found with the overlap of the positions of the sun in a particular location always in the same timeset on a watch. The analema, in turn, describes variations in the durations of the solar day. In some dates, these variations in solar days occur in accordance with the change in velocity of the planet, but at other times, they get along perfectly Conversely, showing in some parts of the solar days year that will gradually reducing their periods as the planet decelerates and also increases periods as the planet accelerates.

Keywords: Kepler's laws; Law of universal gravitation Newton; Analemma; Equation of time; Solar day

Introduction

Due the laws of planetary motion, specifically described by the first law of Kepler and the law of universal gravitation of Newton, the planets have speed variations in its orbit by traffic and also in its rotation.

Because of the speed variations the meridian passage of the sun on your east of apparent westward movement across the planet also occurs with varying speed, and its faster passage when the planet is slower when compared to the time of a clock, and slower when the planet is faster. If we took pictures of where the sun shines in the zenith every day in a same time on the clocks, considering only the variation described speed, the analemma would be a horizontal line, describing only the delays and lates of light in its apparent movement. However, the planet also has an inclination in its imaginary axis, which gives rise to the seasons, specifically in the temperate zone, which makes the analemma also present a variation between north and south and in their form as we know, that resembles the number eight or infinity symbol.

The variations in the analemma chart indicate variations in periods of solar days that in some parts of the year occur concurrently with the change in velocity of the planet, that is, when the planet accelerates, periods of solar days successively decrease and the opposite also. However, the chart also shows that in other parts of the year variations occur completely contrary to variations in the speed of the planet, in which slows down the planet and the sun days have successively decreasing periods and otherwise, too.

In studies of physics, it uses the concept of speed for moving bodies, as stated in relation to the planets, while for the solar day, which is not a mobile; the term is used in reference to the period duration.

Solar days measurements were taken on a few dates for demonstration of paradoxical changes in periods of solar days.

Theoretical

Solar day

According to Oliveira Filho and Saraiva, solar day is "the time interval between two successive passages of the sun across the meridian of the place". Can be measured with a gnomon that can demonstrate the meridian passage of the sun on a date and the next date.

Not to be confused with the 360° rotation around the imaginary axis itself and also according to the authors:

It's 3 m 56 s and longer than the sidereal day because the sun is moving in the opposite direction of diurnal motion, that is, from west to east. This difference is due to the Earth translation movement around the Sun, of about 1 degree (4 minutes) per day (360/y = 360°/ (365.25 days) = 0.9856°/day) (Figure 1).

Thus the solar day is longer than a full rotation of 360° on the planet, because when the sun in its passage east to west, focuses exactly on the meridian passage of a certain locality, only focus again on the same meridian passage or that same Location ~360,9856° and not the exact 360 what would be a complete rotation of the planet around its imaginary axis.

The difference occurs because the time of one complete rotation of the planet space is also moving along its orbit.

The laws of Kepler

Johannes Kepler, German astronomer who lived between 1571 and 1630, although it has not given due importance in the time he published his works, formatted the three laws that describe the motions of the planets, published in different works and not in the same work, that are named today as the three laws of Kepler.

Various trigonometric operations were employed to design the first two. For them, we used data from another important astronomer who preceded him called Tycho Brahe, referring to its observations on the planet Mars.

Working with Brahe, Kepler could conceive the first of its laws, which states that the planets orbit around the sun not in circular orbits,

---

*Corresponding author: Luiz Sampaio Athayde Junior , Professor at School of Accounting Sciences, Federal University of Bahia, Brazil
E-mail: sampaioathayde@yahoo.com.br

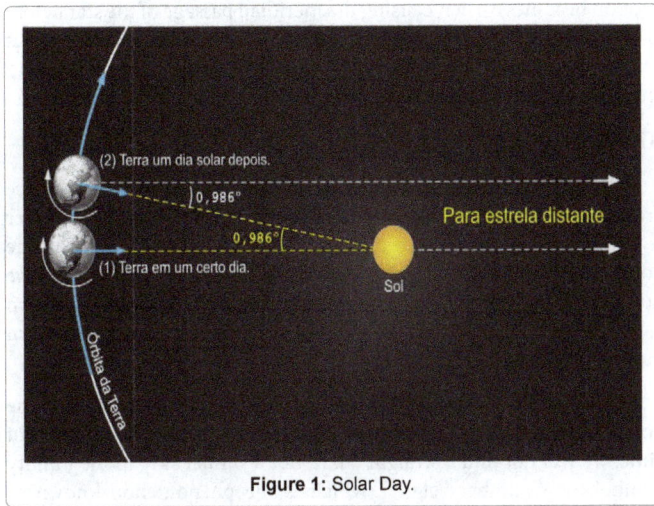

Figure 1: Solar Day.

but in elliptical orbits and the Sun is at one of the focus.

The second teaches us that an imaginary radius of the orbit, joining the sun to the planet takes equal areas in equal times. When the planet is closer to the sun logically has lower imaginary rays than when the planet is farthest.

Kepler also found that when this is the earliest time, position called perihelion or perigee, the planet moves faster in its rotation and its translation and the opposite occurs when the planet is farthest from the sun, called aphelion position or height when travels slower through its orbit and also features slower speeds.

The University of Nebraska in United States of America - USA, provides on its website several simulators of planetary movements, through its Nebraska Astronomy Applet Project - NAAP. We will use here some pictures of the simulator entitled Planetary Orbit Simulator to help us understand Kepler's Laws.

Let´s check the demonstration of the first law in Figure 2.

To improve the understanding of the subject, we can determine, through controls of the simulator, the empty focus, which is in opposition to the sun, the center of the ellipse, besides the semimajor and semimenor, or larger radius and smaller radius of the ellipse, as is done in the figure. When opens the simulator is already selected Mercury, having more eccentric ellipse, which also helps to understand. The simulator also allows increase esssa eccentricity for better visualization of the listed laws, by controlling "eccentricity". Now a graphic demonstration of the second law for the same simulator is shown in Figure 3.

The simulator lets you paint areas covered by the planet at the same time at the moment is slower and faster (when it's out of the sun or closer, respectively). Obviously when slower travel a shorter distance in its orbit, as it were, than when it is faster. Kepler discovered that the painted areas (as shown in the simulator) of two imaginary triangles formed by imaginary rays of the planetary orbits, are equal. This is a very important finding in an attempt to derive the planetary motions by physical causes. If we consider the time in which he lived, we can not fail to recognize how much is an amazing discovery.

Finally, in the year 1619, was published work called Harmonices Mundi by Kepler, in which it is prescribed that the square of the orbital period of the planets is directly proportional to the cube of the average distance of the planet to the Sun, which is known today as your third

law. The simulator present, is hereinafter in this work simply first simulator.

Consider the graphic demonstration of tarceira law of Kepler in the simulator (Figure 4).

Law of Newton gravitation

After finding that the land attracted all objects, starting with the picturesque episode of the falling apple, Isaac Newton further narrowed the ties between physics and astronomy through his Law of Universal Gravitation. He also concluded that the same way force that attracted the objects to the Earth center kept the moon in orbit around them and the planets around the sun. In a single theoretical framework, Newton was able to explain all gravitational phenomena (Figure 5).

The title of his work is PhilosophiaeNaturalis Principia Mathematica. It derived Kepler's Laws of first principles. Thus, many of the foundations of modern physics were created by theoretical

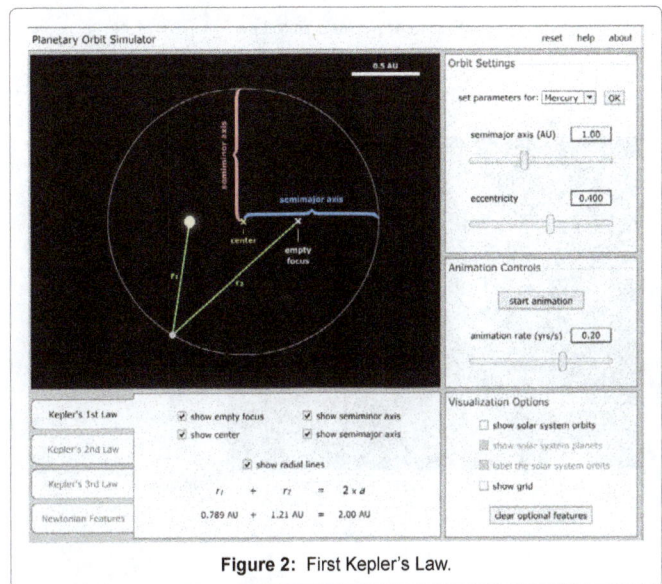

Figure 2: First Kepler's Law.

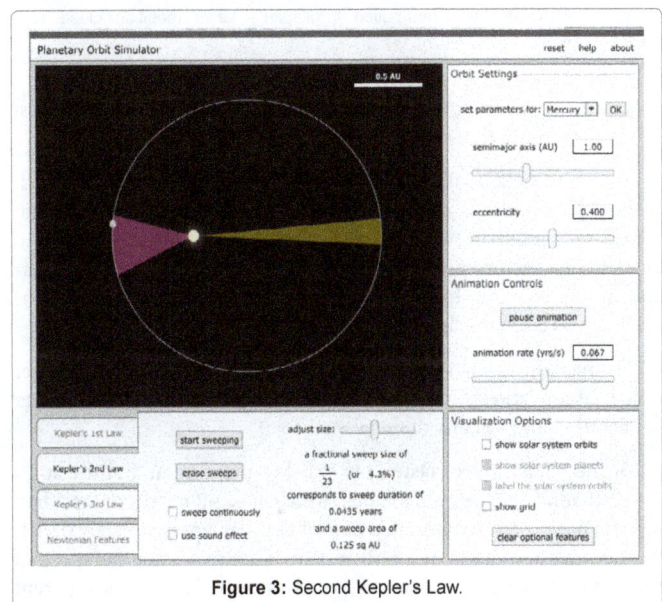

Figure 3: Second Kepler's Law.

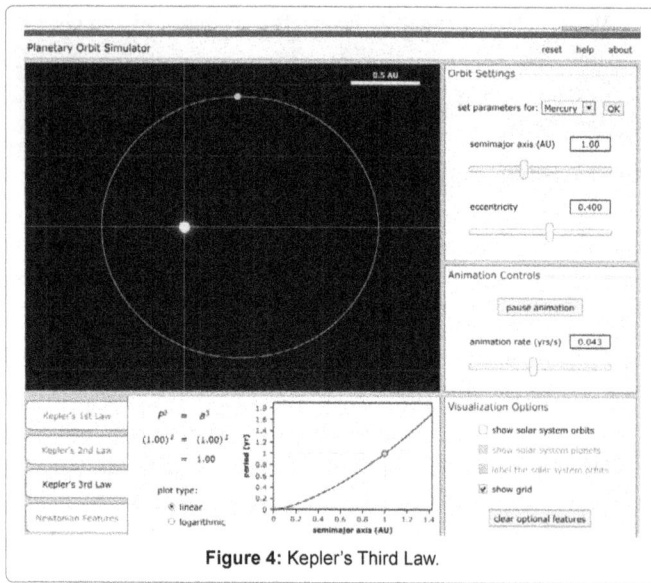

Figure 4: Kepler's Third Law.

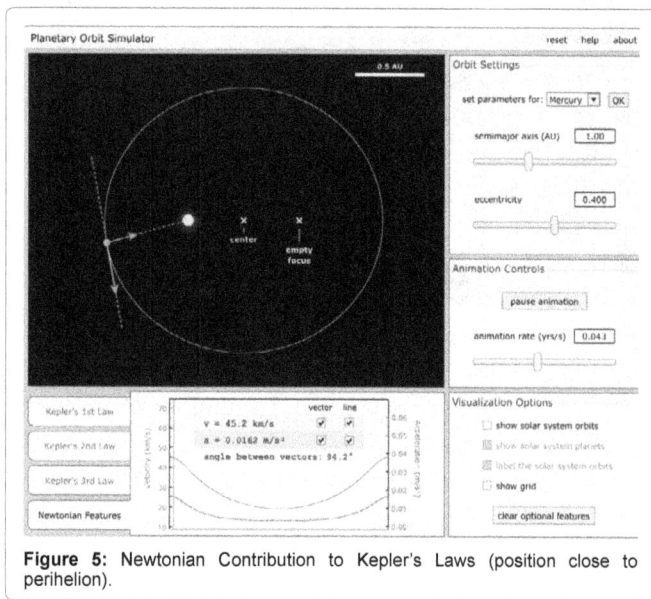

Figure 5: Newtonian Contribution to Kepler's Laws (position close to perihelion).

contribution of Newton. Let´s check the next two images of the first simulator (Figure 6).

When the planet is close to the called perihelion position, which is the time when it is closer to the sun, the incidence of gravity that for Kepler, was unknown, the simulator shows by the application and overlapping vectors of design Larger than when the planet is near aphelia call position. In this, the speed and severity of the conditions are reversed from the first.

In the existing chart at the bottom of the figures is shown another influence of gravity newtoneana, justifying the higher or lower acceleration and velocity of the planet.

If the planet moves faster, our watches mark noon, but the sun's apparent motion is delayed manner, which makes it not occur meridian passage of the star. Around this time of the year, approximately 05/01, the sun rises and sets later for other dates. Similarly, when the speed of the planet is slower, the sun sets and rises early, causing its apparent

motion becomes earlier, causing the meridian passage of the star before noon on the clocks. These speed variations of the planet throughout the year and the slope of the imaginary axis of the planet relative to the orbital plane of about 23.5° originate the analemma phenomenon.

The Analemma

We know that the sun's apparent motion over one day, takes place from east to west, given the planet's rotation is from west to east. Over a year, the apparent motion of the star is between the north and the south, given the slope of the imaginary axis of the planet and so the sun peaking north on the day of the summer solstice for the northern hemisphere and reaches the most southern of the date of the southern summer solstice.

However, if we observe the sun's path between the north and the south, always in the same time on the clock in the one year period of time, we will not find a straight chart, but a similar way to the infinity symbol or a number "eight", which is the phenomenon known as Analemma.

One of the most common forms of his observation is taking pictures of the sun's position in the sky with ten intervals in ten days or every seven for one year always in the same time displayed on a watch. Making an overlay of all the photos you can see the chart clearly demonstrated, or by making an animation between them we can see the outline of the "eight", so to speak, throughout the year.

The reasons for the phenomenon to occur can be explained by the first law of Kepler, and the fact that the imaginary axis of the planet has an inclination of about 23.5°. If the planet had perfect circular orbit, there would be no change in velocity, so the only analemma would be a vertical line on north-south direction. If, added to the hypothesis of perfect circular orbit, there were no imaginary axis inclination, then the analemma would be only a single point, overlapping the images of the sun over a year. In one last assumption, there were no axial inclination but the orbit was elliptical, as indeed is the analemma chart would only be a horizontal line, with variations only between east-west. With the combination of the elliptical orbit (first law of Kepler), which causes variation in the speed of the planet and delays and sun advances regarding our watches, and the slope of the imaginary axis, causing the

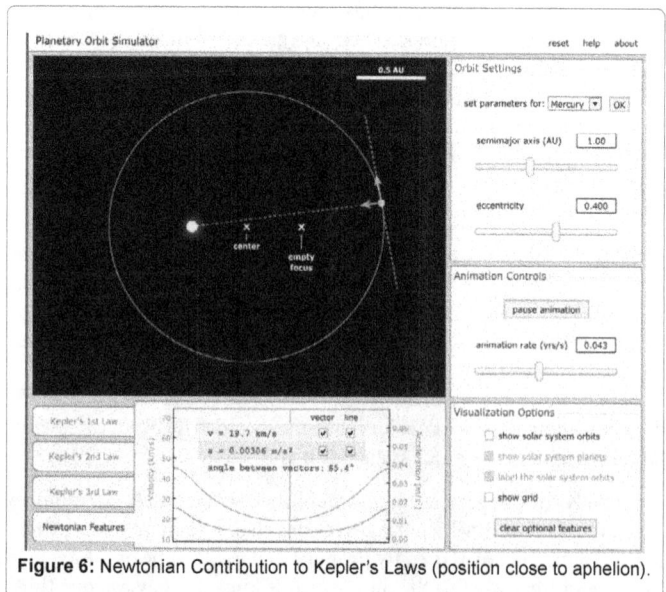

Figure 6: Newtonian Contribution to Kepler's Laws (position close to aphelion).

north-south variation of the analemma During successive positions for one year, the phenomenon is properly configured.

The analemacan also be shown at any location that can be achieved sunlight throughout the year, just to cause both a shadow with a gnomon or a stick placed on the ground or on a wall where the shadow is projected. If the shadow of the points are united tip this rod throughout the year and always in the same time on a clock, the graph into infinity symbol shape will be found. This is a very didactic and requires great simplicity of materials used for demonstration of the phenomenon, especially for students.

In the next parts of this work will bring some additional explanations about the approximate date and chart positions described through images and simulators for teaching purposes, with the primary objective of demonstrating the dates correlated with the position that the sun is on the analemma graph to throughout the year. Analemma image seen in the sky shown in Figure 7.

Simulator of Sun Movings

In order to know in detail the sun's apparent motion will use other simulator NAAP/UNL, entitled Sun Motions Demonstrator, hereinafter called here simply second simulator (these movements are apparent because it is actually not the sun that moves on planet, but the earth revolves around the star).

We set the latitude in the small world map to the right to Salvador/ Bahia. To do so, putting the value of 12.9 degrees south latitude, since the simulator works with only one decimal and decimal fractions.

We can see a little shadow on the side and in front of the spectator represented by a stickman because of analemma position, on the date that is shown. We should mark the "show analemma" controls on the left and below to appreciate the graphic phenomenon throughout the year and "show month labels" to view the position of the sun in the location chosen for the passage of the calendar.

Viewing the occurrence of successive days, however, we must keep track "step by day" clicked for the watch remains frozen at noon, for better observation of the occurrence of the analemma phenomenon over an entire year is possible and including repeat for as many times

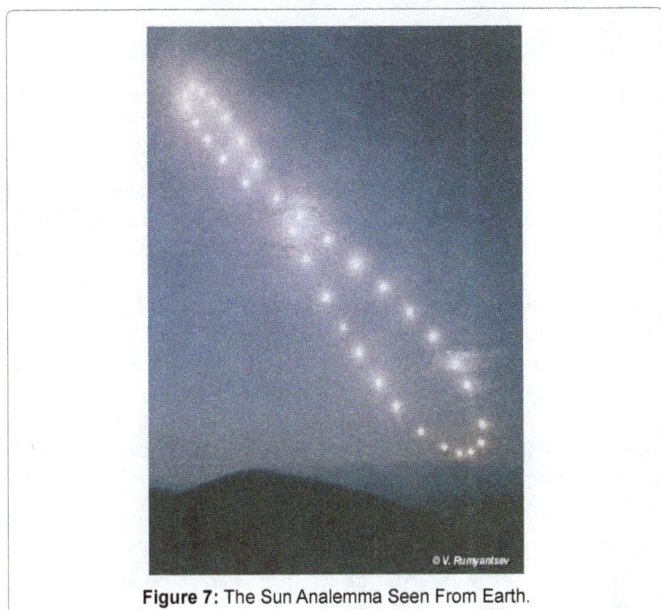

Figure 7: The Sun Analemma Seen From Earth.

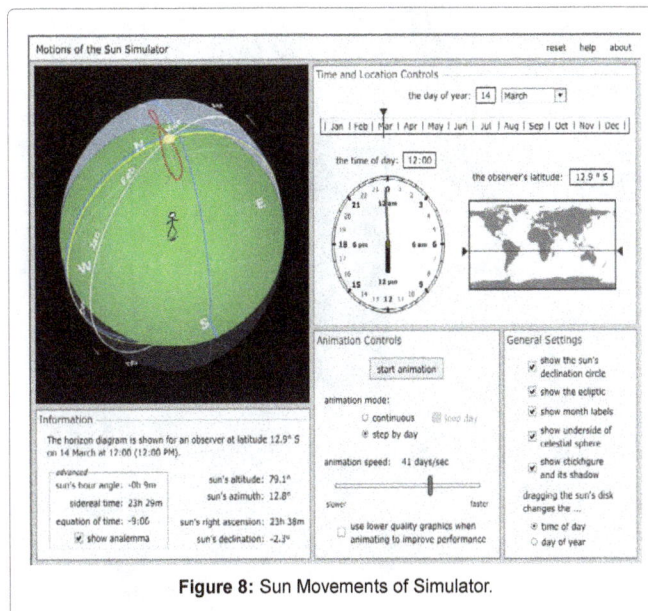

Figure 8: Sun Movements of Simulator.

as desired, simply click the "start animation" always to start or stop the cycle. The simulator also allows, among many other possibilities, with a click made in the calendar arrow, the occurrence of the dates on the contrary, if the desire of those who operates by dragging the pointer with the mouse click (Figure 8).

A Translation with two Simulators

The date of the aphelion, orbital position where the planet is farthest from the sun, by the USNO table, always occurs in the first days of July. In 2012 and in 2013 it was 07/05. It will vary but not much, will always be between July 3rd to 6th until 2020.

However the date of perihelion position in the orbit where the planet is closest to the sun, which in 2012 it was wasjanuary5th and in 2013 it was in was January 2nd, it also will not vary much, will always atay between January 2nd to 5th January 2020.

We can only be sure of these two dates in the first simulator, since it does not have a calendar to show simultaneously month and positions as in the second one. Thus, the positions on other dates will be suggested in the first and not so accurate but without prejudice to any understanding of the content or for attaining the objectives of this work [1].

Obviously our translation described and demonstrated by the images of the simulators will only deadlines of analemma locations and dates of the aphelion and perihelion and not 365 days to prevent the aggrandizement of this text. We also believe that the intervals and the consecussion of the images are sufficient for understanding the occurrence of the translation and its visualization in both simulators.

Starting our translation on the date of 21/12, which is the date of the summer solstice for the southern hemisphere. In the first simulator, is a date before the position of the perihelion (closest planet to the sun, which is ~ 01/05) and the second sun is at the southernmost point of the analemma (Figures 9 and 10) [2].

After 21/12 the next date for the demonstration is ~ 01/05, which is the date of perihelion.

We can see in the simulators that the planet moves along the

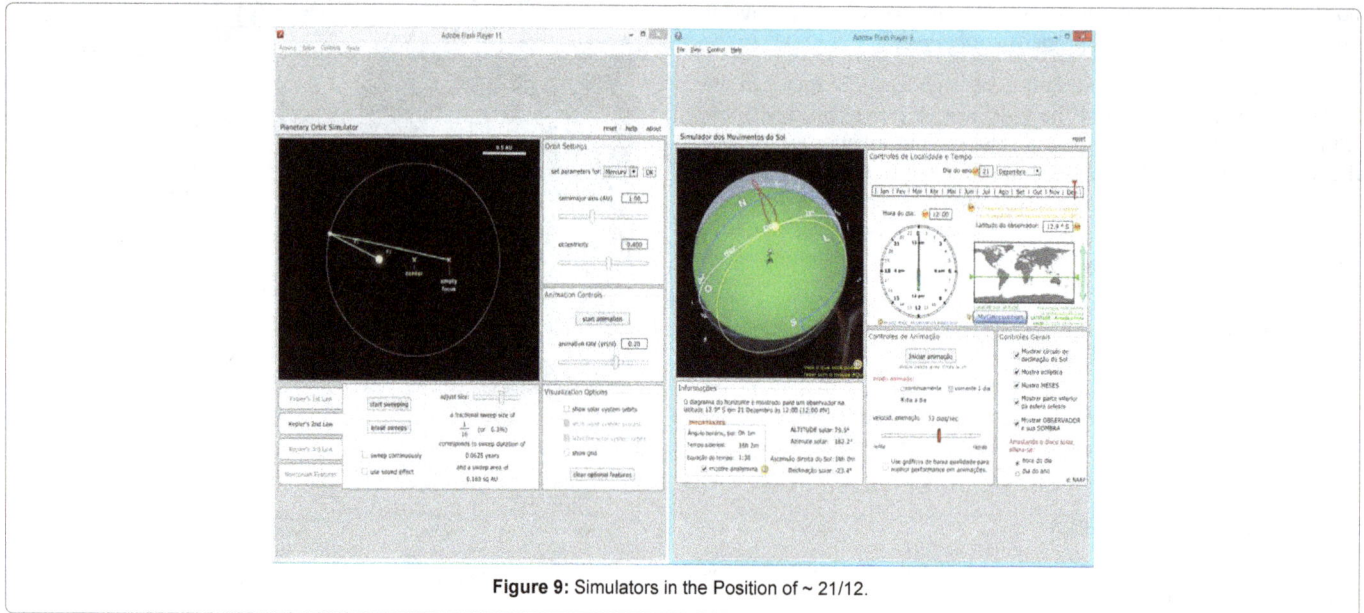

Figure 9: Simulators in the Position of ~ 21/12.

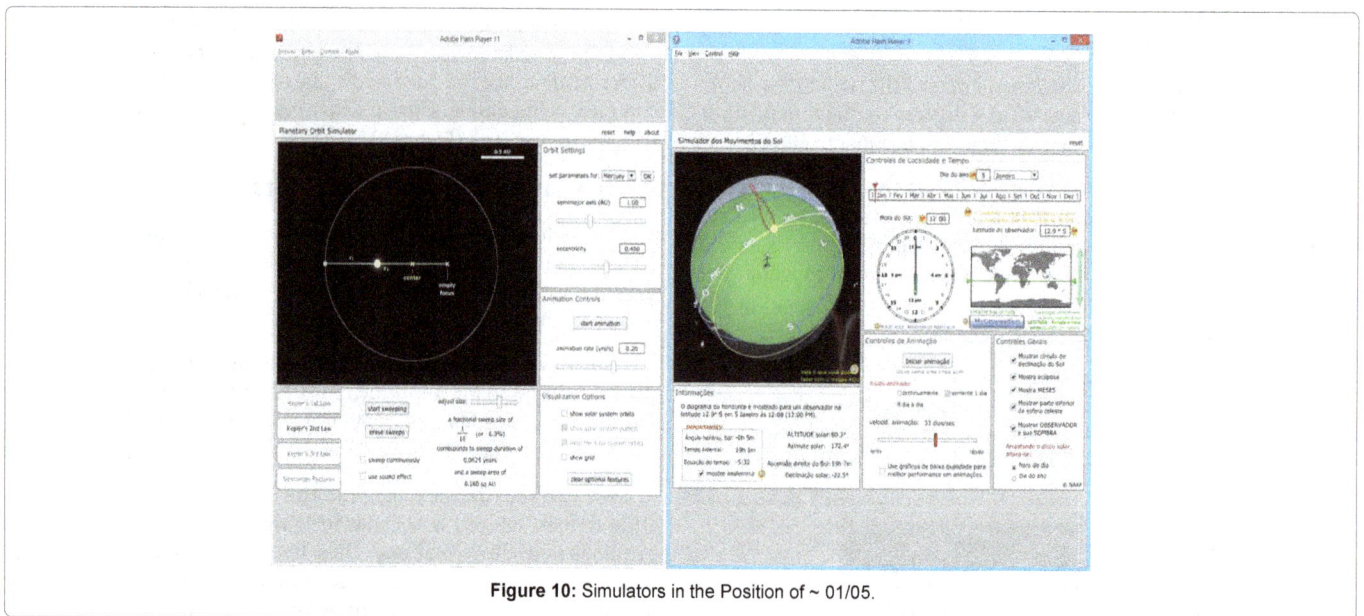

Figure 10: Simulators in the Position of ~ 01/05.

analemma, going towards its lower half in point further east, as the second simulator and is now perfectly pointed to the left in the first, which is one of the dates that we can identify this simulator as cited and reaches the approximate date of ~01/05, which is the date of perihelion.

So if we're on the earliest date the sun, ~01/05, the planet is at its fastest time but the solar day will still showing periods smaller and smaller, ie, will still be with his "speed" increasing to ~02/12 , which is the date on which the solar day has its shorter period, ie, is "faster", as can be seen (Figure 11) [3].

In the second simulator, this date (02/12) is the most eastern part of the larger half of the analemma (a time limit, so) and the first simulator that shows the orbit, at some point after the position where the planet was closest to the sun when it was perfectly on the left of our screen.

Leaving ~02/12, there are successive periods longer solar

day, or decreasing variation of the "speed" of the same and thus the solar day "slow down" to the approximate date of 04/12. On this date, the planet passes through the midpoint in the second simulator, just between the two arches of the "eight" which is the approximate shape shown analemma [3]. The orbit of the planet, or the first simulator, however, although it is a date of "speed" balanced the solar day, you can not be sure of the position, for lack of an attached schedule, if that date is reached point where r1 equals r2 (at which the speed of the planet is average). There are dates on which the period of the solar day is average, namely ~06/21 or northern solstice, ~21/12 or southern solstice and on the dates through the central apex of the analemma ~04/12 and ~08/ 31 (Figure 12).

After ~04/12 to reduce the daylight savings time or a "deceleration" continues only for a "speed" smaller than average because analema followed by the planet is at its maximum point of the lower half west or

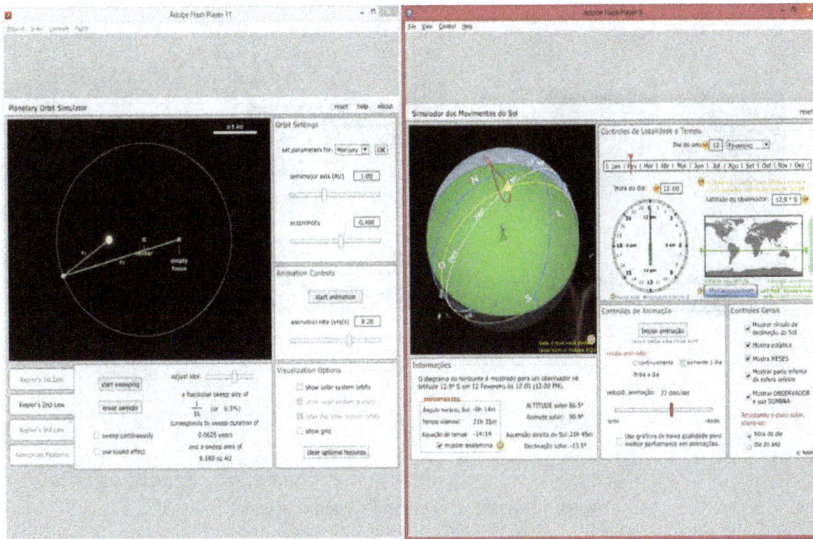

Figure 11: Simulators in the Position of ~ 02/12.

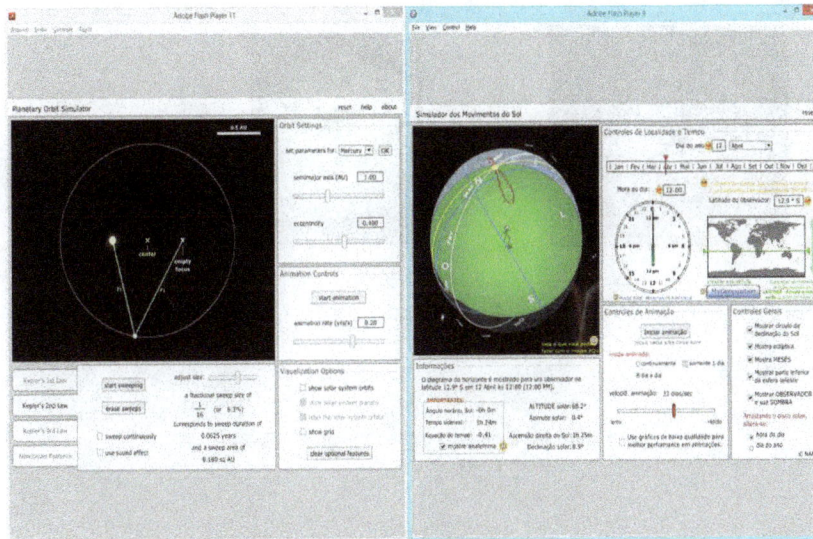

Figure 12: Simulators in the Position of ~ 04/12.

north of it half the approximate date of 05/11 (Figure 13).

The first simulator is in any position before the point entirely right (only come on that date at aphelion, ~07/ 05 and before that still occur the 06/21 deadline) (Figure 14).

Going from ~05/11 solar day reduces its periods, or "accelerates", even though the planet is heading toward aphelion, which is ~07/05. After ~05/11, so the planet reaches the approximate date of the summer solstice in the northern hemisphere, the approximate date of 06/21, where the solar day duration periods are average again, as aforesaid (Figure 15). Remember that from 01/05 to 07/05 the planet is slowing down. On this date the planet is in position in the second simulator date ~06/21 (summer solstice in the northern hemisphere and most northern point of the analemma) and the first, the planet points almost perfectly right.

Continuing with the unlikely reducing periods of solar days, which contradicts the slowdown in the world demonstrated by Kepler and Newtonian gravitation explained by (perihelion to aphelion only), periods of solar days, decrease (or "speed") even after passes through the planet aphelia, ~07/05 to 07/25 approximate date when the sun reach the maximum point on the second simulator east, still in the northern half of analema.

In the first simulation, the date of ~07/05 which is the aphelion, the planet is pointing perfectly to the right of our screen (one of the two dates we can be sure that simulator) and the second already exceeds the most northerly point, at the east toward the lower or upper half of the analemma [4].

On the date of ~ 07/25 the first simulator is in a position after the aphelion (~07/05), she points at up and right. The second shows the

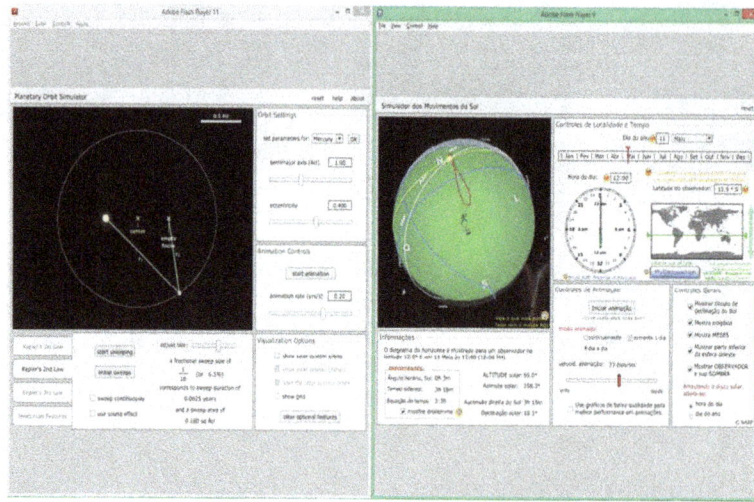

Figure 13: Simulators in the Position of ~ 05/11.

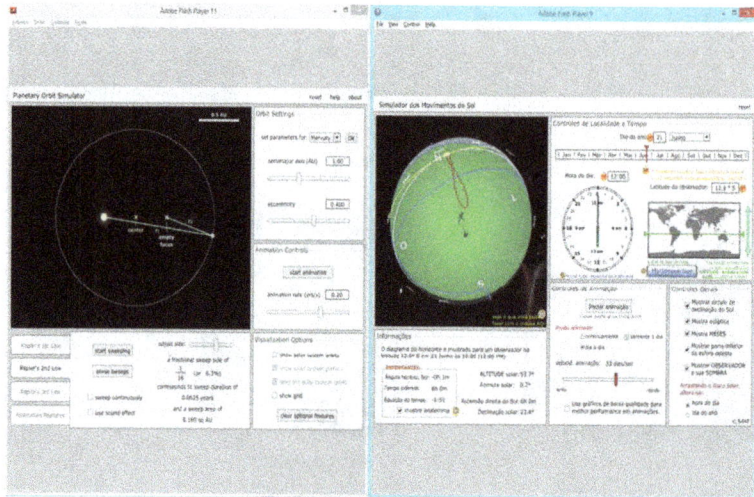

Figure 14: Simulators in the Position of ~ 06/21.

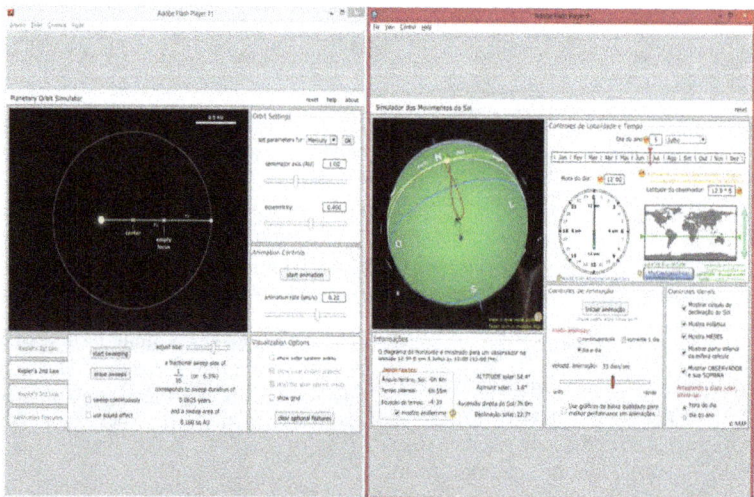

Figure 15: Simulators in the Position of ~ 07/05.

sun through the top half of the analemma in its time further at east (Figure 16).

After only reduced periods of solar day to ~07/25, these increase again increasingly showing a trend of "deceleration" up to the date of the slowest day of the solar year, which is ~11/03, noting that just before he will by the average speed again in ~08/31, when going through the central apex of the analemma (now headed 7back), the second simulator (Figure 17).

This "slowdown" or an increase in the solar day periods is also unlikely because the planet is heading toward perihelion, given the laws of universal gravitation.

In the first simulation, the date of ~08/31 will still be a little more to the left (now on top in our video), but will still be before r1 = r2, because this equality only occurs on the date that the imaginary radius ellipse, (between the planet and the sun) is medium, the exact half between aphelion and perihelion and there is no perfectly gauge the date of occurrence that through the first simulator [5].

After ~08/31 solar days follow as well, contrary to Newtonian gravitation and the first Kepler's Law again, "slowing down" until the date of ~11/03, as the planet is accelerating, with the second simulator at the west position the higher half, which is the date of the slower solar day in the year and the first simulator further pointing to the left (Figure 18).

After the date of ~11/03, the solar day finally begins a series of shorter periods, ie its "acceleration" (now congruently with the planet), and reaches the southern solstice ~21/12, which can be clearly seen in the second simulator, where the speed of the solar day is average and the location is perfectly in the southernmost point of the analemma. At first, this time is when the planet is in the position pointing up and to the left on our screen, before pointing perfectly to the left (because there would be the date of perihelion, ~01/ 05). This concludes the translation with an image identical to that of departure (Figure 19).

Extrapolating a little translation, the reader of the case which is fixed over the first simulator, once used as a delimiter for the positions of the second solstice, it should be recalled that thereafter the solar day following "accelerating", ie showing periods smaller and smaller until you pass the perihelion ~01/05 [6].

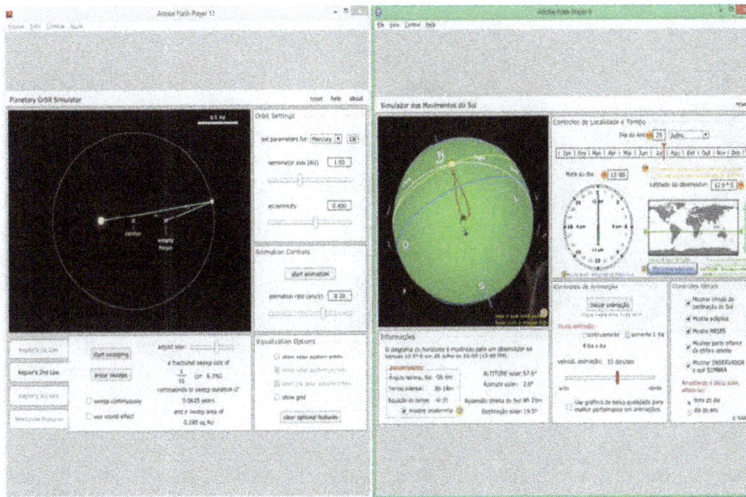

Figure 16: Simulators in the Position of ~ 07/25.

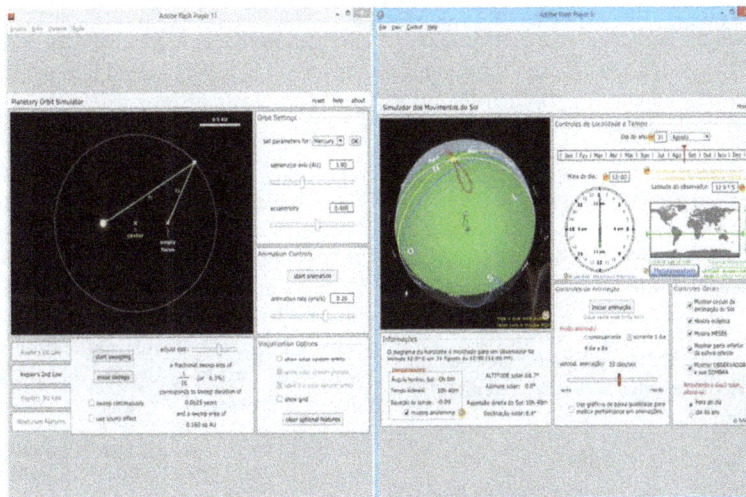

Figure 17: Simulators in the Position of ~ 08/31.

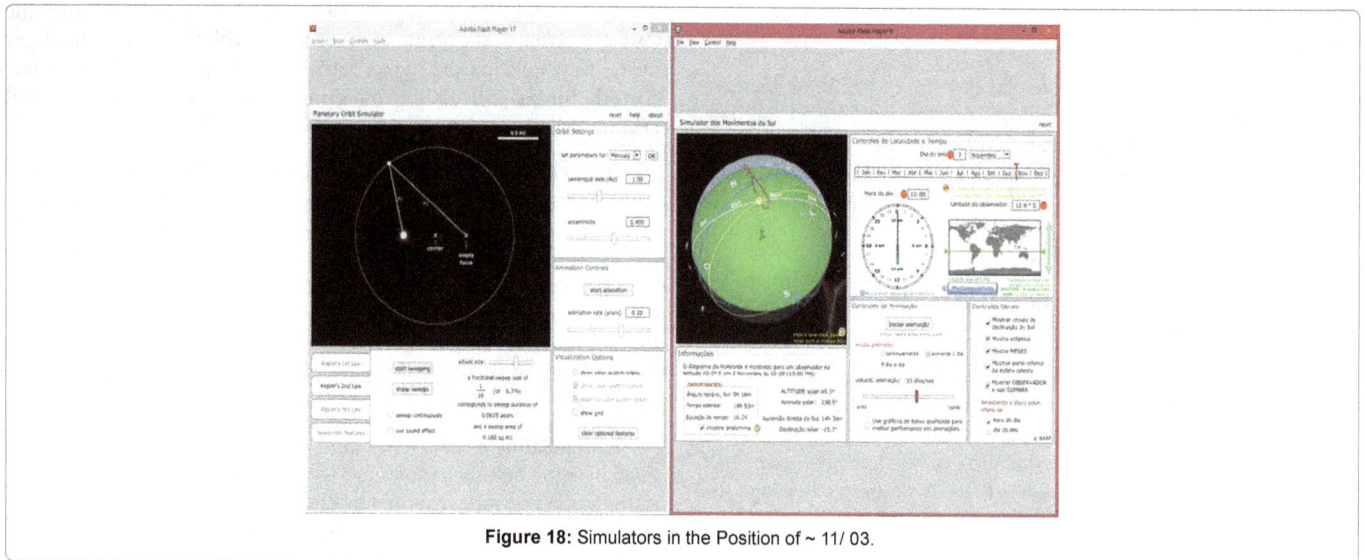

Figure 18: Simulators in the Position of ~ 11/ 03.

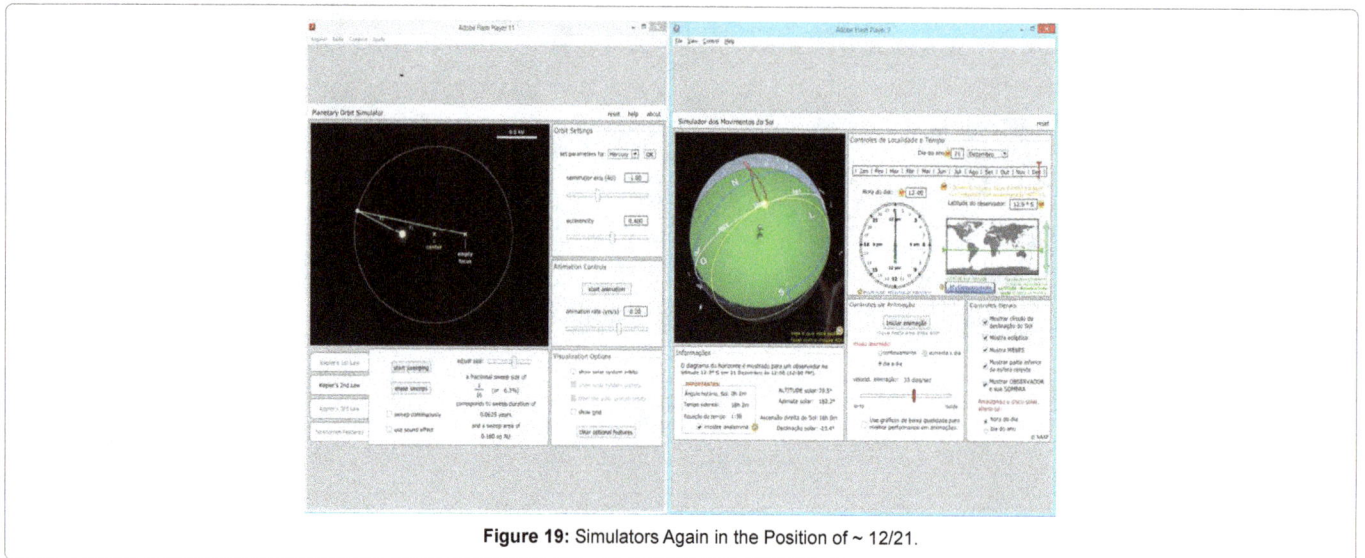

Figure 19: Simulators Again in the Position of ~ 12/21.

In the second simulator this date is after the southern solstice curve and the first simulator is perfectly to the left of our screen (Figure 10) until the date of the fast solar day of translation, ~02/12, the second simulator is demonstrated maximum east, halfway greater analemma, getting a little after perihelion, at the beginning of their "path" to the aphelion (Figure 11).

With this translation described and accompanied by simulators, it demonstrates the coincidence between the solstices and dates aphelion and perihelion. The perihelion occurs ~14 days after Southern solstice, while aphelia occurs ~14 days after the northern solstice (Figures 10 and 15).

When we say that the solar days have decreasing periods or "acceleration" and increasing periods or "deceleration" refer to the analemma while demonstrating chart [7].

For a better understanding, we will make the composition of aanalemma with four elements, two lines and two curves. Although the accuracy of the design is not tasteful, the arrows indicate its path

correctly in each of the four parts, always taking into consideration the analemma as seen in the simulator (from heaven to earth). In analemma vision projected into the sky, to behold the earth, calendar positions would be reversed.

As the sun apparent movement is from east to west, every day the sun shines on a point of the analemma, is considered the same time for observation. Where it is located further at east, to say that the sun "walked a little," that is, at the time of observation he has not advanced much, which can lead us to deduce that the sun is so slow that date or the planet is faster (since it varies its speed as Kepler/Newton). Similarly, on a certain date the sun was farther west, to say that the sun "walked a lot," that is, at the time of observation he advanced a lot, which can lead us to deduce that the sun is then fast or that time the planet is slower.

As the arrows of the four elements that make up the figure indicate the correct direction of the sun's path by the analemma, we can say, at times when it is in the two curves are times when it is going more and more to the east, it is presenting increasingly periods of lower solar days or, we can say with the license of Physicists, the solar days are "accelerating". In the times when it is in two straight lines, are periods of the year when it is going more and more to the west, that is, is presenting more and more periods of the largest solar days or we can say, again with the license of Physical solar days are "slowing down".

It is worthful remembering that the dates on which the sun is in the upper curve, with eastbound ("accelerating") are ~05/11 to ~07/25 (Figures 13-16) and the dates on which it is the bottom curve, also with eastbound ("accelerating") are ~11/03 to ~02/12 (Figures 9-11 and 18). However the dates when it is on the line pointing upwards with westbound ("slowing down") are ~02/12 to ~05/11 (Figures 11-13) and the dates on which it is on the line that points to down with westbound ("slowing down") are from ~07/25 to ~11/03 (Figures 16-18).

To help a little more the understanding and to allow a preview of the date of each paradoxical variations, below there is a figure with two diagrams. The red arrow indicates decrease in speed when it is in reference to the planet and increasing periods of solar days, when compared to the solar day [8]. Blue arrows serve to indicate the exact opposite, acceleration when referring to the planet and reducing periods when referring to the solar day. Of course by the allusion proposal, reducing solar days would be the "acceleration" of the same and increased solar days would be his "slowdown". Hence the use of the same color to these movements when referring to the planet (Figure 20).

Research

Measurements of the solar day periods were carried out with the aid of a fixed gnomon for determining the duration of a solar day and the results were documented in videos.

There are 16 videos that can be viewed on the internet with a simple search the title "Solar Day 01 Speed" to "Speed Solar Day 16". For each measurement there are two videos, one showing and doing the marking of the shade from the sun's position with a fixed gnomon, as described, on a certain date and at a certain time and another the next day showing what time it is when the sun passes the same point marked on the earlier date [9].

Thus, are also added to the titles of videos the words "departure" and "arrival" to signal that one is the day of the shadow position of marker and the following is to demonstrate what time is again passing through the same fixed point in planet.

The number 01 is played, made on 07/12/2012 and its full title is "Solar Day Speed 01/Departure 07/02/2012 14:45 (GMT)." The number 02 is of arrival, made on 07/13/2012 and its full title is "Solar Day Speed 02/Arrival 07/13/2012 14:42 (GMT)."

There is only one exception on the date of 08/27/2012, when

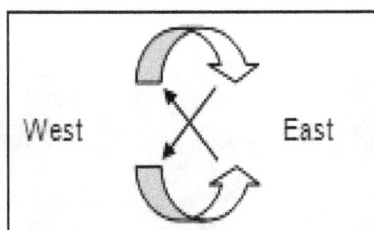

Figure 20: Analemma Composed of Four Elements.

it was made three starting videos (the numbers 03, 4:05) and the corresponding arrival, made the following date, 28/08/2012 leading the numbers 06, 07 and 08. Then, from nine to sixteen is always only one at a time corresponding to the arrival always the next day. We will avoid the transcription of all the titles not to enlargethe text so much [10].

The clock used is also available on the Internet with the official time of Brazil's capital, the city of Brasilia. The clock can be accessed at horariodebrasila.org address. We decided to use the same for a better reliability of the measurements, since they do not have a rigorous accuracy of seconds or fractions thereof, especially because they are spending a few seconds (about 08 or 10 seconds) for the displacement risk for paper made in document the position when a video of "departure" and similar time in the videos of "arrival" to get the brand to the clock on the Internet. The videos are generally for 40 seconds.

The poor accuracy, however, does not alter what we are demonstrating, which is a variation in the length of the solar day at the home of several minutes. The second simulation also shows that over time the outer analema demonstrates a difference in the amount of about 30 minutes, the data above the "analemma show". In ~11/03 hour angle of the sun features 16 minutes and ~02/12 presents -14 minutes. The poor accuracy do not alter the demonstration that is essentially a paradoxical variation of the solar day in relation to planetary motion.

To further ensure the usefulness of the measurements, a video was produced with a digital clock, this time placed next to the drawing board to the measurements, this video that is hosted on the same channel on a popular video site entitled "Method of accuracy", placed between videos numbers 10 and 11 and was published on 11/01/2012.

Recording begins at 14:07:30 and at that moment the shadow of the pointer is pointing precisely to the extent of 9.5 cm on the ruler affixed to the clipboard. The video is about two minutes and the camera goes behind the drawing board to demonstrate that they are all properly attached elements. When you are near the end, as you can see, the clock showing 14:09:30, two minutes flat, the shadow moves to the measure of 9.3 cm, which was approximately one millimeter for every minute.

Thus, it is attested total utility of measurements of this research, it would be impossible to fake a shadow passing the measure as this would only occur if they had passed several minutes. In short, the loss of a few seconds does not invalidate a measurement that only want to show that there is a variation more or less (and paradoxically) in periods of occurrence of solar days.

The measurements documented in the internet videos showed the following results:

- 07/12 and 13//2012 – 1437 min
- 08/27 and 28/2012 – 1440 min
- 10/08 and 09/2012 – 1441 min
- 10/29 and 30/2012 – 1442 min
- 12/03 and 04/2012 – 1441 min
- 01/ 07 and 08/2013 – 1439 min

So that the reader does not need to go back several pages to confront the figures, put down a analemma here with the schedule according to the display used by the simulator (top down to earth and not designed in the sky).

By observing the reader may notice that the first four are presenting increasingly longer periods, that is, are "slowing down", and are in line pointing west and the path going down, see explanations in decomposed analemma in the previous section. However the last two periods are presenting smaller and smaller, ie, are "accelerating", under license of physics and are at the bottom of the analemma curve sector leading the path to the east [11].

We know that the day in summer lasts longer than the night and in winter the opposite occurs with respect to any one hemisphere of the planet. However, these differences are more perceived in temperate and polar zones. In the tropical zone, where the measurements were made, having the coordinates -13.00987 -38.51045, this time lag between days and nights on the dates of the solstices north and south are not very prevalent. For a viewer delays and tropical sun adiantosare more easily perceived than the difference of duration periods of days and nights (Figures 21 and 22).

Conclusion

The ellipse of the earth is almost round, with a low eccentricity, not showing, it is known, major deformities. The radius of the orbit

to the sun is lower in the perihelion and aphelion greater. Perihelion to aphelion they are ever increasing. Aphelion to perihelion they are always decreasing. There are equidistant rays in two movements of "going" and "coming" so to speak, or both halves of the ellipse, if we consider that the aphelion and perihelion are highlighted spots. Certainly there are no discrepancies or completely different radii in different sectors of the orbit, which is in agreement with the previous observation. That's right even if we use the simulator to greatly increase the eccentricity.

We can say that the earth only accelerates toward the perihelion or only slows down toward the aphelion. It happens that the periods of solar days do not exhibit this same behavior, since the dates indicated its variations periods (or speed) are exactly reversed.

We know that there is no disconnect between the speed of translation and rotation, they are related. As one increases, the other also increases, and decreases when, decrease together. The variations of the solar day, which in some dates of the year came with the system

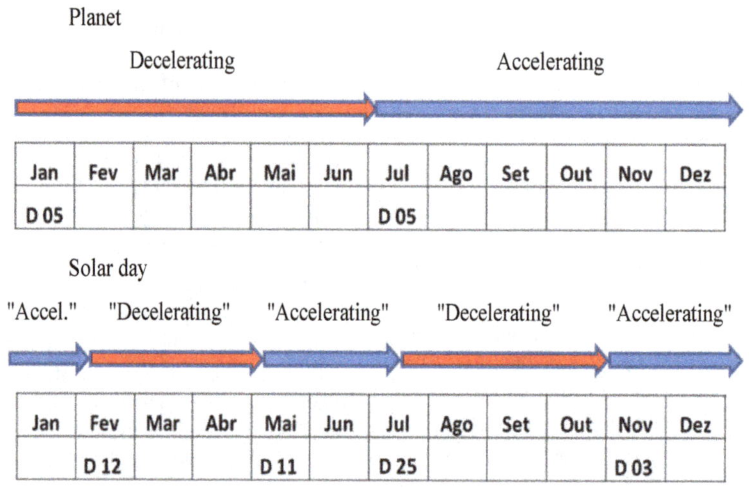

Figure 21: Comparison Between the Dates of Variations.

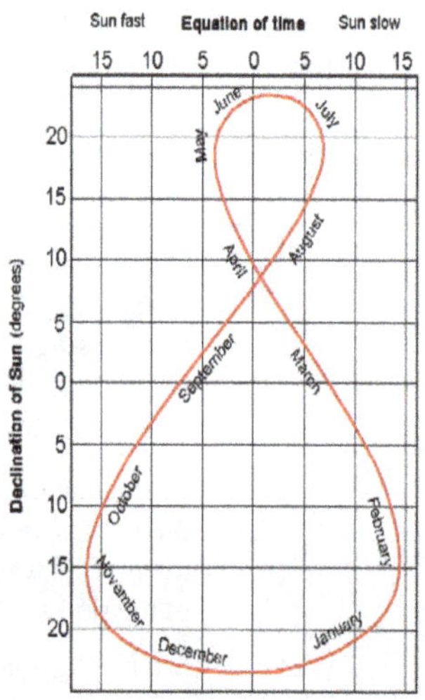

Figure 22: Analemma the Sun Seen From Above With Calendar.

Kepler/Newton, in other time periods behaves exactly the contrary, demonstrated graphically by variations of the analemma itself.

There are paradoxes because that solar days decreases their periods or "speed" between ~05/13 and ~07/25, which are the extreme dates of the analemma in the north (and among them are the northern solstice and aphelion, ~06/ 21/and ~07/05, respectively). Compared the rates of variations in solar days with changes in the world we see that the planet is slowing down to ~07/05. The same paradox occurs because the solar days increase their periods or decelerate between ~07/25 and ~11/03, and between those dates the planet is accelerating, given the law of universal gravitation.

The first date is the maximum position east of the northern analemma and the second is the maximum position west of the south analemma, that is, when it crosses the summit trend of "deceleration".

We know that the analemmais caused by two factors: The change in land speed in orbit (Kepler/Newton) and the slope of the imaginary axis of the planet. The greater the eccentricity of the orbit, the greater the east/west difference analema because of the greater difference in speed at different times during the year. The more inclined the axis, the greater the north/south differences. If there were alternating the speed of the planet and there was only the axial slope, the analemma would have a vertical line format. If there were no axial tilt and varying the speed or he would be a point. If there were no axial tilt and only had varying the speed it would be a horizontal line.

The interesting point is that the analemma describes paradoxes at the speed of variation of the solar day in relation to the variation of the velocity of the planet. The main curiosity is that if we score aphelion and perihelion dates in the analemma, we see that they occur almost perfectly about fourteen days after the two solstices:

Northern Solstice ~06/21/ aphelion ~07/05;

Southern Solstice ~12/21/ perihelion ~01/05.

Another interesting coincidence demonstrated by the analemma is that after aphelion the solar day "accelerates" only for a few days (~07/05 to ~07/25) and then enter the described paradoxical variation (of ~07/25 to ~11/03). Already after perihelion, it also still "accelerate"

for a few days (~11/05 to ~02/12, which is also a paradoxical change), only to "slow" (now according to the speed of variation Planet) of ~02/12 to ~05/13. This is because the dates of the aphelion and perihelion, to limit the acceleration and deceleration of the planet, do not occur in the culminating dates analemma chart.

Another point worth mentioning is that the solar days are "slowing" the straights and "accelerating" in the curves. Straight are in low latitudes while the curves are in the higher latitudes. As we have seen, the dates of the solstices are the curves, as are the high points of the analemma and aphelion and perihelion dates are after them.

Considering the analemma graph condition and taking into account the sun's apparent motion on the planet takes place from east to west, the further east the sun shines at the same time, the slower the solar day is, since the sun " walked a little "on the map. The more west, faster, for similar reasons. The news is that not always the sun is slow because the planet is fast and not only the sun is fast because the planet is slow, given the paradoxical variations identified.

References

1. Niu MCY (1989) Airframe structural design. Practical design information and Junior A, Sampaio L (2014)The Theory of Solar Zenith: a proposal for the seasons in inter-tropical locations. Salvador Edufba.

2. Junior A, Sampaio L (2014) VPPDDSRSKN.

3. GabrielKuhnB. The spatial part of human time.

4. Google Earth.

5. Konstanta (2015) Analemma.

6. (2015) Nebraska Astronomy Applet Project - NAAP. Sun Motions Demonstrator. University of Nebraska-Lincoln UNL.

7. Nebraska Astronomy Applet Project - NAAP. Sun Motions Demonstrator. University of Nebraska-Lincoln - UNL. (Translated and Adapted version).

8. Filho O, de Souza K, Saraiva, Maria de Fátima Oliveira (2014) Time measures.

9. Filho O, de Souza K, Saraiva, Maria de Fátima Oliveira. Motion of the planets: TychoKepler and Galileo.

10. Rumyantsev V. Crimean Astrophysical Obsevatory.Analemma.

11. (USNO) United States Naval Observatory, The United States Naval 3450 Massachusetts Ave, NW, Washington, DC20392-5420.

4

The Research of Supersonic Aircraft Low Sonic Boom Configuration Design and Optimizations

Xuan H*, Cheng S and Fang L

China Academy of Aerospace Aerodynamics, Yungang West Road, Beijing, 100074, P. R. China

Abstract

High noise level of sonic boom is one of the most important reasons that the supersonic transport can't be applied to civil aviation broadly. Sonic boom is a complicated problem relating to aircraft configuration design, aerodynamics, acoustics and so on. The traditional sonic boom minimization theory is an inverse design method with single object, which makes it difficult to be applied in multi-objective optimization, effectively. For the low sonic boom configuration optimization, the sonic boom noise level prediction method based on supersonic linear theory was developed. The sonic boom level of a basic configuration of supersonic business jet was computed and the cause of formation of sonic boom was analyzed, based on which the fuse and wing plane wais optimized to decrease the noise level of sonic boom. Compared with the basic configuration, the sonic boom level of optimized configuration decreased distinctively, with the overpressure decreasing 41% and the A-weighted noise level decreasing 7.55 decibel. The aerodynamic characteristics of optimized configuration were computed. Compared with the basic configuration, the drag decreased obviously at the cruise condition without moment change.

Keywords: Sonic boom; Supersonic transport; Optimization design; Aeroacoustics; Supersonic linear theory

Introduction

The big noise caused by sonic boom when the flight vehicle flying supersonically will not only has an influence on human lives but also bring a destroy to the constructions, especially for the infrasonic boom. The noise level of a Concord civil aircraft flying at the altitude of 50000 ft will be 133 dB, but the noise level of the civil aircraft is only about 90 dB when taking off and landing. Thus, the Concord was forbidden to fly over the continent supersonically because of the high sonic boom level, which played down the economy of Concord. The noise level of next generation supersonic transport is demanded lower than 70 dB, which is comparative with the transonic civil aircraft. Thus mitigating the sonic boom is an exigent problem for next generation supersonic transports development.

The sonic boom of supersonic aircraft is a complex problem which refers to aircraft configuration design, aerodynamics, acoustics and so on. The research of sonic boom can be traced back to 1950's [1-3]. Since then, the sonic boom prediction methods based on supersonic linear theory, geometry acoustics [4-7] and the sonic boom minimization theory [8-12] have been applied to the supersonic transport design. Recently, The CFD (Computational Fluid Dynamics) has been used for sonic boom prediction and physics research more and more. Besides, the modern optimization theories have been applied to the low sonic boom configuration optimization. Compared with the sonic boom minimization theory, the low sonic boom optimization based on modern optimization theory can obtain the configuration not only with the low sonic boom character but also with other good performance such as good aerodynamic character, weight character, structure character and so on.

In this paper the low sonic boom configuration optimization of supersonic business jet were researched. First of all, the sonic boom noise level prediction method based on supersonic linear theory was developed. Then, the sonic boom level of a basic configuration of supersonic business jet was computed and the cause of formation of sonic boom was analyzed, based on which the fuse and wing plane was optimized to decrease the noise level of sonic boom. At last, the

aerodynamic characteristic and the sonic boom pressure near the optimized configuration were computed by CFD.

Prediction Method

Linearized supersonic flow

For a slender axisymmetric body in cylinder coordinates, the over pressure $\Delta p = p-p_0$ can be written as:

$$\Delta p (x - \beta r, r) = p_0 \frac{\gamma M^2 F(x - \beta r)}{(2\beta \gamma)^{1/2}} \qquad (1)$$

$$F(y) = \frac{1}{2\pi} \int_0^y \frac{A''(\xi)}{(y - \xi)^{1/2}} d\xi \qquad (2)$$

Where, M is Mach number, $y = x-\beta r$, $\beta = \sqrt{M^2 - 1}$, A is the cross-sectional area of body by the normal projections of cuts alone planes aligned with the Mach angle. F(y) is also called "Whitham F function". For the asymmetric body, the area A was generalized to equivalent area $A_e(x,\theta)$, which consists two components: a volume component and a lift component. The lift component of the equivalent area is given by:

$$A_L(x,\theta) = \frac{\beta}{\rho u_\infty^2} \int_0^x L(x,\theta) dx \qquad (3)$$

Where is L (x,θ) the component of lift per unit length at axial station x, in the θ direction. In this paper, the uniform distribution of the lift alone the wing was supposed so the eq. (3) can be written as:

*****Corresponding author:** Xuan H, China Academy of Aerospace Aerodynamics, Yungang West Road, Beijing, 100074, PR China
E-mail: haoxuan0722@sohu.com

$$B(x) = \frac{\sqrt{M^2 - 1}W\cos\alpha\cos\theta}{1.4p_v M^2 S}\int_0^x b(x)dx \qquad (4)$$

Where b(x) is the span alone axial station x, W is the weight of the body.

Noise level computation

The over pressure near the aircraft computed by eq.(1) was extrapolated to the ground by waveform parameter method proposed by Thomas [13] in 1972. The narrow band spectrum of over pressure on the ground was obtained by Fast Fourier Transform:

$$P(k) = \frac{1}{N}\sum_{n=0}^{N-1} p(n)e^{-jkn\frac{2\pi}{N}}, k = 0,1,2,...,N-1 \qquad (5)$$

Where $p(n)$ is the over pressure at some time, $p(k)$ is the over pressure at the frequency of k. N is the number of input data, which equals the exponential of 2.

The sound pressure was computed by:

$$SPL = 20\log_{10}\frac{p_e}{p_{ref}} \qquad (6)$$

Where p_e is the effective pressure, p_{ref} is the reference pressure.

Based on narrow band noise level, the 1/3 octave noise level can be obtained, by which the A weighted noise level was computed.

The Basic Supersonic Business Jet Configuration

The sonic boom noise level of a supersonic business jet was analyzed. Figure 1 shows the sketches of the aircraft, and the parameters of this supersonic business jet are listed in Table 1. This configuration was designed without considering the sonic boom characteristic. Figure 2 gives the F function and equivalent area distribution alone axial station at cruise lift coefficient and zero lift condition. It can be seen that both at the cockpit and the wing, the remarkable change of the cross-sectional area of aircraft by the normal projections of cuts alone planes aligned with the Mach angle made the shock wave generation. The lift component of equivalent area enhanced the strength of shock wave further. Figure 3 gives the over pressure of sonic boom below the aircraft at the radius of 5 lengths of the aircraft. Figure 4 gives the over pressure of sonic boom at ground. The over pressure on the ground is 100 pa with the reflect factor 1.9. When the aircraft cruising with Mach 1.6 at the altitude of 14km, the A weighted noise level of the sonic boom over pressure on the ground is 91.3 dB.

Passenger	8-12
Cruise Mach	1.6
Cruise altitude(km)	14
Range(nm)	4000
Maximum takeoff weight(ton)	45
Length(m)	45.2
Spanwise(m)	20

Table 1: Supersonic business jet parameters.

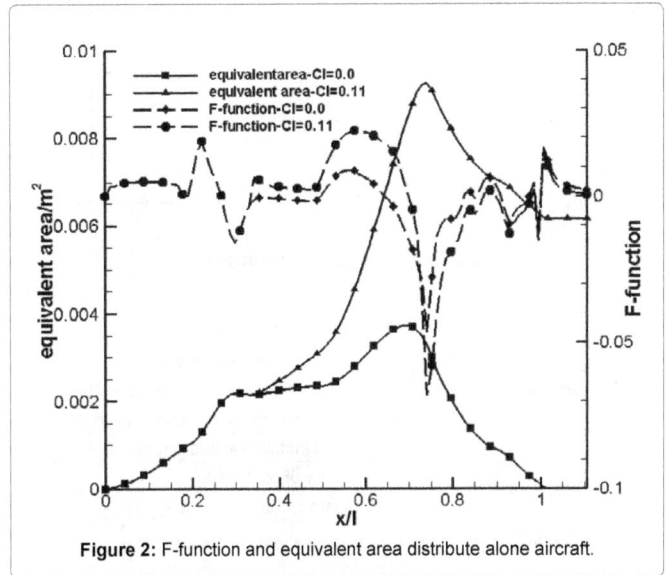

Figure 2: F-function and equivalent area distribute alone aircraft.

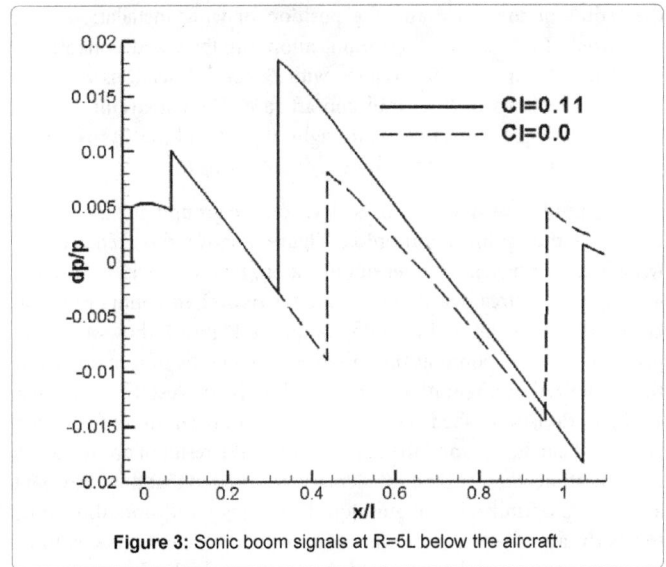

Figure 3: Sonic boom signals at R=5L below the aircraft.

The Low Sonic Boom Optimization of Supersonic Business Jet Configuration

The optimization of aircraft fuse

In order to bate the strength of shock wave led by cockpit, the fuse of aircraft was optimized. Because of the little contribution to equivalent area of the rudder, elevator and engine, only the wing body configuration was optimized with nose and cockpit were designed integrally. Supposing the nose, mid-fuse and aft-fuse were configured

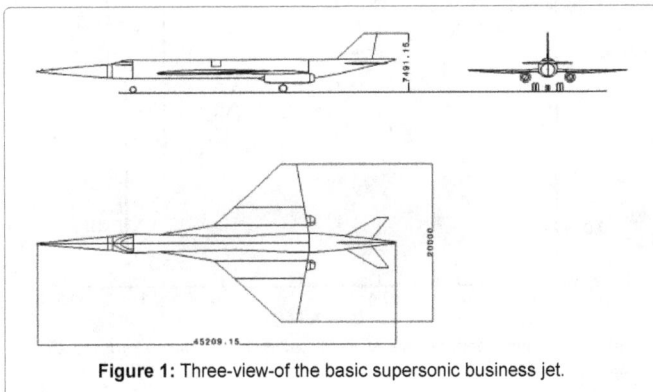

Figure 1: Three-view-of the basic supersonic business jet.

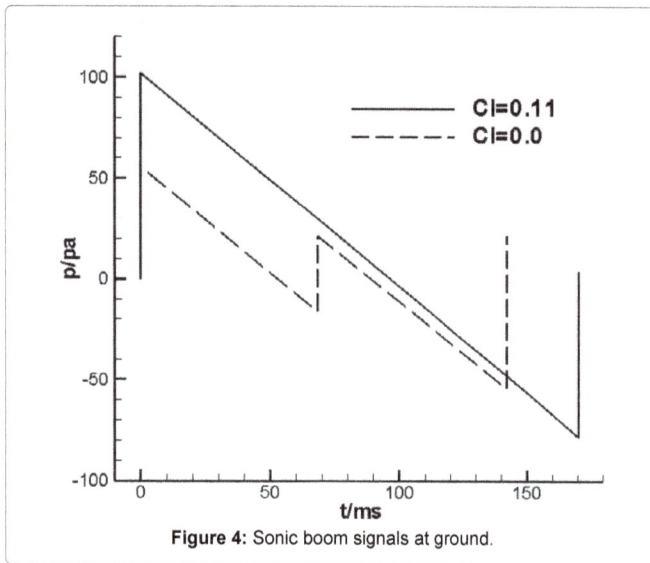

Figure 4: Sonic boom signals at ground.

Variables	Radius of nose sections:d1-d5
	Radius of fuse sections: d6-d13
	Center coordinates of nose sections:y1-y5
	Center coordinates of fuse sections:y6-y13
	Wing installation position:x0, y0
Constraints	View of pilot :a1$\in$(a1$_{min}$,a1$_{max}$)
	Clearance angle: a2$\in$(a2$_{min}$,a2$_{max}$)

Table 2: The optimization variables and constraints of fuse.

by a series of circle section with different radius, the center coordinates, the radius of the circle and the position of wing installation were optimized. Table 2 lists the optimization variables and constraints. There are 28 optimized variables with 5 control sections of nose, 8 control sections of both mid and aft fuse. The constraint of nose declination angle was set to ensure sight of pilot and the aft fuse angle was constrained to meet the requirement of taking off.

The minimum A-weight noise level on the ground at Ma=1.6 was chosen as the optimization object. Figure 5 shows the sketch of the wing body configuration after optimization. Figure 6 gives F-function and equivalent area distribution alone the optimized configuration at cruise lift coefficient and zero lift condition. Figure 7 shows the over pressure of sonic boom at the distance of 5 lengths of aircraft below the optimized configuration. The solid line is the result of the basic configuration, the dashed line is the result of optimization configuration at zero lift coefficients and the dash dot line is the result of optimization configuration at cruise lift coefficients. It can be seen that the equivalent area of the optimized configuration distributes more smoothly alone the body and from the results of zero lift coefficients, the volume component of equivalent area didn't cause notable shock wave, which means that the optimization of fuse and nose was effect. The shock wave was caused by the lift component which can be seen from the results of cruise lift coefficient. Figure 8 gives the over pressure of sonic boom on the ground varying as time. The solid line represents the result of basic configuration and the dashed line represents the result of optimized configuration. The over pressure on the ground was 81 pa with the reflect factor 1.9, and the maximum over pressure decrease 20% after optimization. The A-weight noise level on the ground at Ma = 1.6 and cruise lift coefficients was 87.5dB, with 3.8 dB reduction after optimization.

The optimization of wing plane geometry

Base on the section 4.1, some optimization still needs to be done in order to obtain a better lift distribution, which could decrease the sonic

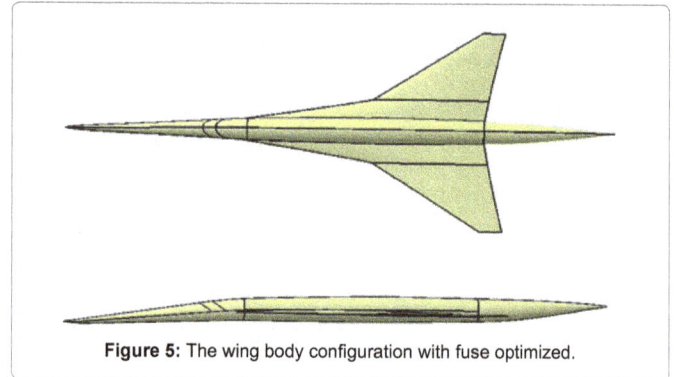

Figure 5: The wing body configuration with fuse optimized.

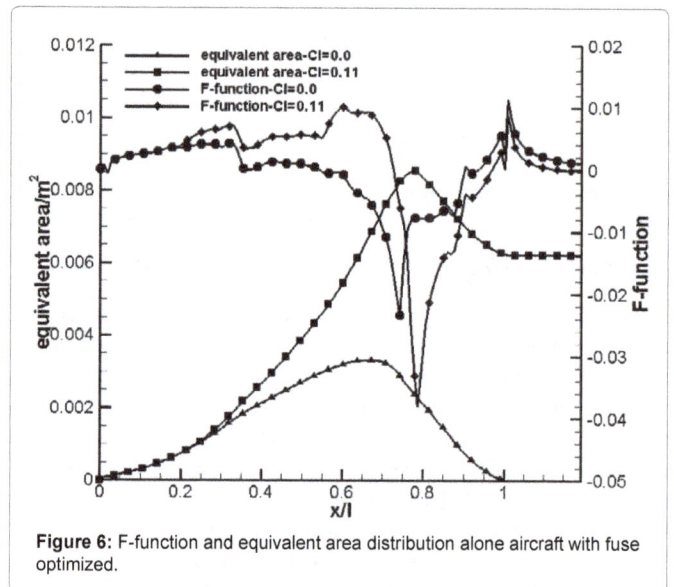

Figure 6: F-function and equivalent area distribution alone aircraft with fuse optimized.

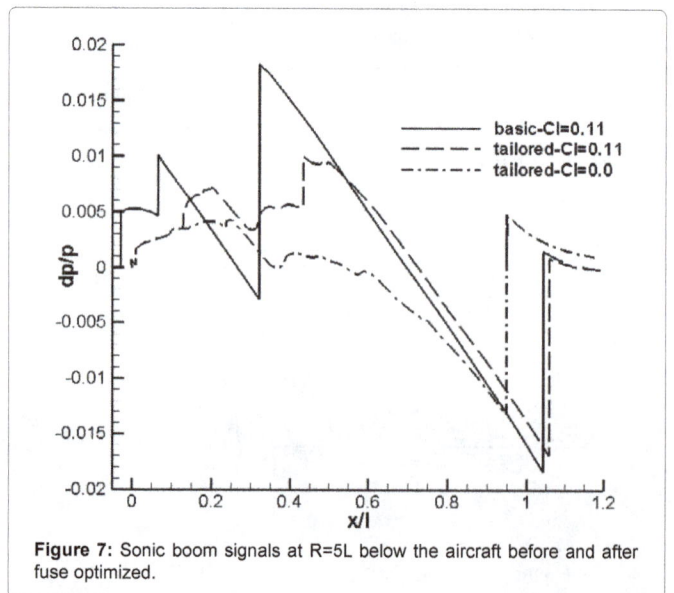

Figure 7: Sonic boom signals at R=5L below the aircraft before and after fuse optimized.

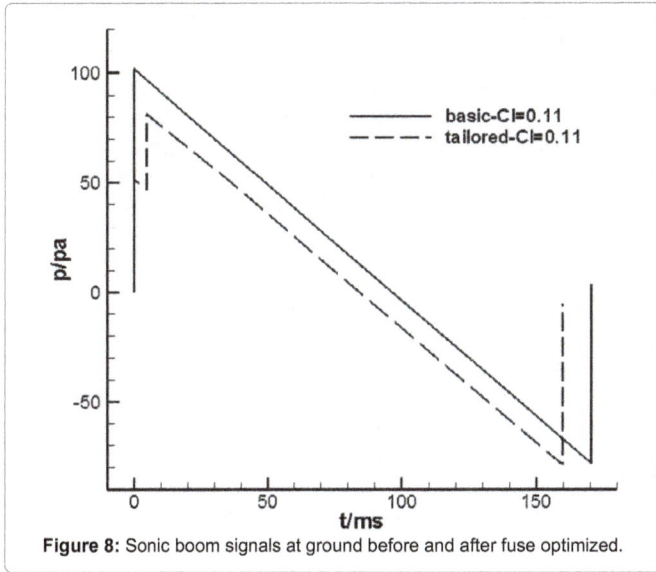

Figure 8: Sonic boom signals at ground before and after fuse optimized.

boom caused by the lift component of the effective area. Supposing the lift distribution was uniform alone the wing, the lift component of equivalent area can be optimized through the optimization of wing geometry.

The genetic algorithm was used to do the optimization. There were 6 optimization variables, including the root chord, tip-root ratio, span, inner wing swept, outer wing swept and KINK position, which are shown in Table 3. The wing area was constrained to ensure the appropriate aerodynamic characteristic. The minimum A-weight noise level on the ground at Ma = 1.6 was chosen as the optimization object. Figure 9 is the sketch of the configuration with wing geometry optimized, and the parameters of wing geometry before and after optimization are shown in Table 4.

Figure 10 shows the F-function and equivalent area distribution alone the optimized configuration at cruise lift coefficient and zero lift condition. Figure 11 shows the over pressure of sonic boom at the distance of 5 lengths of aircraft below the optimized configuration. The solid line is the result of the optimized configuration in section 4.1, the dash dot line is the result of the configuration with wing geometry optimization at zero lift coefficients and the dashed line is the result of the configuration with wing geometry optimization at cruise lift coefficients. It can be seen that the over pressure caused by the lift component of equivalent area decreased obviously after the wing geometry optimized. The shock wave generated at about 20% alone the aircraft was caused the change of the wing geometry, which changed the cross-sectional area of body by the normal projections of cuts alone planes aligned with the Mach angle. Thus, the fuse should be tailored to weaken the shock wave. Figure 12 gives the over pressure of sonic boom on the ground. The solid line is the result the optimized configuration in section 4.1 and the dashed line is the result of the configuration with wing geometry optimization. The over pressure of wing geometry optimization is 60 pa with the reflect factor 1.9, 26% decreased compared with the optimization configuration in section 4.1, and the A-weight sound pressure level is 83.74 dB, 3.8 dB decreased after optimization.

From the results, it can be seen that the shock wave caused by the lift component of the effective area decrease obviously, and the tiny shock wave at the 20% alone the aircraft was caused by the change of

wing geometry, which lead to the volume component of effective area change. The over pressure on the ground is 60 pa with the reflect factor 1.9, 26% decreased compared with the configuration without wing optimization, and the A-weight sound pressure level is 83.74 dB, 3.8 dB decreased after optimization. Compared with the basic configuration, the sonic boom noise of the optimized configuration of fuse and wing

Variables	Optimization Range
Root Chord/m	18-28
Tip-root Ratio	0.07-0.1
Span/m	14.4-17.6
Inner wing Swept/·	68-75
Outer wing Swept/·	50-65
KINK/m	3.0-6.0
Wing Area/m²	130-150

Table 3: The range of optimization variables.

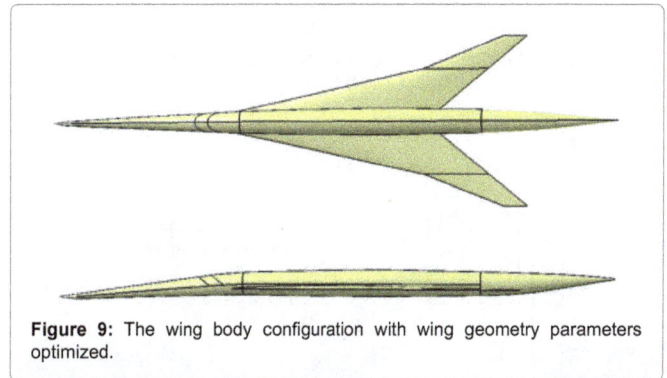

Figure 9: The wing body configuration with wing geometry parameters optimized.

Variables	Basic	Optimization
Chord Root/m	25.36	19.1
Tip-root Ratio	0.13	0.0992
Span /m	20	16.496
Inner wing Swept /·	72	74.95
Outer wing Swept /·	50	63.8
KINK/m	32%	61.94%
Wing Area/m²	164.5	146.12

Table 4: Geometry parameters of the wing.

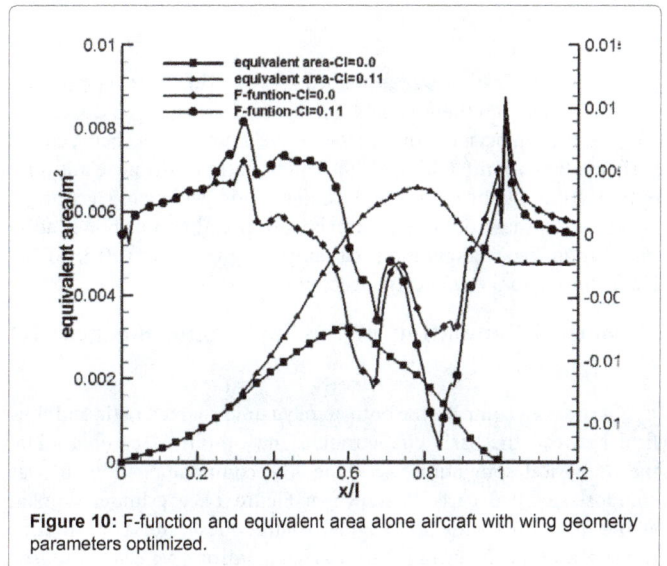

Figure 10: F-function and equivalent area alone aircraft with wing geometry parameters optimized.

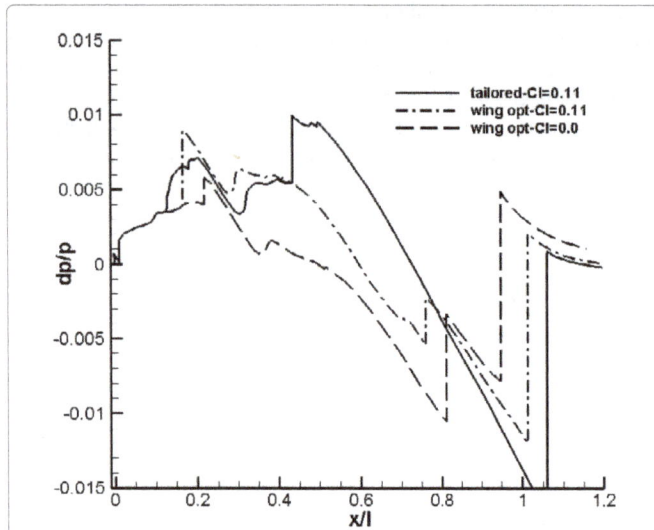

Figure 11: Sonic boom signals at R=5L below the aircraft before and after wing geometry parameters optimized.

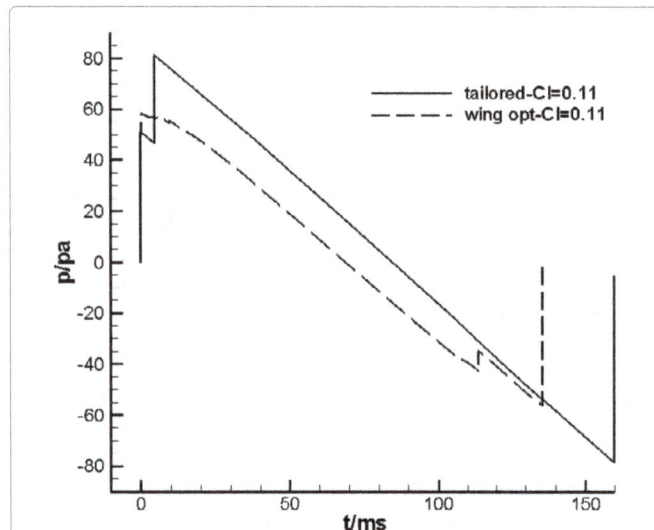

Figure 12: Sonic boom signals at ground before and after wing geometry parameters optimized.

geometry optimization was decreased greatly, with the over pressure decreased 41%, and the A-weight noise level decreased 7.55 dB. Figure 13 gives the 1/3 octave sound pressure level comparisons between the basic configuration (solid line), the configuration with nose and fuse optimization (dashed line) and configuration with wing geometry optimization (dash dot line). It can be seen that the sonic boom noise of the optimized configuration was mitigated greatly from 10Hz to 10k Hz according to the 1/3 octave spectrum.

Numerical Simulation of the Supersonic Business Jet Configuration

In order to compare the both aerodynamic characteristic and flow field between the basic configuration and optimized configuration, the numerical simulation was done. The computation domain was composed of two parts as shown in Figure 14: a cylinder domain at the inner part with non-aligned volume mesh and a mach cone domain around the cylinder with aligned anisotropic cells [14]. The

Euler equation was solved for the basic configuration and optimized configuration.

Figure 15 gives lift coefficients, drag coefficients and pitching moment coefficients comparison between the basic configuration and optimized configuration in section 4.2. It can be seen that the slope of lift coefficients of the optimized configuration decreased because of the increased wing swept. The angle of attack when cruising (Cl = 0.11) varied from 1.65 to 2.6 degree after optimization. The drag decreased after optimization while the pitching moment change was tiny.

Figure 16 shows the pressure distribution around the aircraft at the symmetrical plane. From the flow field results, it can be found that the shock wave of the optimized configuration was weakened remarkably compared with the basic configuration, which meant that the low sonic boom optimization was available.

Conclusion

For a supersonic business jet without considering the sonic boom attenuation, distinct shock waves were induced that made big noise. According to the analysis results by supersonic linearized theory, the fuse, nose and wing geometry were optimized. Compared with the basic configuration, the sonic boom level of optimized configuration

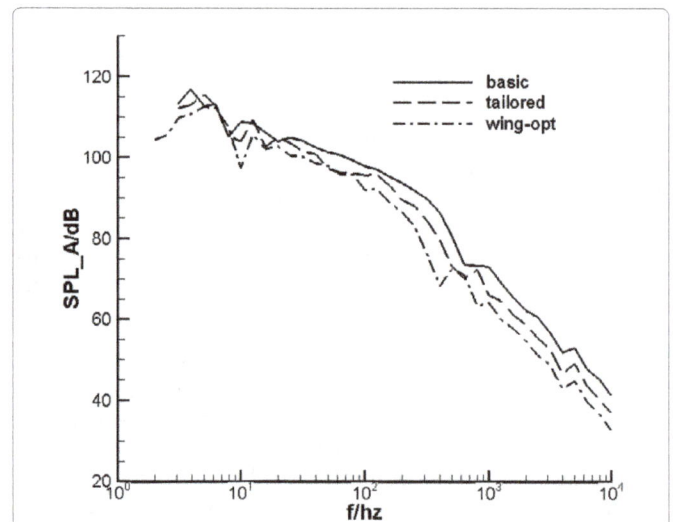

Figure 13: 1/3 octave sound pressure level before and after optimization design.

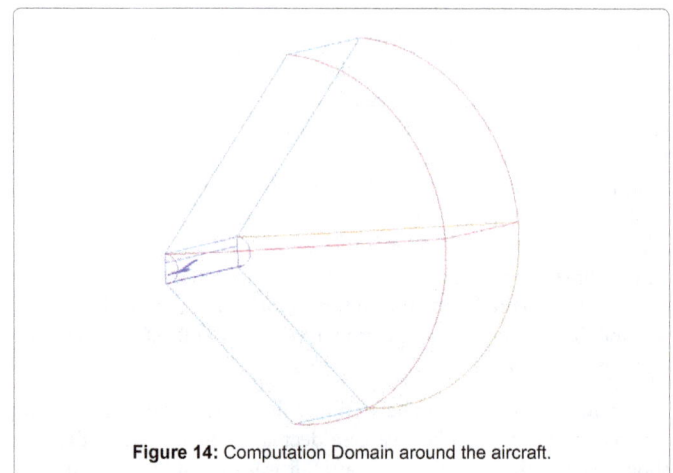

Figure 14: Computation Domain around the aircraft.

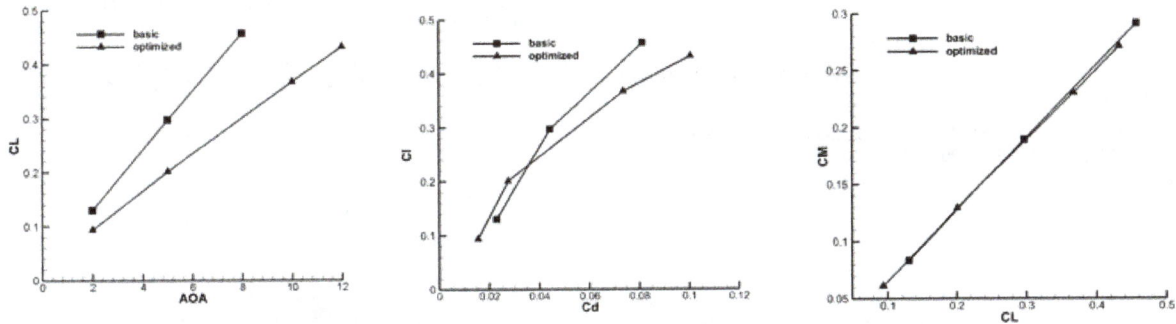

Figure 15: Aerodynamic characteristic comparison between basic and optimized configuration.

(a)   Basic configuration(Ma=1.6,AOA=1.65,φ=0)

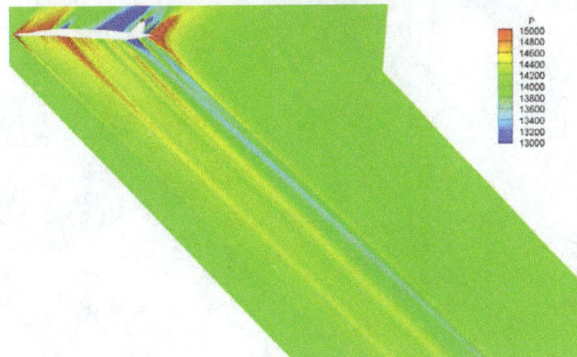

(b)   Optimized configuration(Ma=1.6,AOA=2.6, φ=0)

Figure 16: Pressure contour around the aircraft.

decreased greatly. The maximum over pressure on the ground decreased 41% and the A weighted sound pressure level decreased 7.55dB. The angle of attack when cruising varied from 1.65 to 2.6 degree after optimization. The drag of the optimized configuration reduced but the change of pitching moment was tiny. The shock waves around the aircraft were weakened obviously from the numerical simulation results, which meant that the low sonic boom optimization was available. The low sonic boom optimization in this paper can be applied to multi-object aerodynamic configuration optimization in the future, with the constraints of aerodynamic characteristic and the optimization objective of weighted sonic boom noise level at multi-flight condition.

References

1. Gold T (1952) The double bang of supersonic aircraft. Nature 170: 808-808.

2. Plotkin K (1989) Review of sonic boom theory.

3. Bastin EW (1953) Noise from aircraft at supersonic speeds. Nature171: 214-216.

4. Whitham G (1952) The flow pattern of a supersonic projectile. Commun Pure Appl Math 5: 301-348.

5. McLean FE (1965) Some non-asymptotic effects on the sonic boom of large aircraft. NASA TN D-2877 Hampton Virginia USA: Langley Research Center.

6. McLean FE (1968) Configuration design for specified pressure signature characteristics. Washington DC USA, Second conference on sonic boom research: 37-45.

7. Whitham GB (1956) On the propagation of weak shock waves. J Fluid Mechanics 1: 290-318.

8. Jones LB (1961) Lower bounds for sonic bangs. J Royal Aeronautic Society 65: 1-4.

9. Seebass R (1969) Minimum sonic boom shock strengths and overpressures Nature 221:651-653.

10. Seebass R (1984) Sonic boom theory. J Aircraft 6: 177-184.

11. Darden C (1974) Sonic boom theory: it's status in prediction and minimization. J Aircraft 14: 569-576.

12. George A, Seebass R (1971) Sonic boom minimization including both front and rear shocks. AIAA Journal 9: 2091-2093.

13. Thomas CL (1972) Extrapolation of sonic boom pressure signatures by the waveform parameters method. NASA TN D-6832 Moffett Field California USA: Ames Research Center.

14. Cliff SE, Durston DA, Elmiligui AA. Computational and experimental assessment of models for the first AIAA sonic boom prediction workshop.

5

Analyze the Mode Transition Logic of Automatic Flight Control System using Semi-Formal Approach

Rathina Kumar V*, Nanda M and Jayanthi J

Department of Aerospace, Electronics and System Division, CSIR-National Aerospace Laboratories, Bangalore, India

Abstract

Autopilot system is a highly critical avionics system in modern aircraft as it steers the aircraft automatically. The autopilot is a highly complex system driven by a complex logic and is one of the major reasons for the accidents in automated airliner. The autopilot logic consists of the mode-transition logic which in automated mode steers the aircraft based on the aircraft aerodynamics. In the automated mode the correct and efficient working of the mode-transition is highly critical; hence a high assurance approach is required to analyze the logic for its functionality and performance.

In this paper, we present a semi-formal method based approach to analyze and validate the Mode-Transition Logic (MTL) for an indigenously developed commercial aircraft in the vertical and lateral directions. The MTL is analyzed and validated for its correct, complete, and reliable functionality and operation using Stateflow. The modeled MTL logic is validated for the allowed transitions based on the input combinations against the requirements for functionality and safety. The outcome of the approach shows encouraging results with respect to assurance in functionality, performance and safety in comparison to the conventional manual approach of testing. Similar semi-formal based approach can be used to reduce the design effort in the design and development of complex system designs as compared to the manual analysis.

Keywords: Autopilot; Mode transition logic; Semi-formal methods; Stateflow; Simulink design verifier; Model advisor; RTR; Reactis; Validator; Tester

Introduction

The complex system of Aircraft, such as Digital Autopilot system, Flight management system and Electronics map displays are used to reduce the workload of pilots and gave better information, improving the flight performance and improve situation awareness. In this avionics architecture the autopilot system have more add on compared to flight management system and it never integrated into a single system. The elements of current avionics system are distributed across the mode control panel, the control display unit and primary flight display which include flight mode annunciations. The autopilot has different types of modes such as heading select, heading hold, vertical speed hold. A mechanical, electrical and hydraulic device of the aircraft is used to assist the autopilot operation. The autopilot logic is used to guide an airplane with minimal or no assistance from the pilot [1].

The autopilot is designed by using complicated mode transition logics. The designer spend more time to design the mode transition logic and their safe transitions and the designs are more strengthened by using verification and validation techniques such as assertions, safe states and safe transitions. The incorrect mode transition logic has led to accidents in the past year. The accidents are overcome by improving the Mode transition logic analysis [2,3].

Mode confusion occurs when the pilot believes the current mode is different from actual mode but it's actually in correct mode instead of the correct mode pilot change inappropriate mode. Mode confusion can also occur when pilot does not understanding the behavior of mode transition logic and pilot has poor knowledge about the mode transition logic. Advancement in digital avionics system has accounted for much of the improvement in air safety seen over the last few decades. At same time the growing complexity places in the system and increase the risk of the mode transition logic. To fly commercial fly today, the pilots must be a master in several complex, dynamically interacting system and should know operating at different levels of automation [3].

In safety critical applications become higher safety and functionality assurance for using the formal method based techniques according to the civil aircraft standard RTCA DO178B [4]. This RTCA DO 178B standard is providing the guidelines for design, development, verification and validation of airborne software in safety critical system. A formal verification technique is a mathematically based languages tools and techniques to formally model the design based on the project requirements. These formal methods technique is used to analyze the system behavior for all the possibilities and converge of the system [5,6].

Model based design and development is used to create the model at each and every stage of the software lifecycle and automatic model transformation for example from code to model. It is well defined method and produces more sustainable software and it is give graphical notations and good abstracting details. This technique has been used in automated high speed train and car autopilot and now being used for aerospace domain for demonstrating the functional and safety properties [7,8].

Semi formal method is technique to analyze the system in model level and code level and then design the computerized version of it. The computerized system having the same structure and functions as we expected and it satisfied the requirements according the standard guidelines [5,6].

*Corresponding author: Rathina Kumar V, Department of Aerospace, Electronics and System Division, CSIR-National Aerospace Laboratories, Bangalore, India
E-mail: rathina2020@gmail.com

The amount of software has increased significantly over the last years and therefore, the verification of embedded software has become of fundamental importance. The most commonly used approaches to verify embedded software are based on co-debugging or co-simulation, which have the coverage problem. Formal verification assures complete coverage but is limited to the size of module that can be verified. In this paper, we present a new semi-formal verification approach in order to verify temporal properties of embedded software, based on the combination of simulation and formal verification approaches. The semi-formal verification approach can be used to overcome the drawbacks of both dynamic and static verification. This approach combines the benefits of going deeper and covers exhaustively the state space of the system. The effectiveness of the semi-formal approach provides the foundation to use this approach in validating the functionality & performance of complex aircraft logic [5,6]. Earlier this logic was validated using conventional approach consisting of reviews, tests at code level, system level and aircraft level. This approach is not only very laborious but also has the drawback in detecting the logic flaws earlier in the phase [9-12].

MTL is a discrete event system with states, inputs, and outputs which is usually modeled as a finite state machine where the states represent the modes. MTL system receives inputs and based on the inputs and the current mode, transits to another mode and produces an output. The outputs are used to command the control surfaces of the aircraft appropriately. Each such computation is referred to as a software cycle [13-15]. The mode and their transitions in the current research have been represented in tabular forms. Mode Transition Logic (MTL) is a design module used by the flight director of an autopilot to switch modes of flight control. The possible transition values from a current value of a state variable are specified in the State Transition Matrix (STM). The actual transition of a state variable from the current value to a new value is allowed when the condition(s) given in the Condition Matrix (CM) are satisfied. Control modes are switched based on the events that are received as inputs or commands to the autopilot [16-18].

In this paper, we implement the MTL vertical and lateral modes using Simulink Stateflow for an indigenous civil aircraft. The modes and logic concerning lateral, vertical modes transition, mode possibilities in lateral and vertical modes are validated at the design level for functional, performance, and safety properties. Comparison of the proposed approach with the conventional approach shows the improvement in the process in understanding and validating the complex logic at the model level rather than at the code level

Literature Survey

The safety, functionality and performance validation and verification of the complex system has always been challenge. Complex logic, complex systems are validated using conventional techniques such as simulation, testing and reviews with bottom-up approach. With the technological evolution, the time from concept to certification of a system is reducing as the demand for these systems are increasing. For example the design to certification of Airbus A350 XBW was in a short span of time [19].

In safety critical domain such as automotive, railways, space and aeronautical the new technologies are complex performing multi-functions. For example car autopilot system, intelligent train control system, electrical flight control system and cyber rail [5,6]. To ensure that these systems perform the functionality as per the requirements of time and safety, lot of analysis is to be performed. Mathematical and graphical based approaches are proving to be more effective than textual based approach. Formal and Semi-formal methods are gaining popularity in validating complex system/ software logic.

Development of the autopilot, the array logic based technique has been used to reduce the design effort. It is easily understood and provides a very concise way of specifying a large number of transitions in simple tabular column. Vertical and lateral modes of the autopilot have been designed using mode transition logic and the technique has been extended to cover the navigation and approach modes [2,3]. All the possible mode transitions in the presence of external or internal event and performance criteria are presented in the subsequent sections. The mode and their transitions have been represented in tabular form called as array logic based technique.

One of the mode transition logic was analyzed by using formal method approach in the name of the paper is mode confusion analysis of a flight guidance system using formal methods and the main author is Anjali Joshi. In this paper they used NuSMV and theorem proving technique to analyze the mode confusion logic [6]. But here we are checking all the state of mode transition logic in semi formal approach using matlab simulink/stateflow tool suite.

In the modern autopilot systems having complex components that are detecting and avoiding collisions with other objects and allow aircraft to land in situation where a human cannot see the runway environment. In earlier days several accidents and incidents have been reported because of the autopilot failure. For example in 1983 the Korean Air Lines Flight 007 flying from Anchorage to Seoul. This aircraft deviated more than 200miles from it path (soviet territory) and got shot down killing the crew and the passengers. The aircraft accident analysis reported navigational failure as the cause of the air crash. It was found that the flight was initially in heading mode, later the pilot either forgot to select the inertial navigation system or otherwise the pilot might activated but system got never activated. The autopilot goes to inertial navigation mode when these following two conditions are satisfied. The aircraft path and the predefined path must be close and the distance between these two paths is within 7.5 miles. These two conditions are continuously checked by software it is called guard. The guard is a logical statement which is used for mode transition of aircraft. The mode is changed only the conditions are true [2-3]. In this case the mode could not be activated, which could be due to improper implementation of the MTL logic or pilot error or system error.

Lateral modes and logic concerning lateral mode transition are less complex compared to the prevailing methods for autopilot design [8]. Various mode possibilities of lateral mode transition in an autopilot is mentioned along with specification criteria's that bound these transition and these possible transitions were given a frame work using MATLAB software. System decomposition, abstraction, and distribution lead naturally to sub problems that can be addressed using formal methods and tools, such as mathematical modeling, control law synthesis, and control implementation verification. We classify these methods and tools, which rely heavily on mathematical formulations of the underlying problem.

Approach

The novel approach proposed works on the solid foundation of the conventional approach. The tabulated MTL and MTL simulation blocks are used as the reference for establishing the semi-formal approach. This is done to ensure the correct implementation of the semi-formal methodology for analyzing the complex MTL algorithm. Figure 1 show the technique which is followed. Simulink stateflow is used to implement the MTL logic. The output of the stateflow is compared with the expected output using conventional manual approach. The comparison proves the efficacy of this approach in terms of ease of implementation, ease of understanding, and the improvement in the process.

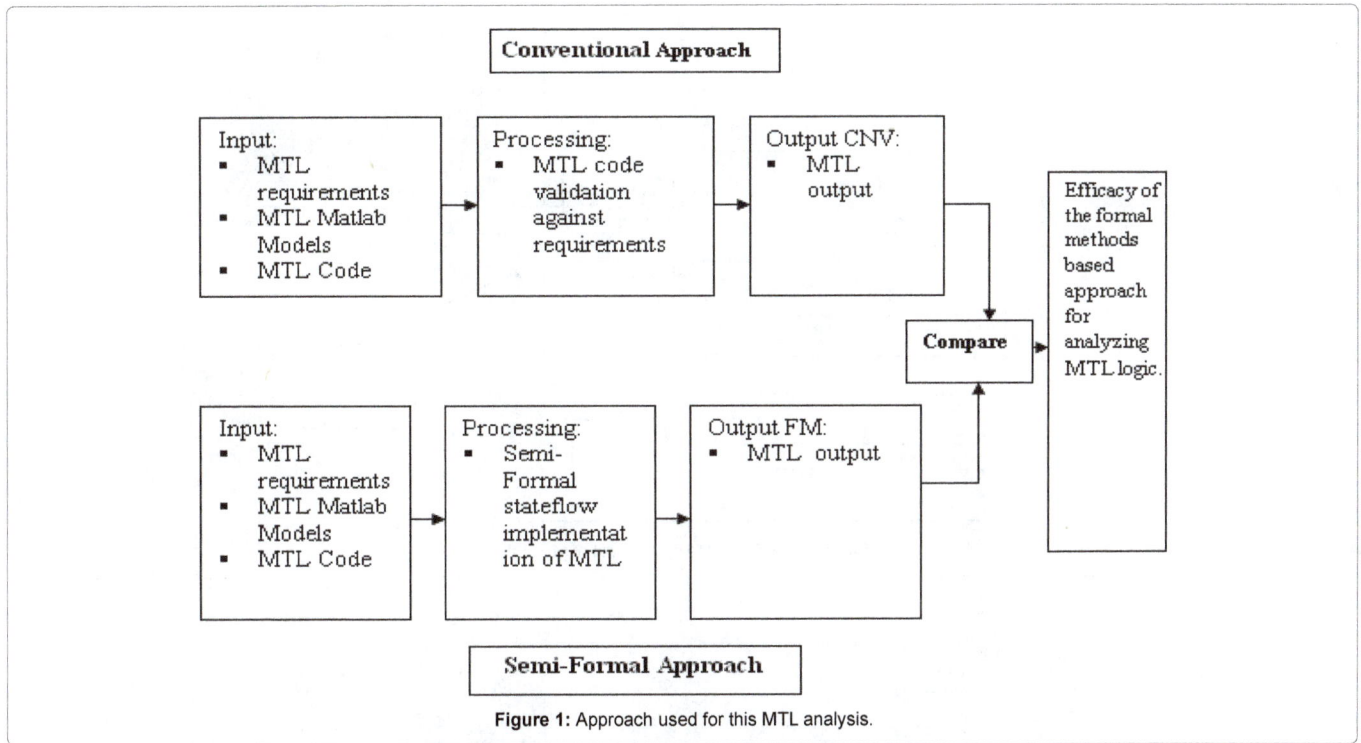

Figure 1: Approach used for this MTL analysis.

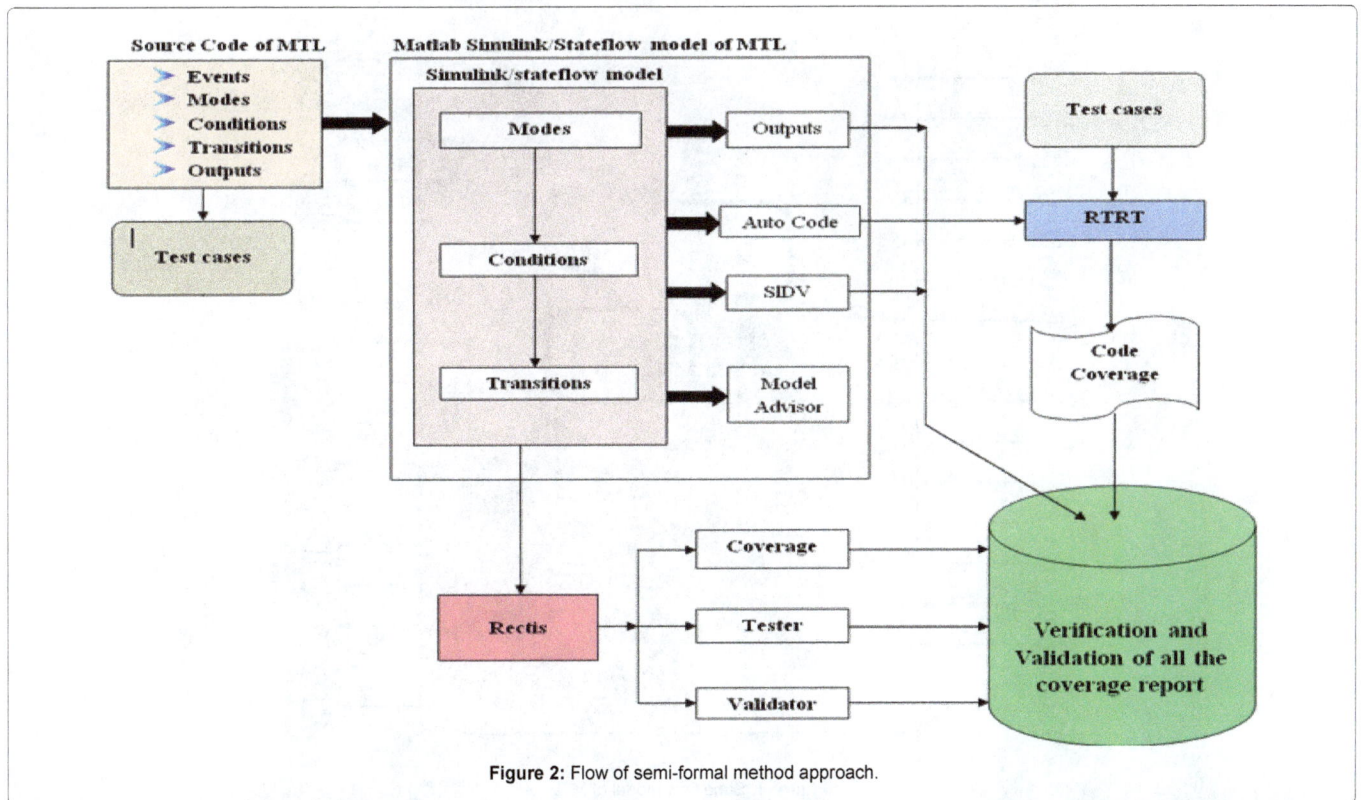

Figure 2: Flow of semi-formal method approach.

The inputs for the semi-formal method based approach are MTL requirements, MTL tables, and software code, as an input. These inputs are translated into the stateflow. The stateflow structure is similar to the code structure to maintain the semantic translation from code to model level. Stateflow standards are followed in order to generate compact auto code. Stateflow model is done according to the MAAB guidelines [20]. The equivalence checking of the MTL requirements is checked for the stateflow by mapping the MTL requirements to the stateflow.

To analyze the robustness of the stateflow design, test-scenarios

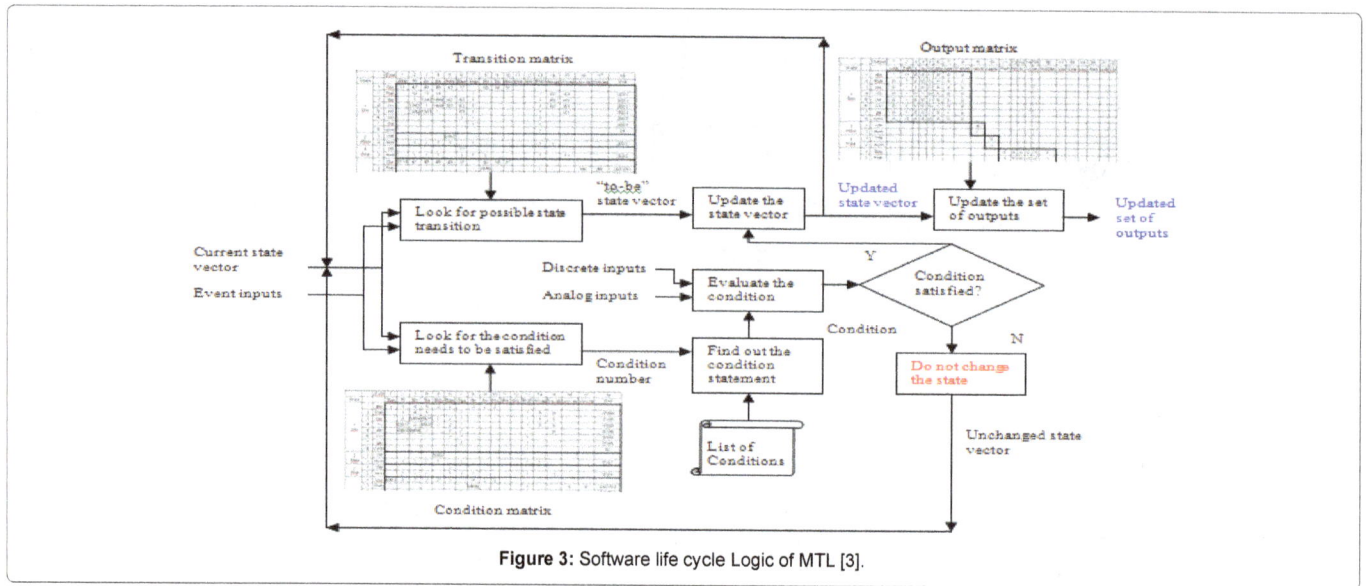

Figure 3: Software life cycle Logic of MTL [3].

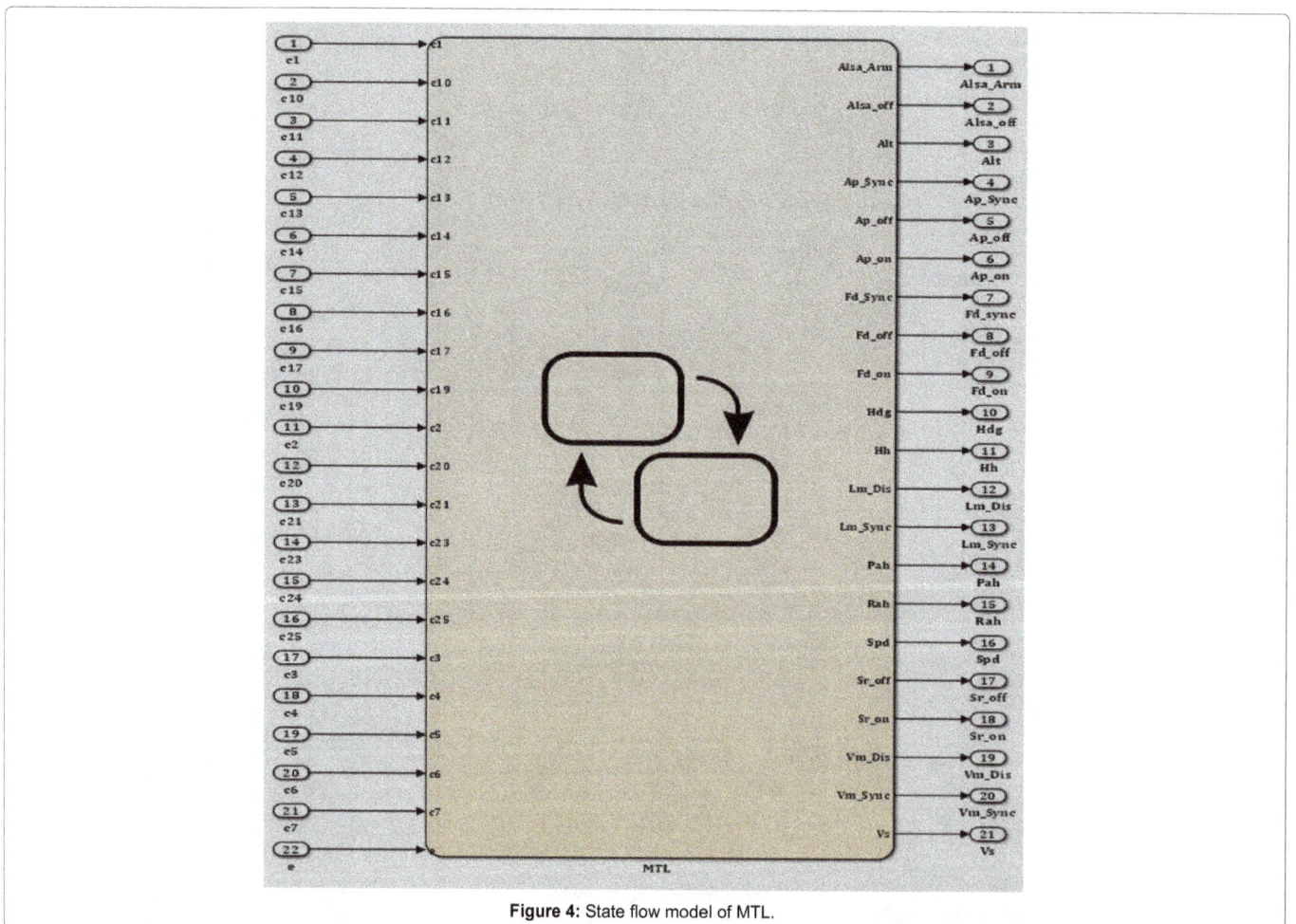

Figure 4: State flow model of MTL.

generated at code level are translated to model-level. The test outputs at the model level are compared to the code level test output. This comparison provides the correct semantic translation of the code to model.

Figure 2 shows the details of the implementation of the approach for the MTL logic. The MTL model using the stateflow is generated based on the manual code structure and the MTL tables for inputs, outputs, events, conditions and transitions. Model inputs are events

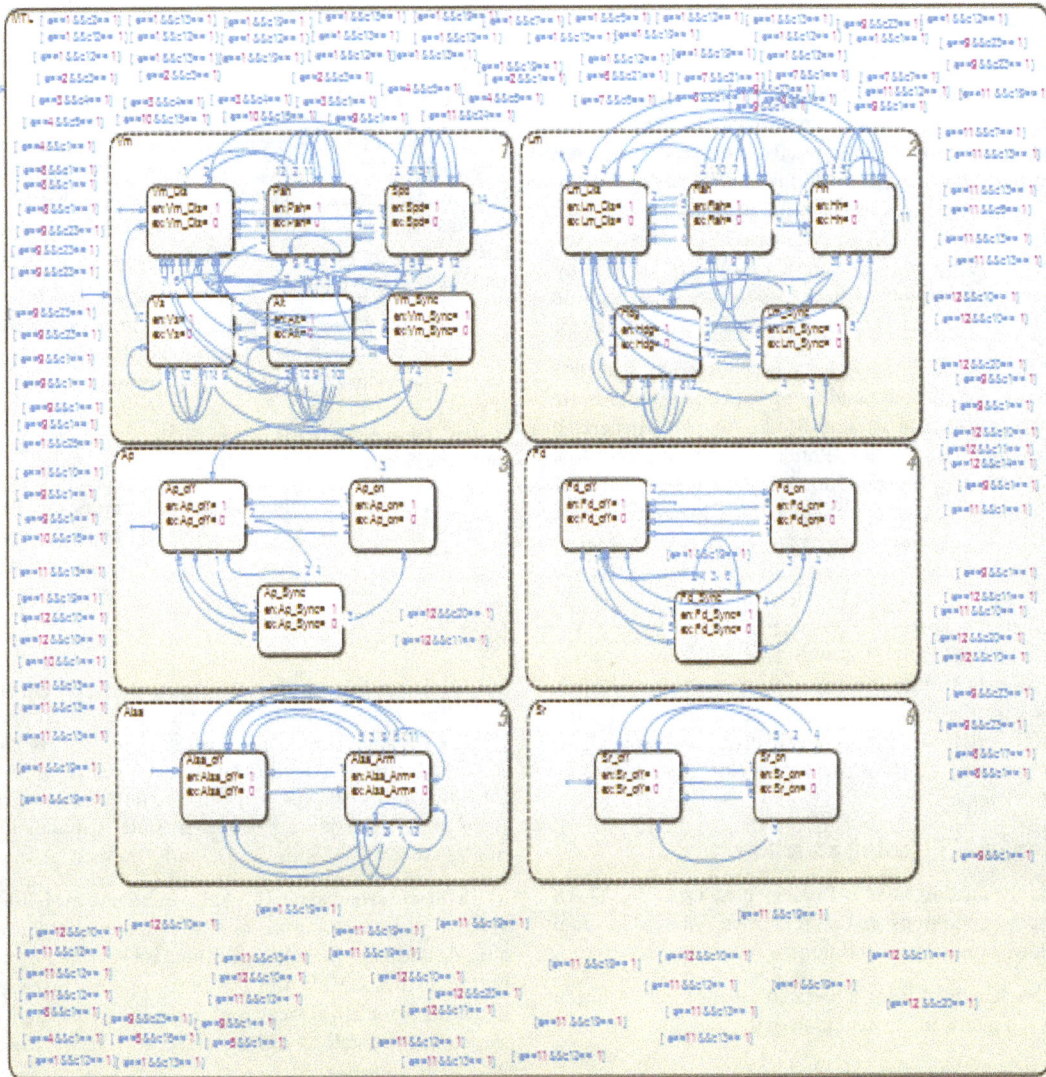

Figure 5: State flow chart of MTL.

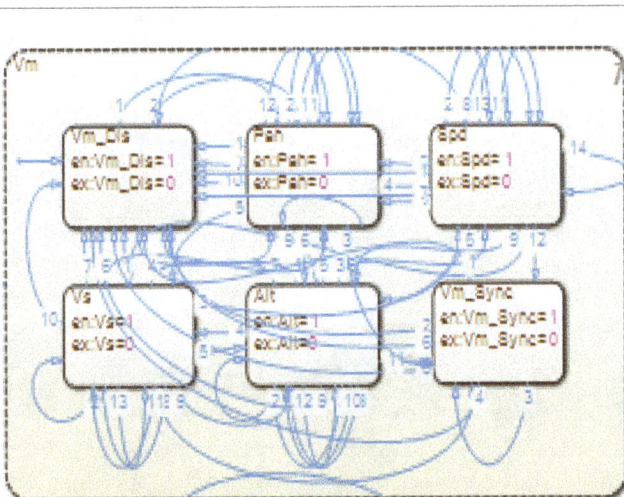

Figure 6: State flow chart of vertical mode.

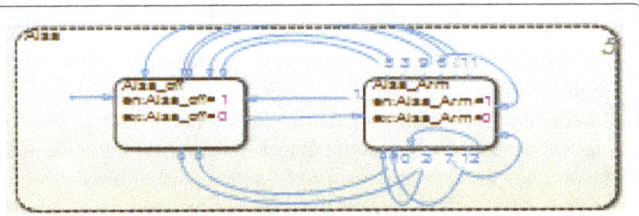

Figure 7: State flow chart of altitude select arm.

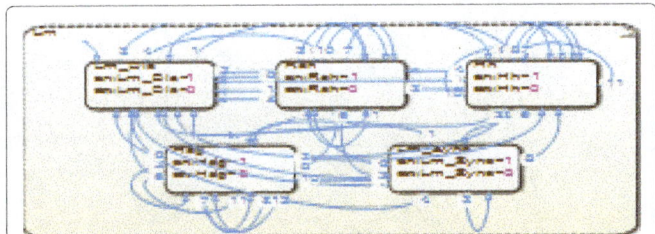

Figure 8: State flow chart of lateral mode.

from ARINC 429, and discrete sources which are acquired from external sensors in the aircraft. State transition logic is generated based on the pre-defined inputs, events, conditions, and allowed transitions.

Validation of the developed stateflow [version 8.1,] is performed at the model level as well as at the code level. At model level, the test scenarios to validate the model are translated from code level test cases. Test cases generated are imported to the matlab workspace and the outputs are analyzed with reference to the expected output [11]. Simulink design verifier [version 3.5,] is also used to generate the test scenarios and capture the test outputs to check the model functionality, performance, safety and stability. These outputs are compared to expected output [12] for the correct implementation of the MTL logic.

Validation at the code level is done by generating the autocode of the MTL stateflow. The in-house and approved code standards are used to generate a compact and safer autocode. Autocode generated from stateflow are imported to RTRT (Rational Test Real Time) tool suite, Version 7.5. Test cases generated for manual code are used for the autocode and analyzed for code functionality, code compactness and coverage. RTRT tool generates the coverage report. In case of safety critical software developed for highest criticality i.e., Level A [3], code is tested for 100% MCDC (modified condition decision coverage) coverage. Model level robustness is validated by importing the autocode into Reactis tool suite [13]. Reactis tool suite provides three types of analyses: Model Coverage Analysis, Reactis Tester and Reactis Validator. Test-cases for the autocode are generated automatically. The coverage report of Reactis Tool suite is compared with the model level and RTRT level test report. All three coverage reports are used in the validation of the MTL logic.

Implementation of MTL using Stateflow

Autopilot mode transition logic is implemented using MATLAB Stateflow. MTL inputs, conditions, and outputs are taken as a Boolean. Transitions are allowed only if the conditions are satisfied.

MTL logic is implemented using a hierarchical approach from top to bottom using a sequential implementation of the logic as:

» Top-model : provides the MTL architecture

» Vertical Mode : provides the logic for entry, exit of vertical modes for the autopilot

» Lateral Model: provides the logic for entry, exit of lateral modes for the autopilot

Figure 3 describes the software logic of the MTL which is to be translated into the stateflow model. Transition matrix guides the change of state on receiving an input from external interfaces. In order to successfully execute a state change on receipt of an input, certain conditions have to be met which is dictated by a condition matrix. Conditions check for values of certain inputs to be within specified

ranges and if the condition is met, state is changed as per the transition matrix. State change also results in an output, as dictated by the output matrix. This output is used to command the flight control surfaces.

Top level model

The top-level model provides the MTL architecture. The control flow of the MTL logic is designed at this level. Figure 4 shows the top-level model of the autopilot MTL. Figure 4 consists of states, events and outputs. The state is a uniquely defined mode variable which can take certain discrete values. The state changes in response to external event and the respective conditions.

During the autopilot, when an event is pressed on the Autopilot Computer and Mode Selection Panel (ACMSP), the control (control matrix) flow checks the pre-defined conditions (condition matrix) for the MTL. If the condition is allowed then the transition (transition matrix) takes place to the allowed state, else the present state is retained and provides the respective output.

Figure 5 depicts different aircraft modes: Vertical Mode (VM), Altitude Select Arm (ALSA), Lateral Mode (LM), Autopilot (AP), Soft Ride (SR) and Flight Director (FD). This is the complete state diagram of MTL logic and it contains all the states, conditions and events as per the given inputs. Each of the state is explained separately in the following sections.

Vertical mode: The basic vertical mode is the (Pitch Altitude Hold) PAH. The transition to the higher level modes like the ALT, SPD, and VS Hold takes place when thecorresponding events conditions are satisfied. Figure 6 shows the state transition chart of vertical modes. The transition to the next state depends on the Event as well as Condition(s). If no event takes place the present state is maintained. The allowed modes are Vm_Dis, Spd, Vs, Alt, and Vm-Sync.

Altitude select arm: The Altitude Select arm (Alsa) is a compound vertical mode allowing the pilot to climb or descend to a pre-selected altitude and hold that altitude. The Alsa is armed by pressing the Alsa button on the ACMSP. Alsa gets engaged only when a specific condition is true, else remains disengaged. Figure 7 shows logic and allowed the state transition state for altitude select arm Alsa. It has only two modes: Alsa_arm, and Alsa_off.

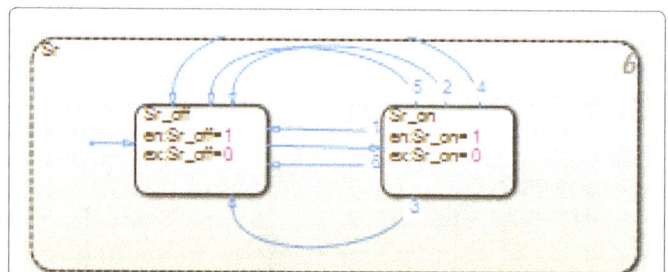

Figure 10: State flow chart of soft ride.

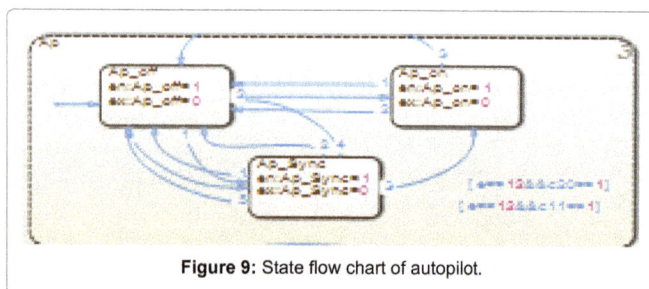

Figure 9: State flow chart of autopilot.

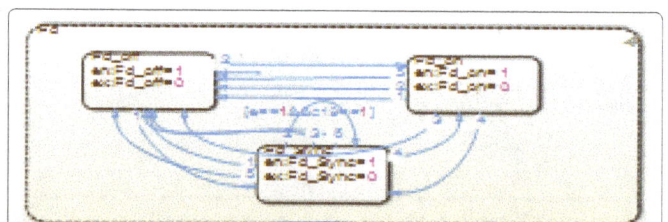

Figure 11: State chart of flight director.

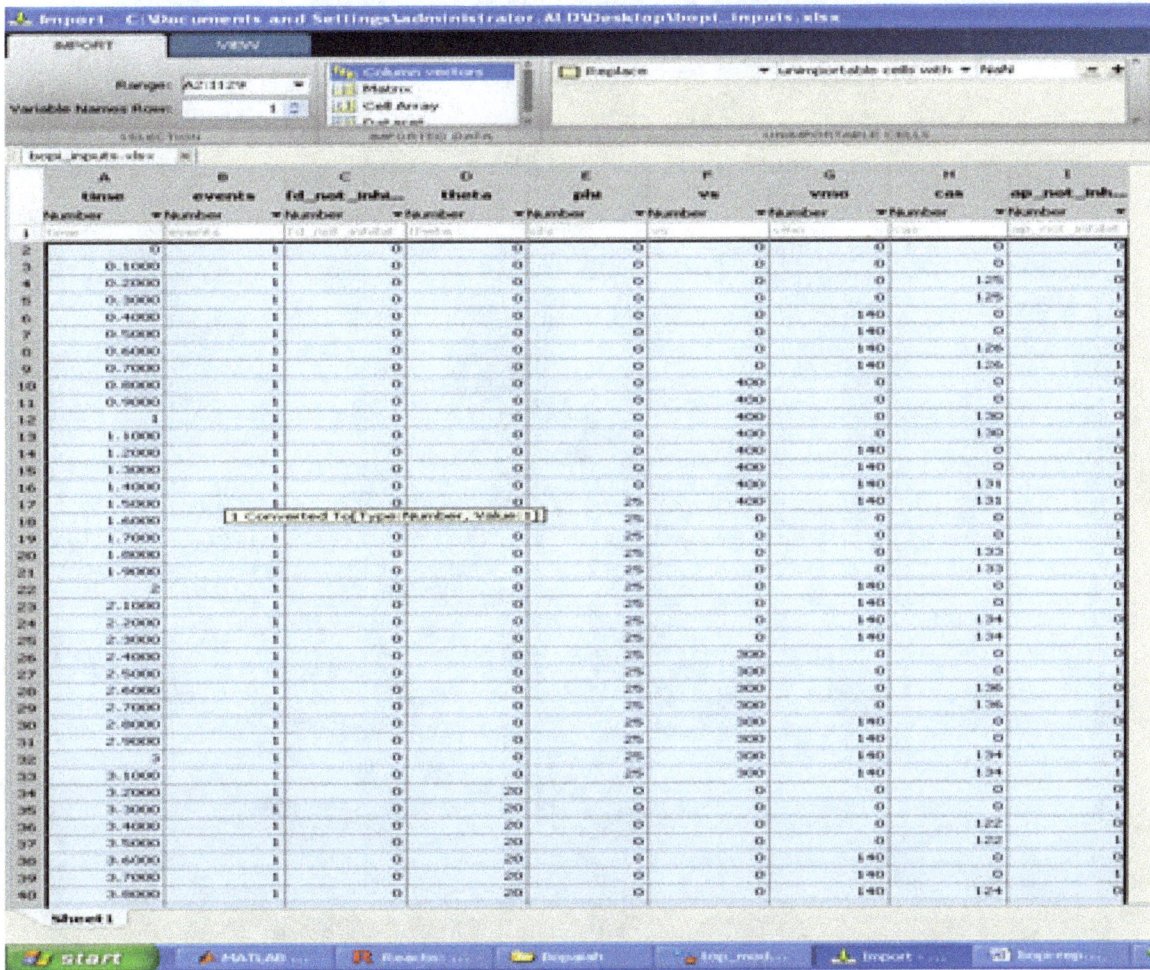

Figure 12: Imported inputs in Matlab environment.

Lateral mode: The basic lateral mode is the (Roll Attitude Hold) RAH. The transition to the higher-level modes like the HH Hold and HDG Select takes place when the corresponding event and conditions are satisfied. If no event takes place the present state is maintained. Figure 8 shows the logic and the allowed state transition chart for lateral mode. The allowed modes are: Lm_Dis, Hh, Hdq, and Lm_Sync.

Auto-pilot: The autopilot gets engaged only when the autopilot button on the ACMSP is pressed where the transition from off-state to on-state takes place, else the mode is in disengaged or off state. Figure 9 shows logic and allowed state transition chart for autopilot.

Soft ride: This mode is selected while encountering turbulence/ gusts in flight. On selecting this mode, any previously held higher mode is dropped and the basic modes (PAH, RAH) are engaged with the inner loop control law gains appropriately reduced to alleviate the effect of external disturbances. Figure 10 shows the logic and allowed state transition chart.

Flight director: Flight Director, FD, function computes the reference commands for the AP function based on the pilot selection of modes through the ACMSP and the aircraft motion parameters obtained from the AHRS and ADCU. The FD also computes the reference commands, which drive the steering bars on the EFIS for Flight Director Guidance. Figure 11 shows the logic and the allowed state transition chart.

MTL Model Design Validation

MTL model design validation was done at the model level, code level and cross validated using third party tool. Validation at model level is performed using the simulink/ stateflow environment. Validation of the autocode generated from the model was done using a third party tool.

Model validation using Simulink/ Stateflow environment

Matlab simulink/stateflow outputs: Simulink stateflow is validated for the functionality and robustness by providing the test cases. The test cases used for the manual code are imported from Excel to Matlab workspace. Figure 12 shows the imported inputs from Excel to Matlab workspace. In the matlab workspace inputs are given in particular timed interval for that particular time and the outputs are displayed in the same workspace. These outputs are used to compare with the experimental results. Figure 13 shows the corresponding outputs as per the imported inputs [20].

Figures 12 and 13 show the inputs and outputs for event 1 i.e., Bap, with 7 inputs and 21 outputs. Time column in the inputs shows the

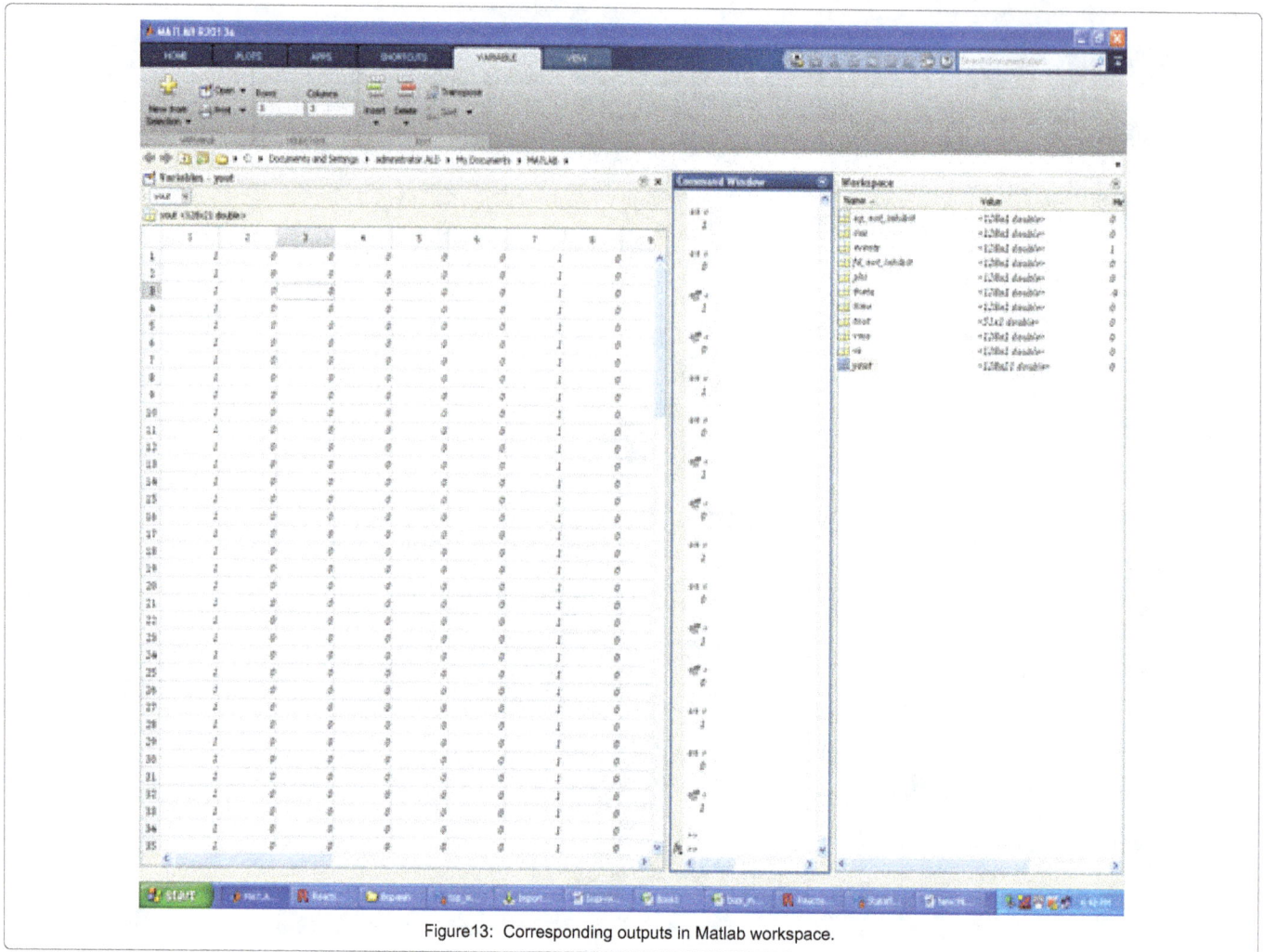

Figure13: Corresponding outputs in Matlab workspace.

specific time at which the corresponding row of inputs loaded into the model [21]. Other MTL modes are tested using similar approach.

Simulink design verifier (SlDV) [20]: Simulink Design Verifier™ uses formal methods to identify hidden design errors in models without extensive simulation runs. It detects blocks in the model that result in integer overflow, dead logic, array access violations, division by zero, and requirement violations. For each error it produces, there is a simulation test case for debugging. The Figure 14 is shows simulink design verifier report for mode transition logic. The logic is executing all the blocks of simulink/stateflow model. Totally 334 test cases have been used to cover the MTL logic.

Figure 15 shows the model advisor report. Model is built in Matlab Simulink/Stateflow Environment according to MAAB standard. Figure 16 shows the model advisor report of MTL. The report shows 10 fail standard and 13 warning because of model is not connected with hardware and internet to check the online resources with math work technical team.

Model checking or model advisor is an automated approach to verify that a model of a (usually concurrent, reactive) finite state system satisfies a formal specification of requirements to the system. In this approach how the system are behaving and generating the test cases automatically to analyze the behavior of the model. Tools that automatically perform model checking are called model checkers.

Rational test real time (RTRT) analysis

Rational Test Real Time is a cross-platform solution for component testing and runtime analysis. It is designed for developers creating complex systems for embedded, real-time and other cross-platform distributed applications. This software helps you debug and correct errors before they go into production code. RTRT resolve software problems during the development phase - allows testing the components. It can analyze the performance and reliability of the applications as run on the host development system. Modeling the system in modeling tool and generate the corresponding C/C++ code. This autocode was imported into the RTRT environment and test cases are written as per the requirements.

Figure 16 shows the 100% coverage of functions, functions and exits, statement blocks, decisions, basic conditions, modified conditions, multiple conditions in percentage. Figure 17 shows the report summary of RTRT unit testing with zero failure. It's totally having 212 test cases to cover the auto code.

Model validation-rectis environment

Reactis offers model-based testing, debugging, and validation for Simulink / Stateflow models. Reactis currently consists of three main

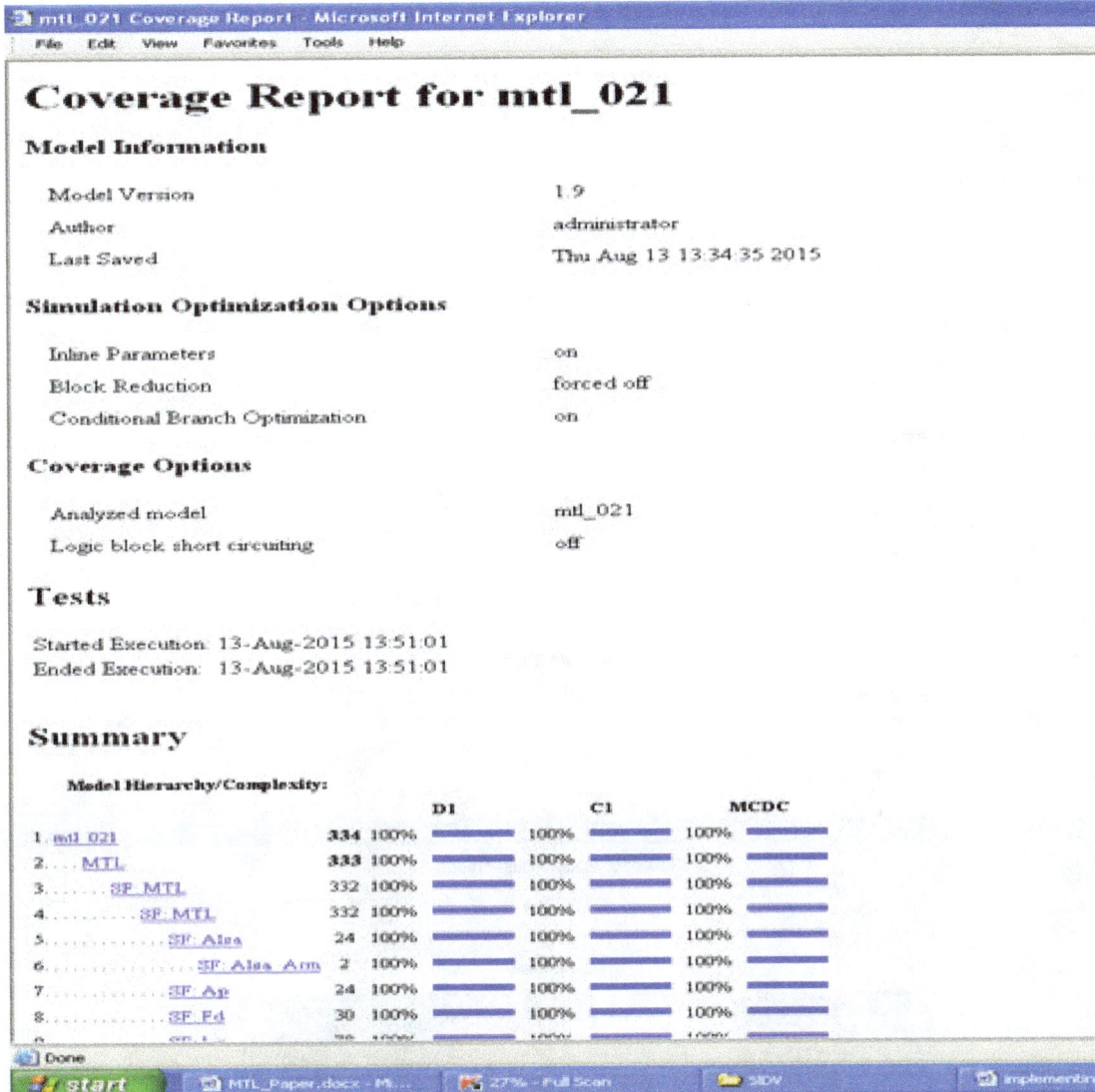

Figure14: Simulink design verifier report.

components: Reactis Tester, Reactis Simulator, and Reactis Validator [21,22]. Reactis Tester automatically generates test suites from Simulink / Stateflow models of embedded control software. The test suites provide comprehensive yet concise coverage of different test-quality metrics. Each test in a test suite consists of a sequence of input vectors as well as the responses to those inputs generated by the model. These tests may be used for a variety of purposes, including [23-25].

- **Implementation conformance.**

The tests may be applied to implementations derived from models to ensure conformance with model behavior.

- **Model testing and debugging.**

The tests may be run on the models themselves to analyze model behavior and to detect runtime errors.

- **Regression testing.**

The tests may be run on a new version of a model to compare its behavior to an older version.

- **Reverse engineering of models from source.**

Tests may be generated from models derived from legacy code in order to check conformance between model and code. Reactis enables to maximize the effectiveness of testing while reducing time and effort.

Reactis coverage: Figure 18 shows the MTL coverage report in Reactis environment. Reactis generated test cases automatically and executed the model and code satisfying the decision, conditions and MC/DC 100% [21]

Reactis validator: Figure 19 is report for reactis validator analysis. Reactis validator analysis is used to validate the simulink model in the reactis environment and it shows the validator coverage report is 99% is true [21].

Reactis tester: Figure 20 shows the report of reactis tester. Reactis tester is testing the model as per the input assigned in the property and generated the coverage report. The report shows the 99% of true in test cases of the MTL logic. Those test cases are covering decision 100%, Condition 100% and MC/DC 98%. Those test cases coverage report

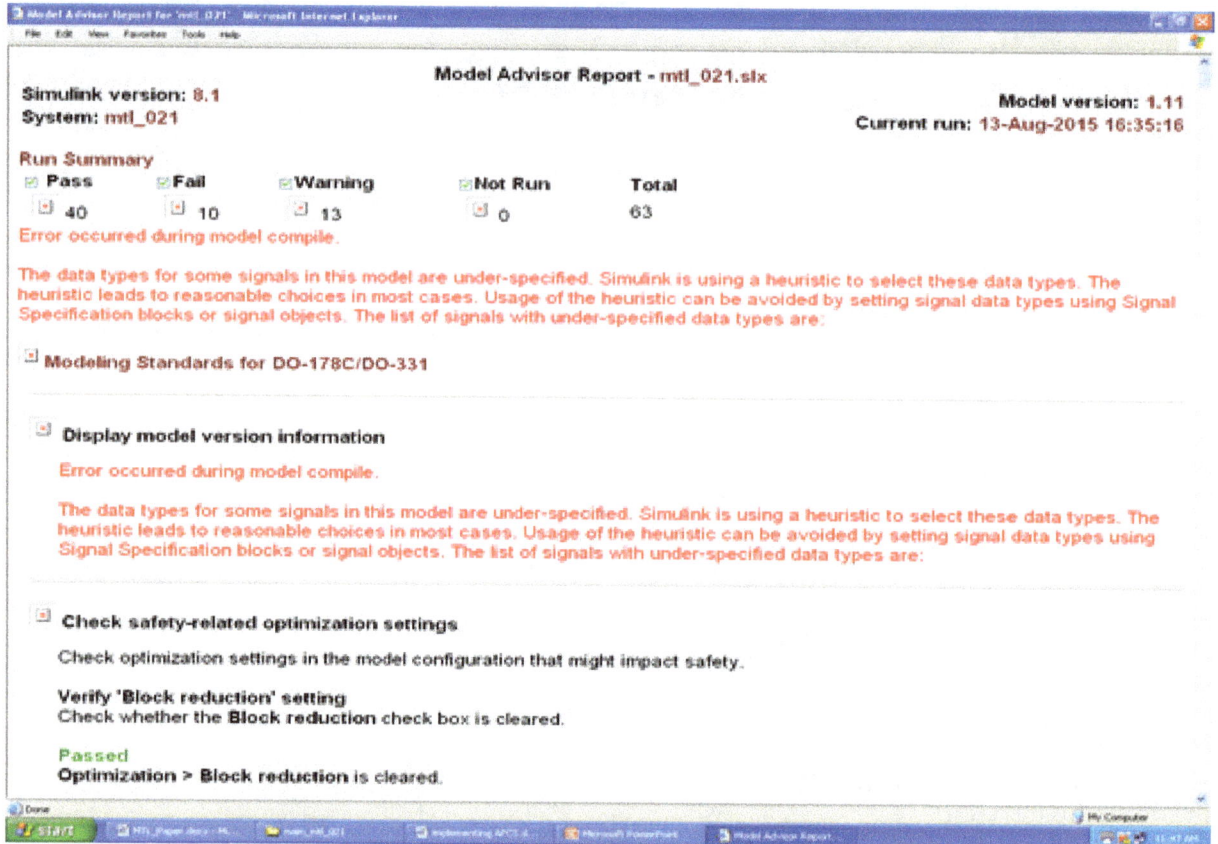

Figure15: Model advisor report.

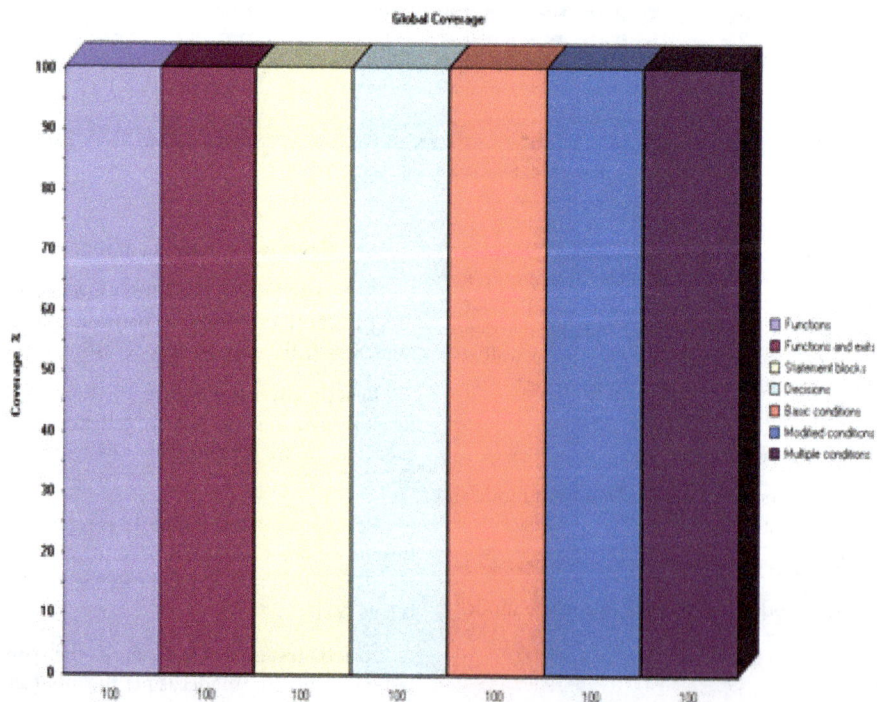

Figure 16: Global coverage of MTL code.

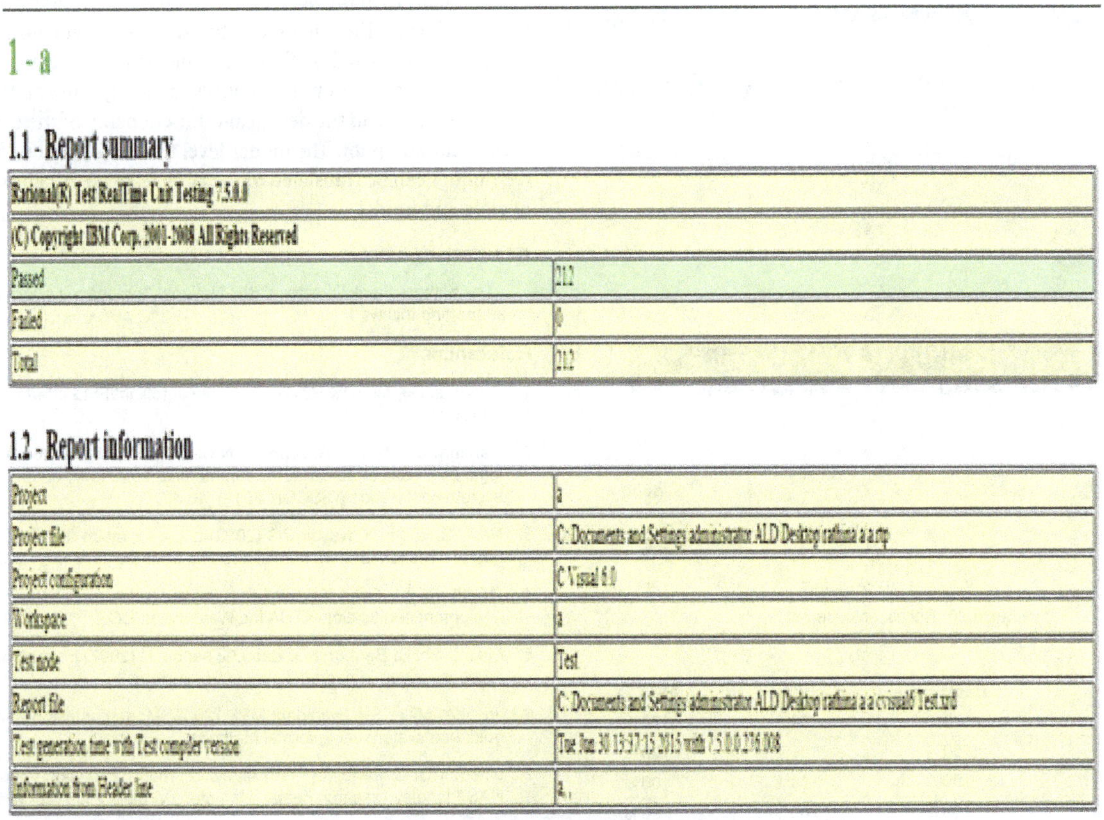

1 - a

1.1 - Report summary

Rational(R) Test RealTime Unit Testing 7.5.0.0	
(C) Copyright IBM Corp. 2001-2008 All Rights Reserved	
Passed	212
Failed	0
Total	212

1.2 - Report information

Project	a
Project file	C:\Documents and Settings\administrator\ALD\Desktop\rathina\a\a.rtp
Project configuration	C Visual 6.0
Workspace	a
Test node	Test
Report file	C:\Documents and Settings\administrator\ALD\Desktop\rathina\a\a cvisual6\Test.xrd
Test generation time with Test compiler version	Tue Jun 30 13:57:15 2015 with 7.5.0.0.276.008
Information from Header line	a,,

Figure17: Report summary of MTL code.

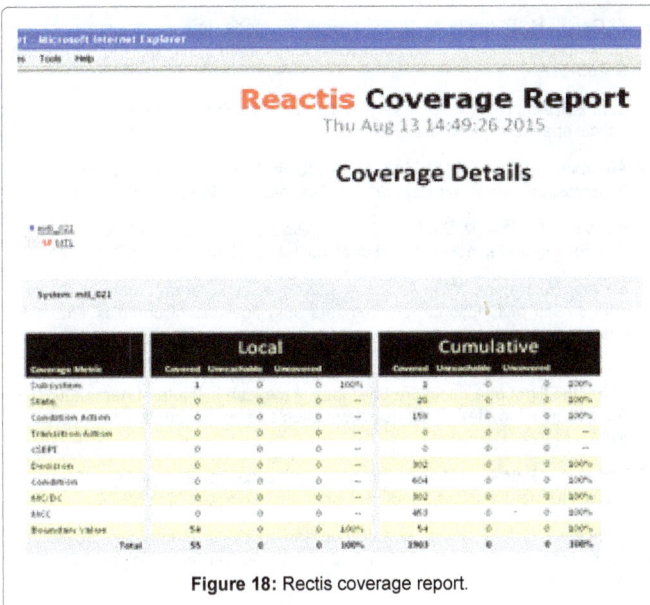

Figure 18: Rectis coverage report.

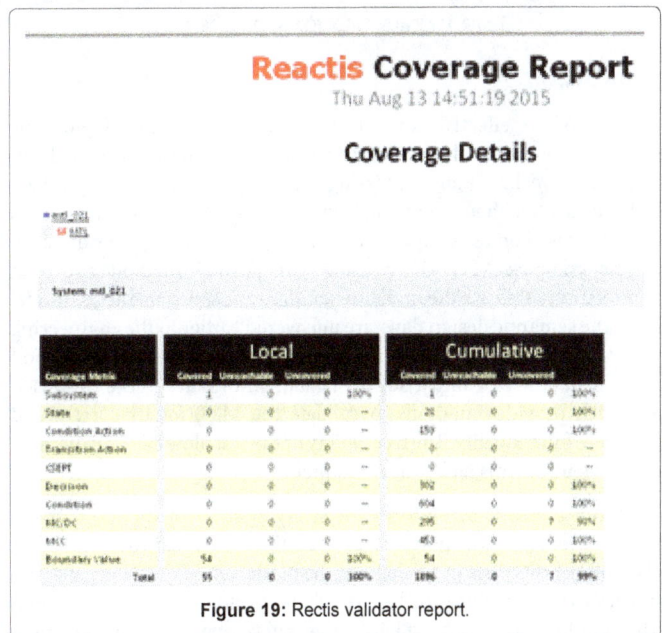

Figure 19: Rectis validator report.

is available in rectis .rst file. Figure 20 shows only the main coverage report of Reactis tester [21].

Result Analysis

Table 1 consolidates the MTL validation result carried out at model and code level. The MTL logic is implemented using stateflow and autocode is generated for the stateflow. The correct and complete implementation is validated using various complementing techniques. At the model level, SlDV and Reactis are used to validate the model as per the requirements. At the code level, RTRT is used to validate the code. The report generated can be used as artifact for the adherence to RTCA DO-178C certification of complex logic.

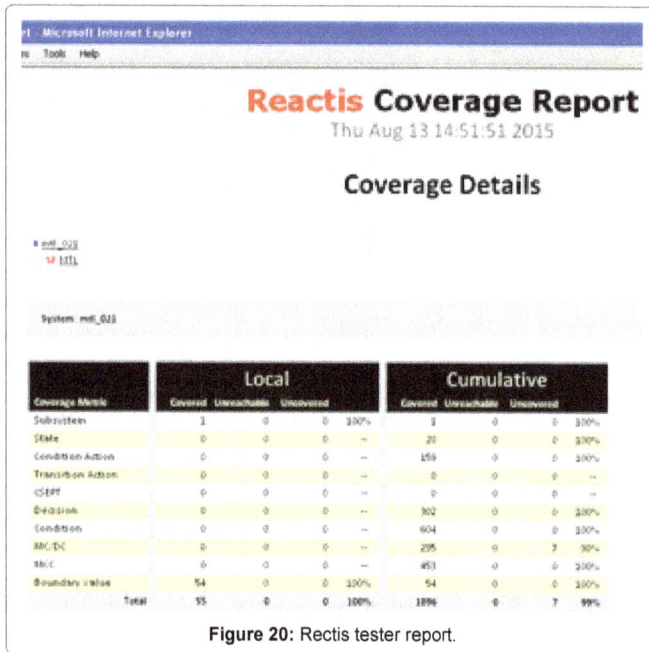

Figure 20: Rectis tester report.

Environment	Decision	Condition	MC/DC
Matlab	100%	100%	100%
(SIDV)	100%	100%	100%
RTRT	100%	100%	100%
Reactis Coverage	100%	100%	100%
Reactis Validator	100%	100%	98%
Reactis Tester	100%	100%	98%

Table 1: Final result analysis comparisons.

Conclusion

One of the effective ways to implement the complex logic is by means of formal methods. Semi-formal method based tools help in visualizing, realizing and validating the complex logic. This approach is more effective than the manual approach where the individual needs to verify based on ones experience. The implementation and validation of the complex MTL logic using semi-formal method approach demonstrates this. As the implementation was carried out at the model level, the semantic design flaws are uncovered earlier in the engineering process. The understanding of such complex logics is better understood using stateflow in comparison to tabular information. The proposed approach provides methods to validate the MTL for its correct and complete functionality. The traceability of the stateflow can be generated to the requirements and code if required.

Work ahead is to perform the safety analysis of the MTL at the model level. This will not only provide the functional robustness but also safety properties of the logic at the model level. The Safety analysis integration will provide the failure behavior information of the system. The modified approach will integrate safety analysis to a integrated model-based formal analysis of complex software designs.

The Simulink stateflow is an effective way to model the complex logic of aircraft mode-transition-logic. In this paper the transitions from one state to another using formal method has been described. All the possible mode transitions in the presence of external event and condition(s) are presented in the stateflow chart, which is easy to understand, analyze, debug and generate source code.

The analysis of the MTL for its functionality & safety is performed at the model level. The outcome of the approach is encouraging approach to adopt semi formal methods in other safety critical application. The proposed approach not only reduces the design effort but also provides higher assurance in the design and functionality of the complex system such as the autopilot. The model-level test cases generated using formal techniques can be translated to the code test cases to ensure traceability of code and model.

Acknowledgment

The authors thank Director, CSIR-National Aerospace Laboratories Bangalore for supporting this work.

References

1. Boorman DJ, Mumaw RJ (2004) A new autoflight/FMS interface guiding design principles.

2. Randhawa P, Mishra A, Jeppu Y, Nayak CG, Murthy N (2012) Mode transition logic of a vertical autopilot for commercial aircrafts IX control instrumentation system conference (CISCON-2012) :16-17.

3. Nair AS, Jeppu Y, Nayak CG (2013) Logic for mode transition of autopilots in lateral direction for commercial aircrafts bonfring. Int J Man Machine Interface.

4. (1992) RTCA DO-178B Software considerations in airborne systems and equipment certification RTCA Inc Washington DC.

5. Meenakshi B, Barman KD, Babu G, sehgal K (2007) Formal safety analysis of mode transition in aircraft flight control system .

6. Joshi A, Miller SP, Heimdahl MPE (2003) Mode confusion analysis of a flight guidance system using formal Methods.

7. Littegen G, Carreno V (1999) Analyzing mode confusion via model checking, NASA langley research center, Hampton, Virginia 23681-2199,USA.

8. Degani A, Heyamann M (2000) Pilot-autopilot interaction: A formal perspective. 8th International Conference on Human – Computer Interaction in Aeronautics :1-11.

9. El-Gendy H, El-Kadhi N (2005) Formal methods: Important experience and comparative analysis. J Computational Methods In Sciences And Engineering 5: 235-247.

10. Lutz RR, Ampo Y (1996) Using formal methods for requirements analysis of critical spacecraft software IEEE 1-6.

11. Ait Ameur Y, Boniol F, Wields V (2007) Towards a wider use of formal methods for aerospace system design and verification. Int J Tools transfer 12:1-7.

12. Kuhn DR, Craigen D, Saaltink M (2003) Practical application of formal methods in modeling and simulation. National Institute of Standards and Technology.

13. Kececi N, Halang WA, Abran A (2002) A semi-formal method to verify correctness of functional requirements specifications of complex embedded system.

14. Gavrilets V, Martinos I, Mettler B, Feron E (2002) Control logic for automated aerobatic flight of a miniature helicopter. American Institute of Aeronautics and Astronautics Inc AIAA Guidance Navigation and control conference and exhibit AIAA.

15. Yadav A, Gaur P (2014) AI-based adaptive control and design of autopilot system for nonlinear UAV. Indian Academy of Sciences 39: 765-783.

16. Aldrich W (2002) Using model coverage analysis to improve the controls development process. AIAA Modeling and Simulation Technologies Conference and Exhibit.

17. Balachandran S, Atkins EM (2015) Flight safety assessment and management during take-off. AIAA.

18. Busser RD, Blacburn MR, Nauman AM Automated model analysis and test generation for flight guidance mode logic. IEEE Xplore Digital Library.

19. Ming Z, Yung L, Yi Q, Xiongwen H, Xiaochuan Z et al. (2012) Model Based Design UAV Autopilot software. Proceeding of the 2012 2nd international conference on Computer and Information Application (ICCIA 2012) published by Atlantis press Paris France.

20. (2015) Mathworks Inc. Simulink/Stateflow.

21. (2015) Mathworks Inc. Model Advisor.

22. Reactis System Inc. Reactis validation tester and analyzer.

23. Cofer D, Whalen M, Miller S (2008) Software model checking for avionics systems.

24. (2015) Mathworks Inc. Simulink Design Verifier.

25. Book: Rational IBM Rational Test Real Time RTRT-User-Guide for version 7.0 G|11-6755-00.

6

Development and Technological Characterization of Multi-functional Aeronautical Coating From Lab-Scale to the Relevant Environment

Mazzola L[1]*, Bruno G[1], Galasso B[1], Quaranta V[1], Albano F[1], Auletta A[1] and Coriand L[2]

[1]CIRA - Italian Aerospace Research Centre, Via Maiorise 1, 81043 Capua, Italy
[2]Fraunhofer Institute for Applied Optics and Precision Engineering, Albert-Einstein-Str. 7, 07745 Jena, Germany

Abstract

Ice adhesion on critical aircraft surfaces is a serious potential hazard that runs the risk of causing accidents. Frozen contaminants cause rough and uneven surfaces which will disturb smooth air flow and greatly degrade the ability of the wing to generate lift and increasing drag.

Amongst icing mitigation systems, passive anti-icing coatings represent a challenge to reduce the ice nucleation and growth, reducing the power consumption of the active de-icing systems and consequently the fuel consumption.

In this work the advanced properties and effectiveness of the new multifunctional coating with ice-phobic and aesthetical properties are described. In particular advanced morphological characterizations based on Atomic Force Microscopy and Laser Scanning Microscopy measurements as well as subsequent Power Spectral Density analysis were performed to evaluate the surface roughness.

Contact angle measurements were executed in order to determine the wettability and surface free energy as well as work of adhesion in flight conditions. In addition, dynamic analysis of the impact of single water droplets on the new multifunctional coating and the classical livery coating were performed in order to demonstrate the different physical behavior during the impingement.

It was also demonstrated that the new multifunctional coating overcome the environmental test similarly to the commercial livery coating in accordance with the aeronautical specification.

Finally, two NACA symmetric airfoils were design and developed using 3D printing technology. The surfaces were coated with a commercial coating in one case and with the new multifunctional coating in the other case. Both airfoils were tested in the Icing Wind Tunnel at different conditions in order to evaluate the effectiveness, in terms of reduction of accreted ice, of the new multifunctional coating respect to the commercial one. Tests demonstrated the reduction of accreted ice of 50% using the new multifunctional coating.

Keywords: Icephobic coating; 3D printing; Icing wind tunnel; Unmanned Air Vehicle (UAV); Roughness; PSD function; FEM analysis; Environmental tests

Introduction

De-icing on the ground is usually done by spraying aircraft with a de-icing fluid based on ethylene glycol, propylene glycol, diethylene glycol as well as urea and acetates which are toxic or has negative effects for aquatic life [1]. The United States Environmental Protection Agency had estimated that before 1990, U.S. airports had discharged approximately 28 million gallons, 50% glycol solutions, of aircraft deicing fluids annually to receiving waters. As of 2000, the discharges have been reduced to 21 million gallons of de-icing and anti-icing fluids (50% concentration) per year to receiving waters with an additional 2 million gallons discharged to publically owned treatment works [2]. The used de-icing fluid during daily operations at the airport can contribute as a source of air, water and soil pollution. Additionally, these operations can adversely affect the climatic conditions, plants, buildings and animals [3,4].

An inflight ice protection system on existing aircrafts heat the vulnerable regions using hot air bleed from the engine compressor [4] and/ or electrothermal solutions. However, the air-bleed reduces the efficiency of the engines and the pipe network for the hot air bleed adds considerable weight and maintenance requirements. Another established approach is to use pneumatic bladders in vulnerable regions which can be inflated to detach ice from these critical locations. There is then a risk of impacting other parts of the aircraft structure or being ingested by the powerplants, damaging fan blades. Due to weight and maintenance

implications, anti-icing and de-icing systems are restricted to specific, albeit critical, regions and so ice can still accumulate at other sections of the lifting surface [5]. The current anti-icing and de-icing system will build up weight, increase fuel consumption and add complexity to the aircraft systems. A Passive Ice Protection System (PIPS) and mixed passive / active methods could significantly reduce aircraft fuel consumption. The aim is to improve the power efficiency and operational effectiveness of the current anti-icing or de-icing system in an environmentally friendly way [6-12].

Studies on passive anti-icing systems such as icephobic coating are actually ongoing because even if, in literature [13-27], there are several formulations with superhydrophobic and icephobic properties, they are not really applicable in aeronautics. Mainly this is due to the following criticisms:

---

*Corresponding author: Mazzola L, CIRA - Italian Aerospace Research Centre, Via Maiorise 1, 81043 Capua, Italy, E-mail: l.mazzola@cira.it

- The new icephobic formulations must be applied on top of the component. Therefore in case of aircraft components, it will be applied over the livery coating altering the aesthetical aspect.

- Difficulties of the production at the industrial scale and following insertion in the actual production chain.

- They must be cost effective.

- They are eco-friendly according to the REACh regulations and VOC free.

- They must overcome all aeronautical specifications (Airbus, Boeing, Bombardier, Embraer, etc...) and the relative standard tests.

For this reason the development of a multifunctional coating with combined aesthetical and icephobic properties represents a product really challenge.

In the previous work Mazzola [28] designed and realized a new polymeric formulation for aerospace applications. In fact the efficacy of the icephobic properties at lab scale (having also aesthetical properties) was demonstrated as well as a complete physical, chemical morphological and mechanical characterizations were performed.

This work describes the advances in the development of the new formulation of this multifunctional coating. In particular, the surface structure at micro and nanoscale was investigated through atomic force microscopy as well as laser scanning microscopy and compared with SEM micrographs. For a complete understanding of the surface roughness and its influence on the wetting and icing behavior respectively, a semi-empirical approach consisting of Power Spectral Density analysis procedure and a subsequently derived universal structural parameter were utilized. This so-called wetting parameter is directly connected to the contact angle and, hence, enables an estimation of the wetting behavior of a specific surface structure. Or, as for this research, the other way around: If different wetting behavior or specific ones occur, then the question about the cause arises. This mainly concerns the roughness of a surface that, besides chemical composition, essentially governs its wetting properties [29,30].

Environmental test was performed in order to evaluate the durability of both coatings in harsh conditions (high percentage of relative humiditiy for a long time).

In addition dynamic tests of water droplet impingement at lab scale were performed in order to study the impact of small supercooled water droplets on the new icephobic surface.

All of these experiments were useful to understand the properties and the real behavior of the new multifunctional icephobic coating. Using all of these experimental data (i.e. the surface free energy and wettability of the new coating), a FEM numerical simulation was performed in order to study the reduction of adhesion between ice and the new multifunctional icephobic coating.

Starting from TRL 2 in the previous work [28], the advances on the characterization of the new multifunctional coating at lab scale, along with the availability of new 3D printing technology and Icing Wind Tunnel facility, allowed to reach to this new multifunctional coating, TRL 5 (relevant environment).

Materials and Methods

Materials

The new ice-phobic coating was obtained starting from the commercial coating used as livery, i.e. a matt grey livery coating. Once that the substrate was scraped with P400 sandpaper and washed, an epoxy-modified polyamide primer using VOC (Volatile Organic Compounds) exempt solvents (solvent based High Solid coating) was applied in order to improve the adhesion of the topcoat, inhibit the corrosion and level the surface.

Above the primer, a water-based 3-component, isocyanate cured polyurethane topcoat (in accordance to Registration, Evaluation, Authorization and Restriction of Chemicals- REACH regulation). The formulation of the topcoat was modified in order to give further functionality, such as ice-phobicity without alter the aesthetical properties and the other mechanical and corrosive properties. Both coatings were deposited through spray process on substrate of CFRP composites (Carbon Fiber Reinforced Polymer) because this type of application is addressed mainly for Unmanned Air Vehicle (UAV) that is usually realized in composite materials.

The temperature of the substrate was 20°C during the spray process. The dimension of the spray nozzle was 1.2 mm, the temperature of the air carrier was 20°C and the air-pressure was 2 bar. The drying of the coatings was realized using a drying chamber at temperature of 120°C.

Physical and chemical characterization

In order to replicate the same thermodynamic conditions of flight, it is necessary to change both temperature and pressure. As known the surface free energy may be defined as the increase in Gibbs free energy of the whole system per unit increase in interfacial area, carried out under conditions of constant temperature and pressure that is [31]:

$$\gamma = \left(\frac{\Delta G}{\Delta A} \right)_{T,p} \tag{1}$$

Contact angle changes in case the component works at ambient pressure and temperature, at flight temperature and ambient pressure, or at temperature and pressure in flight. Aiming at a replication of the flight conditions, a test room, to mount on a classical contact angle measurement instrument, was designed and developed at CIRA [28]. In this work, a new version of the test room is described. In fact the main limitation of the first version of the test room was mainly the supplying and use of pellets of dry ice (Carbon Dioxide). To overcome this limitation the new version of the test room uses a Ranque-Hilsch Vortex Tube – RHVT powered by the line of compressed air. In this way, utilizing this system, it is possible to reach temperature out of the cold side of the RHVT temperature of -70°C.

The new test room was realized in insulator material (in particular top, bottom and two lateral and parallel surfaces), such as polycarbonate with a thickness of 2 mm, whereas the other two lateral and parallel surfaces were realized in aluminum Al2024-T6 with a thickness of 1 mm.

Polycarbonate was chosen because, other than thermal insulating, it is also transparent. This property is needed to capture the liquid drops applied on the surface with the camera of the contact angle instrument. The two lateral and parallel surfaces were realized in Aluminum where one was isolated using Polystyrene whereas the other was used for the heat exchange.

On the external surface of this last Aluminum surface was a cooling chamber of polystyrene mounted, on which the cold side of RHVT is connected. In this way the cold air exits from the RHVT and it come into contact with the Aluminum surface for the heat exchange. The holes present on the sides of the cooling chamber were used to let out the cold air produced by RHVT.

Using this method it is possible to reach temperature of -50°C in the test room. The temperature within the test room was detected using a thermocouple K-type.

The top surface of the test room has a hole with an elastic membrane in order to allow the insertion of the syringe needle to deposit the liquid drop. The syringe employed was Hamilton 600 series – 5 μl.

Supercooled water droplets were realized using a microliter syringe of Hamilton. Once reached the temperature and pressure in the test room, a small quantity of bi-distilled water was sucked up within the capillary of the needle. Afterwards, the needle was inserted through the elastic membrane of the test room and leaved within the room for 30 seconds, in order to reduce the temperature of the water from ambient temperature to near zero Celsius degree. After this time, the water droplet was deposited very slowly on the surface of the sample presents on the bottom of the test room. During the deposition, the water droplet had a further drastically reduction of temperature; Before reaching the surface of the sample, The droplet becomes supercooled. This supercooled property was simulated with a model of the heat exchange surface and phase shift of the water considering the main characteristics of the water droplet. In order to have the same flight altitude pressure, a circuit with a Venturi tube model (ZH-05-DS-06-06-06) was realized [28].

With this new tool it is possible to replicate the pressure and temperature until a flight altitude of about 16000 meters. According to the standard certification (FAR CS-25 – appendix C), the highest probability to have icing phenomena is 5000 meters, where the temperature is about -12/-15°C and pressure is about 0.5 bar. Note that the test room could be mounted on a micrometric sample-holder, which can be moved on three axes and in addition it is possible to tilt the entire test room in order to determine the advancing and receding contact angle as well as the roll-off angle.

Bi-distilled water, methylene iodide and formamide were used to determine surface free energy and its components, together with the adhesion work and other performance indexes, i.e. advancing and receding contact angle, hysteresis, roll-off angle. During the test, 10 drops (with volume smaller than 3 μl) of each liquid were deposited on the sample surface. The surface free energy was calculated according to the Owens-Wendt method [32]. The Owens-Wendt approach is one of the most commonly used methods for calculating the surface free energy of the materials [33]. The principal assumption of the Owens-Wendt method is that the surface free energy is the sum of the two components: dispersion and polar components [34].

The Owens-Wendt model is represented by the geometric mean relationship:

$$\frac{1}{2}(1+\cos\theta)\gamma_L = \left(\gamma_S^D \cdot \gamma_1^D\right)^{\frac{1}{2}} + \left(\gamma_S^P \cdot \gamma_1^P\right)^{\frac{1}{2}} \tag{2}$$

Where θ is the contact angle between the liquid droplet and surface, γ_L is the liquid total surface tension, γ_1^P is the dispersion component of liquid surface tension and γ_1^P is the polar component of liquid surface tension.

The unknown terms in the equation (2) are γ_S^P which is the dispersion component of the solid surface free energy and γ_S^P which is the polar component of the solid surface free energy.

The value of total surface free energy of the solid is obtained using the following equation:

$$\gamma_S = \gamma_S^D + \gamma_S^P \tag{3}$$

The surface tension of liquids and its components are taken from literature [35]. However these values are referred to standard conditions of ambient temperature and pressure at sea level.

Using the Karbanda's equation, it is possible to rescale the values of surface tension from temperature point of view:

$$\gamma_{T_2} = \gamma_{T_1}\left(\frac{T_C - T_2}{T_C - T_1}\right)^{1.12} \tag{4}$$

where γ_{T_1} represents the surface tension of liquid at 20°C while T_C represents the critical temperature of the liquid. The critical temperatures of the three liquids are reported in Table 1:

Once rescaled the values from temperature point of view, it is necessary to rescale these new value for the new pressure, using the Laplace equation:

$$\gamma_{p_2} = \gamma_{p_1}\left(1 - \frac{k.p_2}{200}\right) \tag{5}$$

where K is a constant value which is a function of employed liquids. For the liquids employed in these experiments K equal to 2 was used. The calculation of all performance indexes of the instrument were determined using a software, designed and developed at CIRA. This software has all main models present in literature other than Owens-Wendt.

It is evident that, in the real condition, the behavior and dynamics of the interaction between the supercooled liquid dropled and surface are more complex, because the supercooled water droplets impact with high speed on the aircraft components. The dynamics of interaction, happening in a few milliseconds, are in non-equilibrium condition and there are several phenomena that can occur, i.e. infiltration due to the water hammer pressure, instantaneous freezing or rapid rolling of the droplet.

In any case, the characterization technique developed in this work allows studying the ice-phobic surface in static conditions. This represents the worst condition since the supercooled water droplet has a long time to reach the equilibrium state and to freeze on the surface.

To evaluate the behavior of the liquid droplet during the impact on both commercial coating and on the new multifunctional coating, a piezoelectric system for microdroplets dispensing, Microdrop MD-K-130, was used. The instrument is equipped with a high speed camera in order to evaluate the instants during the impact of the single small droplets. It is highlighted that this system produces microdroplets with a diameter between 20 μm and 150 μm. These values are in accordance with the dimension of microdroplets produced by the spray bar system of the Icing Wind Tunnel facility. Tests were performed analyzing the impact of water microdroplets with a diameter of 70 μm that imping the surfaces tilted of 40° respect to the horizontal line.

FEM Analysis to Evaluate the Adhesion at the Coating-Ice Interface

The aim of FEM analysis was focused to evaluate the maximum value of force that cause the debonding at the coating-ice interface both for the new multifunctional coating and for the commercial one.

Critical Temperature (°C)		
Water	Formamide	Diiodomethane
373.94°C	376.45°C	474.42°C

Table 1: Values of the critical temperature of the three liquids employed.

Virtual Crack Closure Technique analysis (VCCT) was chosen as solver and the software Abaqus was used. In this analysis the experimental values of the work of adhesion (measured previously through contact angle measurement technique) were assigned to coating-ice interface and a displacement was imposed. A pre-existing crack (or debonding) was explicitly included in the finite element mesh. During the simulation, the imposed displacement generates an external work that becomes internal energy released by extending the crack (or debonding). The assumption of the method is that the required work to close or to open the crack over the length Δa (propagation variation of the crack as a result of the increment of applied energy) is equal (Figure 1).

The model is composed by:

1. A composite plate with layup [45, -45, 0, 90]$_{3s}$, 24 plies of 0.186 mm of thickness and dimensions of 100 mm × 100 mm;

2. An ice block of 50 mm × 25 mm × 3 mm.

3. A cohesive zone at the interface with a maximum value of work of adhesion which simulate surface property of the coating.

According to this method, the first line of nodes, at the interface, are not linked in order to create a preexisting crack. Thanks to the imposed displacement, the crack proceeds its propagation until to cover the entire zone of interface. The displacement is constant and tangential to the plate, as shown by yellow arrows in Figure 2a. The bottom of the composite plate is clamped and between ice and plate there are cohesive element for adhesion and a friction coefficient of 0.05.

In the Figure 2b, FEM model is reported with a mesh size of 2 mm. As described above, the initial crack has a depth of 2 mm for all width of the ice block (X-direction).

Environmental Test

Environmental tests in the climatic chamber were performed on the samples with commercial and new multifunctional coatings in order to evaluate the durability. Tests were performed in accordance with the standard AER(EP)M-P-001 and ASTM D2247-15. These standards define the combined environmental conditions of the temperature and humidity and the test duration. The ASTM is a general standard, otherwise the AER(EP)M-P-001 is a standard for aeronautical coating, so the main environmental conditions were taken from this last one.

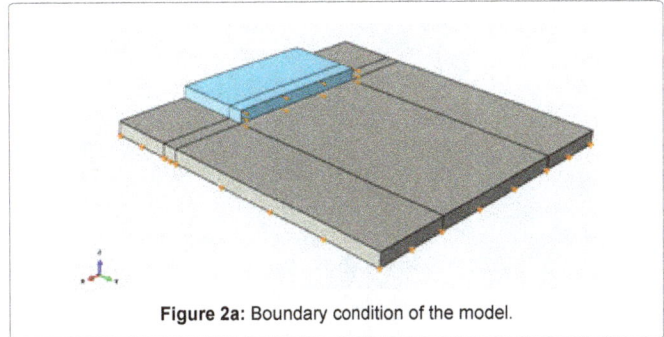

Figure 2a: Boundary condition of the model.

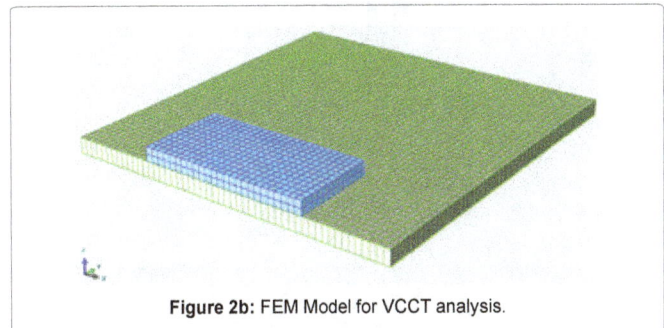

Figure 2b: FEM Model for VCCT analysis.

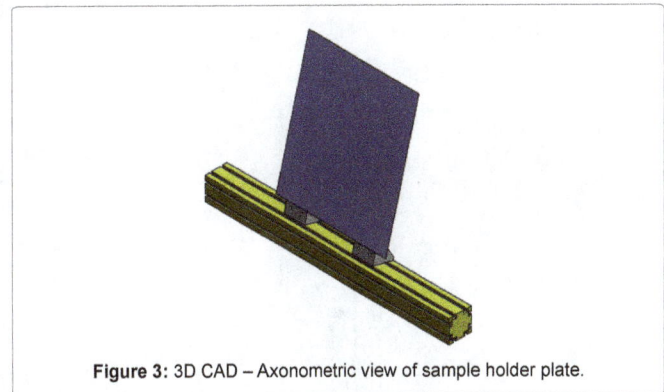

Figure 3: 3D CAD – Axonometric view of sample holder plate.

Tests were performed using the Angelantoni climatic chamber (mod. CH2000). To overcome this test, the coating shouldn't have bubbles, blistering, delaminations and color changes.

For each type of coating, three samples of 150 mm × 80 mm were placed on two sample holder plates tilted of 15° respect to the vertical in accordance with the standard (Figure 3). The process parameters were 38°C of temperature and 100% of relative humidity, test was performed for one month (30 days).

Each week, the test was interrupted for a few minutes in order to oversee and to take a picture of the samples in order to verify if the sample surfaces were degraded, in accordance with the ASTM standard.

The trends of temperature and relative humidity are reported in Figures 4a and 4b.

After the environmental test, several tests on the samples have been carried out with the aim to verify if the environmental condition modified the surface properties of the coating. The first test was the visual inspection of the samples to compare them pre and post-test in order to evaluate blistering or color changes.

The second characterization was performed using optical

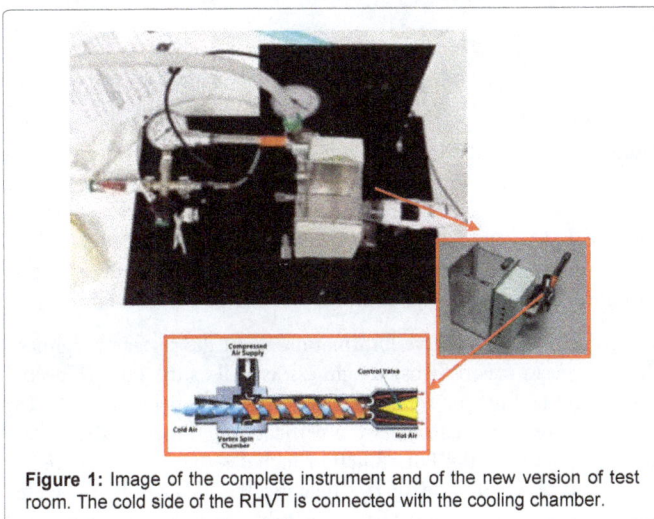

Figure 1: Image of the complete instrument and of the new version of test room. The cold side of the RHVT is connected with the cooling chamber.

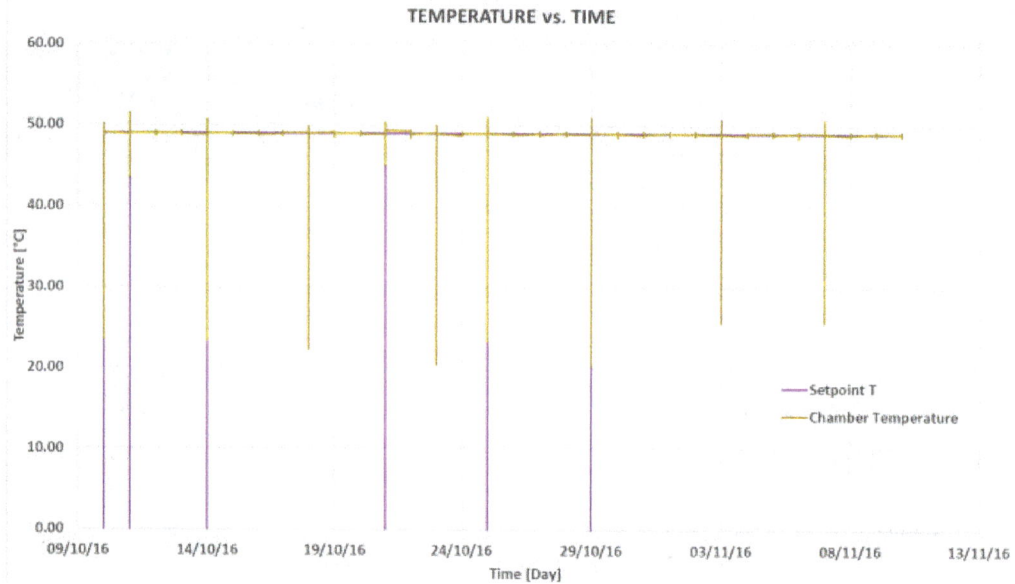

Figure 4a: Test progress of temperature vs. time.

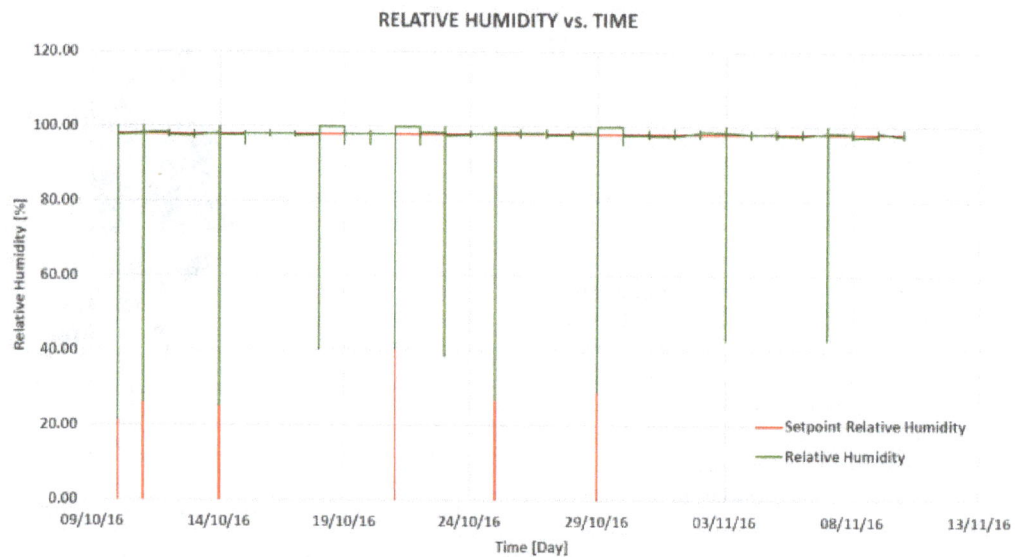

Figure 4b: Test progress of relative humidity vs. time.

microscopy in order to verify uniformity and homogeneity of both coatings and to evaluate in deep blistering or the presence of bubbles. A stereomicroscope Leica MZ12 with camera Leica MC 190 HD was used.

The last test was performed to evaluate the adhesion of the coatings with the substrate. In fact the humidity could infiltrate at the interface between coating and substrate, reducing the adhesion; qualitative adhesion tests were performed in accordance with ASTM D 3359 method B; in particular a cutter PA 2056 (6 teeth – 2.0 mm) and subsequently an adhesive tape PERMACEL #99 were used.

Roughness Analysis

For a robust investigation of the surface morphology and its roughness, atomic force microscopy (AFM) and laser scanning microscopy (LSM) measurements have been performed using a Dimension 3100 AFM from Bruker in the Tapping Mode™ and a LSM 510 from Carl Zeiss Jena GmbH. The vertical resolution of the AFM is limited by the instrumental noise to rms values as low as about 0.04 nm. For the LSM examinations the vertical resolution depends on the objectives (e.g. 100x: rms < 100 nm). For a maximum depth resolution a minimum diameter of the confocal pinhole (10 µm) is used.

To receive quantitative information about the surface roughness, the root mean square (rms) roughness as well as the Power Spectral Density (PSD) function were calculated from the topography data. The rms roughness only considers the vertical dimensions of the surface structure, whereas the PSD function includes the vertical as well as the lateral distribution of surface heights. The PSD can be calculated from surface topography data z(x,y) within a certain scan range L and afterwards simplified to a 2D-isotropic PSD in case of isotropic surface roughness [29-38]:

$$PSD\left(f_x, f_y\right) = \lim_{x \to \infty} \frac{1}{L^2} \left|FT\left\{Z\left(x, y\right)\right\}\right|^2 \qquad (6)$$

Using the PSD functions, it is possible to combine the information resulting from the AFM measurements for the higher spatial frequency range and from the LSM measurements for the lower spatial frequency range to analyze the surface roughness within a wide spatial frequency range (0.01 μm^{-1} ≤ f ≤ 1000 μm^{-1}).

Furthermore, a roughness based structural parameter can be derived from the PSD by means of data reduction procedure. This so-called wetting parameter k_B enables a direct link between the roughness characteristics and the wetting behavior of real surfaces. Thus, it opens up the possibility to separate the influence of roughness properties and chemical material properties on the wettability. For example, previous studies at Fraunhofer IOF [36-38] showed that a κ_B value of at least 0.3 is necessary to reach an extreme wetting behavior like superhydrophobicity. Further information about κ_B as well as an overview of the algorithm for the calculation from PSD data developed at the Fraunhofer IOF are given in [29,30,36].

Design and Production of NACA Airfoils and Deposition of Coatings

In order to evaluate the efficacy of the new multifunctional coating with aesthetical and icephobic properties Icing Wind Tunnel tests were performed. In fact the icephobic and superhydrophobic properties at lab scales in static and dynamic conditions were demonstrated. The following step was to scale up the new formulation of the icephobic coating in relevant environment (reaching Technology Readiness Level 5 – TRL 5).

In order to reach this goal a symmetric NACA 0015 airfoil was design through CAD design using CATIA software. The advantages of utilizing this type of airfoil are mainly due to the absence of the lift (because it is symmetric) and it is enough thick as frontal section area, in order to head-off the supercooled water droplet during the Icing Wind Tunnel tests. The airfoil was designed with respect to achieve a chord length of 100 mm, a wingspan of 150 mm and the height max 15 mm.

Figures 5a and 5b show the CADs of the optimized airfoil and the realized NACA airfoils printed by the 3D printer.

The CAD of NACA airfoil was replicated three times and printed in ABS (Acrylonitrile butadiene styrene) using 3D printer Stratasys Dimension SST-1200ES (Figure 5b).

The three symmetric NACA airfoils were also coated with the commercial coating and the new multifunctional coating. On the third airfoil was applied only the primer to demonstrate the levelling effect of this layer respect to the original roughness of the airfoil (Figure 6a).

Quality control tests were performed in order to confirm the differences in term of contact angles between the two airfoils. As reported in Figure 6b the contact angles were measured at the stagnation point and at the ¼ of chord length. Therefore the two airfoils were ready to be tested in the IWT.

Icing Wind Tunnel Tests

In order to evaluate the efficiency of the icephobic properties of the new multifunctional coating, it was planned to perform comparative tests in an icing facility on both airfoils (equipped with icephobic coating and commercial coating). The aim of the tests was to evaluate the reduction of accreted ice starting from the leading edges without using active de-icing systems.

The tests were performed in the CIRA Icing Wind Tunnel (IWT) facility.

Officially inaugurated in September 2002, the CIRA Icing Wind Tunnel (Figure 7a):

1) It is the largest refrigerated wind tunnel in service;

2) It is the highest speed icing wind tunnel (M=0.7);

3) It is the only facility combining altitude and Temperature simulation;

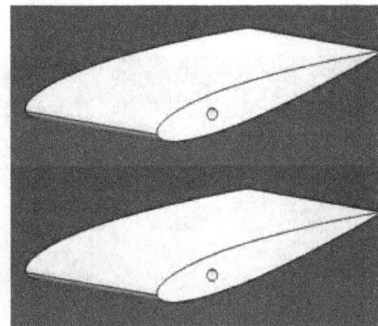

Figure 5a: CAD design of NACA 0015 symmetric air-foil.

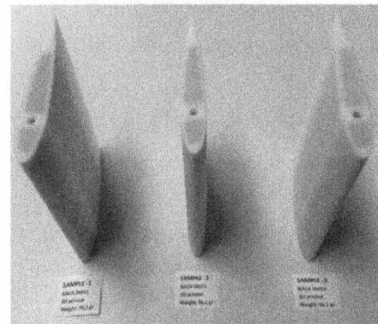

Figure 5b: Three NACA 0015 air-foils printed in ABS using 3D printing technology.

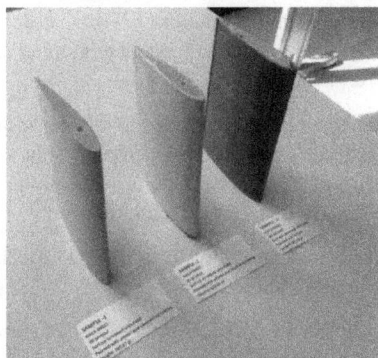

Figure 6a: Three NACA 0015 airfoils coated (from left) with commercial coating, new multifunctional coating and only with primer.

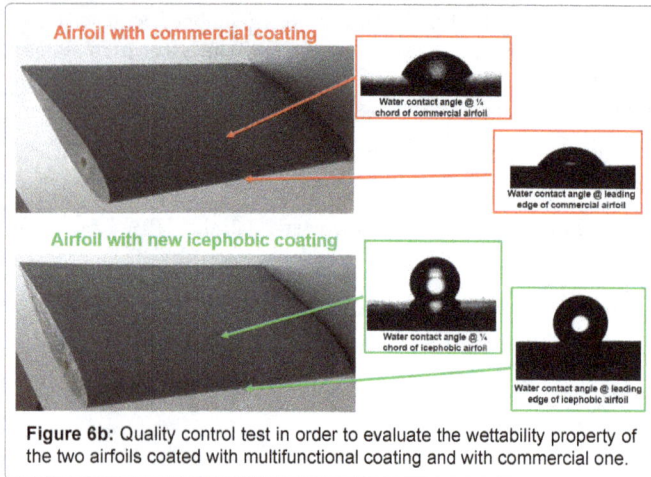

Figure 6b: Quality control test in order to evaluate the wettability property of the two airfoils coated with multifunctional coating and with commercial one.

Figure 7a: CIRA icing wind tunnel.

4) has the largest number of different test sections configurations (4);

5) It has a large number of spray bar nozzles (500) and bars (20);

6) It has the widest operating range for engine flow simulation (1.5-55 kg/s mass flow).

Located in the stilling chamber, the Spray Bar System (SBS) is dedicated to the generation of the icing cloud in all the conditions as prescribed in the current FAR/JAR regulations. Looking at future revisions of the airworthiness regulations, the SBS is also capable to generate Supercooled Large Droplets (SLD) within the range of freezing drizzle conditions.

The extreme values that can be reached in the IWT are the following: temperature simulation down to −40°C, a de-pressurization system for altitude simulation up to 7000 meters (about 23000 feet) and a humidity control down to 70% RH.

The two coated airfoils were installed in the center of the IWT test section as reported in Figure 7b in order to reduce the wall effects, the possible turbulence of the airflow and consequently to have the best distribution of the cloud in terms of cloud uniformity.

Tests were performed according to the FAR-25 appendix C and the test matrix is reported in Table 2.

Where V is the airflow speed, T is the static temperature, h is the altitude, MVD is the water droplet median volumetric diameter, LWC is the Liquid Water Content that measures the concentration of the water in a cloud.

Test 5 was performed in the same tunnel and cloud conditions of Test 1 in order to check the repeatability of the results.

Results

Tests with the new tool were performed both on a classical livery coating and on the new icephobic formulation. In particular ten droplets of each liquid: bidistilled water, di-iodomethane and formamide were deposited on the surfaces in order to determine surface free energy and its components (polar and dispersion components). Tests were performed in standard conditions, i.e. ambient temperature and pressure at sea level, and in simulated flight conditions, i.e. temperature of -12°C and pressure of 0.5 bar. As reported in Figure 8, the great difference between the water contact angle of the commercial coating and the new icephobic formulation is evident.

From Figures 8a and 8b, it is evident that the two samples have the same aesthetical aspect. No differences in terms of color, gloss, brilliance are evident. The two samples seem the same, but, as showed in Figures 8c and 8d, the behavior of the freezed water droplet on the surface is completely different. The commercial coating is hydrophilic, whereas the new formulation has super hydrophobic/icephobic behavior. It is necessary to highlight that this static condition represents the harsh condition in which the coating will work. In fact, as described previously, the coating is under airflow. Therefore the water droplet does not remain on the surface but it rolls away. It is evident that in case of hydrophilic coating, as reported in Figures 8a and 8c, this behavior does not happen; at most it tends to slide away.

In any case, in Figures 8c and 8d the contact angle measurements of the supercooled water droplets applied on both samples are reported. The commercial coating has a water contact angle of about 48°, whereas the new icephobic coating has a water contact angle of about 160°. The improvement of water contact angle was higher than 70%.

Realizing also the contact angle measurements with other two liquids (formamide and diiodomethane), it was possible to determine the surface free energy and other chemical and physical performance indexes.

Figure 7b: The two coated air-foils installed in the centre of the IWT test section.

S. No	V [m/s]	T (°C)	h[ft]	MVD [mm]	LWC [g/m³]	Note
Test 1	95	-25	20000	22	0.23	
Test 2	95	-25	17000	20	0.2	
Test 3	95	-25	5000	19	0.14	repeatability
Test 4	95	-25	sea level	25	0.37	
Test 5	95	-25	20000	22	0.23	

Table 2: ITW process parameters.

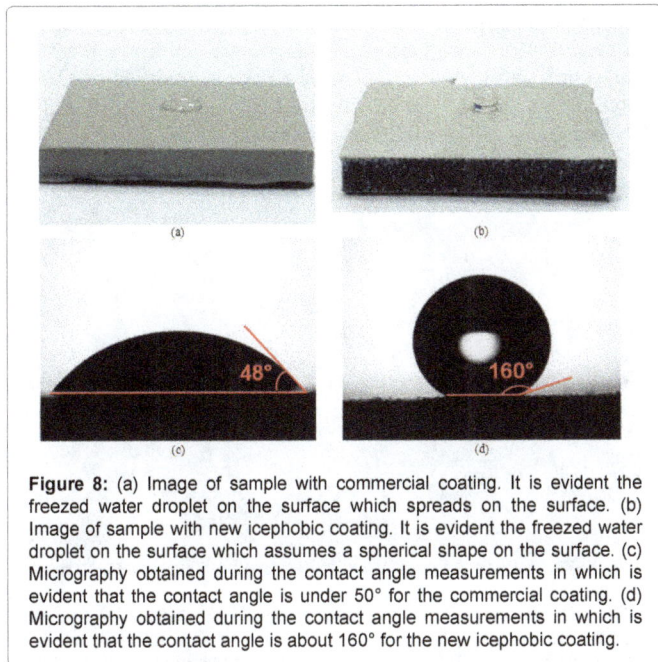

Figure 8: (a) Image of sample with commercial coating. It is evident the freezed water droplet on the surface which spreads on the surface. (b) Image of sample with new icephobic coating. It is evident the freezed water droplet on the surface which assumes a spherical shape on the surface. (c) Micrography obtained during the contact angle measurements in which is evident that the contact angle is under 50° for the commercial coating. (d) Micrography obtained during the contact angle measurements in which is evident that the contact angle is about 160° for the new icephobic coating.

As showed in Figure 9, a drastic reduction of these performance indexes is evident; in fact the surface free energy decrease of the 85%, whereas the dispersion component of the 77% and polar component of 99%. Note that the polar component mainly influences the adhesion between water droplet and the surface, therefore the drastic reduction of this component allows to estimate the reduction of the adhesion of water.

This last statement is also confirmed by the experimental values of work of adhesion. In fact author calculated the work of adhesion of the classical commercial coating and the new icephobic formulation. The reduction of the 92% is evident from Figure 10.

From FEM analysis, a comparative graphic between commercial coating and new multifunctional coating (Figure 11) was obtained. Thanks to the knowledge of the adhesion energies of two coatings, the maximum value of force necessary to debond the ice from the surface coating (in blue the commercial coating and in red new multifunctional coating) was determined. In Figure 11 it is evident that on the surface of the new multifunctional icephobic coating, the required force is about four time lower than the commercial one.

From Figure 12a, it is possible to notice clearly the initial crack in blue, required for the boundary condition of the VCCT method, whereas in Figures 12b and 12c it is evident that two different frames that describe the evolution of the debonding of the ice from the surface coating (in blue the debonding area and in red the adhesive area).

The different behavior of the two coatings is due not only to the different chemical composition of the surface but also to the different morphology of the two coatings. In fact, as showed in Figures 13a-13c at three different magnifications, it is evident that the classical commercial coating shows a smooth surface.

Whereas the new anti-ice formulation shows a surface rougher than the commercial one (Figure 13d). In particular in Figure 13e several micro-features are showed; in addition, improving the magnification on each micro-feature, a nanostructure is evident. A spherical shape roughness is described in Figure 13f.

In addition to the SEM images, Figures 14 and 15 show the topographic images and rms values resulting from AFM and LSM measurements. These results confirm the previous observation: The icephobic sample surface exhibits an obviously higher surface roughness in all investigated measurement fields than the reference

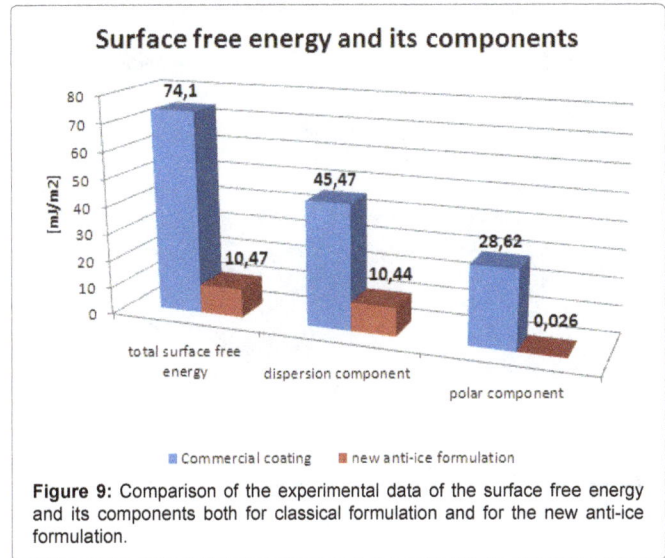

Figure 9: Comparison of the experimental data of the surface free energy and its components both for classical formulation and for the new anti-ice formulation.

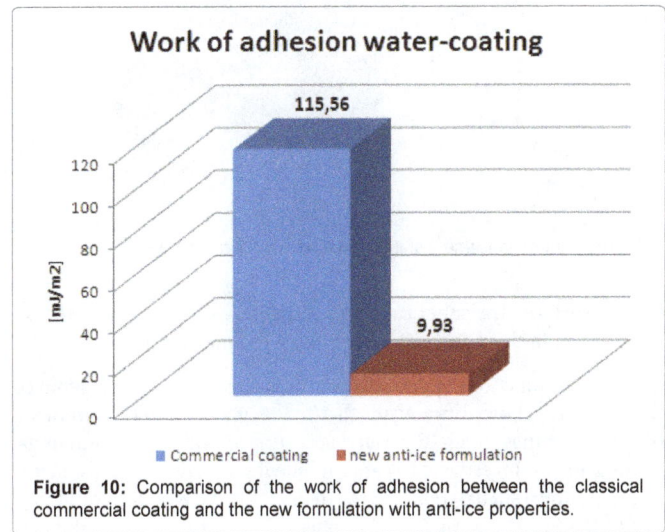

Figure 10: Comparison of the work of adhesion between the classical commercial coating and the new formulation with anti-ice properties.

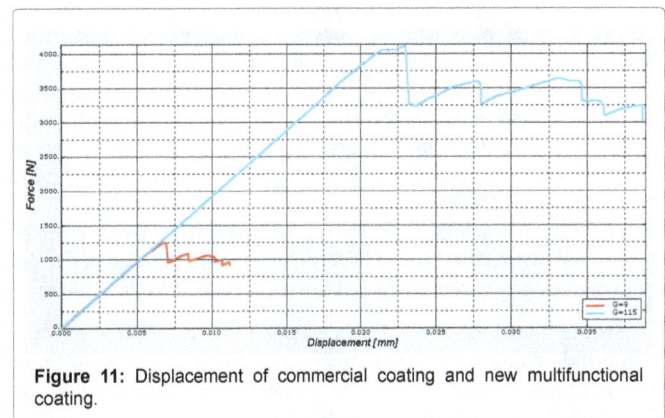

Figure 11: Displacement of commercial coating and new multifunctional coating.

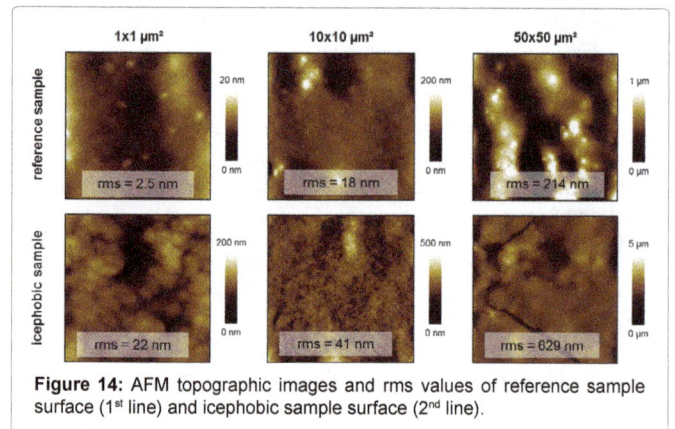

Figure 12a: Detachment before FEM simulation.

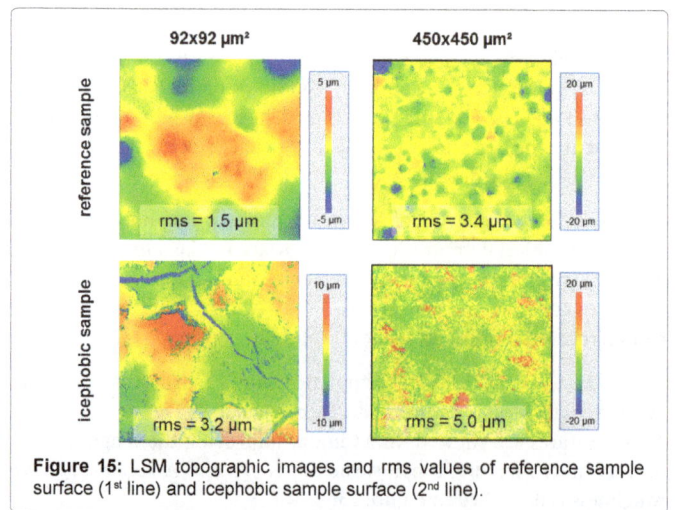

Figure 12b: Advancement of the ice detachment during the FEM simulation.

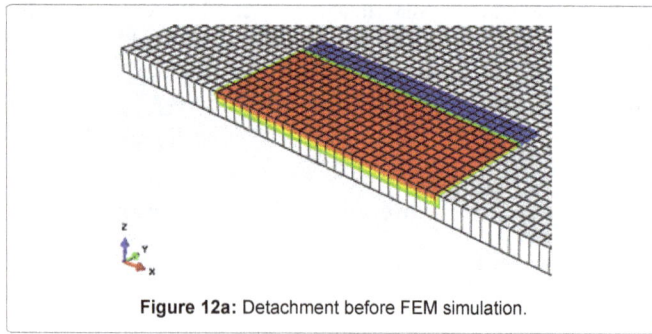

Figure 12c: Advancement of the ice detachment during the FEM simulation.

sample surface. The spherical shape roughness structure (cf. 13f) can also be seen in the small scan area of the AFM measurements.

The PSD functions calculated from the topography data and depicted in Figure 16, shows once again that the icephobic surface exhibits a higher roughness than the reference surface with the commercial coating in the investigated spatial frequency range. Especially in the range around a spatial frequency of 10 μm^{-1}, the PSDs of both functions differ significantly because of the spherical nanoroughness of the new multifunctional coatings.

The wetting parameter K_B was subsequently calculated from the PSDs to 0.14 for the reference sample surface and to 0.30 for the icephobic sample. These values confirm the results of the wetting analysis at the beginning of this section: The commercial coating with a contact angle of about 48° exhibits a surface structure with a K_B value clearly below the necessary criterion for super-hydrophobicity. But for the icephobic one, the necessary criterion for super-hydrophobicity with $\kappa_B = 0.3$ is reached. Thus, the special roughness characteristic of this surface enables the observed hydrophobic properties and hence the improved icing behavior compared to the reference sample surface.

Taking SEM images and PSD analysis into account, it is possible to conclude that a hierarchical structure is obtained, similar to those present in nature such as lotus leaf. This hierarchical structure allows the coating to have stable super hydrophobic properties. In summary,

the observed nanoroughness (spherical shapes) of the icephobic sample surface promotes the entrapment of air or vapor within the micro and nanofeatures and the wetting of such surface is predicted by Cassie-Baxter wetting model. In particular, the droplet touches only the crests of the roughness and air pocket remains entrapped between the droplet and the surface.

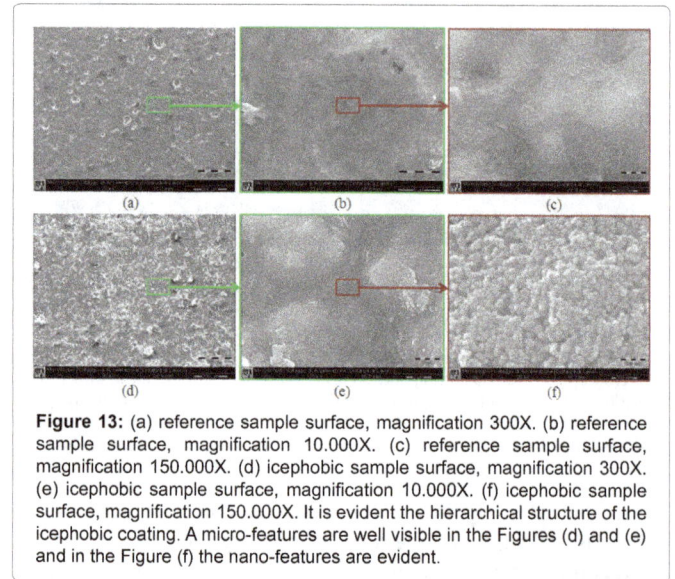

Figure 13: (a) reference sample surface, magnification 300X. (b) reference sample surface, magnification 10.000X. (c) reference sample surface, magnification 150.000X. (d) icephobic sample surface, magnification 300X. (e) icephobic sample surface, magnification 10.000X. (f) icephobic sample surface, magnification 150.000X. It is evident the hierarchical structure of the icephobic coating. A micro-features are well visible in the Figures (d) and (e) and in the Figure (f) the nano-features are evident.

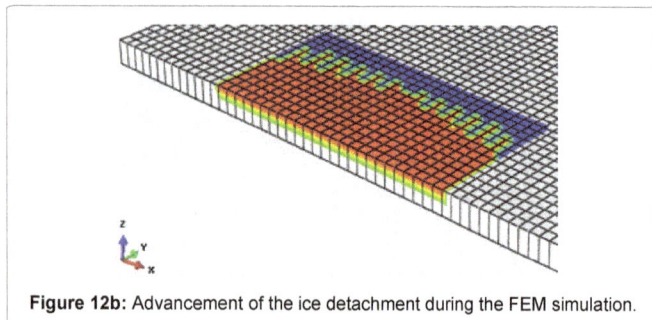

Figure 14: AFM topographic images and rms values of reference sample surface (1st line) and icephobic sample surface (2nd line).

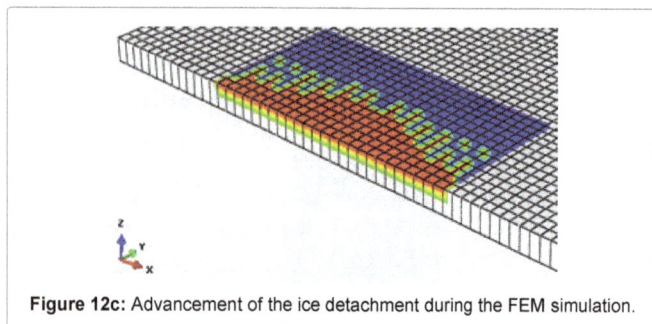

Figure 15: LSM topographic images and rms values of reference sample surface (1st line) and icephobic sample surface (2nd line).

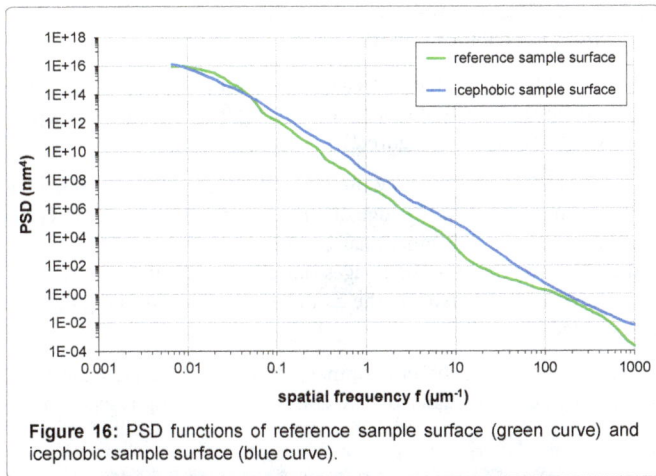

Figure 16: PSD functions of reference sample surface (green curve) and icephobic sample surface (blue curve).

Even if the morphological aspect is extremely important to have super-hydrophobicity and icephobicity, it is also important to preserve it for the entire cycle of flight from one maintenance and other of the aircraft. For this reason, it is important to evaluate the mechanical properties of the new coating. As described in the previous work [28] the main mechanical properties of the new multifunctional coating are the same or higher than the commercial one. The same consideration can be done also for the resistance to the hydraulic fluids [28].

Before scaling up this technology in relevant environment, the dynamic analysis at lab scale of the impingement of the microdroplets on the surface of both coatings was done. Tests showed clearly the different behavior of the surfaces. In fact as described in Figure 17a the water microdroplets that fall down on the commercial coating are well adherent with the surface and passing the time, they pile up until to create a one large macrodroplet. It is interesting to see in Figure 17a the large contact area between water droplet and surface.

On the contrary the new multifunctional coating with icephobic properties showed a complete different behavior. In fact in Figure 17b are reported the same instant times of those described previously for the commercial coating. The single water droplet, that falls down onto the surface, bounces from the surface, therefore there isn't accumulation of droplets. In particular in Figure 17b three different phases of droplet-surface interaction are evident:

The impact

It is evident the super-hydrophobic behavior of the surface analyzing the water droplet during the impact. No water spreading is evident and the droplet remains with a spherical shape.

The sliding

Once that the water droplet touches the surface, it slides on the surface due to the poor adhesion between water and material surface. As described previously, the work of adhesion of the new coating is 92% lower than the commercial one. This is corroborating experimentally from this test.

The sloshing and spinning

Once that the water droplet bounces from the surface, it starts to slosh and to spin far from the surface.

The new multifunctional coating showed also high resistance to harsh conditions as confirmed by the environmental test. In fact,

compared with the commercial one (that is already certified), the behavior of the new coating was the same. In fact the visual inspection before and after the test, demonstrated that the new multifunctional coating preserves the aesthetical aspect and blistering; color changes and delamination are not present (Figures 18a-18d).

For a profound determination of the coating quality after the environmental test, analyses using optical microscopy were performed. As showed in Figures 19a-19d no cracks or in homogeneities are present on the surfaces of both coatings at microscopic level; in fact they are homogenous.

Finally in order to evaluate possible reduction of the adhesion between coating and substrate due to the infiltration of the humidity at the interface, cutting and tape test was performed on both coating.

The test confirmed the results reported in the previous work [28]. The new multifunctional coating showed an hardness higher than the commercial coating, in fact, as shown in the Figures 20a-20d, the blades of cutter failed to damage the surface of the new coating, only a few scratches are evident. On the contrary on the surface of the commercial coating, all scratches are well evident (Figure 20d). In

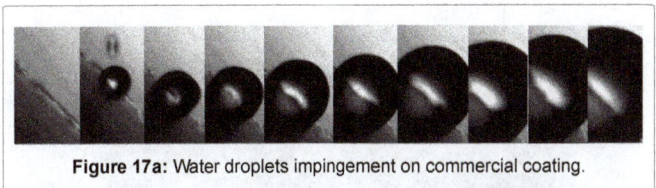

Figure 17a: Water droplets impingement on commercial coating.

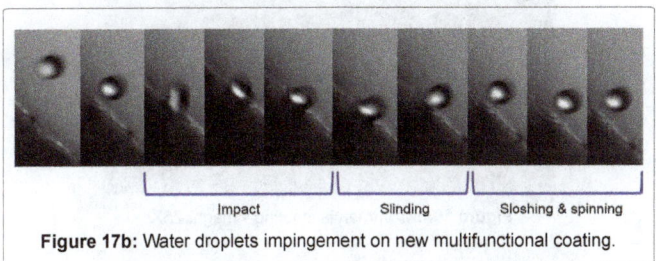

Impact   Slinding   Sloshing & spinning

Figure 17b: Water droplets impingement on new multifunctional coating.

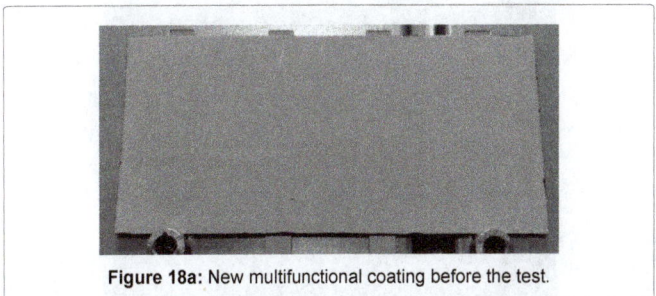

Figure 18a: New multifunctional coating before the test.

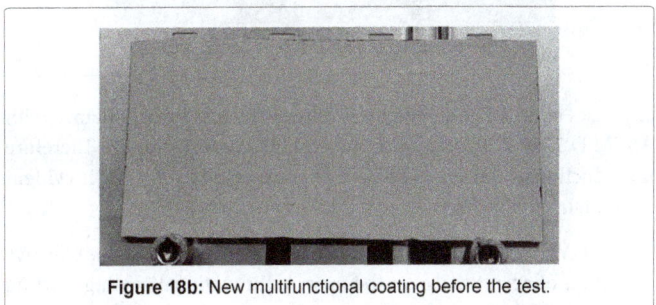

Figure 18b: New multifunctional coating before the test.

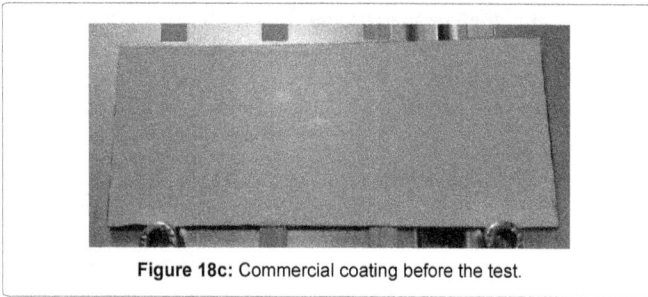

Figure 18c: Commercial coating before the test.

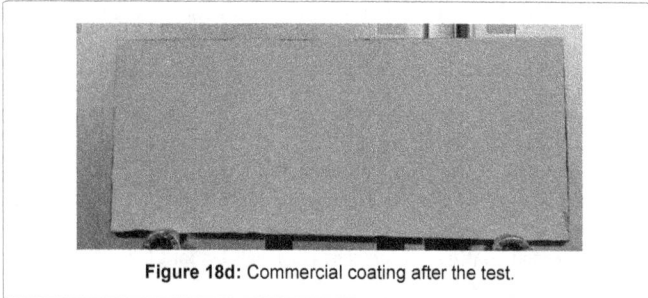

Figure 18d: Commercial coating after the test.

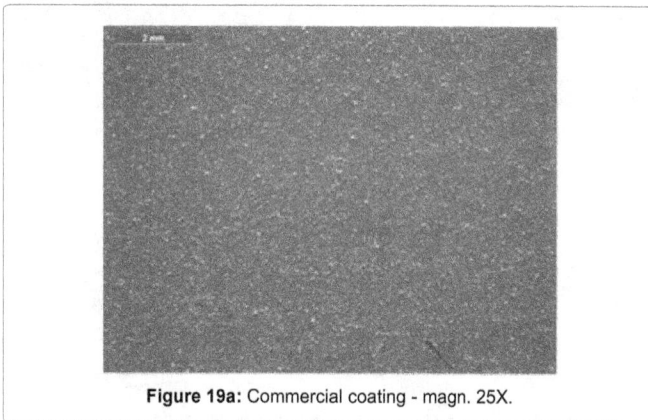

Figure 19a: Commercial coating - magn. 25X.

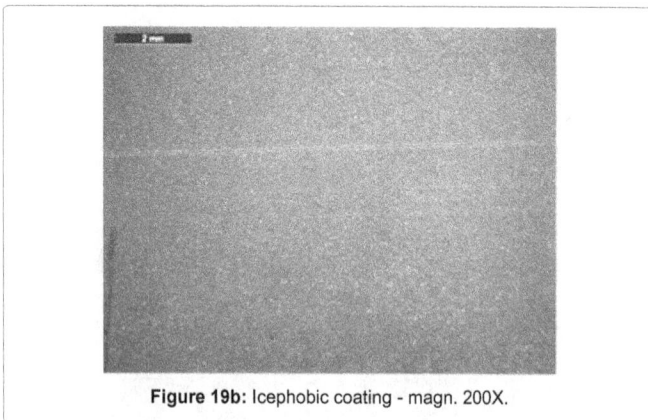

Figure 19b: Icephobic coating - magn. 200X.

any cases after the complete procedure of the test (in accordance with ASTM D 3359-2) the classification for both coatings was 5B. Therefore no reduction of the adhesion between coating and substrate is evident after environmental test.

The IWT tests demonstrated, in relevant environment, the effective reduction of accreted ice on the new multifunctional coating starting

from the leading edge respect to the commercial one. This statement confirms the icephobic properties of the new coating.

The reduction of the accreted ice was measured through the length of compact ice that is accreted starting from the leading edge as reported in the sketch in Figure 21.

Figures 22a and 22b show the original images of the accreted ice in the region of the leading edge obtained respectively on commercial coating and new multifunctional coating. Whereas the Figures 22c and 22d represent the same images but after image analysis in order to optimize and to maximize the contrast with the aim to highlighted the accreted ice.

It is evident that the development of accreted ice (starting from the leading edge) is higher on the commercial coating respect to the

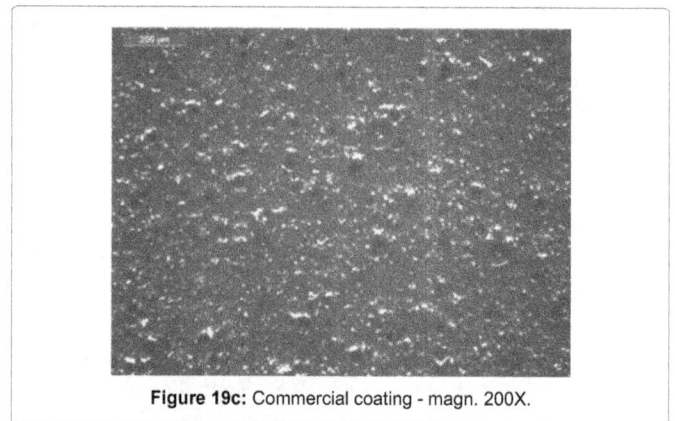

Figure 19c: Commercial coating - magn. 200X.

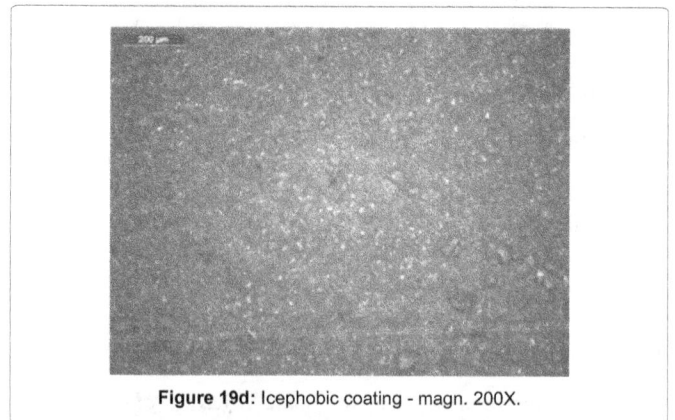

Figure 19d: Icephobic coating - magn. 200X.

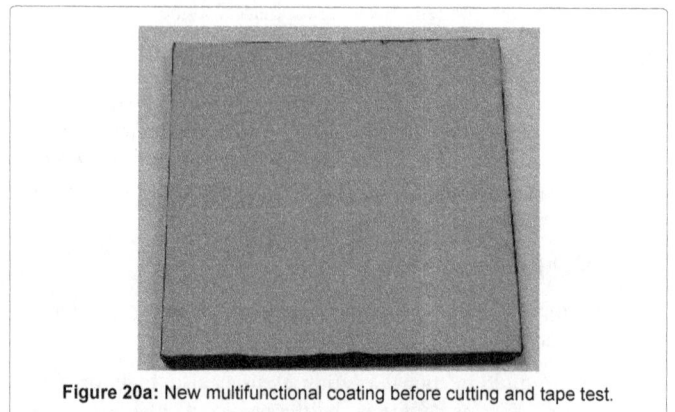

Figure 20a: New multifunctional coating before cutting and tape test.

Figure 20b: Commercial coating before cutting and tape test.

Figure 20c: New multifunctional coating after cutting test.

Figure 20d: Commercial coating after cutting test.

to remove the ice on the two airfoils is lower for the airfoil with new icephobic coating respect to the commercial one. This qualitative result corroborate the weak adhesion between ice and surface demonstrated quantitatively during the lab test (reduction of the work of adhesion between iced water droplet and surface of 92% respect to the commercial coating) and FEM analysis. This means that, combining passive icephobic systems with active de-icing systems, the ice is removed easier than the classical coating. It should also be considered that the wings of the aircrafts are in the airflow consequently, using a small energy from the airflow, the ice detaches from the surface; it go away from the components.

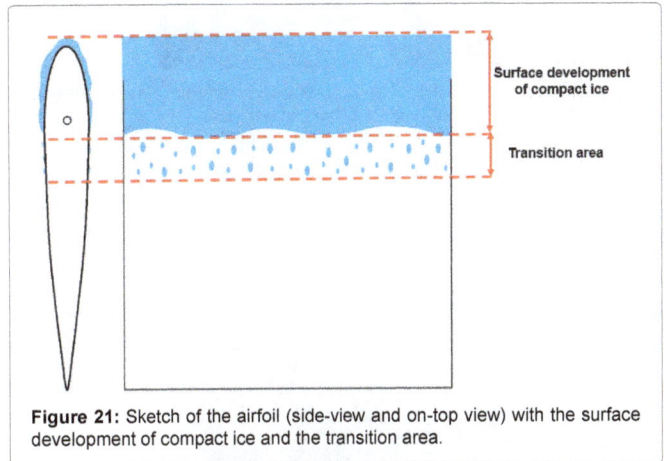

Figure 21: Sketch of the airfoil (side-view and on-top view) with the surface development of compact ice and the transition area.

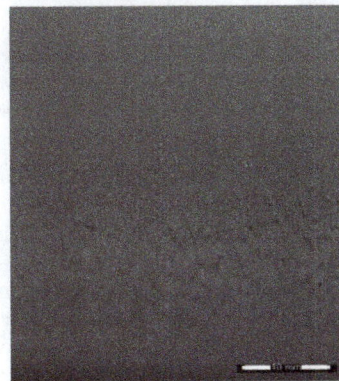

Figure 22a: Accreted ice on the commercial coating (original image).

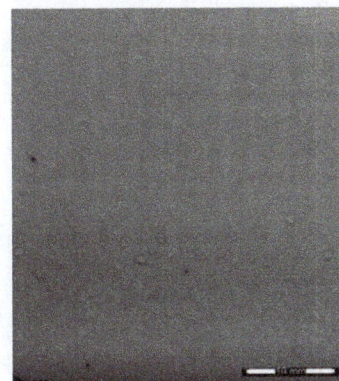

Figure 22b: Accreted ice on the new multifunctional coating (original image).

new multifunctional coating with icephobic property. This confirms the effectiveness of the new formulation and its functional property (icephobicity).

It is also interesting to study the type of accreted ice on the two airfoils, in fact in both cases the accreted ice is rime; in addition, the commercial coating shows also feathers developed on the final part of the compact accreted ice region. These feathers are not present on the new multifunctional coating.

Differences on a development of the accreted ice are mainly due to the icephobic property of the new formulation that in general delays the nucleation and growth of the ice (corroborating the effectiveness of the new coating). In addition, after all tests the manual strength necessary

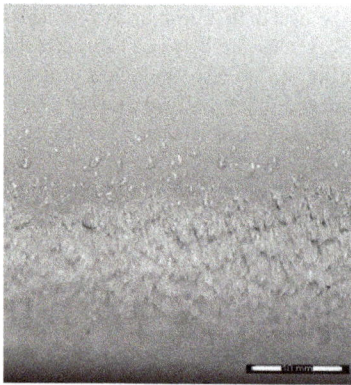

Figure 22c: Accreted ice on the commercial coating (after image analysis through DCE (Differential Contrast Enhancement filter).

Figure 22d: Accreted ice on the new multifunctional coating (after image analysis through DCE (Differential Contrast Enhancement filter).

It is also interesting to analyze the transition region (as reported in the sketch of Figure 21) of the two airfoils in Figures 22a-22d. In fact on the commercial coating the transition area is wider than the new coating formulation and the freezed water droplets are numerically greater and more spread respect to the new multifunctional coating. In fact it is evident that the new coating shows a few droplets in a very short transition region and in addition they are freezed in a most spherical shape. It is highlighted that the points in which these droplets are present, are points in which the surface roughness is changed or it has surface defects and in these points happens that supercooled water droplet start to begin ice. Even if the water droplets have a spherical shape, the Cassie-Baxter condition is no longer maintained, in fact probably the Wenzel regime acts in this points. This is due to the local deterioration of the coating. In fact it is necessary to highlight that the most exposed area of the airfoil to the airflow, and consequently to the water impingement, is the leading edge. Therefore the leading edge is undergone to deterioration phenomena (rain, insects, sand, erosion phenomena) respect to the other part of the airfoil. For this reason the wings of civil aircrafts the leading edges are realized with naked (uncoated) metal. On the contrary the remaining parts of the wings are coated.

These results are obtained for all tests performed in the IWT and they confirm the statements reported above; a further test is represented in Figure 23 where the drastic reduction of the accreted ice on the new multifunctional coating with icephobic property is evident. Anyway to summarize the results of all tests, the bar-chart in Figure 24 is reported.

The value of accreted ice of the test 2 is not available, because during the measurements, the ice was broken and detached from the surface (further attestation of the very low adhesion between the ice and the new multifunctional coating).

Beyond the test 2, it is evident that the surface development of compact accreted ice is higher in case of commercial coating respect to the icephobic one. This statement corroborates the efficiency of the icephobic coating.

In all tests the development of accreted ice of the icephobic coating is 50% lower than the commercial coating.

Test 5 was performed using the same process parameters of test 1 in order to study the repeatability. It is evident a reduction of accreted ice on both coatings (commercial and new coating). This is due to the erosion phenomena at the leading edge and in particular on the stagnation point for both coatings. The accreted ice on new coating of the test 5 was reduced respect to the test 1; this means that the icephobic properties are improved during the utilization.

The reduction of accreted ice was reached also for commercial coating as demonstrate making a comparison between test 1 and test 5. In this case, the deterioration phenomena increase the roughness of the commercial coating giving hydrophobicity and consequently a small quantity of supercooled water droplets remain adherent on the surface.

Another way to determine the effectiveness and efficiency of the new multifunctional coating with icephobic property was to correlate the main IWT process parameters with the accreted ice for both types of airfoils (Figures 25a-25c).

In particular a correlation between altitude, MVD and LWC with accreted ice of two airfoils are plotted. It is clearly evident that the spatial distribution of the experimental points relative to the two airfoils is the same but the points of new multifunctional coating with icephobic property are shifted on the left. This gap gives an estimation of the effectiveness and efficiency of the icephobicity. As bigger the gap, the greater is the icephobicity.

Conclusions

In this work a further development of the new multifunctional aeronautical coating with aesthetic and icephobic properties were described.

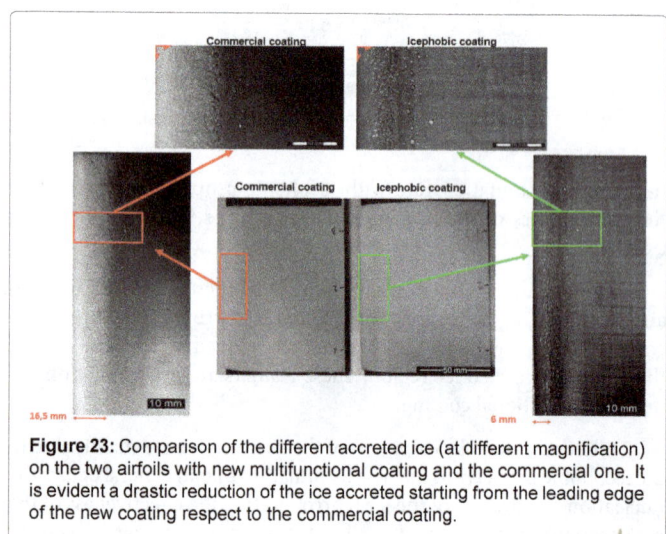

Figure 23: Comparison of the different accreted ice (at different magnification) on the two airfoils with new multifunctional coating and the commercial one. It is evident a drastic reduction of the ice accreted starting from the leading edge of the new coating respect to the commercial coating.

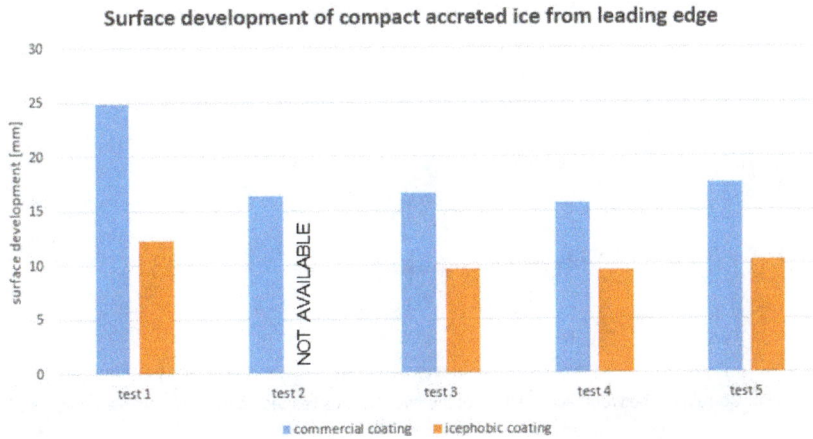

Figure 24: Comparison of the accreted ice on the two coatings for all tests performed in the IWT.

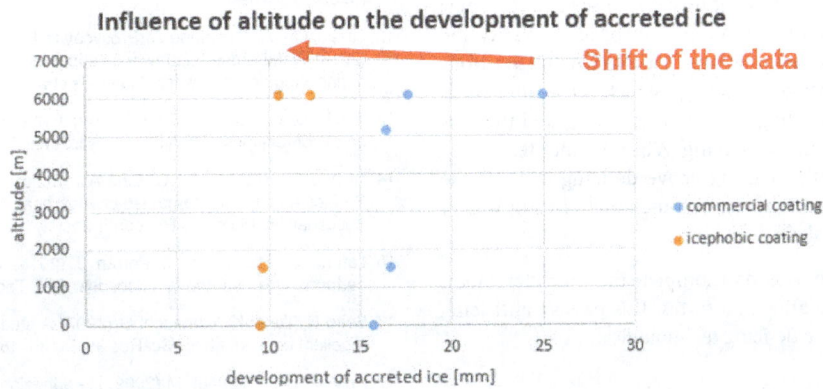

Figure 25a: Correlation between accreted ice of the two coatings respect to the altitude simulated in the IWT.

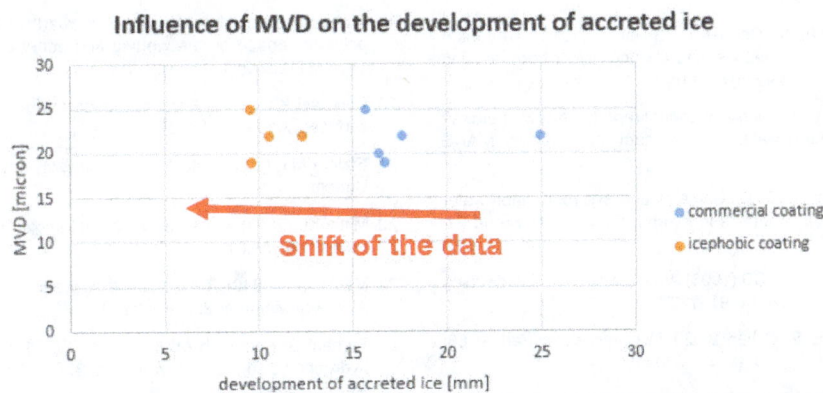

Figure 25b: Correlation between accreted ice of the two coatings respect to the MVD simulated in the IWT.

Once design and realized at lab scale, the new multifunctional coating was tested at lab scale with the aim to check the new icephobic property. Subsequently a deep analysis of the morphology was done with the aim to investigate the reason for the improved icephobicity. It has been shown that spherical nano roughness seen for the novel icephobic coating enables the special functionalities.

The characterization at lab scale reported in this work supplements the results obtained in the previous work, corroborating the high quality of the new coating. In fact the new coating formulation passes all standard tests performed until now. In particular, the new multifunctional coating passes also the test of long-time exposure in high concentration of relative humidity similar to the commercial one.

The different behavior of the liquid droplets during the impingement onto the sample surfaces was described through dynamic characterization at lab scales. It is clearly evident the behavior of the new coating and in particular its super-hydrophobic property. The

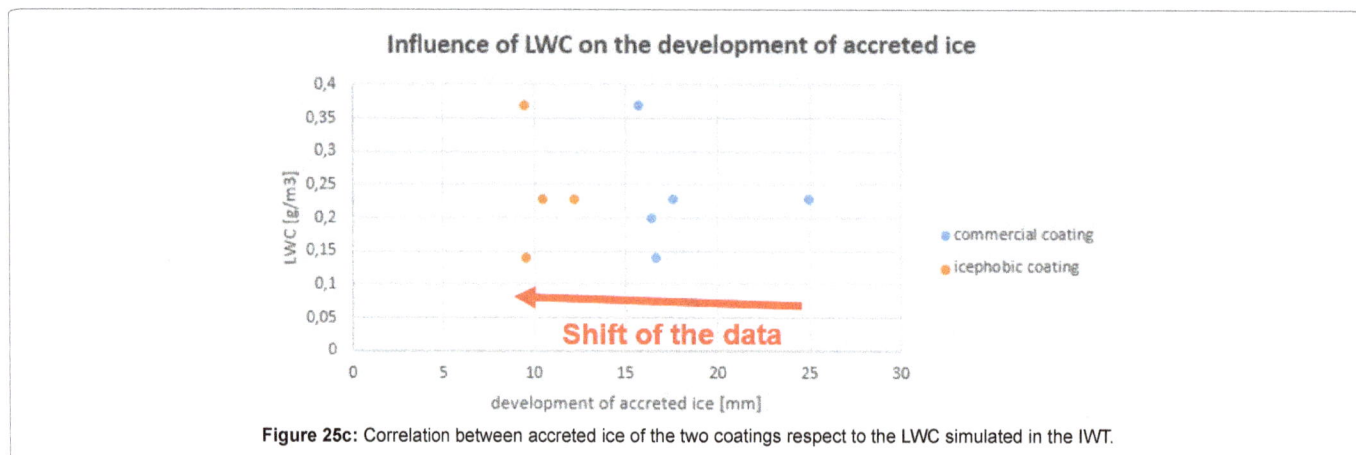

Figure 25c: Correlation between accreted ice of the two coatings respect to the LWC simulated in the IWT.

dynamics of interaction between droplet and surface are completely different respect to the commercial coating.

A further issue reported in this work was also to scale up this new coating formulation in relevant environment in order to verify the icephobic property. For this reason symmetric NACA 0015 airfoils were realized using 3D printing technology which were afterwards coated with the commercial coating (to use as reference) and the new multifunctional coating. Results after Icing Wind Tunnel tests were very encouraging, in fact without using active de-icing systems the reduction of accreted ice on the new multifunctional coating was of 50% respect to the commercial one.

Future activities will be addressed to complete the characterization of the new multifunctional coating and to test this passive anti-icing system with one or more active de-icing technologies.

Acknowledgement

Author would like to thank to Ottavio Minieri for the support during the IWT tests.

References

1. Murphya C, Wallace S, Knight R, Cooper D, Seller T (2015) Treatment performance of an aerated constructed wetland treating glycol from de-icing operations at a UK airport. Ecological Engineering 80: 117–124.

2. Gray L (2013) Water pollution prevention and control report. Review of aircraft deicing and anti-icing fluid storm water runoff control technologies. MANE 6960H01.

3. Sulej M, Palkowska Z, Namieśnik J (2012) Contaminants in airport runoff water in the vicinities of two international airports in Poland. Pol J Environ Stud 21: 725–739.

4. Thomas SK, Cassoni RP, MacArthur CD (1996) Aircraft anti-icing and de-icing techniques and modeling. J Aircraft 41: 1291–1297.

5. Falzon BG, Robinson P, Frenz S, Gilbert B (2015) Composites: Part A, 68: 323–335.

6. Boinovich LB, Emelyanenko AM (2013) Anti-icing potential of super hydrophobic coatings. Mendeleev Commun 23: 3-10

7. Boinovich L, Emelyanenko AM, Korolev VV, Pashinin AS (2014) Effect of wettability on sessile drop freezing: When super hydrophobicity stimulates an extreme freezing delay. Am Chem Soc 30: 1659-1668

8. Makkonen L (2013) Ice adhesion- theory measurements and countermeasures. J Adh Sci and Tech 26: 413-445.

9. Susoff M, Siegmann K, Pfaffenroth C, Hirayama M (2013) Evaluation of icephobic coatings-Screening of different coatings and influence of roughness. App Surf Sci 282: 870-879.

10. Momen G, Farzaneh M (2014) Facile approach in the development of icephobic hierarchically textured coatings as corrosion barrier. App Surf Sci 299: 41-46.

11. Strobl T, Storm S, Kolb M, Haang J, Hornung M (2014) Development of a hybrid ice protection system based on nanostructured hydrophobic surfaces. Proc. 29th Congress of the International Council of the Aeronautical Science St. Petersburg Russia.

12. Lazauskas A, Guobiene A, Prosycevas I, Baltrusaitis V, Grigaliunas V, et al. (2013) Water droplet behavior on super hydrophobic SiO$_2$ nanocomposite films during icing/deicing cycles. Materials char. 82: 9-16.

13. Zhu L, Xue J, Wang Y, Chen Q, Ding J, et al. (2013) Ice-phobic coatings based on silicon-oil-infused polydimethylsiloxane. Appl Mater Inter 5: 4053–4062.

14. Attinger D, Frankiewicz C, Betz AR, Schutzius TM, Ganguly R, et al. (2014) Surface engineering for phase change heat transfer: A review. MRS Energy & Sustainability Materials Research Society 1: 1-40.

15. Laforte C, Blackburn C, Perron J (2015) A review of ice-phobic coating performances over the last decades. SAE Technical paper 01: 2149.

16. Saito H, Takai K, Yamauchi G (1997) A study on ice adhesiveness to water-repellent coating. Mater Sci Res Int 3: 185–189.

17. Kulinich SA, Farzaneh M (2009) Ice adhesion on super-hydrophobic surfaces. App Surf Sci 255: 8153–8157.

18. Kulinich SA (2011) Super hydrophobic surfaces: are they really ice-repellent? Langmuir 27: 25–29.

19. Riosa PF, Dodiukb H, Kenigc S, McCarthyd S, Dotan A (2007) The effect of polymer surface on the wetting and adhesion of liquid systems, J. Adh. Sci. Tech. 21: 227-241.

20. Varanasi KK (2010) Frost formation and ice adhesion on super hydrophobic surfaces. App Phys Let 97: 23.

21. Etzler FM (2003) Contact angle wettability and adhesion. 3 KL Mittal (Ed.) VSP Utrecht.

22. Mittal KL (2013) Advances in contact angle. Wettability & Adhesion. Scrivener Publishing, Willey.

23. Meuler AJ (2010) Relationships between water wettability and ice adhesion. ACS App Mat Inter 2: 3100–3110.

24. Farhadi S, Farzaneh M, Kulinich SA (2011) Anti-icing performance of super hydrophobic surfaces. App Sur Sci 257: 6264–6269.

25. Jung S (2011) Are super hydrophobic surfaces best for icephobicity? Langmuir 27: 3059–3066.

26. Hassan MF, Lee HP, Lim SP (2010) The variation of ice adhesion strength with substrate surface roughness. Meas Sci Tech. 21: 7.

27. Bascom WD, Cottingt RL, Singlete CR (1969) Ice adhesion to hydrophilic and hydrophobic surfaces. J.Adhes 1: 246.

28. Mazzola L (2016) Aeronautical livery coating with icephobic property. J Surface Engineering 32: 733-744.

29. Flemming M, Duparré A (2006) Design and characterization of nanostructured ultrahydrophobic coatings. Appl Opt 45: 1397-1401.

30. Flemming M, Coriand L, Duparré A (2009) Ultra-hydrophobicity through

stochastic surface roughness in: Superhydrophobic Surfaces. CRC Press Boca Raton FL: 19-38.

31. Jones RAL, Richards RW (2006) Polymers at surface and interfaces. Cambridge University Press.

32. Owens DK, Wendt RC (1968) Estimation of the surface free energy of polymers. J Appl Polym Sci 13: 1741.

33. Zenkiewicz M (2007) Methods for the calculation of surface free energy of solids. Pol Test 26: 14-19.

34. Rudawska A, Jacniacka E (2009) Analysis of determining surface free Energy Uncertainty with the Owens-Wendt method. Inter J Adhes Adhes 29: 451-457.

35. Holysz L, Chibowski E, Terpilowsi K (2008) Contact angle wettability and adhesion, Leiden.

36. Duparré A, Coriand L (2013) Assessment criteria for superhydrophobic surfaces with stochastic roughness, in advances in contact angle, wettability and adhesion1. Wiley-Scrivener Beverly MA: 197-201.

37. Coriand L, Mitterhuber M, Duparré A, Tünnermann A (2011) Definition of roughness structures for superhydrophobic and hydrophilic optical coatings on glass. Appl Opt 50: C257-C263.

38. Duparré A, Ferré-Borrull J, Gliech S, Notni G, Steinert J, et al. (2002) Surface characterization techniques for determining rms roughness and power spectral densities of optical components. Appl Opt 41: 154-171.

7

Impact Analysis of Composite Repair Patches of Different Shapes at Low Velocities for Aircraft Composite Structures

Gangadharan S[1]*, Baliga SV[2], Sonawane NH[2], Sathyanarayan P[2] and Shruti Kamdar[3]

[1]Department of Mechanical Engineering, Embry-Riddle Aeronautical University, 600 South Clyde Morris Blvd, Daytona Beach, Florida 32114, USA
[2]Department of Aerospace Engineering, Embry-Riddle Aeronautical University, 600 South Clyde Morris Blvd, Daytona Beach, Florida 32114, USA
[3]Department of Design Engineering, EGA, Honda Engineering North America, 640 Colemans Crossing Blvd, Marysville, OH 43040, USA

Abstract

The area under crack for various aircraft composite structures can be effectively repaired using composite materials. Low velocity impact can cause barely visible damage to the interior structure of laminated composite. These impacts can cause delamination in composite materials. In this study, a finite element analysis was conducted using Abaqus/Explicit and the results of the analysis were compared to the experimental data from literature. E-glass/epoxy composite laminate was subjected to a low velocity impact test. To study the effect of patch repair, a composite patch was applied on a cracked laminate and a low velocity impact was then conducted on this model. The FEA results were validated with the experimental data and an approach to model an ideal composite patch shape was conducted. Different patch shapes like square, rectangle, circle and ellipse were designed and analysed on the crack by keeping the surface area of the patch common. All these patches were compared and an ideal patch shape was found for the model based on stress concentration on the patch. Finally, a parametric study was performed considering the change in impactor speed and impactor material on impact damage. The effectiveness of finite element analysis of low velocity impact on aircraft composite structures is demonstrated.

Keywords: Aircraft; Composite structures; Laminate

Nomenclature

E = Modulus of Elasticity

F = Force

G = Shear Modulus

ν = Poisson's Ratio

ρ = Density

σ = Normal Stress

σM = Von Mises Stress

τ_{max} = Maximum Shear Stress

$\sigma 1$ = Principal Stress

V_f = Fraction by Volume

V_m = Fraction by Weight

Introduction

High strength and high stiffness fiber-reinforced materials like glass/epoxy and carbon/epoxy are significantly used in the aerospace industry and material industry. They are highly flexible and have low elastic modulus. Due to low weight and low coefficient of thermal expansion these composite materials are used substantially. However, one of the biggest concern is that such structures are prone to impact loading while handling loads or when the loads are dropped. Serious damages may be caused by failure as a result of impact in composite structures in a variety of ways. It may cause delamination, matrix cracking or fiber breakage of the material. Low to moderate energies caused typically by impact forms delamination, cracking and fiber breakage. Penetration and shear damage at an excessive amount is caused by high impact energies Abrate [1]. The strength and stiffness of the damaged object, the stress state on the damage and the response of the damaged structure makes the problem complex.

It is a known fact that composite structures after impact can endure a major decrease in tensile strength and compressive strength Sierakowski Robert and Chaturvedi [2]. To study and analyse the damage on a composite structure, several experiments have been conducted. Such experiments are conducted by replicating the real-life situations in controlled environment. For instance, drop weight test is conducted to simulate the dropping of hard tools on composites. This test is generally low-velocity impact test. Damage because of low velocity impact on fiber reinforced composites is thought to be very risky for the most part, in light of the fact that the damage is not detectable to the exposed eye; this kind of damage is called as Barely Visible Impact Damage. A composite's compressive strength can undergo a loss of about 60% with this type of damage.

All in all, there are numerous parameters which characterize the way of the damage in composite structures, for example, delamination in composite structure, caused due to pressure loads. Different parameters which characterize the morphology of the impact incorporates impactor speed, geometric imperatives connected to the framework, impactor shape, and design of the affected structure. In this manner

*Corresponding author: Gangadharan S, Professor, Department of Mechanical Engineering, Embry-Riddle Aeronautical University, 600 South Clyde Morris Blvd, Daytona Beach, FL 32114,USA, E-mail: sathya@erau.edu

investigations of these parameters are critical in comprehension to the effect procedure and the damage brought on by them in the composite structures.

The damage caused by low velocity impact is inevitable. Hence, a repair or fortification of the damaged portion of the structure to restore the basic structural strength and efficiency is required. Applying composite patch repair is one of the latest solutions. Little research into the combined low-velocity impact damage resistance of the patch is available in published literature. The potential for an outwardly unnoticeable mix of the composite damage with likely adhesive damage recommends that low-velocity effect damage in composite repair is ought to be studied about and considered amid design configuration. It is costly and quite complex to conduct and perform physical experiments to evaluate impact damage on composite patches considering the quantity of distinctive parameters to be viewed and internal damages to be examined. Finite Element Analysis (FEA) gives a more financially savvy approach to foresee and survey damage in composite patches, and giving a road to investigate numerous material mixes and designs. FEA can then show the areas where constrained trial testing may be important for acceptance of the damage behaviour as reported by Goodmiller [3]. Patch shape, properties of materials, thickness, orientation, and number of plies in the composite structure, quality of the bond surface, and damage tolerance properties of materials are some of the parameters that are important to feed in for the impact performance of the structure. The mechanics of the damage of the patch is also imperative to study the analysis of the patch performance. To have an appropriate and optimized patch design, it is important to understand the effects of input parameters, damage mechanics, and their interactions.

The aim of this research was to conduct a FEA which studied the damage mechanisms related to a composite patch performance on E-glass/epoxy material under low-velocity impact loading. The results from this analysis and simulation was compared to available experimental data in quantitative terms of stress, energy, displacement and contact force. Abaqus 6.13 was used for this research, which provided modules for composite structures and adhesive properties. Composite patch performance has limited availability of experimental data. Due to that and also because of a few obscure properties of materials needed for damage models, several assumptions were made. These includes assumptions of material strength, adhesive thickness and its properties. In addition to analysis of the patch, the parameters were studied to obtain an optimum composite patch shape for impact damage resistance based upon the stress carrying capacity. Other potential factors such as number of plies and its orientation, patch size, adhesive type, and thermal expansion mismatch were not examined in this study, but should be investigated in future work.

Procedure

Experimental setup

Geofrey and Yatin [4] conducted low velocity impact experimentally. Experimental data from this study was used as reference for FEA analysis in this research. Geofrey et al. conducted a drop weight test to simulate low velocity impact on an E-glass/epoxy composite laminate. The experimental setup had nine layers of E-glass/epoxy laminates with alternating 0° and 90° plies. The dimension of the laminate was 100 mm × 100 mm and its total thickness was 4.04 mm. This plate was subjected to an impact of 20 J under the velocity of 4.472 m/s. E glass fabric, type C of IS: 11273 were used to fabricate composite laminates. An epoxy

matrix based on Lapox L-12 resin and K-5 hardener was selected for making composite panels.

In the next step of the experiment, a cracked laminate was applied a composite patch upon it. The crack was deep up to the third layer of the composite ply while the crack dimensions were varied. The crack dimension for the first case was 5 mm × 5 mm and for the second case it was 5 mm × 7.5 mm. This composite patch had an orientation of 90°. The dimension of the patch used was 10 mm × 10 mm and the thickness of the patch was 1 mm.

Modeling and analysis

Finite element method is a numerical technique that is used to find solutions to a large level and variety of engineering problems which includes stress analysis in dynamic conditions. The three basic steps to perform finite element analysis are, pre-processing, solving and post-processing. In pre-processing, geometric models are made as per the requirement. The modelled geometry is then applied with appropriate meshing. Material properties are assigned to the elements and boundary constraints are applied to the nodes of the element. The next step involves, solving which is the processing of geometric data. After the data is processed the output file is generated.

The third step is post-processing which involves studying the obtained data in the form of stress, strain and force graphs. In this research Abaqus serves as both, pre-processor and post-processor. Abaqus is an interactive 3D modeling software that can be used to model many complex and simple components in engineering. Since, it has very user friendly tool interface and extensive customizing capacity; it is used on a large scale for modeling. Solving and post-processing both the jobs are done in this software. Abaqus software has explicit and implicit finite element program that is used to analyze the responses that are non-linear and dynamic. It has a fully automatic definition of contact areas and a large library of constitutive material models and failure models.

A finite element model of a symmetric, cross ply, laminated composite and impactors were modeled in Abaqus design module. The finite element model consisted of nine separate layers with each layer being 0.44 mm thick and 100 mm × 100 mm in dimension. The orientation of these layers was [0/90/0/90/0/90/0/90/0]. Every layer was attached to each other with a cohesive layer between them having a thickness of 0.1 mm. The total thickness of the composite structure was 4.04 mm. These plies were modeled with SC8R: 8 nodes, quadrilateral, reduced integration, continuum shell element. It had enhanced hourglass control with Hashin damage viscous stabilization factor of 1×10^{-7}.

The material that was modelled was E-glass/epoxy. The material properties of the E-glass/epoxy used in this test is shown in Table 1. Elements are 0.5 mm × 0.5 mm in the center of the mesh and their size increases with the distance from the impact zone. The adhesive layer between every ply is of 0.1 mm thick.

An impactor was modeled, providing impact energy of 20 J and velocity of 4.472 m/s. A friction penalty of 0.5 was provided for the contact between the impactor and the composite layer (Figure 1).

The second part of the test involved creating crack in the composite layer and a patch for the crack. This involved two tests with crack of thickness 1.34 mm and varying thickness. The first composite was modeled with 5 mm × 5 mm crack dimension and the second composite was modeled with 7.5 mm × 5 mm crack dimension. A patch

Property	Units	Value
X_c	Mpa	800
Y_t	Mpa	40
Y_c	Mpa	145
S_L	Mpa	73
S_T	Mpa	54.8
ε_{1t}	%	2.807
ε_{1c}	%	1.754
ε_{2t}	%	0.246
ε_{2c}	%	1.2
G_f^t	N/mm	17.965
G_f^c	N/mm	7.016
G_m^t	N/mm	0.049
G_m^c	N/mm	0.87

Table 1: Properties of E-glass/epoxy.

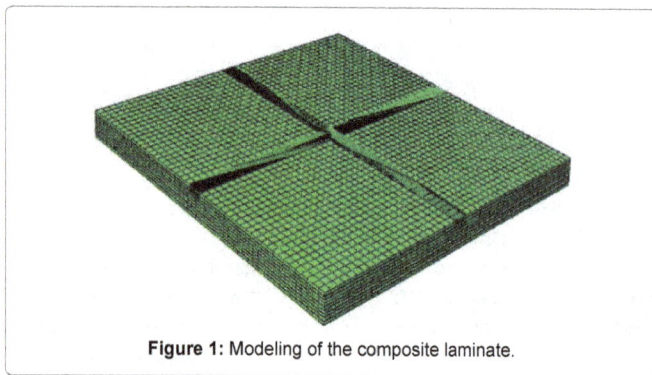

Figure 1: Modeling of the composite laminate.

was modeled for both the conditions. This patch was made of the same E-glass/epoxy element with a single layer having orientation of 90°. The thickness of this patch was modeled to 1 mm and other dimensions were 10 mm × 10 mm. The patch was attached to the composite using the cohesive layer. Both these models were validated comparing with the experimental results and the shape of the patch was changed as per the stress concentration to provide with an ideal shape.

Sensitivity study

A sensitivity study was performed to obtain a good mesh. Meshes that are good enough are ones that produce results with an acceptable level of accuracy, assuming that all other inputs to the model are accurate. Mesh density is a significant metric used to control accuracy (element type and shape also affect accuracy). Assuming no singularities are present, a high-density mesh will produce results with high accuracy (Figure 2).

Validation results

A comparison of experimental results and finite element analysis was done. Both the results showed a good agreement in between the two (Table 2).

The above results were a comparison for nine layer composite laminate without the patch. The experimental tests conducted with the patch also showed good accordance with the finite element test results (Table 3).

Following is the comparison of both the approaches for 5 mm × 5 mm crack on the composite:

Other laminate had a crack of 5 mm × 7.5 mm (Table 4). The results of these laminates are as shown below:

Towards ideal repair patch

The patch shape matters a lot when it comes to repairing of the material. The amount of stress concentration changes with the change of shape of any material. For instance, a shape with more cornered edges may have higher stress concentration when compared to the ones with lesser or no edges. This is good enough to know that the patch shape used in the experimental test may not be an ideal one. To have a better patch shape for the crack, different shapes of nearly same areas were modeled and analyzed. The experimental test which was taken into consideration was the one with the crack length of 5 mm × 7.5 mm (Table 5).

Results and Discussion

Rectangular, circular and elliptical patches were created. Following are the result comparison of all the patches (Table 6).

It is shown in the table above that the maximum displacement is

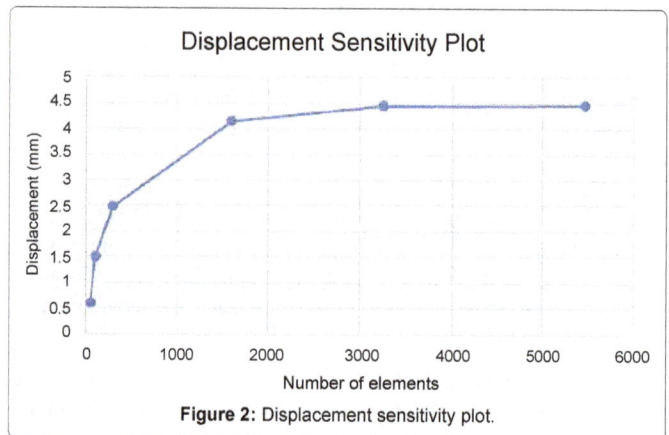

Figure 2: Displacement sensitivity plot.

Parameter	Contact Force	Maximum Displacement	Von Mises Stress
Experimental	5170.4 N	6.283 mm	54.98 MPa
FEA	5468 N	4.472 mm	55.88 MPa
% Difference	5.44	28.82	5.43

Table 2: Comparison of results for composite laminate.

Parameter	Contact Force	Maximum Displacement	Von Mises Stress
Experimental	1097 N	1.42 mm	78.53 MPa
FEA	1579 N	1.20 mm	75.95 MPa
% Difference	30.52	15.49	3.28

Table 3: Comparison of results for the first patch.

Parameter	Contact Force	Maximum Displacement	Von Mises Stress
Experimental	3732 N	0.79 mm	38.42 MPa
FEA	4294 N	0.689 mm	37.75 MPa
% Difference	13.08	12.78	1.74

Table 4: Comparison of results for the second patch.

Shapes	Area (mm²)
Square	100
Rectangle	100.5
Circle	100.1
Ellipse	100.2

Table 5: Areas of different patch shapes.

Patch Shape	Maximum Displacement (mm)
Square	0.689
Rectangle	0.448
Circle	0.447
Ellipse	0.447

Table 6: Maximum displacement comparison.

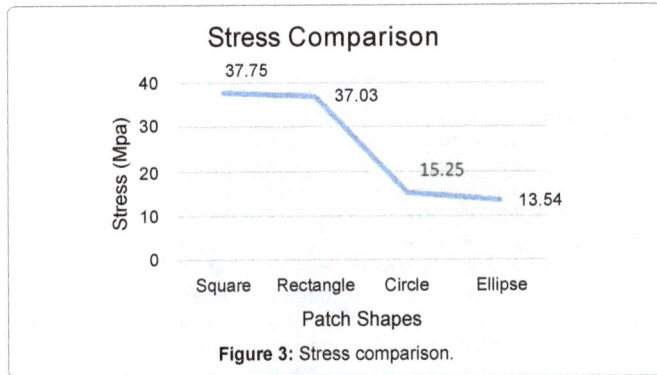

Figure 3: Stress comparison.

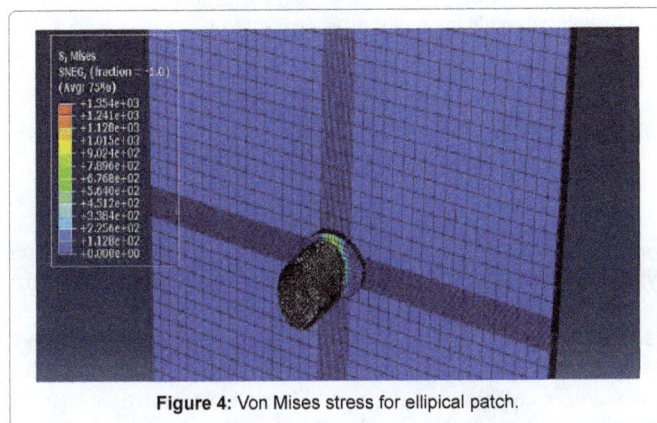

Figure 4: Von Mises stress for ellipical patch.

more in the square patch. The maximum displacement decreases for the remaining patches. The maximum displacement is more or less equal for rectangle, circle and elliptical patches. The figure below shows the stress comparison of all the patches. Elliptical patch shape has the lowest stress of all the patch shapes which is 13.54 MPa (Figures 3 and 4).

Parametric Study

A parametric study was done to understand the effect of uncertain inputs with existing boundary conditions and geometry. Impactor material and impactor velocity are the two topics included in this parametric study. For all the simulations, the thickness of the composite was kept constant throughout the process.

Steel and aluminum projectile were used for parametric study of impactor material. The impactor diameter and velocity was kept the same as that used in the tests. This study was specifically to see the effect of changing material of the impactor on the impact damage. There were differences observed in the impact force with the change in materials. Having the same impact velocity, aluminum and steel had the kinetic energy in a similar ratio. The maximum impact force of steel was found to be 1000.5 N and that of aluminum was 912 N. The figure also shows a plot of impact force vs. time for both aluminum and steel impactor materials.

The next parametric study was conducted for the change in velocity of the impactor. The impactor given in the experimental setup is used the way it is. The velocity is given as 5 m/s and 6 m/s. The maximum damage is high as compared to that of the velocity used in the experiment. Since the velocity is more, the impact damage would be greater too. Impact force of approximately 1500 MPa was calculated for the impactor speed 5 m/s and impact force approximately 1200 MPa was calculated for impactor speed at 6 m/s.

Conclusion

To simulate low velocity impact scenario on a composite material Finite element method can be effectively used. A Finite Element Analysis model of E-glass/epoxy and impactor were successfully modeled and developed to analyze their behaviour during low-velocity impact analysis. The results from the FEM simulations matches and are in good accordance with the experimental data.

The ideal patch shape analysis was done. Keeping the surface area of all the patches as constant. All the different patch shape geometries were compared to each other on the basis of stress concentration. Elliptical patch shape had stress value of 13.54 MPa and displacement 0.447 mm. It was evident from the results that elliptical patch shape is the ideal patch for the model. The stress concentration on the elliptical patch shape was the least as compared to the other patch shape geometries. Also, it is proved that square patch is not the ideal one. After the analysis of the patch, the model was subjected to parametric studies. To understand the difference obtained by change in the material nature of the impactor on the impact damage, two different type materials were used. Aluminum and steel were used as the impact material on the composite for the parametric study. It was found that the impact energy due to aluminum as well as steel impactor increases with time at a similar constant ratio. The impact force was highest for the steel impactor giving 1000.5 N while that for the aluminium impactor was 912 N. The change in velocity of the impactor was also checked in the parametric study. The experimental tests had velocity of the impactor as 4.472 m/s. The increase in velocity of the impactor for the 8 parametric studies gave high values for the maximum impact force. The damage caused by both these velocities gave excessive distortion for the laminate.

Acknowledgement

The authors of this paper would like to thank Embry-Riddle Aeronautical University, for providing the resources for this research.

References

1. Abrate S (1998) Impact on composite structures. Cambridge University Press.

2. Sierakowski-Robert L, Chaturvedi SK (1997) Dynamic loading and characterization of fiber-reinforced composites. John Wiley & Sons Inc., NJ, USA.

3. Goodmiller G (2013) Investigation of composite patch performance under low-velocity impact loading Knoxville.

4. Geofrey R, Yatin B (2010) Low velocity impact analysis on composite patch 16th Conference for Material Conference of Russia.

8

Design and Structural Analysis of Solid Rocket Motor Casing Hardware used in Aerospace Applications

Dinesh Kumar B*, Shishira Nayana B and Shravya Shree D

Department of Aerospace engineering from M.L.R Institute of Technology, Dundigal, Hyderabad

Abstract

Rocket motors are widely used to generate thrust or impulsive force to impart a desired velocity to flight vehicle to transport its payload to the intended destination. The working principle of Rocket motor is mainly Newton's 2nd and 3rd laws. Rocket motors are non-air breathing propulsion class i.e., won't require oxygen from the atmosphere for combustion of the fuel which is stored in the rocket motor. During the operating conditions of the motor hardware, it will be subjected to high temperatures and pressure loads. Structural and thermal design has to carried out for a given input parameters and analysis to be carried out to check the stress levels and temperatures on the hardware. The present paper deals with structural design of motor hardware. The main input parameters considered are the maximum operating pressure and maximum diameter of the Motor hardware. The material properties considered are up to 100°C. Structural analysis and fracture analysis are to carry out after the design of each component of the rocket motor hardware. For design, the motor hardware is considered as a pressure vessel. To compute parameters like thickness some initial assumptions were made. 2D drawing is developed using Auto Cad software and structural analysis is carried out in ANSYS. This software employs finite element analysis techniques to generate the solution. Hence the displacement magnitude, von mises stress and strain developed within the motor is pictorially visualized. Fracture analysis is also carried out on the material.

Keywords: Ansys; Finite element analysis; Fracture analysis; Motor casing; Solid rocket motor; Stress; Strains; Structural analysis

Introduction

Rocket motors are non-air breathing propulsion class i.e., won't require oxygen from the atmosphere for combustion of the fuel which is stored in the rocket motor [1]. A rocket motor is a typical energy transfer system. The chemical energy inside the fuel is converted to the thermal energy by a combustion process. High pressure and high temperature combustion product gases are expanded through a converging-diverging nozzle. By this process "internal energy of the gas is converted into kinetic energy of the exhaust flow and the thrust is produced by the gas pressure on the surfaces exposed to the gas" [2].

Solid propellant rocket motor is the most commonly used compared to other rocket motors due to its relatively simple design, high reliability, ease of manufacture and ready to use on demand etc. Since solid-fuel rockets can remain in storage for long periods, and then reliably launch on short notice, they have been frequently used in military applications such as missiles. Solids are, however, frequently used as strap-on boosters to increase payload capacity or as spin-stabilized add-on upper stages when higher-than-normal velocities are required. Solid Rocket Motor can be used for a wide variety of applications requiring wide range of magnitude of thrust [3].

The design and the construction of the solid rocket motor hardware involve consideration of various stresses acting on the motor hardware due to pressure and thermal loads. For this analysis to be done, selection of material and their properties, motor hardware performance and operating conditions, a few design considerations, etc.., are the parameters required to be studied to obtain the solution.

Literature Review

Solid rockets were invented by the Chinese; the earliest versions were recorded in the 13th century. Hyder Ali, king of Mysore, developed war rockets with an important change: the use of metal cylinders to contain the combustion powder [4]. Rocket propulsion system is a non-air breathing system, in which the propulsive effort or thrust is obtained by variation of the momentum of the system itself. They do not depend on the atmospheric air, either as oxidizer. As its name implies, the propellant of the motor is in the solid state. The oxidizer and the fuel is premixed and is contained and stored directly in the combustion chamber [5]. Since the solid propellant both includes fuel and oxidizer, solid propellant rocket motors can operate in all environmental conditions. In comparison to other types of rockets, solid propellant rocket motors have simple design, are easy to apply and require little or no maintenance. Rocket motor propulsion can be classified based on the type of propellant of rocket propulsion units used in a given vehicle and type of construction, the number by the method of producing thrust. Even though there are many rocket propulsion types only chemical rocket propulsion is widely used [6]. Other rocket propulsions have their drawbacks in the weight consideration and thrust produced. Further advancements in technologies may lead to usage of other rocket propulsion systems in future.

Main Parts of Solid Propellant Rocket Motors

A simple solid rocket motor consists of a Motor casing, Nozzle, propellant grain and igniter (Figure 1).

*Corresponding author:** Dinesh Kumar B, Department of Aerospace engineering from M.L.R Institute of Technology, Dundigal, Hyderabad
E-mail: dineshbajaj113@gmail.com

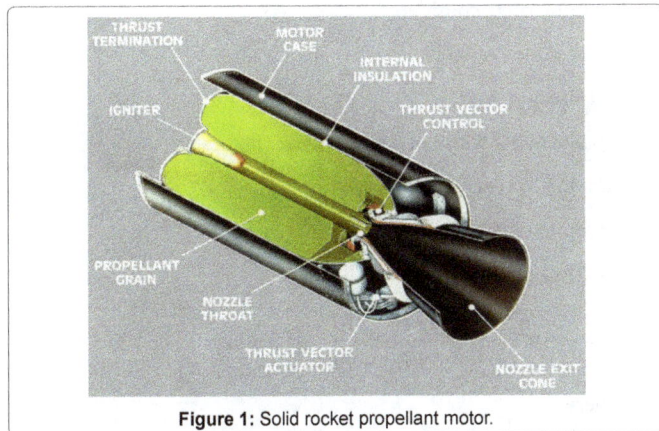

Figure 1: Solid rocket propellant motor.

Motor case

The combustion takes place in the motor case; therefore, sometimes it is referred to as combustion chamber. The case must be capable of withstanding the internal pressure resulting from the motor operation, approximately 3-30 MPa, with a sufficient safety factor. Therefore motor case is usually made either from metal (high-resistance steels or high strength aluminum alloys) or from composite materials (Glass, Kevlar and Carbon).

Insulation

High temperature of the combustion gases, ranging from approximately 2000 to 3500 K, requires the protection of the motor case or other structural subcomponents of the rocket motor. Typical insulator materials have low thermal conductivity, high heat capacity and usually they are capable of ablative cooling. Most commonly used insulation materials are EPDM (Ethylene Propylene Diene Monomer) with addition of reinforcing materials.

Igniter

The ignition system gives the energy to the propellant surface necessary to initiate combustion. Ignition usually starts with an electrical signal. The ignition charge have a high specific energy, they are designed to release either gases or solid particles. Conventional heat releasing compounds are usually pyrotechnic materials, black powder, metal-oxidant formulations and conventional solid rocket propellant.

Nozzle

High temperature, high pressure combustion gases are discharged through the converging-diverging nozzle. By this way, chemical energy of the propellant is converted to kinetic energy and thrust is obtained. The geometry of the nozzle directly determines how much of the total energy is converted to kinetic energy. Therefore nozzle design has a very important role on the performance of a rocket motor.

The grain behaves like a solid mass, burning in a predictable fashion and producing exhaust gases. The nozzle dimensions are calculated to maintain a design chamber pressure, while producing thrust from the exhaust gases. Once ignited, a simple solid rocket motor cannot be shut off, because it contains all the ingredients necessary for combustion within the chamber in which they are burned. More advanced solid rocket motors can not only be throttled but also be extinguished and then re-ignited by controlling the nozzle geometry or through the use of vent ports.

Solid Rocket Motor Hardware Design

Design inputs

Maximum Expected Operating Pressure (MEOP) considered is 150 ksc which normally can be obtained from the ballistic design. The motor maximum diameter considered is 200 mm. Motor hardware design is carried out using pressure vessel code.

Solid rocket motor or pressure vessel consists of the following components

i) Cylindrical Motor casing

ii) Head end and nozzle end domes

iii) Head end and nozzle end flange

iv) Convergent divergent nozzle

v) Head end cover

vi) Bolted Joint between motor- nozzle and motor –head end cover

Material selection criteria

Maraging steel -250 grades (MDN-250) is chosen to minimize the hardware weight due to its high specific strength. Moreover, it is easily available and fabrication technology is well established. This material has been widely being used in space and defense programs.

The detailed data on the chemical composition and mechanical properties are as given below.

Chemical Composition (Wt. %):

C: 0.03 Max.

Mn: 0.1 Max.

Si: 0.1 Max.

Ni: 17 - 19

Mo: 4.6 - 5.2

Co: 7 - 8.5

Ti: 0.3 - 0.5

Mechanical Properties of maraging steel:

UTS (MPa)   :   1750 MPa

Y.S. (MPa)   :   1680 MPa

Elongation(%) : 15

K_{IC} fracture toughness: 80 MPa-m$^{\frac{1}{2}}$

Young's modulus : 210 GPa

Factor of safety selection criteria: As per AVP –32 and MIL standards, the following safety factors have been chosen.

On Ultimate tensile Strength (UTS) : 1.5

On Yield Strength : 1.33

Allowable Ultimate tensile stress: 175/1.5 = 116.6 kgf/mm^2

Allowable Yield stress   : 126 kgf/mm^2

Lowest of above two allowable stresses is on UTS (116.6 N/mm^2) is taken for design.

Design inputs

In brief, the following inputs have been taken for motors hardware design.

i) Material	=	Maraging steel (250 grade)
ii) Motor outer diameter (D) =		200 mm
iii) MEOP (P)	=	150 kgf/cm^2
iv) UTS	=	175kgf/mm^2
v) Yield strength	=	168kgf/mm^2
vi) Factor of safety (F.S.)	=	1.5 on UTS
vii) Weld efficiency (E)	=	90%
viii) Biaxial gain	=	1.1(10%)
vii) Mismatch factor	=	1.15 (5%)

Design Calculations

Thickness of motor casing:

The cylindrical shell thickness is calculated by using conventional formula from ASME Pressure vessel code. It is given by

$$t = \frac{P * D * Mismatch\ Factor}{biaxial\ gain * 2 * (SE - 0.6P)}$$

The Cylindrical shell is assumed to be made from the sheet which is rolled and weld method. In this case, both mismatch factor and weld efficiency (E) have to consider.

Allowable strength(S)	:	UTS/F.S.
Calculated thickness (t)	:	1.506 mm

Head end and nozzle end dome:

Various domes shapes viz. ellipsoidal, hemispherical, tori spherical etc are used in pressure vessel. The shape of the dome considered is Tori-spherical type because of easy fabrication compared to other shapes. This is assumed to fabricate from the forged rod of 200 mm diameter. After iterations, the following parameters are chosen to get optimum parameters

Crown Radius (L)	=	150 mm
Knuckle radius (r)	=	15% of 'L' = 22.5 mm
L/r	=	6.6 mm
M	=	0.25*(3+ √(L/r)
$t = \dfrac{P * L * M}{2.0 * (SE - 0.1P)}$	=	1.39
Thickness (t)	=	1.347 mm

Head end flange design and bolted joint between motor–head end cover

Schneider's approach is used to calculate the flange thickness and to finalize the size and number of bolts 12.9 grade Socket head bolts are considered for design.

Head end opening diameter:		333 mm
MEOP	:	1.5 kg/mm^2

Factor of safety considered:		1.5
No. of studs assumed	:	24 No. s
Bolt size and class	:	M6 x 1.0, 12.9 class

N = No. of bolts/mm of stud circle circumference

= 24/468.1 = 0.0512

Circumferential pitch (d) = π x 149/(24 x 10) = 2.43

A = Minimum required area/bolt

$$A = \frac{P\ R_m^2(1.0 + l / b_{max})}{2.0 * \sigma_{bolt} * N(R_m + l)} = 35.9\ mm^2$$

Where

σ_{bolt} = yield strength of the bolt = 108 kg/mm^2

M6 x 1.0 bolt having cross sectional area 36.6 mm^2

Thickness of flange can be calculated by

$$t = 1.1\ R_m \left[\frac{3.0 * P * l}{\sigma_f\ (1 - N * d)\ (R_m + l)} \right]^{0.5}$$

t = 7.22 mm

Thickness chosen: 8.0 mm;

Max. stress on each bolt due to pressure load and preload

Stress due to pre load on each bolt = 0.4 σ_{yield_bolt} = 36.0 kg/mm^2

Stress due to pressure load on each bolt   = 11.3 kg/mm^2

Total stress on each bolt = 36.0+11.3 = 47.3 kg/mm^2

Factor of safety available on each bolt: 1.90 on yield strength

Head end cover

Head end cover thickness can be same as flange but it is a flat plate, the thickness has to be calculated by using flat plate closure with bolted joint formula and higher one has to be finalized. Flat plate with bolted joint formula given in pressure vessel code is

$$t = d\sqrt{\frac{CP}{SE} + \frac{1.9\ whg}{SEd^3}}$$

Where d = Diameter up to the centre of 'O' ring = 156 mm

C =   0.3

P =   1.5 kgf/ mm^2

E =   1

W =   π/4 d^{2}p = 35668 kgf

h$_g$ = 10.5 mm

$$t = 80\sqrt{\frac{0.3 * 1.5 * 2.0}{175 * 1.0} + \frac{1.9 * 7540 * 6.0 * 2.0}{175\,x\,1.0\,x\,(80)^3}}\ \ t = 11.09\ mm$$

Nozzle end flange design and bolted joint between motor – nozzle

Schneider's approach is used to calculate the flange thickness and to finalize the size and number of bolts 12.9 grade Socket head bolts are considered for design.

Head end opening diameter: 667 mm

MEOP: 1.5 kg/mm^2

Factor of safety considered: 1.5

No. of studs assumed: 24 No. s

Bolt size and class: M6 x 1.0, 12.9 class

N = No. of bolts/mm of stud circle circumference

= 24/468.1 = 0.0512

Circumferential pitch (d) = π x 149/(24 x 10) = 2.43

A = Minimum required area/bolt

$$A = \frac{P\,R_m^2(1.0 + l/b_{max})}{2.0 * \sigma_{bolt} * N(R_m + l)} = 35.9 \text{ mm}^2$$

Where

σ_{bolt} = yield strength of the bolt = 108 kg/mm^2

M6 x 1.0 bolt having cross sectional area 36.6 mm^2

Thickness of flange can be calculated by

$$t = 1.1\, R_m \left[\frac{3.0 * P * l}{\sigma_f\,(1 - N * d)\,(R_m + l)} \right]^{0.5}$$

t = 7.22 mm

Thickness chosen: 8.0 mm;

Max. Stress on each bolt due to pressure load and preload

Stress due to pre load on each bolt=0.4 $_{yield\_bolt}$ = 36.0 kg/mm^2

Stress due to pressure load on each bolt = 11.3 kg/mm^2

Total stress on each bolt = 36.0+11.3 = 47.3 kg/mm^2

Factor of safety available on each bolt: 1.90 on yield strength

Nozzle design

Nozzle is a convergent – divergent conical nozzle. It is designed with an area ratio of 9. This is planned to fabricate from the forged rod of 200mm diameter.

Nozzle throat diameter: 35 mm

Nozzle exit diameter: 105 mm

Convergent diameter: 200 mm

Area of the divergent ratio: 9

Nozzle Convergent Thickness:

Convergent thickness is designed by using conical shell formula

$$t = \frac{P * d}{2\cos\alpha((UTS/F.S) * \eta_{weld} - 0.6P)}$$

α = half of the included angle of the cone = 55 °

By formula thickness (t): 1.511 mm

Nozzle throat metal backup thickness: The pressure at nozzle throat section will be approximately 0.54P$_c$. Therefore, at the throat metal back up

P$_t$ = 0.54P$_c$

$$t = \frac{P * d}{2((UTS/F.S) * \eta_{weld} - 0.6P)}$$

Where

Throat back up diameter : 35 mm

Pressure at throat : 81 ksc

Thickness by formula : 0.32 mm

Thickness chosen : 1.0 mm (from fabrication point of view)

Nozzle divergent thickness: Nozzle divergent thickness is with conical shell formula

$$t = \frac{P * d}{2\cos\alpha((UTS/F.S) * \eta_{weld} - 0.6P)}$$

α = half of the included angle of the cone = 14 °

Thickness obtained is 0.7 mm. The thickness obtained is very small; however from the manufacturing point of view a minimum thickness of 1.0 mm is selected (Figures 2 and 3).

Structural analysis of solid rocket motor hardware

The operating maximum internal pressure of motor is 150 ksc. Finite Element analysis of above motor is done for internal pressure.

Material properties:

Following material properties are considered in this analysis

Material: Maraging steel (MDN-250)

UTS: 175 Kgf/mm^2

YS: 168 Kgf/mm^2

E: 19000 Kg/mm^2

ν : 0.3

Weld zone properties:

Weld efficiency: 90%

UTS: 120 Kg/mm^2

YS: 108 Kg/mm^2

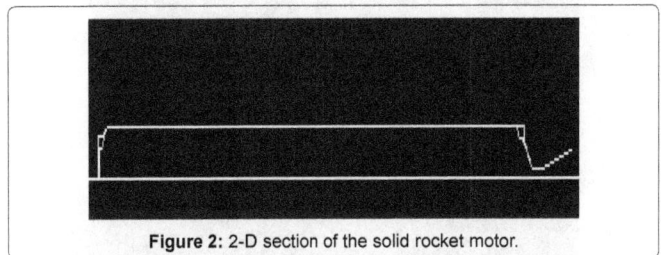

Figure 2: 2-D section of the solid rocket motor.

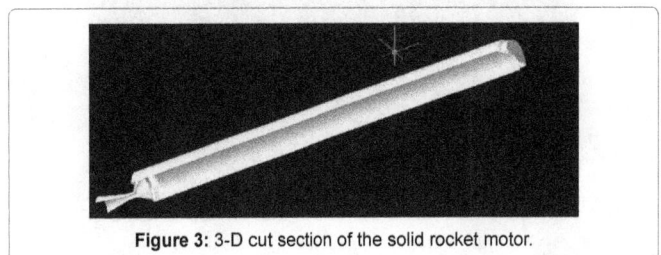

Figure 3: 3-D cut section of the solid rocket motor.

Allowable stress:

Factor of safety on UTS is 1.5 and factor of safety on YS is 1.33

σ_{all} = minimum of (σuts/1.5, σy.s/1.3)

= min (116.6, 126.3)

In Parent material = 116.6 Kg/mm^2

In weld zone = 80 Kg/mm^2

Stress Analysis of Motor Casing

Axisymmetric Finite Element model of motor casing is prepared by using 8-nodded axisymmetric element, (plane 183). Stress analysis of model is carried out for internal pressure of 150KSc by fixing upper bulkhead portion. FE model with boundary condition is shown in the Figures. Stresses at different location are given in Table 1 and locations are shown in Figures. Stress pattern is also shown in Figures 4-10.

Flange joint is modeled by arresting relative nodal displacement in appropriate direction. This will not give accurate result for bolt stresses, but our aim is to find out the stresses in casing only and this model will fulfill our requirements.

Fracture Analysis of Motor Case

The traditional design approach of rocket motor case is generally based on thin shell theory wherein the ultimate or yield strength of the material enters the design along with the chosen safety factors. However, it is well known that the strength-based design carries with it an inherent assumption that the material is free from flaws and perfect fabrication [7]. But in practice, the material may contain voids, cracks

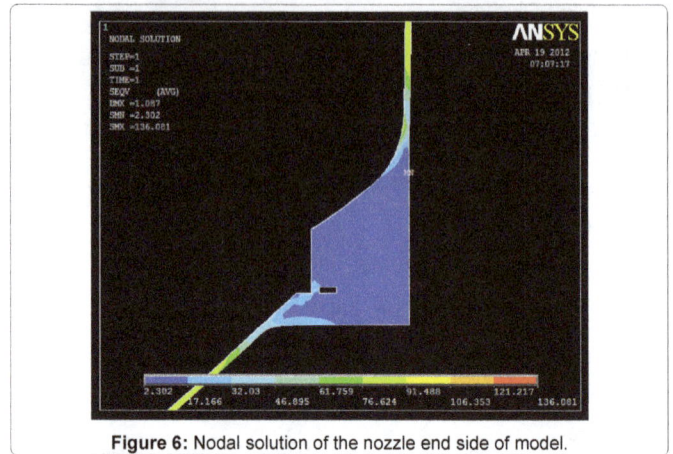

Figure 4: Showing the elements of the motor model.

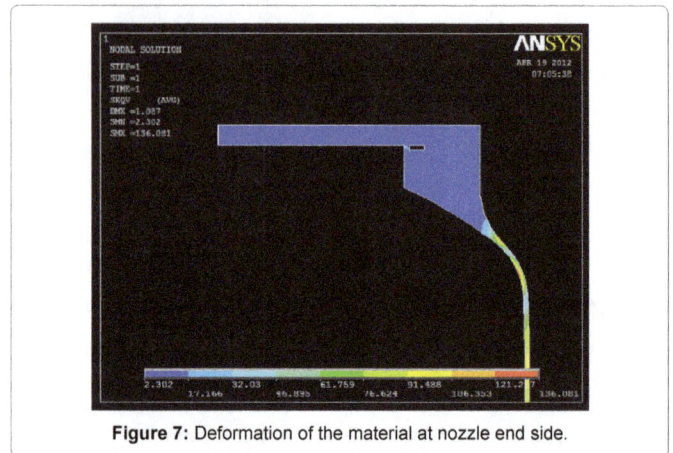

Figure 5: Nodal solution of the total model.

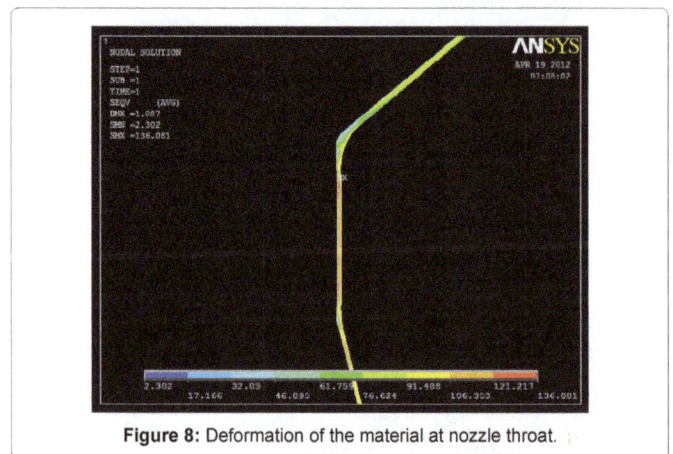

Figure 6: Nodal solution of the nozzle end side of model.

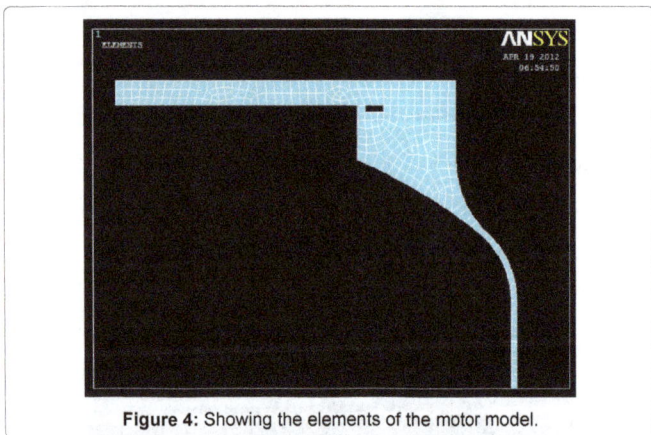

Figure 7: Deformation of the material at nozzle end side.

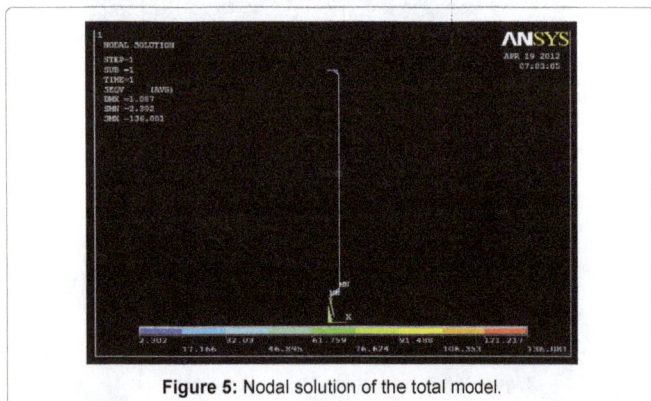

Figure 8: Deformation of the material at nozzle throat.

or some induced cracks in weldment and during weld repair process. These flaws can act on sites for crack initiation under certain conditions of operation and leads to catastrophic failure of the motor case.

Fracture analysis makes it possible to determine the critical crack size dimensions in selected motor case thickness for known fracture toughness of the material, which leads to failure. Based on analysis, the designer can specify the critical flaw sizes to take care in the raw material and inspection stage and selection of the material based on its fracture toughness value. For the present analysis, the following design inputs have been considered Figure 11.

Location	Von misses	F.S. on UTS
		Von misses
A	44.77	1.95
B	94.23	0.93
C	132.77	0.66
D	80.32	1.1
E	77.73	1.13
F	77.73	1.13
G	164.87	0.53
H	121.3	0.72
I	78.12	1.12

Table 1: Stresses at different locations.

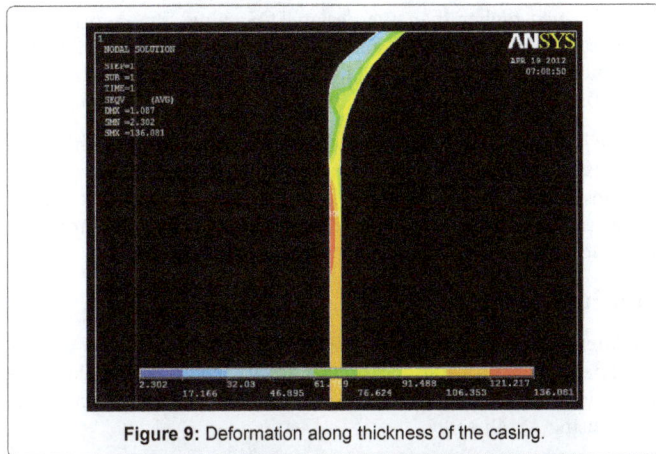

Figure 9: Deformation along thickness of the casing.

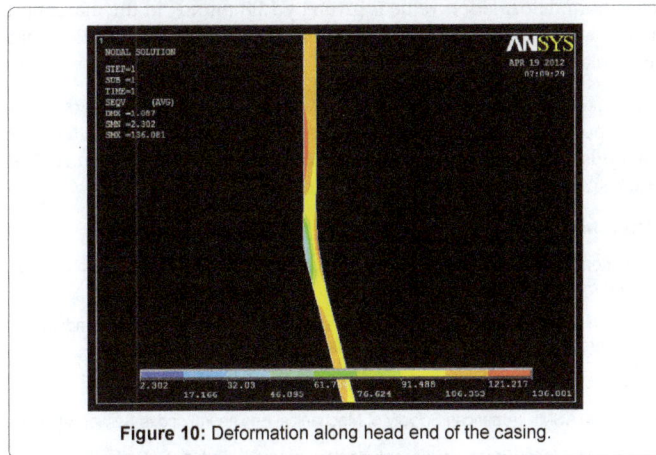

Figure 10: Deformation along head end of the casing.

Design inputs

Material: Maraging steel-250 grade

Yield strength of the material (σ_{ys}) : 168 kgf/mm²

Plane strain fracture toughness on parent metal (K_{IC}) : 90 MPa√m

Plane strain fracture toughness on weldment (K_{IC}) : 75 MPa√m

: 241.5 kg/mm³/²

Plane stress fracture toughness on parent metal (K_C) : 120-140MPa√m

The thickness of the motor case hardware (t) : 1.5 mm.

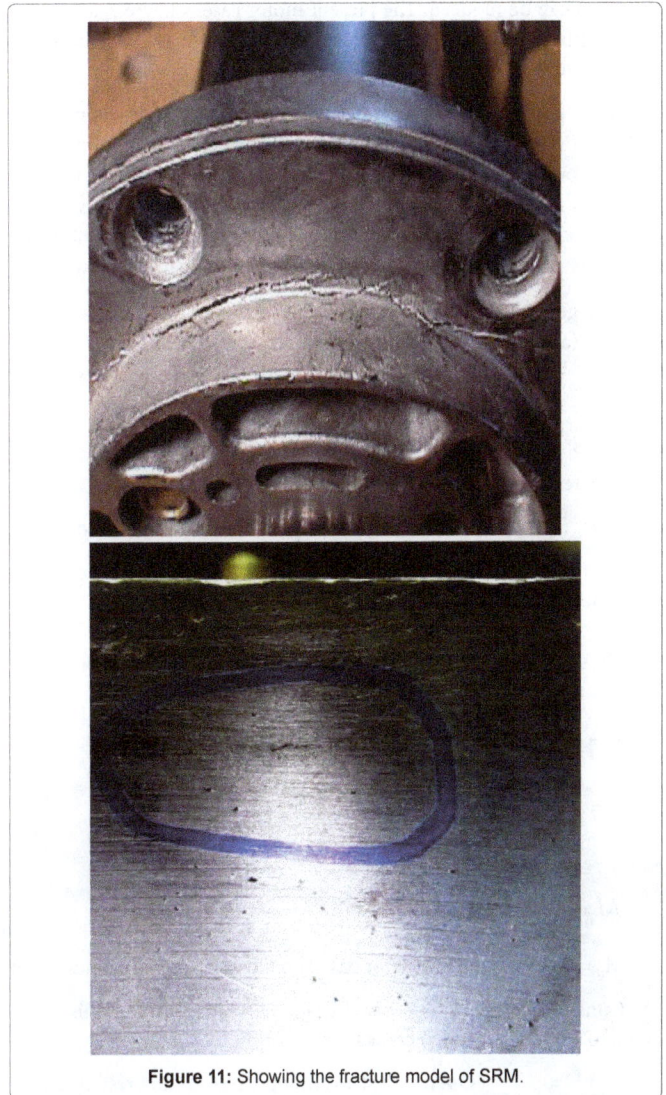

Figure 11: Showing the fracture model of SRM.

MEOP (P): 150 ksc

Motor outer diameter (D): 200 mm

The state of stress exists in thin shells is plane stress condition, whereas in thick shells it is a plane strain condition. The minimum thickness to achieve the plane strain conditions is given by

$$t = 2.5\left(\frac{K_{IC}}{\sigma_{YS}}\right) = 8.1 \, mm$$

The present motor case thickness is 1.5 mm. So, the plane stress fracture toughness ($K_{c\text{-}max}$) is relevant rather than the plane-strain fracture toughness. However, the plain strain condition is considered for this analysis due to the following reasons.

i) The plain strain fracture toughness(K_{IC}) is a material property like UTS, yield strength etc. and it is independent of thickness of material and flaw geometry for high strength materials. Whereas

ii) The plane stress fracture toughness (K_C) is not independent of material property but it depends on thickness of material, flaw geometry and its dimensions.

iii) "K_{IC}" of the given material will be lower than "K_C". Consideration of K_{IC} for design is more conservative and the number of allowable crack sizes also can be reduced. The present motor case is having number of weld joints. The welded motor case fracture toughness will be lower than parent metal fracture toughness.

Hence, the plane strain fracture toughness of weld ment of 75 MPa√m (assumed) is taken for analysis. The next assumption is that the flaw present is in such an orientation that it would tend to open up under the tangential stress. Elliptical surface flaws have been considered, as they tend to propagate at a faster rate than the embedded flaws. The stress intensity factor is more in elliptical surface flaws.

The general relation among K_{IC}, applied membrane stress(σ) and flaw size (a) for elliptical surface flaws is given by

$$K_I = 1.12 \, M_K \, \sigma \, \sqrt{\frac{a\,\pi}{Q}}$$

Where

K_I = Stress intensity for a given flaw size and stress ratio (σ/σ_{ys})

Q = Flaw shape parameter which is a function flaw aspect ratio (a/2c) and stress ratio

$$Q = \left[\varphi^2 - 0.212 \left(\frac{\sigma}{\sigma_{YS}} \right)^2 \right]$$

φ = elliptical integral of the second kind which is given by

$$\phi = \int_0^{\pi/2} \left[1 - \left(\frac{c^2 - a^2}{c^2} \right) \sin^2 \theta \right]^{1/2} d\theta$$

$$M_K = 1.0 + 1.2 \left(\frac{a}{t} - 0.5 \right)$$

M_k = Back correction factor based on flaws depth to thickness ratio

Constant 1.12 = Front free surface correction factor, this value equal to 1.0 for embedded cracks.

σ = Operating stress (hoop stress in the present case)

a = Crack depth (half of minor axis of ellipse)

c = Semi-crack length (semi major axis of ellipse)

Any meaningful fracture analysis is possible when industry data on minimum detectable flaw size with a specified degree of reliability and confidence levels is available. The minimum detectable flaw size invariably has to be validated at shop floor level and is a function of many factors viz, material properties, thickness in question, NDT methods used vis-à-vis state of art, NDT inspector's skill etc. For the sake of present analysis a set of flaw sizes have been assumed based on past industry experience on maraging steel plates and forgings.

Analysis:

Hoop stress due to internal pressure (σ) = PD/2t

= (1.5*178)/(2.0*1.5)

=100 kg/mm^2

Stress ratio (σ/σ_{YS}) = 0.69

Comments:

Fracture analysis has been carried out for different flaw depths

ranging from 0.5 mm to 1.2 mm and flaw length of 2 mm to 5 mm. It is seen from the results that up to crack depth of 1.2 mm and crack width of 2.5 mm for all the cases, the stress intensity levels are within limits and the factor of safety available is fairly good.

From the crack depth 1.5and crack width 3, the flaws detected are **critical** where stress intensity levels are high and FOS is not satisfied. However, these critical cracks can be detected through F_notch and G_notch NDT methods.

Results and Discussion

On observing all the results we got from ANSYS it is clear that the design can withstand the stresses produced because of the pressure of 150 Kg-f/cm^2. Since the stresses developed are lower than the UTS of the material i.e., 175 Kg-f/mm^2 = 1716.164 N/mm^2 and the factor of safety FOS obtained higher than the design FOS value. The deformations of the casing are also known and plotted in the analysis portion. The maximum deformation occurred in the casing is at divergent portion of the nozzle and its value is 1.087 mm.

We have also found the various allowable crack lengths which can be withstand by casing and also the critical crack lengths which cannot be withstand by the material. By knowing these critical crack lengths we can estimate the allowable crack lengths in imperfections during the manufacturing process.

Conclusion

After structural analysis of the Rocket motor casing it is clear that the stress developed at various locations of the casing are within allowable limit and a factor of safety of (FOS) 1.5 is obtained by designing the casing using ASME codes.

The deformation of the casing is also known using ANSYS and the maximum deformation value is given by 1.08 mm. And the maximum deformation occurred is located at the end portion of the divergent section of the nozzle.

It is seen that all crack sizes (and aspect ratios) considered, the corresponding critical crack sizes are well above the NDT detectable methods. The thickness of 1.5 mm motor casing with less than critical crack sizes will not affect based on fracture.

References

1. Sutton GP, Biblarz O (2001) Rocket propulsion elements. (7thedn), John Wiley and Sons New York.

2. Singh S (2013) Solid rocket motor for experimental sounding rockets advances in aerospace science and applications 3: 199-208.

3. (1970) NASA SP-8025 Solid rocket motor metal cases (N72-18785).

4. (2004) ASME (American Society of Mechanical Engineers) codes Section VIII Div.

5. Dennis Moss Pressure vessel design manual (3rdedn), Gulf professional publishing Burlington USA.

6. Davanas GE, Jensen DW, Netzer (1996) (eds.) Solid rocket motor design chapter 4 of tactical missile propulsion progress in aeronautics and astronautics AIAA 170: 323-379.

7. Beena AP, Sundaresan MK, Nageswara Rao B (1995) Destructive tests of 15CDV6 steel rocket motor cases and their application to lightweight design. Int J of Pressure Vessel and Piping: 313-320.

9

Fuel Leak Detection on Large Transport Airplanes

Behbahani-Pour MJ* and Radice G

Division of Aerospace Sciences, School of Engineering, University of Glasgow, Glasgow G12 8QQ, UK

Abstract

Fuel leaking from the tanks can be ignited by different sources, with catastrophic consequences for the flight; therefore it is important to detect any fuel leakage before the departure of the aircraft. Currently, there are no fuel leak detection systems installed on commercial aircrafts, to detect fuel tank leakage, while only a small number of more recent aircraft, have a fuel monitoring system, that generates a fuel leak-warning message in cockpit in the case of fuel imbalance between the tanks. The approach proposed in this paper requires the fuel vent ports on the wings to be replaced with fuel vent valves, which can be controlled to be in open or close position. The fuel vent valve will be in close position, when certain conditions are fulfilled (all the related fuel valves closed, pumps not operating, etc.), the fuel tank ullage area is then pressurized to 4 psi and the rate of change of the pressure is measured over a period. Several experiments have been conducted and, the result show that a continuous fuel leak of one liter per minute can be detected. Further experiments show that if the fuel tank is pressurized to higher pressures, a fuel leak can be detected sooner.

Keywords: Airplanes; Ignition sources; Pressures

Nomenclature

A/C = Aircraft

ADIRUS = Air Data Inertial Reference Unit

ASM = Air Separator Module

CMD = Command

CTR = Center

Diff. Press. = Differential pressure

EEC = Engine Electronic Controller

FTPC = Fuel Tank Pressure Controller

FADEC = Full Authority Digital Electronic Control

FWC = Flight Warning Computer

FCMC = Fuel Control Management Computer

LGCIU = Landing Gear Control Interface Unit

LH = Left Hand

LP = Low Pressure

M = Torque Motor

OBIGGS = Onboard Inert Gas Generation System

PS = Pressure Sensor

P/B = Pushbutton

Press = Pressure

SNR = Position Sensor

SOV = Shut-Off Valve

S/W = Switch

VLV = Valve

RVDT = Rotary Variable Differential Transformer

TK = Tank

Temp. SNR = Temperature Sensor

XFR = Transfer

Introduction

There have been several incidents of fuel tank explosion on commercial transport airplanes, which in some cases, have resulted in loss of lives and aircraft. Experts predict that fuel tank explosions on commercial airplanes could escalate due to increased number of electrically operated components inside fuel tanks, such as booster pumps, quantity probes, level sensors, fuel valves, etc. The U.S Federal Aviation Administration (FAA) states that in order to reduce the risk of fuel tank explosion on large transport aircrafts, any internal or external ignition source inside fuel tanks must be eliminated. When heat or a spark of sufficient energy comes into contact with fuel and oxygen, the result will be fire or explosion. Fuel leakage onto hot surfaces such as engine, APU, brakes, wheels, electrical fuel pump, hydraulic pumps, etc., may also result into fire and consequently fuel tank explosion.

The aim of this paper is to present a technique, which can be used on large transport airplanes, to detect fuel leak in fuel tanks before aircraft departure. Fuel leaks could lead to fuel tank explosion, and consequently could result in loss of aircraft and lives. Clearly, it is very important for any fuel tank leakage to be identified and repaired before takeoff. In August 2007, a China Airlines Boeing 737-800, registered B18616, took off from Taiwan International Airport scheduled Flight 120 at 08:23, and landed at Naha Airport at 10:27. It was destroyed by fire, after a fuel tank explosion due to fuel leak [1]. The purpose of minimizing fuel leak is clearly to increase flight safety, by reducing

*Corresponding author: Behbahani-pour MJ, Division of Aerospace Sciences, School of Engineering, University of Glasgow, Glasgow G12 8QQ, UK
E-mail: behbaa@yahoo.co.uk

the possibility of fuel tank explosions but it also avoids a reduction in aircraft range and prevents flight cancellations [2]. Fuel tanks may leak due to several reasons: excessive flexing of the tank structure causing cracking and overstraining of the sealant; sealant ageing; accidental damage; incorrectly set rivets during initial manufacture and more [3]. Additionally, there are several fuel booster pumps and fuel valves, located inside the fuel tanks, and therefore incorrect fuel component installation or defective seals can result into fuel leakage, and possible risk of fuel tank explosions. Fuel valves and pipelines are designed to be 100% leak proof, however loose joints or damaged seals may nevertheless lead to leaks from such pipes. For safety reason nitrogen is used to pressurize the fuel tanks and fuel system, however ambient air can be used to perform leak check on water/pneumatic/hydraulic/ oil pipelines and associated systems, using the proposed technique. In the proposed technique, in addition it can detect leak in the water / oil / air / hydraulic pipelines, leak proof valves and components.

The methodology presented in this paper, requires pressurizing the fuel tank with nitrogen, and measuring the rate of change of pressure over a period. Nitrogen is used because it is an inert gas, and thus reduces the risk of fire ignition and hence fuel tank explosion. In all aircraft, the fuel tanks are vented into the atmosphere, but in the proposed approach, the fuel vent ports must be closed. Therefore, the fuel vent ports on the wings are replaced with fuel solenoid or torque motor operated fuel vent valves, which can be commanded to open or close position, depending on the operational requirements. To carry out a fuel leak detection test, the tank is then pressurized to 4psi. Higher pressure would allow a leak to be detected sooner however, this may not be always possible as in some large transport airplanes, such as the Airbus A340-300 series, and there are several carbon disks, which act as overpressure protectors. The overpressure protector installed in the surge tank, breaks to relieve excessive pressure inside the fuel tank. Overpressure protectors are also installed in the fuel pipe between the center fuel tank and the inner fuel tank, and are designed to break between 5-7 psi [4]. Some fuel tanks such as the additional center fuel tank (bladder type) used on the Airbus A320s to increase the range of aircraft, have no over pressure protector, and hence these fuel tanks can sustain higher pressurization values.

Several experiments have been conducted by designing two sample fuel tanks from steel, installed with the pressure gauge to measure fuel tank ullage pressure, fuel pipelines, fuel flow-meter, fuel vent valves, fuel leak valve, fuel inlet and outlet valves, etc. and a flow-meter was used to measure the rate of fuel leak. During the experiment, the fuel tank is pressurized with nitrogen at different pressures, and then a fuel leak at different flow rates is produced, and the pressure drop of the ullage of the tank is measured. The result of the experiment was that firstly, if the size of fuel is increased, then the system detected the fuel leak quicker. Secondly, if the fuel tank is pressurized to higher pressure, then the system can detect a fuel leak in less time. The same technique could be used to detect leakage in air /water /oil /hydraulic pipelines, airtight fuel and pneumatic valves. In most aircraft, very hot, engine bleed air, is used to deice the wing leading edges (slats) and this air is routed via pneumatic ducts to the slats. Any hot bleed air leakage could result into a loss of engine thrust and in order to compensate, the fuel controller increases the fuel flow to the engine, resulting in higher operational costs. Therefore, it is important, to also perform a leak check on the pneumatic ducts before aircraft departure. In such case, ambient air instead of nitrogen can be used. The fuel leak detection test must be performed while the aircraft is on the ground so that in case of fuel leakage, the system can generate appropriate warnings in the cockpit and corrective maintenance action can be taken before take-off.

Following the TWA flight 800 crash in 1996, the US National Transportation Safety Board (NSTB) attributed the accident to the explosion of the center fuel tank. This accident led to the introduction of new regulations by the Federal Airworthiness Authority (FAA) to reduce ignition sources inside fuel tank. Currently, most modern aircrafts have center fuel tank inert system, which uses nitrogen generated by Air Separator Module (ASM) [5]. To perform a leak check on the fuel tanks, the current system can be modified by installation of an air compressor that supplies pressurized, ambient air to ASM when the engine is not operating, and to install fuel vent valves on fuel vent ports in order to isolate the fuel tank from the atmosphere. Therefore, the proposed design can be implemented at little cost, while greatly increasing flight safety.

Accidents Due to Fuel Leak

It is very difficult for a pilot to determine if there is a leak in the fuel tanks. For example, it was a passenger who saved the lives of 300 passengers, on a flight from Chicago to Tokyo. He spotted an outflow from the wing tip and when the captain was informed he confirmed the leak, and diverted the flight to San Francisco [6]. There have been several other narrow escapes, such as the case of the Japan Airlines flight from Boston to Tokyo on January 8, 2013, a Boeing 787 Dreamliner experienced a heavy fuel leak from left wing fuel tank before departure [7]. On the Air Canada flight ACA216 on November 6, 2003, the pilots did not receive any abnormal engine indications or any other warning, but following indications from air traffic control, they shut down the engine and made an emergency landing; the investigation revealed that the LP inlet fuel line to the fuel/oil heat exchanger on engine 2 was detached thus resulting in fuel leakage. If the pilot had not been informed by the tower, there could have been an explosion or fire [8]. Air Transat, Flight 236 was flying over the Atlantic Ocean on August 24, 2001, when the pilot received a fuel imbalance message but assumed it was a false warning. The pilots followed the standard procedures, and opened the fuel cross-feed valve and switched OFF the right wing pumps in order to feed the right engine from the left wing fuel tanks. However, due to a three inches crack on the fuel tube, a leakage developed on the engine and consequently both engines flamed out during flight, leading to an emergency landing. The leak was due to a maintenance error however, the pilot also made a mistake in assuming a false fuel imbalance warning. After the incident, the investigation recommended that the flight manual be amended, to ensure that fuel is not transferred by opening the cross-feed valve if there is suspicion of a possible leak [9].

Fuel leaks can however lead to disastrous consequences as happened to the China Airlines, B737-800 destroyed by fire at Naha airport in Japan on August 20, 2007, after a fuel tank explosion due to fuel leakage caused by a bolt piercing the wing [1,10] as shown in Figure 1. The leakage had started shortly after the landing, when the slats were fully retracted inside the wing, as shown in Figure 1, and within five minutes, approximately 60 pounds of fuel leaked from the right main tank and got in contact with hot engine surfaces resulting in fuel tank explosion and the aircraft being engulfed in flames [1,11].

British Airways Boeing 777-236 departed London Heathrow enroute to Harare in Zimbabwe, on 10 June 2004. Immediately after takeoff, a fuel leak was detected and emergency landing initiated. The UK Air Accident Investigation Branch (AAIB) revealed that the fuel leak was caused by fuel escaping through the center fuel tank purge door inside the left main landing gear, shown in Figure 2. The aircraft had recently undergone a scheduled, major maintenance check, during which the center fuel tank purge door was removed for maintenance,

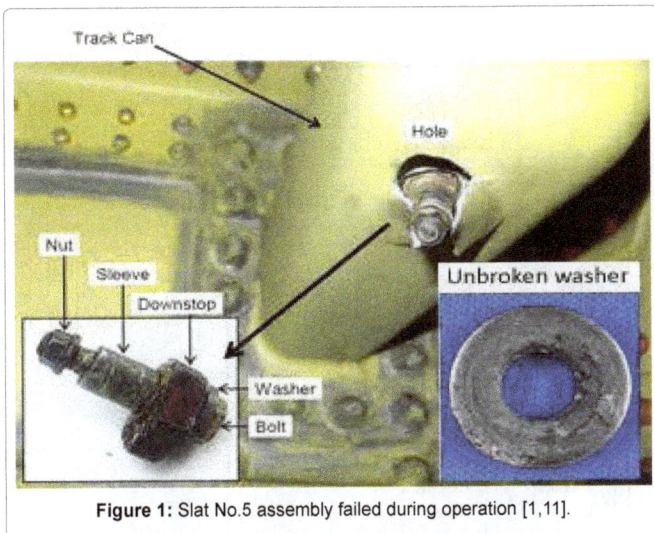

Figure 1: Slat No.5 assembly failed during operation [1,11].

Figure 2: Center fuel tank purge door circled in red was left open by maintenance personnel [12].

but not re-installed after the completion of the task [12]. The most modern airplane Boeing 787 Dream Liner suffered fuel leak on runway, and the passenger alerted the cabin crew. The incident was captured on a video available on You Tube [13].

A UK military accident investigation concluded that the fatal accident of Royal Air Force BAe Nimrod crash on the 2 September 2006, in Kandahar, Afghanistan, which resulted in the death of 14 servicemen, was caused by fractured fuel pipes. As the aircraft was being refueled in mid-air, the leaking fuel from fractured fuel pipe caught fire, possibly due to a spark from faulty wiring [14-17].

Current Operational Fuel Leak Procedures

Currently, there is no onboard system that can detect a fuel leak from the tanks. The Airbus A320 Flight Crew Operation Manual (FCOM) states that "a fuel leak may be evidenced by either passenger observation (fuel spray from engine or wing tip) or total fuel quantity decreasing at an abnormal rate or if there is a fuel unbalance" [15]. Hence in case of fuel leak, the pilot must land as soon as possible. Operational procedures state that a fuel imbalance between left and right wing fuel tank, greater than 500 KG, indicates a fuel leak. This approach however, is of somewhat limited effectiveness, because, fuel imbalance between the two tanks can occur due to many factors: defective fuel quantity probe harness, contaminated fuel probes, leaks from fuel booster pump check valves; unbalanced fuel flow from fuel

booster pump; fuel cross-feed valve or fuel inlet valve left in open position; broken fuel delivery pipes, etc.

Several methods are used during aircraft maintenance checks to perform leak test on the fuel tank. A common approach is to totally defuel the tanks, and then manually applying special materials to the internal surfaces of the fuel tank. Compressed air is then applied to the outside of the tank where a leak may be suspected [4]. The disadvantage of this method is that it requires maintenance personnel to enter inside the fuel tanks, increasing the risk of fuel tank explosion if safety advices are not strictly observed. To enter a fuel tank, the fuel tank must be totally dry so it has to be purged with dry air for at least 16 hours, all tools must be free from static charge to avoid sparks, safety shoes to be used, etc., before compressed air is applied to suspected areas. However, if the fuel leak location is not known, then this process is very time consuming. In some cases, such as the fuel trim tank inside horizontal stabilizer there is no space for personnel to access the inside of the tanks.

Other leak detection methods include defueling the aircraft, sealing off the fuel vent ports by installing NACA vent dummy door to the fuel vent ports and then using a manometer, to create a negative pressure of not more than 2.9 psi (to prevent operation of overpressure burst disks). Negative pressure is applied via the dummy door to fuel tanks and then applying a dye solution to the external surface of the fuel allows the identification of the location of leak [4]. The disadvantages of this method are that it again requires personnel to enter fuel tanks, as well as being time consuming. In addition, it requires specialized equipment, and qualified staff to perform the leak test.

Fuel leak test on the additional center, flexible bladder type tank, on Airbus A319/320 requires the tank vent port, fuel drain pipe and fuel transfer pipes must be sealed off in order to prevent air to vent outside the tank during the test. The fuel tank is the pressurized with dry air up to 20.3 psi to detect any leaks [16]. The disadvantage of this procedure is that again it requires special equipment and certified staff.

Fuel Tank Leak Detection System

In order to perform a fuel leak test, nitrogen in the atmosphere is extracted through an Air Separator Module (ASM) and, injected into the appropriate fuel tank. The pressure level inside the tank is then monitored to ascertain the presence of any leaks. In order to prevent nitrogen escaping through the fuel vent ports on the wings, the fuel tank must be isolated from any vent system as shown in Figure 3. The Fuel Tank Pressure Controller (FTPC) will control the fuel leak detection process, and contain the logic software instructions. It receives position feedback from the valves, fuel pressure sensors, etc., and performs the leak test and displays the result in the cockpit for the flight crew or aircraft maintenance engineers.

In the proposed approach, the fuel tanks are pressurized with nitrogen, and each fuel tank has a fuel vent valve, which closes during the test. The system is able to determine which fuel tank is leaking and leading to corrective actions by maintenance personnel. Location of the leak can be then determined either through visual observation or by applying a soap solution to exterior surface of the tank. The main benefits of this approach are that it does not require specialized equipment or qualified personnel and can be performed at the terminal with fuel inside the tanks.

The FTPC is a controller, which stores the program in its main memory module and determines what actions are required. On large transport airplanes such as Airbus A330/A340 series, the fuel is stored in

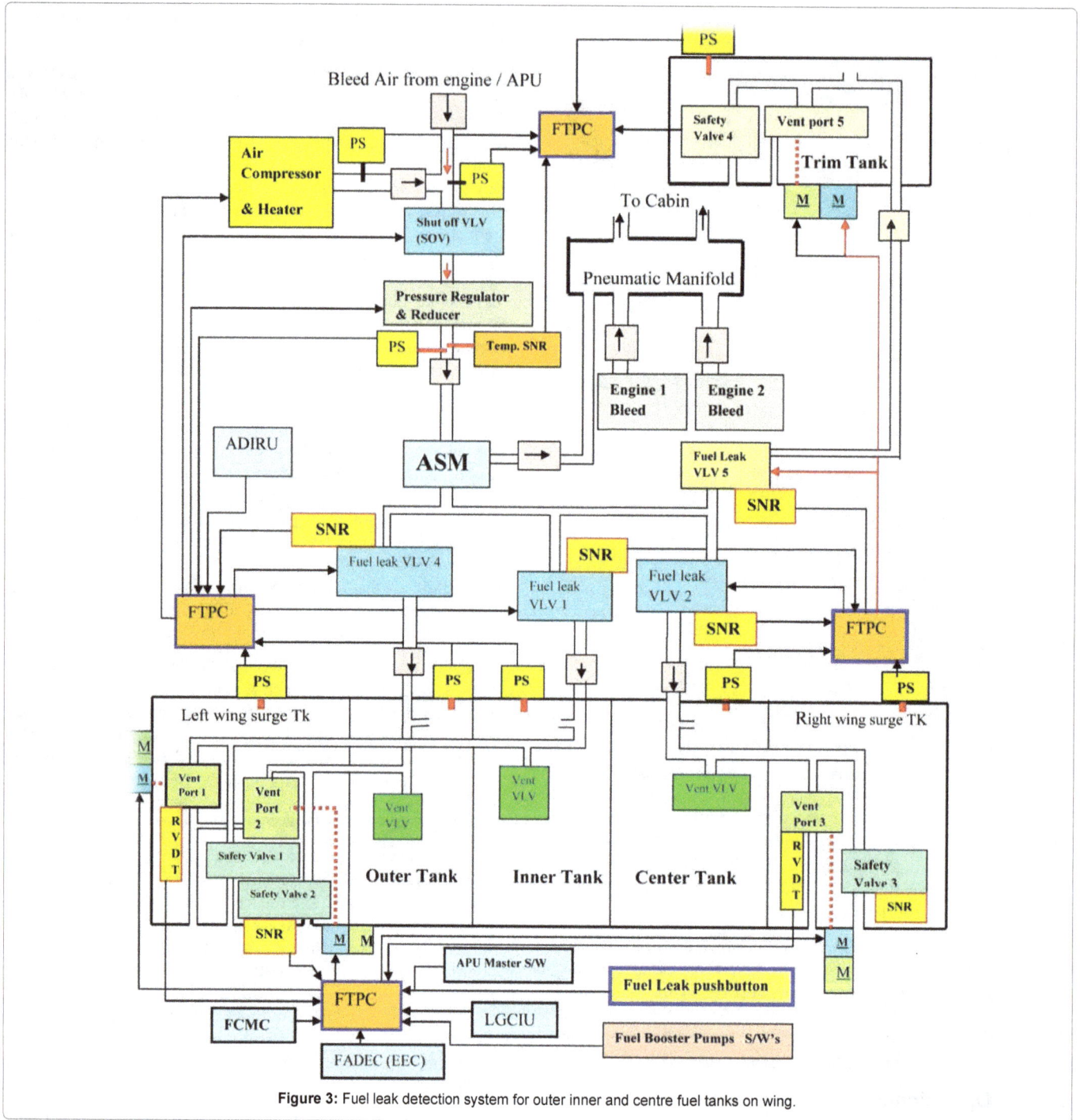

Figure 3: Fuel leak detection system for outer inner and centre fuel tanks on wing.

the left and right wings, centre tank in the forward cargo compartment, and possibly additional centre tanks in the aft cargo compartment, and horizontal stabilizer. Fuel tank in horizontal stabilizer is called trim tank. The wing tanks are divided into inner fuel tanks, and outer fuel tank located towards the wing tip. In the Airbus fleet, fuel is transferred from the centre tank to the inner tanks in the wings so that fuel is left in the wing tanks as long as possible, to avoid excessive wing bending during flight. Additionally, fuel is transferred forward or aft during flight in order to control the position of the centre of mass.

Currently, all transport airplanes have one fuel vent port in the

left wing and one fuel vent port in the right wing, and one in the trim fuel tank (horizontal stabilizer). The fuel vent ports in both wings are interconnected to each other so that all wing fuel tanks are linked together, and are permanently open to the atmosphere. However, in the proposed method, each fuel tank has its own fuel vent (valve) port, which is controlled by the FTPC. The reasons behind each fuel tank having its own fuel vent port, are to prevent nitrogen escaping into the atmosphere during fuel tank pressurization, and to easily determine which fuel tank may be leaking. We have used the Airbus A330 fuel system for the proposed leak detection design but believe that it can easily be applied to any other aircraft fuel system. In the proposed methodology, the FTPC can be programmed to give the operator the choice to select on which fuel tank the test is to be performed. The fuel leak detection test can be done on individual tanks or all tanks at the same time, regardless of their location or type. The fuel leak test is performed while the aircraft is on ground, and the personnel have to press and hold fuel leak test switch for seven seconds, this is to prevent inadvertent operation of the leak detection system, as part of safety feature, to allow the crew to de-select the system, in case it was incorrectly selected.

Fuel leak test starts when certain conditions are met :all the related fuel valves closed, no fuel transfer, no refuelling / no defueling, no jettison, and fuel booster pumps not operating, etc. as shown in Figure 4. Additionally, the fuel tank should not be full to its maximum capacity, because, if the fuel tank is filled with fuel to maximum, then there is no or very little ullage area in the tank. With the aircraft on ground and engines and APU not operating, there will be no bleed air supply, therefore, the air compressor combined with electrical heater, extracts air from atmospheric, pressurizing it to above 40 psi and heating it. The air supply to ASM is heated to 180F plus or minus 10F, in order to increase the efficiency of ASM then air is supplied to the Air Separator Module (ASM) via the shut-off valve. The ASM supplies nitrogen to the fuel tanks via fuel leak valves, as shown in Figure 3, until the nitrogen pressure reaches the appropriate threshold of 4psi. Each fuel tank is equipped with two high fuel level sensors located on the top of the tank; and two low fuel level sensors located at the bottom of the tank, in such case, as part of fuel leak test condition, the high fuel level sensors should not sense wet state condition which is maximum fuel quantity. Each fuel tank has its own separate fuel vent port controlled by FTPC, this is done in order to identify which fuel tank on aircraft is leaking fuel (failed the test), so to assist the engineer in trouble shooting while reducing time and manpower.

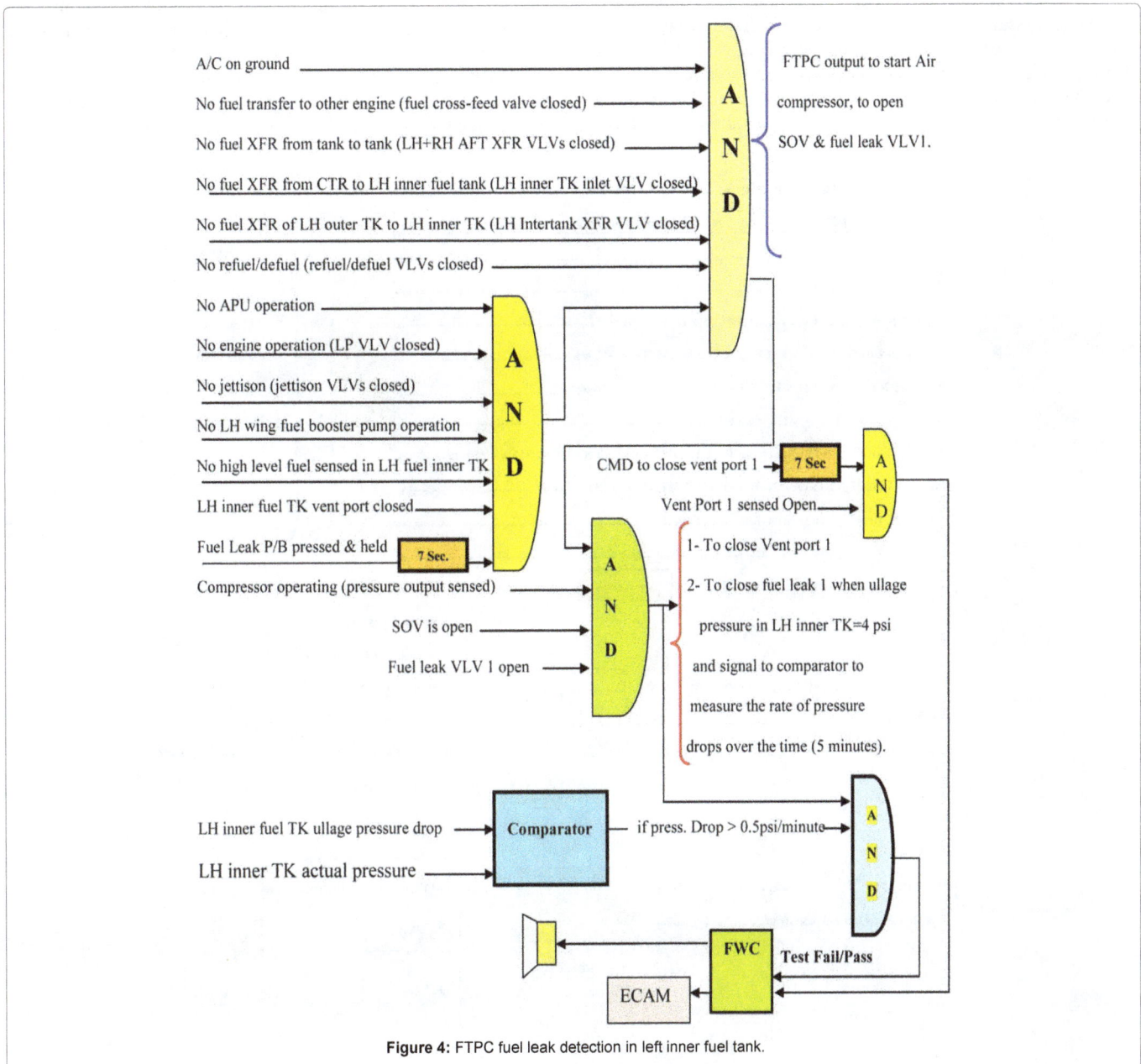

Figure 4: FTPC fuel leak detection in left inner fuel tank.

The FTPC receives the fuel tank ullage pressure data and measures the rate of change of pressure. With the fuel tanks pressurized to 4 psi if there is no pressure drop after five minutes then fuel leak test is satisfactory and a text message appear in the cockpit for crew attention and they can switch off the leak test system. However, if the leak test is not satisfactory, the FTPC will send signal to Flight Warning Computer (FWC) to generate oral and text warning messages on ECAM screens and the maintenance personnel have to take corrective action to rectify the defect before the aircraft departure. In order to perform a fuel leak test on the centre fuel tank, then certain conditions must be fulfilled, as shown in Figure 5.

Safety Feature

In case of the proposed system design malfunctions, such as over pressurization of the fuel tank during fuel leak test, the FTPC will command the shut-off valve to close, to stop the tank pressurization, in order to prevent damage to the wing structure. In case of excessive differential pressure, the safety valve installed in the surge tank opens. In the proposed approach, there is one safety valve, which connects all the wings tanks to the atmosphere as well as one safety valve for the trim tank. The FTPC has an ambient pressure transducer to determine the ambient atmospheric pressure, while as a back-up the FTPC receives ambient pressure data from air data computers known as Air Data Inertial Reference Unit (ADIRU). The FTPC, compares the internal pressure of the fuel tank with ambient atmospheric pressure and if the maximum differential pressure between the tank ullage pressure and ambient air pressure exceeds the aircraft manufacturer structural limit, then FTPC will command shut-off valve (SOV) to close, and hence stop fuel tank leak test. A fuel tank can withstand certain amount of mechanical stress, without any failure, due to the differential pressure between the fuel tank and ambient air pressure. However, modern aircrafts such as Boeing 787 Dreamliner using composite materials can tolerate much higher structural stress than metal fuel tanks, due to higher flexibility higher resistance to mechanical fatigue [18].

In addition, for safety reasons, the fuel vent valve is controlled by two torque motors, so that in case of one torque motor fails, the second will take over. In case, the fuel vent port valve fail in close position, the position sensor senses the position of the valve, and the controller compares the command output with actual valve position, thus generate warning in the cockpit. The fuel vent port valve can be moved to open position manually, in order to dispatch aircraft on time. In addition, the fuel vent valve can be a solenoid-operated valve, with two coils, which can move the fuel vent port to fully open or close position. The position sensors or RVDT provides position feedback to FTPC.

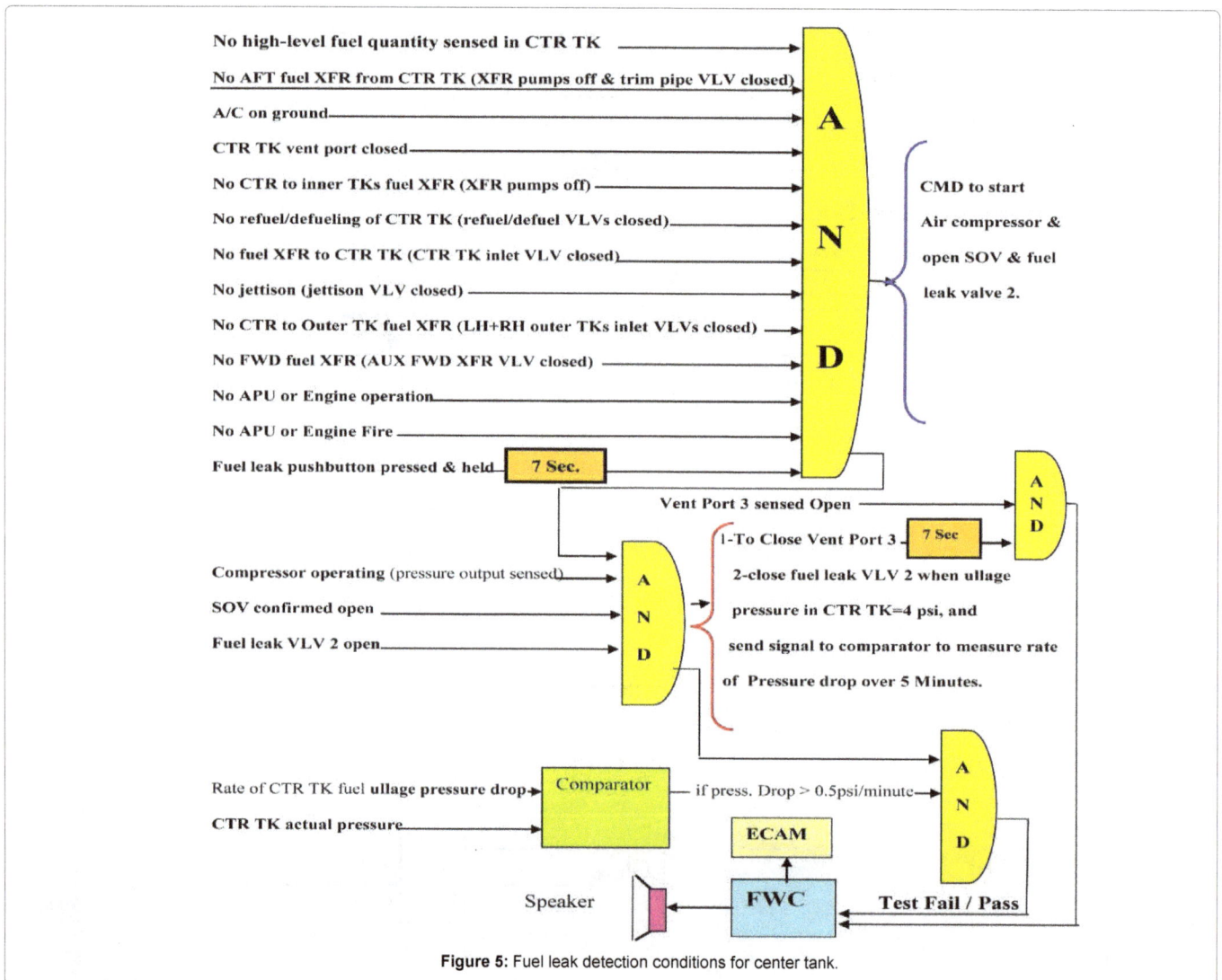

Figure 5: Fuel leak detection conditions for center tank.

Safety Valve

The safety valve design, shown in Figure 6, consists of a diaphragm held by spring tension, and in case of positive differential pressure of more than XXXX psi, the valve moves down (opens) and allows the fuel tank to vent into the atmosphere. The safety valve proximity sensor-2 will sense the proximity of the valve and send a signal to FTPC to indicate that there is a positive maximum differential pressure, and the FTPC will command shut-off valve (SOV) and to stop the pressurization of the tank. In case of negative differential pressure (atmospheric pressure greater than fuel tank pressure), the safety valve moves upward (opens) and allows air to enter the fuel tank. As the valve moves upward, the safety valve proximity sensor-1 will sense it, send signal to FTPC, which in turn, will command the pressure regulator to open more, to increase air supply to the ASMs in order to supply more nitrogen flow to the fuel tank, and reduce the differential pressure to be within the limit.

Fuel Tank Leak Detection Experiments

The hypothesis that underpins this work is that if a vessel is leak proof, then when pressurized with air or nitrogen, it should not experience any pressure drop. During the experiment, the ambient pressure of day was 1013 mb and ambient temperature of 30°C.

Figures 7- 9, present the experimental setup used to validate the fuel leak detection system. Two galvanized steel tanks with a capacity of 62 liters are equipped with several ports for pipeline connections. All pipelines had washer seals and sealant applied to all joints to ensure that the tank is air-tight. A flow-meter is screwed on the pipelines, and there are also several shut-off valves (SOV). All valves are opened and closed manually during the experiment. A Bourdon tube type pressure gauge is screwed on the top of the tanks, to measure the tank ullage pressure. Initially, SOV 2 and SOV 3 are fully closed, and the vent valve 1 opened fully, to allow trapped air to escape while 27 liters of fuel are transferred to fuel tank#1 through SOV 1. When the fuel has been transferred vent valve 1 is closed manually. Dry nitrogen stored under high pressure is then connected to the inlet of SOV 1 through a flexible hose. SOV 5 is opened very slowly, to allow nitrogen to flow into the tank to a pressure of 2 psi. The fuel leak simulation valve is then opened to simulate fuel leak of half of liter per minute, as shown on the flow-meter with the leaking fuel being collected in a container. During the

fuel leak experiment, the pressure gauge was monitored to assess any pressure changes. At the same time, when the fuel leak was simulated, a timer was used to determine when the pressure on the gauge started to change.

As shown, in the Figure 10, it takes just over three minutes for the pressure to change. Different fuel leak flow rates were generated, by varying the opening of SOV. With a fuel flow leak rate of 1 liter per minute it takes just over two minutes before the pressure gauge registers a change in ullage pressure. Clearly as the fuel leak rate is increased, any pressure change is detected more rapidly. A further experiment, in which the fuel tank #1 is subjected to pressures above 4 psi, is conducted. The purpose of this experiment is to determine the effect of pressurization on leak detection time. Therefore, tank #1 pressurized to 6 psi, then with all the valves closed, SOV 3, fuel-cross feed valve, fuel tank vent 2 is opened manually and fuel is transferred from fuel tank # 1 via SOV 3, flow-meter, and fuel-cross feed valve to fuel tank #2. It was observed that if fuel is transferred from one tank to another, it affects the pressure of fuel tank ullage area; therefore, it is important that during fuel leak test, there should be no fuel transfer between the fuel tanks. Therefore, this condition is reflected in the proposed approach, as part of requirement logics, that there should be no fuel transfer, no defueling and no refueling of the fuel tank during the leak test.

With all the valves closed, the nitrogen bottle pressure regulator is manually adjusted to read 4 psi and SOV 1 opened slowly to allow nitrogen to pressurize fuel tank # 1. SOV 1 closed when the pressure gauge showed 4 psi. Then SOV 3 is opened to let fuel to inlet of the flow-meter, and the fuel leak simulation valve is slowly opened to create a fuel leak of one liter per minute. It was noted, as shown in Figure 11, that it takes around 1 minute for a pressure change to be observed on the pressure gauge. This experiment was then repeated by increasing the ullage pressure by 2 psi. It can be seen that an increase of 2 psi results in a decrease of approximately 10 seconds in the time to detect a fuel leak.

During the experiment, the ambient pressure of day was 1013 mb and ambient temperature of 30°C.

Analysis and conclusion of the experiments show that higher

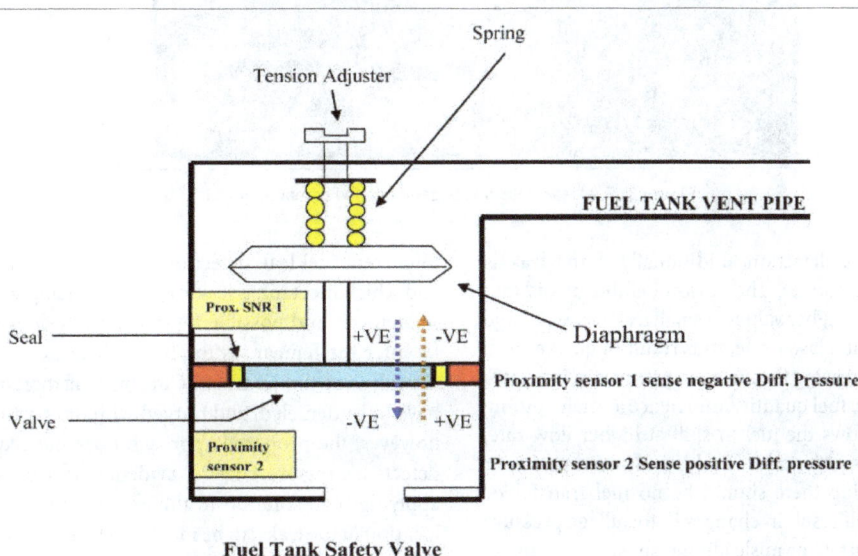

Fuel Tank Safety Valve

Figure 6: Fuel tank safety valve.

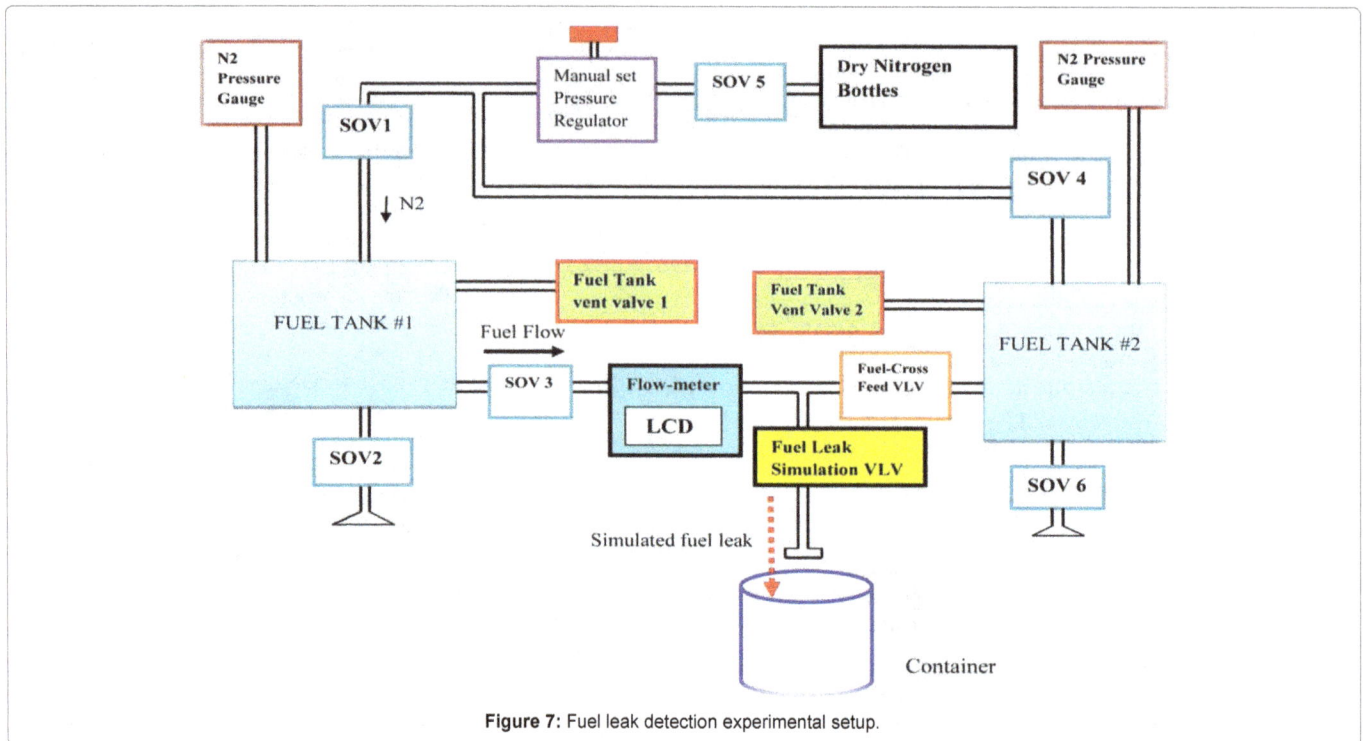

Figure 7: Fuel leak detection experimental setup.

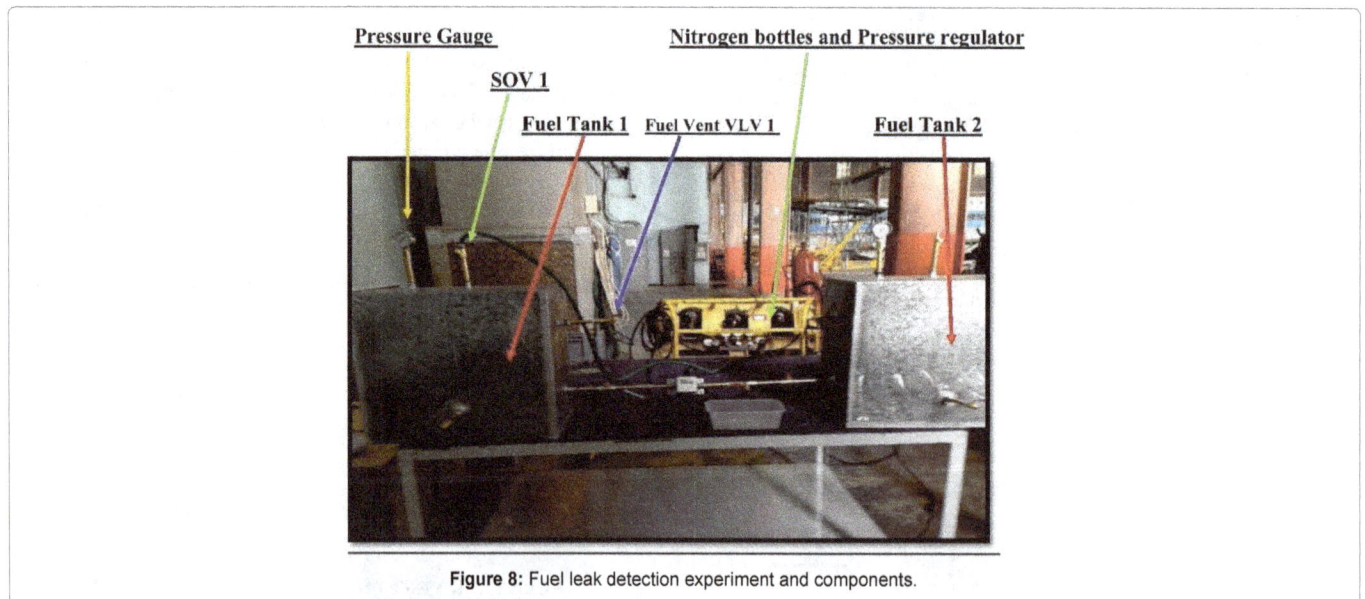

Figure 8: Fuel leak detection experiment and components.

pressures lead to quicker leak detection, additionally if the leak is increased, then it is identified sooner. The reason behind this is that when the tank is pressurized to higher values, a small leak causes a large drop in tank ullage area, and at a faster rate, thus result in quicker leak detection time. Because by increasing the nitrogen pressure, it increases the head pressure acting on the fuel quantity, and hence a small opining as in case of tank leakage, allows the fuel to spill at higher flow rate, and ullage area to increase at faster rate. Analysis of the experiment also showed that during testing there should be no fuel transfer or defueling/refuelling, as this will result in changes in the ullage pressure and hence could lead to inaccurate or misleading results.

The proposed approach presents many advantages when compared to current fuel leak detection approaches. It can quickly determine if and which fuel tank is leaking, thus reducing trouble shooting time and manpower, and possible flight delays. It does not require the aircraft to enter the hangar, as the fuel leak test can be performed while the aircraft is at the terminal. Currently, if there is a fuel leak, the tanks have to be defueled, and purged with dry air to get rid of fuel vapours; however, the proposed approach does not require all this. If the test detects a pressure change, evidence of a possible fuel leak, then by applying soap solution to the external surfaces of the tank, the exact location of the leak can be identified through bubble formation. Finally, the proposed approach does not require maintenance personnel to enter fuel increasing workplace safety.

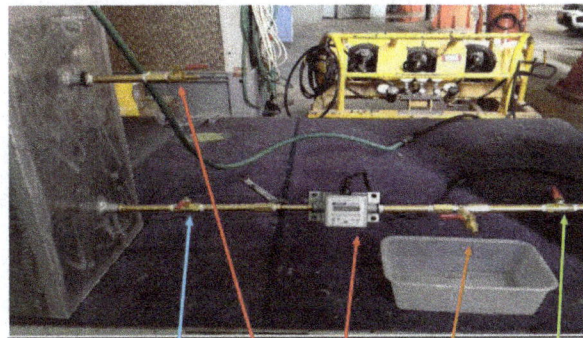

Figure 9: Fuel leak detection experiment.

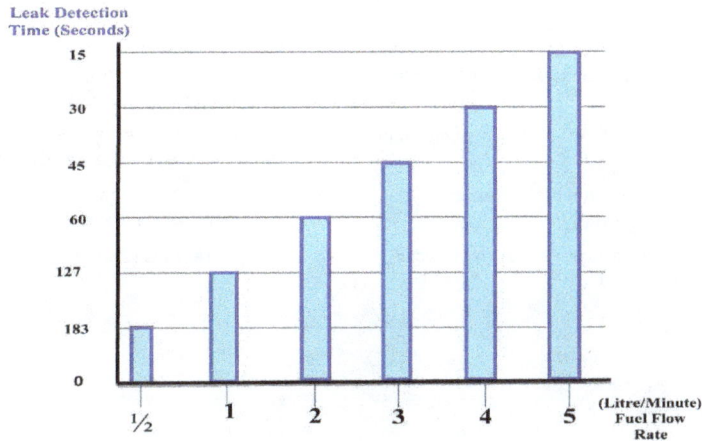

Figure 10: Leak detection time as a function of leakage flow rate.

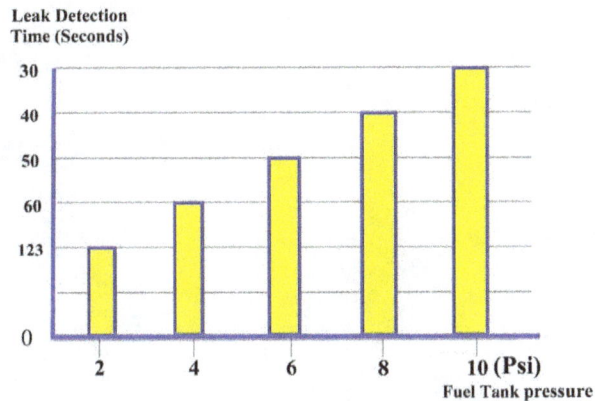

Figure 11: Leak detection time as a function of ullage pressure.

Conclusion

- Fuel tank leak on large transport airplanes is obviously a very serious concern. Fuel may leak onto hot brakes or hot surfaces and ignite. Even if the fuel leakage does not present ignition possibilities, it will nevertheless lead to a reduction in the aircraft range and possible emergency landing prior to destination arrival.

- Currently, there is no on-board system that can accurately detect a fuel leak and maintenance procedures are often hazardous, time consuming and staff intensive. Therefore, the authors have proposed a system, which can quickly detect any fuel tank leak before the aircraft departure, which requires low manpower and does not necessitate empty fuel tanks.

* The underlying hypothesis is that if a vessel is 100% leak proof, then when it is pressurized there should be no pressure drop. Any leakage will allow air to escape from the vessel, leading to the pressure inside the vessel to drop. The purpose of the experiments carried out was to prove that this hypothesis can be used to detect a fuel leak by measuring the pressure drop in tank ullage area. The sample fuel tank was therefore pressurized with nitrogen at different values, and fuel leak at different flow rates was introduced. Pressure changes in the ullage area of the tank were then recorded. The results of the experiment show that larger fuel leaks are detected more quickly and that higher pressures in the ullage area lead to a quicker determination of any fuel leak. In the experiment, a basic bourdon tube pressure gauge was used to detect the changes in the ullage pressure but we believe that more sensitive pressure measuring devices could detect very small fuel leak, of the order of just a few drops per minute.

References

1. JTSB (2009) Aircraft accident investigation report. China airlines B18616 Rept AA2009-7, Tokyo.

2. Langton R, Clark C, Hewitt M, Richards L (2009) Aircraft fuel systems John Wiley & Sons Ltd UK.

3. Sheridan NG (2015) Fuel tank technology AGARD Rept. 771. Research group for aerospace research and development, France.

4. Airbus (2001) A340 Aircraft maintenance manual. Airbus manufacturer technical publications department, France 28: 13-14.

5. MacIsaac B, Langton R (2011) Gas turbine propulsion systems. Wiley Publication, USA.

6. Leak L (2014) Airman spots jet liner fuel leak. We never would have made into Japan.

7. Moir I, Allan S (2015) Aircraft systems: Mechanical, electrical and avionics subsystems integration (3rdedn). Wiley Publications, UK.

8. Transportation Safety Board of Canada (2014) Maintenance error- in-flight fuel leak air canada airbus A330-300 C-GHKX Vancouver International Airport British Columbia. Government of Canada.

9. Accident Investigation Final Report (2014) All engines-out landing due to fuel exhaustion air transat airbus A330-243 marks C-GITS Lajes, Azones, Purtugal, Rept. 2/2007. Aviation Accidents Prevention and Investigation Department, Portugal.

10. Barbara C (2009) A week of drama. Flight Global, UK.

11. FAA (2015) Accident overview history of flight. US Department of transportation, Washington.

12. Air Accidents Investigation Branch (2007) Report on the serious incident to Boeing 777-236, G-YMME on departure from London Heathrow airport on 10 June 2004. Department of Transport, UK.

13. Saive H (2008) Boeing 787 spills fuel on runway as passengers alert crew.

14. Hoyle C (2007) UK Nimrods face new safety review following crash report. Flight Global, UK.

15. Airbus A320 (2015) A320 flight crew operating manual. Airbus Support Division Technical Publications Department, France, Vol. 3, 2015, pp.8-9.

16. Airbus A319/A320 (2010) Aircraft trouble shooting manual. Airbus Support Division Technical Publications Department, France 28: 219-220.

17. Wastnage J (2006) RAF sources leak September fatal Afghan Nimrod investigation findings, showing fractured fuel pipes. Flight Global, UK.

18. Wanger M, Norris G (2009) Boeing 787 Dreamliner. Zenith Press, USA.

10

Aerodynamic Load Estimation of Helicopter Rotor in Hovering Flight

Reddy MVR*

Department of Aeronautical Engineering, Nitte Meenakshi Institute of Technology, India

Abstract

The aerodynamic characteristics of a helicopter rotor blade are highly dependent on the wake induced flow. The rotor blades, which are rotating wings, shed vortices which trail in helical path along the axis of rotation forming a wake. The velocity field induced by the vortex system influences the blade loading. The free wake model analysis is computationally intensive and numerically unstable in very low speed applications.

For the aerodynamic analysis of a hovering rotor, Miller and Reddy have proposed simple wake models in which the spiral vortex wake is replaced by a system of vortex rings or vortex line elements in conjunction with lifting line theory to overcome some of the problems encountered in free-wake models.

In the Proposed model of the rotor, the same lifting line model of the rotor is retained, but a further simplification of the vortex wake is proposed. This method requires relatively very small computing time compared to the above models, but without necessarily sacrificing the accuracy in estimating the aerodynamic blade loading.

Keywords: Helicopter; Rotor blade; Wake; Vortices

Introduction

The helicopter is an aircraft that uses rotating wings to provide lift, propulsion and control. The rotor lift is obtained as a reaction force by accelerating the air downward of the rotor. The blades of the rotor are maintained in uniform rotational motion, usually by the shaft torque from an engine. The lift and drag forces on these rotating wings produce the torque, thrust and other forces and moments of the rotor [1-4].

Development of lift on the rotor

In its simplest model the rotor can be considered as a rotating device capable of producing lift by drawing air from above, increasing its axial velocity and made to pass through the rotor (Figure 1). The change in the velocity of the slipstream caused by the rotor is called the induced velocity and this change results in the change of momentum of the slipstream so that a reaction to this change occurs on the rotor which is the thrust.

The blade is a rotating wing and as such it can be modeled by a vortex system consisting of bound vortices within the wing and free vortices coming off from the trailing edge (Figure 2). These free vortices are carried downstream by the axial velocity of the slipstream so that a helical vortex sheet formation takes place within the slipstream (Figure 3). This wake is to be modeled for the estimation of aerodynamic characteristics.

Modeling of the wake

The slipstream below the rotor has significant influence on the performance of the blade. Therefore the wake has to be modeled so that one can get a quantitative estimate of lift and torque produced by the blades, the most accurate method of calculating the performance of the blade is to model the blade as a lifting surface with its span-wise and chord-wise vortex system and the free vortex sheet emanating from the trailing edge in the form of a helical vortex system. This is called a free wake (Figure 4) analysis and calls for large computing time.

The Proposed Method

It may be observed from earlier studies that several models such as free spiral vortex wake model (Figure 4), vortex ring model used by Miller (Figure 5), have been proposed to predict the load distribution of helicopter rotor blades. These models even after simplification require extensive computing time. This is the purpose of this present study to explore a model that would retain the accuracy of computation and to introduce further reduction and simplicity in computing effort.

Description of the proposed model

In the proposed model, the blade is treated as a rotating wing of finite span. Each blade is divided into a number of segments 'n' in the

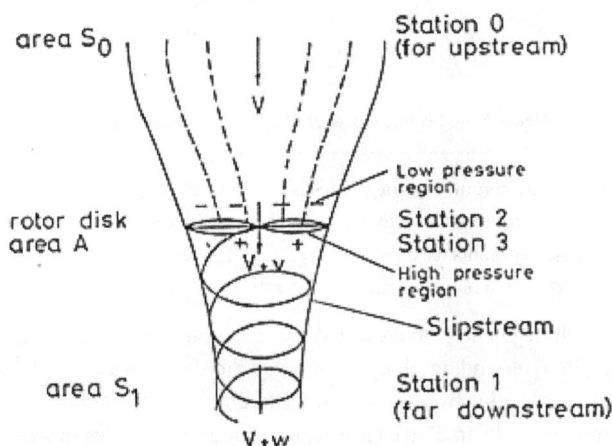

Figure 1: Slipstream formed by a vertical climbing rotor.

*Corresponding author: Reddy MVR, Department of Aeronautical Engineering, Nitte Meenakshi Institute of Technology, India
E-mail: mulavrreddy@yahoo.com

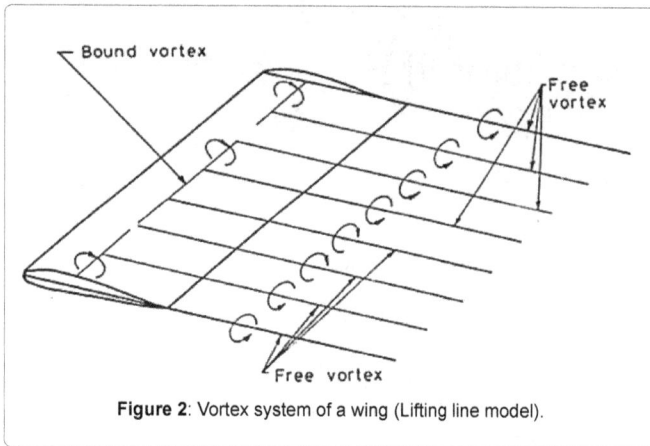

Figure 2: Vortex system of a wing (Lifting line model).

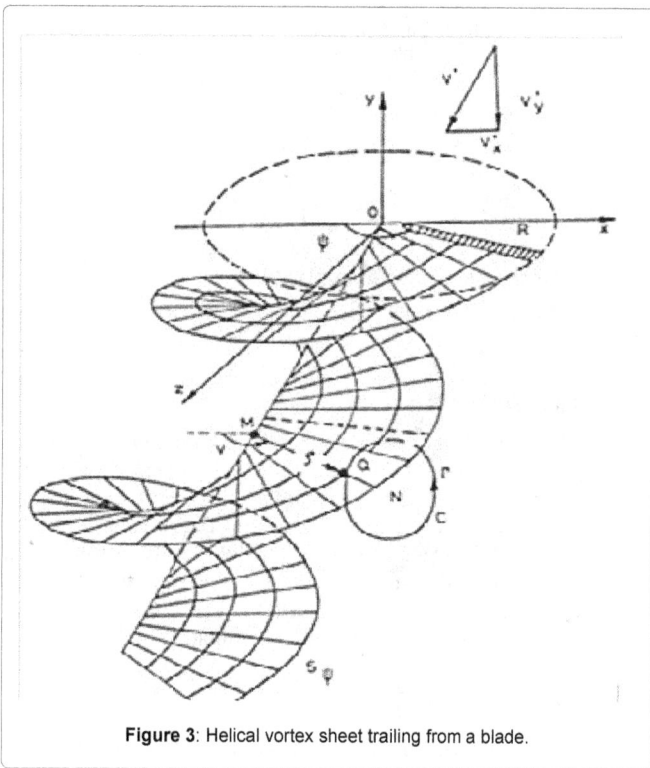

Figure 3: Helical vortex sheet trailing from a blade.

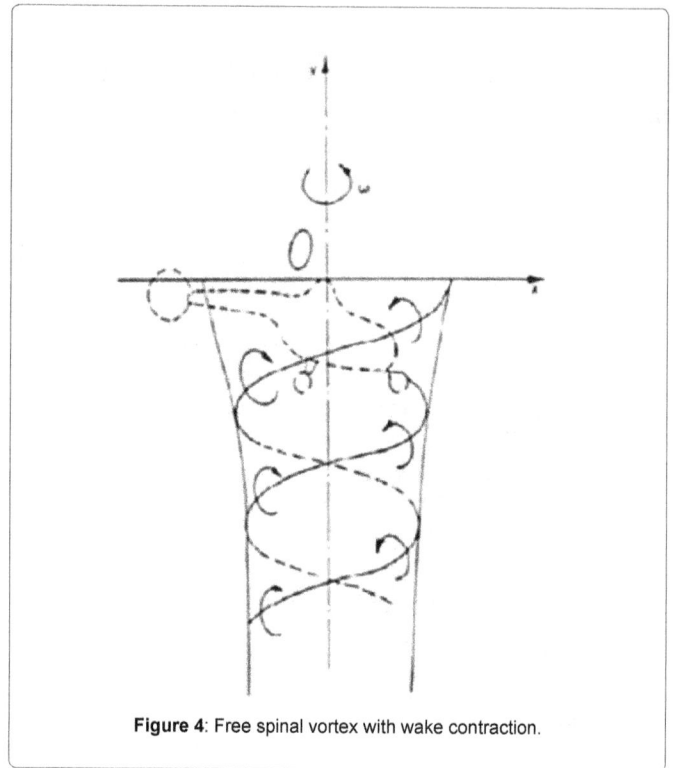

Figure 4: Free spinal vortex with wake contraction.

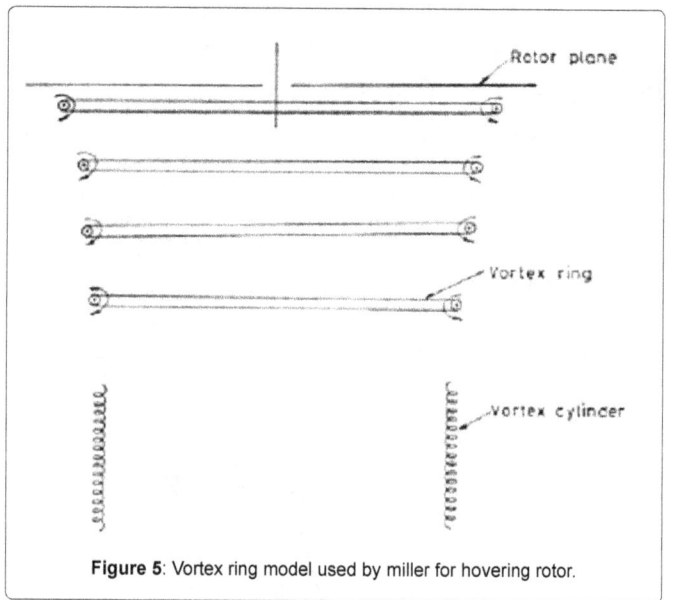

Figure 5: Vortex ring model used by miller for hovering rotor.

span-wise direction (Figure 6). Each segment is represented by a bound vortex placed along the span at the quarter-chord point and chord-wise vortices coming off from the ends of the bound vortex towards the trailing edge and leaving at the trailing edge as helical vortices. According to the present model those helical vortices forming the wake are replaced by two ring vortices kept below the rotor plane at a suitable location and the chord-wise vortices are connected to the ring vortices by two straight vortex filaments parallel the axis as shown in Figure 7. The radius of the vortex ring and its axial position are pre-determined. This system therefore forms the complete vortex system as it satisfies Kelvin and Helmholtz laws of vortex motion. The velocity induced by the vortex system is obtained by the application of Biot-Savart's law to the vortex line. The contributions to induced velocities at any control point stem from all the blades of the rotor and the wake.

Formulation of the problem

The coordination system of the rotor blade at an instant is shown in

Figure 8. The rotation of the blades is in the clock-wise direction about the x-axis, with angular velocity Ω. Each of the blades is divided into 'n' span-wise segments as shown in Figure 8. The relative velocity of flow with respect to a blade section is shown in Figure 9.

Each segment is represented by a bound vortex of unknown strength Γ_j extending along its quarter-chord line with chord-wise vortices extending from its ends to the trailing edge. Each of the vortex starting from the trailing edge is replaced by a single vortex ring of the same strength placed below and parallel to the rotor plane but with modified radius.

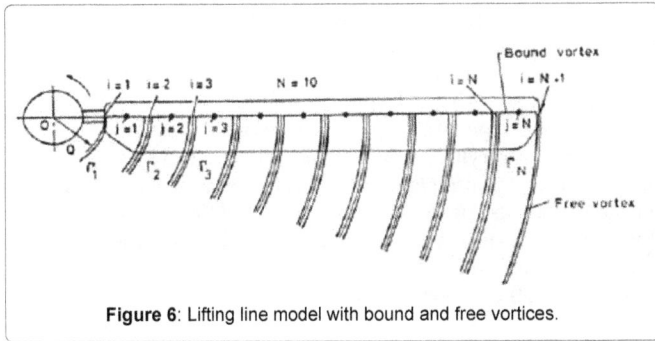

Figure 6: Lifting line model with bound and free vortices.

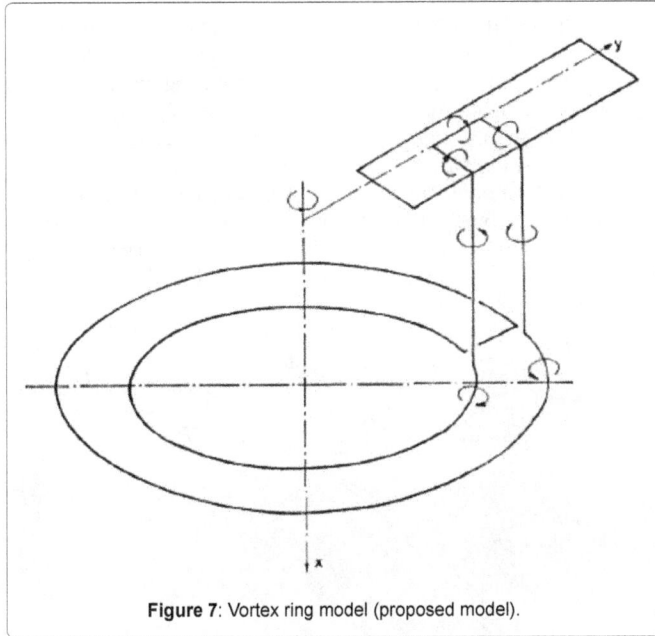

Figure 8: paneling of blades and helical horse shoe vortex representation of panel.

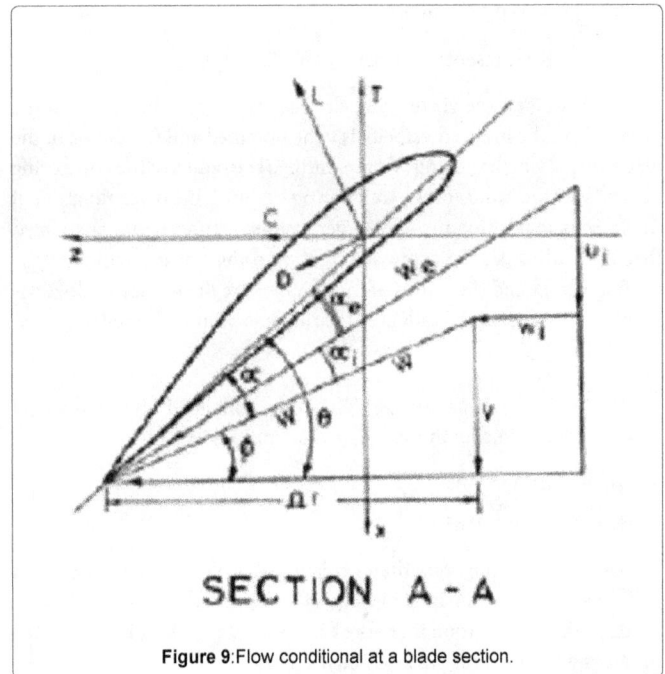

Figure 7: Vortex ring model (proposed model).

The induced velocity is calculated at the mid- point of each segment along the quarter-chord line due to the vortex system of the blades and the wake-vortex rings by using Biot-Savart's law

$$w_i = \frac{\Gamma}{4\pi}\left(\int \frac{\vec{ds} \times \vec{r}}{r^3}\right)$$

Where, $\vec{ds}$ is elemental length of vortex line, and $\vec{r}$ the position vector directed from the element to the field point (i) where w_i is required.

The Prandtl's kinematic flow condition is fulfilled at the mid-point of the segments of the reference blade in the form,

$$\alpha = \alpha_e + \alpha_i$$

Where, α is the geometrical angle of attack, which can also be expressed as

$$\alpha = \theta - \phi$$

$$\alpha_e = \frac{2\Gamma}{Wcc'l}$$

$$\alpha_i = \frac{v_n}{W}$$

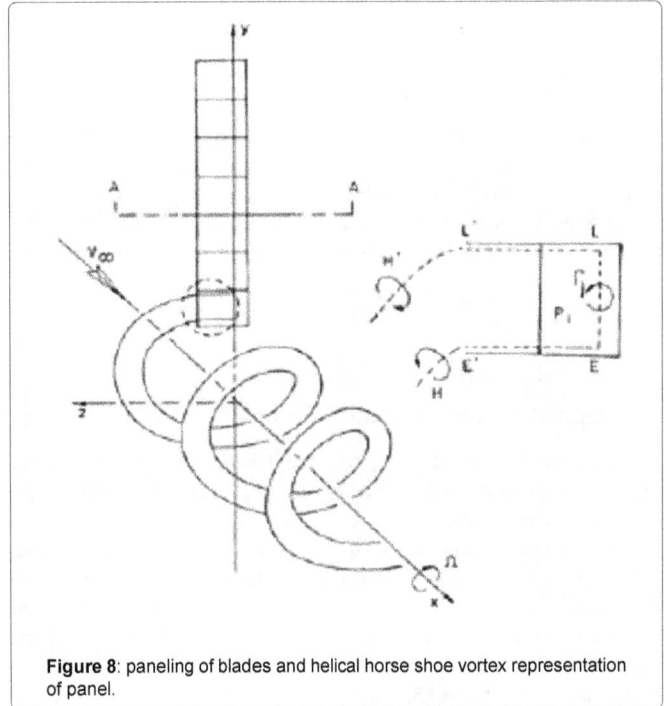

SECTION A - A

Figure 9: Flow conditional at a blade section.

$$v_n = u\cos\phi - w\sin\phi$$

These equations may be expressed in non-dimensional form for the control point P_i as

$$\alpha_e = \frac{2\gamma}{\overline{W}\overline{c}c_i'}$$

$$\alpha_i = \left(\eta\sum u_{ij}\gamma_j - \lambda\sum w_{ij}\gamma_j\right)\Big/\overline{W}^2$$

The kinetic boundary condition may therefore be expressed as

$$\theta_i - \phi_i = (\alpha_e + \alpha_i)_I \; \text{Or}$$

$$\theta_i - \phi_i = \left(\frac{2\gamma_i}{W_i c_i c_i'}\right) + \frac{\left(\eta_i \sum \overline{u}_{i,j}\gamma_j - \lambda \sum \overline{w}_{i,j}\gamma_j\right)}{\overline{W}_i^2} \qquad\qquad i = 1, 2.........n$$

Solving this set of simultaneous equations, one can obtain the non-dimensional circulation distribution 'γ' along the blade. From this circulation distribution using Kutta-Joukowski theorem one can calculate the lift, induced drag for each segment of the blade and hence the thrust and torque coefficients can be obtained in the following form:

$$C_T = \frac{2}{\pi}\int(\eta + \overline{w})\gamma d\eta$$

$$C_Q = \frac{2}{\pi}\int(\lambda + \overline{u})\eta\gamma\, d\eta$$

Determination of the size and location of vortex ring

The axial location of the vortex ring depends to a great extent on the geometry of the blade, particularly the chord, and the setting angle. The location was therefore expressed by a combination of these two factors by an expression namely, 0.75 c sin θ, which represents the axial displacement of a fluid particle as it moves in the chord-wise direction from the quarter-chord point to the trailing edge. Also, the induced velocity is influenced very much by the number of blades 'B' of the rotor. Therefore it was postulated that the axial location of the vortex ring can be expressed by the combination as

$$X_c = (K + B)\, 0.75\, c \sin\theta$$

Where, K represents a constant to be determined.

The radius of the vortex ring also depends upon the radius of the point where the induced velocity is to be obtained and the radius of the blade vortex which the ring is representing. The radius of the vortex ring depends on the radius of the trailing vortex. Since the determination of the free-wake position involves larger computer time, it was postulated that this radius can be expressed by the product of the radius of the trailing vortex and the radius of the point where the induced velocity is required. Therefore, the radius of the ring has been expressed as

$$y_{ring} = K_1 y_c y_j$$

Where the value of K_1 has been obtained by numerical experimentation using the above method and is given by

$$K_1 = 1.0 \text{ for } y_c \leq 0.81$$

$$K_1 = 0.1 \text{ for } y_c > 0.81$$

Using these equations the circulation distribution was calculated and was found to be in good agreement with the results of other models. This formulation therefore has been adopted for different kinds of rotors with different blade geometries.

Results and Discussion

The circulation distribution obtained after solving the system of simultaneous equations shows that γ depends on the number of segments taken on each blade. Therefore, the computations are performed for determining the minimum number of segments required for estimating the circulation distribution reasonably well.

Comparison of the proposed method with other theoretical and experimental results

In order to test the validity of the proposed method, the results of the proposed method are compared with the available experimental and theoretical results. In the present case the comparison is made for a two bladed rotor of rectangular plan form with 11° twist which is due to Miller [1,2] for hovering conditions (Figure 10). For the geometrical data of the rotor, computation of the span-wise circulation distribution for a rotor with varying number of blades is obtained by the proposed method. This is shown plotted in Figure 11 for hovering. Also, carried out computations for comparison of Thrust/Torque coefficients by different models (Figure 12). It may be observed from these figures that the present simple method predictions follow closely the experimental trend and also the theoretical results.

Conclusions and Recommendations

The simplified ring vortex wake model predicts the circulation distribution along the blade which is in close agreement with Miller's experimental and theoretical results for hovering condition. The comparison was made for a setting angle of 30°. This can be generalized for any other setting angle. The results of this model agree well with the results of spiral vortex model for climbing condition also. It may be noted that no other simpler model has been found to exist in the literature for climbing condition. Further, the model is able to predict the aerodynamic characteristics of rotor without sacrificing the

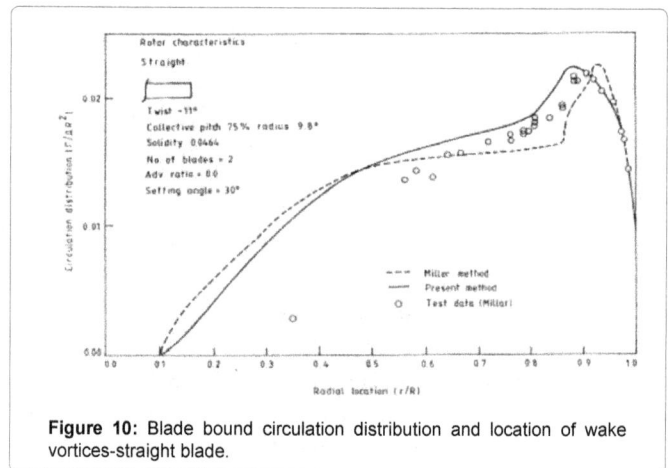

Figure 10: Blade bound circulation distribution and location of wake vortices-straight blade.

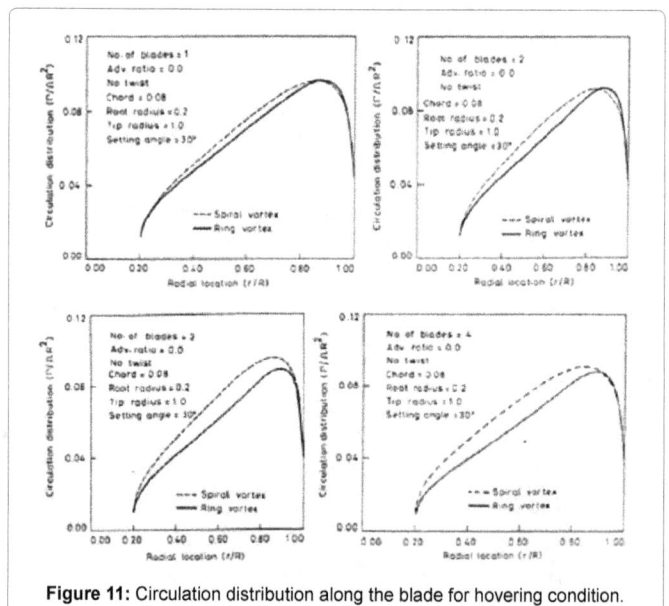

Figure 11: Circulation distribution along the blade for hovering condition.

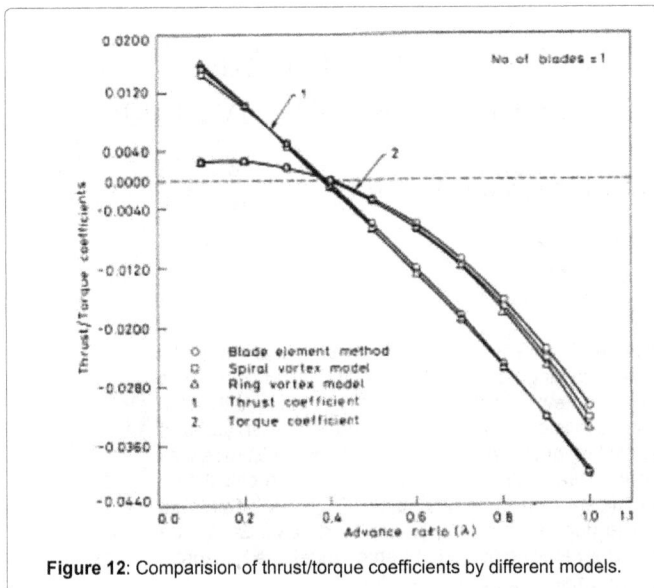

Figure 12: Comparision of thrust/torque coefficients by different models.

accuracy in spite of gross simplification of the model.

This model predicts the aerodynamic characteristics of the rotor as accurately as the spiral vortex wake model by taking considerably smaller computing time which is less than 1/15th of that of other models.

The method is very simple as it involves comparatively smaller mathematical equations and take smaller amount of computer time, at the same time without sacrificing the accuracy of the results.

Therefore, this method can be recommended to calculate the aerodynamic characteristics of the rotor in hovering flight and for vertical climbing flight and also it can be extended for forward flight.

Nomenclature

C_T: Thrust coefficient

C_Q: Torque coefficient

R_t: Blade tip radius

U: Induced velocity in the axial direction

V_c: Vertical climbing flight

v_n: Induced velocity component perpendicular to W

W: Resultant velocity of fluid relative to blade

w: Induced velocity in the circumferential direction

C: Blade section chord

c_l' : Lift curve slope of Blade section

y_c, y_i: Radial coordinate of control point where induced velocity is calculated

y_j: Radial location of chord-wise vortex in a blade segment

Greek Symbols

α: Geometric angle of attack

$α_e$: Effective angle of attack

$α_i$: Induced angle of attack

θ: Setting angle of attack

φ: Angle of the relative velocity of air at a blade section with respect to the plane of rotation

Γ: Circulation strength of vortex filament

η: Non-dimensional radial location (y/R_T)

Ω: Angular velocity of rotor

λ: Climb inflow ratio (Advance ratio)

Non-dimensional quantities

$$\bar{c} = \frac{c}{R_T}$$

$$\bar{u} = \frac{u}{\Omega R_T}$$

$$\bar{w} = \frac{w}{\Omega R_T}$$

$$\bar{W} = \frac{W}{\Omega R_T}$$

$$\gamma = \frac{\Gamma}{\Omega R_T c}$$

References

1. Miller RH (1982) Vortex theory for hovering rotors. AIAA Journal 2: 1754-1756.

2. Miller RH (1982) A simplified approach to the free-wake analysis of a hovering rotor Vertica 6: 89-85.

3. Reddy KR (1980) The Vortex flow field generated by a hovering helicopter 7 Australian hydraulic and fluid Mechanics conference Brisbane p. 553-556.

4. Reddy KR (1986) The effect of rotor wake geometry variation on hover induced power estimation for a UH-1H IROQUOIS HELICOPTER ARL-AERO-TM-384.

11

Determination of the Earth's Magnetic Field Gradients from Satellites Measurements and Their Inversion over the Kursk Magnetic Anomaly

Kis KI[1], Taylor PT[2*] and Wittmann G[3]

[1]Geophysics and Space Sciences Department, Loránd Eötvös University, 1117 Budapest Pázmány Péter Sétány 1/c. Hungary
[2]Planetary Geodynamics Laboratory, NASA/GSFC, Greenbelt, MD 20771, USA
[3]MOL Hungarian Gas and Oil Company, 1117 Budapest, Október Huszonharmadika u. 18. Hungary

Abstract

We computed magnetic field gradients at satellite altitude, over Europe with emphasis on the Kursk Magnetic Anomaly (KMA). They were calculated using the CHAMP satellite total magnetic anomalies. Our computations were done to determine how the magnetic field observations data from the new ESA/Swarm satellites could be utilized to determine the structure of the magnetization of the Earth's crust, especially in the region of the KMA. Ten years of CHAMP data were used to simulate the Swarm data. An initial east magnetic anomaly gradient map of Europe was computed and subsequently the North, East and Vertical magnetic gradients for the KMA region were calculated. The vertical gradient of the KMA was also determined using Hilbert transforms. Inversion of the total KMA was derived using Simplex and Simulated Annealing algorithms. The depths of the upper and lower boundaries are calculated downward from the 324 km elevation of the satellite. Our resulting inversion depth model is a horizontal quadrangle. The maximum errors are determined by the model parameter errors.

Keywords: Earth's magnetic field; Gradients

Introduction

Measuring magnetic gradient field anomalies have been used to map regions of differing crustal magnetization, both induced and remanent. To record the gradients over large segments of Europe and particularly the Kursh Magnetic Anomaly (KMA) satellite data are essential. Ten years of the CHAMP mission are used in this study but the more recent three satellite Swarm mission, launched in November 2013 will be able to determine the field gradients directly. Our study should be considered as a base line from which Swarm and future mission will serve as a reference.

Source of the Kursk Magnetic Anomaly

The exploitation of iron deposits is of a great economic importance. The Kursk Magnetic Anomaly is located 400 km south of Moscow it is one of the largest magnetic anomaly on Earth. The Proterozoic iron ore deposits are located in a NNW-SSE oriented complex syncline which is superimposed on the Voronezh Bulge anteclise. The Voronezh Bulge is bordered by the Ryazan-Saratov and Pripyat-Dnieper-Donetsk aulacogens Shchipansky and Bogdanova [1]. The extent of the anomaly is approximately 190,000 km^2 and its crustal depth is between 0.5 - 3.0 km. According to Heiland [2] the anomaly was discovered by Smirnov in 1874. The discussion of the anomaly is described by the early works of Lasareff [3], Haalck [4] and more recent investigations by Lapina [5], Taylor and Frawley [6], and Rotanova [7]. Due to its extent and large magnetization it can be detected by satellite measurements (Magsat, Taylor and Frawley 1987; The resultant magnetization is 3 Am^{-1} (Taylor and Frawley [6]. Our inversion computations are based on their value. The direction of the resultant magnetization is 47° East declination 67° down inclination. The Kursk magnetic anomaly was computed 324 km altitude from CHAMP measurements Figure 1 Kis et al. [8].

Development of the Iron-Ore Deposits

The texture of iron ore bodies can be banded iron formation (BIF) and/or granular iron formation (GIF), Bekker et al. [9]. These marine sediments were formed in the Archean and Paleoproterozoic. Their maximum age is 2.6 ca. Ga. The BIF were dominant in the Archean and

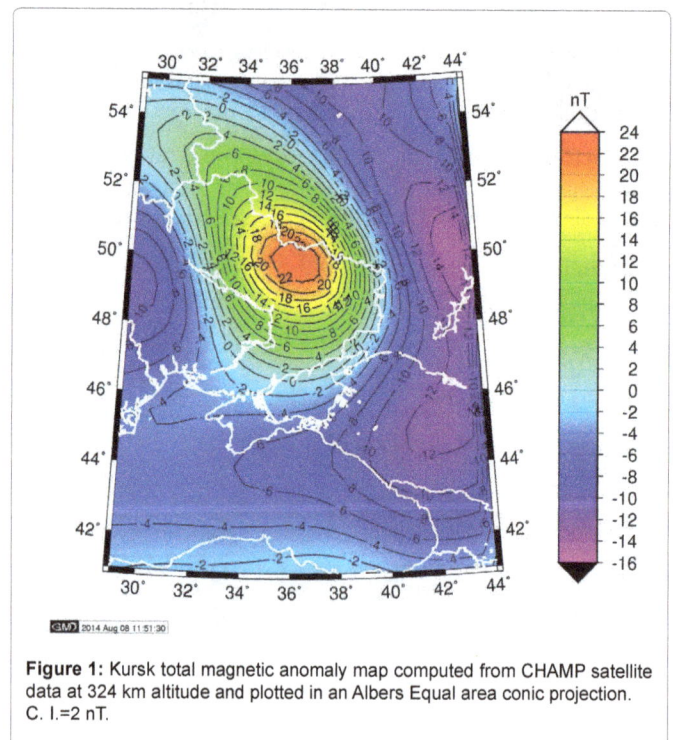

Figure 1: Kursk total magnetic anomaly map computed from CHAMP satellite data at 324 km altitude and plotted in an Albers Equal area conic projection. C. I.=2 nT.

*Corresponding author: Taylor PT, Planetary Geodynamics Laboratory, NASA/GSFC, Greenbelt, MD 20771, USA, E-mail: patrick.t.taylor@ nasa.gov

Determination of the Earth's Magnetic Field Gradients from Satellites Measurements and Their Inversion...

95

earliest Paleoproterozoic while GIF were in the later Paleoproterozoic. A summary of these geological processes which formed the Kursk iron-ore structures are given in Voskresenskaya [10], Alexandrov [11], Shchipansky and Bogdanova [1], Bekker et al. [9], Kovács and Pálfy [12].

These deposits contain considerable qualities of iron in siliceous banded structures. One theory is that the iron originated from marine and submarine volcanic activities or the possible from mantle plumes near the sea bottom. The precipitation of iron depends on the redox conditions of the environment, since the reduced Fe (II) remains as a liquid while the oxide Fe (III) precipitates in anoxic setting. The formation of iron deposits was prevented by the Great Oxidation Event ca. 2.4 Ga. The reductive environment of the hydrosphere ended and the oxygen content increased. This event was due to the decline of volcanic activities. The production of BIF's may have reached a maximum at *ca* 1.85 Ga, since the magmatic activity was greater at this time. The composition of BIFs and GIFs indicates that it is probably of submarine origin.

Three possible oxidative processes are summarized by Bekker et al. [9]. The first description: the origin of the oxygen is from a photosynthetic process of cyanobacteria in a thin oxidative zone in the upper layer of the sea.. Under this layer there are some anoxic water columns where Fe (III) precipitates. According to the second version the iron oxidizing bacteria live in an anoxic environment. These protobacteria are able to absorb water and carbon dioxide. The third process: ultraviolet photons oxidize the Fe (II) liquid to Fe (III) solids. Bekker et al. [9] state that the third process is less probable.

Determining CHAMP Magnetic Anomaly Gradients

The gradients of the magnetic field can be either determined directly from the two side by side lower orbiting SWARM satellites. The gradients of the magnetic field can also be computed from the CHAMP anomalies.

Previously we Taylor et al. [13] calculated the horizontal gradients over the Kursk magnetic anomaly. In this study the vertical, north and east gradients will be computed.

CHAMP magnetic anomalies over the European region and the KMA were calculated at 324 km altitude. These anomalies are plotted in a spherical polar coordinate system Kis et al. [8] (Figure 2).

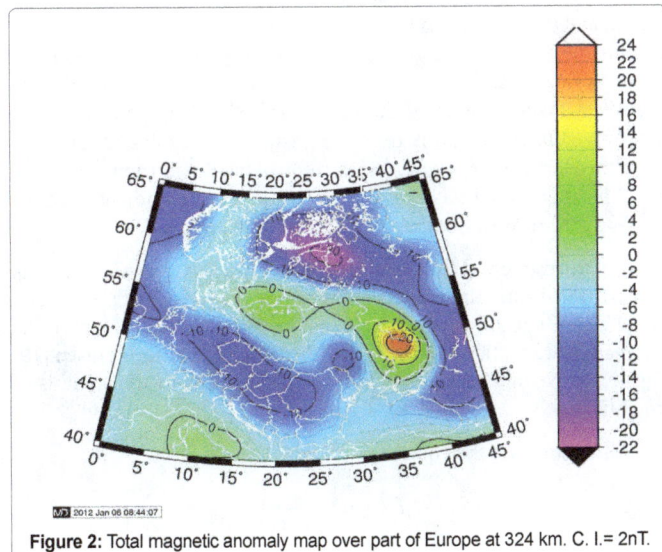

Figure 2: Total magnetic anomaly map over part of Europe at 324 km. C. I.= 2nT.

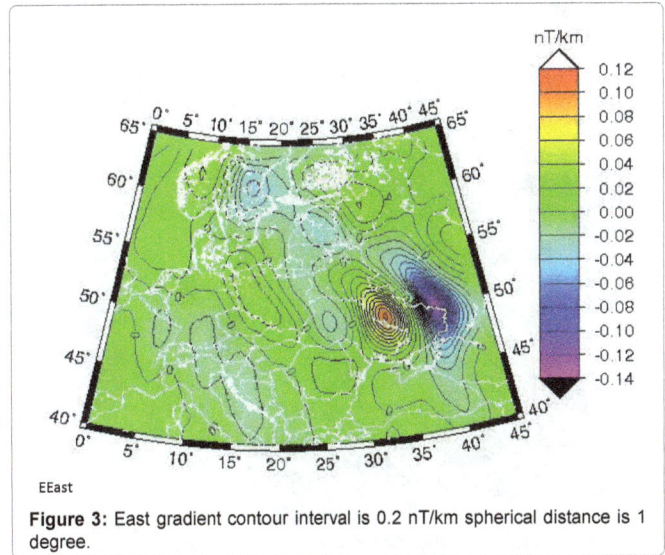

Figure 3: East gradient contour interval is 0.2 nT/km spherical distance is 1 degree.

The numerical horizontal gradients are calculated (from the anomaly data plotted in Figure 2) when the data are in the same latitude but different longitude (the longitude distances will be 1, 2, and 4 degrees, respectively). The spherical distance Δ between two data values is given by the spherical triangle cosine theorem:

$$\cos \Delta = \cos \vartheta \cos \vartheta + \sin \vartheta \sin \vartheta \cos (\lambda_2 - \lambda_1) \tag{1}$$

Where $_{d=k\,cm^1}$ is the polar distance, λ_2 and λ_1 are the two longitudes, respectively. The distance of these two data is given by the equation:

$$d = R \cos^{-1} \Delta \tag{2}$$

Where R is the Earth's spherical radius, 6371.2 km + 324 km. The Eastern gradients are approximated by the simple equation of

$$\frac{T(R, \vartheta, \lambda_2) - T(R, \vartheta, \lambda_1)}{d} \tag{3}$$

where T indicates the total magnetic anomaly. The East gradients determined by this method are shown in Figs. 3, 4 and 5. The longitude distance, λ_2-λ_1 was 1° the East gradients are shown in Figure 3. In this case the d distance varies between 88.86 and 49.75 km due to the meridional convergence. The longitudinal distance is 2° in Figure 4, and distance d is between 177.71 and 99.51 km, while the distance d is changed between 355.42 and 199.02 km. It can be seen from Figures 3-5 that the appropriate spherical distances will be 1° or 2° for the gradient determination for the SWARM anomalies.

The second method for computing the gradients is based on the application of the transfer and weight functions of the *x*-, *y*- and *z*- axes Kis and Puszta, [14]. The transfer functions of the above mentioned gradients are given by Blakely [15]:

$$S_{dx}(f_x, f_y) = j2\pi f_x, \quad S_{dy}(f_x, f_y) = j2\pi f_y, \quad S_{dz}(f_x, f_y) = 2\pi \left(f_x^2 + f_y^2\right)^{1/2} \tag{4}$$

Where S_{dx}, S_{dy} and S_{dz} are the transfer functions of the *x*-, *y*- and *z*- gradients; f_x and f_y are the spatial frequencies, j is the imaginary unit. Gaussian low-pass window of:

$$S^w(f_x, f_y) = e^{-k^2\left(f_x^2 + f_y^2\right)} \tag{5}$$

Is applied for the above mentioned transfer functions, where k is the appropriate parameter of the windowed transfer functions. The weight functions of the windowed transfer functions are:

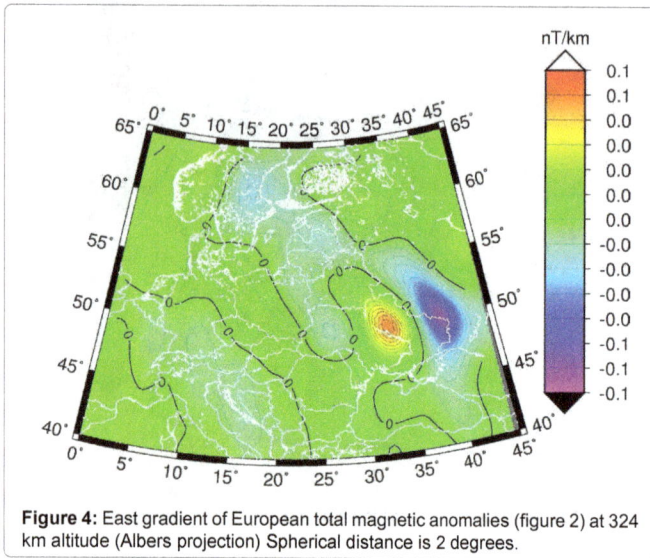

Figure 4: East gradient of European total magnetic anomalies (figure 2) at 324 km altitude (Albers projection) Spherical distance is 2 degrees.

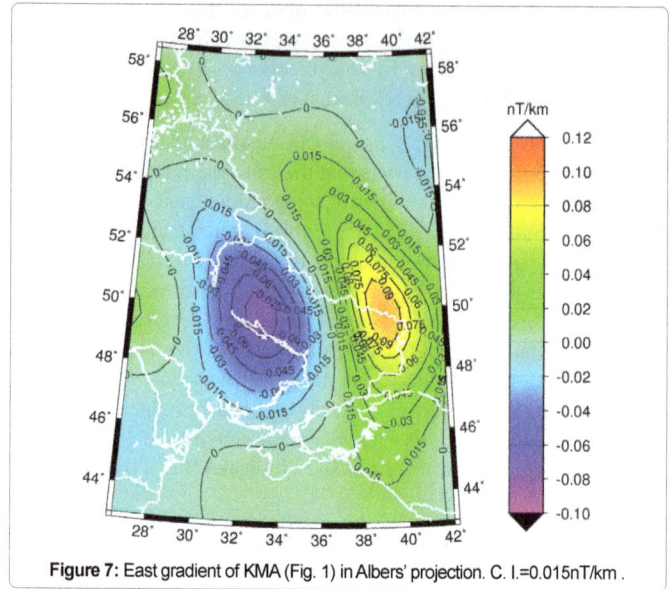

Figure 5: East Gradient of European total magnetic anomalies (Figure 2) spherical distance is 4 degrees.

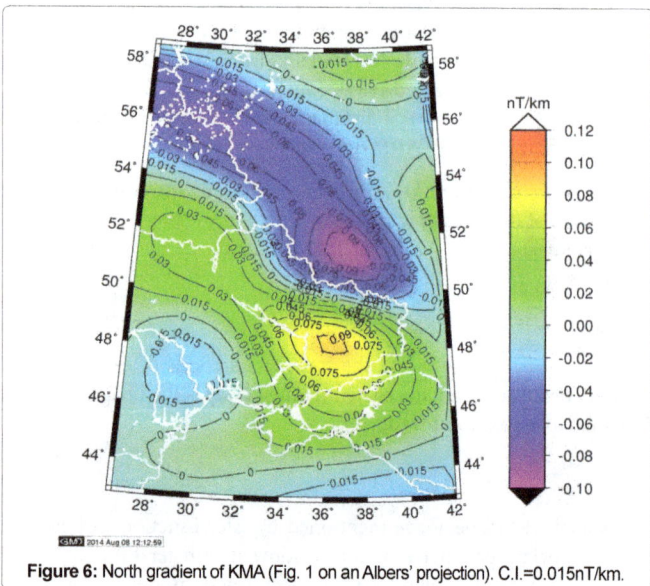

Figure 6: North gradient of KMA (Fig. 1 on an Albers' projection). C.I.=0.015nT/km.

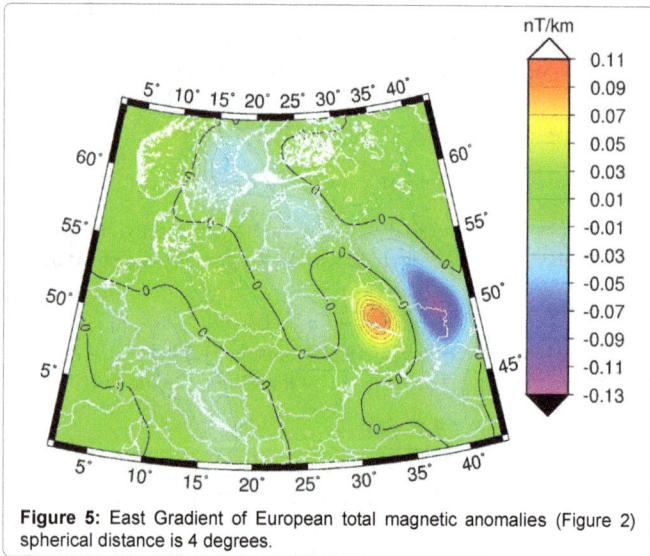

Figure 7: East gradient of KMA (Fig. 1) in Albers' projection. C. I.=0.015nT/km .

$$S_{dx}^{w}(x,y) = -\frac{2\pi^2}{k^4} x e^{-\frac{\pi^2}{k^2}(x^2+y^2)} \tag{6}$$

$$S_{dy}^{w}(x,y) = -\frac{2\pi^2}{k^4} y e^{-\frac{\pi^2}{k^2}(x^2+y^2)} \tag{7}$$

$$S_{dz}^{w}(x,y) = \frac{\pi^{5/2}}{k^3} e^{-\frac{\pi^2}{k^2}(x^2+y^2)} M\left(-\frac{1}{2}, 1, \frac{\pi^2(x^2+y^2)}{k^2}\right) \tag{8}$$

Where M is the confluent hypergeometric function (Slater, 1976). The development of the (6) – (8) functions is given by Kis and Puszta [14].

If the gradients are computed by the above method than the CHAMP anomalies should be transformed from the spherical polar coordinate system to the Cartesian coordinate system. The zero of the Descartes coordinate system is placed at an altitude of 324 km and at latitude 48.75° and longitude 36.25°. The transformation is given by Kis et al. [16]. The windowed gradients are computed in the Cartesian coordinate system and the next step is their transformation into the spherical polar coordinate system. These gradients are shown in the Figures 6-8 in an Albers' projection. These gradients emphasize the depth variation or change in magnetization of the source or both. These results are not determined unambiguously from the gradients. The gradients in Fig. 6 show the Northeast- Southwest elongation of the Kursk anomalies while Fig. 7 illustrates the East-West variation. The vertical gradient, Figure 8 displays a Northwest-Southeast directional asymmetry.

The third method for calculating the vertical gradient is using the Hilbert transform. The Hilbert transform was named by G. H. Hardy (1932), the English mathematician, out of deference to D. Hilbert, the German mathematician. Nabighian [17,18] Nabighian and Hansen [19] Guspi and Novara [20] applied Hilbert transforms to the analysis of potential fields.

Let us consider the equation:

$$F\left\{\frac{\partial T}{\partial z}\right\} = -\frac{jf_x}{\left(f_x^2+f_y^2\right)^{1/2}} F\left\{\frac{\partial T}{\partial x}\right\} - \frac{jf_y}{\left(f_x^2+f_y^2\right)^{1/2}} F\left\{\frac{\partial T}{\partial y}\right\} \tag{9}$$

Given by Nabighian [18], where F is the Fourier transform. This equation can be expressed in a simpler form:

Where H_1 and H_2 are the Hilbert transform, that is:

$$F\left\{\frac{\partial T}{\partial z}\right\} = H_1 F\left\{\frac{\partial T}{\partial x}\right\} + H_2 F\left\{\frac{\partial T}{\partial y}\right\} \tag{10}$$

$$H_1 = -\frac{jf_x}{\left(f_x^2 + f_y^2\right)^{1/2}} \quad \text{and} \quad H_2 = -\frac{jf_y}{\left(f_x^2 + f_y^2\right)^{1/2}} \tag{11}$$

The inverse Fourier transform of Equation (9) is:

$$F^{-1}\left\{\frac{-jf_x}{\left(f_x^2 + f_y^2\right)^{1/2}}\right\} = \frac{1}{2\pi}\frac{x}{\left(x^2 + y^2\right)^{3/2}} \tag{12}$$

and

$$F^{-1}\left\{\frac{-jf_y}{\left(f_x^2 + f_y^2\right)^{1/2}}\right\} = \frac{1}{2\pi}\frac{y}{\left(x^2 + y^2\right)^{3/2}} \tag{13}$$

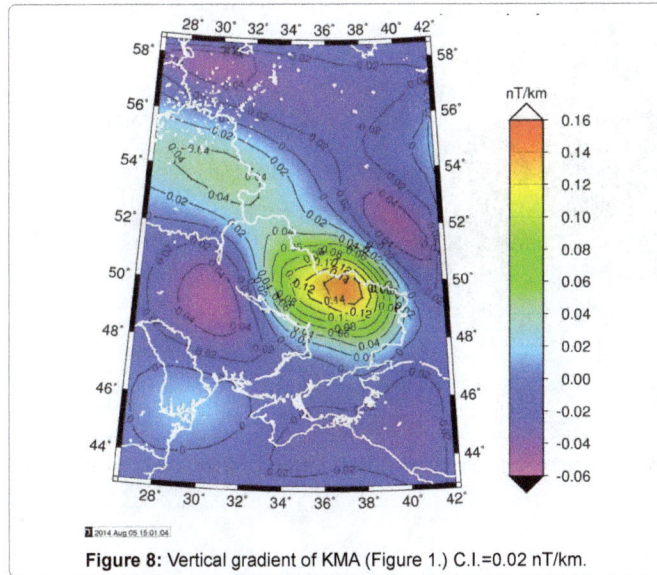

Figure 8: Vertical gradient of KMA (Figure 1.) C.I.=0.02 nT/km.

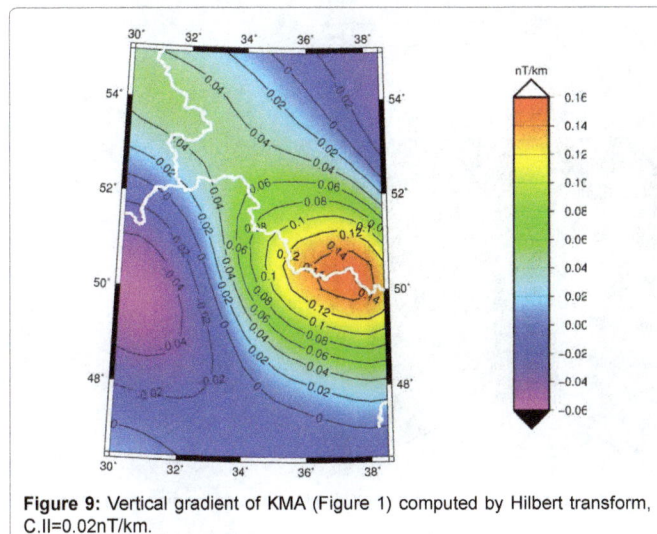

Figure 9: Vertical gradient of KMA (Figure 1) computed by Hilbert transform, C.II=0.02nT/km.

These inverse Fourier transforms are given by Nabighian [18] in his Appendix A. The CHAMP magnetic anomaly field (Fig. 1) is transformed from the spherical polar coordinate system to the Descartes coordinate system. Now the Hilbert transform is computed in the spherical polar coordinate system and the vertical gradients given in Figures 8 and 9, they illustrate virtually the same results.

Inversion of the Kursk Magnetic Anomaly

The Bayesian inference was applied to the inversion of the Kursk Magnetic Anomaly. The Bayesian inference is widely used in the inversion procedures and is summarized by Box and Tiao [21,22], Tarantola [23], Duijndam [23,24], Menke [25], Gregory [26], Kis et al. [16] and Kis et al. [8].

The Kursk Magnetic Anomaly are shown in the spherical polar coordinate system (Figures 10-13). The inversion procedure is applied in the Descartes coordinate system so the earlier mentioned transformation should be used. The result is given in the Descartes coordinate system that was computed from the CHAMP data.

The forward model from the inversion is a quadrangle with horizontal upper and lower levels we computed the anomaly using Plouff's [27] method. This is an idealized model of the Kursk source [28-33]. We used 3 Am^{-1} average magnetization given by Taylor and Frawley [6] and the average direction of magnetization inclination 47°down and 67°East declination given by Bhattacharyya.

The basic equation of the Bayesian inference is:

$$p(\mathbf{m} \mid \mathbf{d}) = p(\mathbf{d} \mid \mathbf{m}) p(\mathbf{m}), \tag{14}$$

Where p ($\mathbf{m}|\mathbf{d}$) is the p$^{a\ posteriori}$ conditional probability density, p ($\mathbf{d}|\mathbf{m}$) is the *likelihood* conditional probability density, $p(\mathbf{m})$ is the *pa priori* probability density. The vector $\mathbf{m}$ is the estimated parameters of the forward model and the vector $\mathbf{d}$ the measured CHAMP magnetic anomalies. The $p^{a\ posteriori}$ conditional probability density for Gaussian multivariate distribution can be expressed in the following form:

$$p^{a\ posteriori} \propto exp\left(-\frac{1}{2}\left(\mathbf{m} - \mathbf{m}^{a\ priori}\right)^T \mathbf{C}_m^{-1}\left(\mathbf{m} - \mathbf{m}^{a\ priori}\right)\right)$$
$$exp\left(-\frac{1}{2}\left(\mathbf{d}^{measured}(x,y) - \mathbf{T}^{calculated}(x,y,\mathbf{m})\right)^T \mathbf{C}_D^{-1}\left(\mathbf{d}^{measured}(x,y) - \mathbf{T}^{calculated}(x,y,\mathbf{m})\right)\right), \tag{15}$$

where vector $\mathbf{m}^{a\ priory}$ is the parameters estimated by the interpreter, $\mathbf{C}_m$ is the covariance matrix of the estimated parameters, vector $\mathbf{d}^{measured}$ is the measured CHAMP anomalies, $\mathbf{T}^{calculated}(x,y,\mathbf{m})$ is the calculated values at the coordinate (x,y), calculated for the $\mathbf{m}$ parameters, $\mathbf{C}_D$ is the covariance matrix of the CHAMP measured anomalies, superscript T the transposed vector.

The $p^{a\ posteriori}$ conditional probability density for a Laplacian distribution is:

$$p^{a\ posteriori} \propto exp\left(-\frac{\left|\mathbf{m} - \mathbf{m}^{a\ priori}\right|}{\mathbf{C}_m^{\frac{1}{2}}}\right) exp\left(-\frac{\left|\mathbf{d}^{measured}(x,y) - \mathbf{T}^{calculated}(x,y,\mathbf{m})\right|}{\mathbf{C}_D^{\frac{1}{2}}}\right). \tag{16}$$

We want to maximize the $p^{a\ posteriori}$ probability density given by the equations (15) and (16) as a function of the parameter $\mathbf{m}$. This is equivalent to minimizing the sum of exponent of the equations (15) and (16). The functions $E(\mathbf{m})$ which will be minimized for multivariate Gaussian parameter distribution are:

$$E(\mathbf{m}) = (\mathbf{m} - \mathbf{m}^{a\ priori})^T \mathbf{C}_m^{-1}\left(\mathbf{m} - \mathbf{m}^{a\ priori}\right) + (\mathbf{d}^{measured}(x,y)$$
$$- \mathbf{T}^{calculated}(x,y,\mathbf{m}))^T \mathbf{C}_D^{-1}\left(\mathbf{d}^{measured}(x,y) - \mathbf{T}^{calculated}(x,y,\mathbf{m})\right) \tag{17}$$

For the multivariate Laplacian parameter distribution:

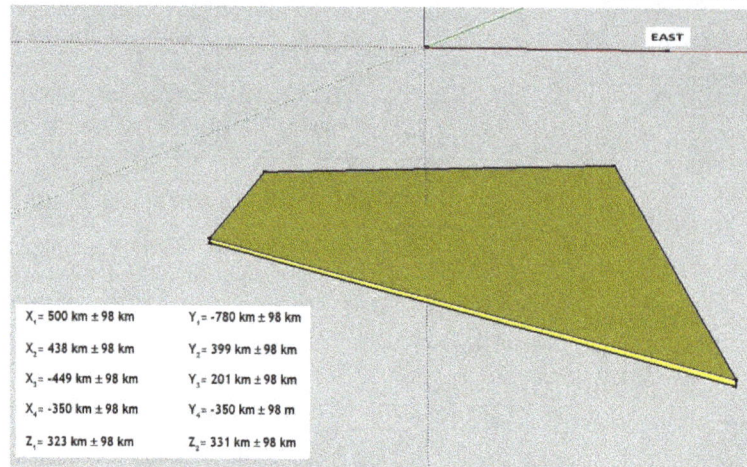

$X_1 = 500$ km $\pm$ 98 km	$Y_1 = -780$ km $\pm$ 98 km
$X_2 = 438$ km $\pm$ 98 km	$Y_2 = 399$ km $\pm$ 98 km
$X_3 = -449$ km $\pm$ 98 km	$Y_3 = 201$ km $\pm$ 98 km
$X_4 = -350$ km $\pm$ 98 km	$Y_4 = -350$ km $\pm$ 98 m
$Z_1 = 323$ km $\pm$ 98 km	$Z_2 = 331$ km $\pm$ 98 km

Figure 10: Computed model in Cartesian coordinates data have a Gaussian distribution with the minimum problem being solved using the simplex method. The table shows the determined parameter values and their maximum error.

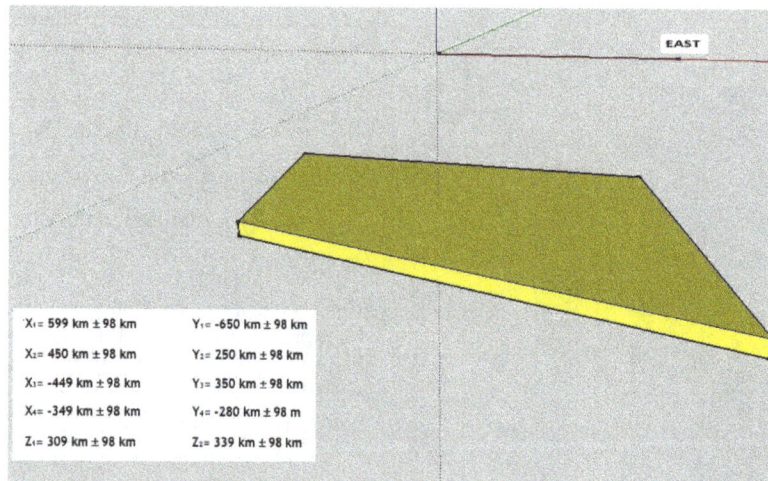

$X_1 = 599$ km $\pm$ 98 km	$Y_1 = -650$ km $\pm$ 98 km
$X_2 = 450$ km $\pm$ 98 km	$Y_2 = 250$ km $\pm$ 98 km
$X_3 = -449$ km $\pm$ 98 km	$Y_3 = 350$ km $\pm$ 98 km
$X_4 = -349$ km $\pm$ 98 km	$Y_4 = -280$ km $\pm$ 98 m
$Z_1 = 309$ km $\pm$ 98 km	$Z_2 = 339$ km $\pm$ 98 km

Figure 11: Computed model in Cartesian coordinates, the data have Gaussian distribution with the minimum problem being solved using the simulated annealing method. Table shows the calculated parameter values and their maximum error.

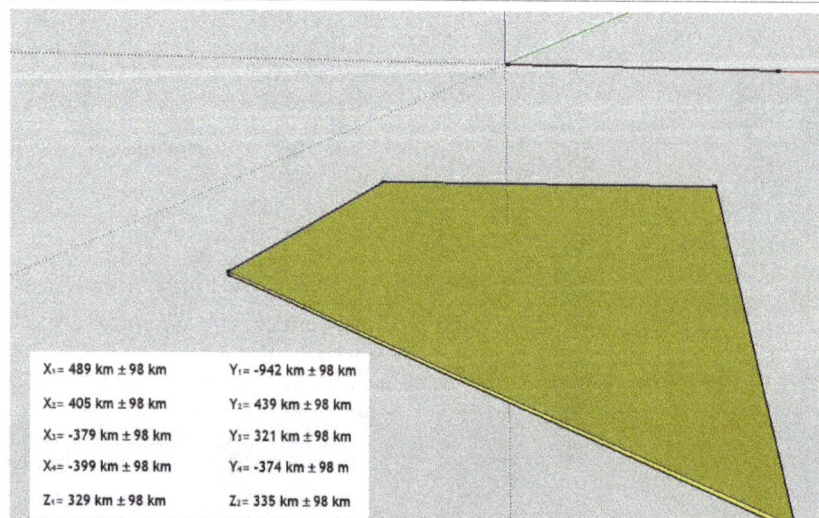

$X_1 = 489$ km $\pm$ 98 km	$Y_1 = -942$ km $\pm$ 98 km
$X_2 = 405$ km $\pm$ 98 km	$Y_2 = 439$ km $\pm$ 98 km
$X_3 = -379$ km $\pm$ 98 km	$Y_3 = 321$ km $\pm$ 98 km
$X_4 = -399$ km $\pm$ 98 km	$Y_4 = -374$ km $\pm$ 98 m
$Z_1 = 329$ km $\pm$ 98 km	$Z_2 = 335$ km $\pm$ 98 km

Figure 12: Computed model in Cartesian coordinates, , the data have a Laplace distribution with the minimum problem being solved by the simplex method. The results are given in tabular form with the determined parameter values and their maximum error.

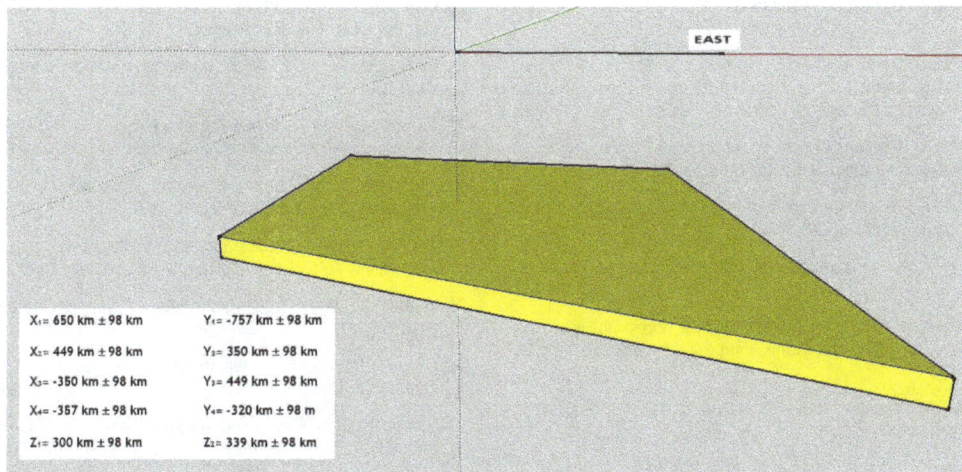

Figure 13: Computed model in Cartesian coordinates, the data have a Laplace distribution with the minimum problem being solved by simulated annealing method. The tale in the figure show the calculated parameter values and their maximu, error.

	Parameter	Gaussian distribution model parameter	Laplacian distribution model parameter
simplex method	x_1	500 km ± 98 km	489 km ± 98 km
	x_2	438 km ± 98 km	405 km ± 98 km
	x_3	-449 km ± 98 km	-379 km ± 98 km
	x_4	-350 km ± 98 km	-399 km ± 98 km
	y_1	-780 km ± 98 km	-942 km ± 98 km
	y_2	399 km ± 98 km	439 km ± 98 km
	y_3	201 km ± 98 km	321 km ± 98 km
	y_4	-350 km ± 98 km	-374 km ± 98 km
	z_1	323 km ± 98 km	329 km ± 98 km
	z_2	331 km ± 98 km	335 km ± 98 km
simulated annealing method	x_1	599 km ± 98 km	650 km ± 98 km
	x_2	450 km ± 98 km	449 km ± 98 km
	x_3	-449 km ± 98 km	-350 km ± 98 km
	x_4	-349 km ± 98 km	-357 km ± 98 km
	y_1	-650 km ± 98 km	-757 km ± 98 km
	y_2	250 km ± 98 km	350 km ± 98 km
	y_3	350 km ± 98 km	449 km ± 98 km
	y_4	-280 km ± 98 km	-320 km ± 98 km
	z_1	309 km ± 98 km	300 km ± 98 km
	z_2	339 km ± 98 km	339 km ± 98 km

Table 1: Computed models in Cartesian coordinates, the data have Gaussian and Laplace distributions. The optimum problems are solved by simplex and simulated annealing methods.

$$E\left(\mathbf{m}\right) = \frac{\left|\mathbf{m} - \mathbf{m}^{a\,priori}\right|}{\mathbf{C}_{\mathbf{m}}^{1/2}} + \frac{\left|\mathbf{d}^{mért}\left(x, y\right) - \mathbf{T}^{számított}\left(x, y, \mathbf{m}\right)\right|}{\mathbf{C}_{\mathbf{D}}^{1/2}}. \quad (18)$$

The $E(\mathbf{m})$ functions equations (17) and (18) will be minimized using the regularization suggested by Tikhonov and Arsenin (1977) is:

$$E\left(\mathbf{m}\right) = \left(\mathbf{m} - \mathbf{m}^{a\,priori}\right)^T \mathbf{C}_m^{-1}\left(\mathbf{m} - \mathbf{m}^{a\,priori}\right) + \left(\mathbf{d}^{measured}\left(x, y\right)\right.$$

$$\left. -\mathbf{T}^{calculated}\left(x, y, \mathbf{m}\right)\right)^T \mathbf{C}_D^{-1}\left(\mathbf{d}^{measured}\left(x, y\right) - \mathbf{T}^{calculated}\left(x, y, \mathbf{m}\right)\right) + \lambda(\mathbf{m}_{i+1} - \mathbf{m}_i)^2 \quad (19)$$

and

$$E\left(\mathbf{m}\right) = \frac{\left|\mathbf{m} - \mathbf{m}^{a\,priori}\right|}{\mathbf{C}_{\mathbf{m}}^{1/2}}, + \frac{\left|\mathbf{d}^{mért}\left(x, y\right) - \mathbf{T}^{számított}\left(x, y, \mathbf{m}\right)\right|}{\mathbf{C}_{\mathbf{D}}^{1/2}} + \left|\mathbf{m}_{i+1} - \mathbf{m}_i\right| \quad (20)$$

Where λ is the regularization parameter. According to the earlier investigations Kis et al. [8] the appropriate value of λ is between 1 – 10.

The minimum problem is solved by the simplex Walsh [28] and simulated annealing Kirkpatrick et al. [29] methods. In the present investigation the *a priori* covariance matrix is a diagonal one whose variances is $10\ \text{km}^2$, the *likelihood* covariance matrix is also diagonal one whose variances is $2\ \text{nT}^2$. The determined models are visualized in the model values are presented in Table 1 for the Descartes coordinate system. The maximum errors are determined by the model parameter errors.

Conclusion

The three methods for the determination of the gradients of the satellite magnetic anomaly data (at an altitude of 324 km) over the KMA are presented in spherical polar coordinates. All of them resolve the anomalies and they show the complex structure of its source. Both the windowed vertical gradient and the vertical gradient by the Hilbert transform give very similar results. All of the gradient methods can be applied. Three inversions (for Gauss distribution and the parameters determined by the simplex method; Gaussian distribution and the parameters determined by simulated annealing method; Laplacian distribution and the parameters determined by simulated annealing method) give comparable result within their maximum errors.

References

1. Shchipansky AA, Bogdanova SV (1996) The Sarmatian crustal segment precambrian correlation between Voronezh Massif and the Ukraine shield across the Dnieper-Donetsk aulacogen. Tectonophysics 286: 109-125.

2. Heiland CA (1946) Geophysical exploration. Prentice Hall Company.

3. Lasareff P (1923) The anomalies of terrestrial magnetism and gravity in the Kursk Government, Russia. Terrestrial Magnetism: 123-124.

4. Haalck H (1929) Zur Frage der Erklärung der Kursker magnetischen und gravimetrischen Anomalie. Gerlands Beiträge zur Geophysik 22: 241-255.

5. Lapina MI (1960) On certain results obtained from the study of vertical gradients of a magnetic field in the area of the Kursk magnetic anomaly. Bulletin Academy of Sciences USSR Geophysics Series: 390-395.

6. Taylor PT, Frawley JJ (1987) Magsat anomaly data over the Kursk region USSR Physics of the Earth and Planetary Interiors 45: 275-265.

7. Rotanova NM, Kharitonov AL, Frunze AKH, Filippov SV, Abramova D (2005) Anomalous magnetic fields measured on the CHAMP satellite for the territory of the Kursk Magnetic anomaly. Geomagnetism and Aeronomy 45: 671-678.

8. Kis KI, Taylor PT, Wittmann G, Toronyi B, Puszta S (2012) Interpretation of the total magnetic field anomalies measured by the CHAMP satellite over a part of

Europe and the Pannonian basin. Acta Geodaetica et Geophysica Hungarica 47: 130-140.

9. Bekker A, Slack JF, Planavsky N, Krapez B, Hofmann A et al. (2010) Iron formation the sedimentary product of a complex interplay among mantle tectonic oceanic and biospheric processes. Economic Geology 105: 467-508.

10. Voskresenskaya MN (1965) Relations between Archean and Proterozoic rocks in Kursk magnetic anomaly. International Geology Review 11: 454-460.

11. Alexandrov EA (1973) The Precambrian banded iron formations of the Soviet Union. Economic Geology 68: 1035-1062.

12. Kovács Zs, Pálfy J (2014) A rozsdamentes Föld talányos bányakincse. Természet Világa 145: 156-160.

13. Taylor PT, Kis KI, Wittmann G (2014) Satellite-altitude horizontal magnetic gradient anomalies used to define the Kursk magnetic anomaly. J Applied Geophysics 109: 133-139.

14. Kis KI, Puszta S (2006) Application of field derivatives for locating Sarmatian graves. J Applied Geophysics 60: 13-26.

15. Blakely R (1995) Potential theory in gravity and magnetic applications. Cambridge University Press.

16. Kis KI, Taylor PT, Wittmann G, Toronyi B, Puszta S (2011) Inversion of magnetic measurements of the CHAMP satellite over the Pannonian Basin. J Applied Geophysics 75: 412-418.

17. Nabighian MN (1972) The analytical signal of two-dimensional magnetic bodies with polygonal cross-section: its properties and use for automated anomaly interpretation. Geophysics 37: 507-517.

18. Nabighian MN (1984) Towards a three-dimensional automatic interpretation of potential field data via generalized Hilbert transform: Fundamental relations. Geophysics 49: 780-786

19. Nabighian MN, Hansen RO (2001) Unification of euler and werner deconvolution in three dimensions via generalized Hilbert transform. Geophysics 66: 1805-1810.

20. Guspi F, Novara I (2012) Generalized Hilbert transforms of the effect of single magnetic sources. Geophysics 77: J7-J14.

21. Box GEP, Tiao GC (1973) Bayesian inference in statistical analysis. Addison-Wesley.

22. Tarantola A (1987) Inverse problem theory methods for data fitting and model parameter estimation. Elsevier Amsterdam, Oxford -New York -Tokyo.

23. Duijndam AJW (1988a) Bayesian estimation in seismic inversion Part I: Principles. Geophysical Prospecting 36: 878-898.

24. Duijndam AJW (1988b) Bayesian estimation in seismic inversion Part II: Uncertainty Analysis. Geophysical Prospecting 36: 899-918

25. Menke W (1989) Geophysical data analysis: Discrete inverse theory. Elsevier Inc., Academic Press, San Diego-New York-Boston-Sydney-Tokyo-Toronto.

26. Gregory PC (2005) Bayesian logical data analysis for the physical sciences. Cambridge University Press.

27. Plouff D (1976) Gravity and magnetic fields of polygonal prisms and application to magnetic terrain correction. Geophysics 41: 727-741.

28. Walsh GR (1975) Methods of optimization. John Wiley & Sons, London.

29. Kirkpatrick S, Gelatt Jr CD, Vecchi NP (1983) Optimization by simulated annealing. Science 220: 671-680.

30. Bhattacharayya BK (1980) A generalized multi-body model for inversion of magnetic anomalies. Geophysics 45: 255-270.

31. Hardy GH (1932) On Hilbert transforms. Quart J Math (Oxford) 3: 102-112.

32. Slater LJ (1970) Confluent hypergeometric functions. In: Abramowitz M, Stegun IA, (eds.) Szerk Handbook of mathematical functions with formulas, graphs and mathematical tables. US Department of Commerce, National Bureau of Standards Applied Mathematics. Series 55: 503-535.

33. Tikhonov AN, Arsenin VY (1977) Solutions of ill-posed problems. John Wiley & Sons, New York-Toronto-London-Sydney.

12

Enhancing the Performance of an Axial Compressor Cascade using Vortex Generators

Diaa AM*, El-Dosoky MF and Ahmed MA

Department of Mechanical Engineering, Assiut University, Egypt

Abstract

Axial flow compressors have a limited operation range due to the difficulty of controlling the secondary flows. Therefore, a new design of vortex generators is considered in the current investigation to control the secondary flow losses and consequently enhance the compressor's performance. Different sets of curved side vortex generators with varying configurations are studied to find their effect on the secondary flow losses. Numerical simulations of a three-dimensional compressible turbulent flow have been performed to explore the effect of vortex generators on the reduction of secondary flow losses. Based on the simulation results, the pressure, velocity, and streamline contours are presented to track the development of secondary flows in the compressor cascade. Thus, the total pressure loss and static pressure rise coefficients, blade deflection angles, and diffusion factors are estimated. Results indicate that vortex generators have a significant impact on secondary flow losses such as reducing the corner vortices, and improving the location of separation lines which are moving toward the trailing edge. At the cascade design point, it is found that vortex generators have a significant effect on the reduction of normalized total pressure loss which is evaluated to be up to 20.7%. Using vortex generators do not lead to a significant change in flow deflection and accordingly the off-design conditions will still be far from reached.

Keywords: Secondary flows; Axial compressors; Vortex generators; Turbulence

Introduction

The importance of axial compressors due to its relevance to gas turbine applications has motivated many re-searchers toward enhancing its overall performance. Controlling the secondary flow phenomena associated with the flow in compressor cascades will significantly improve the aerodynamic performance of compressors. This is because secondary flows are extracting energy from the fluid and increasing the flow instability. Endwall boundary layer separation, horseshoe vortex, corner vortex, tip vortex, endwall crossflow, and passage vortex are secondary flow components in the cascade. Many researchers investigated the impact of three-dimensional blades and endwall boundary layer separation as well as flow separation in corners of blade passages on the development of secondary flows [1-5]. To control the secondary flows, both passive and active methods have been applied to reduce or overcome the effects of secondary flows in axial compressors. It was found that the passive control methods remain the preferable techniques because of their simplicity and cost effectiveness [6,7]. Numerous types of passive flow control devices were investigated such as slotted blading in linear cascades [8], vane and plow vortex generators placed on several positions [9], counter rotating and co-rotating rectangular, triangular, and parabolic vane type vortex generators [10-12], cavity as a control of shock wave interactions with the turbulent boundary layer [13], low profile vortex generators to reduce the boundary layer thickness [14], and doublet vortex generators [15]. An excellent comprehensive review of boundary-layer flow-separation control by a passive method and their applications had been compiled Lin JC [16,17].

There are numerous other reported studies on the control of separation in turbulent boundary layers using low profile vortex generators. In such devices, the mechanism of flow control is to energize the low momentum layers near a solid surface without adding extra energy through the momentum transfer from the outer (free-stream) flow to the near wall region. Yet, this leads to an overwhelmed stronger adverse pressure gradient and hence avoids or delay the flow

separation. In case of turbulent flow over a flat plate, experimental results indicated that the vane and wheeler type of vortex generators can efficiently reduce the flow separation. Using the vortex generator height (h/δ) of 0.1 to 0.4 was efficient with much reduce in the drag effect [18]. It was reported that flow control by means of vane-type vortex generator arrays is robust with respect to changes in the pressure gradient and changes of the separation point [19]. In addition, the van type with height (h/δ) of 0.8 attained the largest pressure recovery [20]. McCormick [21] experimentally compared between two passive methods for controlling shock induced separation on a turbulent flat plate boundary layer. A doublet wedge type vortex generator with h/δ=0.36 was used versus passive cavity (porous wall with a shallow Cavity underneath). He reported that the low-profile vortex generators were found to be significantly suppressing the shock induced separation and improve the boundary layer characteristics downstream the shock whereas the mass-averaged total pressure loss increased. In case of turbulent flow over backward facing ramp [22], wheeler doublet and wishbone type vortex generators were used to control flow separation. They concluded that both wheeler doublet and wishbone type achieved the best effect in separation control when their heights (h/δ) varied in the range of 0.1 to 0.2.

A theoretical investigation was conducted of three-dimensional turbulent flow provoked in a boundary layer by an array of low-profile vortex generators on the surface [23]. Triangular type vortex generators,

---

*Corresponding author: Diaa AM, Department of Mechanical Engineering, Assiut University, Egypt, E-mail: ahmeddiaa@aun.edu

with various span wise spacing were considered. He concluded several suggestions for vortex generator designs such spanwise packing, enlarged vortex generator length, and suitably of non-smooth spanwise profile. Recently, experimental and numerical studies of the effect of two vortex generator types with different configurations on the performance of the compressor cascade were conducted by Hergt [24]. They reported that at the cascade design point, the total pressure losses were reduced by about 9% with the vortex generator. Moreover, vortex genera-tors have a significant effect on the cascade deflection and a remarkable enhancement of the cascade stall range. However, the static pressure rise due to inserting a vortex generator was nearly unaffected.

To conclude, many types of low profile vortex generator devices were widely investigated for different applications. However, application on compressor cascade is still limited, and the optimal design and position of vortex generators to control the development of secondary flows are not fully established yet. Therefore, the objective of the present research is to investigate the influence of vortex generators on the development of secondary flows and flow separation zones of compressor cascade. Therefore, different sets of vortex generators with varying configurations are numerically studied. Based on the numerical results, pressure, velocity, and streamline contours are presented in order to track the development of the secondary flow losses. Furthermore, the total pressure loss coefficient, static pressure rises, blade deflection angles, and diffusion factors are estimated and discussed.

Computational Work

Compressor cascade

In the present work, a linear high speed compressor cascade that was reported by the research group of Hergt et al. [24] is used. Their compressor cascade was designed by "MTU Aero Engines". The design parameters and the operating conditions of the cascade are summarized in Table 1. In addition, the inlet and staggered angles of the compressor cascade are shown in Figure 1.

Computational domain and boundary conditions

The computational domain used in the present work is depicted in Figure 2. The non-slip boundary condition is applied at the walls representing the top boundary, the bottom boundary (Endwall), and the blade surfaces demonstrating the suction and pressure sides including the leading and trailing edges. Periodic boundary conditions are applied on the domain sides. The pressure outlet boundary condition is defined at the outlet plane. The fully developed flow is adopted at the inlet with an average Mach number of 0.66 and inlet angle ($\beta 1$) of 132°. Turbulence intensity is set to be 1% at the inlet and 3% at the exit. The blade is tested under the design operating conditions.

Numerical solution

To investigate the effect of inserting vortex generators with varying dimensions on secondary flow losses, the Reynolds-average Navier–Stokes with fully coupled turbulence model equations are numerically solved using the commercial flow solver Fluent-14. The three dimensional multiblock grid is constructed using a structured mesh of H-O-H topology. Five different numbers of grids are selected in order to investigate the effect of grid size on the computed results. Figures 3a and 3b present the effect of grid size on the mass flow averaged integrated velocity, and the velocity contours at (exit plane). Based on these figures, it is found that there is no grid dependency after 0.8 million cells. Therefore, the present simulation is performed using 1.2 million cells to get free grid independent results and to reduce the

computational time and with minimum y+<1 near the walls, which is considered to capture and resolve the boundary layer at the blade surfaces and enwalls.

Computational investigations for curved side vortex generators (CSVGs) with and without a rounded nose are performed to find their effects on the development of secondary flows. The dimensions of studied sets of vortex generators named A, B, C, D, E, and F are summarized in Table 2. All tested CSVGs are shown in Figure 4, h/δ is varied from 0.1 to 0.5 for each set which is defined as low profile vortex generators as reported by [16].

Mach number at inlet	M1 = 0.66
Inlet flow angle	β1=132°
Turning angle	Δβ = 38°
Stagger Angle	βst =105.2°
Blade chord	c = 40 mm
Blade span	L = 40 mm
Pitch to chord ratio	s/c = 0.55
End-wall boundary layer thickness at inlet	δ = 4 mm
Maximum blade thickness	t = 2.6 mm
Relative maximum camber	n/c = 0.446

Table 1: Compressor cascade design parameters and operating conditions.

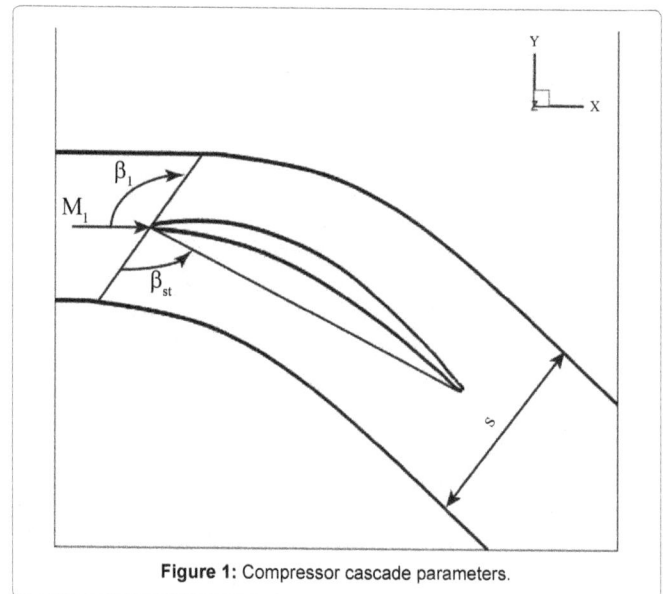

Figure 1: Compressor cascade parameters.

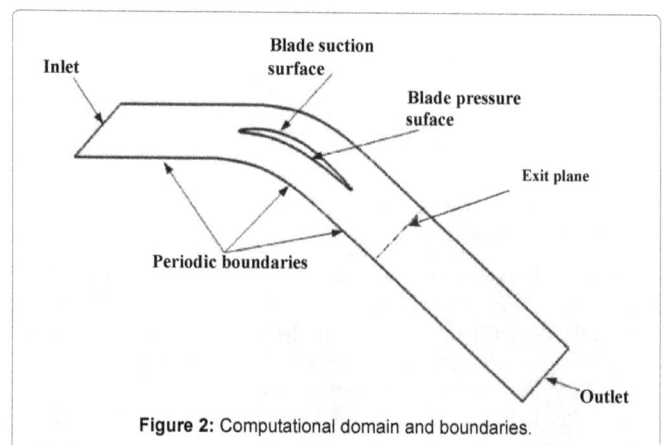

Figure 2: Computational domain and boundaries.

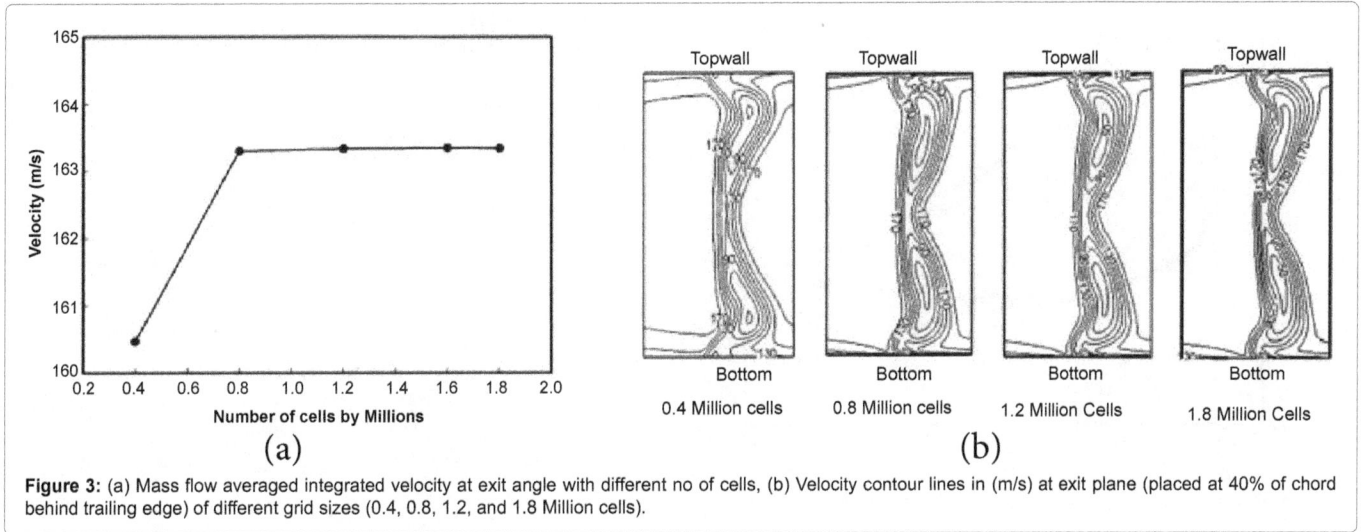

Figure 3: (a) Mass flow averaged integrated velocity at exit angle with different no of cells, (b) Velocity contour lines in (m/s) at exit plane (placed at 40% of chord behind trailing edge) of different grid sizes (0.4, 0.8, 1.2, and 1.8 Million cells).

	e/h	w/h				h/δ						
	6	3	4	5	6	0.1	0.15	0.2	0.25	0.3	0.4	0.5
Set A	X	X	-	-	-	X	X	X	X	X	X	X
Set B	X	-	X	-	-	X	X	X	X	X	X	X
Set C	X	-	-	X	-	X	X	X	X	X	X	X
Set D	X	-	-	-	X	X	X	X	X	X	X	X

	e/h				w/h		h/δ		
Set E	4	5	6	8	6		0.4		
	X	X	X	X	X		X		

	e/h		w/h		h/δ	r/δ			
Set F	5		6		0.4	0.25	0.5	0.8	1
	X		X		X	X	X	X	X

Table 2: Different parameters values for CSVG used in sets A, B, C, D, E and F.

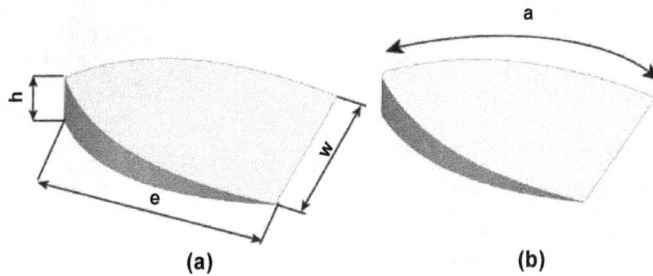

Figure 4: Curved sides' vortex generator geometry, (a) without round nose, (b) with round nose.

Results and Discussion

This section is classified into six subsections. The first subsection demonstrates the validation of computational results by comparing with the available experimental data and numerical results. The second one depicts the influence of vortex generators on the development of secondary flows. The third subsection describes the influence of vortex generators on the total pressure loss coefficient. The influence of vortex generators on cascade deflection is shown in the fourth subsection. The fifth subsection illustrates the influence of vortex generators on the static pressure rise coefficient. In the last subsection, the influence of vortex generators on the diffusion factor is presented.

Validation

The numerical results are validated, first, by comparing between the isentropic Mach, Mis, distribution at mid span calculated from the numerical and the experimental results of Hergt et al. [24] as shown in Figure 5. Comparisons indicate that a good agreement exists between the present simulation and the experimental results where the maximum deviation is found to be less than 3.4%.

Second, the total pressure loss distribution is compared with the results obtained by Hergt [24] at the same plane (exit plane) as shown in Figure 6. Based on comparison, a reasonable agreement is observed where the maximum deviation is less than 4% except near the endwall. It reaches about +9%. Hergt et al. [24] reported that the predicted total pressure loss near the endwall is undepreciated the measured value, consequently the difference between the present results and measured by Hergt [24] is expected to be much less than 9%. Thereby the numerical method can predict the time-averaged blade loading with

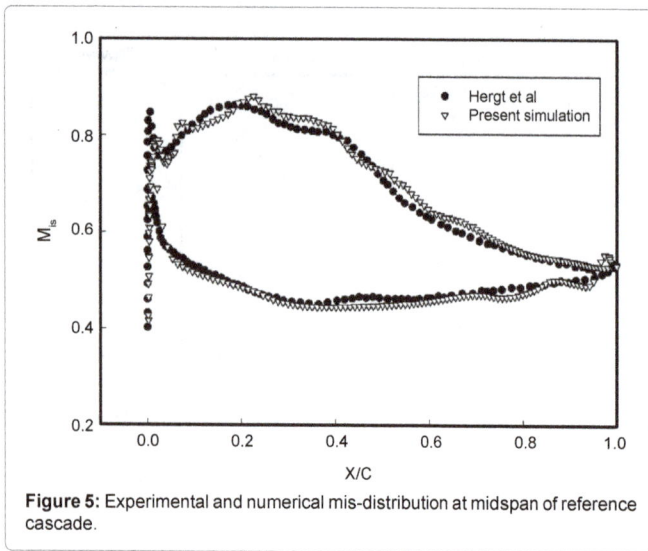

Figure 5: Experimental and numerical mis-distribution at midspan of reference cascade.

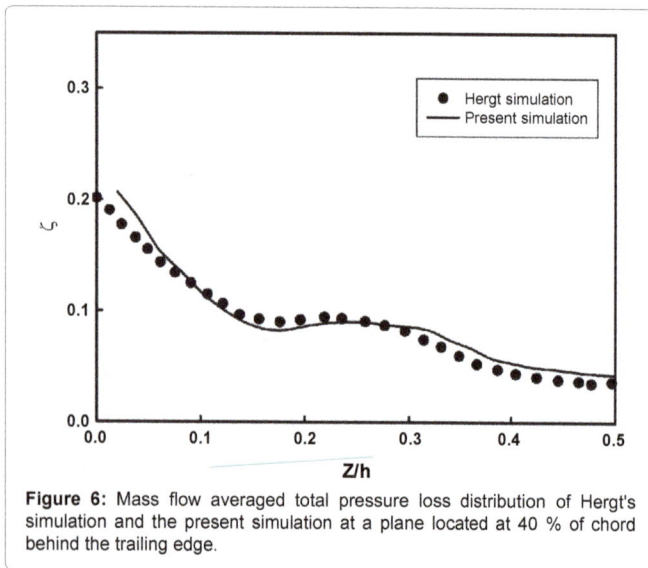

Figure 6: Mass flow averaged total pressure loss distribution of Hergt's simulation and the present simulation at a plane located at 40 % of chord behind the trailing edge.

adequate engineering accuracy as appropriate boundary conditions are applied.

Influence of vortex generators on the development of secondary flows

Streamline patterns are used to display the separation lines and the translations that occur to their positions on the suction surface for the different sets of vortex generators. Figures 7-9 illustrate the streamline contours on the suction side for different configurations of vortex generators of A, C, D, E, and F. Figure 7 shows the streamline contours for the reference case and set A where h/δ varies between 0.1 and 0.5 in step of 0.1. Based on Figure 7, comparisons between the reference set and other sets indicate that at h/δ=0.1, streamlines show how separation lines move at the suction surface. The cross flow from the endwall moves toward the leading edge where a new separation line and the formation of a separation bubble are observed. This occurs in the position between the cross flow and the corner separation while it moves slightly downstream. Increasing h/δ to 0.2 indicates that streamlines move to the corner separation downstream, and endwall cross flows are still developed and formed towards the leading edge.

Furthermore, for h/δ=0.3, streamlines show a movement of the corner separation towards the trailing edge and a separation line is noticed near the corner separation region. Thus, causing the endwall cross flow to be deflected in the downstream direction. At h/ δ=0.4, streamlines show

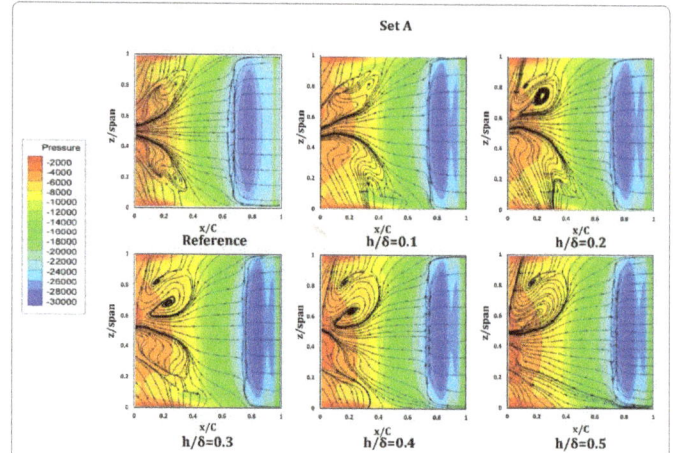

Figure 7: Calculated static pressure contours and streamlines on blade suction side of reference case and set A with h/δ ranged from 0.1 to 0.5.

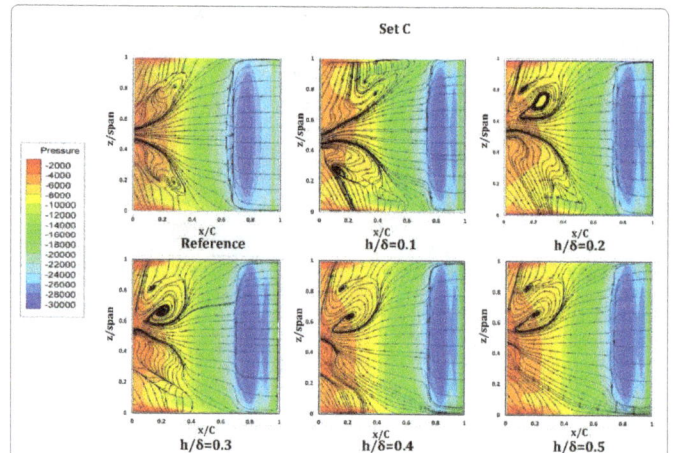

Figure 8: Calculated static pressure contours and streamlines on blade suction side of reference case and set C with h/δ ranged from 0.1 to 0.5.

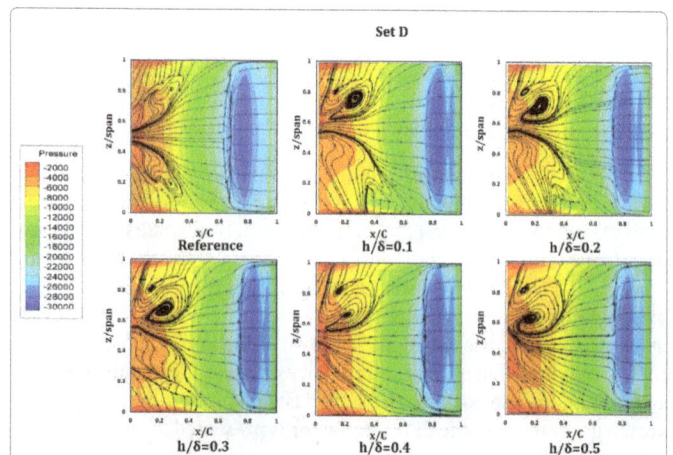

Figure 9: Calculated static pressure contours and streamlines on blade suction side of reference case and set D with h/δ ranged from 0.1 to 0.5.

a noticeable displacement of the corner separation downstream, and so leading the endwall cross flow is to be translated in the downstream direction as well. In addition, the growth of previous formed separation lines occurs. Finally, at h/ δ=0.5, streamlines show regression of the corner separation towards the trailing edge, and at the same time the endwall cross flow deflects in the downstream direction, as well as the separation line near the corner region disappears. The changes in streamlines for set B is like that cause by A, so it is not reported.

Figure 8 shows the streamline contours for the reference case and set C where h/δ varies between 0.1 and 0.5 in interval of 0.1. Comparisons between the reference set and other sets indicate that at h/δ=0.1, streamlines show a translation of corner separation in the downstream direction towards the trailing edge. As the endwall cross flow propagates towards the leading edge, a separation bubble is noted between the endwall cross flow and the corner separation. However, at h/δ=0.2, no significant changes in streak lines are noted. At h/δ=0.3, streak lines also show a translation of the corner separation towards the trialing edge, and a formation of a separation line between the endwall cross flow and the corner separation. At h/δ=0.4, streak lines show a significant movement of the corner separation downstream towards the trailing edge, and the endwall cross flow is deflected in a downstream direction. However, the separation line formed in the previous case disappears. Finally, at h/δ=0.5, streak lines show that the corner separation is at 0.85 of chord (downstream) which is significantly translated, and the endwall cross flow is deflected in the downstream direction as well.

Figure 9 shows the streamline contours for the reference case and set D where h/δ varies between 0.1 and 0.5 in intervals of 0.1. Comparisons between the reference set and other sets indicate that at h/δ=0.1, the cross flow is slightly reduced meanwhile it propagates in the span direction, while separation lines do not display significant changes. Though, at h/δ=0.2, no significant changes in streak lines are noted. At h/δ=0.3, streamlines show a translation of the corner separation towards the trialing edge, a formation of laminar separation line between the endwall cross flows, a corner separation, and a turbulent reattachment line. At h/δ=0.4, streak lines are significantly moved from the corner separation region towards the trailing edge, and the endwall cross flow is deflected in the downstream direction. However, the separation line formed in the previous case disappeared. Finally, at h/δ=0.5, streak lines show that the corner separation is at 0.85 of chord (downstream) which is significantly translated, and the endwall cross flow is also deflected in the downstream direction.

To conclude, for all those sets, the pressure side effects the development of the passage vortex and its propagation which in turn affects the flow structure on the blade suction side. As a result of inserting the vortex generators, the passage vortex is lifted from the endwall and swept downstream which travels the separation line on the blade suction surface towards the trailing edge of sets A, C, and D. Increasing h/δ results in the downstream movement of this separation line as noted in the previous figures. For set E, increasing e/h leads to the increase of the downstream movement of the separation line on the blade suction surface. A similar trend is observed by increasing r/δ as set F.

Influence of vortex generators on total pressure loss coefficient

Total pressure loss coefficient (TPLC) is usually used as an indicator for losses that take place in a cascade. Reducing pressure losses tends to increase the cascade efficiency and consequently enhances the compressor performance. TPLC (ζ^*) refers to local mass averaged total pressure loss coefficient (TPLC) and it can be defined as the following:

$$\zeta^* = \frac{p_{t1} - p_{t2}(y,z)}{p_{t1} - p_{s1}} \tag{1}$$

where: Pt1 is the total pressure at the inlet, and Pt2 (y,z) is the total pressure at the exit plane while, Ps1 is the static pressure at the inlet. TPLC is estimated as a function of (y, z) based on the calculated values at a plane downstream of the trailing edge at the distance of 0.45 of the chord (C). This plane is defined as the exit plane where variables computed at this plane take the index 2.

Total pressure loss coefficient contours: TPLC contours for set A, B, C, and D shows gradual change in the TPLC by increasing the dimensions of the VG. The reported results are from set D where the changes are very noticeable. Figure 10 shows the total pressure loss contours for set D. For set D, the total pressure loss contours are depicted for different values of h/δ varied from 0.1 to 0.5 in intervals of 0.1 along with the contours for the reference case. As reported earlier, the reference case is related to the flow in the compressor cascade without the vortex generator. Comparisons between the TPLC for reference case and those for different values of h/δ show two different trends. Firstly, a slight reduction in TPLC values are observed for h/δ ranged from 0.1 to 0.25 (values in-between are not reported in figures). Secondly, further increase of h/δ beyond 0.3 results in a significant reduction in TPLC. This is most likely because of increasing h/δ which strengthens the generated vortices and consequently increases the mixing of high momentum fluids with low ones. This process enriches the boundary layer velocity and therefore exhibits more resistance to separate from the suction side. Furthermore, increasing h/δ leads to an increase of the mean stream wise momentum of the boundary layer till h/δ=0.4, nevertheless beyond this value further increase in h/δ has no effect in TPLC reduction. This is due to energizing the lower part of the boundary layer velocity profile, since it is more crucial to reduce the tendency of the flow to separate.

Figure 11 shows the TPLC for the reference case and set E where e/h=4, 5, 6, and 8, while h/δ=0.4, and w/h=6. As shown in Fig. 11, the reduction in TPLC increases from e/h=4 to 5. Further increase in e/h from 5 to 8 leads to a decrease in TPLC. The reason is that the increase of e/h leads to a reduction in bluntness factor, and consequently a reduction in stretching rate and vortex strength. This trend can be attributed to when the parameter "e" increases the BF decreases, on the other hand as "e" increases, "a" increases which can increase or

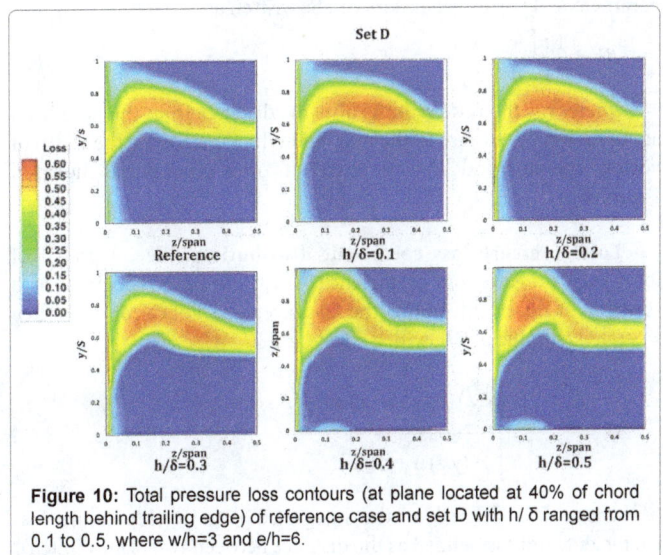

Figure 10: Total pressure loss contours (at plane located at 40% of chord length behind trailing edge) of reference case and set D with h/ δ ranged from 0.1 to 0.5, where w/h=3 and e/h=6.

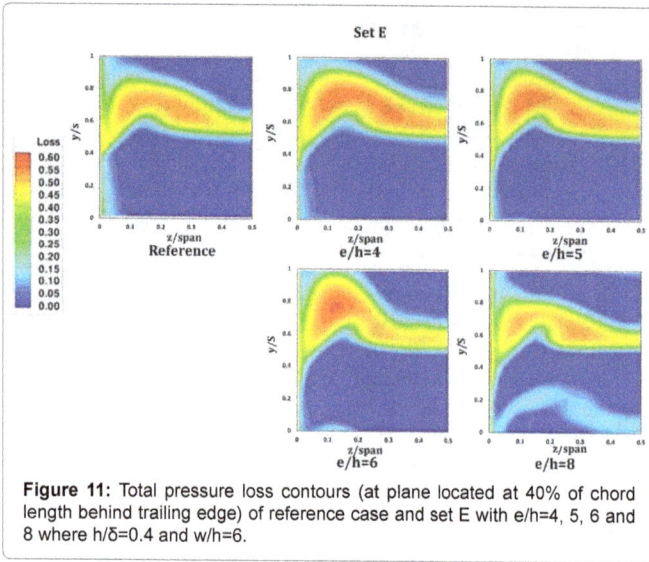

Figure 11: Total pressure loss contours (at plane located at 40% of chord length behind trailing edge) of reference case and set E with e/h=4, 5, 6 and 8 where h/δ=0.4 and w/h=6.

decrease the BF. Therefore, the results show that the reduction gained in TPLC increases by increasing "e/h" up to 5. As e/h becomes larger than 5, the reduction in TPLC decreases due to the weakness of the generated vortices and consequently the reduction in its stretching rate.

Figure 12 shows the TPLC for the reference case and set F with r/δ = 0.25, 0.5, 0.75, and 1.0, where h/δ=0.4, w/h=6, and e/h= 5. It was found that a significant reduction in TPLC occurs with increasing r/δ from 0.25 to 0.5. Further increase of r/δ beyond 0.5 results in a decrease of TPLC. The reason is most probably due to the increase of the bluntness factor which is a correlation on the effect of wing shape on the horseshoe vortex stretching rate developed as reported by Fleming [25]. In rounded nose vortex generator, the parameter "e" affects the blunt nose shape which controls the vortices strength through the bluntness factor "BF" [26,27]. Moreover, as the vortex generator nose radius increases, the time-mean vortex size and strength increase as reported [25]. These strong generated vortices enhance the performance of the compressor cascade till r/δ≈0.5. Further increase in r/δ leads to stronger vortices with larger vortex stretching rate which in turn reduces the enhancement degree [28].

The relationship between Bluntness Factor (BF) and geometrical parameters of vortex generators can be written as:

$$BF = \frac{1}{2}\frac{r}{e}\left[\frac{w}{a}+\frac{a}{e}\right] \qquad (2)$$

where: r is the nose radius, "a" is the distance from the nose head along the CSVG surface to the maximum thickness "w" is the width of vortex generator, and "e" is the length of vortex generator as shown in Figure 4.

Total pressure loss coefficient distribution: The integrated of local total pressure loss coefficient (ζ) distribution is calculated as a function of Z/L by integrating the mass averaged TPLC from y = 0 to y = S for each value of Z/L with a constant Δ Z.

$$\zeta = \frac{\displaystyle\int_{z=0}^{h}\int_{y=0}^{t}\zeta\,(y,z)\,\rho_2(y,z)\,u_2(y,z)dydz}{\displaystyle\int_{z=0}^{h}\int_{y=0}^{t}\rho_2(y,z)\,u_2(y,z)dydz} \qquad (3)$$

where: ζ is defined as the integrated local total pressure loss coefficient, and s is defined as the distance between two blades (pitch).

For sets A, B, C, D, E, and F, the total pressure loss coefficient distribution at the exit plane from the endwall till the midspan is calculated. It was found that for all investigated sets, the mass average of TPLC at Z/L = 0 is approaching unity. The results of sets A, B and C is not reported here. Total pressure loss distribution of sets D, E, and F is reported as shown in Figures 13-15. Figure 13 shows the total pressure loss distribution for set D, at Z/L<0.24, the TPLC is slightly less than the corresponding TPLC for the reference case except at h/δ =0.4, 0.5 where the TPLC is slightly higher than those of the reference case. Whereas at Z/L > 0.32, the TPLC is slightly less than the corresponding TPLC for the reference case except at h/δ=0.1, and 0.15 where the TPLC is slightly higher than those of the reference case.

Figure 14 shows that the TPLC distribution is reduced compared to the reference case along the span for e/h=4, 5, while it's slightly reduced for e/h=8. For e/h=6 it is reduced at Z/L< 0.06, and increased until Z/L=0.3, then reduced again. The highest reduction in TPLC occurs at e/h=5. Figure 15 shows that the TPLC is reduced for all values of r/δ from 0.25 to 1. The highest reduction in TPLC occurs with r/δ=0.5 which is the highest of all sets.

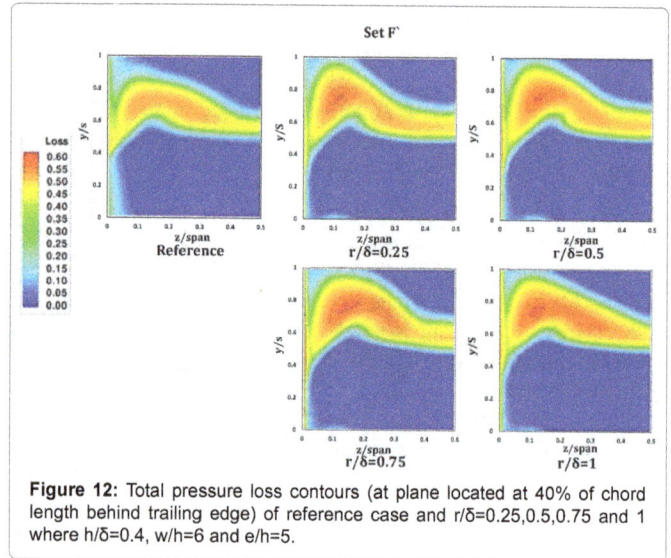

Figure 12: Total pressure loss contours (at plane located at 40% of chord length behind trailing edge) of reference case and r/δ=0.25,0.5,0.75 and 1 where h/δ=0.4, w/h=6 and e/h=5.

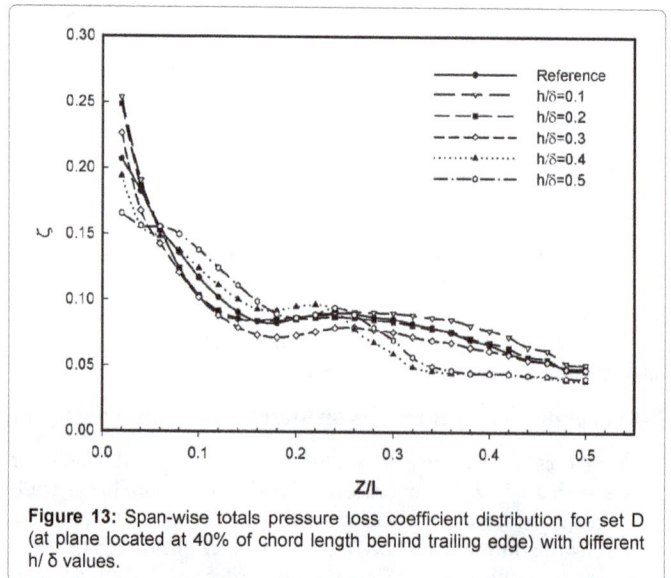

Figure 13: Span-wise totals pressure loss coefficient distribution for set D (at plane located at 40% of chord length behind trailing edge) with different h/ δ values.

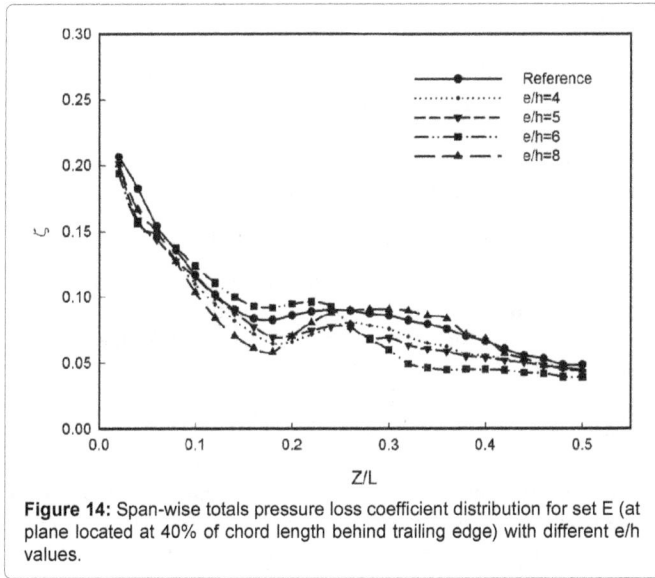

Figure 14: Span-wise totals pressure loss coefficient distribution for set E (at plane located at 40% of chord length behind trailing edge) with different e/h values.

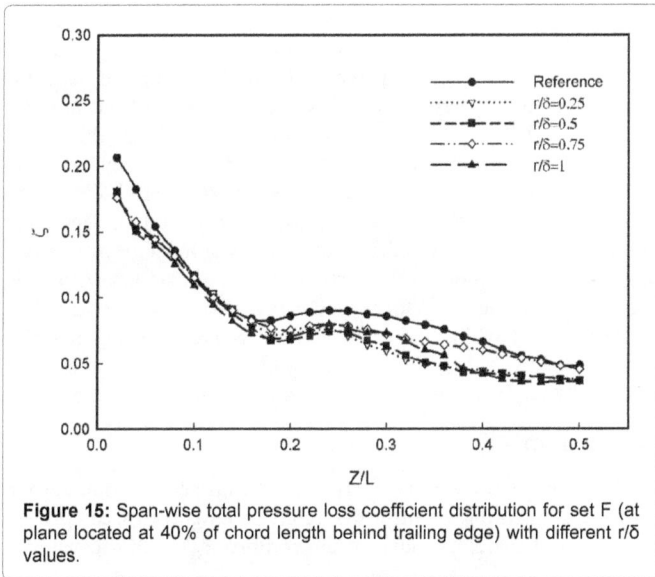

Figure 15: Span-wise total pressure loss coefficient distribution for set F (at plane located at 40% of chord length behind trailing edge) with different r/δ values.

Normalized total pressure loss coefficient: Normalized TPLC is defined as $\zeta_n = \Delta\zeta/\zeta\text{Ref}$ and it is calculated for all sets by integrating the mass-averaged TPLC value across the exit plane and subtracting it from the integrated mass-averaged TPLC for the reference case. Figure 16 presents the variation of ζ_n versus h/δ for sets A, B, C, and D. Based on Fig. 16, for set A the ζ_n reduces by 2.3% at h/δ=0.1. As h/δ increases, the reduction in ζ_n increases from 3% at h/δ=0.15 to about 5.3% at h/δ=0.4 then decreases to about 3.7% at h/δ=0.5. A Similar trend is observed for set B where ζ_n is reduced by 1.9% at h/δ=0.1. As h/δ increases, the reduction in ζ_n increase from 3.25% at h/δ=0.15 to about 6.67% at h/δ=0.4 then decreases 5.82% at h/δ=0.5. For set C, the ζ_n is reduced by 1.7% at h/δ=0.1. As h/δ increases, the reduction in ζ_n increase from 3.73% at h/δ=0.15 to about 7% at h/δ=0.4 then decreases to 5.1% at h/δ=0.5. Finally for set D, the ζ_n is reduced by 1% at h/δ=0.1. As h/δ increases, the reduction in ζ_n increases from 3.4% at h/δ=0.15 to about 12.6% at h/δ=0.4 then reduced to 5.1% at h/δ=0.5. Figure 17 presents the variation of ζ_n versus e/h for set E. As shown in Fig. 17, ζ_n is about 13.4% at e/h=4. As e/h increases, ζ_n reaches to 18.6% at e/h=5 and then reduces to about 5.73% at e/h=8. The effect of varying

the vortex generator nose radius r/δ on the normalized total pressure loss coefficient is shown in Fig. 18. Based on this Figure 18, ζ_n is about to 18.62% at r/δ=0. Increasing r/δ to 0.5 results in an increase of ζ_n to 20.7%, while further increase in r/δ to 1.0 leads to a decrease in ζ_n to 15.8%. In conclusion, the normalized total pressure loss coefficient can be reached to up 20.7% for set F with r/δ=0.5. This value of 20.7% can be considered the best achievable reduction in a normalized total pressure loss coefficient.

Influence of vortex generators on cascade deflection

Cascade deflection (ε) can be calculated from the following equation:

$$\varepsilon = \beta_1 - \beta_2 \qquad (4)$$

where: β1 is the inlet blade angle and β2 is the outlet angle. Figure 19 shows the variation of the percentage of deflection angle Δε/εRef versus h/δ for sets A, B, C, and D. Based on Fig.19, deflection is slightly enhanced for all sets between h/δ=0.1-0.5. However, for set D a significant reduction is observed between h/δ=0.4 and 0.5. As shown in Fig. 19 set E has a slight change in deflection occurring by varying e/h. It is noted that deflection increases by increasing the value of e/h. For set F, as shown in Figure 19 deflection barely changes it's values range from 0 to -0.5% with r/δ =0.25, 0.5, 0.75, and 1. These results are

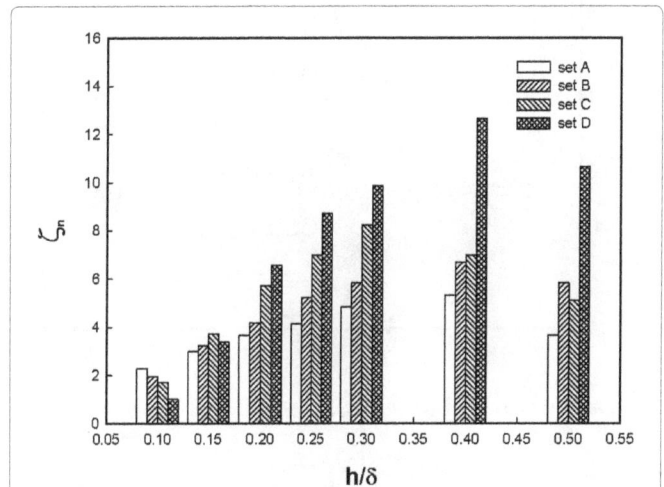

Figure 16: Normalized total pressure loss coefficient for sets A, B, C and D.

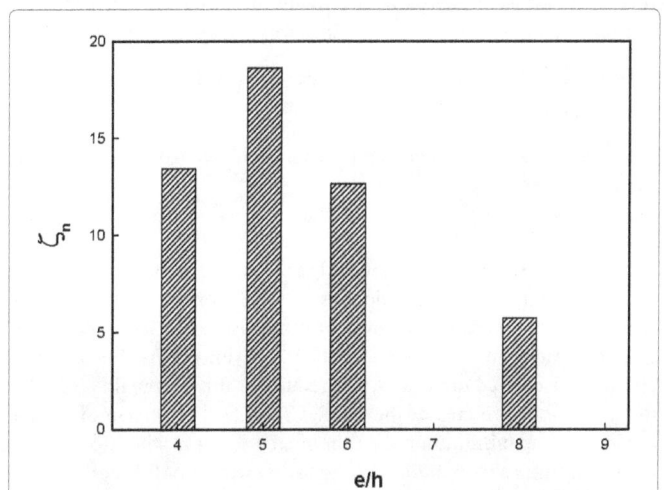

Figure 17: Normalized total pressure loss coefficient for set E.

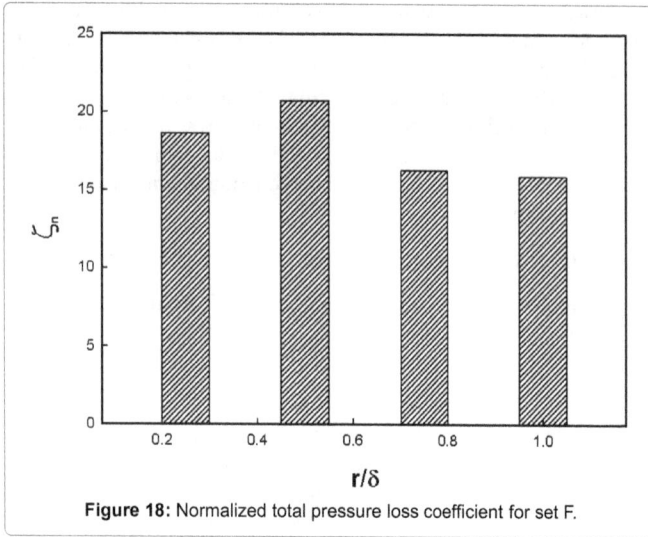

Figure 18: Normalized total pressure loss coefficient for set F.

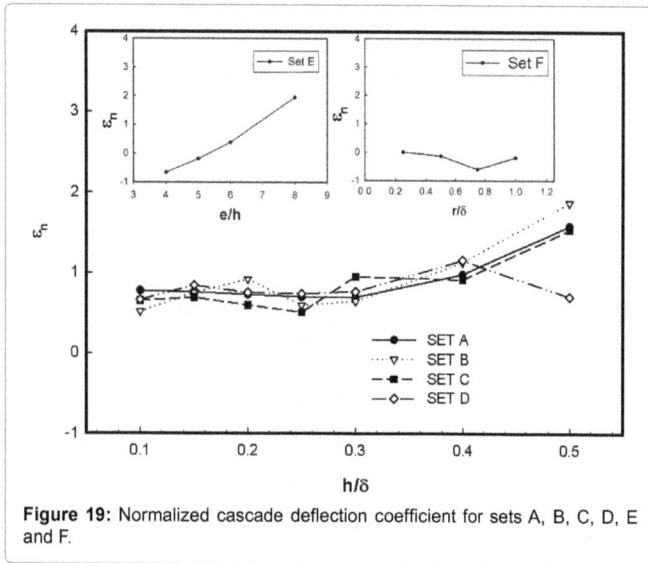

Figure 19: Normalized cascade deflection coefficient for sets A, B, C, D, E and F.

important for multistage compressors, since the change in deflection angle causes an off design operation for the next stages. However, using vortex generators does not lead to a significant change in deflection and consequently the off-design conditions will still be far from reached.

Influence of vortex generators on static pressure rise coefficient

Static pressure rise coefficient can be calculated from the following:

$$Cp = \frac{P_{s2} - P_{s1}}{P_{t1} - P_{s1}} \qquad (5)$$

where: Pt1 is the total pressure at the inlet; Ps1 is the static pressure at the inlet; and Ps2is the static pressure at the exit. Fig. 20 shows the variation of the normalized static pressure rise coefficient Cpn=ΔCp/CpRef at mid span versus h/δ for sets A, B, C, and D. As shown in Fig. 20 Cpn is decreased for all sets, this reduction almost remains constant for set A, while it increases for sets B, C, and D. The percent of change in Cpn has the minimum reduction of 2.7% at set D with h/δ=0.1 and the maximum reduction of -11.6% at set D with h/δ=0.5. For set E, as shown in Figure 20, increasing e/h leads to a decrease in the reduction of Cpn from -12 to -8% as e/h =4, 5, 6, and 8.

For set F, as shown in Figure 20, the reduction in normalized static pressure rise coefficient is almost constant with various r/δ values. This reduction with the increase in total pressure means that there is a considerable increase in the dynamic pressure when using the vortex generators. To benefit from the increase in the dynamic pressure, optimization study can be carried out on the suction blade profile to recover part of the dynamic pressure into static pressure.

Influence of vortex generators on diffusion factor

Blade loading is assessed by the diffusion factor (DF) which relates to the peak velocity on the suction surface of the blade to the velocity at the trailing edge. The diffusion factor can be defined by the following equation [28]:

$$DF = 1 - \frac{v_2}{v_1} + \frac{|\Delta v_\theta|}{2\sigma v_1} \qquad (6)$$

where: v1 is the inlet relative velocity at inlet plane, v2 is the outlet relative velocity at exit plane, Δvθ is the difference of tangential components of inlet and outlet velocity, and σ is the solidity.

Figure 21 shows the variation of the normalized difference of the diffusion factor DFn (=ΔDF/DFRef) for sets A, B, C, and D versus h/δ at midspan. The diffusion factor is used as an indicator of probability of occurrence in separation. Based on Figure 21, as h/δ increases, the diffusion factor is decreased for all sets except for set C, at h/δ=0.4, and set D at h/δ=0.5 where the diffusion factor is slightly increased. Reduction of the normalized diffusion factor as a percentage varies from 0.5% for set C with h/δ=0.4 to 5.5% for set C with h/δ=0.5. The reduction in the diffusion factor causes a reduction in separation occurrence which may be reflected in the delay or for some sets leading to eliminate separation on the suction surface. For set E, as shown in Figure 21, the diffusion factor reduction is almost constant at -2%. This is also similar to set F, as shown in Figure 21, where the reduction is also constant at -2%. Values of DF in excess of 0.6 are thought to indicate blade stall [24]. In addition, the DF for the reference case is about 0.35.

Conclusion

New designs of vortex generators are considered in the current investigation in order to control the secondary flow losses in compressor cascades and therefore enhance the compressor's performance. Six

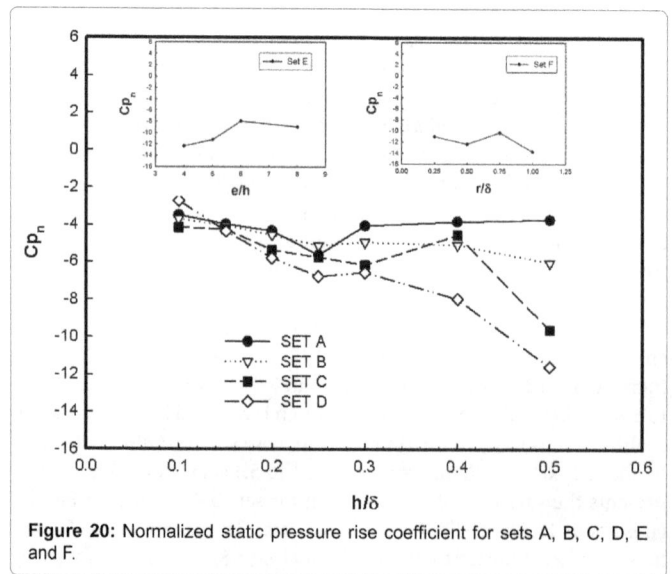

Figure 20: Normalized static pressure rise coefficient for sets A, B, C, D, E and F.

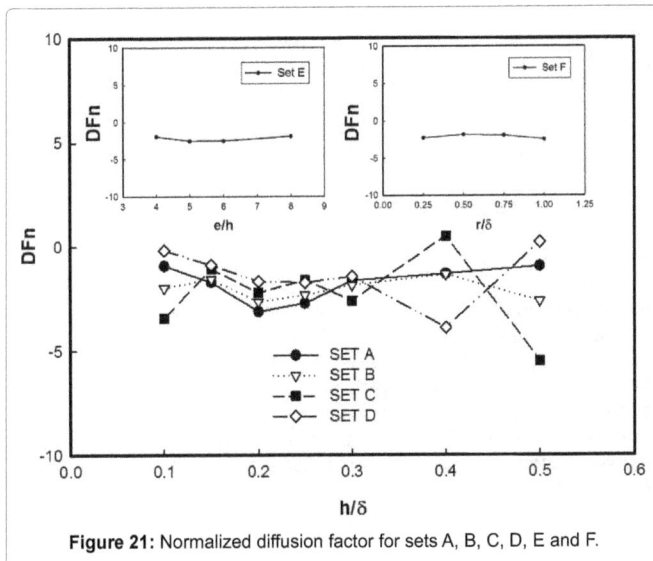

Figure 21: Normalized diffusion factor for sets A, B, C, D, E and F.

different sets of vortex generators with varying geometrical parameters and designs are numerically investigated. The flow in the compressor cascade is a 3-d compressible turbulent flow with an inlet Mach number of 0.66 and an inlet angle (β1) of 132°. Numerical simulations are performed using a numerical solver Fluent-14 of the flow through the compressor cascade without and with vortex generators placed on the endwall region near the leading edge of the cascade blade. Considering the current study, some important observations can be made. First, vortex generators have a significant impact on secondary flow losses such as improving the location of separation lines by its moving toward the trailing edge, and reducing the corner vortices. Second, a significant reduction in the normalized total pressure loss of up to 20.7% is accomplished using the curved surface vortex generator with a rounded nose of r/δ=0.5, w/h=6, e/h=5, and h/δ=0.4. Third, using vortex generators results in reduction of the diffusion factor of about -2.0%. This will increase the safe range of operating conditions without stall. Finally applying vortex generators has insignificant effect on the change of deflection and consequently the off design conditions will still be far from reached. However, there is a reduction of static pressure rise which leads to an increase of dynamic pressure. To benefit from the increase in the dynamic pressure, a further investigation is needed to recover part of the dynamic pressure into static pressure.

References

1. Gbadebo SA, Cumpsty NA, Hynes TP (2005) Three-dimensional separations in axial compressors. J Turbomachinery 127: 331-339.

2. Dorfner C, Hergt A, Nicke E, Moenig R (2011) Advanced non axisymmetric end wall contouring for axial compressors by generating an aerodynamic separator Part I: principal cascade design and compressor application. J Turbomachinery 133: 021026.

3. Hergt A, Dorfner C, Steinert W, Nicke E, Schreiber HA (2011) Advanced non axisymmetric end wall contouring for axial compressors by generating an aerodynamic separator Part II: experimental and numerical cascade investigation. J Turbomachinery 133: 021027.

4. Horlock JH, Louis JF, Percival PME, Lakshminarayana B (1966) Wall stall in compressor cascades. ASME J Fluids Eng 88: 637- 648.

5. Greitzer EM (1980) Review-axial compressor stall phenomena. ASME J Fluids Eng 102: 134-151.

6. Gad-el-Hak M (1996) Modern developments in flow control. Applied Mechanics Reviews 49: 365-379.

7. Lin JC, Howard FG, Bushnell DM (1990) Investigation of several passive and active methods for turbulent flow separation control. AIAA 21st Fluid Dynamics Plasma Dynamics and Laser Conference, June 18–20 1990 Seattle WA AIAA: 90-1598.

8. Mdouki R, Gahmousse A (2013) Effects of slotted blading on secondary flow in highly loaded compressor cascade. J Engineering Science and Technology 8: 540-556.

9. Lin JC (1999) Control of turbulent boundary-layer separation using micro-vortex generators. 30th AIAA Fluid Dynamics Conference Norfolk VA.

10. Kuya Y, Takeda K, Zhang X, Beeton S, Pandaleon T (2009) Flow separation control on a race car wing With vortex generators in ground effect. ASME J Fluids Eng 131: 121102.

11. Wijdeven T, Katz J (2013) Automotive application of vortex generators in ground effect. ASME J Fluids Eng 136: 021102.

12. Katz J, Morey F (2008) Aerodynamics of large-scale vortex generator in ground effect. ASME J Fluids Eng 071101.

13. Seo J I, Kim SD, Song DJ (2002) A numerical study on passive control of shock wave/turbulent boundary layer in a supersonic compressor cascade. The International J Rotating Machinery 8: 423-430.

14. Sahin F, Arts T (2012) Experimental investigations on the effects of low profile vortex generators in a compressor cascade. 9th National Congress on Theoretical and Applied Mechanics Brussels.

15. McCormick DC (1992) Shock–boundary layer interaction control with low-profile vortex generators and passive cavity. 30th AIAA Aerospace Sciences Meeting and Exhibit Reno NV AIAA Paper 31: 91-96.

16. Lin JC (2002) Review of research on low-profile vortex generators to control boundary-layer separation. Progress in Aerospace Sciences 38: 389-420.

17. Lu FK, Li Q, Shih Y, Pierce AJ, Liu C (2011) Review of micro vortex generators in high-speed flow. 49th AIAA Aerospace Sciences Meeting including the New Horizons Forum and Aerospace Exposition Orlando Florida.

18. Lin JC, Howard FG (1989) Turbulent flow separation control through passive techniques. AIAA 2nd Shear Flow Conference Tempe AZ AIAA: 13-16.

19. Ola L, Kristian A, Henrik A (2010) On the robustness of separation control by stream-wise vortices. European Journal of Mechanics B/Fluids 29: 9-17.

20. Lin JC, Selby GV, Howard FG (1991) Exploratory study of vortex-generating devices for turbulent flow separation control. 29th Aerospace Sciences Meeting Reno Nevada and AIAA Paper.

21. McCormick DC (1993) Shock/boundary layer interaction control with vortex generators and passive control. AIAA J 93: 91-96.

22. Lin JC (1999) Control of turbulent boundary layer separation using microvortex generators. AIAA Paper.

23. Smith FT (1994) Theoretical prediction and design for vortex generators in turbulent boundary layers. J Fluid Mech 270: 91-131.

24. Hergt A, Meyer R, Engle K (2012) Effects of vortex generator application on the performance of a compressor cascade. J Turbomachinery 135: 021026.

25. Fleming JL, Simpson RL, Cowling JE, Devenport WJ (1993) An experimental study of a turbulent wing-body junction and wake flow. Experiments in Fluids 14: 366-378.

26. Rood EP (1984) The governing influence of the nose radius on the unsteady effects of large scale flowstructure in the turbulent wing and plate junction flow. ASME Forum on Unsteady Flow 15: 7-9.

27. RD M (1984) Effect on a wing nose shape on the flow in a wing/body junction. Aerosp J 88: 456-460.

28. Falck N (2008) Axial flow compressor mean line design. Master thesis Lund University Sweden.

13

Computational Study of Ailerons in Cross Flows Ground Effects and Biplanes Configurations

Alsarheed M* and Sedaghat A

Department of Mechanical Engineering, Australian College of Kuwait Mishref, Kuwait

Abstract

An aileron, part of the trailing edge of a fixed-wing airplane, is used to control aircraft's movement around its longitudinal axis (roll). Ailerons have significant impacts on airfoil surface pressure and its lift and drag coefficients. Both panel and finite volume methods were used on a NACA 2412 airfoil with a 20% aileron in a cross flow. The aerodynamic performance of ailerons alone, in a biplane configuration, and in the ground effects has been computationally investigated using both the panel and the finite volume methods. Several parameters were analyzed including the effects of the attack angle of the airfoil, aileron deflection angle, the ride height from the ground, and the characteristics of biplanes. Results of both computational methods are presented and discussed for the aforementioned configurations of NACA 2412 airfoil with aileron.

Keywords: Aileron; Biplane; Finite volume; Ground effect; NACA 2412; Panel method

Introduction

From flying a kite to operating a hypersonic vehicle, achieving and maintaining a high lift is always critical to a successful performance of a flying craft. Some of the advantages of an efficient lift system include improved maneuverability, higher wing payload capacity, longer range for given gross weight, and lower takeoff and landing speeds. Wing components such as trailing edge flaps play a major role in altering lift in conventional fixed-wing aircrafts. They control lift and extend speed range by upward or downward movements causing an increase or decrease in lift. Also, the flap deflection can reduce efficiency by causing or increasing drag, which sometimes is difficult to reduce due to flow with high Reynolds number. Part of the trailing edge assembly is a flap called ailerons, which is a French word for 'little wing' that is used to control the aircraft in its longitudinal axis (roll). This movement is called 'rolling' or 'banking' and it alters the flight path due to the titling of the lift vector. One of the main components that impacts lift is the airfoil. An airfoil is a cross sectional part designed to generate lift when it is subject to an air flow. Ailerons are not just limited to aircraft applications. They are also used in high speed levitation trains to control and maintain the hovering in track.

The airfoil configuration has been widely studied. There are several studies that optimize the airfoil configuration and assess the maximum airfoil lift capabilities [1,2]. Other literatures study the efficiency of the airfoil and the effects of different parameters on it [3]. The camber and the Reynolds number effects have also been studied by many including Levy [4]. Other work done by Birch [5] and Rinoie [6] include calculations of unsteady loading on the airfoil and the effect of trailing edge flaps on flow vortex shapes. Also, the performance of a wing in ground effect and the flow field characteristics has been studied in [7,8]. In addition, the lift distribution between biplane wings and other biplane characteristics have been looked at in [9]. The objective of this paper is to investigate the performance and the ground effects on the ailerons of a biplane configuration. Both panel and finite volume methods were used on a NACA 2412 airfoil with a 20% aileron in a cross flow. Numerical measurements of surface pressure distributions were also obtained to determine the lift and drag coefficients for various configurations.

Airfoils Characteristics

For every action, there is an equal and opposite reaction. According to Newton's third law [10,11] an airfoil generates lift by diverting the motion of flow over its surface in a downward direction, resulting in an equal upward reaction. Typical airfoil geometry is shown in Figure 1.

The main parameters of an airfoil, which play a key role in its aerodynamic performance, are the angle of attack, the chord length, and the mean camber line. The chord is the straight line across the airfoil and it is used to measure the airfoil length. The mean camber is the line halved the airfoil thickness and it is used to measure airfoil curvature.

Common airfoil shapes have been characterized by the National Advisory Committee for Aeronautics (NACA) [12]. Each airfoil shape is defined by this system by a series of digits that correspond to non-dimensional airfoil properties. The number of digits series varies, but in this paper the focus is on four digit airfoils (NACA 2412).

A NACA 2412 airfoil has a maximum camber that is 2% of the chord length with a maximum thickness of 12% of the chord length, located 4/10 of the chord length away from the leading edge.

Computational Modeling

A NACA 2412 airfoil configuration with a 20% aileron was used for the panel and finite volume methods, and it is shown in Figure 2a. The chord length of the airfoil is designated by c, the velocity of the uniform flow is denoted by U, δ_f is the aileron deflection angle, and α represents the angle of attack. The flow is assumed to be two-dimensional steady and incompressible Laminar flow.

---

*Corresponding author: Alsarheed M, Department of Mechanical Engineering, Australian College of Kuwait Mishref, Kuwait, E-mail: m.alsarheed@ ack.edu.kw

Panel method

The airfoil was analyzed using a vortex panel method, where the airfoil was approximated by a finite number of panels. As shown in Figure 2b, the airfoil surface is divided into 70 panels. There are more panels near the leading and trailing edge to accommodate for the rapid changes in flow near these two ends. Each panel had local, uniform and distributed vortices. Since the velocities can be singular at the center of the vortex, and to avoid singularities, these vortices had infinitesimal strength $\gamma_0 ds_0$, where γ_0 is the vortex strength per unit length, and ds_0 represents the length of a small segment of the airfoil.

The governing equation for the stream function due to all such infinitesimal vortices at a point in space may be given by:

$$\psi = \oint \frac{\gamma_0}{2\pi} \ln\left(\overline{r} - \overline{r}_0\right) \, ds_0 \tag{1}$$

where $\overline{r}_0$ is a point in the x-y plane with coordinates x_0 and y_0 and the distance between the position of the vortex, and the point where the velocity is evaluated with coordinates x and y is denoted by ⌐⌐. The x-y plane is originated at the leading edge of airfoil with x-axis along chord length and y perpendicular to it.

Equation (1) is integrated over all vortices on the airfoil surface and the stream function associated with the free stream is given by $u_\infty y - v_\infty x$, where u_∞ and v_∞ are the x- and y- components of the free stream velocity. Taken these effects into account, the stream function at any point $\overline{r}$ in space is given by

$$\psi = u_\infty y - v_\infty x - \frac{1}{2\pi} \oint \gamma_0 \ln\left(\left|\overline{r} - \overline{r}_0\right|\right) \, ds_0 \tag{2}$$

Since the body itself is a streamline, the stream function value at

all the points on the airfoil will be constant and the value is denoted by C. Also, for all points $\overline{r}$ on the airfoil surface, equation (2) becomes:

$$u_\infty y - v_\infty x - \frac{1}{2\pi} \oint \gamma_0 \ln\left(\left|\overline{r} - \overline{r}_0\right|\right) \, ds_0 - C = 0 \tag{3}$$

The Kutta condition is an aerodynamics principle and it is applicable to the trailing edges of the airfoils. This condition states that both the pressure above and below the trailing edge of the airfoil must be equal. It also requires that the flow be smooth and must leave the trailing edge in the same direction at both upper and lower edges and it is expressed as follows:

$$\gamma_{upper} = -\gamma_{lower} \tag{4}$$

For a finite number of N panels, there are equal N unknown vorticity strengths, denoted by $\gamma_{0,j}$, as shown in Figure 2(b). To approximate the line integral over the entire airfoil, it can be integrated as several line integrals over N panels, taking each value of γ_0 as constant and so equation (3) becomes:

$$u_\infty y_i - v_\infty x_i - \sum_{j=1}^{N} A_{i,j} \gamma_{0j} - C = 0 \tag{5}$$

$$A_{i,j} = \frac{1}{2\pi} \int \ln\left(\left|\overline{r}_i - \overline{r}_0\right|\right) ds_0$$

Equation (5) can also be solved over the entire airfoil as several line integrals over the panels for the unknown vorticity strengths.

Finite volume method

The numerical simulations were conducted using FEMLAB software with a Reynolds number Re = 3 × 10⁶. The Reynolds number Re is defined as

$$Re = Uc/\upsilon \tag{6}$$

Where U is the velocity of a uniform flow, c is the chord length of the airfoil, and υ denotes the kinematic viscosity

The governing non-dimensional equations for the fluid flow are described by the following continuity and Navier-Stokes equations:

$$\nabla.u = 0$$

$$\rho \frac{\partial u}{\partial t} - \nabla.\upsilon\left(\nabla u + (\nabla u)^T\right) + \rho(u.\nabla)u + \nabla p = 0 \tag{7}$$

Where ρ is the density, u represents the velocity vector, and p is the pressure.

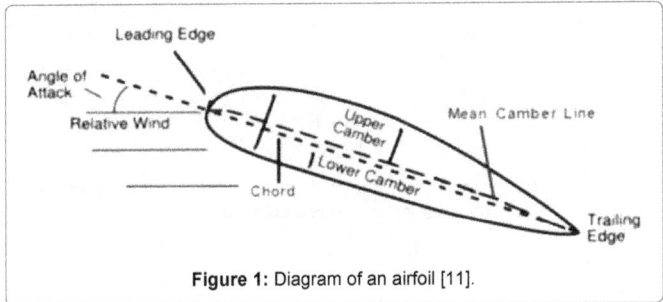

Figure 1: Diagram of an airfoil [11].

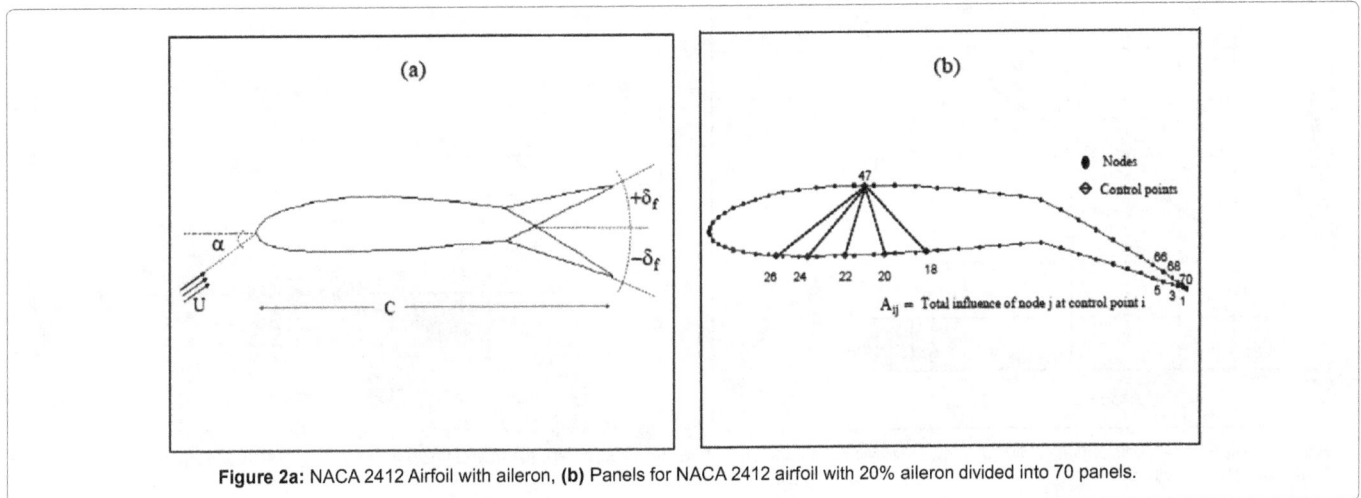

Figure 2a: NACA 2412 Airfoil with aileron, **(b)** Panels for NACA 2412 airfoil with 20% aileron divided into 70 panels.

Results and Discussions

Panel method

Aileron effects: For the zero aileron deflection case ($\delta = 0$), the experimental results of NACA 2412 airfoil with data from Panel code and FEMLAB are compared in Figure 3. The results show a good agreement in all cases and validate the accuracy of the panel method.

Effects of a 20% aileron applied to the NACA 2412 airfoil on the value of lift force and moment coefficients are shown in Figure 4. Figure 4a shows the use of ailerons can change the lift over 50%. Also, by increasing the deflection angle, the drag coefficient increases rapidly. The moment coefficients for different aileron deflection angles are shown in Figure 4b and are based on 25% airfoil chord length. For a constant aileron deflection angle, the moment coefficient stays approximately constant for different attack angles.

Figure 5 shows variation of pressure coefficients for different deflection angles and an incidence angle of $\alpha = 3°$. When the aileron is upward, the pressure coefficient has a considerable change approximately at the middle of the airfoil. For $\delta_f \neq 0$ around 0.8 airfoil length, there are two pressure peaks at the lower and upper surfaces of the airfoil. As it is shown in the figure, the deviation of these pressure peaks from the initial case ($\delta = 0$) is larger when the aileron is downward.

Ailerons in ground effects: Fundamental geometric parameters such as ground clearance, wing camber and its thickness ratio can have impact on the performance of a wing in presence of a solid wall (ground). In Panel method, a wing in ground effect is equivalent to a two-wing case, where the second wing is the mirror image of the first, as shown in Figure 6a, and the wing-body combinations are solved simultaneously. The effect of ride height on the lift coefficient is

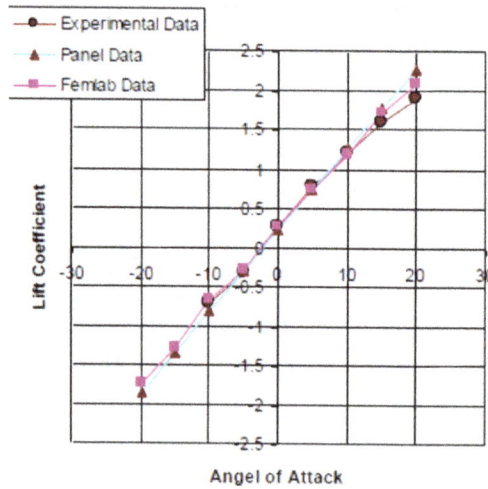

Figure 3: Comparison of lift coefficient between Panel, FEMLAB and experimental data for a NACA 2412 airfoil.

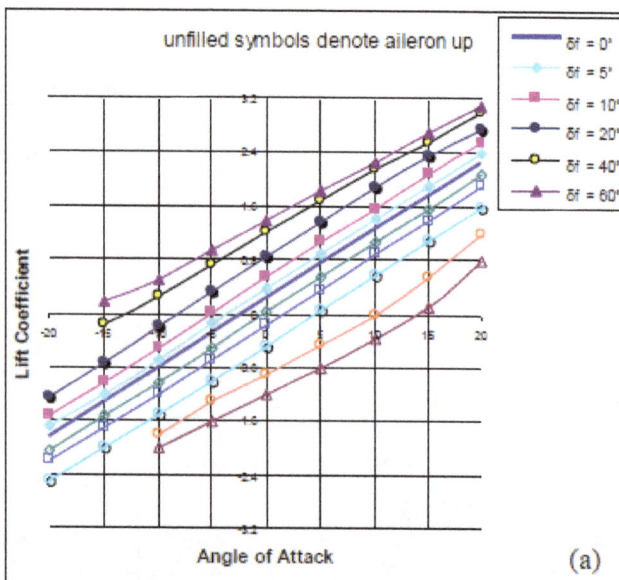

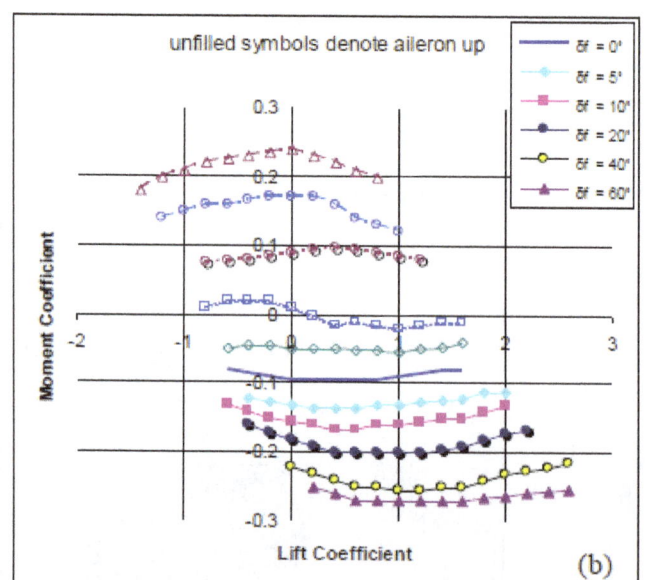

Figure 4: (a) Force Coefficients for a NACA 2412 airfoil with aileron in cross flow (Panel Method), **(b)** Moment Coefficients for a NACA 2412 airfoil with aileron in cross flow (Panel Method).

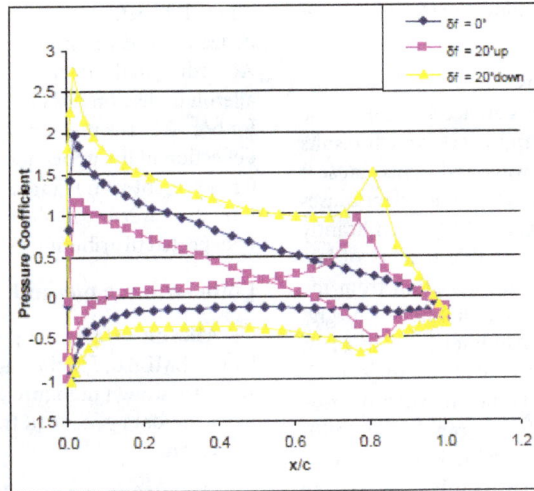

Figure 5: Surface pressure distributions for δ = 0, +20, −20 f and α = 3° (Panel method).

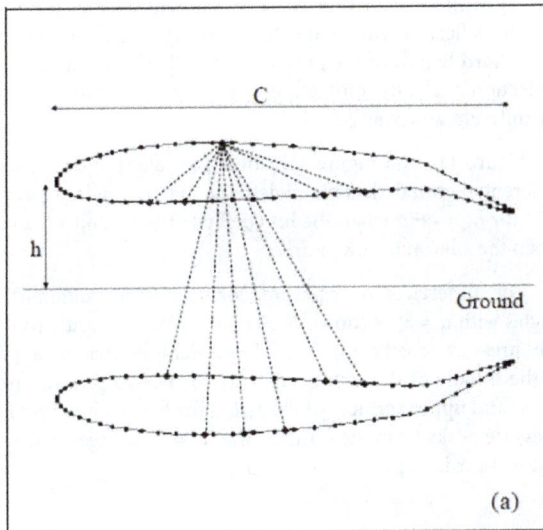

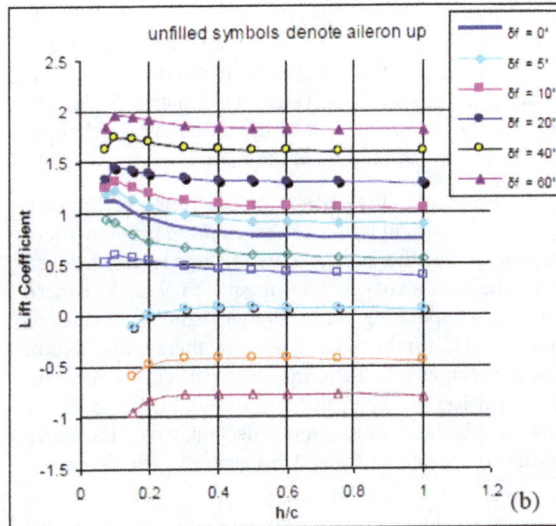

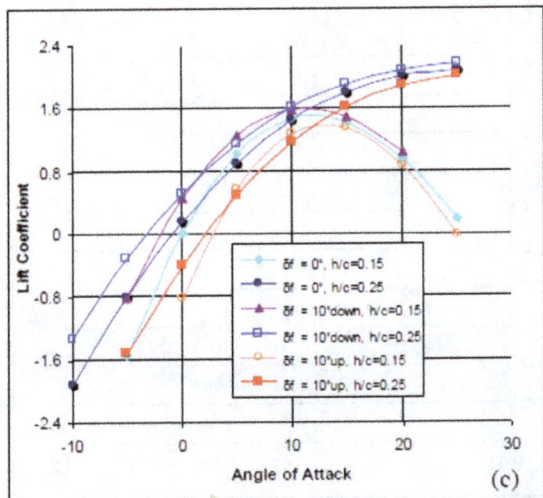

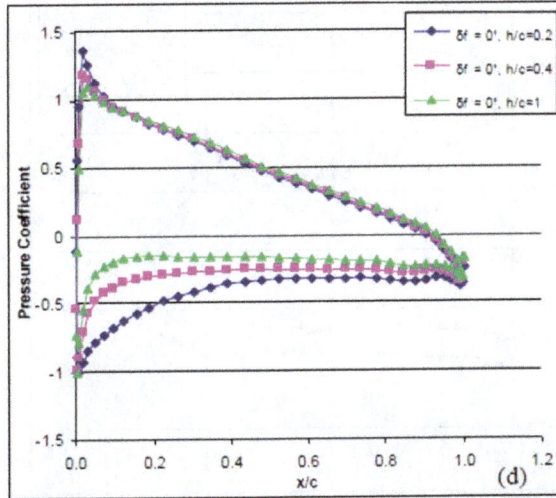

Figure 6: (a) Ground effect (Panel method) **(b)** Lift coefficients in ground effect for different ride heights (Panel method). **(c)** Lift coefficients in ground effect for different configurations (Panel method), **(d)** Pressure coefficients for a NACA 2412 airfoil without aileron (Panel method).

presented in Figure 6b. The tests were performed at an angle of attack of 3°. At a close proximity of the wing to the ground, the effect of higher down force coefficients is clearly noticeable.

The physical effect of the ground is to constrain the airflow on the suction (lower) surface of the wing. This causes an acceleration of the flow compared with the case away from ground effects which results in a greater suction on the lower surface of wing and hence cause a higher lift force. As the ride height decreases, the ground effect causes the flow to be accelerated to a higher degree, generating a significantly higher lift force compared with free stream. At ride heights of less than $0.1c$, there is a gradual and then significant deviation from the earlier trend of ever-increasing lift force with reduction in ride height. Eventually, the lift force falls off to reach a maximum C_L of 2 at a ride height of $0.1c$, as shown in Figure 6(b). The effect of attack angle on C_L in ground effect is shown in Figure 6c. As it is shown, when the ride height is less than $0.15c$, the lift coefficient decreases rapidly for $\alpha > 10°$ as also observed in ref. [13]. In Figure 6d, C_p distribution of a NACA 2412 without aileron is shown in ground effect. As it is expected, the pressure coefficient in this case only changes at the lower surface of the airfoil for different ride heights.

Figures 7a and 7b show the pressure coefficient (C_p) distribution when the aileron deflection of ± 20° is occurred in the airfoil. For $\delta_f = -20°$, the pressure coefficient of the lower surface changes gradually due to the distance between airfoil and ground. It also decreases when the airfoil is closer to the ground. However, as it is shown in Figure 7b, when $\delta_f = +20°$ the pressure coefficient of the lower surface changes rapidly and reaches 8 at the connection of the aileron.

Ailerons in biplanes: The effect of ailerons in biplane configurations on lift coefficient (C_L) is examined using panel method. For a biplane configuration, the surfaces of two airfoils are divided into several N panels. The number of equations is equal to N+2 since there are N panels and (corresponding vortex strength) and two values of stream function C_1 and C_2 for the two airfoils. Also, there are two Kutta conditions, one for each airfoil. Then, the system of equations is set from influence of vortices on each panel and solved. Figure 8a shows the configuration of a biplane with aileron deflection, where the upper and lower airfoils are labeled as 1 and 2 respectively. The difference

between aileron deflection in upper and lower airfoils and their effect on the lift coefficient are shown in Figure 8b and 8c. For a high aileron deflection and attack angle, the lift coefficient (C_L) can reach as high as 4. According to the data shown in Figure 8(c), the choice of the airfoil for aileron deflection does not have significant effect on the lift coefficient for h/c >0.5. The pressure coefficient and the difference between aileron deflection of the upper and lower airfoils for $\alpha = 3°$, $\delta_f = -20°$ and h/c = 0.5 is presented in Figure 8d. The results show that when the deflection is on the lower airfoil, the pressure coefficient of the upper airfoil has an unexpected distribution among the surfaces of the airfoil.

Finite-volume method

Aileron effects: The results of NACA 2412 airfoil with data from FEMLAB, for the case where the aileron deflection is equal to zero, are shown in Figure 9. The experimental results and the results from the data generated from FEMLAB for NACA 2412 airfoil are in agreement.

Figure 10 shows the effects of a 20% aileron applied to the NACA 2412 airfoil on the value of lift and moment coefficients. From Figure 10a, it can be seen that the use of ailerons can change the lift over 50% but by increasing the deflection angle, the drag coefficient increase rapidly which is not desired. The moment coefficients for different aileron deflection angles are shown in Figure 10b and are based on a 25% chord length of the airfoil. As it is shown, for a constant aileron deflection angle, the moment coefficient stays approximately constant for different attack angles [14].

Figure 11a and Figure 11b show the values of drag coefficient for different aileron deflection angles. The maximum drag can reach up to 0.24 for $\delta f = -60°$ when the lift coefficient is around 2.5, and it occurs when the aileron is downward.

The difference of pressure coefficients for different deflection angles with $\alpha = 3°$ is shown in Figure 12. When the aileron is upward, the pressure coefficient has a considerable change approximately at the middle of the airfoil. Also, there are two pressure peaks at the lower and upper surfaces of the airfoil for $\delta_f \neq 0$. The deviation of these pressure peaks from the normal case ($\delta_f = 0$) is larger when the aileron is downward.

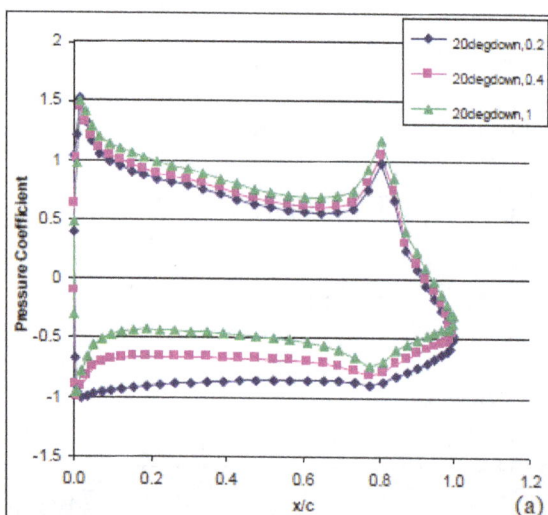

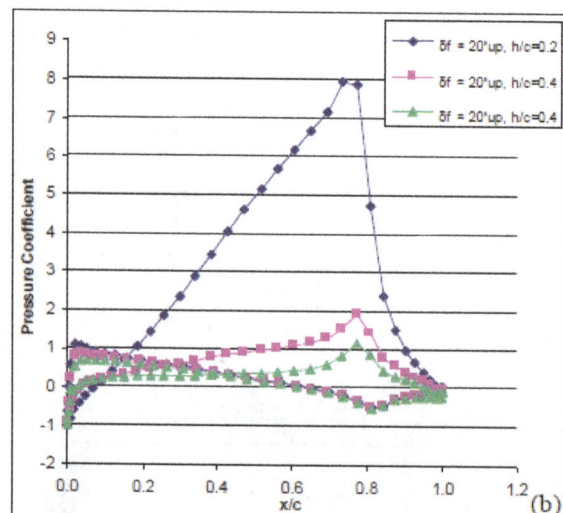

Figure 7: (a) Pressure coefficient distribution for a NACA 2412 airfoil with aileron in ground effect, $\alpha = 3°$ and $\delta f = -20°$, (Panel method) **(b)** Pressure coefficient distribution for a NACA 2412 airfoil with aileron in ground effect, $\alpha = 3°$ and $\delta f = +20°$, (Panel method).

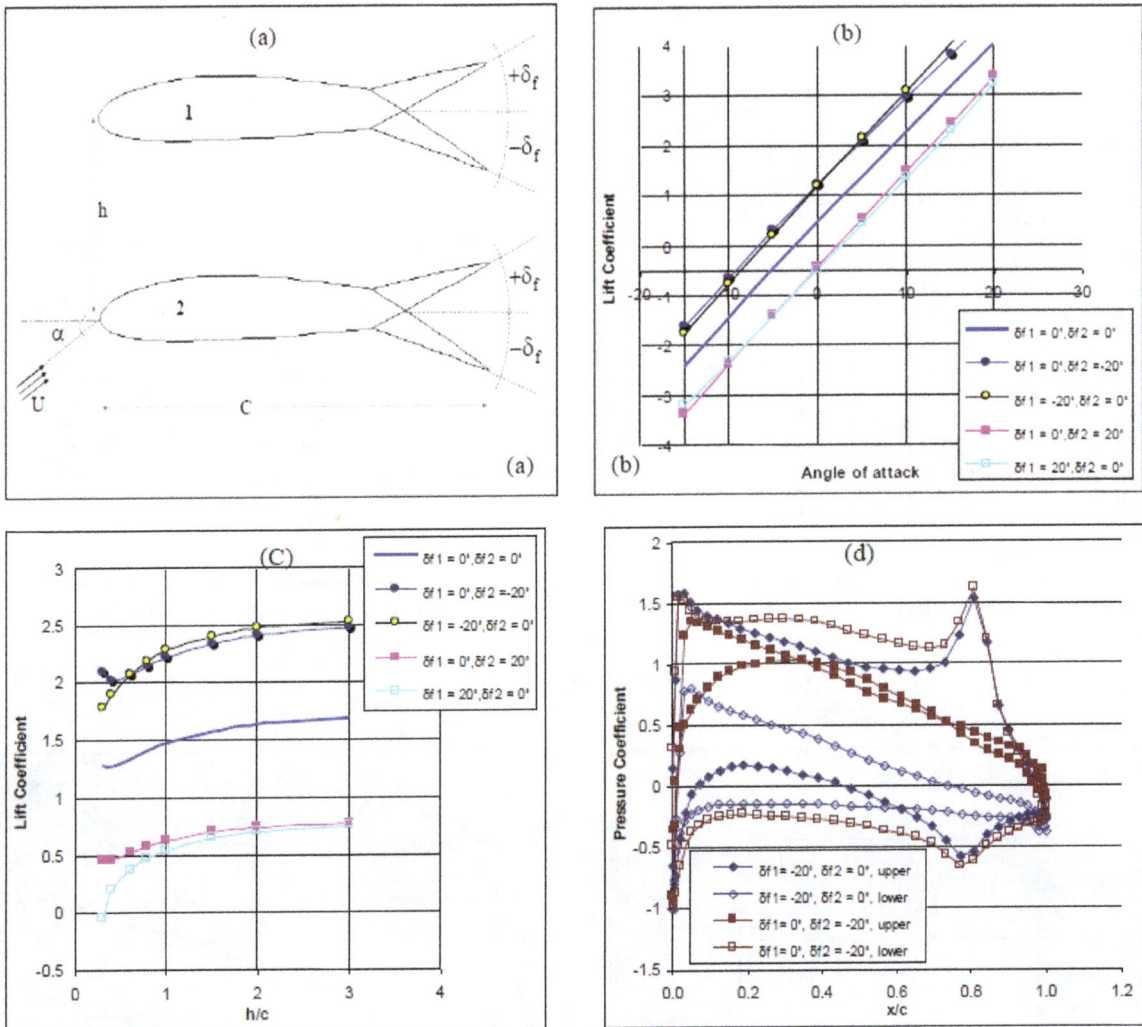

Figure 8: (a) Biplane configuration of NACA 2412 Airfoil with aileron (Panel method) **(b)** Lift coefficients in biplane with aileron deflection due to angle of attack (Panel method). **(c)** Lift coefficients in biplane with aileron deflection due to distance between the airfoils (Panel method) **(d)** Pressure coefficients in biplanes with aileron deflection in different configurations (Panel method).

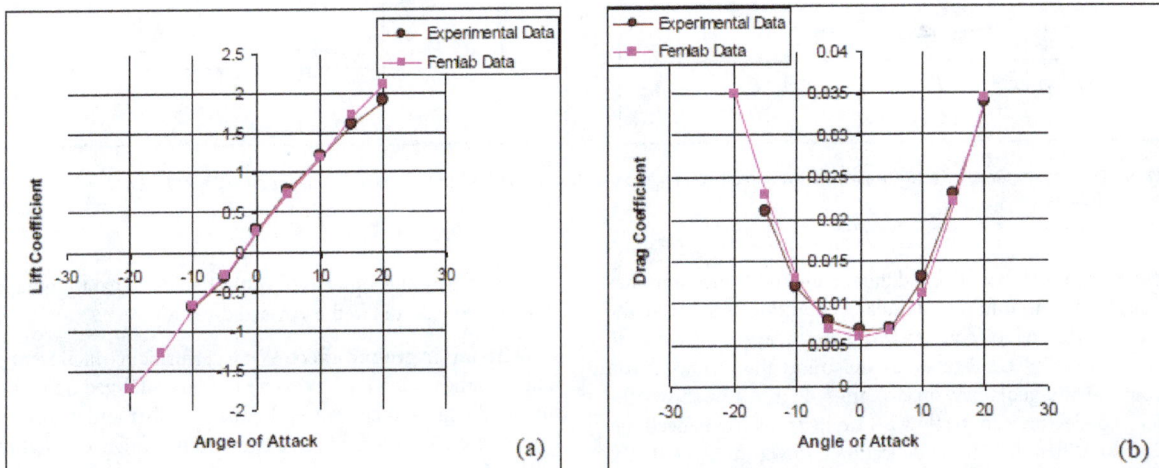

Figure 9: (a) Lift coefficients-experimental and numerical data for a NACA 2412 airfoil **(b)** Drag coefficients-experimental and numerical data for a NACA 2412 airfoil.

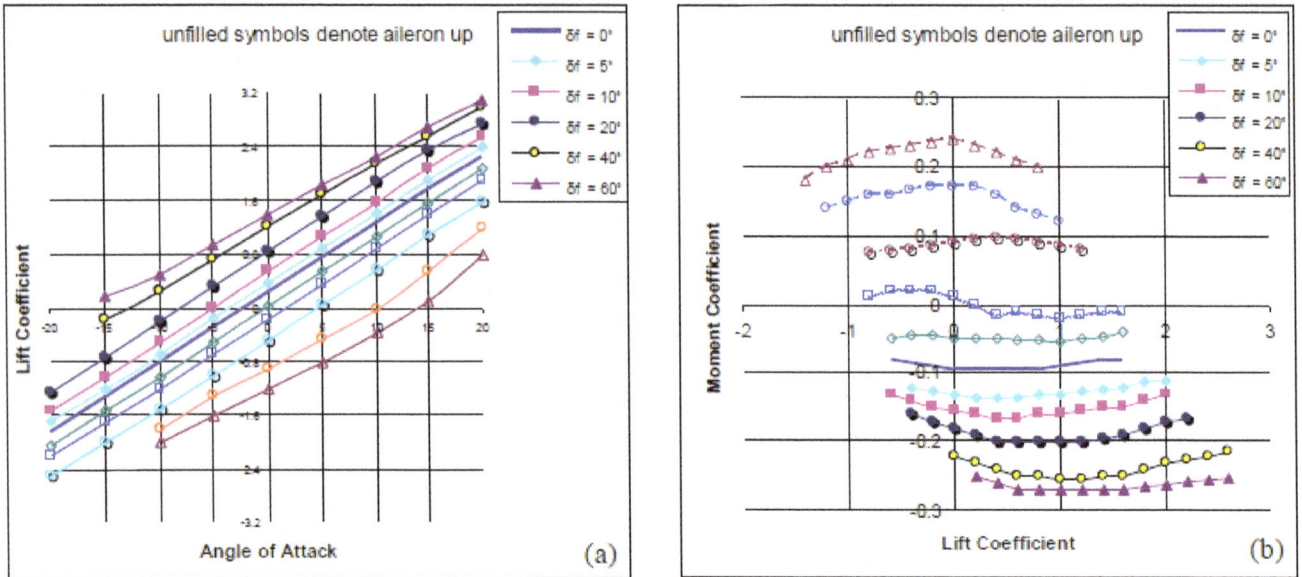

Figure 10: (a) Lift coefficients for a NACA 2412 airfoil with aileron in cross flow (FEMLAB) **(b)** Moment coefficients for a NACA 2412 airfoil with aileron in cross flow (FEMLAB).

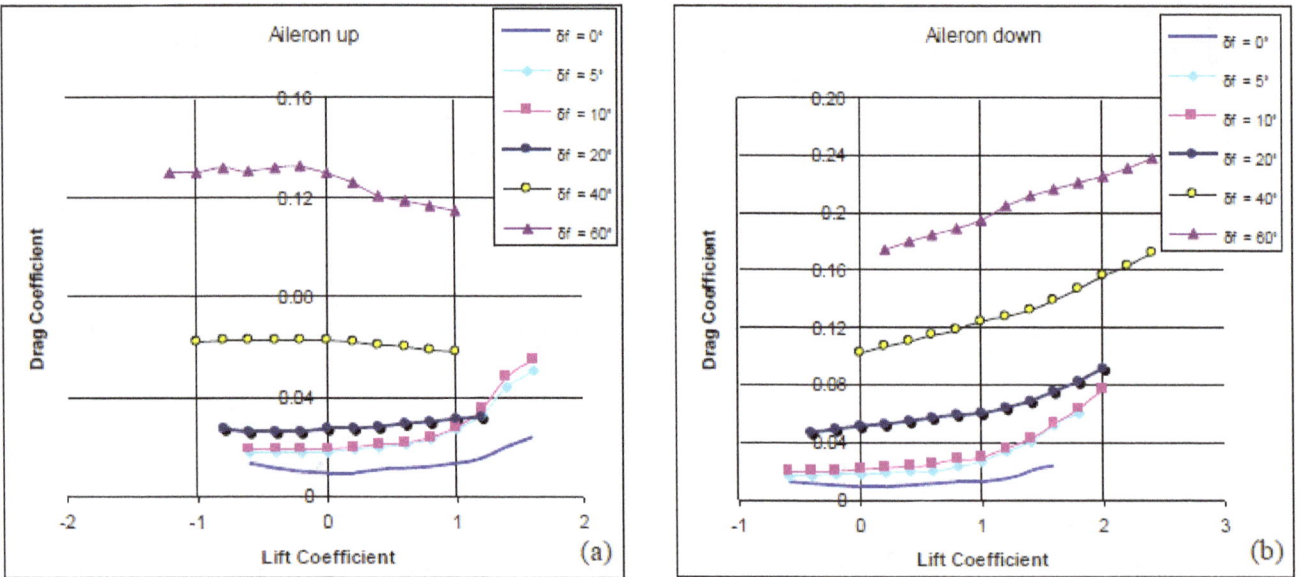

Figure 11: (a) Drag coefficients for a NACA 2412 airfoil with aileron up in cross flow (FEMLAB) **(b)** Drag coefficients for a NACA 2412 airfoil with aileron down in cross flow (FEMLAB).

The flow separation caused by different aileron deflections has also been analyzed. The outflow boundary condition is used at the downstream and the velocity at upstream boundary is selected to be the uniform velocity U. Figure 13a illustrates the airfoil-aileron configuration and the geometry surface grids. Figure 13b shows the effect of aileron deflections on the flow field at three aileron deflections, δ_f = -20, 0 and 20. When the aileron deflection is downward ($\delta_f < 0$), the flow separation occurs earlier with more flow in the upward direction. Then, the flow reconnects again near the trailing edge when the attack angle is very small and forms a small separation region. When the

aileron deflection is upward ($\delta_f > 0$), the flow separation happens later with less flow occurs in the upward direction.

Ailerons in ground effect: Wing performance in presence of a solid wall (ground) is affected by some fundamental geometric parameters such as the ground clearance, the wing camber and its thickness ratio. Figure 14 shows flow field velocity contours of ailerons in ground effect.

When the wing is in close proximity to the ground, the effect of higher lift force coefficients is noticeable. Figure 15a shows the effect of ride height on the lift coefficient at an incidence of 3°. The physical

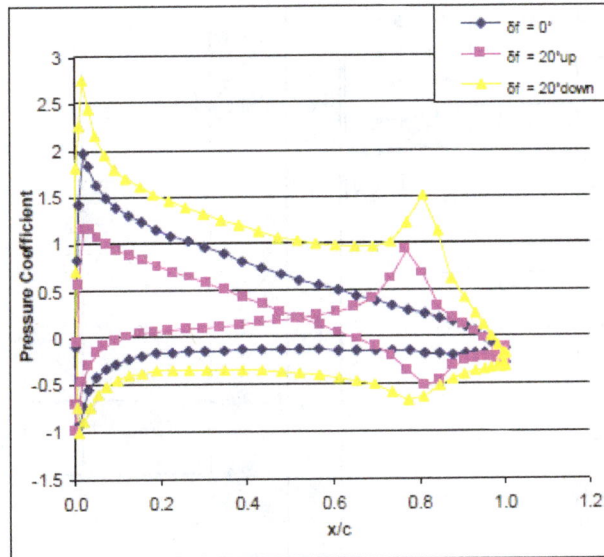

Figure 12: Surface pressure distributions for δf = 0°, +20°, -20° and α = 3° (FEMLAB).

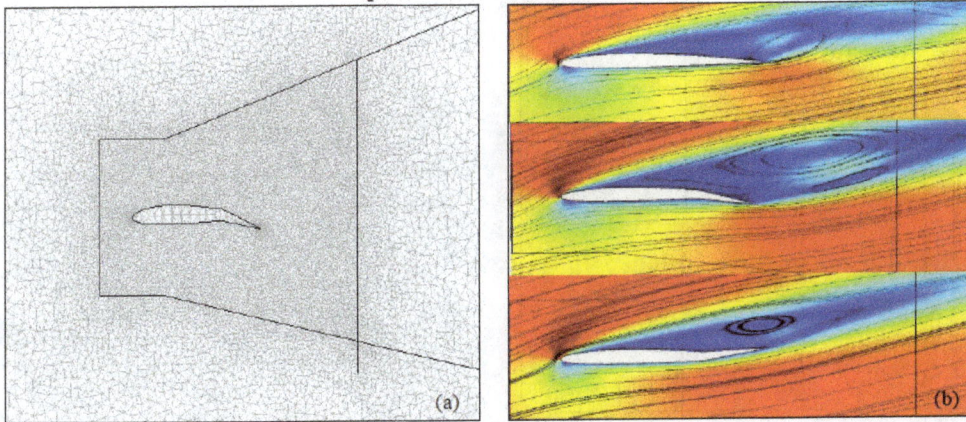

Figure 13: (a) Surface geometry and grids of a NACA2412 airfoil with 20% aileron, δf = -20° **(b)** Flow field velocity contours show separation patterns for different aileron deflections δf = 0, -20, 20, α = 20˚.

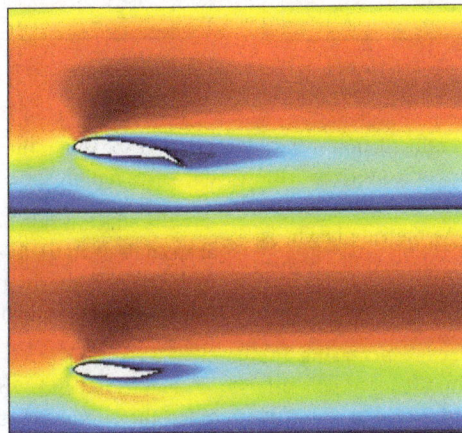

Figure 14: Flow field velocity contours for NACA 2412 airfoil-aileron configuration in ground effect δf = ± 20°, α = 5°.

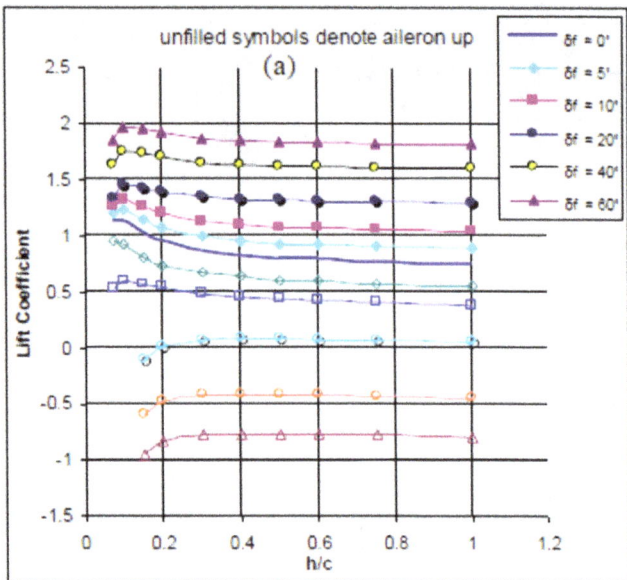

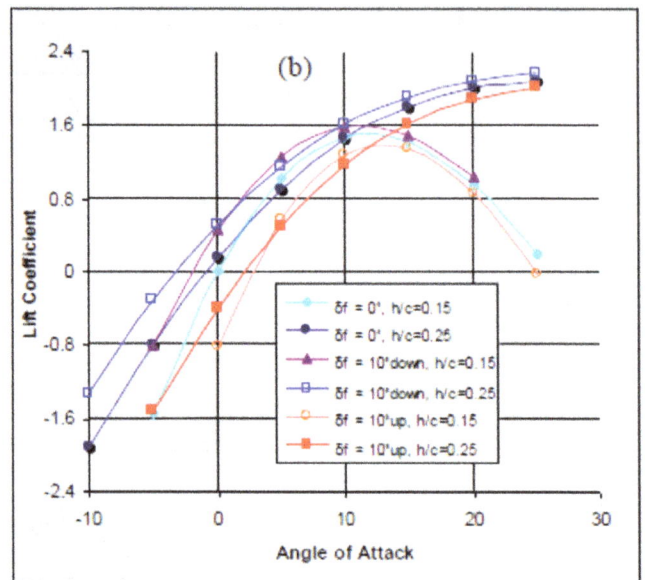

Figure 15: **(a)** Lift coefficients in ground effect for different ride heights (FEMLAB) **(b)** Lift coefficients in ground effect for different angles of attack (FEMLAB).

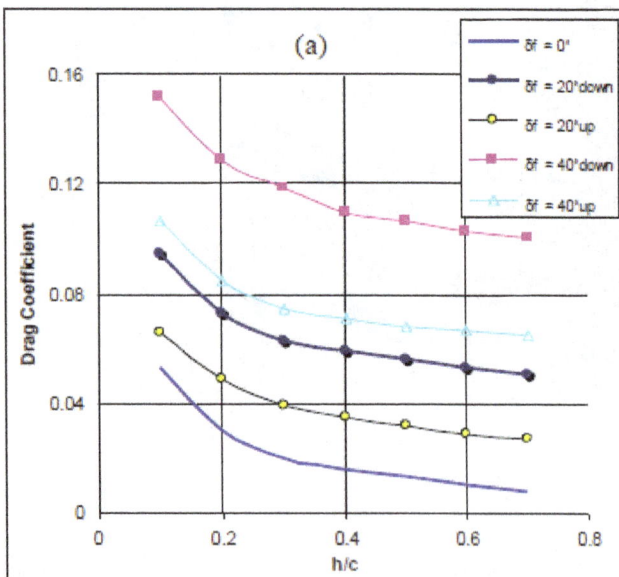

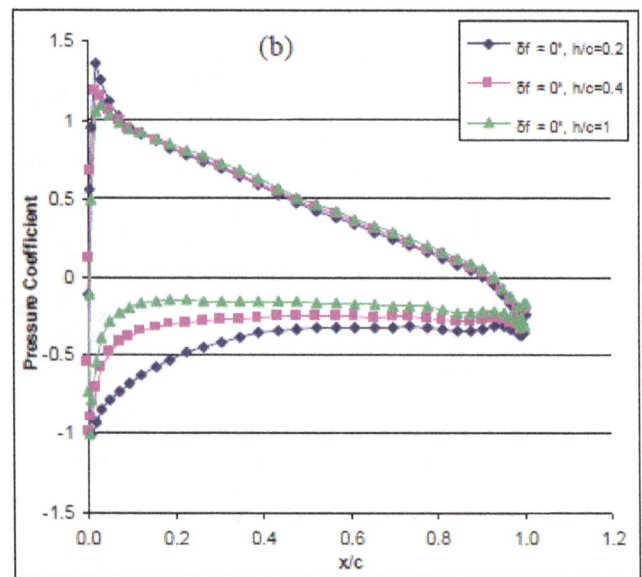

Figure 16: **(a)** Drag coefficients in ground effect for different ride heights (FEMLAB) **(b)** Pressure coefficients for a NACA 2412 airfoil without aileron (FEMLAB).

effect of the ground is to constrain the airflow over the suction surface of the wing. This causes an acceleration of the flow compared with the case out of ground effect and results in a greater suction on the suction surface and, hence, a higher lift force. As the ride height is reduced, the ground effect causes the flow to be accelerated to a higher degree, generating a significantly higher lift force compared with free stream. At ride heights of less than $0.1c$, there is a gradual and then significant deviation from the earlier trend of ever-increasing lift force with reduction in ride height. Indeed, the lift force falls off, to reach a maximum C_L of 2, at a ride height of $0.1c$. The closer it is to the ground, the more lift force reduction is seen. The effect of attack angle on the lift coefficient (C_L) in ground effect is shown in

Figure 15b. When the ride height is less than $0.15c$, the lift coefficient decreases rapidly for $\alpha > 10°$.

The effect of ride height on drag coefficient is shown in Figure 16a at an angle of attack fixed to 3°. As it is expected, when the airfoil is close to the ground, the drag coefficient is high and can reach up to 0.16 for a ride height of 0.1 and $\delta_f = -40°$. In Figure 16b, pressure coefficient (C_p) distribution of a NACA 2412 without aileron is shown in ground effect.

Figure 17a and 17b show the pressure coefficient (C_p) distribution when the aileron takes deflection angle of $\pm 20°$. For $\delta_f = -20°$ case, the pressure of the lower surface changes gradually due to the distance between airfoil and ground and decreases when the airfoil is closer to

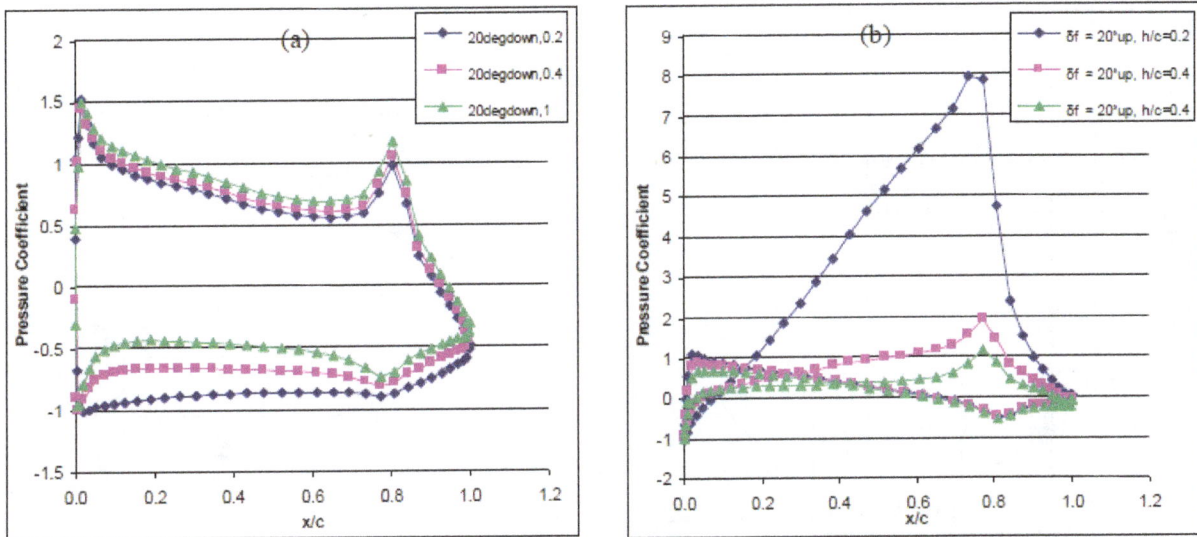

Figure 17: Pressure distribution for a NACA 2412 airfoil with aileron in ground effect, α = 3° (FEMLAB).

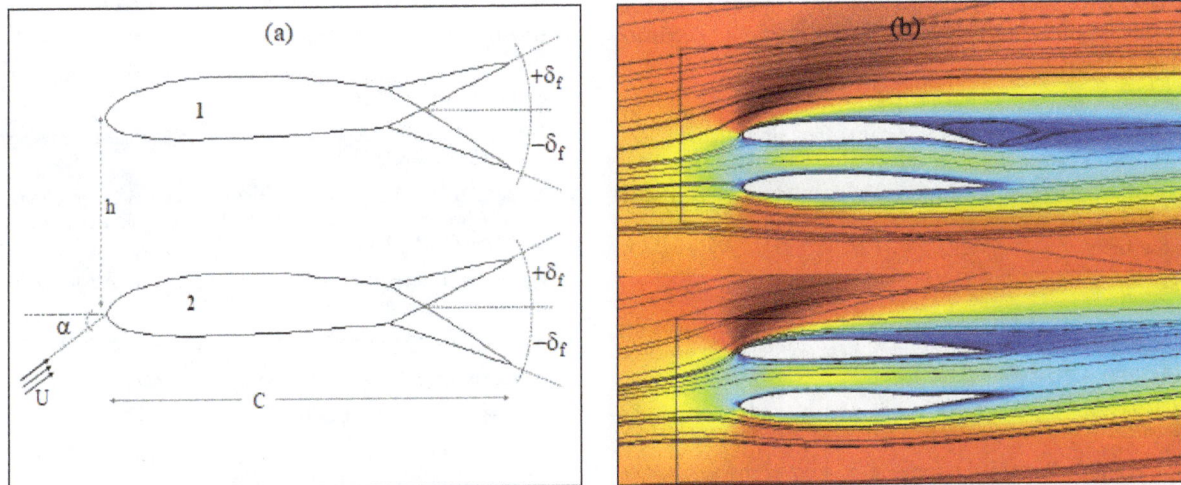

Figure 18: (a) Configuration of biplane (left) and (b) flow field velocity contours (right) for a biplane with aileron deflection.

the ground; but as it is shown in Figure 17b, when $\delta_f = +20°$ the pressure of the lower surface changes rapidly and reaches 8 at the aileron joint.

Ailerons in biplanes: In this section, the effect of ailerons in biplane configurations on the lift coefficient (C_L) is examined. Figure 18a and 18b show the configuration and flow field for a biplane with aileron deflection. The difference between aileron deflection in upper and lower airfoils and their effect on the lift coefficient are shown in Figure 19a and 19b. For a high aileron deflection and attack angle, C_L can reach as high as 4. Additionally, Figure 19b shows that the choice of the airfoil for aileron deflection has little or no significant effect on lift coefficient for h/c >0.5.

Conclusions

The flow field over an airfoil with 20% aileron has been numerically examined using panel and finite volume methods. Many parameters play important roles in the efficiency of the airfoil. The effects of the

attack angle of the airfoil, the flap length, flap deflection angle and Reynolds number are investigated in this paper. Also, viscous flows over the flapping airfoil with different flap combinations are computed. The results of both Panel and the Volume numerical methods are consistent are in agreement with the experimental results. The results also indicate that the existence of an aileron has significant impacts on the lift and moment of the airfoil. The use of ailerons in ground effect and biplane configurations relevant to levitated high speed train has been studied as well. The following conclusions are drawn:

- For the airfoil with aileron when the attack angle is increased, the lift is also increased by two folds.

- When the aileron is upward, the pressure coefficient has a considerable change approximately at the middle of the airfoil. There are two pressure peaks at the lower and upper surfaces of the airfoil for $\delta_f \neq 0$. The deviation of these pressure peaks from the normal case ($\delta_f = 0$) is larger when the aileron is downward.

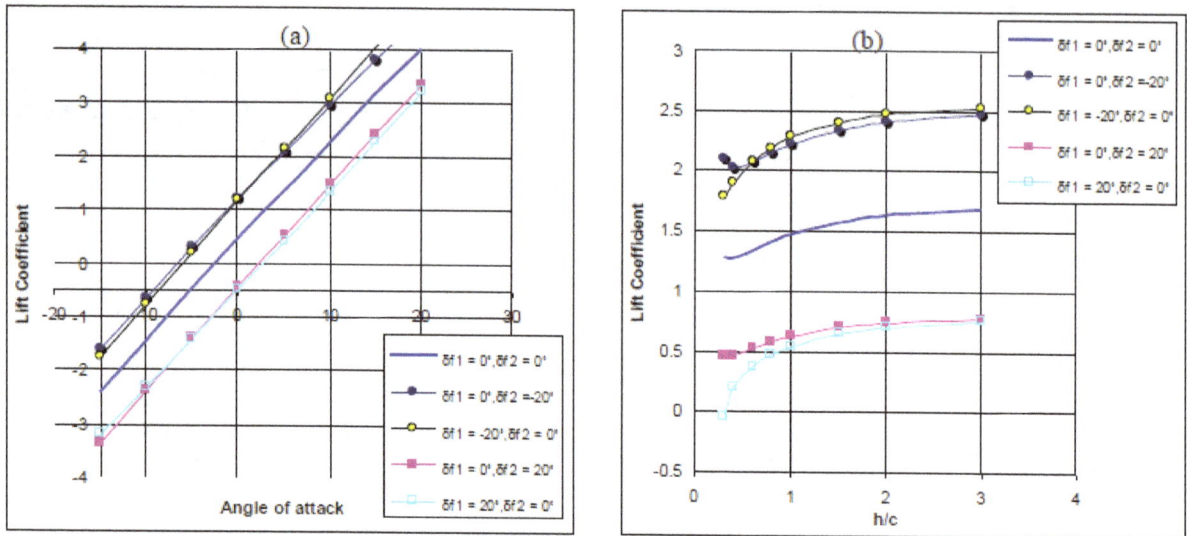

Figure 19: Lift coefficients (using FEMLAB) in biplane with aileron deflection due to: **(a)** angle of attack; **(b)** distance between the airfoils.

- When the airfoil-aileron configuration is close to the ground surface, force characteristics increase significantly and when the ride height is less than $0.15c$, the lift coefficient decreases rapidly for $\alpha >10°$.

- For the biplane case, the choice of the airfoil for aileron deflection has little or no important effect on lift coefficient for $h/c >0.5$. However, it is shown that when the deflection is on the lower airfoil, the pressure coefficient of the upper airfoil has an unexpected distribution among the surfaces of the airfoil.

References

1. Valarezo WO, Dominik CJ, Mcghee RJ, Goodman WL, Paschal KB (1991) Multi-element airfoil optimization for maximum lift at high Reynolds numbers. Technical Papers 2: 969-976.

2. Besnard E (1998) Two-dimensional aircraft high lift system design and optimization. Aerospace Sciences Meeting and Exhibit 36th Reno NV.

3. Sommerer A (2000) Numerical optimization of adaptive transonic airfoils with variable camber. International Congress of Aeronautical Sciences 22nd Harrogate United Kingdom International Organization .

4. Levy D (2007) Reynolds number effects on the performance of a hinged slotted flap. 45th AIAA Aerospace Sciences Meeting Reno NV United States.

5. Birch D (2005) Effect of trailing-edge flap on a tip vortex. J Aircraft 42 : 442-447.

6. Rinoie K (2003) Benefit and performance of various vortex flap configurations. J Aircraft 40: 1215-1218.

7. Zerihan J, Zhang Xin (2000) Aerodynamics of a Single Element Wing in Ground Effect. AIAA Journal.

8. Zhang X, Zerihan J (2003) Aerodynamics of a double-Element Wing in Ground effect. Advances in Computational Methods in Fluid Dynamics ASME New York 1994 FED41: 1007-1016.

9. Maruyama D (2006) Aerodynamic analyses of airfoil configurations of biplane type supersonic transport. J stage 72: 2132-2139.

10. Anderson, David, Eberhardt S (1988) Understanding Flight. 2nd edn. Prentice Hall.

11. US Centennial of Flight Commission. Airfoil Diagram.

12. Alexander Greg NACA Airfoil Series.

13. Pedreiro N (1999) Aileron effectiveness at high angles of attack Interaction with forebody blowing. J Aircraft 36: 981-986.

14. Li, XQ (2004) Computation of a wing-body combination with aileron.

NASA Environmentally Responsible Aviation Project Develops Next-Generation Low-Emissions Combustor Technologies (Phase I)

Clarence T Chang[1], Chi-Ming Lee[1]*, John T Herbon[2] and Stephen K Kramer[3]

[1]*NASA Glenn Research Center, Cleveland, OH 44135, USA*
[2]*General Electric Company, Cincinnati, OH 45215, USA*
[3]*Pratt and Whitney, East Hartford, CT 06108, USA*

Abstract

NASA's Environmentally Responsible Aviation (ERA) Project is working with industry to develop the fuel flexible combustor technologies for a new generation of low-emissions engine targeted for the 2020 timeframe. These new combustors will reduce Nitrogen Oxide (NOx) emissions to half of current state-of-the-art (SOA) combustors, while simultaneously reducing noise and fuel burn. The purpose of the low NOx fuel-flexible combustor research is to advance the Technology Readiness Level (TRL) and Integration Readiness Level (IRL) of a low NOx, fuel flexible combustor to the point where it can be integrated in the next generation of aircraft. To reduce project risk and optimize research benefit NASA chose to found two Phase 1 contracts. The first Phase 1 contracts went to engine manufactures and were awarded to: General Electric Company, and Pratt & Whitney Company. The second Phase 1 contracts went to fuel injector manufactures Goodrich Corporation, Parker Hannifin Corporation, and Woodward Fuel System Technology. In 2012, two sector combustors were tested at NASA's ASCR. The results indicated 75% NOx emission reduction below the 2004 CAEP/6 regulation level.

Keywords: Environmentally responsible aviation; Nitrogen oxide; Technology readiness level

Introduction

NASA's Environmentally Responsive Aviation Project (ERA) is working with the industry to develop the combustor technologies for a new generation of low-emissions engines targeted for the 2020 timeframe. These new combustors will reduce Nitrogen Oxide (NOx) emissions to half of current State-of-the-Art (SOA) combustors, while simultaneously reduce particulate emissions.

NASA has been driving the NOx reduction effort in the aviation industry over the last four decades [1], resulting in approximately 50% reduction every generation of about 15 years (Figure 1). The initial concerns were health issues such as ground-level NOx and its contribution to photochemical smog. As a result, a series of increasingly stringent NOx emission standards by the International Civil Aviation Organization's (ICAO) Committee on Aviation Environmental Protection (CAEP) have, over the years, regulated aviation emissions

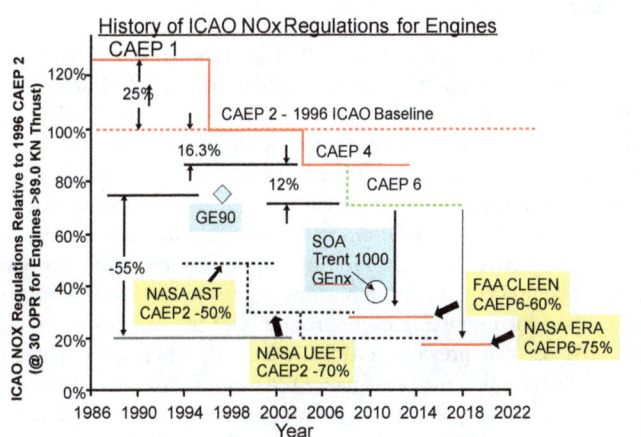

Figure 1: History of ICAO NOx regulation for engines and NASA program goals.

below 3,000-foot altitude. These standards cover the take-off, climb, descent, and taxing/ground idle phases of the engine operation, the so-called Landing and Take-Off (LTO) cycle, in a prorated fashion. With forward-thinking research, the aviation propulsion industry has turned these past NASA-sponsored combustor concepts into flight hardware. Notable among these were NASA's Experimental Clean Combustor Program (ECC, 1974-79) and its follow-on Energy Efficient Engine Program (E[3], 1980-84) that resulted in combustor concepts that were introduced into service a decade later in the GE90 and V2500 engines. Collaborative work during NASA's High Speed Research Program (1989-94), the Advanced Subsonic Transport Program (1994-99) and its followed-on Ultra Efficient Engine Technology Program (UEET, 2000-04) generated valuable insights into the designs of P&W's TALON combustor series and GE Aviation's TAPS combustor for the GEnx engine.

This continuing NOx-reduction effort is even more difficult under ERA than it was under previous programs. After four decades, the existing NOx level is already pretty low and there's not much room from which to squeeze further improvements (Figure 1). At the same time, ERA's system-level goal also includes a 50% fuel burn reduction for the platform (Table 1). While much of these savings may be taken up by airframe drag reduction, the contribution required from engine efficiency improvement means increasing the engine Overall Pressure Ratio (OPR) to about 55 from the State of the Art (SOA) 45. This increased combustor pressure and temperature also increases the NOx

*Corresponding author: Chi-Ming Lee, NASA Glenn Research Center, Cleveland, OH 44135, USA, E-mail: chi-ming.lee-1@nasa.gov

Technology Benefits	Technology Generations (Technology Resdiness Level = 4-6)		
	N+1 (2015)	N+2 (2020)	N+3 (2025)
Noise (cum. margin rel. to Stage 4)	-32 dB	-42 dB	-71 dB
LTO Nox Emissions (rel. to CAEP/6)	-0.6	-0.75	-0.8
Cruise Nox Emissions (rel. to 2005 best in class)	-0.55	-0.7	-0.8
Aircraft Fuel Consumption (rel. to 2005 best in class)	-0.33	-0.5	-0.6

Table 1: NASA's Subsonic Transport System Level Metrics [2].

formation rate. Thus, ERA's NOx-reduction effort fights on two fronts. On top of these, cruise-level NOx reduction is specifically mentioned [2], and the program itself also called for the technology concept to be fuel-flexible, capable of operating on 50% blend of non-petroleum-sourced hydrocarbon alternative fuels.

NOx reduction needs to be considered from the conceptual stage of combustor technology development. Fuel-air mixture preparation before burning starts affects what a combustor emits. The fuel from the fuel injectors sprays in as liquid, and it needs to vaporize and mix with the air before burning can occur. A very non-uniform mixture (with some pockets being too fuel-rich and some too fuel-lean) can lead to unacceptable levels of Carbon Monoxide (CO), unburned hydrocarbons, and soot due to quenching or inadequate residence times to achieve complete burnout. In contrast, some near stoichiometric pockets of fuel-air mixtures will burn very hot and produce NOx very quickly. Since NOx emission level is the time integral of the nitrogen-oxide's formation rate, the latter being an exponential function of the air temperature, NOx emission level correlates well to the fuel injector's ability to prepare the fuel-air mixture. Mixing the fuel as quickly and uniformly as possible before burning starts is a key technology for clean burning. The difficulty is in doing it during the available time, which must decrease with increasing temperature and pressure due to risk of auto-ignition. These issues need to be considered in totality from the onset.

Technology trade-off is a required practice. A solution of opportunity often also can present a challenge. Fuel-air can mix faster if the fuel can be introduced through smaller holes in fuel injectors to speed up breakup and vaporization. However, fuel heats up going through the fuel passage. Eventually, some components in the fuel reacts with the dissolved oxygen and breakdown into a gummy substance which in time turns into carbon buildup (coking) that blocks the fuel passage. Increasing the OPR increases the air temperature and speeds up coking. The availability of alternative hydrocarbon fuels that don't coke easily enables the use of smaller injection passages to speed up fuel-air preparation. However, every fuel injector also has its own combustion dynamics characteristic in which fluid dynamics interact with the combustion process. When the time scale and phase match, they can interact with the combustor acoustics to set up instabilities or limit-cycle behavior that can result in severe pressure oscillations, or disrupt the normal flame stabilization process. Thus, a balanced design of a fuel injector that mixes quickly, resists coke formation, burns stably, and still operates over a wide range of power conditions is key in bringing a new generation of cleaner-burning combustors on line.

Enabling technologies makes possible new design opportunities. Higher OPR combustion will need combustor liners able to withstand higher temperatures. Ceramic Matrix Composite (CMC) liner materials and Environmental Barrier Coatings (EBC) are complementary enabling technologies to the new injectors. A CMC liner can withstand higher temperatures than a traditional super alloy metal liner, while needing less cooling air. This capability allows the extra air to be used in the fuel injector to increase fuel-air mixing, which in turn provides a more uniform mixture with fewer hot spots such that the liner needs less air for cooling. EBCs protect the CMC surface from oxidation as well as allowing the CMC liner to operate cooler and extend the liner's life.

Phase I of ERA's Low NOx Fuel Flexible Combustor Integration task pursues this technology development via two levels of activities. The fist type of combustor activities entails a screening of potential combustor system-level concepts that includes designs on fuel injector, thermal liner, dynamics, and operability as a whole system. This task assembles the best technology sufficiently mature and currently available for demonstration in a multi-cup arc-sector form. The second type of activities narrows the scope to Lean Direct Injection (LDI) injectors and their enabling technologies that may be required for design concepts that are designed for operation above 50 atm. These injector activities are confined to flametube-level mid-power testing.

Ultra Low-Emissions Combustor Concept Developments

The combustor activity engages industry partners General Electric (GE) and Pratt & Whitney (P&W) to develop combustor concepts that can achieve the 75% LTO NOx reduction below CAEP/6 standard. These two cost-shared contracts leverage from past NASA-sponsored works and industry partners' internally-developed technology. They cover the full set of combustor challenges with full-sized injectors, liners, as well as the challenge to manage combustor system-level dynamics. This activity starts from flame tube (Technology Readiness Level (TRL) 3), through sector combustor form (TRL4) and full-annular combustor (TRL 5), and potentially can go to an engine core demonstration (TRL 6). Both programs also need to demonstrate that their designs can burn a 50%/50% alternative fuel to jet fuel blend. Both Phase I proof of concept designs met the emission goal in multi-cup arc-sector combustors, although they took different development paths and design paths.

GE N+2 advanced low-NOx combustor technology (ERA Phase I)

GE's concept design started with the legacy Twin Annular Premixing Swirler (TAPS) design that was developed via multiple technology and commercial programs, including GEnx and LEAP (Figure 2) [3], and advance the capabilities of this technology to meet the aggressive N+2 NOx and performance goals. The engine architecture, scale, and cycle were set by an engine-aircraft system analysis, pointing to a conceptual Hybrid Wing Body (HWB) aircraft and engine that could meet the key N+2 objectives for NOx, fuel burn, and noise reduction. The basic concept behind GE's N+2 combustor design is to increase the fraction of air used for premixing in the front end of the combustor beyond the 70% used in previous TAPS designs [3], while simultaneously adding features that further enhance the fuel-air mixedness. Increased premixing air can present a significant challenge to both operability (efficiency and combustion dynamics) as well as durability (less cooling air for the combustor dome and liner). To meet durability challenges, high temperature Ceramic Matrix Composite (CMC) materials with

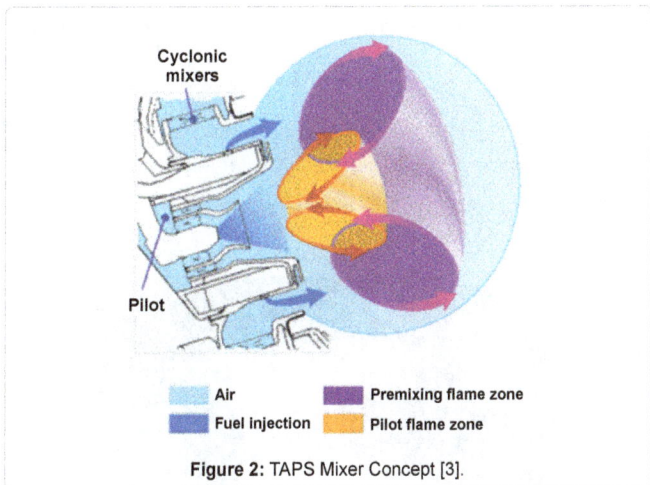

Figure 2: TAPS Mixer Concept [3].

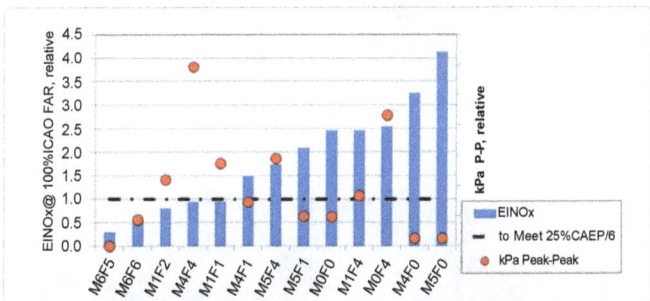

Figure 3: Flametube EINOx emissions and combustion dynamics (100% ICAO fuel/air ratio) at max rig conditions 538°C (1000°F) and 1728 kPa (250 psia).The dynamic pressure data represent the maximum peak-to-peak amplitude recorded during steady-state operation at these specific conditions, and serves as a relative indicator of potential dynamics concerns for each concept. A normalized EINOx=1 indicates the target EINOx value required at these conditions and fuel/air ratio to meet 25% CAEP/6 LTO NOx in the full combustor.

advanced cooling are utilized for the combustion liners. The new combustor design concepts were benchmarked against data from previous successful development programs. A series of combustion tests ultimately provided the opportunity to down select and further optimize the designs, leading up to the testing of one final configuration in a new 5-cups sector at NASA.

The combustor development program began with an extensive CFD effort to identify and optimize a suite of main mixer/swirler and main fuel injector concepts that could increase fuel-air mixedness while maintaining the required operability across the range of engine cycle conditions. Main stage swirler concepts included multiple designs intended to increase turbulence for fuel-air mixing, while simultaneously avoiding generation of such excessive turbulence that the swirl number was detrimentally decreased (impacting flame stability) or transporting coherent turbulence downstream into the flame front (impacting combustion dynamics). Concepts included both co-rotating and counter-rotating vanes. Seven different swirler concepts were down selected and manufactured for the initial flame tube testing, denoted here as M1-M7 (Figure 3).

The CFD effort also explored options for the number and sizing of the main stage fuel injection orifices. Main stage fuel injection concepts included varying both the radial and axial location of fuel injection (relative to the mixer exit), as well as varying the number of fuel injection points. Other concepts explored means for increasing

jet penetration into the main stage air flow either mechanically or aerodynamically. Seven different main injection concepts were down selected and manufactured for the initial flame tube testing, denoted F0-F6 (Figure 3). In all of the concept fuel nozzles, a GEnx-style pilot was scaled and utilized for the N+2 combustor due to its proven operational capability.

Flame tube testing: In the first combustion screening tests, 13 fuel/air mixer concepts were evaluated in a single-cup flame tube (FT) rig at GE Aviation. The test facility was able to achieve pressure and temperature conditions up to 538°C (250 psia) and 1728 kPa (1000°F). While this is significantly lower than the take-off and climb cycle points important to the LTO NOx evaluation, the conditions were high enough to enable fully-staged operation (fuel splits similar to the takeoff design point) and perform a relative assessment of the NOx performance of the different designs. NOx emissions and dynamic pressures were measured over a range of temperature & pressure conditions, fuel/air ratios, and pilot/main fuel splits. Comparative data at the 100% ICAO fuel/air ratio (FAR) is shown in Figure 3.

The expected performance of each concept relative to the LTO NOx goal is a critical assessment, and provides a quantitative target for acceptability of any given design. Rig data at this stage of the program was limited to low T_3/P_3 conditions (538°C and 1728 kPa) in the flame tube geometry. LTO NOx was therefore estimated from the flame tube data–corrected for T_3, P_3, and flametube-to-engine combustor correlation factors based on legacy programs. Those calculations provide a target EINOx level for the 100% ICAO fuel-to-air ratio, as measured at maximum flame tube conditions, which would be required to meet the 25% CAEP/6 objective in the eventual sector test.

The flame tube testing provided a fairly clear comparison of the performance of the various concepts. The configurations were ranked based on NOx emissions, the most important factor in the down select. Efficiency calculations, based on CO and unburned hydrocarbon measurements, were also evaluated and used to compare concepts. Quantitative efficiency measurements are considered less reliable in the flame tube vs. an actual combustor due to the differences in flame geometry and recirculation zones; however, these measurements highlight a potential challenge that must be met as combustor designs continue to get leaner and more premixed. At the lower cruise fuel/air ratios, the lowest-NOx designs also tend to have efficiencies that fall off faster as fuel/air ratio decreases. Finally, dynamic pressure data identified two concepts with elevated concerns for combustion dynamic sensitivities, specifically the M0/M1 mixer family and F4 fuel nozzle design.

Based on flame tube data, 3 concepts were chosen for further design and testing. The M6F6 concept was chosen due to its ultra-low NOx performance and a slightly better efficiency than the M6F5. The M1F2 concept provided the next-best NOx performance, with better relative efficiency than M6F6 but somewhat higher dynamic pressure signatures. Finally, the M4F1 concept was chosen for its fairly good NOx performance, but especially its improved dynamics and slightly better efficiency than the M1F2 concept. These 3 designs provided a range of mixer and fuel injection strategies going into the next round of screening tests.

Tunable combustor acoustics testing: The 3 concepts down selected from flame tube testing were further evaluated in a similar flame tube rig with tunable acoustic boundary conditions. This rig allows a more detailed mapping of the relative acoustic sensitivities of the designs. The dynamics data provide relative comparisons of the

TCA/HTP Configs	FT Normalized EINOx	FT P4' p-p Ranking 1=Best	FT Cruise Eff. Ranking 1=Best	TCA P4' p-p Relative to max limit	HTP Normalized EINOx	HTP A/I margin Relative to limit
M6F6	0.512	1	3	-	0.343	-
M1F2	0.8	3	2	>		
M4F2					1.07	-
M4F1	1.509	2	1	<	0.72	+

Table 2: Summary of flametube results at 100% ICAO FAR for the top concepts.

operability limits of the three tested configurations, and delineate the differences and features of these designs. At 1000°F, the M4F1 and M6F6 configurations show acceptable acoustics throughout the desired FAR36 and main/pilot split range of the nozzles. The M1F2 design, with its more aggressive mixer, encounters the dynamic pressure boundary limit at lower FAR36 and Main fuel flow split, making it the less attractive design from an acoustics point of view.

High Temperature/Pressure flame tube testing: In the next round of testing, new engine-style fuel nozzles were manufactured to advance the concepts into the form that would eventually be tested in the 5-cups sector. These final single-cup flame tube tests were designed to validate the concepts at high T_3, P_3 conditions near the 100% ICAO cycle point, including high power emissions measurements and evaluation of durability risk due to autoignition. Three concept nozzles were manufactured, F1, F2, and F6. The F1 and F2 nozzle concepts were slightly modified to improve durability as well as achieve an expected further reduction in NOx emissions. Both nozzles were tested with the M4 mixer due to its lower acoustics sensitivity compared to M1.

Area-averaged NOx data for all three tested configurations at 85% and near-100% ICAO generated the final ranking of the concepts for with respect to LTO NOx. The less aggressive mixing of the M4F2 resulted in predictably worse NOx than the M4F1. Similar to the initial flame tube measurements, the M6F6 configuration exhibits the best NOx performance but worse CO (and therefore efficiency) compared to the M4 designs.

Autoignition margin data: The high temperature/pressure flame tube rig also was utilized to collect autoignition data for all 3 configurations. Autoignition boundaries were mapped at various combustor inlet conditions up to the maximum facility capabilities. Autoignition data were reduced using GE design tools, and a relative risk is calculated for the different designs at 100% ICAO N+2 conditions on Jet-A fuel. Of the three designs, only the M4F1 meets the criteria for acceptable operational margins. The risk level can be reassessed for the specific alternative fuels of interest for use in future NASA testing at the Advanced Subsonic Combustion Rig (ASCR).

Conclusions for flame tube testing: Data from the 3 flame tube test campaigns is summarized in Table 2, and leads to the down select of one design for the 5-cup sector. An LTO NOx reassessment, based on the high temperature & pressure flame tube data, indicated that all 3 designs could likely meet the 25% CAEP/6 NOx target. This assessment uses a correction for flame tube vs. sector emissions data. Among the 3 designs tested in all 3 rigs, the M4F1 design resulted in the best balance between NOx emissions performance, combustion efficiency, autoignition margin, and combustion dynamics; and was selected for the sector test.

Sector testing: A major part of the combustor development program was the design and manufacturing of a new 5-cup sector rig for operation at the NASA ASCR facility (Figure 4). The combustor design utilizes high temperature CMC liner materials in order to reduce cooling air requirements and enable the high mixer air flow

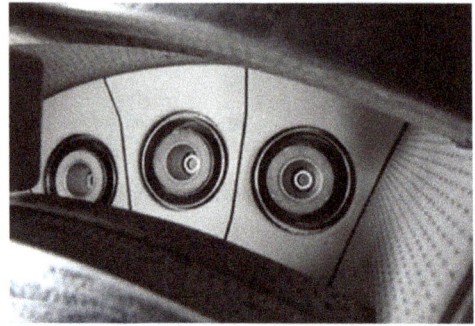

Figure 4: GE N+2 5-cup CMC combustor sector rig.

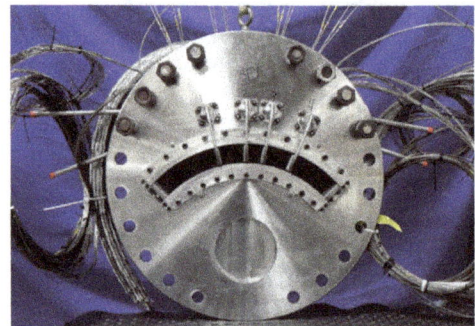

Figure 5: GE N+2 sector rig-emissions rake layout.

split. Mechanical and thermal analyses were performed, and the cooling design and mechanical construction were optimized to ensure viability of the hardware up to the takeoff conditions of the engine cycle.

The combustor rig has 4 emissions rakes, each with 4 sample elements. Rakes are located within Cups 2, 3, and 4 and are spaced in different locations relative to the cup centerline in order to capture a comprehensive averaged sample when all 16 sample points are ganged together (Figure 5). Generally, data is taken at a fixed rig T_3, P_3, and dP/P_3 while the overall fuel-to-air ratio is swept over the range of interest.

Combustor emissions data is presented in Figures 6-8. Data for 7% ICAO is shown in Figure 6. Additional single points taken on 2 other test days are shown to confirm fairly good repeatability of the data. Data for 30% ICAO is shown in Figure 7. High pressure data at the maximum main fuel split, used to determine NOx at the 85 and 100% ICAO points, is shown in Figure 8.

In general, NOx emissions results in the sector tests were in line with expectations based on correlations (low power, pilot-only points) and the high temperature/pressure flame tube data (high power, fully staged operation). Table 3 summarizes the LTO NOx data for the ICAO points. The facility was unable to deliver T3 temperatures high enough to run the 85% and 100% ICAO points at the exact T_3/

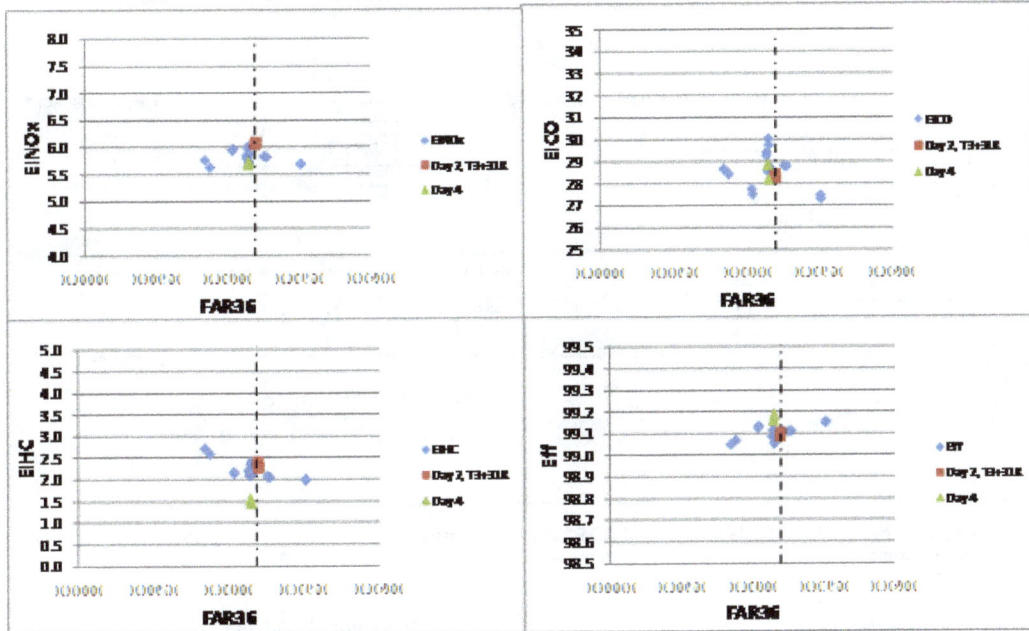

Figure 6: Sector rig emissions data (EINOx, EICO, EIHC, and combustion efficiency) at the 7% ICAO point, plotted vs. the fuel/air ratio based on sampled emissions. Repeated points taken on 2 additional test days are shown for repeatability. The vertical line indicates the target 7% ICAO cycle fuel/air ratio.

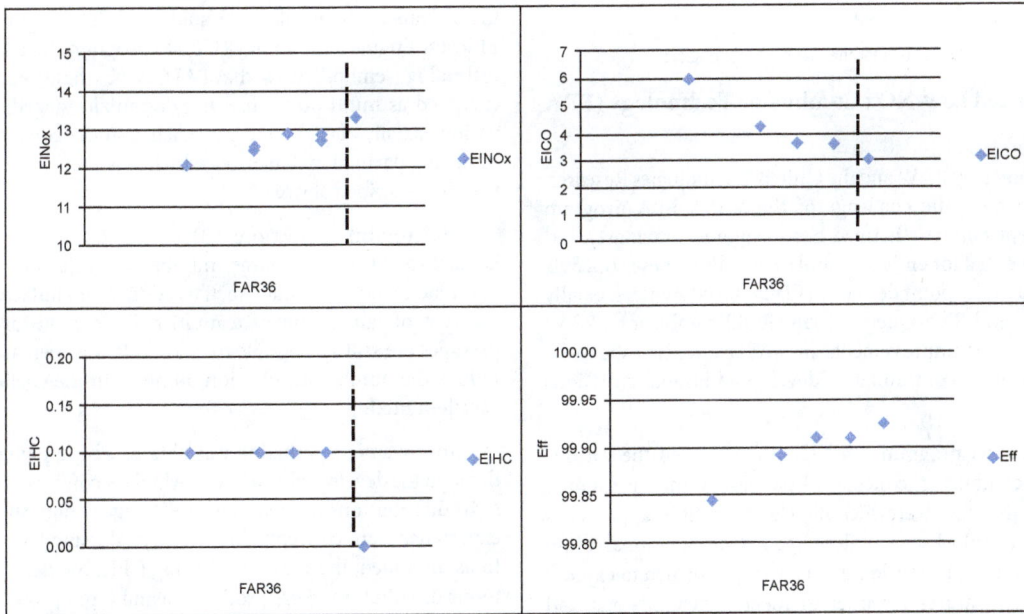

Figure 7: Sector rig emissions data (EINOx, EICO, EIHC, and combustion efficiency) at the 30% ICAO point. The vertical line represents the target fuel/air ratio.

P_3/flow/FAR (fuel-to-air-ratio) conditions. The 85% ICAO point is taken directly from the data in Figure 8 at the appropriate mixer flame temperature. For the 100% ICAO point, the data in Figure 8 was curve fit and extrapolated to the appropriate flame temperature. The standard humidity correction, based on the measured dew point in the combustor inlet air, was applied to the data in Figures 6-8 to arrive at the final EINOx values in Table 3.

Cruise NOx emissions and efficiency were also measured. The data

resulted in a 60-70% reduction in EINOx over the previous state-of-the-art, with better than 99.9% efficiency.

GE Phase I Conclusions: The GE combustor delivered 19% CAEP/6 NOx, surpassing the N+2 goal of 25% CAEP/6, with good combustion efficiencies and acceptable dynamic pressures for this stage of development. Further development of this technology will focus on thermal and mechanical durability, manufacturability, and optimization of the design to balance combustion efficiency and dynamics vs. LTO NOx capability.

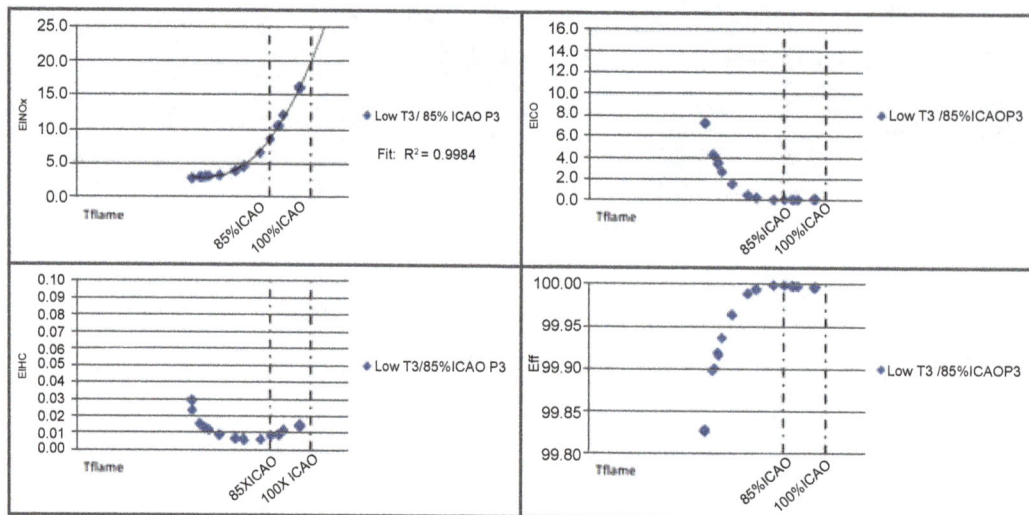

Figure 8: High pressure, fully staged emissions data at the 85% ICAO P3 and maximum facility T3 for this air flow rate, for determination of the 85 and 100% ICAO NOx values. Vertical dashed lines represent the target cycle flame temperature for 85% and 100% ICAO.

M4F1 in the Sector				
% ICAO	Time [min]	EINOx	dp/Foo	% CAEP/6
100	0.7	17.6		
85	2.2	7.9	20.6	18.9
30	4	11.8		
7	26	5.2		

Table 3: LTO NOx results for the GE N+2 5-cup sector.

P&W N+2 Advanced Low-NOx Combustor Technology (ERA Phase I)

The approach taken by P&W and the United Technologies Research Center (UTRC) to meet the challenge of the NASA ERA program revisited the concept families that had been explored previously and determine their potential for emissions and operability. These concepts include lean-staged multi-point designs, radially staged swirlers, axially staged combustors, and Rich-Quench-Lean (RQL) combustors. P&W has achieved significant improvements in TALON X Rich-Quench-Lean Technology and is continuing to develop additional emissions capability.

For the NASA ERA program, P&W/UTRC reviewed the various staged and lean combustor concepts. Roadblocks that prevented their adoption in the past were identified and possible approaches to address those roadblocks were developed. Computational Fluid Dynamics (CFD) and single nozzle rig tests were performed to explore and understand key features in order to meet both emissions and operability requirements. Exploration of the designs continued into the rig design phase. A multi-sector rig of the preferred design was designed and fabricated, then tested at UTRC and at NASA.

In conjunction with the NASA ERA program, Pratt continues to develop the TALON X combustor technology. The basic technology was developed with support from NASA under the UEET program, and is the combustor for the P&W Geared TurboFan engine on upcoming Airbus, Bombardier and Mitsubishi aircraft, meeting all program metrics. Work performed on the TALON X as part of the program includes further reductions in smoke, LTO NOx, and cruise NOx.

Staged/Lean Combustor Concepts: Over the past 40 years, PW/

UTRC has explored and developed various combustor concepts to reduce NOx. For example, an axially staged combustion systems was developed for the V2500 in the 1990's. More directly related to the current NASA program, PW explored and developed several concepts for the NASA sponsored High Speed Civil Transport (HSCT) Program, the Advanced Subsonic Technology (AST) program and the Ultra Efficient Engine Technology (UEET) program. In addition to the RQL technology embodied as the TALON X, these technologies can be classified as multi-point, fuel-nozzle radially staged, multi-dome and various axially staged concepts. Each of these concepts was identified as having particular challenges which was then addressed through the process described above.

Multi-point injection: The multi-point injection concept is challenged by the large number of injection points. Modern manufacturing techniques such as additive manufacture have reduced the cost of fabrication of a multi-point combustor. The many fuel passages are still susceptible to coking. It was seen as advantageous to reduce the number of injection points with the challenge to maintain excellent mixing.

Improvement in mixing was obtained by applying lessons learned through the development of PW TALON X combustors. Swirl cups with swirl distributions in the range of PW legacy high shear swirler design experience were conceptualized, analyzed using PW experience based tools, modified, then improved using CFD. Swirlers with desired fuel-to-air distributions were produced using rapid prototyping techniques and tested in the UTRC spray facility, which has the capability to measure velocities of gas and liquid phases, droplet size, and spray uniformity. Tests were then performed of the concepts in the Advanced Aeroengine Combustor (AAC) single nozzle test facility at UTRC. This facility has moderate pressure capability (up to 1037 kPa (150 psi)) and the test section has optical access as well as extraction probes that can traverse axially and radially. Emissions results from these single nozzle rig tests of the multipoint design showed acceptable levels of NOx, with estimated NOx at cruise conditions to be significantly less than 5 EI, an indication that the concept could meet the program goals.

Radially staged swirlers: The challenge identified for the radially staged swirler was to aerodynamically separate the pilot from the main in such a fashion to ensure good low-power efficiency, yet permit sufficient mixing for stability over the range of operation. In addition, the radially-staged swirler concept is challenged to obtain uniform mixing in a large swirler. The mechanical packaging of this concept is also difficult due to the combined size of the multiple swirlers. Coking is an issue here, as in all lean staged systems, due to the many fuel injection points and the need to distribute these points in such a fashion to allow for uniform mixing with air.

Once again, improvements to this concept were derived from P&W legacy swirler design experience. Swirler mixing and pilot stability were improved over previous UEET concepts with the application of lessons learned during TALON X development. Various swirler combinations and swirl distributions were conceptualized and analyzed, modified, then improved using insight gained from CFD analyses. Spray tests were made of key designs, and results used to further modify the designs. Tests of the final concept were then performed in the UTRC AAC rig and then, as part of another NASA contract examining potential Low-NOx concepts for supersonic engines, the P&W radially staged swirler concept was tested at the NASA CE-5 single nozzle test facility. NOx results were significantly less than 5 EI, once again experimentally demonstrating the capability to meet program emissions goals. Importantly, low power efficiency was also very good, showing that the necessary separation and stabilization of the pilot flame had been achieved.

Axial controlled stoichiometry: The challenge identified for the axially staged combustor was to implement it in a simpler fashion than the version developed for the V2500. Axially staged combustors characteristically have good separation of the pilot with accompanying positive stability and efficiency. Packaging and mixing of the main stage is traditionally the concern in such systems. Coking again is an issue.

For the system envisioned for this effort, the pilot stage was kept simple, using experience gained in developing TALON combustors. Various concepts for the mains were conceptualized, analyzed, and explored with CFD. The level of mixedness required to achieve low levels of NOx was a key question that was explored through a series of idealized UTRC AAC rig tests. Mixer designs were then created that achieved the desired level of mixedness. The resulting mixers were then evaluated in the UTRC AAC rig to determine if they achieved the desired level of emissions. NOx results were the lowest observed of the PW configurations tested, providing significant developmental margin.

Testing the ACS combustor in the NASA ASCR facility: The performance, strengths and challenges of each of the concepts were reviewed. Each of the concepts had demonstrated the potential to meet the program goals for NOx. Each of the concepts has challenges with complexity, packaging, and coking. Approaches were conceptualized for each of the concepts to overcome these challenges.

The P&W team chose the Axially Controlled Stoichiometry (ACS) concept for testing at NASA. The arrangement of the separation of the pilot and the main provides for efficiency and stability at low power, and stability at all operating conditions. Mixing of the pilot and main is controllable according to PW experience. P&W has experience in the design and manufacture of axially staged combustion systems due to the V2500 design. The ASC distributes the heat release axially, reducing susceptibility to acoustics. Finally, the NOx emissions were the lowest tested, providing the most margin for development of any of the concepts.

The ACS concept was then implemented in a 3-sector arc rig that

was tested first at UTRC, then at NASA in the ASCR facility (Figure 9). Results between the two series of tests were consistent. Acoustic issues were only experienced at off-design conditions. Efficiency was above 99.9% at all fully staged high power points. Due to the conventional pilot zone, high efficiency was also achieved at idle and approach conditions. Cruise NOx levels were 2 EI and below (Figure 10). Emissions were measured idle, take-off, climb, and approach, and a NOx EPAP of 88% below CAEP/6 was calculated using an N+2 cycles based on an advanced Geared Turbo Fan. Performance with respect to all CAEP regulated emissions are shown in Figure 11.

P&W ERA phase I conclusion: The next stages in the development of the concept include exploring approaches to make the concept more product-ready and simplify the packaging. The concept must fit into the envelope available in the current and planned Geared Turbo Fan engines. Further, the concept must be verified and matured in a full annular design. The evaluation and testing of the Phase I effort provided an excellent basis for the continued development of this concept.

Continued development of the TALON X combustor: In conjunction with the NASA ERA program, P&W continues to develop the TALON X combustor. In particular, swirler and front end modifications were explored that continue to reduce the low smoke

Figure 9: Axially Controlled Stoichiometry Combustor Rig Installed at NASA ASCR Facility.

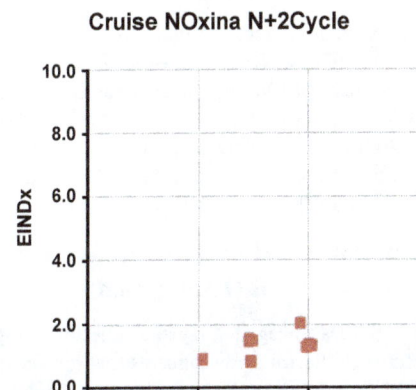

Figure 10: PW ACS configuration achieved less than 2 EI NOx at typical cruise conditions of an advanced N+2 GTF Cycle.

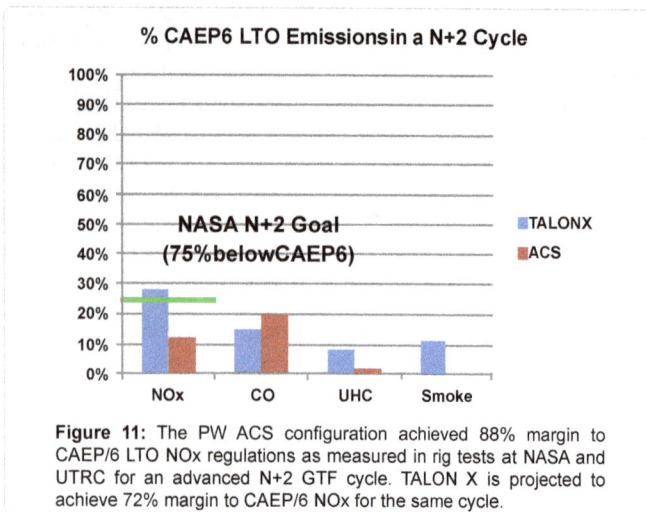

Figure 11: The PW ACS configuration achieved 88% margin to CAEP/6 LTO NOx regulations as measured in rig tests at NASA and UTRC for an advanced N+2 GTF cycle. TALON X is projected to achieve 72% margin to CAEP/6 NOx for the same cycle.

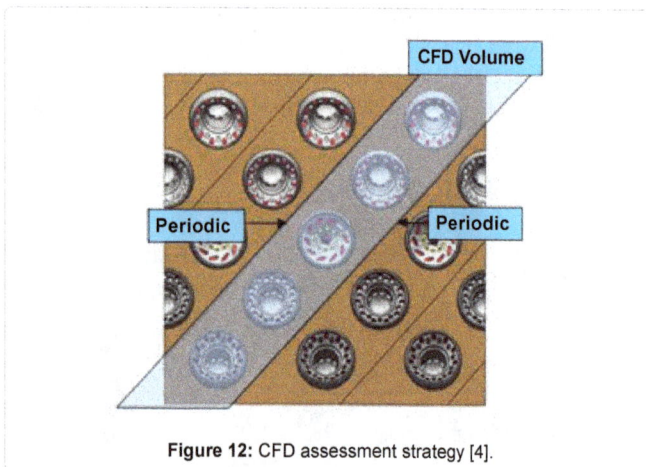

Figure 12: CFD assessment strategy [4].

and NOx levels demonstrated in the TALON X combustors which were developed and tested in the Gear Turbo Fan family of engines.

A design of experiments approach was followed to further improve the swirlers. They were analyzed, evaluated using CFD, and tested in the spray facility at UTRC in order to determine critical uniformity parameters. A five sector rig was designed and fabricated that was shown to duplicate the results of the full annular tests and the engine emissions. Tests were performed of the chosen concepts.

Results indicated smoke levels below those for the swirlers currently used in the engines. NOx levels were also improved. When results are projected, using P&W experienced based methods, to the same advanced GTF cycle used for the staged results; a NOx EPAP of 72% margin to CAEP/6 is calculated, closely approximating the N+2 program goal of 75% reduction to CAEP/6. Thus, the TALON X remains a viable option for the current generation of aero engines.

Ultra-Low-Emissions High-Pressure Lean-Direct-Injection Injector Concept Developments

The injector activities engaged three fuel-injector companies to develop lean-direct-fuel injector array concepts that can accommodate faster-burning fuel blends at the more aggressive higher-pressure engine cycle conditions envisioned for the future. Goodrich, Woodward FST, and Parker Hannifin are on contract to design and develop these more aggressive ideas, but their scopes are limited

to the injector and its dynamic behavior to keep the cost low. These concepts have to demonstrate being able to burn the more aggressive 80%/20% alternative fuel to jet fuel blends. Some of these concepts will incorporate fuel-flow control features to provide fueling precision while others instability control. While these industry partners will be able to verify their performances at low power, they also will verify their higher-power performance at NASA's test facility. All three of these efforts utilized CFD to provide design screening.

The LDI concept is a natural fit for ultra-high-pressure operation. While a majority of ERA's fuel reduction goal can be reached through airframe drag reduction or increasing propulsive efficiency, improving the thermodynamic cycle efficiency by raising the compression ratio also is considered. Both of the previously mentioned engine cycle concepts considered in the combustor development activity raised the maximum combustor pressure from the current 45 bar to over 50. This means that the combustor inlet air temperature will be hotter, and the liquid fuel sprayed into the combustor will heat up quicker and ignite sooner than current designs. Since good emission characteristics heavily depend upon mixing the fuel and air well before the burning starts, certain amount of premixing is used in current designs. However, at the elevated air temperatures, such a feature may not be available as auto-ignition or flash-back of the flame into the fuel-air mixing nozzles can pose a serious hardware damage hazard. LDI averts this issue by directly injecting the fuel into the flaming zone, mixing quickly to keep the residence time of the non-uniform mixture to a minimum.

LDI typically manifest itself in arrays of smaller nozzles. Quick fuel-air mixing can be done by increasing air flow turbulence where the fuel is injected, but this is done sacrificing air pressure across the fuel nozzle which no longer will be available for expansion in the turbine for work extraction. The alternative is to use an array of small nozzles to reduce the physical distance that the fuel needs to travel to achieve uniform mixing. All three of the design concepts in this current program fall into this form. For comparison, fuel-air mixing nozzles in current engines are from 2 to 4 inches in diameter. In the current program a typical size is about 1 inch.

All three LDI fuel injector designs accommodate to using alternative hydrocarbon fuels. All alternative fuels are composed of saturated hydrocarbons that do not have sulfur. They are not as easy to coke as currently used petroleum distillate fuels. As a result, the designers can use smaller fuel orifices to make the liquid fuel jets smaller to speed up break up and vaporization. This will help greatly as these new fuels also generally have faster kinetics and will start to burn earlier than the current distillate fuel, resulting in flames that can be much closer to the fuel injector. While the combustor programs mentioned in the earlier portion of this paper are designed to accommodate using 50%/50% blend of alternative with distillate fuels, these advanced LDI injectors are designed to take advantage of up to 80%/20% mixture of alternative fuels. Using these fuels potentially generates much less particulate emissions as well as having much lower probability of coking inside fuel nozzles.

Goodrich multi-point combustion system

Goodrich (currently UTC Aerospace Systems) designed their modular LDI array around discrete-jet-based airblast fuel nozzles (Figure 12) [4]. Intense mixing turbulence level is created very close to where the fuel is injected through strong shear layers. The resulting rapid breakup and vaporization promotes a more homogeneous and burnable mixture quickly, especially at low power conditions.

The multi-stage array (Figure 13) is distributed with the row of

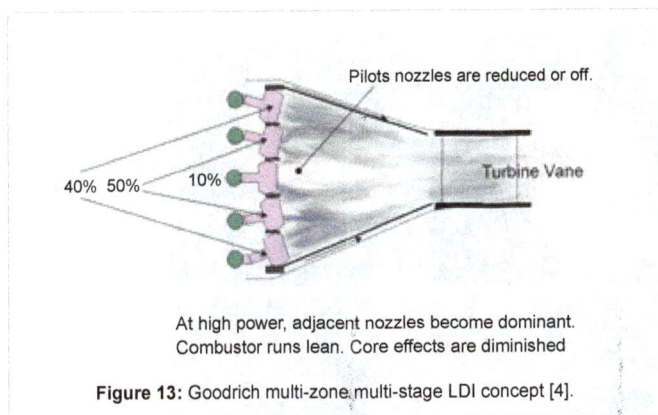

Figure 13: Goodrich multi-zone multi-stage LDI concept [4].

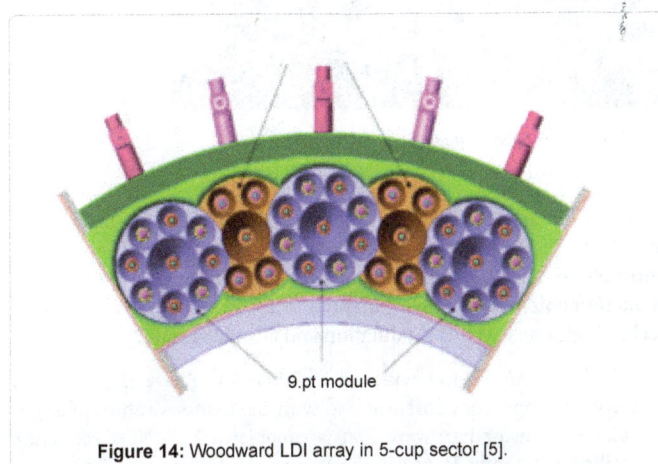

Figure 14: Woodward LDI array in 5-cup sector [5].

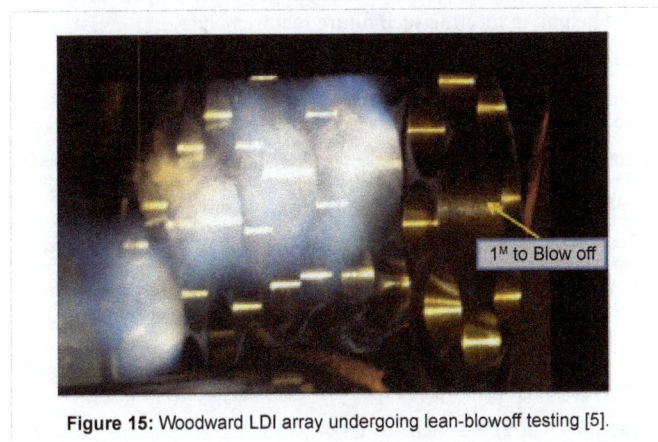

Figure 15: Woodward LDI array undergoing lean-blowoff testing [5].

slightly recessed pilot nozzles that will be fueled independently to provide ignition stability. While it may burn a little hotter and generate more NOx than the main burners, it only flows 10% of the air flow. The separately fueled stages will be scheduled to generate the minimum emissions and broadest operating range [4]. This design concept has gone through mid-power testing with Jet-A at NASA.

Woodward multi-point LDI

Woodward FST based their designed concept on mixed types of fuel nozzles aided by simple swirl-generated shearing action. Their LDI array module consists of multi-staged series of smaller and leaner-fueled nozzles surrounding a central larger pilot that provides the low-

power operations. The modules may be stacked as a multi-cup sector (Figure 14) [5]. The recessed pilot provides a zoned effect and shields the pilot from its neighbors. At low powers, evenly-distributed fuel-air mixture may be too lean to burn. As a result, multiple fueling stages are used to distribute the fuel so that burnable mixtures at very low overall equivalence ratios are achievable. Of particular concern is the staging process to provide stable light off and ultra-low- power operation. Figure 15 shows a 5-cup arc sector undergoing lean-blowout testing. A more precise fuel-staging system that can reduce fuel flow drift is under development for demonstration on this modular injector concept.

Two variations of this type of design were tested at NASA. One variation successfully ran through the complete mid-power test matrix using mixtures of up to 100% alternative fuel.

Parker 3-Zone lean direct injection

Lastly, Parker introduced a variation of their multi-staged 3-zone injector module from their work under NASA's UEET Program. The concept originated from the idea of retrofitting current combustor injectors with better performing new hardware (Figure 16). The module is composed of miniature mixing cups each fueled by a pressure-swirl nozzle. The cups are formed using Parker's platelet technology and introduce intense turbulence where the fuel injection takes place. The three zones are formed by canting the side nozzles away from the middle layer to provide some relief from inter-injector interaction. Multiple fuel stages are used to shift fuel spatially to provide the leanest and acceptably-stable burnable fuel-air mixture (Figure 17). Figure 18 shows the flame distribution during lean-blowout. This array has gone through mid-power test conditions.

Part of this particular task also includes a high frequency fuel actuator development that is capable of being integrated into the 3-zone fuel module to provide spatial and temporal fuel redistribution (Not shown.)

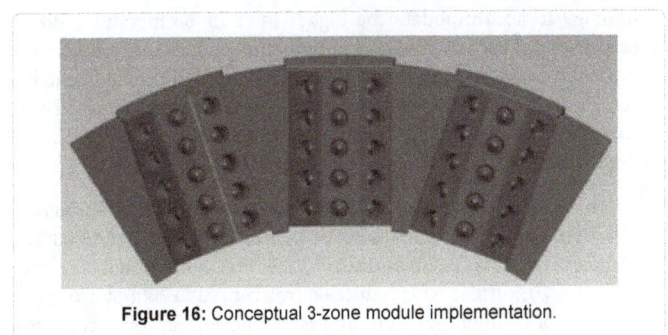

Figure 16: Conceptual 3-zone module implementation.

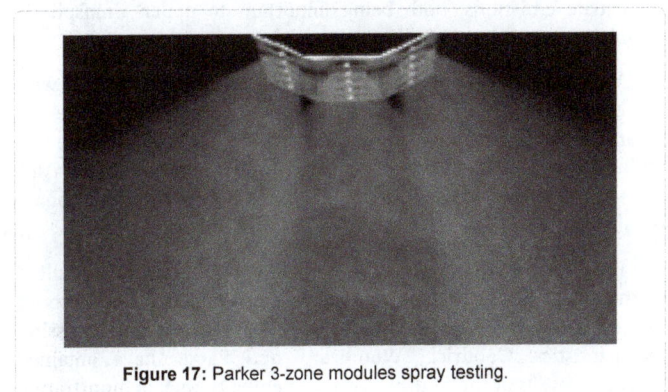

Figure 17: Parker 3-zone modules spray testing.

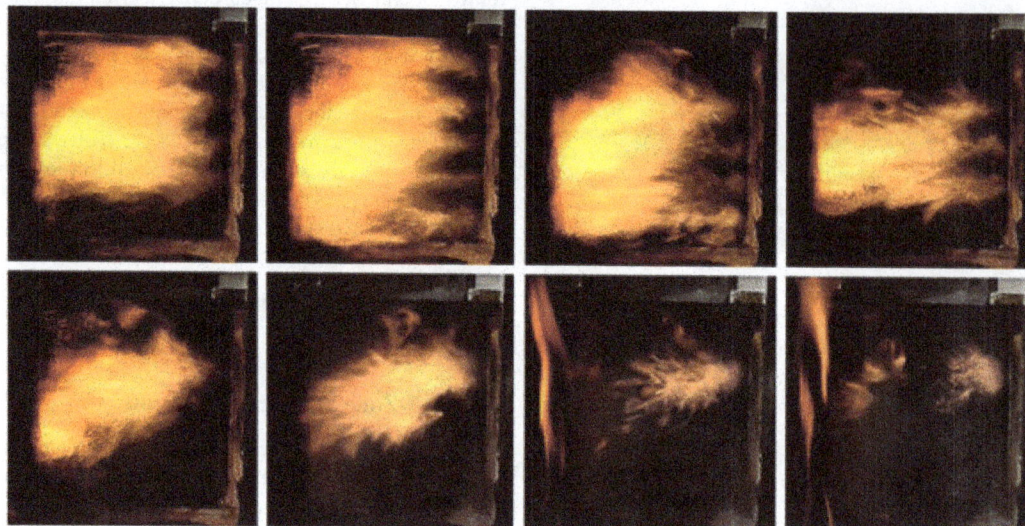

Figure 18: Parker 3-Zone injector module during lean-blow off testing.

All three of these injector designs under this second activity have achieved scaled NOx reduction levels for the two lower-power ICAO test points and have achieved their desired lowered lean-blowoff levels.

Summary and Next step-ERA Phase 2

NASA ERA's combustor development project has reached the goal of demonstrating technology for the (N+2) generation of engines. The Phase I concept screening work (2009-2011) with GE and P&W resulted in two distinctively different designs, although they followed very different design paths. GE stayed with their more familiar TAPS-type platform and focused in improving their fuel-air mixer and staging operation. GE's design included a CMC liner as an enabling technology to accommodate the higher inlet air temperature. P&W screened several very diverse concepts, including an improved version of their rich-burn TALON combustor, but settled on the final axially-staged ACS concept with vastly improved performance characteristics compared to two decades ago. Both sector designs were tested in NASA facility over the range of their designed operating conditions.

Both of these combustor concepts from GE and P&W surpassed the N+2 goal of 25% CAEP/6 LTO NOx level, with good combustion efficiencies and dynamic pressures at TRL 4 level. Both design concepts also met the 70% cruise NOx reduction (referenced to 2005 stat-of-the-art combustor technology). Both of these advanced lean-burn concepts not only reduce the cruise-level NOx, they also produce vastly reduced particulate emissions, both being important to upper atmospheric chemistry.

A progressive competition strategy will be used in Phase II with down-selection of sources from the initial phase to determine the second phase contractors. The competition for the second phase will build on the results of the initial phase, and the award criteria for the second phase includes successful completion of the initial phase requirements.

The more focused injector (sub-component) investigations resulted in three LDI arrayed-injector concepts that evaluated the effects of injector recess and design mixtures on operational and emission characteristics. Goodrich, Woodward, and Parker have obtained excellent results at the mid-pressure flametube testing, demonstrating

ignition, flame propagation, improved lean blowout capabilities, and NOx reduction. NASA plans to continue testing these design concepts at NASA's intermediate pressure rig to assess their NOx reduction potential and their potential to accommodate other enabling technologies such as fuel modulation and advanced staging techniques.

NASA's ERA Project will end in 2015. The Phase II combustor concept development effort will end with the demonstration of a full-annular combustor hardware that is capable of 75% NOx reduction capability. Currently ICAO is discussing whether to regulate cruise NOx in the future. The combustor technologies developed under ERA will be helpful in meeting such future requirements.

References

1. Chang C, Tacina K, Lee C, Bulzan D, Hicks Y, et al. (2013) NASA Glenn Combustion Research for Aeronautical Propulsion. J Aerosp Eng 26: 251-259.

2. Collier F (2012) NASA Aeronautics - Environmentally Responsible Aviation Project - Solutions for Environmental Challenges Facing Aviation. AIAA-2012-0936, 50th AIAA Aerospace Sciences Meeting, Nashville, TN, USA.

3. Foust MJ, Thomsen D, Stickles R, Cooper C, Dodds W (2012) Development of the GE Aviation Low Emissions TAPS Combustor for Next Generation Aircraft Engines. AIAA-2012-0936, 50th AIAA Aerospace Sciences Meeting, Nashville, TN, USA.

4. Prociw A, Ryon J, Goeke J (2012) Low NOx Combustion Concepts in Support of the NASA Environmentally Responsible Aircraft Program. GT2012-68426, Proceedings of ASME Turbo Expo 2012, Copenhagen, Denmark.

5. Lee Philip (2012) Research and Development of Low-Emissions Combustor Concepts and Associated Fuel Control Valves: Gen-2 Lean Direct Injection System (LDI-2).

15

NACA653218 Airfoil Aerodynamic Properties

Abdelghany ES[1], Abdellatif OE[2], Elhariry G[3] and Khalil EE[4]*

[1]*Institute of Aviation Engineering, Cairo, Egypt*
[2]*Department of Mechanical Engineering, Benha University, Egypt*
[3]*Department of Mechanical Engineering, Cairo University, Egypt*
[4]*Department of Mechanical Engineering, Cairo University, Egypt*

Abstract

In this research we have obtained the drag and lift coefficients, velocity, pressure and path lines contours using CFD which can also be determined by using wind tunnel experimental test. This process is relatively difficult and surely price more than CFD technique cost for the same problem solution. Thus we have gone through analytical method then it can be validated by experimental testing. A CFD procedure is described for determination aerodynamic characteristics of subsonic NACA65$_3$218 airfoil. Firstly, the airfoil model shape, boundary conditions and meshes were all formed in GAMBIT® 2.3.16 as a pre-processor. The second step in a CFD model should be to examine the effect of the mesh size on the solution results. In order to save time take case for a grid with around 100000 cells. The third step is validation of the CFD NACA65$_3$218 airfoil shape model by different turbulence models with available experimental data for the same model and operation conditions. The temperature of free stream is 288.2 K, which is the same as the environmental temperature. At the given temperature, the density of the air is ρ=1.225kg/m³, the pressure is 101325 Pa and the viscosity is μ=1.7894×10⁻⁵ kg/m s. A segregate, implicit solver is utilized (FLUENT® processor) estimate were prepared for angles of attack variety from -5 to 16°. The Spalart-Allmaras turbulence model is more accurate than standard $k - \varepsilon$ model, RNG $k - \varepsilon$ model and standard model k–ε models. For lift coefficient, it is found maximum error by Spalart-Allmaras model about 12% lower than other turbulence models. For drag coefficient, it is found maximum error by Spalart-Allmaras model about 25% lower than other turbulence models. For pitching moment coefficient, it is found maximum error by Spalart-Allmaras model about 30% lower than other turbulence models.

Introduction and Literature Review

The CFD grow to be instrument for developing, sustaining, optimizing, innovating, verifying and, particularly here, for validating steps. The CFD has become a widely used tool for aerodynamic applications. On Aerodynamics, the four main forces which act on the aircraft during the flight are Lift, Drag, Thrust and Weight. Drag is one of the most critical phenomena amongst all and is the opposing force of aircraft's forward motion, [1, 2]. A class of body exists, however for which a wing profile is not symmetrical (or when there is a nonzero angle of incidence); a velocity difference is upheld between upper and lower surfaces. This creates a pressure difference and a circulation around the wing: lift is generated, [3].

Airfoil is famous aerodynamic shape that used in aeronautical applications. When the aerofoil is in motion through air, the air is passing above and below the wing. The wing's upper surface is shaped so the air velocity increases. The air pressure above the wing decreases. The wing's lower surface is shaped so the air velocity decreases. The air pressure above the wing increases. Lift of a wing is produced by high pressure on the lower surface and low pressure on the upper surface. And when the force of gravity is lower than the force of lift, the airplane is able to fly [4-6].

From Figure 1, at the front of the airfoil, the leading edge is the point has the maximum curvature. At the rear of the airfoil, the trailing edge is defined the point of maximum curvature. A straight line connecting the leading and trailing edges points of the airfoil is the chord line. AOA is the angle between the direction of air velocity and a chord line on the wing [7]. AOA increases when the nose of the wing pitches up, and lift increases. Drag increases also, but not the same as lift. The drag force, lift force, pitching moment equations are shown in equations (1), (2), (3).

$$D = \tfrac{1}{2} \, \rho \times V^2 \times S \times C_D \qquad (1)$$

$$L = \tfrac{1}{2} \, \rho \times V^2 \times S \times C_L \qquad (2)$$

$$M = \tfrac{1}{2} \, \rho \times V^2 \times S \times C \times C_M \qquad (3)$$

The investigation of the 2D subsonic flow over a NACA 0012 airfoil at different AOA and running at a Reynolds number of 3000000 is considered by [8]. In this project, the steps of computational solution are consisting of three stages as shown in Figure 2. The project starts from preprocessing step of geometry design and grid generation. The model geometry and the grid are generated by GAMBIT® 2.3.16. The second step was solving equation of motion by FLUENT solver using Finite Volume Approach. Finally is the post-processing step where the aerodynamics properties of NACA65$_3$218airfoil. Then the drag, lift, pressure contours, pitching moment coefficient, path lines and velocity contours around aerofoil at all AOA are determined by CFD package.

Governing Equations

The air flow is modeled as 2-D compressible viscous flow. Thus the governing equations are the continuity equation together with x- y and z governing equations for a compressible flow. Turbulence is modeled by the Spalart-Allmaras model. The complete system of equations is presented here in differential form, FLUENT® Documentation [9] and [10]. The governing equations in this model are:

*Corresponding author: Khalil EE, Professor, Department of Mechanical Engineering, Cairo University, Egypt, E-mail: khalile1@asme.org

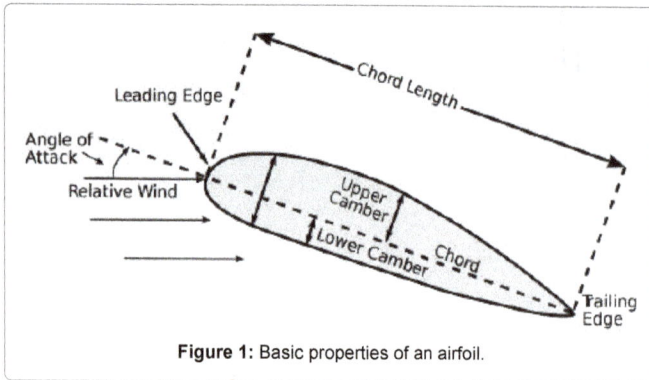

Figure 1: Basic properties of an airfoil.

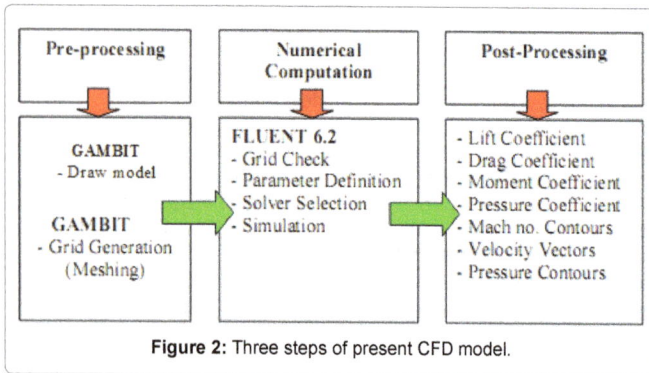

Figure 2: Three steps of present CFD model.

Continuity equation in vector form:

$$\frac{\partial \rho}{\partial t} + \nabla \cdot (\rho \vec{V}) = 0$$

Momentum equation in vector form:

$$\nabla \cdot (\rho . \vec{v} \vec{v}) = -\nabla p + \nabla \cdot (\overline{\overline{\tau}}) + \vec{F}$$

Energy Conservation Equation:

The energy equation is applied on the control volume that is primarily derived from the first law of thermodynamics. The energy equation may be written in the differential form as:

$$\frac{\partial (\rho E)}{\partial t} + \nabla \cdot (\vec{v}(\rho E + p)) = \nabla \cdot \left[k_{eff} \nabla T - \sum_j h_j \vec{J}_j + \left(\overline{\overline{\tau}}_{eff} . \vec{v} \right) \right] + S_h$$

Numerical model of NACA653218 airfoil

Figure 3, the numerical model of NACA65$_3$218airfoil is shown below:

Boundary conditions: The flow field, temperature, pressure and Mach number in the numerical model of NACA65$_3$218airfoil considered are solved by FLUENT®, with the following boundary conditions:

Pressure far field: At Pressure far field boundary, Reynolds number (as shown in Figure 1) was Re=3x10^6, same with the reliable experimental numbers from [11,12], to validate the present CFD simulation. The free stream temperature is 288.2 K, which is the same as the environmental temperature. The density of the air at the given temperature is ρ=1.225kg/m3, the pressure is 101325 Pa and the viscosity is μ=1.7894×10^{-5} kg/m s. A segregated, implicit solver is utilized (ANSYS FLUENT® processor) calculation were done for varies angles

of attack range from -5 to 16°. The airfoil shape, boundary conditions and meshes are created in GAMBIT® 2.3.16 as a pre-processor.

NACA65$_3$218airfoil: The NACA65$_3$218airfoil is considered adiabatic and no slide wall, as shown in Figure 3.

Drawing of NACA65$_3$218airfoil: According to airfoil database [13], scatter drawing of an aerofoil in this problem was a 6-digit NACA series, NACA65$_3$218airfoil. The airfoil is drawn using GAMBIT® 2.3.16, as shown in Figure 4.

Grid Creation

Grids near the airfoil wall boundary must be dense enough and computed fields must be large enough to satisfy far field boundary conditions to obtain accurate aerodynamics properties such as drag, lift, and pitching moment on airfoil. However, extreme grids will cost too much computing resources and increase computing time. Thus, the compromise is that grids far from the airfoil wall boundary are scattered and grids near the airfoil wall boundary are intense. Figure 3, is shown Computed fields where right part are two rectangles, where AB=ED=GC=20c, and AG=GE=12.5c. Left part is a half circle with a center at G, and a radius of 12.5c.

It is meshed each of 3 faces individually to get our final mesh. Figure 5(a), shows all grids in computed fields. By performing the command "Grid Check" in FLUENT, it is known that total number of grids is 100000 quadrilateral cells, the volume of the smallest grid 4.1224x10^{-10} m3, and the volume of the largest grid 9.3985x10^{-1} m3. Figure 5(b), shows the grids surrounding the airfoil.

Solver

FLUENT® package is used to calculate the flow field and properties

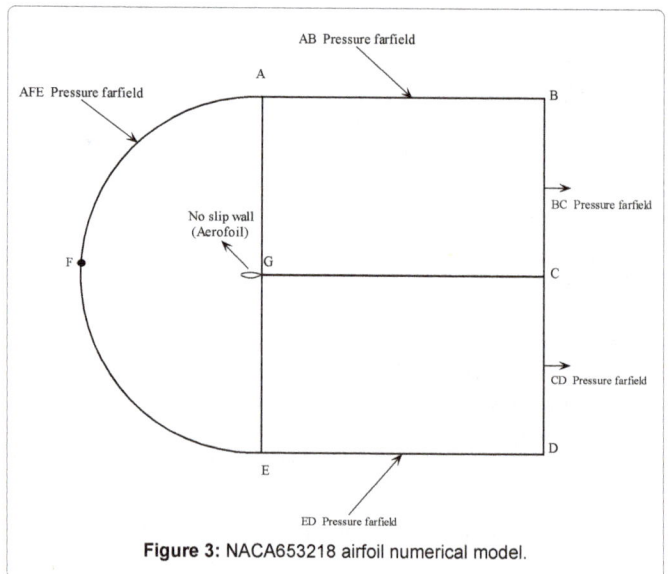

Figure 3: NACA653218 airfoil numerical model.

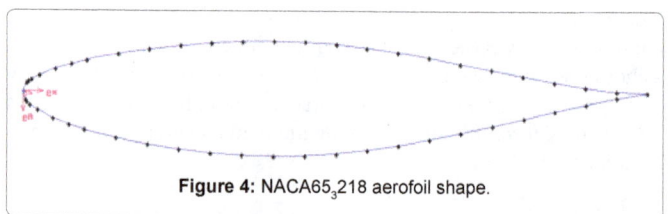

Figure 4: NACA65$_3$218 aerofoil shape.

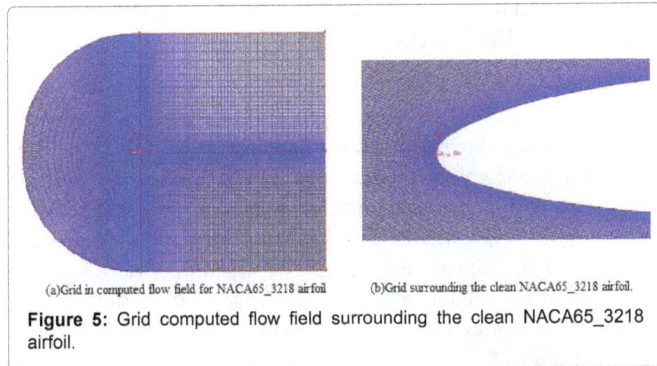

(a)Grid in computed flow field for NACA65_3218 airfoil.   (b)Grid surrounding the clean NACA65_3218 airfoil.

Figure 5: Grid computed flow field surrounding the clean NACA65_3218 airfoil.

through the different configurations. Simulations as Velocity contours, Pressure contours, drag and lift coefficients values by the same package.

Convergence criterion

Continuity equation, linear momentum equations and turbulence model Spalart- Allmaras equations are calculated for the mesh control volumes to a residual of 10^{-3}, while energy equation is calculated to a residual of 10^{-6}.

Grid dependency check

It should be to investigate the grid size effect on the solution results for the first step in performing a CFD simulation. Generally, a numerical solution to be further accurate as more cells are used, but using added cells also increases the essential computational time and computer memory. The appropriate number of cells is determined by increasing the number of cells until the grid is satisfactorily fine so that further refinement does not vary the outcomes. To check the independency of the outcomes to cell number, seven types of grids are produced. The results of these seven grids are seen in Table 1, at stall AOA (15°).

Figure 6 explains the effect of number of mesh cells in aerodynamic lift coefficient at stall AOA (15°).

Figure 7 explains the effect of number of mesh cells in aerodynamic drag coefficient at stall AOA (15°).

Figure 8 explains the effect of number of mesh cells in aerodynamic moment coefficient of at stall AOA (15°).

In order to save time when running the computations, the grid with the smallest number of cells displaying an independent solution should be used for the calculations. This is seen to be the case for a grid with around 100000 cells.

Verification of numerical model

A similar Numerical Model NACA65$_3$218airfoil of the same previously-mentioned grid size and type was developed, for verifying numerical model with experiment and numerical models measurements. The NACA65$_3$218airfoil model was used to verify the work done by [11-13]. To apply the same boundary conditions at pressure far field, temperatures are 288.2K, velocities are 43 m/s, pressure is 101325 Pa. The density of the air at the given temperature is ρ=1.225kg/m^3, the viscosity is μ=1.7894×10^{-5} kg/m s at Re=3x10^6. The NACA65$_3$218airfoil is considered zero heat flux wall and no slide wall. Compare the outcomes of the numerical model by standard $k - \varepsilon$ model, RNG $k - \varepsilon$ model, the standard model $k-\omega$ model and Spalart-Allmaras turbulence model to those of the numerical and experimental models measurements. The results show good agreement

of lift, drag and moment coefficients with the corresponding values in the experimental and numerical models measurements. Figure 9; see the coefficient of lift (C_L) with AOA from -4 degree to stall angle of attack 16 degree of numerical models and experimental studies, plotted on the same axes and scale for comparison. By Spalart-Allmaras, it is found maximum error model about 12% but for standard $k - \varepsilon$ model, RNG $k - \varepsilon$ model, the standard model $k-\omega$ model it is found maximum error increase from Spalart-Allmaras model maximum error and reach in $k-\omega$ model about 60%.

Figure 10; see the coefficient of drag (C_D) with AOA from -4° to stall angle of attack 16° of numerical models and experimental studies, plotted on the same axes and scale for comparison. By Spalart-Allmaras model, it is found maximum error about 25% but for standard $k - \varepsilon$ model, RNG $k - \varepsilon$ model, the standard model $k-\omega$ models, it is found maximum error increase from Spalart-Allmaras model maximum error and reach in $k-\omega$ model about 300%.

Figure 11 see the coefficient of pitching moment (C_m) with AOA from -4° to stall AOA 16° of numerical models and experimental studies,

Number of Cell	7200	22500	56250	90000	97500	105000	200000
C_L	0.613	1.1739	1.439	1.5272	1.4939	1.497	1.517
C_D	0.187	0.0884	0.05107	0.05308	0.05615	0.0593	0.056
C_m	0.068	0.0362	0.03693	0.04511	0.04391	0.0431	0.043

Table 1: Effect of grid size of main aerodynamic flow parameters.

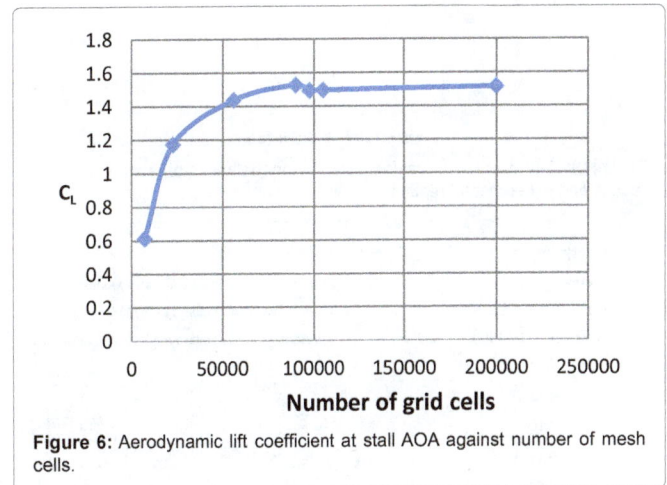

Figure 6: Aerodynamic lift coefficient at stall AOA against number of mesh cells.

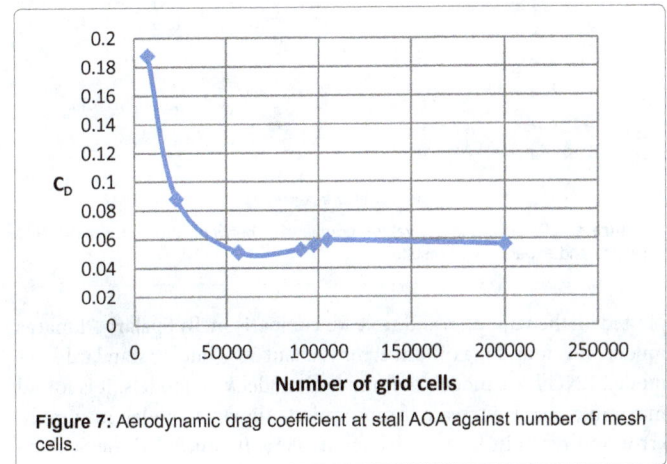

Figure 7: Aerodynamic drag coefficient at stall AOA against number of mesh cells.

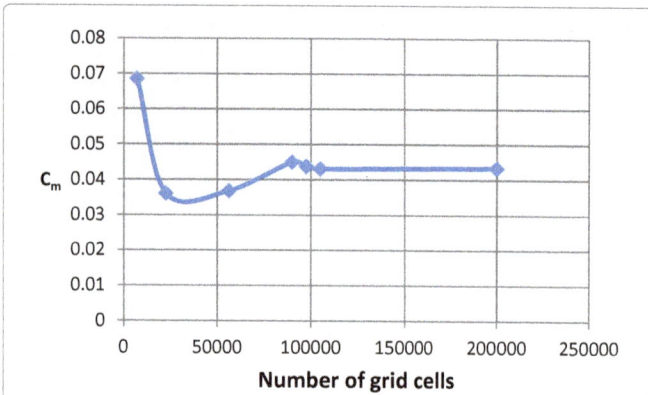

Figure 8: Aerodynamic moment coefficient at stall AOA against number of mesh cells.

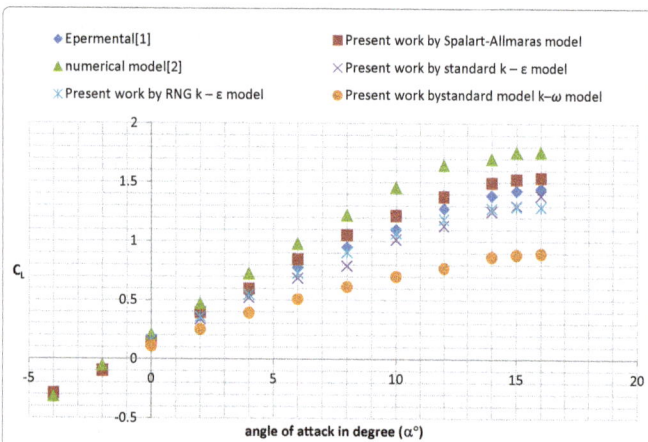

Figure 9: Lift coefficient values comparison between present numerical results and experimental results.

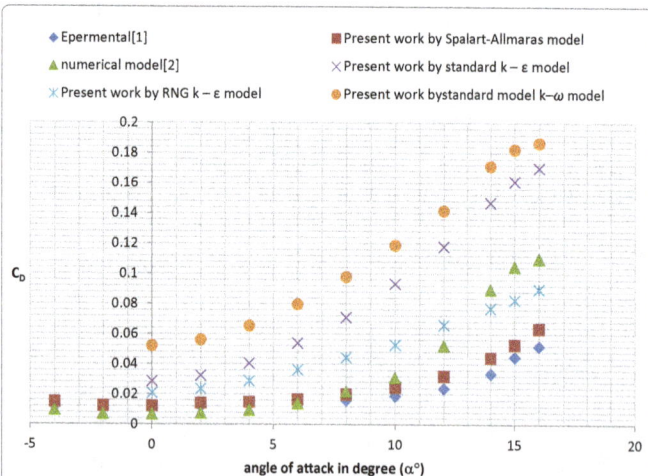

Figure 10: Drag coefficient values comparison between present numerical results and experimental results.

plotted on the same axes and scale for comparison. By Spalart-Allmaras model , it is found maximum error is about 30% but for standard $k - \varepsilon$ model, RNG $k - \varepsilon$ model, the standard model $k-\omega$ models, it is found maximum error increase from Spalart-Allmaras model maximum error and reach in $k-\omega$ model about 500%. It concluded the Spalart-

Allmaras model more Accurate than standard $k - \varepsilon$ model, RNG $k - \varepsilon$ model, the standard model $k-\omega$ models.

Figure 12 show the velocity contours explain the flow development from α= -4° to α=16°. The range values of all figures shows maximum value of velocity about 130 m/s obtained for α=16°. At α= -4° it is shown that the low velocity area value around leading edge is small, and it starts to disappear with increasing the AOA then start building up from α=8° raises steadily up to approximately α=16°on pressure side of aerofoil. From around =10° the separation is clearly seen and reattaches to the suction side at trailing edge of the airfoil and the separation area raises until it arrive at about 50% of the suction side of the airfoil at α=16°.

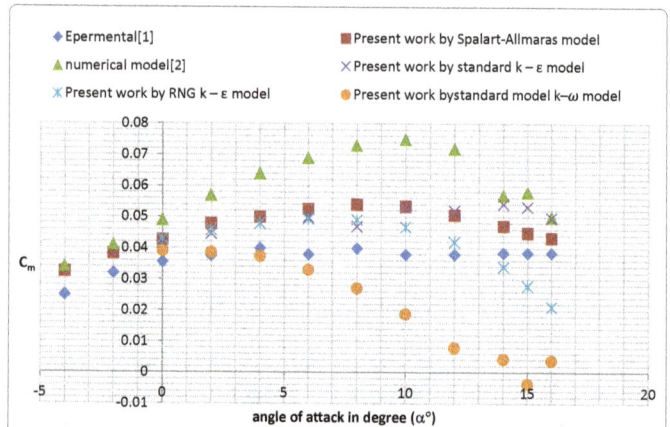

Figure 11: Numerical results of Cm in comparison to corresponding experimental results.

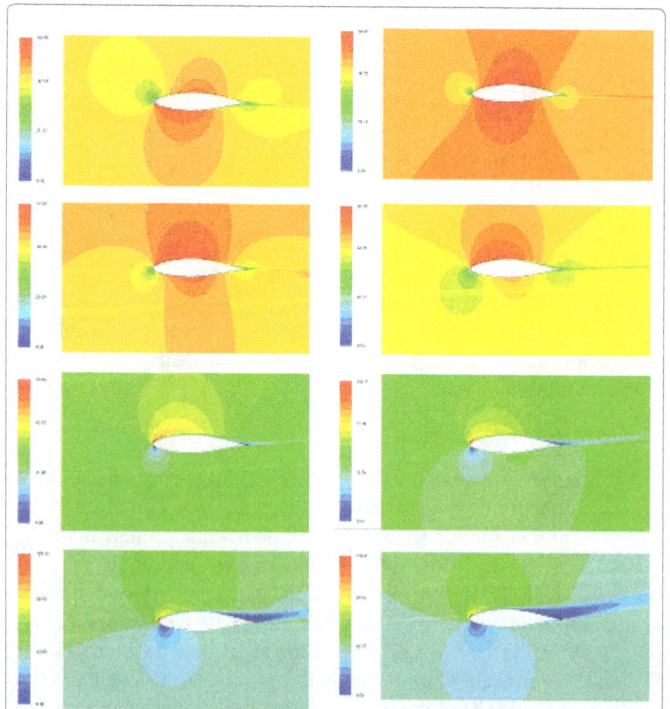

Figure 12: Velocity contours around the leading edge for clean airfoil case. 1st row: α =-4 °(left) and α =-2° (right), 2nd row: α =2° (left) and α =4° (right), 3rd row: α =8° (left) and α =10° (right), 4th row: α =14 ° (left) and α =16° (right). Values are in m/s at Re=3*10⁶.

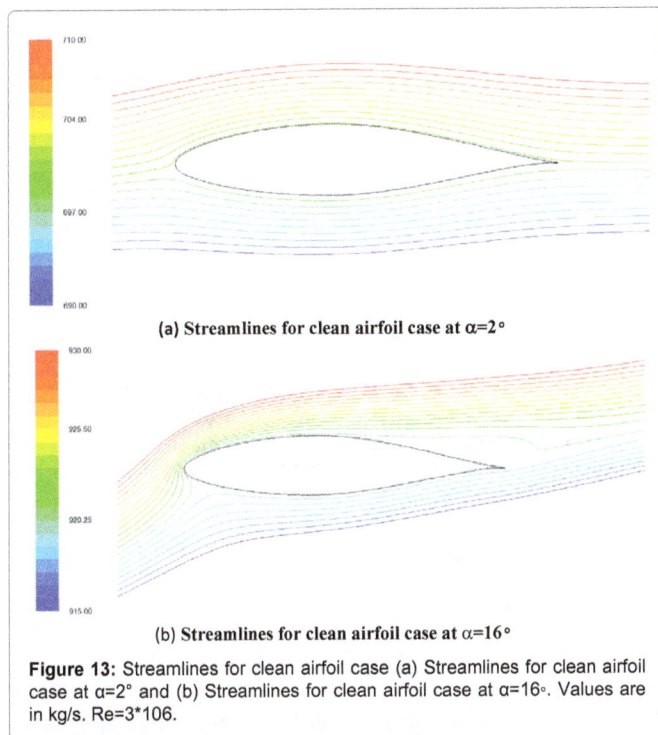

(a) Streamlines for clean airfoil case at α=2°

(b) Streamlines for clean airfoil case at α=16°

Figure 13: Streamlines for clean airfoil case (a) Streamlines for clean airfoil case at α=2° and (b) Streamlines for clean airfoil case at α=16°. Values are in kg/s. Re=3*106.

The reason for the streamline unit kg/s not kg/m.s is the 2D geometry. The missing m in the denominator denotes per unit depth. It is evident from Figure 13(a) that the flow at α=2° to is rather smooth and well attached to the surface of the airfoil. It is evident from Figure 13(b) that the flow at α=16° a separation bubble starts to form at the trailing edge and moving upstream for bigger angles of attack.

Conclusions

- By using CFD to calculate performance of numerical model NACA65$_3$218airfoil, huge amount of time and money can be saved before testing the wing in the wind tunnel. Calculations show that trends of numerically-simulated curves are in excellent agreement with trends of experimentally-obtained ones.

- The Spalart-Allmaras turbulence model was found to be more Accurate than standard $k - \varepsilon$ model, RNG $k - \varepsilon$ model, the standard model $k - \omega$ models.

- Lift coefficient increases with increases AOA. After stall AOA about 16 degree, Lift coefficient decreases.

- Drag coefficient increases with increases AOA.

Nomenclature

List of symbols

C Chord length

C_L Lift coefficient

C_D Drag coefficient

C_M Moment coefficient

D Drag force

E Total energy of a fluid

P Particle constant

h Enthalpy

M Mach Number, Pitching moment

S Reference area

P Pressure value

Re Reynolds number, Re = ρ U Lc / μ

t Time

T Temperature

u Instantaneous x direction velocity

v Instantaneous y direction velocity

w Instantaneous z direction velocity

x, y, z Cartesian coordinate components

Greek Letters

α Angle of attack

ε Turbulence dissipation rate

μ Dynamic viscosity

μ_t Turbulent viscosity

ρ Density

$\rho\vec{g}$ Gravity body forces

List of Abbreviations

AOA Angle of attack

CFD Computational Fluid Dynamics.

References

1. Versteeg H, Malalasekera W (1995) An introduction to computational fluid dynamics: The Finite Volume Method-Longman.

2. (2005) FLUENT Documentation. © Fluent Inc.

3. Glauert H (1983) The element of aerofoil and airscrew theory.

4. Roskam J (1985) Airplane design roskam aviation and engineering corporation Rt4 Ottawa Kansas.

5. http://www.larc.nasa.gov

6. Jenkinson LR (1999) Civil jet aircraft design Arnold London.

7. Houghton EL, Carpenter PW (2003) Aerodynamics for engineering students (5thedn) Oxford Great Britain Butterworth-Heinemann.

8. Karna S, Saumil B, Utsav B, Ankit PA (2014) CFD analysis of an aerofoil. Intl J Engineering Research 3 :154-158.

9. (2013) FLUENT Documentation. © Fluent Inc.

10. Abbott IH, Von Doenhoff AE (1958) Theory of wing section. Dover New York 634-635.

11. Abbott IH, Von Doenhoff AE, Stivers LS (1945) Summary of airfoil data. NACA Report No. 824: 222-223.

12. Aerodynamic coefficients for NACA 653-218 airfoil.

16

Novel Methods for Multi-hop Sensor Positioning on the Surface of a Solar System Body

Lászlo Bacsárdi[1], Árpád Huszák[1], Aliz Szeile[1] and Tamás Haidegger[2]*

[1]*Budapest University of Technology and Economics, Budapest, Hungary*
[2]*Óbuda University, Budapest, Hungary*

Abstract

Fundamental methods used in space research and exploration are continuously changing and developing, depending on the major scientific target of the community. Currently, we use space probes and expensive rovers to discover and analyze the surface of solar system bodies. However, these rovers could be replaced with thousands of cheaper sensors which are organized into a sensor network. In order to monitor the surface and the atmosphere of a solar system body, positioning accuracy and energy efficiency are key determining factors in such a network. In this article, a mobile sensor network capable of measuring and forwarding data on the surface of a distant planet is investigated, and several problems of spatial positioning are addressed.

Keywords: Sensor positioning; Solar system exploration; Multi-hop sensor network

Introduction

Sensor-based networks offer advantages compared to the currently used methods in space research. Recently, the NASA and European Space Agency (ESA) tests different sensor-based solutions like Seamless IP Diversity based Generalized Mobility Architecture (SIGMA) or Disruption Tolerant Networking tools [1,2]. Deploying sensors on the surface of distant planets will allow remote monitoring of non-easily accessible areas in preparation for human or robotic missions. For such kind of network, the mobility of sensing devices is an important element, due to the potential valuable scientific results gained from different sites, as opposed to static landers. In present days, expensive and sensitive multifunctional rovers were sent to analyze other planets. However, it is possible that in the future, thousands of cheap sensors will be placed on the surface of distant solar planet.

Sensor networks capable of distant planetary missions have been investigated by different groups. Our aim was to propose a mobile sensor network architecture that can localize its elements with minimal required resources. We used a recursive technique for the position estimation process, which technique extends the accessible coverage area using only three high performance devices with accurate positions. The location information of these super nodes serves as reference points for the recursive positioning of sensors [3]. At the same time, the super nodes play a gateway role between the deployed sensors and satellite(s).

We analyzed the performance of the recursive positioning algorithms in a custom implemented simulator. In our simulations, the surface and environmental characteristics—that influences the mobility and communication of mobile device—were also taken into account, e.g., dunes, holes, electromagnetic storm.

In this work, we identified different key questions of movement and positioning on the surface of a distant planet. We developed a simulation framework to analyze these questions. The article is organized as follows. Related Works describes the state-of-art of related works, including wireless sensor network deployment, data gathering methods and positioning techniques. Optimization of Multi-Hop Sensor Positioning gives a detailed description of the proposed model and the design goals of the architecture are introduced. The results are discussed in Results and Discussion.

Related Works

Wireless Sensor Networks (WSNs) will play a critical role in different environments, e.g., personalized healthcare, autonomous vehicles, distributed scientific measurements, meteorology, as well in space and planet exploration. The simple and cheap sensor devices have the ability to remotely monitor non-easily accessible areas, and can be used even to ensure the safety of actual human or robotic missions [4]. Due to their low cost and dimensions (even millimeter scale is possible), high number of these devices can be dispersed at the investigated area, to monitor atmospheric, terrestrial, electromagnetic features, and forward the collected data through their communication interface.

The efficient deployment of sensors is very important for the successful completion of sensing tasks. A sensor may move independently from others, or in a group, but usually uniform dispersion is preferred to minimize the uncovered area in the monitored environment. Different strategies were investigated in the literature, proposing movement control methods of the devices [5]. Most of these strategies [6-8] assume that the environment is sufficiently known and under control. However, in unknown or hostile environment—such as distant planets or disaster areas—sensor deployment cannot be performed manually. In these cases, the devices are scattered over great distances (e.g., airplane, space capsule), therefore the actual landing position cannot be precisely controlled due to the existence of wind or other obstacles. Zou and Chakrabarty proposed a centralized approach, where a powerful "cluster head" collects the sensor location, and determine the target location of the mobile sensors [9]. In certain deployment environments, the centralized approach is not always acceptable, because it may suffer from the problem of single point

***Corresponding author:** Haidegger T, Óbuda University, Budapest, Hungary
E-mail: haidegger@irob.uni-obuda.hu

failure. In the case of special conditions, self-controlled methods are preferred.

Wang et al. investigated how to maximize the sensor coverage given less time, range of movement and message complexity [5]. The first step of their distributed self-deployment protocols is to discover the existence of coverage holes in the target area (areas not covered by any sensor) based on Voronoi diagrams [10,11]. After discovering a coverage hole, their proposed protocols calculate the target positions of these sensors. They introduced three movement-assisted sensor deployment protocols, VEC (VECtor-based), VOR (VORonoi-based) and Minimax, based on the principle of sensor movement from densely deployed areas to sparsely deployed areas. Common assumption of all the proposed control protocols is that the sensors have perfect positioning and navigation capability. Another alternative solution is, if the mobile sensors are proceeding on a determined path [12]. In this case, their current position can be estimated based on the elapsed time and movement velocity. Moreover, the future positions can be also predicted, so delivery of collected measurement data can be forecasted more efficiently.

The required energy for wireless transmission depends on the distance of the devices. The relation between the energy consumption and the distance (d) of devices is da, where a is between 2–5 depending on the wireless propagation conditions. The distance between the transmitter and the receiver is one of the most important parameters from the consumption point of view. Therefore, energy efficient network operation can be applied, if the data transmission is triggered when the distance between the source sensor and the receiver (central) sensor is the smallest. In the case of multi-hop sensor network, the devices close to the central equipment will consume more energy, because the data will travel through these sensors towards the data collector equipment. Assuming a sensor network deployed on a distant planet, the central data collector device will serve as a gateway, which forwards the collected records to the satellites, as it is illustrated in Figure 1.

In most of the cases, the collected measurement data is useful only if the accurate position of data gathering is known. Therefore, we can categorize the related works according to data gathering or positioning point of view.

Data gathering

Two types of data collector devices can be used to collect sensor data: fixed or mobile. In the first case, the sensors must move to the vicinity of the collector to transfer the measured data, while in the second case, the central device is also moving in order to collect the

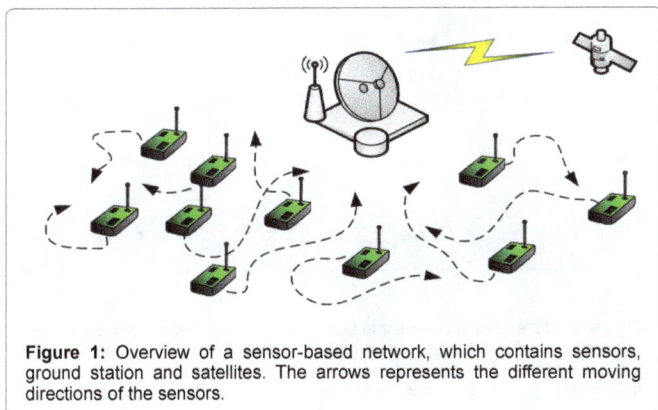

Figure 1: Overview of a sensor-based network, which contains sensors, ground station and satellites. The arrows represents the different moving directions of the sensors.

measurement records, consuming less energy for wireless transmission due to shorter distances, but the fuel economy of the motion must also be taken into account. The motion of data collector device can be random, predictable or controlled.

MULE devices (Mobile Ubiquitous LAN Extension) [13] visit randomly the sensors. When a MULE device arrives in close range to a sensor, it picks up the measured data, buffers it, and delivers the records to the data center when the communication becomes possible with it. A similar solution is used by the SENMA (SEnsor Networks with Mobile Agents) architecture [14] for low power and large scale sensor networks. Mobile agents in SENMA are powerful hardware units, both in their communication and processing capability, and in their ability to traverse the sensor network. Mobile agents can be aerial or ground vehicles equipped with sophisticated terminals and power generators, or specially designed light nodes that can hop around in the network.

In the case of predictable movement of the collector device, it can move on a pre-defined path as Chakrabarti et al. propose [15]. As the observer is assumed to traverse the same path repeatedly, the data is pulled by the receiver device periodically by waking up the nodes when it is close to them. Since the sensor nodes only transmit when the receiver device approaches them, the power requirements can be significantly reduced.

The third possibility is the controlled receiver device movement. Controlled movement generally consists of mobile devices in the network and moving to specified destinations with defined mobility patterns for specific objectives. The receiver node can adapt its route to achieve more efficient coverage management [16-18], energy composition reduction [19,20], transport layer parameters' improvement [21,22], etc. Controlling the device's movements offers new possibilities to improve the data collection efficiency, or optimize the network topology for different type of objectives as the above referenced works show [23].

Sensor positioning

Localization is one of the most significant challenges for mobile sensor networks. In order to gather sensor data in a spatial context or for proper navigation through a sensing region, the position must be precisely known. Mobile sensors must frequently estimate their coordinates, which takes time and energy, and consumes other resources needed by the sensing application. Different types of position estimation method exist, but all of them are based on measurement of radio signal propagation features. While receiving a radio signal, some of its properties, such as arrival time, signal strength, and direction, are captured by the receivers. In the second phase, certain signal parameters, such as TOA (Time of Arrival), TDOA (Time difference of Arrival), RSS (Received Signal Strength) and AOA (Angle of Arrival) are extracted from the captured values. The three most popular categories of methods for position estimation are time based, angle based and received signal strength based method.

With TOA [24], the distance between the transmitting node and the receiving node is deduced from the transmission time delays and the corresponding speed of signal. The main drawback of this approach is that it is difficult to precisely record the arrival time of radio signals, since they travel close to the speed of light. TDOA localization [25] improves upon the TOA approach by eliminating the need to know when the signal was transmitted. Several time-synchronized nodes receive a signal, and look at the difference in arrival times. For each TDOA measurement, the transmitter must lie on a hyperboloid with a constant range difference between the two measuring units. The

equation of the hyperboloid is given by

$$R_{i,j} = \sqrt{(x_i - x)^2 + (y_i - y)^2 + (z_i - z)^2} - \sqrt{(x_j - x)^2 + (y_j - y)^2 + (z_j - z)^2} \quad (1)$$

where (xi, yi, zi) and (xj, yj, zj) represent the fixed receivers i and j; and (x, y, z) represent the coordinate of the target.

The AOA method [26] determines the angular separation between two beacons, or a single beacon and a fixed axis. This method requires special antennas. Using RSS-based technique [27], the distance is estimated based on the wireless signal attenuation from the transmitting node to the receiving node. Empirical mathematical models are used to calculate the distance according to signal propagation. As the final step, the calculation of the coordinates is done using triangulation (AOA) or trilateration (TOA, TDOA, RSS) [23].

Positioning systems assume that reference points exist in the network with precisely known coordinates. Recursive positioning [3] is an alternative solution that can increase system coverage iteratively, as nodes with newly estimated positions join the reference set. In hostile environment, where only few high performance centralized devices can be deployed, recursive positioning can extend the sensor network coverage area. Our model—proposed for monitoring distant planets with wireless sensors—utilizes the benefits of this method.

Optimization of Multi-Hop Sensor Positioning

Several questions arise regarding monitoring the environment of a distant planet. In the sensor network planning process, the monitored data types, the network topology and the sensor movement strategy must be defined. In this article, we focus on a set of related questions, discussed in details in this Section. Other issues of reliability (including the effects of hardware or software errors on a WSN) are out of the scope of this work.

Measurable data on a planet

On the surface of a planet different types of data can be monitored and forwarded. We must define how important the measured data is. In the case of water discovery or soil pattern monitoring, it is not a problem if the collected data reach the command center on the Earth later since the measurements remain valid. However, if temperature is measured, it could be important to arrive in time, because the collected data could lose its actuality. Moreover, we must define that the measuring process is periodic, continual or event controlled. In our model, we chose a sensor network assuming that the collected data are valid for period of time and the measurements are periodic.

Satellite system

A satellite system makes it possible to forward the measured data to the communication station. If more satellites are used, permanent coverage can be ensured, but on the other hand the deployed network will be too expensive or even unrealizable. Therefore, we suppose to use a satellite system, keeping the number of satellites as low as possible in the proposed model.

Supersensors in the network

In our model, we used some special sensors, named supersensors. The supersensors collect data from the other sensors and forward to the Earth via satellites. Their actuation is more expensive, because the communication with satellites needs more energy that must be produced using bigger solar cells, compared to other regular sensors. We assume that these sensors are able to precisely determinate their own position (e.g., through GPS) and serve as base points for the iterative position estimation process of regular sensors. The supersensors move in formation with the other, more simple measuring sensors too, so they ensure to be at service, if there is data to be forwarded.

Sensors motion

The sensors are able to move and take measurement at different locations. In order to model the movement of the sensor equipment, we used both random and fixed path motion. The sensors get a random motion direction first, then a protecting zone border is defined, which determines the limit of the y coordinate displacement.

Due to the iterative positioning technique that is used to estimate the coordinates of the equipment, the sensors must stay in groups and without moving away from each other. If a sensor reaches the predefined border limit, the movement direction must be changed in order to keep the sensors in group. During the angle adjustment, y coordinate of the motion changes, so the chance of leaving the border will be smaller. In case of crossing the border, the connection will be probably lost with the other sensors, as it is shown in Figure 2. The left original spots sign the starting point of the sensors. The highlighted points sign supersensors that are able to communicate with satellites and determinate their accurate position. In Figure 2, we can follow the movement of a sensor. In Step 2, the sensor reaches the protecting zone border, thus its y coordinate changes (it will become -y) and begins to move to a new direction. The x coordinate is not modified during the movement, so the sensors will move until reaching the line of point D. Using a large number of sensors, we can ensure that measurements will be performed uniformly on the determined territory of the lane from the starting point to the final destination (D).

Sensor communication

The measuring sensors can communicate with each other (multi-hop network) forwarding the monitored data sensor to sensor. The other solution is if the communication works only between a sensor and supersensor (single-hop network). In the first case, the data can reach to the supersensors in multiple steps, while in the second case, the sensors must wait until they arrive within the range of a supersensor. In our model the supersensors move in formation with the regular sensors, so we chose the second case due to its simplicity.

Sensor positioning

Simple sensor devices have limited energy production and complexity, so we cannot use GNSS-based (Global Navigation Satellite System) navigation. Instead of the satellite system, we use sensor based

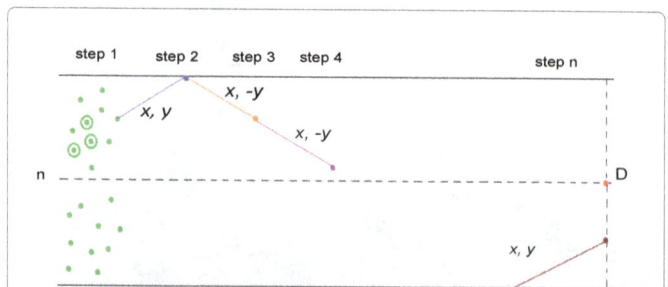

Figure 2: Movement of a sensor in our model. The left original spots sign the starting point of the sensors. The highlighted points sign supersensors. x and y are the actual coordinates of the sensor. The line represents the border of the protecting zone. The sensor will move until reaching the line of point D.

triangulation method for position determination as it is shown in Figure 3. Utilizing the position information of three sensors, we are able to estimate the position of a fourth sensor, if this sensor is within the range of the three reference sensors [28].

Figure 3 shows a scenario, where B1, B2 and B3 points are the known positions of the three sensors and sensor A (in the middle) has the unknown position. In order to calculate its position we use circle engraving. If the cover of sensors B1, B2 and B3 are bigger than the distance from sensor A, then the method is adaptable. Circles, with d1, d2 and d3 radius and B1, B2 and B3 center, define the position of A. In this case, the B1, B2 and B3 are reference points. The reference point is the position of a sensor, which is used by the positioning algorithm. After determining the position of A sensor, it will become a reference point for determining still unknown sensor positions.

Using this technique recursively and assuming that there are no lost sensors, the positions of all sensors within the group will be known. If there are some lost sensors, the determination process will be harder because probability of finding three reference sensors will be lower.

At the beginning of our algorithm, the first reference points are the supersensors. These positions are always precisely known, because they can communicate with the satellites.

We use the following formulas for the calculation:

$$A(x) = p(x) \pm \frac{h(y2 - y1)}{dist(B1, B2)} \qquad (2)$$

$$A(y) = p(y) \pm \frac{h(x2 - x1)}{dist(B1, B2)} \qquad (3)$$

where x, y are the coordinates of sensors, p(x) and p(y) are the coordinates of point P, and function dist(B1,B2) returns the distance between B1 and B2. The accuracy of the method was analyzed by using different optimization algorithms as it will be discussed later.

Optimization algorithms

We developed three different algorithms for optimization of positioning and we compared them from different viewpoints.

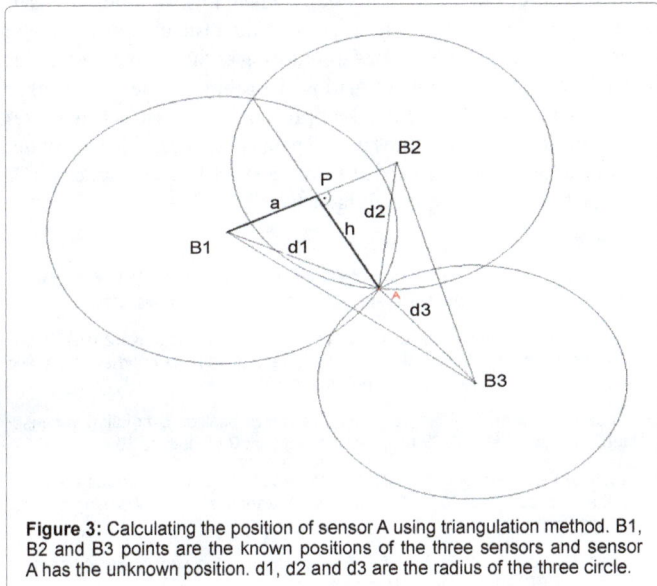

Figure 3: Calculating the position of sensor A using triangulation method. B1, B2 and B3 points are the known positions of the three sensors and sensor A has the unknown position. d1, d2 and d3 are the radius of the three circle.

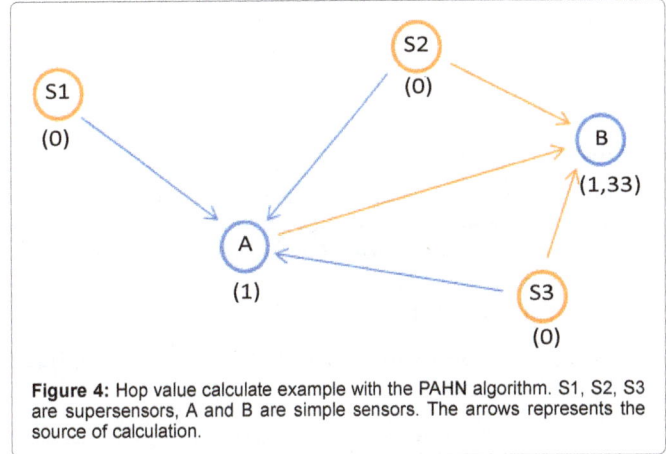

Figure 4: Hop value calculate example with the PAHN algorithm. S1, S2, S3 are supersensors, A and B are simple sensors. The arrows represents the source of calculation.

Basic position algorithm without optimization (BPA): In case of the initial algorithm, there is no optimization method. In order to calculate the position of a sensor, three parent sensors are chosen from the list of potential reference devices. The algorithm examines the potential reference sensors in the sensor group, and selects the first one that is capable to serve as a reference. After the first three hits, the examination stops, and the triangulation process can be started.

BPA is not an effective solution, because if the algorithm chooses three sensors, which estimated position is not accurate enough due to accumulated errors, the new calculation will contain all of these errors, too. This means that the calculated coordinates will be different from the real position. Actually, if the three basic supersensors can be selected as parents, then the calculated value will contain lower error, since the position of supersensors are precisely known.

Positioning algorithm based on hop number (PAHN): The PAHN algorithm differs from the BPA in the method of choosing the parents. In PAHN, the sensors use a hop value that contains how many steps of positioning lead to the calculated coordinates. Hop values of supersensors are zero, because the supernode positions are precisely known, and their coordinates are not estimated. Each simple sensor position is estimated from the position of three parent nodes, therefore the average of the hop values of the parents are calculated and set as a hop value of the currently estimated sensor.

For example, if we want to calculate the hop value of sensor A, and the parents of this sensor are supersensors S1, S2, and S3, the hop value will be the following:

$$(0+0+0)/3 + 1 = 1. \qquad (4)$$

If sensor B is chosen which parents are sensor A and two supersensors, then the hop value will be calculated as follows:

$$(0+0+1)/3 + 1 = 4/3. \qquad (5)$$

This means that the hop value shows how accurate calculation can be performed in the position estimation. If the hop value is small, then the calculation value is probably more accurate. Therefore a sensor chooses its parents on the basis of smaller hop values in order to estimate the position as accurately as possible. An example is illustrated in Figure 4.

Extended basic positioning algorithm (EBPA): We extended the previously described BPA with heuristic optimization. During the positioning process, the sensor sets all combination of reference sensors and calculates the coordinates for each one. The sensor position is calculated as the average of the estimated coordinates based on the examined reference sensor combination.

Results and Discussion

In order to analyze the performance of the proposed solutions, a simulation tool was implemented in C#. The tool is capable to set the surface of the planet, the number of the sensors, the communication range, the starting point and the final destination, etc. The parameters used in the simulation are listed in Table 1.

Analyzing the error during the positioning process applying the three algorithms

In the research, we were interested in answering the following question: How much error will appear using these positioning algorithms? The error means the difference (in meters) between the real coordinates (controlled by the simulator itself) and the calculated coordinates, using the positioning algorithm. The averaged errors are presented in Figure 5.

According to Figure 5, the BPA algorithm works with an average error of 3.4 meter. The PAHN algorithm generates less error (around 3.1 meter) and the error of the EBPA is between these two values with 3.2 meter. This means that the entire two improved algorithm shows better results than the original one. There is the less error distance with less lagging sensors, because the lane border is stepped over with lower probability and the chance of data loss is reduced as well. According to the simulation results, the most effective algorithm is PAHN. In the case of PAHN, the calculations were always performed with the most accurate coordinate values, therefore the estimation of the position was also more precise compared to the other algorithms.

Name of parameter	Default value [unit]
Number of sensor	200 [p]
Size of the area	400 [m²]
Maximum value of displacement	4 [m]
Range of sensor	30 [m]
Distance of point D	400 [m]
Pixel/meter	0.5 [pixel/m]
Change of RGB color	1 [m]

Table 1: Except otherwise indicated, we used the above described parameters.

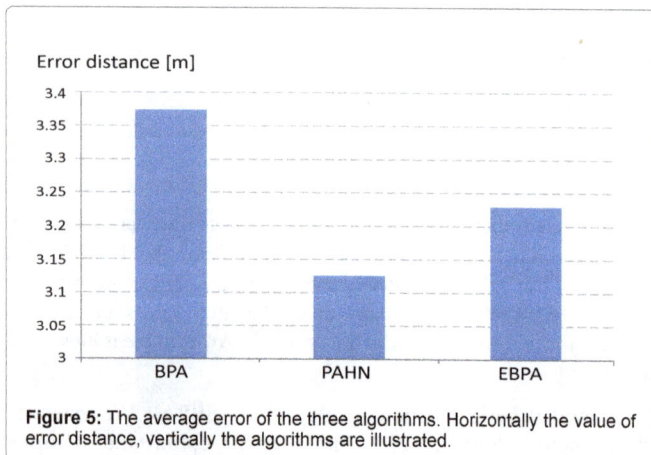

Figure 5: The average error of the three algorithms. Horizontally the value of error distance, vertically the algorithms are illustrated.

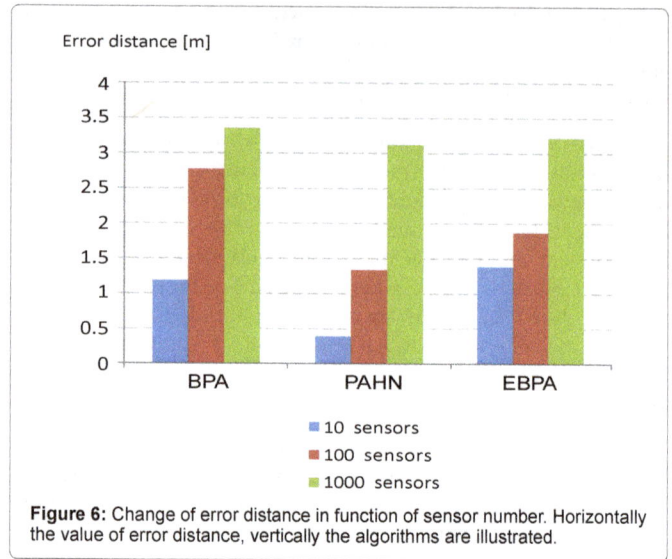

Figure 6: Change of error distance in function of sensor number. Horizontally the value of error distance, vertically the algorithms are illustrated.

Change of error distance in function of sensor number

In this analysis, we were interested in the changes of the averaged error value in the function of the number of sensors. The results are interpreted for all of three algorithms, therefore we illustrated the results in one graph for the sake of transparency.

The measurement results are shown in Figure 6. In the case of BPA algorithm, if sensor number increases, the three supersensors are selected with lower probability as a reference sensor. Hence, the calculated sensor coordinates are less accurate. The EBPA algorithm shows better results than BPA. As we can see, the PAHN algorithm is the most effective one.

Discussion

The new algorithms, especially the PAHN can be effectively used in unexplored territories without any infrastructure. If three initial reference points are determined, the sensor positions can be estimated recursively expending the area of the covered region where the measurements can be performed. The proposed algorithms can be used in future exploration missions, where high number of sensor devices is used to monitor the surface of the distant solar body. We focused on distant planet surface monitoring as the main object of our research work, our position estimation algorithms can be used in other environment, too, where the lack of positioning infrastructure makes general positioning algorithms unacceptable. However, further studies of localization errors are needed before a related space application is developed, and a space system is launched.

References

1. Atiquzzaman M, William I (2007) SIGMA for Space Sensor Web Networks. In Proc. of ESTO AIST Sensor Web Technology Meeting, Unites States.

2. Jenkins A, Kuzminsky S, Gifford KK, Pitts RL, Nichols K (2010) Delay/Disruption-Tolerant Networking: Flight test results from the international space station. In. Proc of. Aerospace Conference, IEEE.

3. Albowicz J, Chen A, Zhang L (2001) Recursive position estimation in sensor networks. In Profc of IEEE Int. Conf. on Network Protocols, IEEE.

4. National Research Council (2011) Vision and Voyages for Planetary Science in the Decade 2013–2022, The National Academies Press, Washington, DC, United States.

5. Wang G, Cao G, Porta T (2004) Movement-assisted sensor deployment. In Proc. of IEEE INFOCOM'04, Hong Kong.

6. Dhillon S, Chakrabarty K, Iyengar S (2002) Sensor placement for grid coverage under imprecise detections. In Proc. of International Conference on Information Fusion, United States.

7. Meguerdichian S, Koushanfar F, Potkonjak M, Srivastav MB (2001) Coverage Problems in Wireless Ad-hoc Sensor Network. In Proc. of IEEE INFOCOM, Alaska.

8. Meguerdichian S, Koushanfar F, Qu G, Potkonjak M (2001) Exposure in Wireless Ad-Hoc Sensor Networks. In. Proc of Mobicom.

9. Zou Y, Chakrabarty K (2003) Sensor deployment and target localization based on virtual forces. In Proc. of INFOCOM.

10. Du D, Hwang F, Fortune S (1992) Voronoi diagrams and Delaunay triangulations, Euclidean Geometry and Computers.

11. Aurenhammer F (1991) Voronoi Diagrams—A Survey of a Fundamental Geometric Data Structure. ACM Computing Surveys.

12. Butler Z, Rus D (2003) Event-based Motion Control for Mobile Sensor Networks. IEEE Pervasive Computing 2: 34–42.

13. Shah RC, Roy S, Jain S, Brunette W (2003) Data MULEs: Modeling a three-tier architecture for sparse sensor networks. In Proc. of IEEE Workshop on Sensor Network Protocols and Applications (SNPA), pp. 30–41, Anchorage, Alaska, USA, 2003.

14. Tong L, Zhao Q, Adireddy S (2003) Sensor networks with mobile agents. In IEEE Military Communications Conference, 22: 688–693, Boston, MA, United States.

15. Chakrabarti A, Sabharwal A, Aazhang B (2003) Using predictable observer mobility for power efficient design of sensor networks, In Proc. of IPSN'03, Second International Workshop on Information Processing in Sensor Networks, Palo Alto, CA, USA.

16. Cortes J, Martinez S, Karatas T, Bullo F (2004) Coverage Control for Mobile Sensing Networks. In IEEE Transactions on Robotics and Automation 20: 243-255.

17. Butler Z, Rus D (2004) Controlling Mobile Sensors for Monitoring Events with Coverage Constraints. In Proc. of IEEE ICRA 2: 1568-1573.

18. Bisnik N, Abouzeid A, Isler V (2007) Stochastic Event Capture Using Mobile Sensors Subject to a Quality Metric. In IEEE Transactions on Robotics 23: 676-692.

19. Rao R, Kesidis G (2004) Purposeful Mobility for Relaying and Surveillance in Mobile Ad Hoc Sensor Networks. In IEEE Transactions on Mobile Computing 3: 225-232.

20. Wang W, Srinivasan V, Chua KC (2005) Using Mobile Relays to Prolong the Lifetime of Wireless Sensor Networks. In Proc. of ACM MOBICOM.

21. Somasundara AA, Ramamoorthy A Srivastava MB (2007) Mobile Element Scheduling with Dynamic Deadlines. In IEEE Transactions on Mobile Computing 6: 395-410.

22. Basu A, Boshes B, Mukherjee S, Ramanathan S (2004) Network Deformation: Traffic-Aware Algorithms for Dynamically Reducing End-to-end Delay in Multi-hop Wireless Networks. In Proc of ACM Mobi Com.

23. Priyantha NB, Balakrishnan H, Demaine ED, Teller S (2005) Mobile-assisted localization in wireless sensor networks. In Proc. of the IEEE 24th Annual Joint Conference of the IEEE Computer and Communications Societies (INFOCOM).

24. Kusy B, Sallai J, Balogh G, Ledeczi A, Protopopescu A, et al. (2007) Radio interferometric tracking of mobile wireless nodes. In: Proc of Mobi Sys.

25. Niculescu D, Nath B (2003) Ad hoc positioning system (APS) using AOA. In Proc. of the Twenty-Second Annual Joint Conference of the IEEE Computer and Communications Societies, INFOCOM.

26. Lee H, Wicke M, Kusy B, Guibas L (2008) Localization of mobile users using trajectory matching In Proc. MELT 2008, the first ACM international workshop on Mobile entity localization and tracking in GPS-less environments.

27. Amundson I, Koutsoukos XD (2009) A Survey on Localization for Mobile Wireless Sensor Networks.

28. Szeile A, Bacsardi L, Huszak A (2013) Analyzing Sensor Based Positioning on the Surface of a Distant Planet. In Proc. of 64th International Astronautical Congress. Beijing, China.

17

Cargo Compartment Fire Extinguishing System

Behbahani-Pour MJ* and Radice G

Division of Aerospace Sciences, School of Engineering, University of Glasgow, Glasgow G12 8QQ, UK

Abstract

In all large passenger transport airplanes, halon fire bottles are used to extinguish fire in the cargo compartments. Halon as a fire-extinguishing agent, contributes to the destruction of stratospheric ozone in the atmosphere and it is banned in many countries. FAA considers halon 1301 as an effective firefighting agent due to its low toxicity and noncorrosive properties but because it damages the ozone layer, it has been phased out of production. However, it is still widely used on commercial aircraft until a suitable replacement is found. In this paper we will present an alternative approach to using halon 1301 as a fire fighting paradigm. In the proposed method, nitrogen is first extracted from the atmosphere by using the onboard air separator module it is then cooled, and pressurized into the cargo compartments to suppress any fire. Several methodologies can be used to increase the flow rate from the air separator module, to extinguish fire in cargo compartment.

Keywords: Fire; Cargo

Nomenclature

A/C = Aircraft

ADIRUS = Air Data Inertial Reference Unit

ASM = Air Separator Module

AVEC = Avionic Equipment Ventilation Computer

CMD = Command

ISO = Isolation

FCV = Flow Control Valve

FWC = Flight Warning Computer

FWD = Forward

LGCIU = Landing Gear Control Interface Unit

OBIGGS = Onboard Inert Gas Generation System

OVBD = Overboard

Pr. SNR = Pressure Sensor

P/B = Pushbutton

PSCU = Proximity Switch Control Unit

PRV = Pressure Relief Valve

SNR = Sensor

SOV = Shut-Off Valve

S/W = Switch

SDCU = Smoke Detector Control Unit

VLV = Valve

VC = Ventilation Controller

T SNR = Temperature Sensor

Introduction

In recent years, environmental concerns have moved towards removing halon as a fire extinguishing agent in transport airplanes. The European commission ruled that halon 1301 and 1211, must not be installed onboard new airplanes by 2018 and by 2040 halon must be removed and replaced in all operating aircraft. The International Civil Aviation Organization (ICAO) adapted a resolution that requires a halon replacement agent for lavatory extinguishers, and hand held halon extinguishers by 2011 and 2016, respectively. There is however an increasing concern for aircraft manufacturers, to find an effective halon replacement. Nitrogen is in fact suitable to suppress class A - petroleum products, cellulosic materials, and polymers – class B – flammable or combustible liquids – and class C - energized electrical equipment – fires [1]. A possible alternative to halon is nitrogen due its inertness and its suitability to suppress fires [2].

Pressurization with nitrogen has been proposed and tested as a technique for suppressing fires in gastight spaces. Laboratory fires of several representative class A and B fuels in a 270-litre confined combustion chamber have been extinguished by 0.5 atm pressure of nitrogen [3]. The approach proposed in this paper, it requires nitrogen to be extracted from the air supply by using the air separator module (ASM). High-pressure air fed into the separator; where some gas molecules such as oxygen and carbon dioxide travel more quickly through the hollow fibers; while nitrogen cannot pass through the fibers and is extracted from a different output port.

The Onboard Inert Gas Generation System (OBIGGS) generates nitrogen-enriched air, which is currently used to inert the air inside the center fuel tank to reduce any fire risk. As civil aviation authorities require all airplanes that carry more than thirty passengers, to have an OBIGGS installed nitrogen extracted from the OBIGGS could be used to extinguish fire in cargo compartment. The main challenge with this approach is the low flow rate and hence the long time it may take to inert the cargo compartment of a large transport airplane [1]. In the

*Corresponding author: Behbahani-Pour MJ, Division of Aerospace Sciences, School of Engineering, University of Glasgow, Glasgow G12 8QQ, UK E-mail: behbaa@yahoo.co.uk

proposed approach, we will describe different methodologies that can be used to increase the flow rate from the ASM and hence overcome this limitation. Nitrogen from ASM could also be used as a backup if the nominal fire extinguishing system onboard the aircraft is not sufficient to extinguish the fire. Several experiments to determine the effectiveness of nitrogen to extinguishing fire in confined and ventilated spaces were performed. The purpose of experiments was to assess the suitability of nitrogen as a fire suppressant agent and to determine the necessary pressure, time and size of fire that can be extinguished.

Current Cargo Fire Extinguishing System

In Airbus A330 series, each cargo compartment has four smoke detectors when two smoke detectors sense smoke, the Smoke Detection Control Unit (SDCU) generate warnings (master warning light and continuous repetitive chimes) and in case of smoke detection in forward cargo compartment, the smoke legend on forward agent pushbutton illuminate. Pilot as soon as he selects FWD Agent pushbutton, the fire bottle 1 supplies halon to forward cargo. If the fire has not extinguished then pilot select AFT agent pushbutton and halon from fire bottle 2 discharge via a flow metering unit which restrict the agent flow, so that the agent is slowly discharged into cargo bay and it takes 240 minutes to completely discharge the agent. This causes a slow and continuous flow of halon into the cargo compartment to keep the fire out. When pressure in the bottle decreases, the bottle pressure switch causes advisory message to show in the cockpit.

One of the disadvantage with current cargo fire extinguishing system in airplanes is that spurious cargo smoke warning results in pilot to discharge the halon fire bottles unnecessary, and result in depletion of extinguishing agent. In addition, in case of genuine fire, then there is no or not enough extinguishing agent left to use.

Other disadvantage with current system is that the fire bottle can automatically discharge outside aircraft, due to thermal expansion due to high internal temperature rise. In such condition when the fire

bottle discharge their agent overboard during the flight than there will be no halon agent left to extinguish the fire or smoke in the cargo compartments so in proposed technique, inert gas nitrogen can be generated at any time and for any duration used to extinguish the fire. In addition, to discharge the content of a halon bottle, it requires an explosive cartridge to be fired. The fire bottle cartridge is a life item, and normally has a five years life. In case of defective cartridge or expired cartridge, then the halon agent cannot flow to extinguish fire. As shown in Figure 1, fire bottles have a thermal relief valve to discharge the agent overboard in case excessive internal pressure due to temperature rises.

FAA considers Halon 1301 as an effective firefighting agent with low toxicity, noncorrosive but because it damages the ozone layer, it is phased out of production however widely used on commercial aircrafts until a suitable replacement is developed [4]. The U.S Army conducted investigation of combustion properties of fuel containing halon 1301 in engines, and the result showed that the combustion products derived from halon were so corrosive that it severely damaged the engine in less than 50 hours of operation [5]. Currently, on all transport airplanes, halon agent during fire is applied to outside the engine and not to inside of the engine because it is corrosive. Halon 1301decomposes to form hydrogen fluoride and hydrogen bromide which if it does not get diluted in air, then 15 minutes' exposure becomes lethal also even small concentration can produce corrosion on metals, therefore halon should not be used on metal fires because of exothermic reactions between halons and metals [6].

Spurious Smoke Warnings

Spurious cargo smoke warnings are false warnings, and can result in flight crew to discharge the cargo fire bottle, extinguishing agent unnecessary. This situation reduces the flight safety, that in case of real fire during flight, then there is no or not enough fire extinguishing agent left to extinguish the real fire. Consequently, fire increases its intensity, thus produce more toxic fumes that can enter the cockpit. Fumes can impair judgment and affect performance of the flight crew.

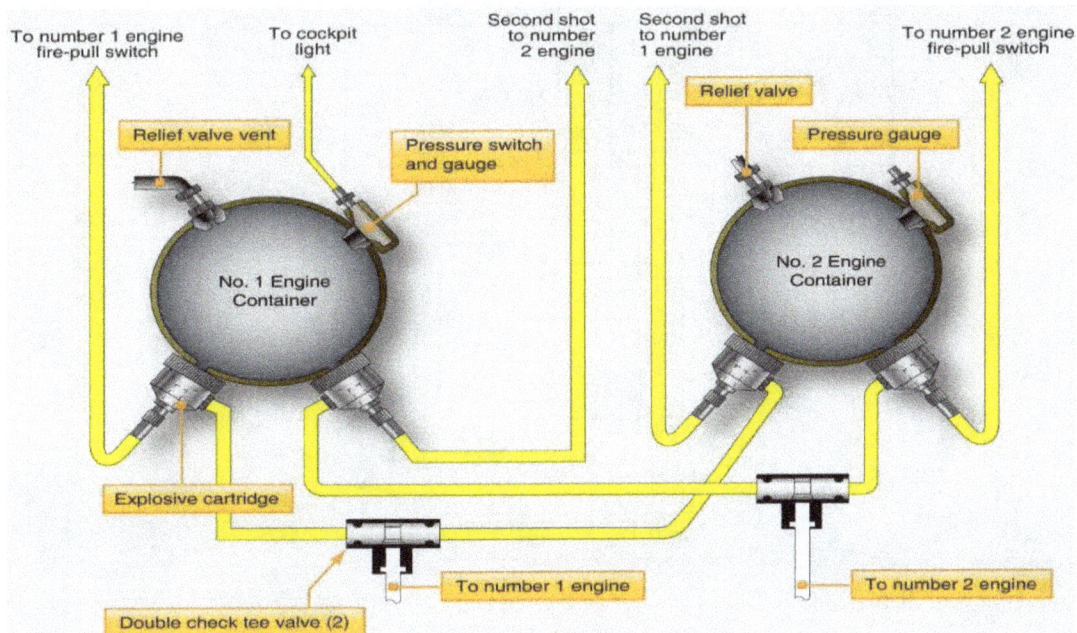

Figure 1: Figure showing current halon engine fire bottle extinguishing system [4].

However, in proposed design, nitrogen is available all the time during the flight, and can be used to extinguish fire.

Spurious cargo smoke warning can be triggered by contaminated air circulation from various sources. For example, smoke warning can be triggered by extreme condensation in the cargo compartment. Cargo smoke warning can be generated by cargos such as tropical fruits, and vegetables etc. These cargos cause humidity levels to increase in unventilated compartments with the result that large quantity of condensation can form, the droplets of which appear to the smoke detector to be like smoke particles. In addition, cargo smoke warning can be triggered when aircraft operating in hot and humid conditions. Cargo smoke warning can be triggered fumes from ground equipment operating in the near vicinity of the cargo-loading door. Smoke warning can be triggered by oil vapor in the bleed air. The oil vapor appears to the detector to be smoke particles and thus generate alarm. The oil vapor is typically the result of oil spillage in the bleed air ducts from the APU. Smoke warnings can be triggered by ingestion of anti-ice/de-ice fluids into the intakes of either the engines or APU. Smoke warning can be triggered by ingestion of engine cleaning agent into the air conditioning system. Most of the smoke warnings been reported by airlines to Airbus, have been due to spurious smoke warnings. Only on very rare occasions, some of the cargo warning were triggered by a real fire [7].

One source of contamination is the APU; the APU bleed-air supply contamination can be due to either APU internal oil leakage or re-ingestion of oil, hydraulic or de-icing fluids. External fumes near the APU air intake, from exhaust gas fumes of main engines of any aircraft nearby or ground power units, can be ingested by the APU, and lead to smell in cabin degraded. Hydraulic fluid leaks from the aircraft main landing gear bay, can be directed along outside of the aircraft fuselage, up to the APU air intake. Hydraulic fluid can be ingested, while APU is running, and causes the aircraft air condition system contamination, and result in associated smell in the cabin. APU external fuel or oil leakage may be re-ingested if the APU inlet seal or APU doors seal degraded [8].

Current Cargo Fire/Ventilation Systems

In modern airplanes such as Airbus A330/340 series, the cargo fire extinguishing system, each cargo compartment has four smoke detectors and when two smoke detectors sense smoke, SDCU generate cargo smoke warnings in the cockpit. In case of smoke warning in forward cargo compartment, pilot selects FWD Agent pushbutton, and the circuit fires the halon-extinguishing bottle number one and halon flow into forward cargo. If the fire has not extinguished then pilot select AFT agent pushbutton and halon from fire bottle number two discharges via a flow metering unit which restrict the agent flow so that the agent is slowly discharged in the compartment, and takes 240 minutes to completely discharge the agent in the second bottle.

In addition, forward and AFT cargo compartments are ventilated and heated, as shown in Figure 2. Air from air condition system (cabin air) enters the cargo compartment through inlet isolation valve, then the extract fan, extract air from the compartment through the outlet isolation valve, and dump it overboard. Cargo ventilation controller controls the cargo ventilation system. Ventilation controller stops the extractor fan if the isolation valve is not open [9].

Some aircraft have cargo heating as well as cargo ventilation system, as shown in Figure 2. The cargo heating system is optional and it is to be selected manually by pilot. For example, in Airbus A320 series,

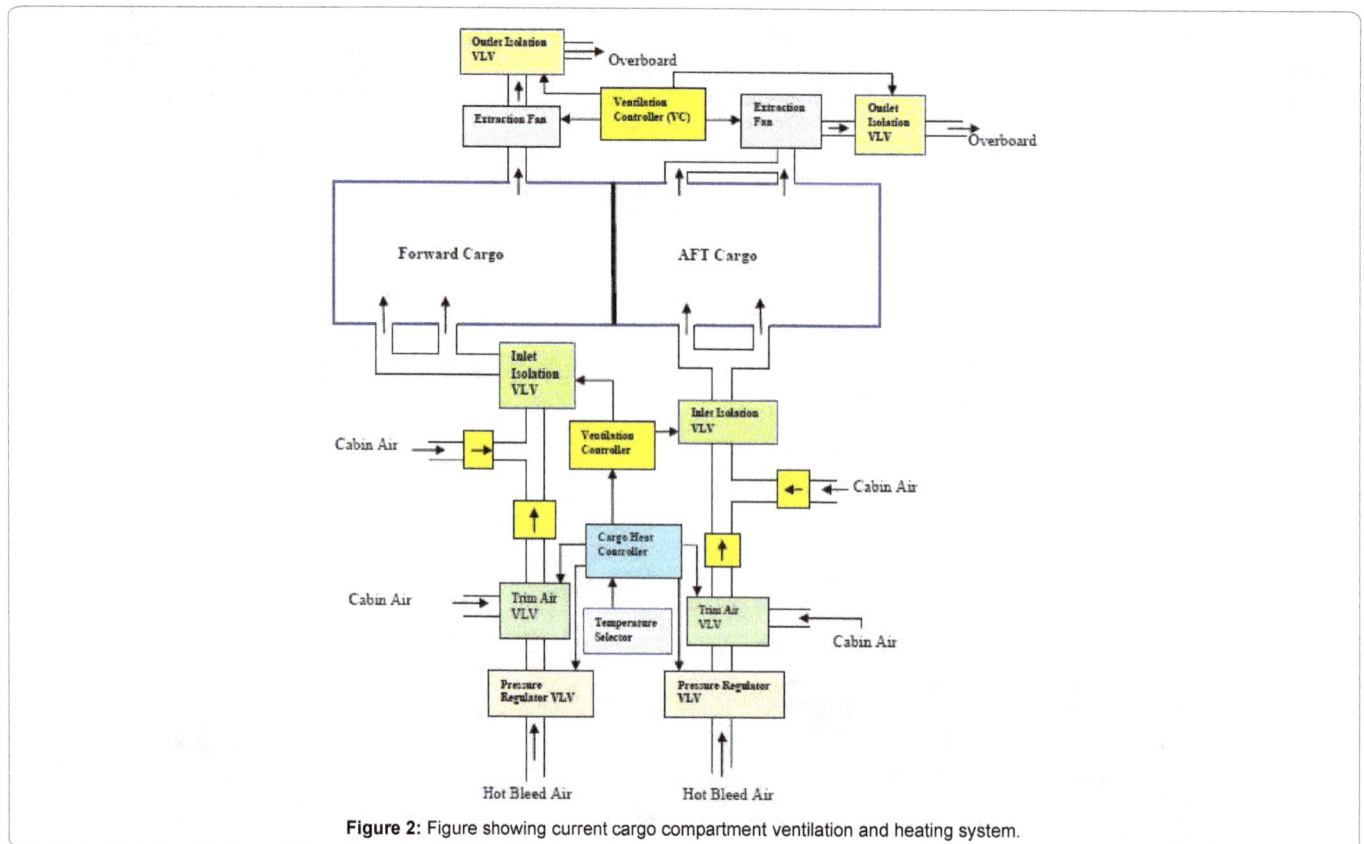

Figure 2: Figure showing current cargo compartment ventilation and heating system.

the cargo heat controller controls the trim air valve and the pressure regulating valve. The trim valve is a stepper motor which modulate the amount of hot air going to the cargo according to temperature selection by pilot. Hot bleed air enters the pressure regulator valve, where its pressure regulated to the required value then mixed with cool cabin air by trim air valve, to achieve required temperature. Heat controller send signal to ventilation controller to open the isolation valves for heated air to enter the cargo compartment. Cargo heat controller controls the pressure regulator valve and trim air valve. The cargo heating stops; if the isolation valves are closed, or ventilation extract fan stopped or in case of duct overheat [10]. In some aircrafts such as Airbus A330/340 series, avionics equipment cooling air, after passing through the equipment, it gets warm, and during flight, it is routed to forward cargo compartment through under-floor valve. The avionics equipment cooling controlled by AVEC. On the ground, the avionics equipment cooling is dumped overboard via overboard extract valve.

Proposed Cargo Fire Extinguishing System

In the proposed designed, the components are nitrogen processor, shut-off valves, a flow control valve, air compressor, heater, three air separator modules in parallel, nitrogen and standby pump, a heat exchanger (cooler 1), and the sensors to measure pressure, temperature, and flow rate. Nitrogen processor contains the logic, and gives command to increase the amount of air supply to ASMs and to pressurize nitrogen to higher values because experiment shows the increasing the nitrogen pressure result in quicker time to extinguish fire. Nitrogen processor send a signal to cargo ventilation and cargo heating system to stop supplying conditioned airflow to the cargo, in order to stop fresh oxygen reaching the fire. Nitrogen processor can

be programmed to supply nitrogen automatically to the cargo when the smoke detectors sense smoke in particular cargo bay, or it can be programmed that pilot manual selection required (by selecting a pushbutton). Air compressor used to increase the pressure of air supply to ASM inlet, in order to increase the nitrogen flow rate. Three ASMs used to increase the nitrogen flow rate to the cargo compartment. Cooler 1 is a heat exchanger, bleed air supply to the ASM is maintained to 180F plus or minus 10F, hence the nitrogen flow from ASMs will be at the same temperature, it is cooled in order to increase the nitrogen density, and reduce the temperature of the fire.

With reference to Figures 3 and 4, the smoke detectors sense smoke in the FWD cargo compartment and send warning signal to SDCU, which in turn send signal to FWC in order to generate oral and visual warnings to the pilots. SDCU send signal to ventilation controller, which causes the FWD cargo inlet and outlet isolation valves to close and the extractor fan to stop. When the cargo ventilation isolation valves are closed, the ventilation controller send such data to nitrogen processor and used as part of the logics. The processor will send signal to AVEC to close under floor valve to stop avionics equipment cooling air to be discharged in the FWD cargo compartment, and to open the overboard extract valve so that the avionics equipment cooling air is dumped overboard during FWD cargo smoke/fire condition. This action will prevent oxygen supply to reach the fire inside cargo compartment while nitrogen supplied into the cargo compartment. Nitrogen processor send signal to the cargo heat controller to stop the cargo heating system. The FCV opens more (high flow mode), in order to increase the bleed-air supply inlet to ASMs and consequently increase the nitrogen output.

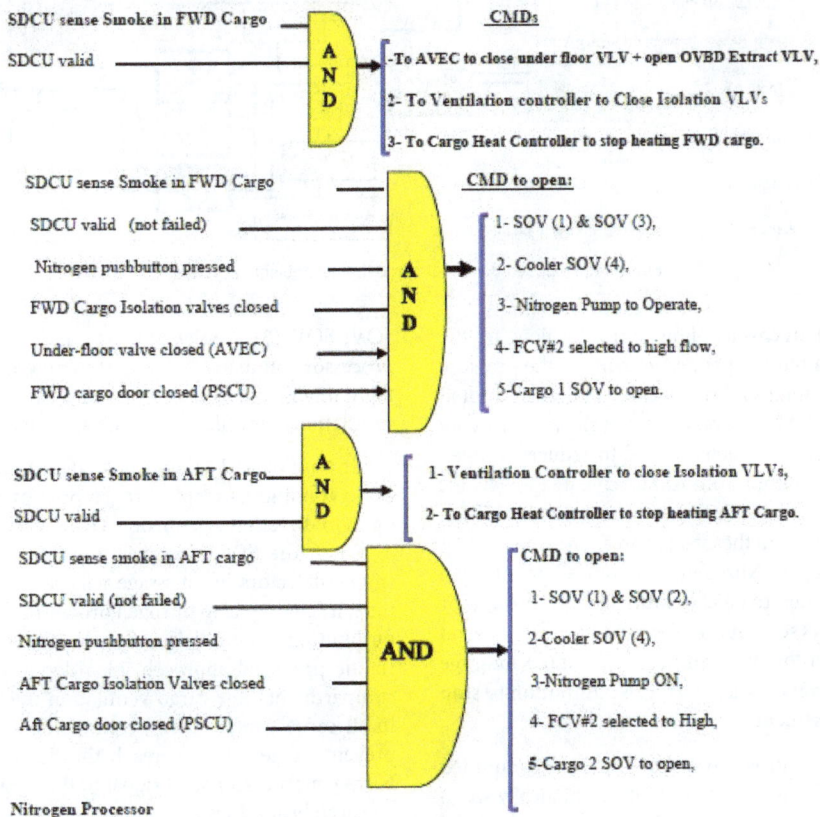

Figure 3: Proposed design of cargo fire extinguishing system for forward and AFT compartments.

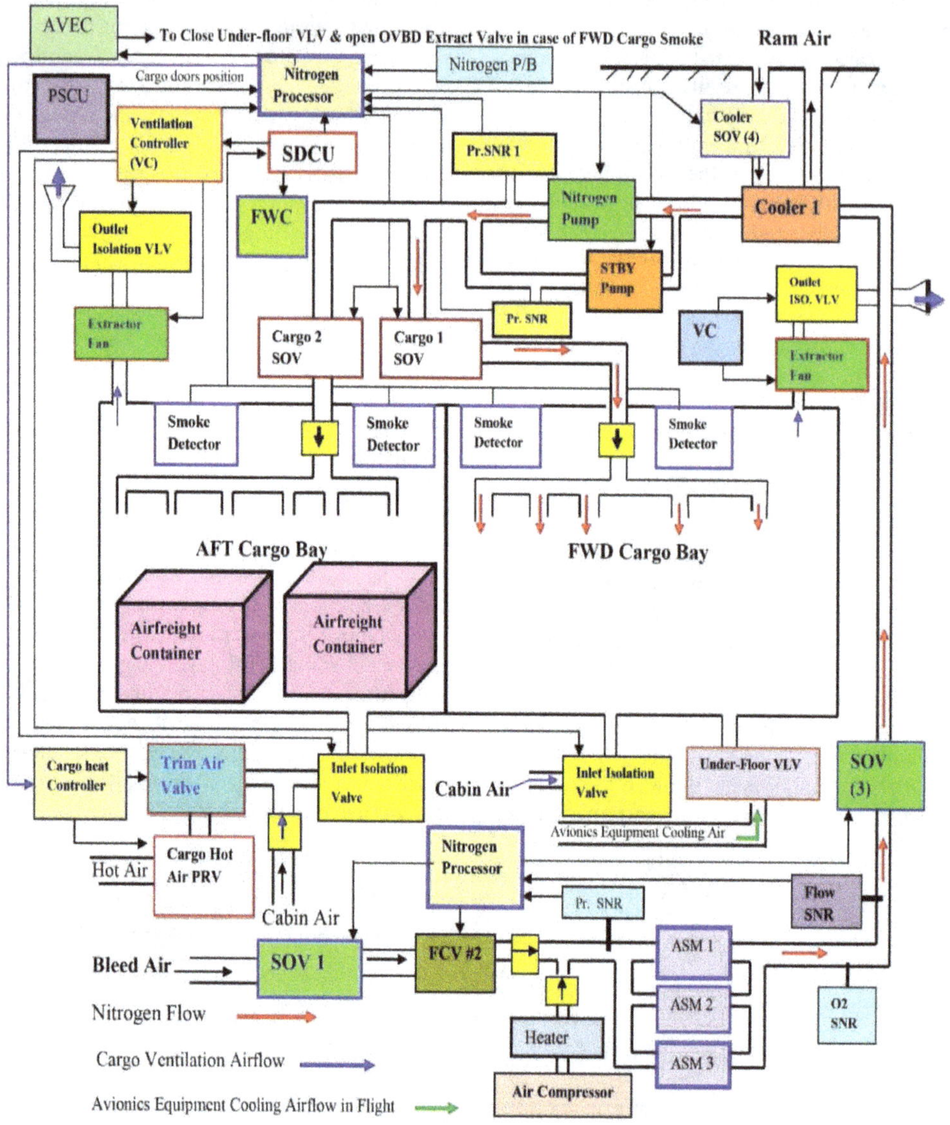

Figure 4: Proposed design of cargo fire extinguishing system.

Moreover, the SOV (3) opens to allow nitrogen flow to the nitrogen pump. The nitrogen pump operates to increase the pressure of nitrogen, in order to extinguish fire quicker, and then routed to forward-cargo, via cargo 1 SOV. Cooler SOV (4) opens to cool the nitrogen in order to increase its density, and to reduce nitrogen temperature because nitrogen output from ASMs regulated to 94C for better operational performance of ASM's. When nitrogen cooled, it can reduce the temperature of the fire in the affected area. The Cooler SOV (4) uses ram air to cool nitrogen. Nitrogen processor allow nitrogen to flow into FWD cargo, in order, to quickly extinguish the fire and or smoke. When Smoke detectors sense no smoke condition, SDCU send no smoke condition signal to the nitrogen processor, and a message appears in the cockpit. Pilot can de-select nitrogen pushbutton to stop nitrogen flow into the compartment.

In the proposed technique, nitrogen processor can be programmed to supply nitrogen to affected cargo compartment automatically when at least two smoke detectors sense smoke in the cargo compartment. When the system is de-selected, nitrogen processor closes, cargo 1

SOV, SOV (3), Cooler SOV (4) and stop nitrogen pump. Nitrogen processor command AVEC to re-open under floor valve (if aircraft in flight mode) and to close overboard extract valve. It also send signal to ventilation controller to re-open the applicable isolation valves.

In case of AFT cargo smoke detection, similar actions to forward cargo condition, except nitrogen processor command cargo 2 SOV to open to direct nitrogen from ASMs to AFT cargo as shown in Figure 4. When the SDCU receives no smoke signal from the associated smoke detectors, no message will be shown in cockpit. At the same time, it send signal to nitrogen processor. If pilot, de-select the nitrogen pushbutton, then nitrogen flow to affected cargo compartment stops. In the proposed approach, in order to extinguish fire in the cargo compartments, the cargo ventilation inlet and outlet isolation valves in aft cargo compartment, commanded to close position in order to prevent oxygen flow to reach the fire in the affected compartment. Nitrogen processor send signal to the cargo heat controller to stop the aft cargo heating system, if it was selected on.

PSCU senses FWD and AFT cargo doors positions and send such

data to nitrogen processor, and used as part logic in the proposed design for cargo smoke extinguishing system. Nitrogen applied to extinguish fire in forward cargo bay, when the FWD cargo door is close, similarly in aft cargo compartment, the aft cargo door have to be closed. This is to ensure that aircraft is in flight mode, and to isolate the fire with fresh oxygen supply from ambient atmosphere.

Challenges

The main challenge with the proposed approach is the limited flow rate from the ASM, and hence the long time it may take for the cargo compartment to be rendered inert. Different approaches can be used to increase the nitrogen flow rate from the ASM. Increasing the pressure of the bleed air supply to the ASM, will increase the nitrogen flow output and decrease oxygen concentration. As shown in tests carried out by FAA and Airbus [11]. Higher bleed air pressure can be supplied to ASM either by taking the bleed air supply from the high pressure compressor stage of the engine, or using a compressor to pressurize the airflow to ASM. If engine compressor bleed air is not readily available, then an alternative could be to use an air compressor to increase pressure supply to the ASM inlet. The compressor type can be rotating either impellers or positive displacement.

An alternative approach that does not require pressurization of nitrogen would be to reduce the size of the cargo compartment, by diving it into smaller sections. This can be achieved by using airfreight containers in the cargo compartments. Cargo compartment can be of different sizes. Air cargo containers are usually made from light aluminum, as shown in Figure 5, however, lightweight carbon reinforced fiber and polypropylene containers have recently become. In either case, the airfreight container is to be installed with a smoke detector, a nitrogen port, and a mechanical pressure relief valve.

In the proposed technique, when the airfreight container is placed inside the cargo compartment, it will have to be connected to a nitrogen flexible hose, and electrical receptacle, to power the smoke detector inside the container. Should the smoke be detected; then the nitrogen processor energizes the associated container solenoid valve, and nitrogen from the ASMs will be routed to the effected container. When the pressure inside the container reaches a predetermined threshold of 30 psi depending upon the size and type of the container, the relief valve opens to vent excess nitrogen and to prevent over pressurization of the container. As safety feature, if the container is not electrically to the aircraft, then a message will appear in the cockpit to alert the crew. This is done by means of a proximity sensor in the nitrogen hose connector.

Figure 5: Figure showing light aluminium air-freight container.

Nitrogen is used in some trucks and ships to transport fresh fruit, vegetables, flowers, plants etc. Nitrogen preserves freshness and increases the fruits and vegetable storage life.

ASM performance can be increased if the ASM fibers are made from nanoparticles materials. Several experiment tests conducted involving membranes based gas separation, involving the hollow-fiber membranes made from three different types of fibers, such as pure polycarbonate, polypropylene, and polycarbonate added with silica nanoparticles. Gas permeabilities of O_2, N_2 and CO_2 were measured to determine the gas separation properties of obtained hollow fibers, and the test results showed that nanoparticles increased the absorption of gas separation, and improved selectivity [13].

For ASM efficiency, the inlet air pressure has to be high (typically 40 psi or more). It is possible to operate at lower pressures but more ASMs are required and this increases weight with each ASM weighting approximately 27kg. For illustration, if the air-supply inlet pressure to ASM is at 15psi, then ten ASMs are required, but if the inlet air pressure is at 56psi, then only two ASMs required to provide the required NEA capacity [12]. A single ASM at altitude of 32,000 feet, if the ASM inlet pressure is 32 psi, then ASM nitrogen flow rate will be equal to 8 cubic feet per minute (SCFM). At 40,000 feet, if ASM inlet pressure is 40 psi, then ASM nitrogen flow rate will increase to 10 cubic feet per minute [13]. In proposed design, air compressor used to increase the pressure of air supply inlet to 65 psi to ASMs.

The proposed design, as shown in Figure 6, allows the airfreight containers to be cooled by means of the aircraft air condition system, in order to meet critical storage needs of the temperature sensitive cargo such as medical or fresh produce. Should smoke be detected, the air supply to the container is stopped so to ensure a higher effectiveness of the nitrogen.

The proposed airfreight containers can be filled with inert nitrogen gas, to 2 to 3 psi. In such case, each container will be inerted with nitrogen, one at a time, when certain conditions fulfilled (cargo door closed and engine operation). It means that as soon as aircraft starts moving, during taxi phase, the containers will be inerted slightly with nitrogen, the containers will be installed with a pressure sensor, which send data to the N2 processor. The proposed containers will be installed with a pressure relief valve, to dump excessive pressure. Experiments conducted and result showed that nitrogen of 2-psi extinguishes fire effectively. In the proposed technique, nitrogen temperature from ASM is controlled to suitable values.

By increasing the number of ASMs, it will increase the nitrogen flow rate. Experiments have been conducted, result showed that nitrogen flow of ASM approximately doubled for the two-membrane configuration [1]. In addition, increasing the size of the Air Separator Module, the nitrogen flow increases.

Another technique, which can be used, is that storing nitrogen from ASM in a cylinder, as a backup source, which can be used in case of fire/smoke, to increase the nitrogen flow to the affected area. The cylinder pressure continuously monitored, and if it falls below certain value, sensed by the pressure sensor, and the system, supply air to ASM to generate nitrogen, then pressurized and stored in the cylinder bottle. Once the pressure in the cylinder reaches the limit, the N2 SOV closes by command from N2 processor, as shown in Figure 7, the refilling of the nitrogen bottle done automatically, if bleed air supply is not available, the air compressor operates to supply pressurized air inlet to the ASMs.

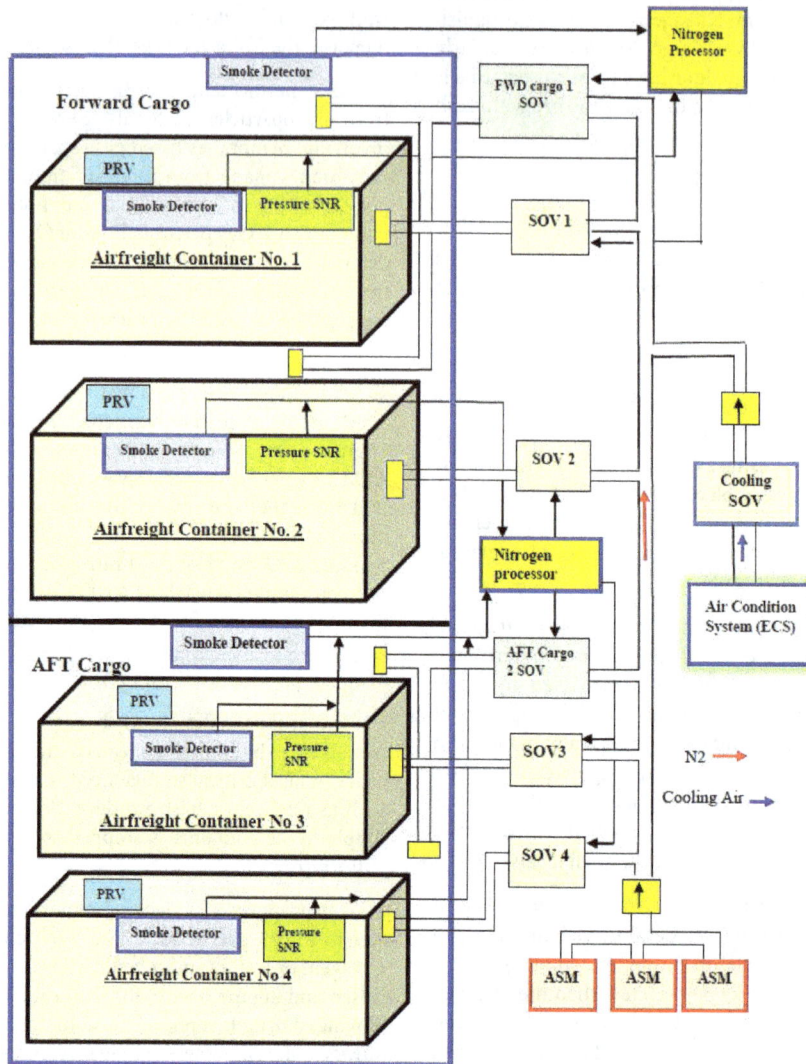

Figure 6: Proposed technique of using individual containers with nitrogen connection.

Nitrogen Experiments

Experiments were performed to evaluate the effectiveness of nitrogen in extinguishing fire in a compartment determine if the size of fire affects the nitrogen fire extinguishing ability, thirdly, determine the time it takes for the fire to be extinguished and determine the minimum pressure for nitrogen to be effective in extinguishing the fire. The first experiment was performed using a transparent, 50 cm, cubic box. The top side of the box his equipped with a port for the nitrogen input connection, as shown in Figure 8.

The experiment is at first performed with only one candle and carried out using one lit candle, and then the number of candles was increased. Experiment conducted initially with the test box having no ventilation. The nitrogen shut-off valve was opened manually, in order to allow nitrogen stored in the bottle to flow to the nitrogen pressure regulator, and then the pressure regulator was manually adjusted in order to control the nitrogen pressure. Experiment was first conducted with the test box having no ventilation to atmosphere, in order to analysis the effectiveness of nitrogen. Then the same experiments conducted with the test box having a ventilation port of 12.73 mm, this

is done in order to establish if nitrogen can effectively extinguish fire in ventilated container (test box), as shown in Figures 9-12.

A series of experiments conducted, aimed at gathering information by using dry nitrogen under different pressure values to extinguish different size of fire. Table 1 shows the nitrogen pressure and time taken to extinguish different sizes of fire inside the test box during the experiments. The analysis of the experiment research showed that increasing nitrogen pressure, resulted in quicker extinguishing time. Figure 13 shows increasing nitrogen pressure, results in extinguishing the fire of the same intensity quicker. This is because nitrogen under higher pressure, quickly decrease the oxygen concentration in the air for the fire already in the process of combustion. In principle, there are two approaches to fire suppression: either decreasing the oxygen concentration or inerting the combustible environment. For fire to occur, it requires oxygen, and by supplying nitrogen at higher pressure, nitrogen pushed away the oxygen from fire at faster rate, hence inhibited combustion process sooner. However, increasing the fire size, it took slightly few seconds longer time. For illustration, five lit candles were extinguished in 0.51 seconds by using nitrogen at 10 psi, however, increasing fire size to 30-lit candles took 2.54 seconds to extinguish

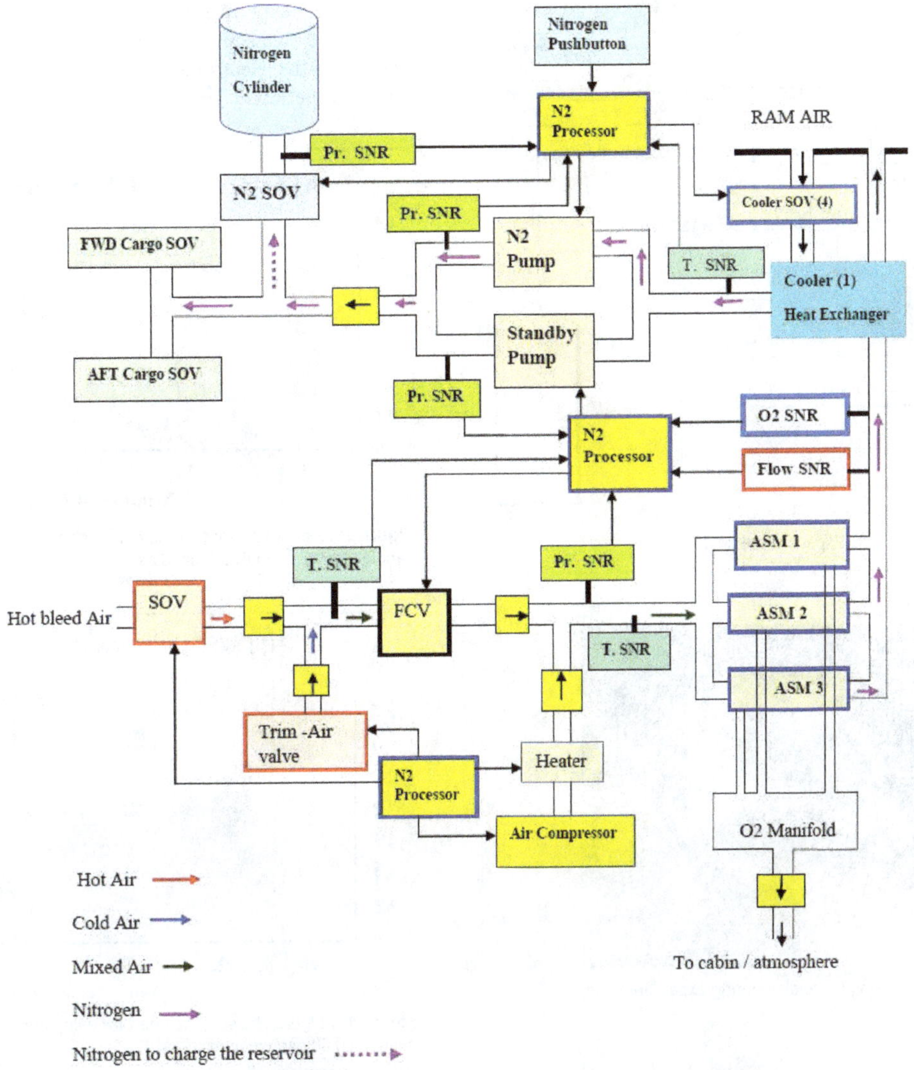

Figure 7: Nitrogen from ASMs stored in the cylinder, and used as backup during cargo fire.

when nitrogen supplied at 10 psi, as shown in Figure 14.

The experimental study of this research shows that nitrogen did not require a cleanup after the fire event experiment, as the result of gas release. Nitrogen discharge did not create a fogging effect hence vision wasn't compromised or obscured, unlike most fire-extinguishing agents. Nitrogen did not damage protected sensitive equipment. It can be concluded that dry nitrogen is non-conductive and can be used in environments where sensitive electronic devices are present.

Figure 11 shows one to five lit candles were extinguished in the same time when supplied with same nitrogen pressure (5 psi). Figure 14 shows dry nitrogen was used at different pressure to extinguish different size of fire, and time to extinguish fire was recorded. Based upon the experimental research, the analysis showed that minimum of 2 psi was sufficient to extinguish fire effectively, but it took few seconds more time. In addition, fire size, did not have effect on nitrogen to extinguish the fire. The experiment had limitation due to difficulties in building a transparent test box to represent actual aircraft cargo size.

Future Work

Although, FAA have conducted extensive ground and flight tests on using nitrogen from ASM to inert center fuel tank compartment on Airbus A320, as published in FAA document: DOT/FAA/AR-03/58. It states that the results of the tests indicated that the concept of the using nitrogen from ASM found effective to inert fuel tank compartment, except bleed air consumption was greater than expected. Therefore, it would be advantageous, to perform an experimental study using dry nitrogen from ASM to extinguish fire in a typical cargo size compartment, which is beyond the scope of this dissertation. The actual aircraft cargo compartment can vary from one aircraft type to another. The experiment had limitation due to difficulties in building a transparent test box to represent actual aircraft cargo size. The theoretical and practical achievement of this research can certainly be considered as an important milestone in the road-map to perform more research in future.

Conclusion

The experimental study and theoretical achievement of this research shows that nitrogen of 2 psi was sufficient to extinguish fire (57 lit-candles) in less than five seconds. In addition, the experimental

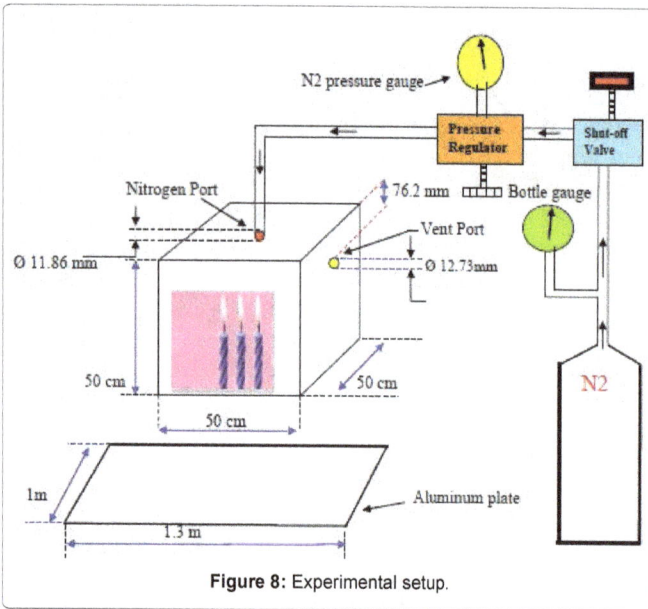

Figure 8: Experimental setup.

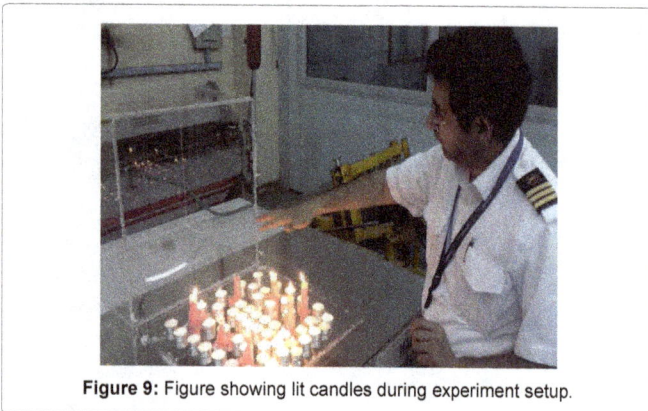

Figure 9: Figure showing lit candles during experiment setup.

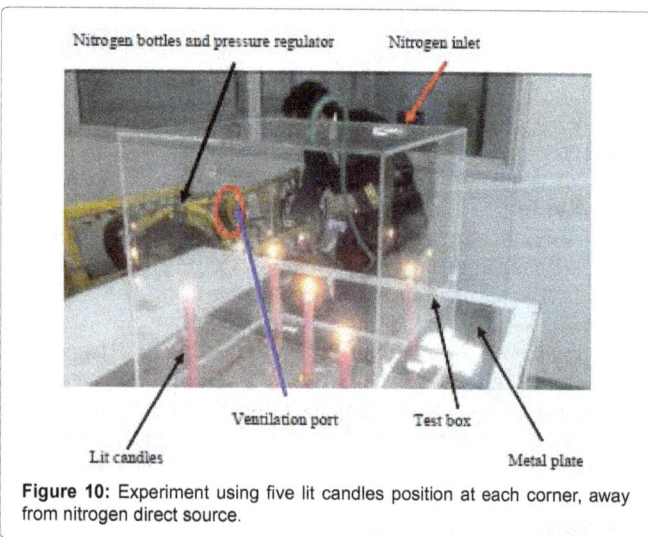

Figure 10: Experiment using five lit candles position at each corner, away from nitrogen direct source.

research showed that increasing the nitrogen pressure, resulted in faster fire extinguishing time because nitrogen at higher pressure, depleted the fire from oxygen at faster rate, and inhibited combustion process

quicker. Nitrogen discharge did not require cleanup after extinguishing the fire during experiment, unlike halon or other extinguishing agent if the affected area is not cleaned, it can result into metal corrosion or damage sensitive electronic devices. In addition, it can be concluded from the experiment research that nitrogen is non-conductive and

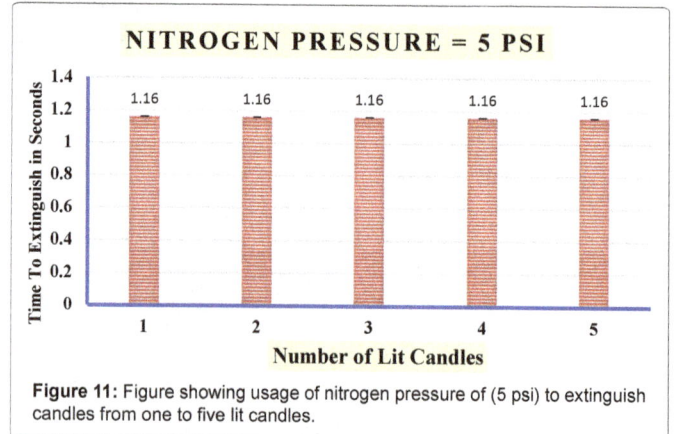

Figure 11: Figure showing usage of nitrogen pressure of (5 psi) to extinguish candles from one to five lit candles.

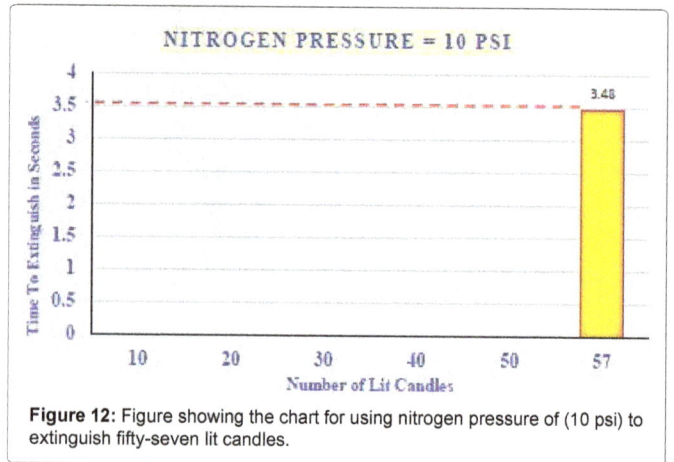

Figure 12: Figure showing the chart for using nitrogen pressure of (10 psi) to extinguish fifty-seven lit candles.

Number of lit candles	Nitrogen pressure (psi)	Extinguishing time (sec)
1 to 5	5	1.16
5	2	3.73
5	4	3.20
5	5	2.70
5	6	2.60
5	8	1.80
5	10	0.51
10	2	4.22
10	5	1.32
10	10	1.28
20	2	4.32
20	5	2.30
20	10	2.22
30	2	4.50
30	5	3.35
30	10	2.54
50	5	3.45
57	10	3.48

Table 1: Table showing the nitrogen pressure and time taken to extinguish different fire intensity inside the test box.

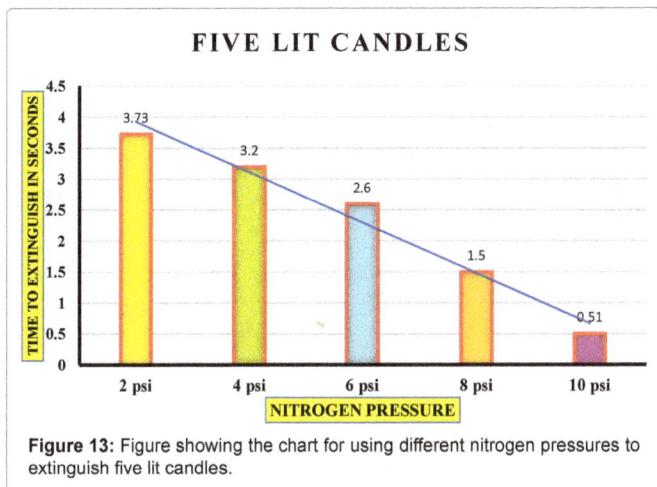

Figure 13: Figure showing the chart for using different nitrogen pressures to extinguish five lit candles.

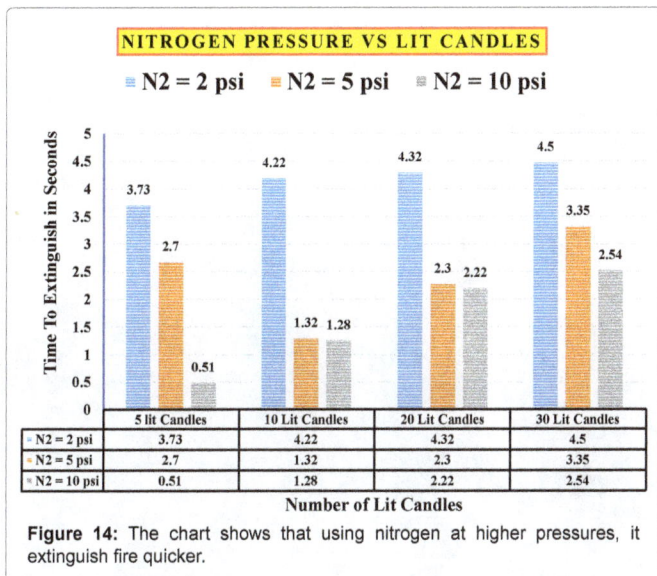

	5 lit Candles	10 Lit Candles	20 Lit Candles	30 Lit Candles
N2 = 2 psi	3.73	4.22	4.32	4.5
N2 = 5 psi	2.7	1.32	2.3	3.35
N2 = 10 psi	0.51	1.28	2.22	2.54

Figure 14: The chart shows that using nitrogen at higher pressures, it extinguish fire quicker.

therefore, it can be used on aircraft environment where sensitive devices are present. Therefore, such research can certainly be considered as important milestone in the road to carry out further experiments in larger space, using dry nitrogen. Many technical challenges have been faced; and overcame in the proposed approach. This is starting point, that some new research paths can be opened, as regard to increasing flow rate from gas separation modules. The final goal will be to generate nitrogen from atmosphere at increased flow rate to extinguish fire in cargo compartment on commercial airplanes.

How the proposed design is going to affect the relevant stakeholders? The aircraft manufacturing company and the shareholders will want the best return on their investments. In the competitive market, and fluctuating oil prices, most airlines forced to reduce fares; however, safety concepts are at very core of any organization. The proposed technique have advantage over the existing cargo fire extinguishing cargo using halon, because nitrogen is noncorrosive when it reacts with moisture and it does not poses threat to the environment (zero ozone layer depletion/zero global warning). Nitrogen effectively extinguish fire, and FAA considers nitrogen inert gas as a suitable alternate to halon. The proposed design increases flight safety, and consequently reduces future aircraft accidents. This in turn, enhance the image of the airline, and wins more customers thus result in business growth and

increases profitability.

How the proposed design is could be implemented in practice? Using dry nitrogen from gas separation modules is already approved by civil aviation authorities such FAA and EASA for use on all commercial airplanes, and meets the civil aviation airworthiness requirements. Nitrogen is used to inert center fuel tank known as On Board Inert Gas Generating System (OBIGGS). FAA requires a decrease in the flammability of center fuel tanks, and the 2008 ruling affects all future fixed-wing airplane design with (passenger capacity greater than 30). All new Airbus and Boeing aircrafts with center fuel tanks are required to have OBIGGS. In order to implement the proposed design, it requires a processor, several shut-off valves, and sensors, additional ASMs to be accommodated in existing OBIGGS. Alternatively, nitrogen from existing OBIGGS, instead of inerting center fuel tank, is diverted to extinguish cargo compartment. In such scenario, priority is given to extinguish fire onboard of airplane. This arrangement can be done by reprogramming the OBIGGS processor, and bears no additional cost.

The main challenge of flow rate limitation of using nitrogen from ASM to extinguish fire in cargo compartment can be overcome by using several methodologies. The ASM flow rate can be increased by several methodology. The flow rate of ASM can be increased by increasing the size of ASM or increasing the number of ASMs. In addition, the ASM flow rate can be increased by increasing the bleed air pressure supply inlet to the ASM. This can be achieved by taking bleed air of higher pressure, from engine high compressor stage, instead of intermediate compressor stage. In addition, the bleed air pressure can be increased by using an air compressor, which can be driven by engine accessory gearbox or an electrical motor or Ram Air Turbine (RAT).

Other methods include using airfreight containers, which are installed by a smoke detector, a PRV, and nitrogen supply connection. Most airlines use airfreight containers. In such arrangement, nitrogen from ASMs can be directed, only to the affected container under fire. The advantages of using airfreight containers are that small pieces of cargo can be wrapped into a single unit, so it saves time to load/off-load items, and reduce turn-around-time. On long haul flights, using more than one airline, the containers could be transported from plane to plane with amazing speed and ease. In addition, it minimizes the aircraft center of gravity to shift during flight. Another technique is to inert each airfreight container with nitrogen, one at a time, when certain conditions fulfilled (cargo door closed and engine operation). Nitrogen temperature can be controlled to desired temperatures; this will help to carry temperature sensitive items. In such arrangement, reduces the risk of ignition/fire, because for fire to occur, it requires oxygen. Nitrogen used in some trucks and ships to transport fresh fruit, vegetables, flowers, plants etc. Nitrogen preserves freshness and increases the fruits and vegetable storage life.

Alternate method, includes storing nitrogen from ASM in a cylinder bottle, which can be used in case of cargo fire/smoke, to increase the nitrogen flow to the affected area. In proposed technique, the nitrogen cylinder pressure is continuously monitored, and if it falls below certain value, sensed by the pressure sensor, the system automatically top it up, hence, no need for personnel to fill the cylinder.

References

1. FAA (2012) Options to the use of halons for aircraft fire suppression systems-2012 Updates. US Department of Transportation, USA.

2. Hasessler M, Walter P (1989) Fire fundamentals and control. Marcel Dekker Inc. USA: 123-125.

3. National Academy of Sciences (1975) Committee on Fire Research Directory of Fire Research in the USA, (1971-1973), (7thedn), USA. p: 105-118.

4. FAA (2016) Fire Extinguishing System (Chapter 17) US Department of Transportation, USA, p: 17-12.

5. National Research Council (1997) Committee on aviation fuels with improved fire safety. A Proceeding of National Research Council, National Academy Press, USA: 134-135.

6. Raymond F (1998) Principles of fire protection chemistry and physics. (3rdedn), Jones and Bartlett Publishers UK.

7. Airbus France (1990) Airbus Technical Digest (FAST) Number 10, July.

8. Airbus Technical Magazine (FAST # 52) A clean APU Airbus France.

9. Airbus A340 (2003) Aircraft Maintenance Manual 36: 7-22.

10. Airbus A319/320 (2010) Aircraft Maintenance Manual.

11. Burns M (2016) FAA flight-testing of the FAA onboard inert gas generation system on an airbus A320. US Department of Transportation. USA.

12. Massey EA, Das A, Joshi P, Mahesh S (2016) On board inert gas generation system. USA Patent: US 20130341465 A1.

13. Moghadassi A, Marjani A, Shirazian S, Moradi, S (2011) Gas separation properties of hollow-fiber membranes of polypropylene and polycarbonate by melt-spinning method. Asian J Chemistry 23: 1922-1924.

18

Simulation of Sloshing in Rigid Rectangular Tank and a Typical Aircraft Drop Tank

Shreeharsha HV*, Shivakumar SG and Mallikarjun SG

Department of Aeronautical Engineering, MVJ College of Engineering, Bangalore, India

Abstract

In this research, the liquid sloshing behavior in a 3D rectangular tank was simulated and validated by applying peak acceleration load using computational fluid dynamics technique. The application of this sloshing phenomenon was carried out on a typical Aircraft Drop tank with and without baffle plates for 7 g peak acceleration. The structural integrity of the drop tank has been taken into consideration during the cruise flight condition. Further, an optimized design of a drop tank has been modeled. The comparison of computed results for 3D rectangular tank case with experimental results showed that the numerical technique is capable of simulating hydrostatic pressure loads exerted on tank walls. Similarly, the necessary sloshing loads in the form of hydrodynamic pressure generated on the tank walls have been estimated for different cases of a typical aircraft drop tank. The kinematic profiles of liquids were observed at different instances for various cases. Computational results indicated that there is a reduction in the peak pressure on aft side of the tank with the use of baffle plates.

Keywords: Liquid sloshing; Volume of fluid; CFD; Drop tank

Introduction

Sloshing describes the free surface oscillations of a fluid in a partially filled tank. For slosh, the liquid must have a free surface, thus posing a dynamic fluid structure interaction problem. This motion of fluid plays a major role in the cargo slosh in ship, spacecraft, rocket tanks, transportation of liquid and oil storage tanks, water oscillations in storage tanks due to Earthquakes and in reactors of boiling water. The behavior of excitation mainly depends on the frequency, shape and motion of the fluid inside the tank [1]. Liquid sloshing in storage containers is due to Earthquakes is of great importance which causes unavoidable problems and failure of the systems.

Fuel sloshing in drop tank or external auxiliary fuel storage tank of aircraft may adversely affect the structural integrity of the drop tank and may even lead to its complete failure. These oscillations result from lateral and longitudinal displacements or angular motions of the aircraft [2]. It is to be noted that fuel tanks should be designed to carry fuel without any sort of leakage and should possess necessary strength for combined action of different loads and stresses. The drop tank's centre of gravity should not be altered because of motion of the fluid due adverse effects of flutter and aerodynamic characteristics of aircraft. Classical storage tanks are gasoline tankers and some of missiles rocket storage tanks of horizontal cylindrical shape, sloshing modes in these tanks generated along the axis of the cylinder [2]. In satellites and in some launch vehicles spherical storage tanks are used because of their high volume to weight ratio [3,4]. Toroidal tanks are of different types which are mainly preferred in some spacecraft and launch vehicles because they can fit around engines and some different parts, in boiling nuclear reactors these toroidal tanks act as supporting pools for condensation of steam released [5]. All the aircraft fuel tanks must last the aircraft life.

The sloshing effect in liquid propellants usually interacts with elastic vehicle structural dynamics and control system dynamics or coupled together [6]. The oscillating frequencies are always nearer to rigid body frequencies than the elastic body frequencies. The control of such dynamic behavior is accomplished either by implementation of baffle plates inside the tanks or by modifying the tank geometry itself or by introducing flexible baffle plates for damping the sloshing [7,8].

In space vehicle storage tanks (for missile under thrust), the lateral sloshing is the severe case, hence to avoid that truncated - cone type ring baffles are usually preferred [9,10]. Hence for the design of fuel tank, it is necessary to estimate sloshing loads generated on the fuel tank. Due to the complex nature of theoretical approaches used for solving such Fluid Structure Interaction (FSI) problems, the use of numerical technique becomes essential.

Geometric Modeling

The sloshing method applied initially to 3D rigid rectangular tank which is 1.008 m long, 0.196 m wide, 0.300 m height and has thickness of 0.03 m [11].

Modeling of a typical aircraft drop tank is carried out based on the optimization study. To reduce the aircraft drag especially wave drag Figure 1. Finess ratio of the drop tank should be high. From the

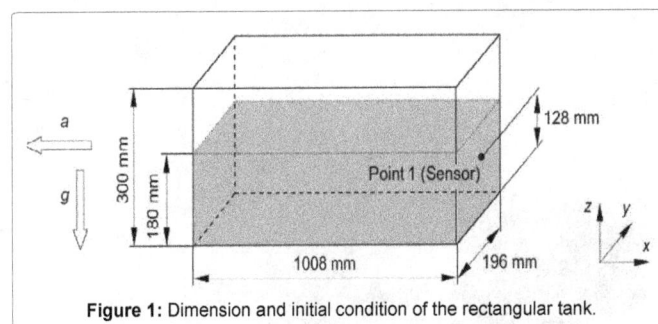

Figure 1: Dimension and initial condition of the rectangular tank.

*Corresponding author: Shreeharsha HV, Department of Aeronautical Engineering, MVJ College of Engineering, Bangalore, India
E-mail: shreeharsha.hv@gmail.com

Sears-Haack equation [12,13] shape of drop tank has been evaluated by considering the cylinder circular area and is given by the equation 1.

$$A(x) = \frac{16V}{3L\pi}[4x(1-x)]^{(3/2)}$$ (1)

Where, A is cross-sectional area of Sears Haack body, V is volume, L is length and x is the ratio of the distance from the nose to the whole body length. From the above formulation, the maximum length and diameter of the drop tank are optimized to their maximum extent using iteration process for high subsonic flow condition. The dimensions obtained are 4.395 m length and 0.64 m diameter, and dimension also reduced the wave drag formation in the total drag [14]. The dimension of drop tank is shown in Figure 2.

In order to reduce the sloshing effects inside the drop tank, vertical, circular baffle plates are placed inside in it and have three holes. Furthermore, the drop tank is modeled with two baffle plates to observe the effects of sloshing. These two vertical baffle plates are positioned at center of the drop tank separated by 750mm. as shown in the Figure 3.

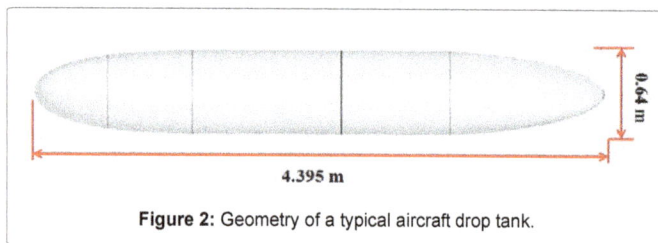

Figure 2: Geometry of a typical aircraft drop tank.

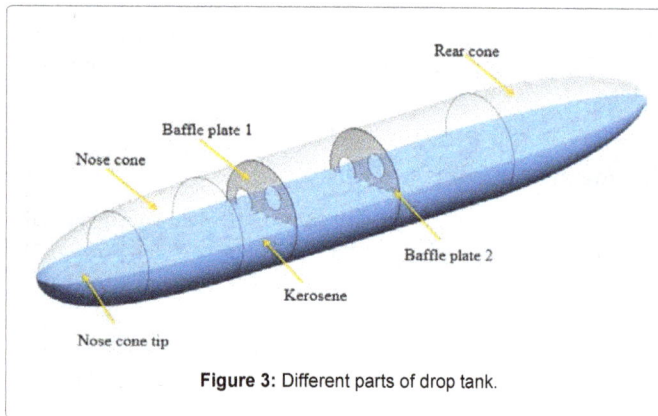

Figure 3: Different parts of drop tank.

Numerical Slosh Modeling

Computational Fluid Dynamics (CFD) analysis is a numerical method based on Finite Volume (FV) technique. It solves for conservation of mass, momentum and energy using continuity, momentum and energy equations respectively [15]. The sloshing basically deals with two immiscible fluids (water and air), it is solved using multiphase Volume of Fluid (VOF) model using ANSYS FLUENT™ to capture the interface of two fluids. The standard k-ε turbulence model [16] has been used for all the computations. The transient simulation has been carried out using explicit time stepping method with a time step size of 0.01 ms up to 0.5 s to maintain numerical stability, to avoid divergence of the solution and to control global Courant number under 250.

Rectangular tank

The fine mesh is generated on rigid rectangular tank and this tank is filled with 60% of water (up to a height of 0.180 m) as shown in Figure 1. The tank is subjected to 30 g peak acceleration at 40 ms [17] in X direction and acceleration due to gravity in vertical downward direction by invoking user defined functions. The material properties for air and water are given below.

1. Air: Density = 1.225 kg/ m³; Dynamic viscosity = 1.784×10^{-5} kg/ms.

2. Water: Density = 1000 kg/m³; Dynamic viscosity = 0.001 kg/ms.

The air and water are treated as primary and secondary phases respectively. The phase interaction between these two fluids is considered and constant surface tension between these two liquids is taken as 0.072 N/m and this tank is maintained at atmospheric pressure. The total tank is treated as rigid and is made of aluminum solid material.

A typical aircraft drop tank

From the state of art of the grid convergence, the medium grid has been chosen for all computations of different cases. The four cases namely: drop tank without baffle, drop tank with one baffle, drop tank with two baffles, and drop tank with two baffles; 180° rotation of second baffle. In all the cases the drop is filled with 50% of kerosene by volume (up to a height of 0.32 m) as shown in Figure 3. This drop tank is excited to 7 g peak acceleration in X direction and subjected to acceleration due to gravity in vertical downward direction (-ve Y direction) by source input. The material properties for air and kerosene are as follows.

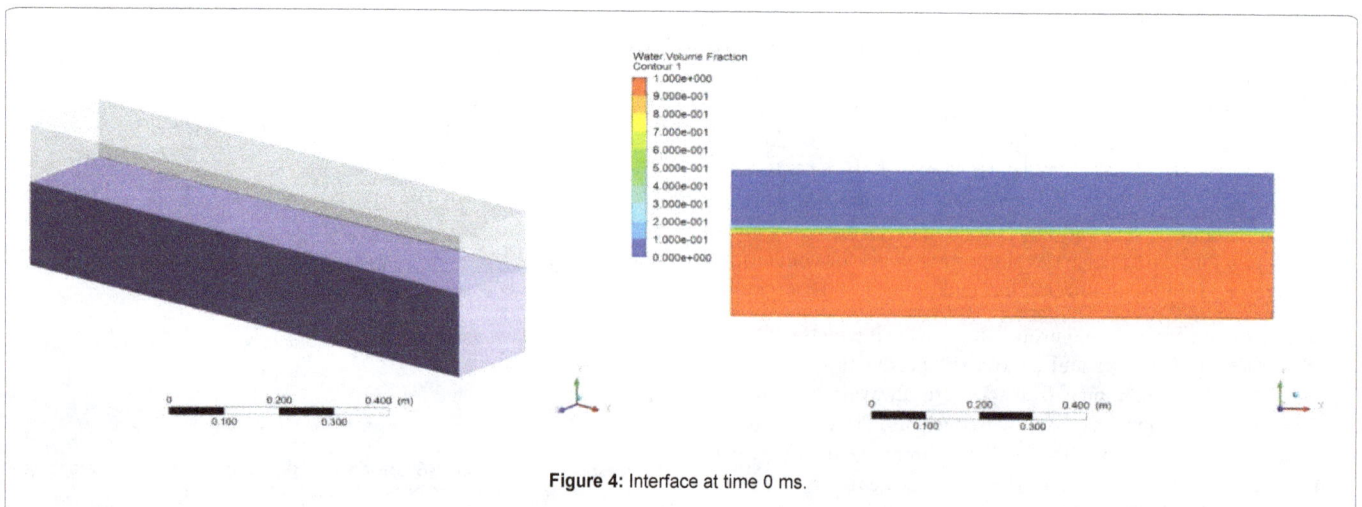

Figure 4: Interface at time 0 ms.

1.   Air: Density = 1.225 kg/ m³; Dynamic viscosity = 1.784×10^{-5} kg/ms.

2.   Kerosene; Density =780 kg/m³; Dynamic viscosity = 0.0024 kg/ms.

The air and kerosene are treated as primary and secondary phases respectively. The surface tension between air and kerosene is taken as 0.025 N/m and the tank is pressurized.

Results and Discussion

The first section of this chapter gives the detailed description about validation of rigid rectangular tank without baffle for 30 g peak horizontal acceleration for available experimental results. The second section includes application of sloshing phenomenon to a typical aircraft drop tank. This application part has been explained in four cases; without baffle, with one baffle, with two baffles and in two baffles' case 180⁰ rotation is being applied to second baffle, and the effect of sloshing on out walls and baffles of aircraft drop tank has been presented for peak 7 g acceleration input. The results obtained from the simulation are satisfactory for both validation and application.

Validation of three-dimensional rectangular tank

Kinematic profiles of water: The contours of water phase at each time instants have been captured and visualized which are shown in Figures 4 - 8. The right side of images are taken by slicing the tank the at the middle section and the lower value (zero) represents empty

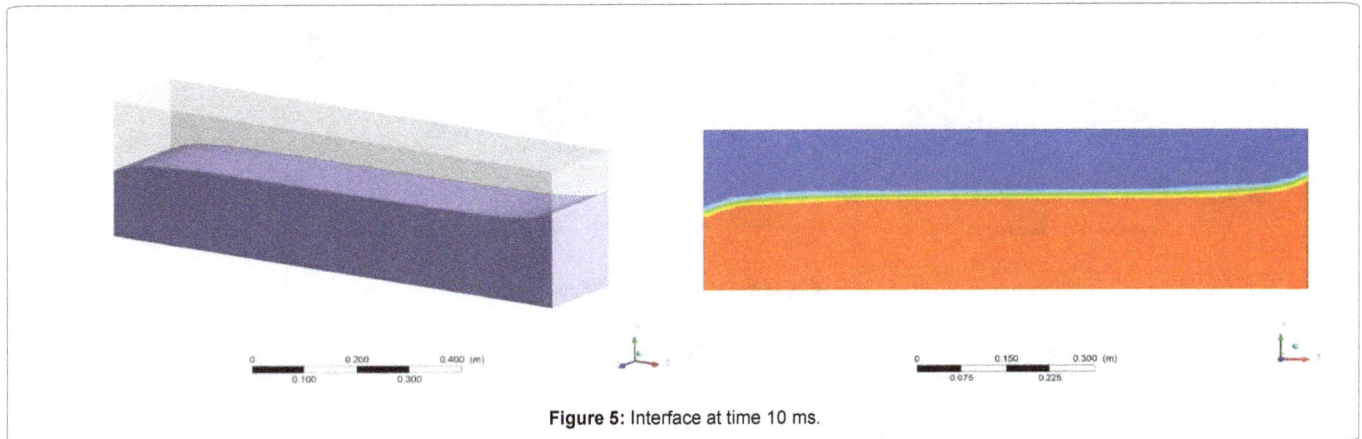

Figure 5: Interface at time 10 ms.

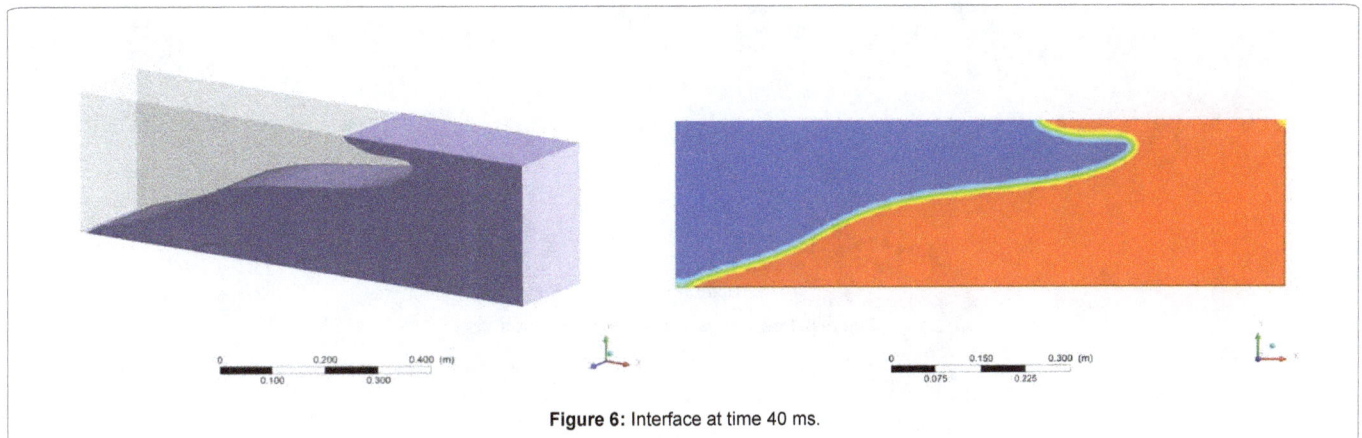

Figure 6: Interface at time 40 ms.

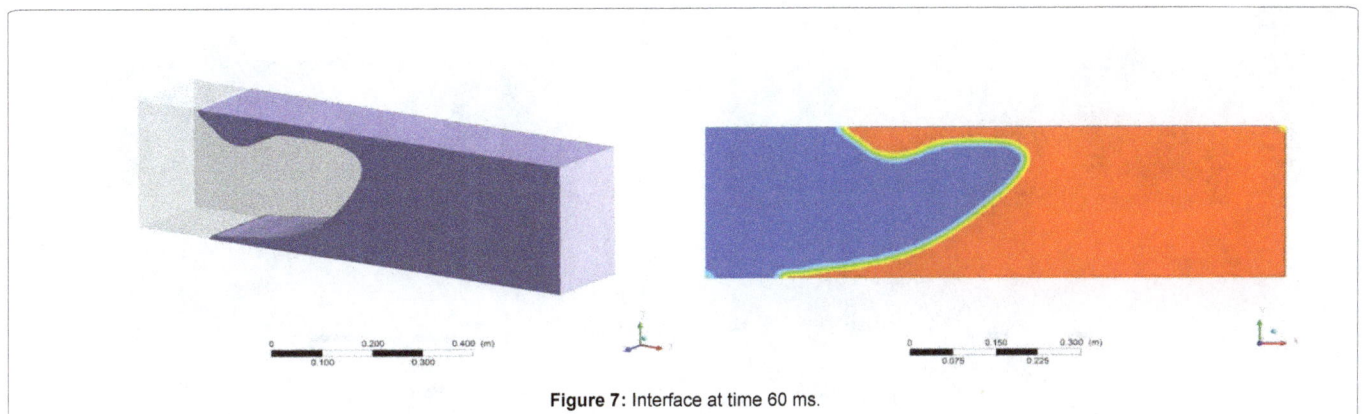

Figure 7: Interface at time 60 ms.

space, higher value(one) shows the cells which are filled with water and between zero and one shows the interface and/or free surface zone (Figures 4-8).

Pressure variation on tank wall: The hydrostatic pressure at bottom of the tank is initially high. As the solution progresses, the pressure further increases and reaches top, and suddenly falls off as shown in Figure 9. This instantaneous rise in the pressure value is because of acceleration input. The effect of this acceleration on tank wall has

been observed. This pressure is measured at a point where the pressure sensor is located as mentioned in [12]. It is evident from the obtained numerical simulation results and experimental values that both results were showing excellent agreement. This gives an insight of sloshing phenomenon when the tank is being accelerated. The 3.30243% error is estimated which is well below the maximum limits.

The hydrostatic pressure variation at each time step is shown in the Figures 10-13. These plots are taken from the front view of the tank. The

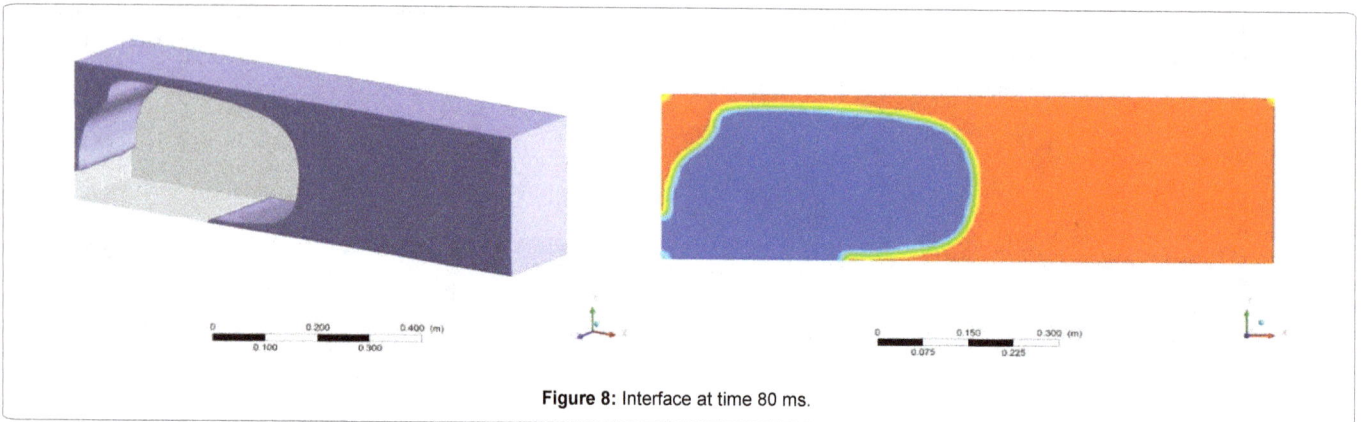

Figure 8: Interface at time 80 ms.

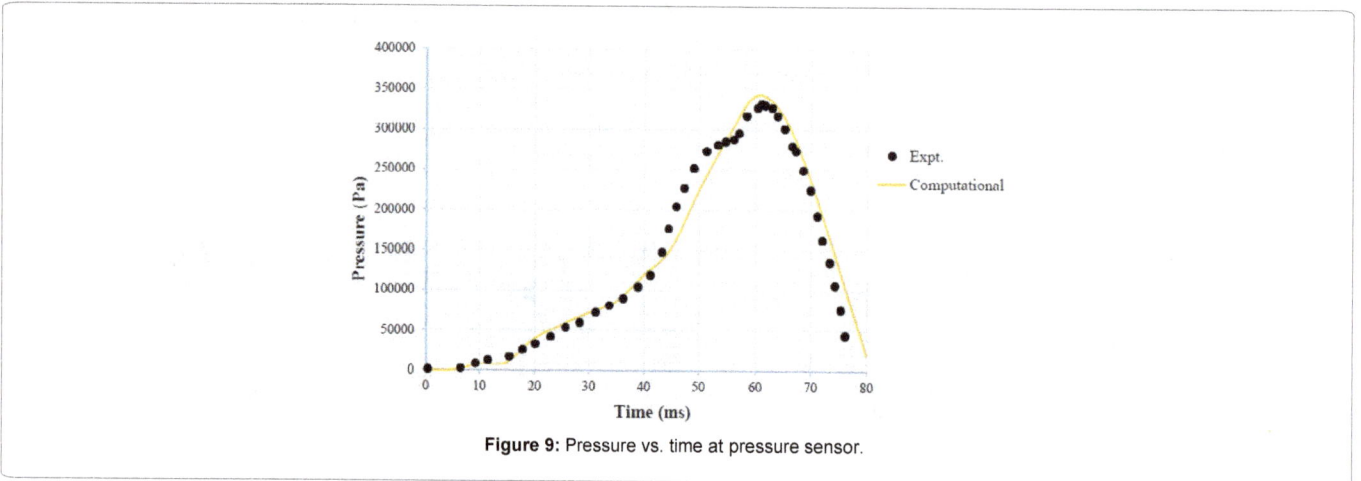

Figure 9: Pressure vs. time at pressure sensor.

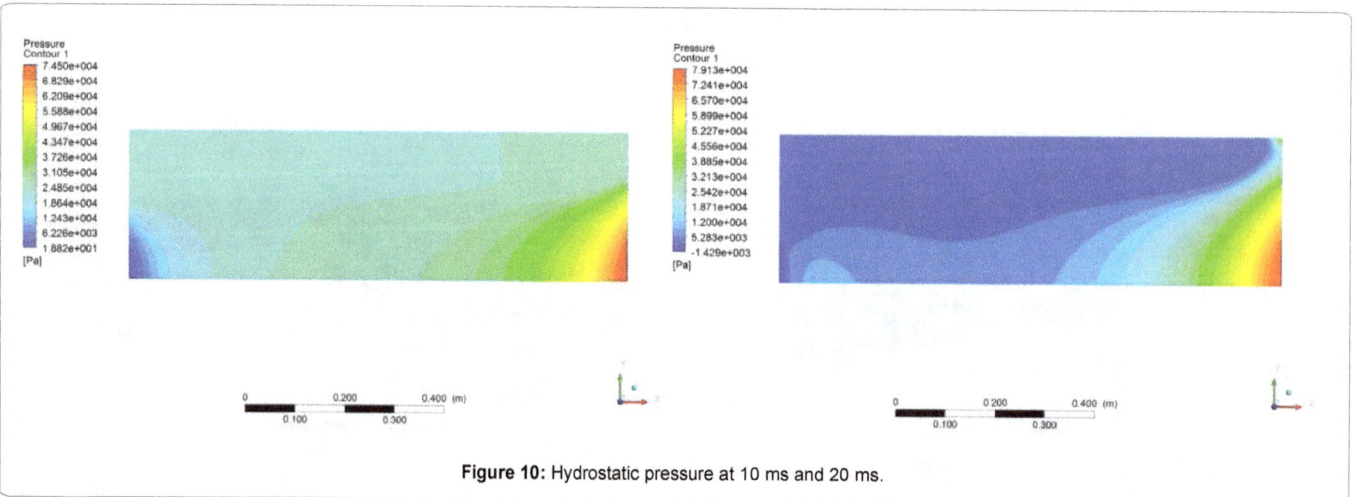

Figure 10: Hydrostatic pressure at 10 ms and 20 ms.

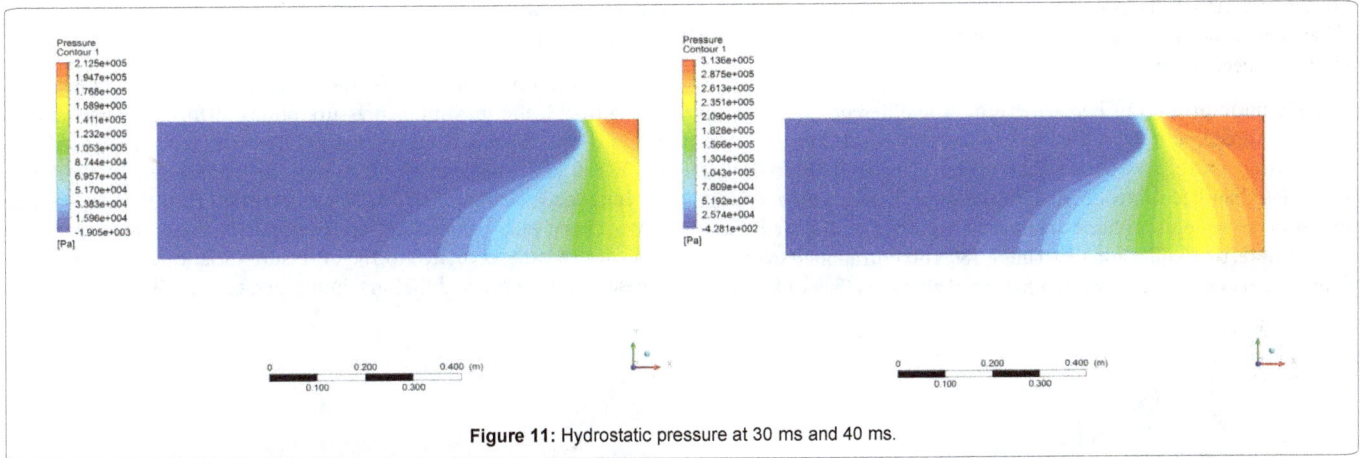

Figure 11: Hydrostatic pressure at 30 ms and 40 ms.

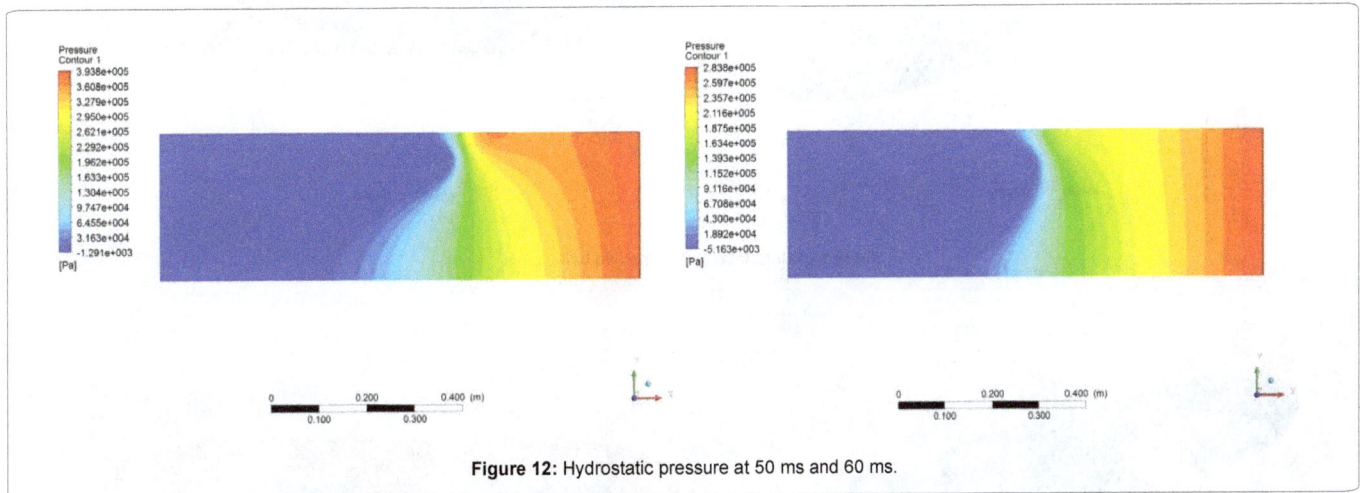

Figure 12: Hydrostatic pressure at 50 ms and 60 ms.

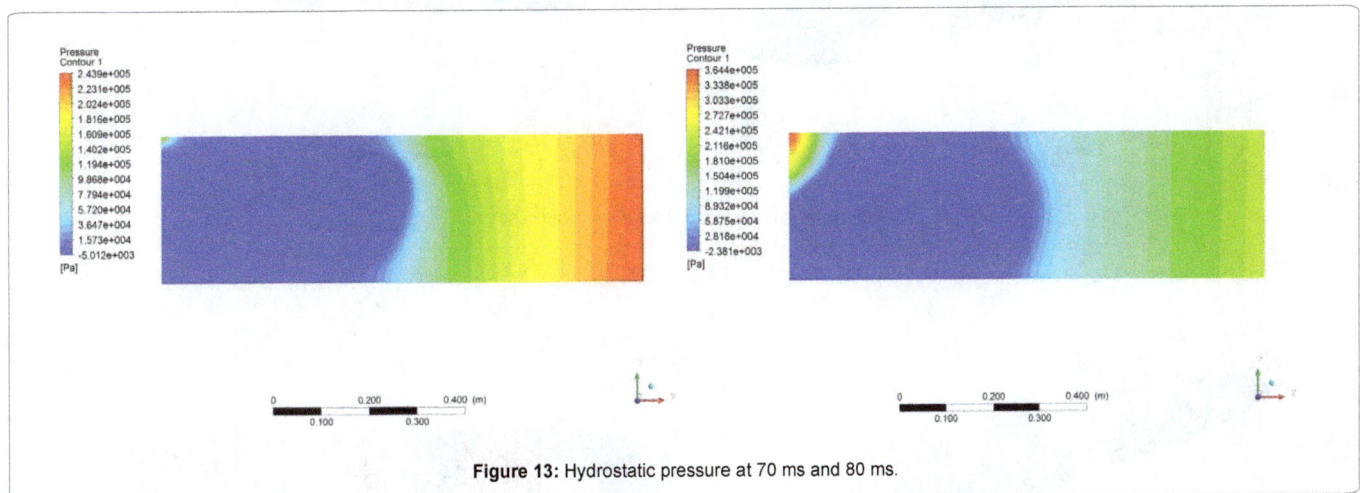

Figure 13: Hydrostatic pressure at 70 ms and 80 ms.

small increment in pressure at the beginning of the solution is observed and when the tank is accelerated at 40 ms, there is rapid rise in the pressure is visualized.

Application of sloshing phoenomenon to a typical drop tank

The sloshing phenomenon is applied to a typical aircraft drop tank. Since fighter aircraft run on white kerosene fuel which is also known as aircraft turbine fuel (ATF), the results obtained using kerosene fuel

is depicted in this analysis. The F-16 fighter aircraft can go up to 7.33 g with 80% fuel by volume [16]. The slosh effects have been observed in the form of pressure variation at aft section of drop tank because most of the fuel moves towards rear end section. When the aircraft accelerates this rear section part experiences higher pressure variations. This typical aircraft drop tank is simulated with and without baffle plates. The similar procedure employed in the validation case to simulate sloshing, is being applied for all the afore mentioned cases of drop tank. From the transient simulation the pressure variation on aft section and behavior of kerosene fluid before and after the application of 7 g acceleration input have been observed.

Kinematic profiles of kerosene: From the different cases of drop tank, it is noted that most of the fuel had moved towards the aft section thus putting more pressure in that region. To avoid this effect, a baffle plate had been introduced into the second case, which showed further decrease in the volume of kerosene impacted on the aft section. Furthermore, two baffles and two baffles; 180° rotation applied to second plate cases had also been observed for visualizing the effect of holes. At different instances of time for different cases, kinematic profiles have been shown in Figures 14-17.

Pressure variation on tank wall: From the pressure vs. time graph in Figure 18, the maximum pressure observed for 'no baffle' case is 158.752 kPa, for 'one baffle' case is 142.851 kPa, and 'two baffles' case is 139.830 kPa. This shows pressure on the rear section is reduced or sloshing loads acting on the walls are decreased with the use of baffle plates. Since each baffle plates are comprise of three holes, in the 4th case, the 2nd baffle is rotated to 180°. Pressure variation in rear section is measured. It is observed that maximum pressure is 139.843 kPa which

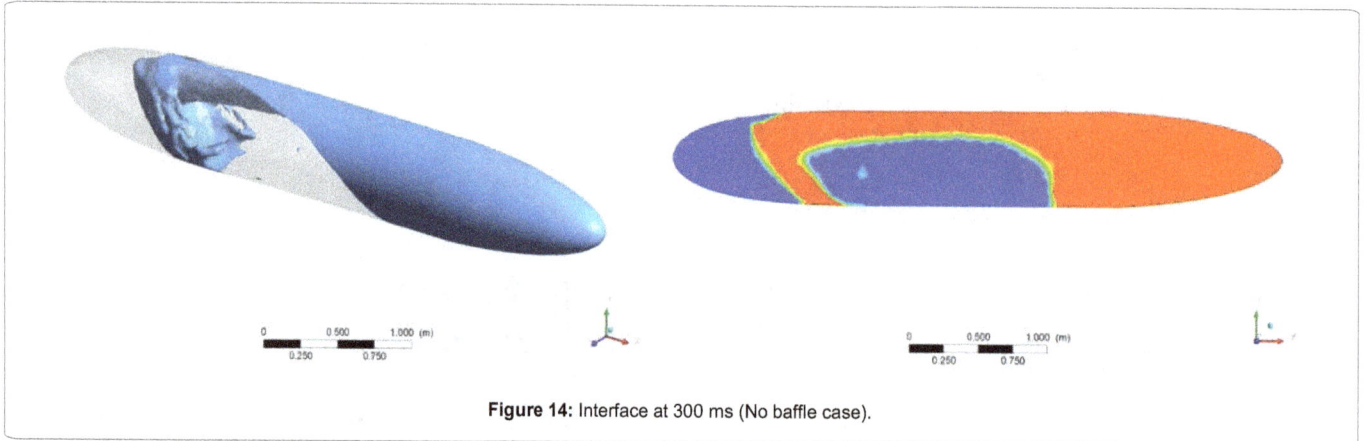

Figure 14: Interface at 300 ms (No baffle case).

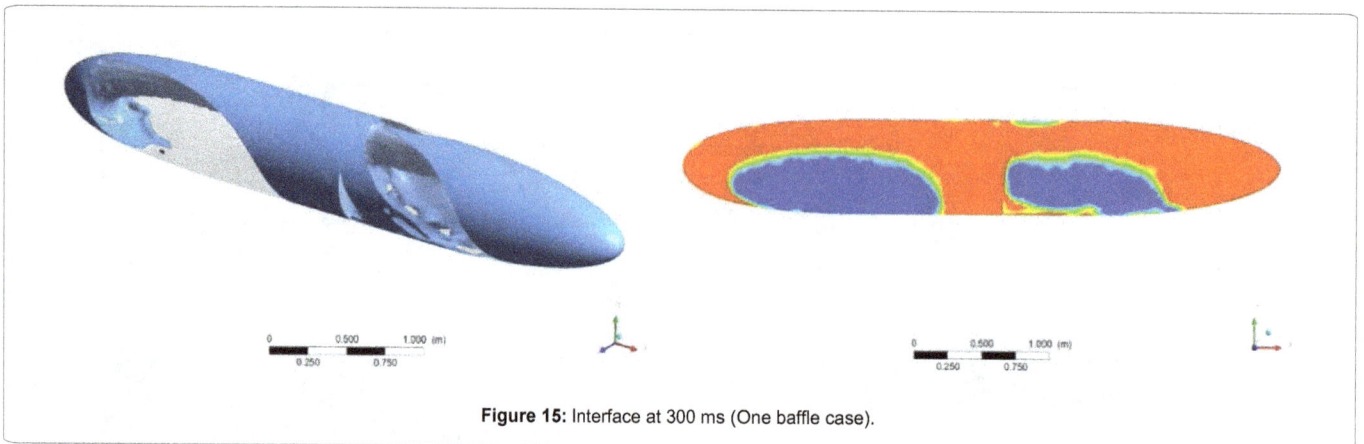

Figure 15: Interface at 300 ms (One baffle case).

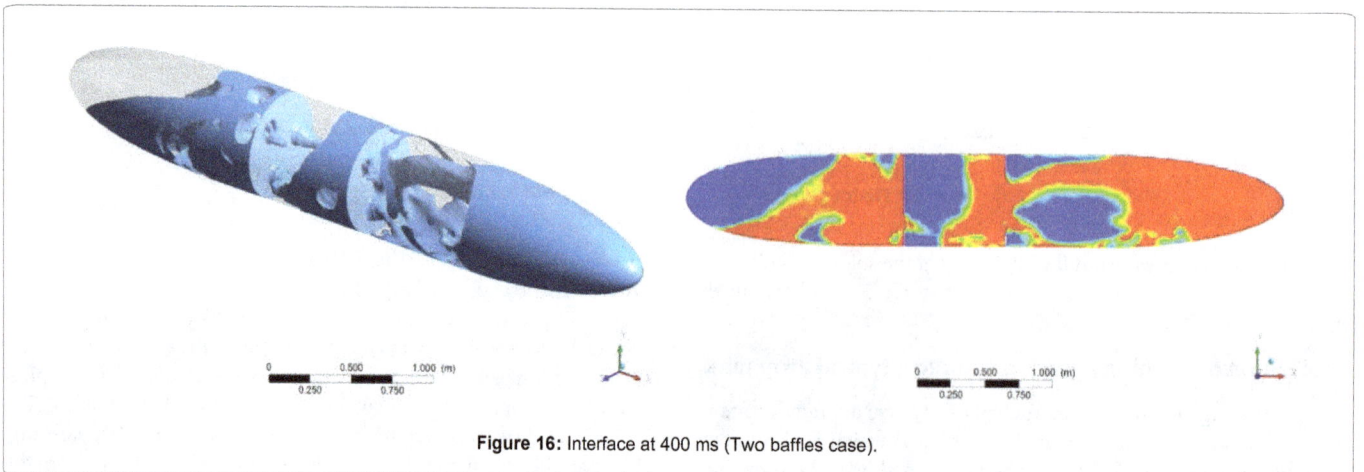

Figure 16: Interface at 400 ms (Two baffles case).

is higher than two baffle case pressure. This is because on 2nd baffle of 'two baffles' case there is only one hole immersed in the kerosene, when 180° rotation is applied to the 2nd baffle plate, top two holes will get immersed in kerosene which allow more amount of kerosene to pass through when the tank is accelerated. Thus, the pressure on the rear cone section of drop tank increases. This Figure 18 also gives a proposal on design and development and placement of baffle plates with holes.

Apart from the rear section of drop tank, the baffle plates which are immersed in the fluid also experience the high pressure due to this accelerating effect of fluid. From the design point of view, it is necessary to establish proper connection between baffle plates and outer wall of the tank. The pressure contour plots for different cases are shown in Figures 19-22.

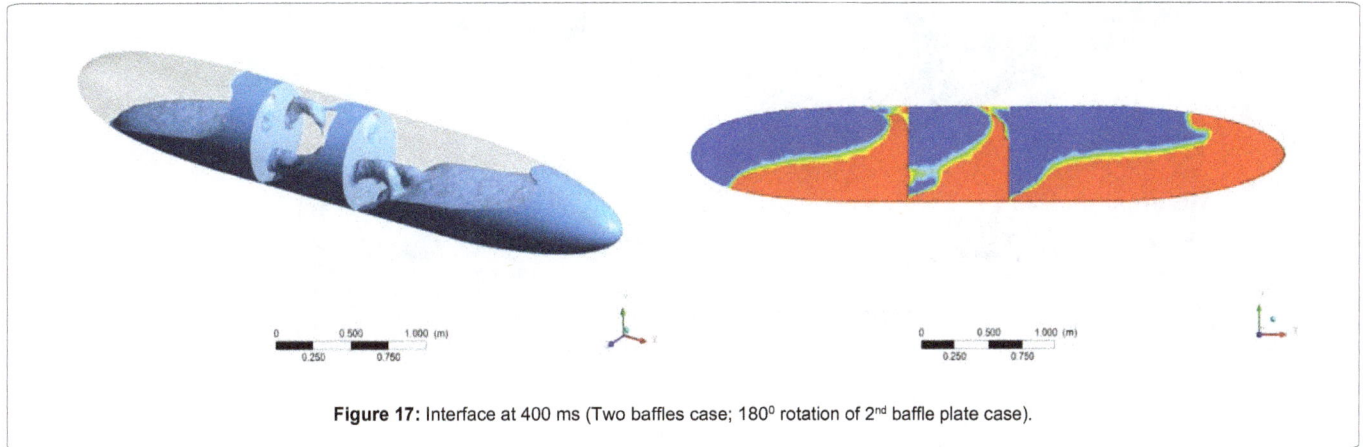

Figure 17: Interface at 400 ms (Two baffles case; 180° rotation of 2nd baffle plate case).

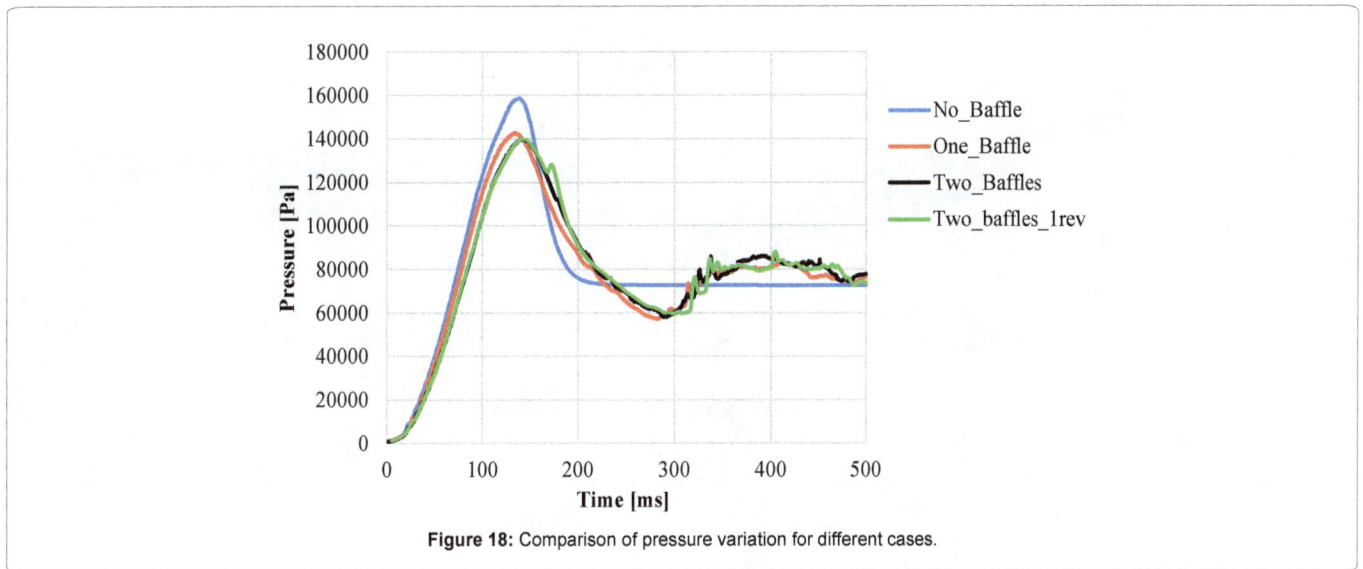

Figure 18: Comparison of pressure variation for different cases.

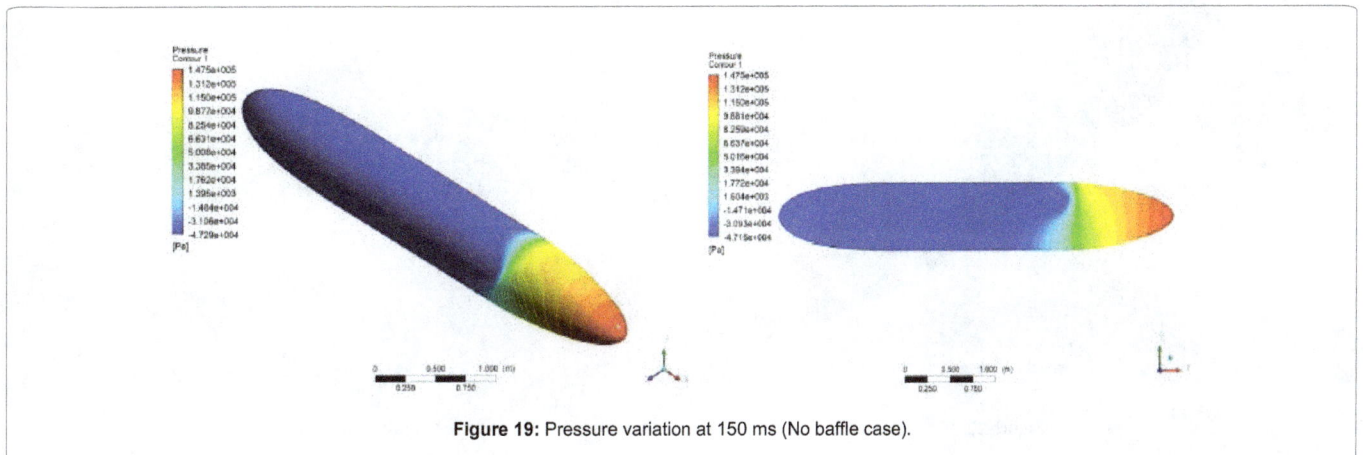

Figure 19: Pressure variation at 150 ms (No baffle case).

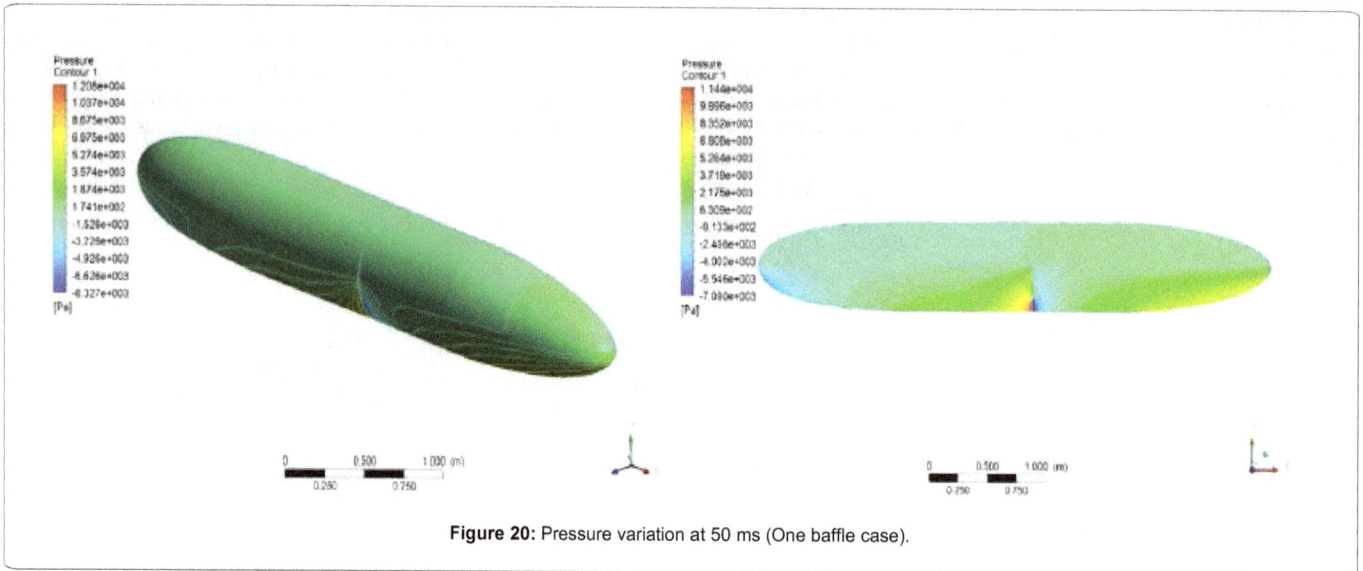

Figure 20: Pressure variation at 50 ms (One baffle case).

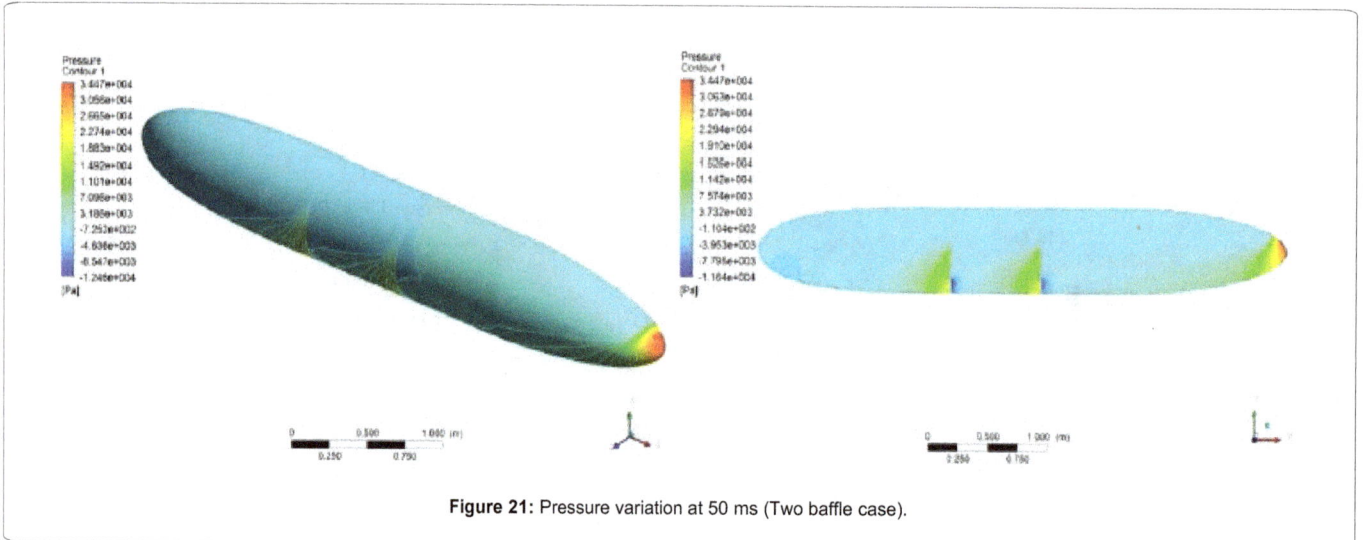

Figure 21: Pressure variation at 50 ms (Two baffle case).

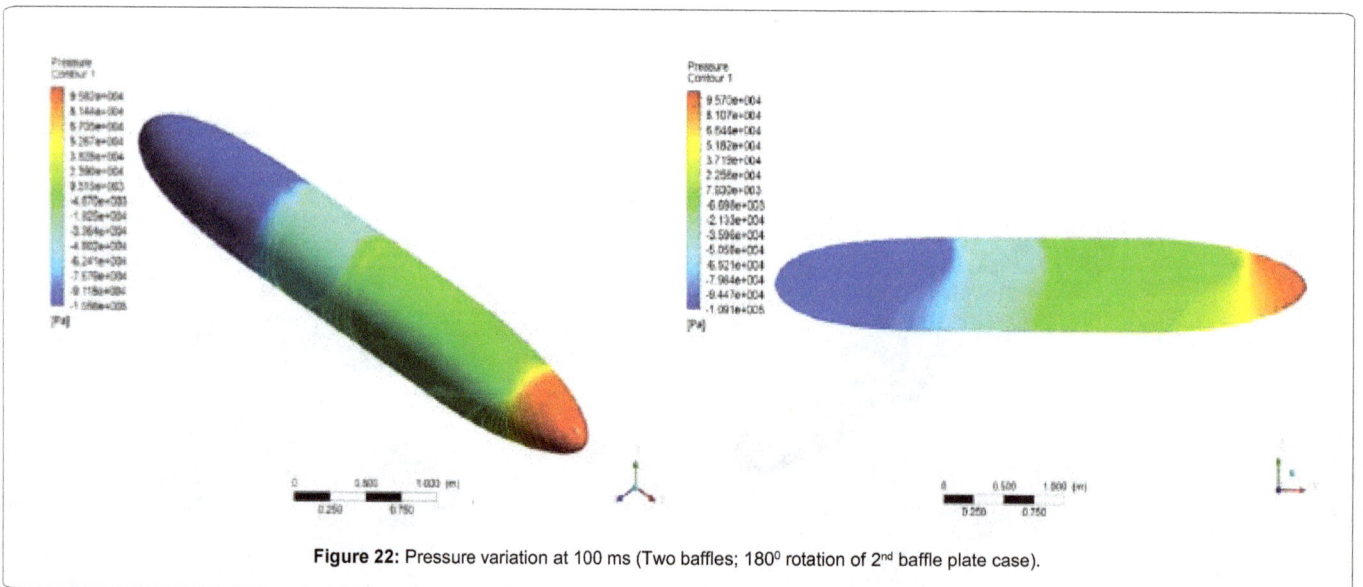

Figure 22: Pressure variation at 100 ms (Two baffles; 180⁰ rotation of 2ⁿᵈ baffle plate case).

Conclusion

In this research, validation of water sloshing in 3D rigid rectangular tank and an application of sloshing using kerosene to a typical aircraft drop tank or external fuel storage tank has been carried out with the help of explicit transient simulation computational fluid dynamics approach. The rectangular tank is filled with 60% of water and is subjected to 30 g peak acceleration at 40ms. The pressure variation results obtained from simulation are compared with experimental results. Numerical results are showing good agreement experimental data. It is noticed that the maximum pressure is achieved after the application of acceleration input. From the kinematic profiles, the free surfaces of water and kerosene have been clearly observed and from the pressure contours, increments in the static pressure at different time intervals have been visualized.

An optimized design of a typical aircraft drop tank has been modeled by taking drag effects into consideration. To this drop tank, sloshing phenomenon has applied and investigated for with and without baffles. It is evident from the explicit transient simulation method conducted at different time interval that pressure at the rear cone section increases significantly without baffle plates and reduces drastically with the use of baffle plates which acts as damper. The pressure variation for 180^0 rotation of one baffle plate (fourth case) actually increased the pressure on rear section side, this is because of large amount of fluid is moving rearwards due to the presence of two holes which are immersed under kerosene. This variation has been visualized from volume fractions. Drop tank with one baffle reduces the peak pressure by 11.25% as compared with no baffle case, drop tank with two baffles reduces the peak pressure by 2% as compared with one baffle case, and two baffle of one baffle reversed condition the peak pressure is reduced by 2.1%. Hence drop tank with one baffle with three holes can be implemented for the design of drop tank.

References:

1. Paniglahy PK, Saha UK, Maity D (2009) Experimental studies on sloshing behavior due to horizontal movement of liquids in baffled tanks. Ocean Engineering 36: 213-222.

2. Abramson HN (1966) Dynamic behavior of liquids in moving containers. NASA SP- 106.

3. Budiansky B (1960) Sloshing of liquids in circular canals and spherical tanks. J Aerospace Sciences 27: 161-173.

4. Riley JD, Trembath NW (1961) Sloshing of liquids in spherical tanks. J Aerospace Sciences 28: 245-246.

5. Mc Carty JL, Leonard HW, Walton WC (1960) Experimental investigation of the natural frequencies of liquids in toroidal tanks. NASA TN D-531.

6. Bauer HF (1964) Fluid oscillations in the containers of a space vehicle and their influence on stability. NASA TR R-187.

7. Bugg F (1973) Evaluation of flexible ring baffles for damping liquid oscillations. NASA TM X-64730.

8. Schwind R, Scotti R, Skogh J (1966) Analysis of flexible baffles for damping tank sloshing. AIAA Paper 66-97.

9. Bauer HF (1962) The damping factor provided by flat annular ring baffles for free surface oscillations.

10. Vesenjak M, Mullerschon H, Hummel A, Ren Z (2004) Simulation of fuel sloshing comparative study. LS-DYNA Anwenderforum, Bamberg.

11. Nihal R, Chaithanya AV, Kuldeep SR, Pathanjali RJ, Muralidhar M (2015) CFD studies of a supersonic drop tank for a generic fighter aircraft. 17th Annual CFD symposium, Bengaluru.

12. Waddington MJ (2015) Development of an interactive wave drags capability for the open VSP parametric geometry tool. California Polytechnic State University San Luis Obispo.

13. Raouf AI (2005) Liquid sloshing dynamics, theory and applications. Cambridge University Press, UK.

14. Singal V, Bajaj J, Awalgaonkar N, Tibdewal S (2014) CFD analysis of a kerosene fuel tank to reduce liquid sloshing. Procedia Engineering. 69: 1365-1371.

15. Hou L, Li F, Wu C (2012) A numerical study of liquid sloshing in a two-dimensional tank under external excitations. J Marine Sci Appl 11: 305-310.

16. Petrescu FI, Petrescu RV (2012) The Aviation History, USA.

17. Khezzar AC, Seibi A, Goharzadeh D (2009) Water sloshing in rectangular tanks. An experimental investigation and numerical simulation. Int J Engineering (IJE) 3: 174-184.

19

Failure Detection in an Annular Combustion Chamber with Experimental and Numerical Methods

Hennecke C, Von der Haar H* and Dinkelacker F

Institute of Technical Combustion (ITV), Leibniz University Hanover, Welfengarten 1A, 30167 Hanover, Germany

Abstract

The inspection of aircraft engines is a complex and time-consuming process, usually requiring the disassembly of the engine or extensive baroscopic examinations. Thus, a method is to be developed in order to evaluate the state of the jet engine prior disassembling by analysing the state and structure of the exhaust jet. This could be done for instance with an analysis of the density, temperature, velocity or concentration distribution in a cross section through the exhaust jet. Assumptions are that failures inside the engine influence the exhaust jet in a measurable way and by means of numerical simulations it is possible to evaluate the flow and combustion process beginning in the engines' interior through the exhaust gas jet. A generic study on a pilot scale annular combustion chamber is the basis to show the feasibility of this approach. The combustion chamber consists of eight premixed swirl burners. One of the burners has the option to be operated independently from the others on defined varied operation points. This simulates a failure state of a burner which allows an investigation of the correlation between defined failure states and resulting pattern in the exhaust jet. Detailed techniques are applied to evaluate the approach and the detection limits. Firstly, extensive numerical simulations of possible failure states were conducted to compare with measurements. Secondly, the particle image velocimetry (PIV) technique was used to measure the velocity field in the measurement plane downstream of the combustion chamber. Local reductions of the thermal power lower the acceleration of the exhaust gas and influence the velocity field. Thirdly, species concentrations were measured with a movable probe, evaluated by Fourier-transform infrared (FTIR) spectroscopy. Species concentration information can indicate combustor failures like locally mistuned air-fuel-ratio, serious defects on the swirl vanes, or oil leakages.

Keywords: Combustion chamber defect detection; Aircraft engines; Turbulent combustion; Particle Image Velocimetry (PIV); Fourier-Transform Infrared Spectroscopy (FTIR); Numerical simulation

Introduction and Methodical Approach

Defects in combustion chambers of aircraft engines might reduce the performance and represent a security risk. A methodology is sought to enable a diagnosis during operation of the aircraft engine on the wing. This would allow decisions on the timing and extent of regeneration measures at an early stage. Thus, operation cycles may be lengthened or shortened depending on demand. Furthermore, the necessary regeneration process could be planned and prepared in detail before the disassembly of the engine, leading to a faster and more efficient repair procedure.

The approach is based on the assumption that defects in the combustion chamber have a significant influence on the internal flow, which induce detectable pattern in the exhaust jet (Figure 1). The distribution of velocity, density, and temperature or gas composition provides information about the status of the combustion and flow processes inside the combustion chamber and may be measured with spatial resolution by appropriate measurement techniques. Due to the high mechanical and thermal load behind an aircraft engine, optical measurement techniques are preferable for real application. A further assumption is that numerical simulation methods can describe the linking process between the failure source inside the combustion chamber and the measured texture of the exhaust jet. In that case, damages at the real engine can be detected with appropriate exhaust jet diagnostics without disassembling the engine. The feasibility of this approach has already been shown in first steps [1–4].

Since measurements on real aircraft engines are complex, experimental studies are carried out on an atmospheric annular combustion chamber in the laboratory. This allows for a detailed investigation whether combustor errors can be localized by analysing

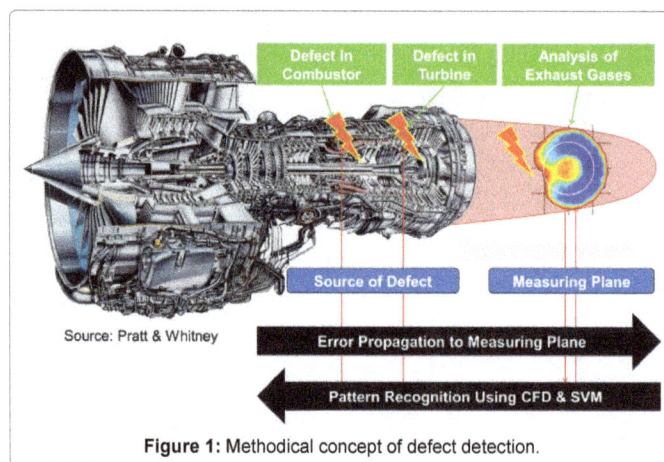

Figure 1: Methodical concept of defect detection.

patterns in the exhaust jet, as this combustion chamber is easily accessible for measurement techniques, and well defined operating points can be compared with numerical simulations.

Following this methodical concept, it is the aim of this work

---

*Corresponding author: Von der Haar H, Institute of Technical Combustion (ITV), Leibniz University Hanover, Welfengarten 1A, 30167 Hanover, Germany
E-mail: vonderhaar@itv.uni-hannover.de

to prove, if influences of the faulty burner are detectable in a cross section in the exhaust gas jet, both with numerical simulation and with experimental methods. Particularly it is to examine which deviations can be measured and which are hidden from the mixing process between the failure location and the measurement plane, dependably. This gives an indication of the minimal observable failure which can be detected sufficiently. Moreover, it has to be studied if a certain measured quantity is appropriate to distinguish different types of failures or if more than one quantity is required to prevent confusion.

Model Combustion Chamber and Investigated Operation Points

For the verification of the approach, a simplified annular combustion chamber has been developed, as shown in Figure 2. The combustion chamber consists of an array of eight premixed-operated methane-air swirl-burners, which are arranged on a circle with a diameter of 210 mm. Every burner (right picture) is composed out of a tube with an inner diameter of 28 mm in which a swirl generator consisting of five blades and a turbulence grid are mounted. The combustion chamber (left picture) is outwardly and inwardly confined by cylindrical quartz glass rings for optical access. The outer diameter is 310 mm and the inner diameter is 105 mm, whereas the height is 200 mm. The inner glass ring is cooled by an air flow from the inside.

For the investigations of defects in a combustion chamber, the test-rig provides the option to operate the chamber with defined failures. One of the eight burners can be varied in its operating condition in the chamber independently from the other ones.

The current work focuses on two types of errors in aircraft engines. In study A, the thermal power of the variable burner is reduced while

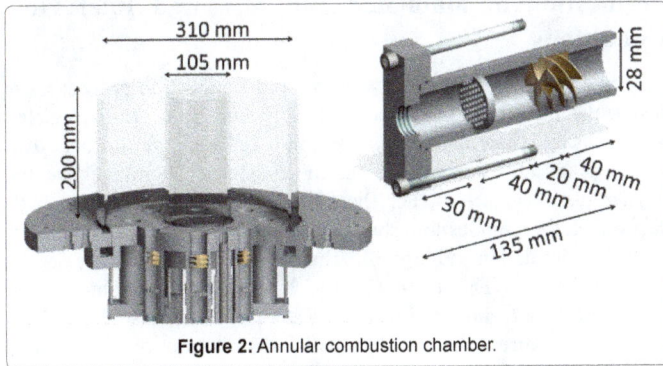

Figure 3: Schematic top view of the annular combustion chamber: Positions of the swirl burners (colored circles, manipulated burner in red) and the locations of the measurements of the species concentrations (crossed).

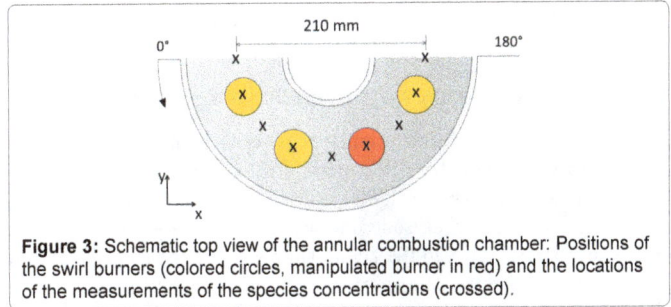

Figure 4: Setup of the computational fluid dynamics simulations of the combustion chamber.

holding the other burners constant on the thermal power of 15 kW. Here, all burners are operated with stoichiometric air-fuel-ratio λ=1.0. The power of one burner is reduced stepwise down to 10 kW (Table 1). With this study, it is investigated which of these rather small variations are still detectable.

In study B, one burner is varied in its air-fuel-ratio, either being reduced to λ_{Single}=0.8 (fuel-rich case) or being increased to λ_{Single}=1.2 (fuel-lean case) while the other seven burners are operated with λ=1.0. Here the thermal power is hold constant at 15 kW for each burner. Table 2 summarizes the operation conditions.

Figure 3 shows the top view of the combustion chamber. The position of the manipulated burner is marked in red. The following evaluations are done for this view: The crosses in the graphic illustrate the positions of the FTIR measurements, the angles are used in the line plots below in this paper.

Numerical Setup of the Computational Fluid Dynamics - Simulations

The connection between the sources of a failure inside the combustor and the flow field further downstream is numerically calculated with the method of the three-dimensional computational fluid dynamics (CFD) with additional models for the reaction processes. The Reynolds-averaged Navier-Stokes simulation (RANS) approach is used. For the calculation, the annular combustion chamber is resolved with a numerical mesh of approximately 18 million tetrahedral cells. Flow and turbulence are determined using the k-ω SST model [5]. The reaction process is modelled with a partially premixed combustion model, consisting of the extended coherent flame model for the premixed regime [6] combined with the presumed probability density function (pre-PDF) as a mixture fraction approach [7–9].

The geometry of the annular combustion chamber is represented by a 180° segment. Periodic boundary conditions ensure accurate transitions on the cut planes, to save computational cost (Figure 4). The third burner can be manipulated individually to represent the failing burner.

For the reference case, each burner is fuelled by a premixed inlet mass flow of 5,471 g/s at a temperature of 300 K, equivalent to a

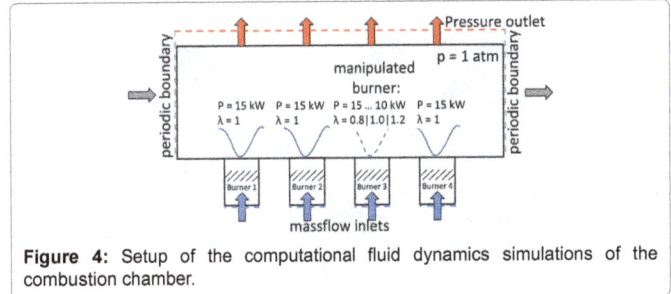

Figure 2: Annular combustion chamber.

Cases	P_{Array}	λ_{Array}	P_{Single}	λ_{Single}
Ref_15 kW	7 × 15 kW	1.0	15 kW	1.0
Fail_14 kW	7 × 15 kW	1.0	14 kW	1.0
Fail_13 kW	7 × 15 kW	1.0	13 kW	1.0
Fail_12 kW	7 × 15 kW	1.0	12 kW	1.0
Fail_11 kW	7 × 15 kW	1.0	11 kW	1.0
Fail_10 kW	7 × 15 kW	1.0	10 kW	1.0

Table 1: Investigated operating points for study A: Variable burner operates with reduced load.

Case	P_{Array}	λ_{Array}	P_{Single}	λ_{Single}
Ref_15 kW	7 × 15 kW	1.0	15 kW	1.0
Fail_λ 0.8	7 × 15 kW	1.0	15 kW	0.8
Fail_λ 1.2	7 × 15 kW	1.0	15 kW	1.2

Table 2: Investigated operating points for study B: Variable burner operates with changed air-fuel-ratio.

thermal power of 15 kW of methane-air-mixture at stoichiometry and atmospheric conditions. The pressure outlet is set to 101330 hPa.

Experimental Setup of the Particle Image Velocimetry

The flow on the model combustors' exit is analysed using the particle image velocimetry (PIV) method with a stereoscopic approach [7,9]. One aspect is to prove, if the manipulated burner has a detectable impact on the exhaust jet. A second aspect is to verify the hypothesis that a correlation between the failure source inside the combustor and the measurable failure pattern on the exit plane is given by applying the computational fluid dynamics (CFD) technique. The comparison

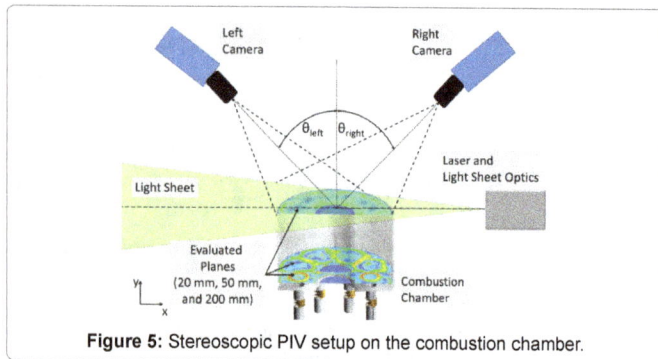

Figure 5: Stereoscopic PIV setup on the combustion chamber.

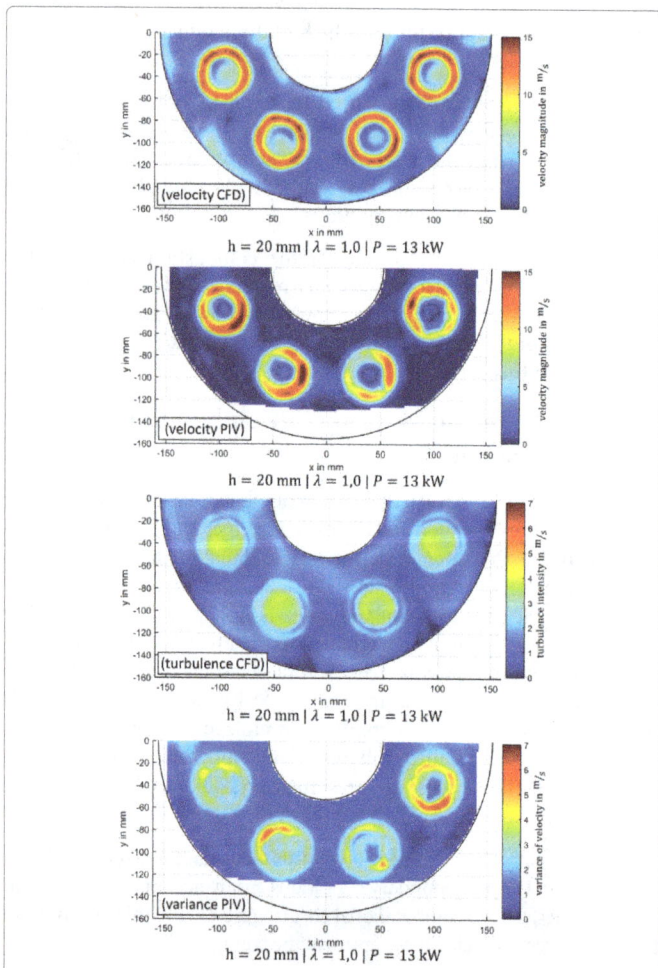

Figure 6: Inflow conditions near the burners' exit (20 mm above). Mean velocity and turbulence intensity, simulated and measured.

between measurement and simulation is of importance to see, if the method of CFD simulation would be applicable also in complex aircraft engine geometries, to locate failures inside the engine, that are identified through the measured pattern in the exhaust gas jet.

PIV is a minimally invasive flow measurement technique that allows a two- or three-dimensional flow field to be acquired by tracking the movement of small particles that are added to the fluid. Within a light-sheet plane illuminated by a pulsed laser the particles are imaged in consecutive pictures with defined time-interval. With the cross-correlation method the velocity field can be determined [10,11].

Due to the highly three-dimensional flow field of the swirl burners flow, a stereoscopic PIV setup was applied (Figure 5).

The measurement plane is aligned horizontally. As light source the Litron Nano L 135-15 PIV double pulse laser with a wavelength of 532 nm and a pulse energy of maximum 135 mJ per cavity was used. A light sheet is formed with a suitable light sheet optic consisting of three lenses. The distance between the measurement plane and the optics is 1000 mm. As tracer, inert TiO_2 particles are mixed to the flow, following the flow through the flame. For the PIV measurements two 14-bit CCD cameras with a resolution of 1600×1200 pixels (pixel size: $7.4\ \mu m \times 7.4\ \mu m$, sensor size: $11.8 \times 8.9\ mm^2$), equipped with 50 mm lenses with a fixed focal length of f / 1.4 are used with laser line filters for 532 nm ± 3 nm. The lens and the camera are connected via manual tilt adapters according to the Scheimpflug criterion [12,13], whereby the camera axes are inclined by the angle θ=50° against the normal of the measuring plane. The cameras are located approximately 830 mm above the measuring plane. The time interval between the pictures was 60 μs for the 200-mm evaluation plane and 20 μs for the 20-mm plane cutting the flame zone.

Verification of Simulated and Measured Inlet Flow Conditions

The simulation domain of the flow and combustion process starts inside the burner 40 mm prior the turbulence grid. The calculated and measured flow conditions are compared near the burners' exit which practically was possible at the height of 20 mm above the burner plane (Figure 5). For the following results, generally a half-section of the combustion chamber is shown in the measurements and the simulation (Figure 6). The single variable burner is the lower right one (Figure 3). For the comparison, all eight burners are operated in similar way, here exemplary the case Fail_13kW is shown. Figure 6 shows the spatial distribution of the velocity magnitude in the 20-mm plane from simulation (first image) and from measurement (second image). The structure of the flow field is sufficiently similar, although the local velocity distributions in the measured cases vary to some extend close to the burner exit. Also, the calculated (third image) and measured turbulence intensity (fourth image) is sufficiently comparable, showing the turbulence intensity fluctuation to be in the range of 4 to 5 m/s.

Failure Detection for Study A by Evaluation of the Velocity Field

The spatial distribution of the velocity magnitude is shown in Figure 7 for the simulation and in Figure 8 for the PIV measurement at the exit plane of the combustion chamber, located 200 mm downstream of the burner plane.

In study A the faulty burner is operated with varied thermal power. In the top case the variable burner is operating identically to the other burners (reference case). The calculated three-dimensional flow field shows a periodic flow pattern with velocity magnitudes in the range of 3 to 4 m/s in the middle section of the ring, with 5 to 6 m/s near the inner ring and with periodic flow pattern near the outer ring.

The measured flow field shows a rather similar velocity distribution concerning the flow pattern. The velocity magnitudes vary more than in the calculation, with a range of 2 to 3 m/s in the middle section and 5 to 6 m/s near the inner and the outer ring.

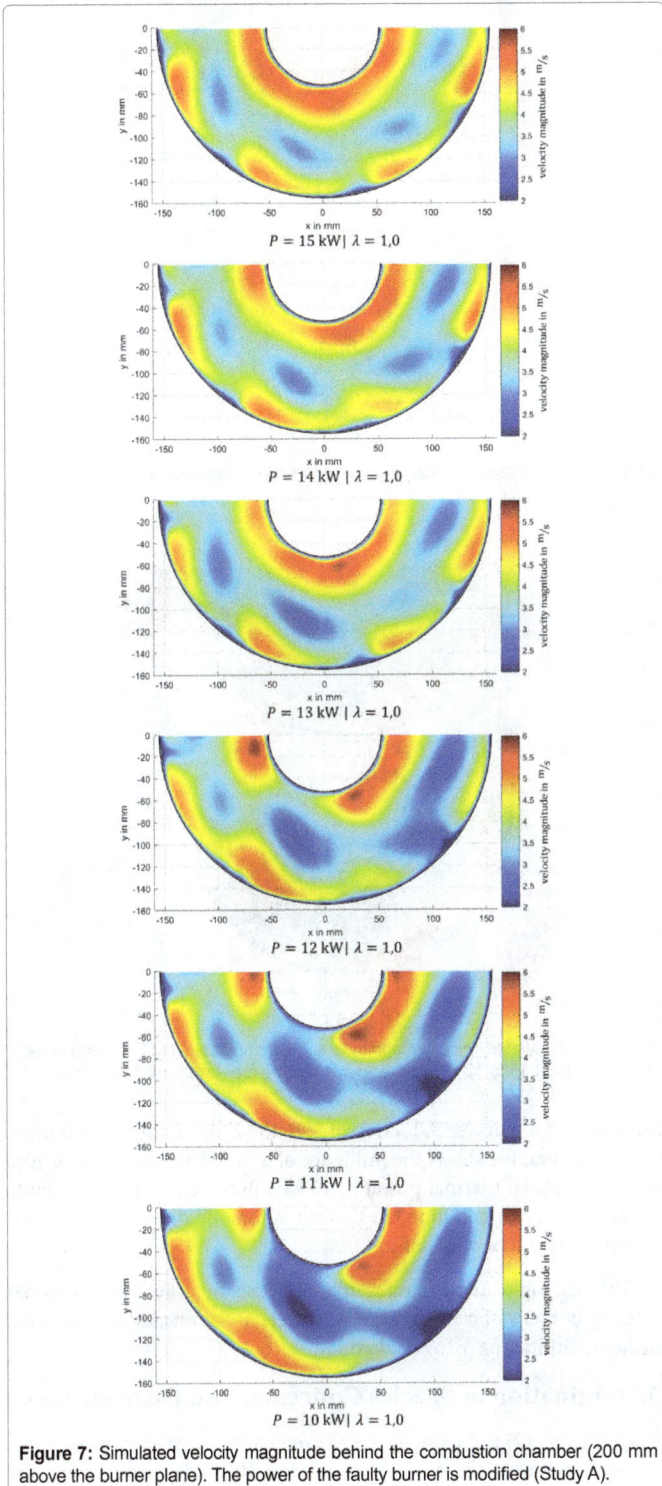

Figure 7: Simulated velocity magnitude behind the combustion chamber (200 mm above the burner plane). The power of the faulty burner is modified (Study A).

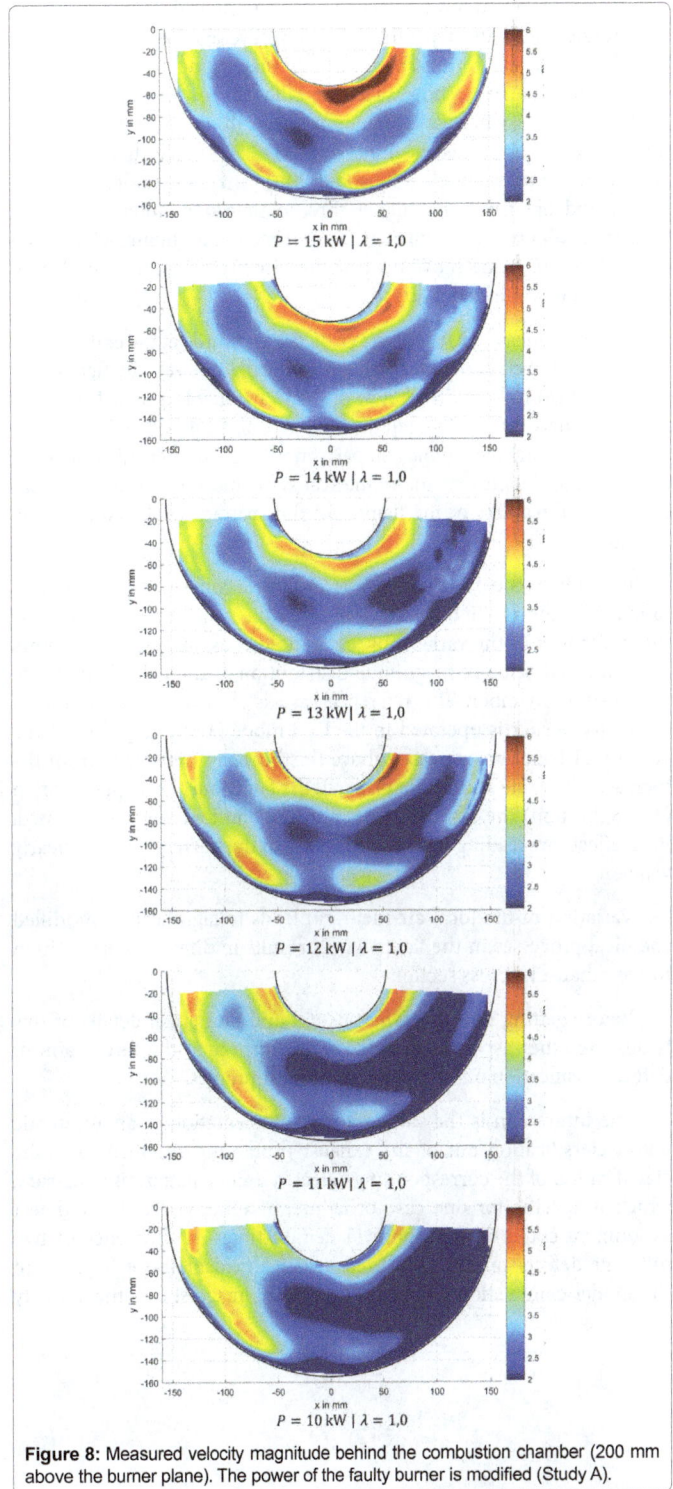

Figure 8: Measured velocity magnitude behind the combustion chamber (200 mm above the burner plane). The power of the faulty burner is modified (Study A).

A significant deviation of the velocity field is detectable, if the faulty burner is operating with 13 kW thermal power instead of the 15-kW reference power (third row). For the 12 kW and lower cases the deviation is clearly visible in the outer part of the velocity field - both for the simulation and the experimental results. The influence of the deviating burner extends the range of the initial 45° segment on the burner plane and the whole flow field is affected increasingly with growing deviation of the faulty burner.

In Figure 9 circumferential profiles (being determined along the middle ring with a width of 7.5 mm, averaged over 1000 single measurements) of the measured velocity magnitude at the exit of the combustion chamber show that the influence of the faulty burner is clearly visible when its power is reduced. It should be noted that the variable burner is positioned at 112.5°, while the observed velocity minimum is moved by about 20° to the 135° position. This is understandable from the highly three-dimensional interacting flow field of the different swirl burners (swirl direction is counter-clockwise) which leads to some movement of the velocity minima behind each burner in circumferential direction.

From the measurement of the velocity magnitude at the exit plane of the combustion chamber it can be concluded that the velocity field is still dominated from the swirling flow of the individual burners. Their flow field is strongly interacting, but the influence of the modified burner can be detected clearly. From the comparison with the numerical simulation it can be concluded that the numerical simulation is able to represent the essential features of the flame and flow pattern in the combustion chamber.

In study B the faulty burner is modified concerning its air-fuel-ratio while the other burners operate in the reference state and the thermal power of the variable burner is hold constant. Figure 10 shows the simulated velocity magnitude distribution at the exit plane of the combustion chamber. The reference case is shown in the middle. If the faulty burner is operated in the lean mode ($\lambda=1.2$, top image) the velocity distribution increases above this burner. This results from the increased flow rate at constant power. If the faulty burner is operated rich ($\lambda=0.8$, bottom) the air-fuel-rate is reduced and less fuel is consumed. Both effects are leading to lower velocities in the burnt part, as is clearly visible.

Variation of the local air-fuel-ratio leads to significantly modified chemical processes in the flame which results in different composition in the exhaust jet cross section.

Before getting into the experimental and numerical details of this issue, one other aspect is discussed with respect to the future aim of defect classification out of measured exhaust profiles.

The future aim is the subsequent implementation of an automatic defect classification out of the exhaust profile measurement. For the classification of the corresponding defect, a unique pattern is necessary, which is specific for one case of failure, as otherwise a faulty defect assignment could occur. Figure 11 demonstrates the influence of two different defects on the simulated velocity field at the exit plane of the model combustion chamber. The upper image shows the velocity

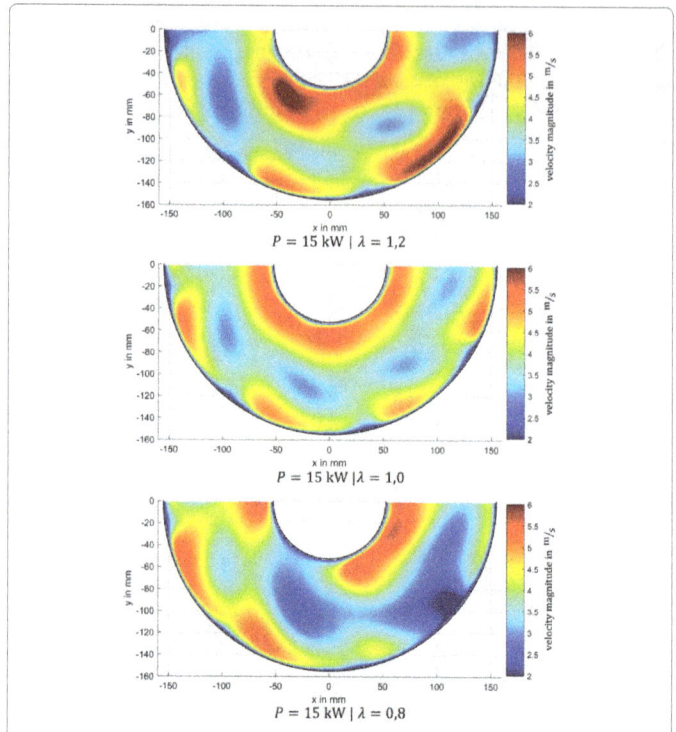

Figure 10: Simulated velocity magnitude at the combustor exit plane for the cases of study B.

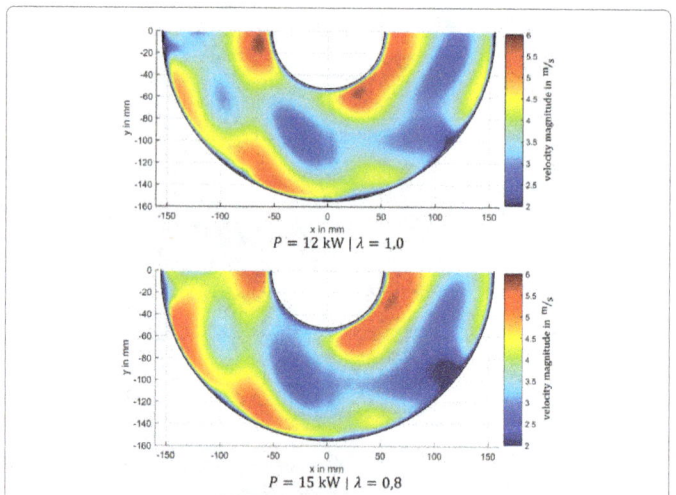

Figure 11: Simulated velocity magnitude at the combustor exit plane for a power reduction (from 15 to 12 kW, top) and a varied air-fuel-ratio ($\lambda=0.8$, bottom).

distribution for a power reduction of 20% to 12 kW of the faulty burner. The second graphic shows the influence of a varied air-fuel-ratio of 20% to $\lambda=0.8$ at initial thermal power of 15 kW. Both defects lead to a quite similar velocity distribution, so that an automatic defect classification will probably not be successful.

This example demonstrates that the use of only one measured quantity is not sufficient for a robust defect classification. Therefore, further quantities have to be taken into account.

Determination of Species Concentration Distributions

In Figure 12, the calculated CO concentration is shown for these two defect cases. The CO concentration depends strongly on the

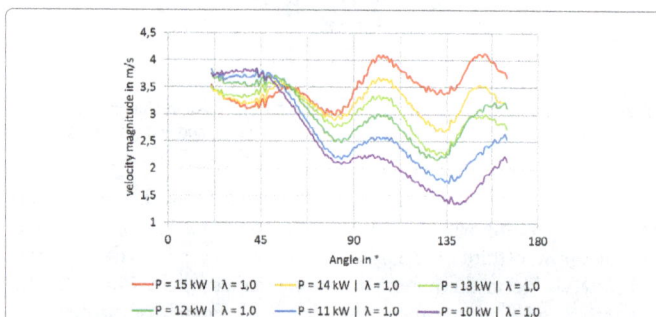

Figure 9: Influence of reduced power of the variable burner positioned at 112.5°, measured radial velocity magnitude (on diameter 205 mm) at the combustor exit plane.

stoichiometric condition in the flame and large amounts of CO due to the incomplete combustion in the fuel-rich case. With this additional information, a differentiation between the two different defects shown in Figure 11 is possible. Figure 12 shows also the spatial distribution of the exhaust flow deriving from the manipulated burner.

In Figure 13, the CO_2 concentration is shown for all three cases of the study B. Also, this quantity can give clear additional information to the velocity magnitude, indicating the influence of the burner failure with respect to varied stoichiometric ratio. The spatial distribution of the exhaust gas coming from the varied single burner is clearly visible.

In the experimental study, additionally species concentration measurements are done at the exit plane of the combustion chamber.

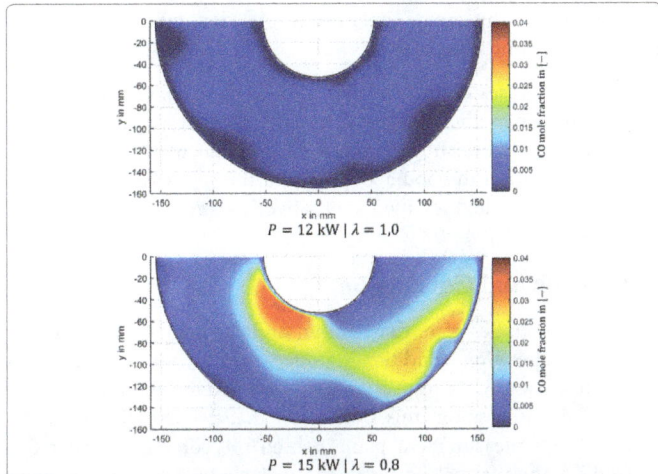

Figure 14: Circumferential profile of calculated (left) and measured CO_2 mole fraction (right) at the combustor exit plane for the cases of study B.

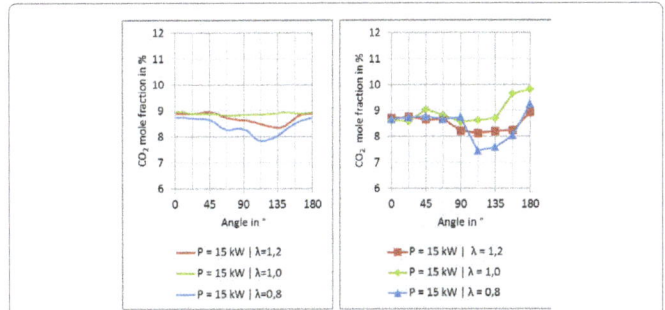

Figure 15: Circumferential profile of calculated (left) and measured CO mole fraction (right) at the combustor exit plane for the cases of study B.

For this purpose, a fourier-transform infrared (FTIR) spectrometer has been used. The FTIR measurement system allows the determination of the concentration of the gas components CO, CO_2, H_2O, NO, NO_2 simultaneously by extracting samples of the exhaust gases. The samples are extracted in the measurement half plane at nine specific positions by a moveable and heat resistant exhaust probe, as shown in Figure 3. The measurement points are arranged in a semi-circle of 210 mm diameter. The recording rate of the used FTIR system is 1 1/s. The measurement period for every position is 45 s and the averaged results are shown in the following. Exemplary measured concentration profiles are discussed in the following for varied air-fuel-ratio of study B.

Figure 14 the measured CO_2 concentrations on the nine measurement points as well as the calculated CO_2 concentration along the circumferential middle line is shown. In the reference case, all burners are operated at air-fuel-ratio λ= 1.0 (green). In the failure cases the faulty burner is operated either in the lean mode (λ=1.2, red) or in the rich mode (λ=0.8, blue). For the reference case the simulation shows a constant CO_2 concentration of 8.8%, while the measurements show the same CO_2 concentration in most of the measurement points, while for the two right positions the CO_2 concentration increased to 8.7% and 9.7%. The reason for this unexpected inhomogeneity is not clear. For the fuel, lean condition at λ=1.2 both the simulation and the FTIR measurements show a reduction of the CO_2 concentration. Between 90° and 157° similar concentration profiles can be found, where the CO_2 concentration is in the same order: 8.4% (simulation) and 8.2% (experiment). The results confirm that the failure can be detected with both methods clearly.

Also, the dispersion of the failure pattern can be analysed. Assuming the initial width of the failure region to be 45° (one out of eight burners), the detected width of the failure region reaches about 75° for the measurement and 60° for the simulation (defined as full width at half maximum, FWHM, of the spatial variation in the concentration

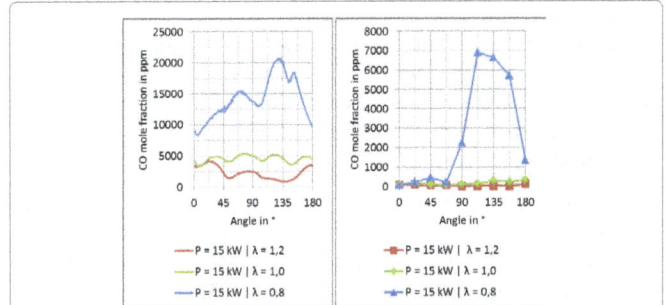

Figure 12: Simulated CO mole fraction at the combustor exit plane for a power reduction (from 15 to 12 kW, top) and a varied air-fuel-ratio (λ=0.8, bottom).

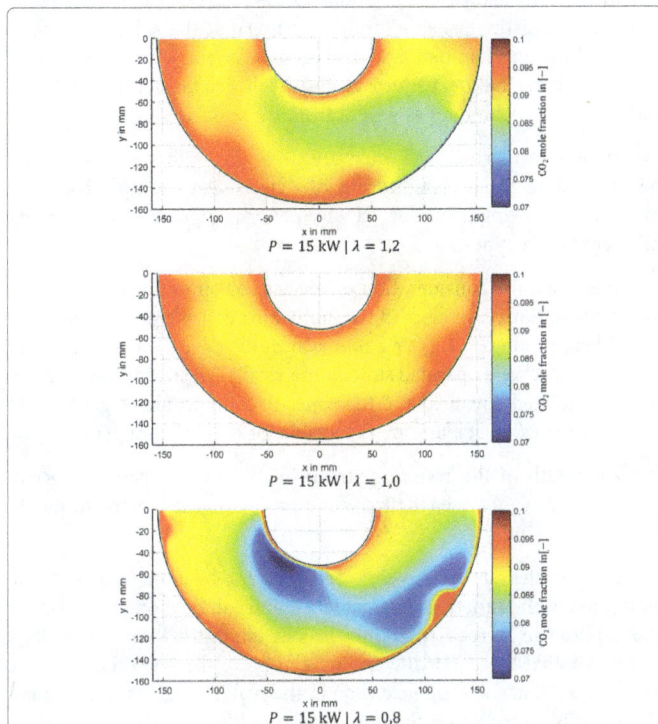

Figure 13: Simulated CO_2 mole fraction at the combustor exit plane for the cases of study B.

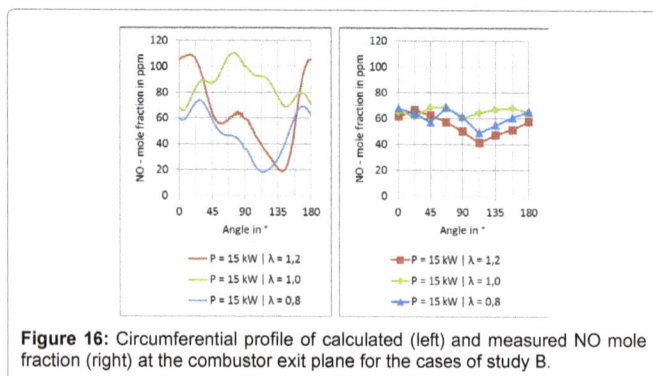

Figure 16: Circumferential profile of calculated (left) and measured NO mole fraction (right) at the combustor exit plane for the cases of study B.

profile) at the measurement distance of 200 mm above the burner plane. This increase of the failure width in the exhaust path is reasonable, and the rather similar increase of this width between simulation and experiment confirms that the simulation method is applicable for the numerical prediction of the correlation between failure source and detection plane.

For the fuel-rich condition at $\lambda=0.8$ both the experiment and the simulation profiles of the CO_2 concentration deviation are clearly visible with a full width of 60° for the measurement and of 75° for the simulation, where the simulation shows a broader dispersion region with low deviation.

In Figure 15 the CO concentration profiles are compared. Here the comparability between measurement and simulation is lower than for the CO_2 concentration. The simulations show significantly increased CO levels compared to the measurement also for the stoichiometric and the fuel lean case, while the measured values are near zero here, as can be expected for nearly complete combustion in these cases. For the fuel, rich condition both simulation and experiment show a strongly increased CO level, as expected for this partially reacting case. The calculations show considerable higher absolute values, which are not reasonable. Here, the reaction model for this intermediate species is definitely not realistic.

However, the relative CO levels are well predicted for the simulation in correspondence with the measured values, if the significant increase for the fuel rich burning faulty burner is regarded. If the CO profile is compared here, the simulation predicts an increased dispersion of the CO molecules over a broader circumferential range. The FWHM values of the spatial distribution are approximately 90° for the simulation and 70° for the experiment.

As a third example the calculated and measured NO concentration profiles are shown in Figure 16. The simulations show the same NO levels as measured, but a stronger reduction for the lean and the rich case. For the homogeneous reference case the measured NO concentration is constantly approximately around 65 ppm. Under lean and rich conditions the measured NO concentration decreases above the position of the manipulated burner. This trend follows the expectation of thermal NO_x formation, as the flame temperature is reduced in both cases and as thermal NO_x depends strongly on flame temperature. The local reduction from 65 ppm to 50 ppm (fuel-rich) or 43 ppm (fuel-lean) is significantly detectable. The FWHM values of the reduced NO zones are between 45° and 60°. This region is narrower than has been detected for the other concentrations. This effect might originate from the non-linear thermal NO formation process as a function of local temperature, which leads to the smaller gap. The reference case of the calculated NO concentration profiles shows a

quite inhomogeneous behaviour. However, the calculated mean values are reasonable: 85 ppm for the $\lambda=1.0$ reference case (measurement: 65 ppm). The strong reduction for both the lean and the rich case at 112.5° is qualitatively in good agreement. The calculated FWHM values are between 70° and 80°.

Not shown here are measured and calculated profiles of NO_2 and H_2O. Also, here the influence of the modified burner is detectable both from the numerical simulation and from the FTIR measurement.

The analysis of the circumferential species concentration profiles for the examples of the faulty burner shows, that the failure has a detectable influence on the concentration profile at the exit of the combustion chamber. The width of the influenced area at the exit is relatively broadened 60° up to 90° in comparison to an initial burner segment of 45°. This can be explained by turbulent mixing leading to a broadened mixing zone.

For these cases both simulation as well as species measurements may be applicable since the trends for both the fuel-rich as well as the fuel-lean cases are distinguishable. The reaction model predicts the CO_2 concentration in good accordance to the species measurements, but is not fully suited for the quantitative CO and NO concentration calculation.

Discussion and Conclusion

Detailed studies were done on an annular combustion chamber where the operation condition of one swirl burner can be varied in a defined way with respect to the other seven swirl burners. The two studies discussed in this work show that defined failures can be detected in a measurement plane behind the combustion chamber. In study A the thermal power of the variable burner in the annular combustion chamber was reduced from the references state (15 kW) in 1 kW steps down to 10 kW, representing a deficient fuel supply. For the investigated cases deviations in the range of 1 kW are already leading to a distinguishable modified pattern at the exhaust plane of the combustion chamber. For deviations above 2 kW (10% to 15% deviations) the failure pattern is clearly detectable for the different measurement techniques as well as from the simulation.

Only considering one quantity like the velocity field might be misleading for a classification of defects. An example has been presented, showing a similar velocity pattern originating from rather different failure types.

Therefore, the consideration of species concentrations is proposed as a useful additive. In study B the faulty burner was operated either in a fuel-lean mode ($\lambda=1.2$) or a fuel-rich mode ($\lambda=0.8$) while the other seven burners were operated stoichiometrically. The resulting deviation in the simulated flow field and the concentration profiles e. g. CO_2, CO or NO at the exhaust plane are detectable for these failures as well.

The width of the resulting streak in the velocity field or species concentrations increases in size while proceeding downstream due to turbulent mixing.

The studies done on this annular combustion chamber are simplified to the situation in a real jet engine combustion chamber. For the application in real jet engines it is of interest if the methodology can be applied there. For the approximation of the potential it can be emphasised that the residence time in the real jet engines' combustion chamber is even lower than in this simplified model combustion chamber. The lateral turbulent mixing between the jet engine flames can be expected to be lower than in the investigated annular combustion chamber. Calculated temperature and velocity profiles at the exit of a real jet engine show correspondingly even more compact structures of the individual burners in the combustor-turbine intersection.

In future work the turbulent dispersion process of failure structures in the exhaust gas path will be investigated in more detail. This is essential to determine the limit of detectable failure structures behind the aircraft engine. It can already be stated that the turbulent dispersion of inhomogeneous gas phase pattern in the subsequently following turbine is lower than might be anticipated [14], as the large scale turbulent mixing is suppressed from the stator and rotor blades.

These fundamental studies are leading the path to the application of the method of exhaust gas pattern detection to evaluate the internal state of aircraft engines without the need to disassemble the jet engine.

Acknowledgements

This project is funded by the German Research Foundation (DFG) within the framework of the Collaborative Research Centre SFB 871, subproject A4 "Influence of combustion chamber defects on the exhaust jet".

References

1. Adamczuk RR, Buske C, Roehle I, Hennecke C, Dinkelacker F, et al. (2013) Impact of defects and damage in aircraft engines on the exhaust Jet. In: Proceedings of ASME Turbo Expo 2013. ASME, San Antonio, Texas.

2. Vonder-Haar H, Hartmann U, Hennecke C, Dinkelacker F (2016) Defect detection in an annular swirl-burner-array by optical measuring exhaust gases. In: Proceedings of ASME Turbo Expo 2016 ASME.

3. Hennecke C, Frieling D, Dinkelacker F (2013) Influence of combustion chamber disturbances to patterns in the hot gas path. Presented at the European Combustion Meeting 2013 , Lund, Schweden.

4. Hennecke C, Maronna T, Dinkelacker F (2013) Detektion von Triebwerks-brennkammerfehlern durch Analyse von Temperatur- und Speziesprofilen der Heißgase einer Modell-Brennkammer. In: VDI-Berichte 2161, Duisburg: 472-480.

5. Menter FR (1994) Two-equation eddy-viscosity turbulence models for engineering applications. AIAA Journal. 32 1598–1605.

6. Poinsot T, Veynante D (2011) Theoretical and numerical combustion. Cerfacs, Toulouse.

7. Jones WP, Whitelaw JH (1982) Calculation methods for reacting turbulent flows: A review. Combustion and Flame. 48: 1-26.

8. Gerlinger P (2005) Numerische Verbrennungssimulation: Effiziente numerische Simulation turbulenter Verbrennung. Springer, Berlin.

9. ANSYS (2017) Fluent theory guide.

10. Adrian RJ (1991) Particle-imaging techniques for experimental fluid mechanics. Ann Rev Fluid Mech 23: 261-304.

11. Raffel M, Willert CE, Wereley ST, Kompenhans J (2007) Particle image velocitimetry: A practical guide with 42 tables. Springer, Berlin.

12. Scheimpflug T (1904) Method of distorting plane images by means of lenses or mirrors. Patents US751347 A.

13. Prasad AK, Jensen K (1995) Scheimpflug stereocamera for particle image velocimetry in liquid flows. Applied Optics 34: 7092 .

14. Hauptmann T, Aschenbruck J, Christ P, Hennecke C, Dinkelacker F, et al. (2015) Influence of combustion chamber defects on the forced response behavior of turbine blades. Proceedings of the 14th International Symposium on Unsteady Aerodynamics, Aeroacoustics & Aeroelasticity of Turbomachines. 2: 9-19.

20

Services for Space Mission Support Within The ESA Space Situational Awareness Space Weather Service Network

Donder ED[1]*, Crosby N[1], Kruglanski M[1], Andries J[2], Devos A[2], Perry C[3], Borries C[4], Martini D[5], Glover A[6] and Luntama JP[6]

[1]Royal Belgian Institute for Space Aeronomy, Uccle, Belgium
[2]Royal Observatory of Belgium, Uccle, Belgium
[3]STFC RAL Space, UK
[4]Institute of Communications and Navigation, German Aerospace Centre, Neustrelitz, Germany
[5]Norwegian Center for Space Weather/Tromsø Geophysical Observatory, Norway
[6]ESA SSA Programme Office European Space Operation Centre Darmstadt, Germany

Abstract

Spacecraft operations are by nature complex and every satellite's operational environment poses a range of potential risks, often a unique combination for a given orbit. The implications of interruptions of operations, data transfer and service provision, are serious, both in terms of cost and capability, thus it is imperative to mitigate against all operational risks to the fullest extent possible.

In the frame of its Space Situational Awareness (SSA) programme, the European Space Agency (ESA) is establishing a Space Weather Service Network to support end-users, in a wide range of affected sectors, in mitigating the effects of space weather on their systems, reducing costs and improving reliability. This service network is currently in a test and validation phase and encourages user engagement and feedback.

The network is organised around five Expert Service Centres (ESCs) focusing on Solar Weather, Heliospheric Weather, Space Radiation Environment, Ionospheric Weather and Geomagnetic Conditions. Each ESC is connecting different expert groups, federating their space weather products, and ensuring the quality and consistency of the provided information. The service network also includes a central Data Centre and the SSA Space Weather Coordination Centre (SSCC).

In this presentation we give an overview of the current status of the network (http://swe.ssa.esa.int/), the targeted end-user groups and Expert Service Centres with a focus on the space community.

Keywords: Space weather; Space situational awareness; Service network

Introduction

The objective of the ESA Space Situational Awareness (SSA) programme is to support Europe's independent utilisation of, and access to, space through the provision of timely and accurate information, data and services regarding the space environment, and particularly regarding hazards to infrastructure in orbit and on the ground.

The SSA programme will, ultimately, enable Europe to autonomously detect, predict and assess the risk to life and property due to man-made space objects (remnant debris, re-entries, in-orbit explosions and release events, in-orbit collisions, disruption of missions and satellite-based service capabilities), potential impacts of Near-Earth Objects (NEOs), and the effects of space weather phenomena on space- and ground-based infrastructure.

The programme is currently in its Period Two [1] and is active in three main areas:

- Survey and tracking of objects in Earth orbit.

- Monitoring and forecasting space weather.

- Watching for NEOs.

Within this framework, the focus in the Space Weather Segment (SWE) of the SSA Programme is on developing a system capable of providing space weather services to end users. The system under development is based on a federated architecture where the service provision will be carried out by Expert Service Centres (ESCs) in the

Programme Member States. These collaborative centres bring together European expertise and assets. Five ESCs focusing on Solar Weather, Heliospheric Weather, Space Radiation Environment, Ionospheric Weather and Geomagnetic Conditions are being established. Space weather services from the ESA SSA system are being made available to the end users because of a framework set to incorporate service level agreements, leading to both a substantial extension of products available and improved reliability of provision. In parallel to the provision of the current services, the ESCs are introducing new, innovative services and developing the necessary processes in order to provide these in a reliable manner. Consequently, the SWE network of services is expected to grow substantially within SSA Period 2, particularly within the context of the ongoing Expert Service Centre Definition and Development (P2-SWE-I) activity taking place during 2015-2017. The overall structure of the network is shown in Figure 1.

For the development of the SWE service network, two distinct

*Corresponding author: Donder ED, Royal Belgian Institute for Space Aeronomy, Uccle, Belgium, E-mail: erwin.dedonder@aeronomie.be

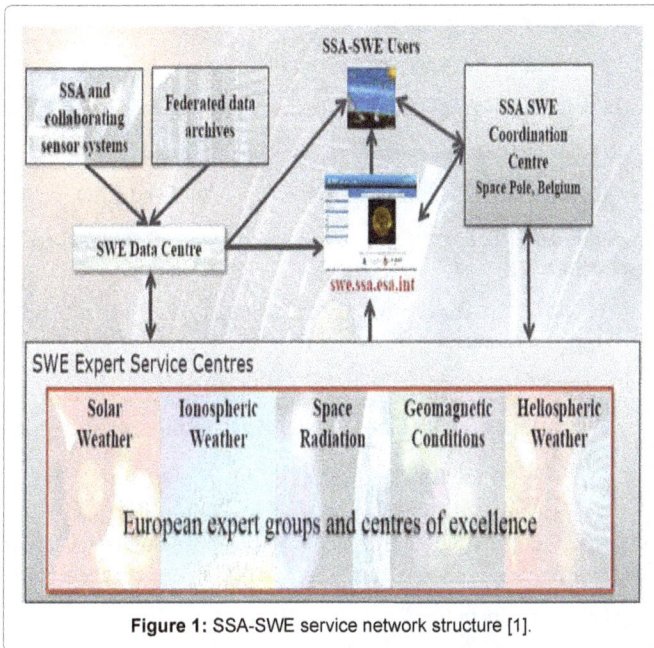

Figure 1: SSA-SWE service network structure [1].

approaches are followed. The first approach is a bottom-up approach where existing space weather assets in Europe are deployed and further developed into a set of representative and essential precursor services. The deployment and operation of this set of precursor services allows to gather feedback from the user community, to make early assessments of the network and to provide key inputs for its strategic service development roadmaps. The second approach is a top-down approach starting with the definition of the customer team also organises visits to the SSCC premises upon invitation and/or request (helpdesk.swe@ssa.esa.int).

In what follows, for each ESC their mission statement, contributing expert groups, provided services and corresponding user types and provided products are described. The SSCC tailoring of some products for space mission support is also described (Figure 1).

Requirements [2] from which system requirements [3] have been derived in succession. The top down design process benefits from the experience and feedback gained from the pre-operation of the SWE service network set up by the first approach. The second approach has defined eight service domains — Spacecraft Design (SCD), Spacecraft Operation (SCO), Human Space Flight (SCH), Launch Operation (LAU), Transionospheric radio link (TIO), SSA Space Surveillance and Tracking (SST), Non-space systems operation (NSO), and General data service (GEN) — for which 39 services have been identified (Refer Appendix).

Since space weather is a rapidly evolving domain, with users gaining more in-depth knowledge of how space weather affects their systems and new user groups becoming aware of the potential impact of space weather, it is important to establish a close relationship with the different potential user communities. Such a relationship enables user awareness and ensures its progress from awareness, to agreement, and from agreement to adoption of evidence-based practices. In the approach to establish a close relationship with the end-users of space weather services, the SSA Space Weather Coordination Centre (SSCC) — which provides the user helpdesk and first line user support for the SWE network conducts three main actions:

- Bringing the network to the users to increase its awareness,

- Identifying key persons within the different user communities

- Setting up end user meetings to learn the user requirements and to translate them into a user tailored space weather service.

Members of the SSCC team attend workshops, conferences etc. to promote the network and to meet end users, and the team also organises visits to the SSCC premises upon invitation and/or request (helpdesk.swe@ssa.esa.int).

In what follows, for each ESC their mission statement, contributing expert groups, provided services and corresponding user types and provided products are described. The SSCC tailoring of some products for space mission support is also described.

The Expert Service Centre Solar Weather

Mission statement

The mission of the Solar Weather Expert Service Centre (S-ESC) is to provide and develop the functionalities, capabilities and expertise in the domain of Solar Weather that are needed within the ESA SSA SWE Network to achieve as a collaborative enterprise its mission of demonstrating and assessing the influences of Space Weather and informing and supporting end-users through the provision of accurate, reliable and timely products and (pre) operational services, tailored to their requirements [4-15].

The S-ESC thus provides, implements and supports the Solar Weather products and capabilities of the ESA SSA SWE. This includes the observation, monitoring, interpretation, modelling and forecasting of Solar Weather conditions with an emphasis on Solar (sub)-surface and Solar coronal features, events and processes that drive Space Weather in our Solar System.

Expert groups

The current network of Expert Groups contributing to the S-ESC network is:

- Solar Influence Data analysis Centre (SIDC), Royal Observatory of Belgium (ROB), Belgium.

- Kanzelhöhe Observatory (KSO), University of Graz (UniGraz), Austria.

- Research Center for Astronomy and Applied Mathematics, Academy of Athens, Greece.

- Osservatorio Astrofisico di Catania (CAO), Istituto Nazionale di Astrofisica (INAF), Italy.

- Institute of 4D Technologies, University of Applied Sciences North Western Switzerland.

Services and user types

The functionalities and responsibilities of the S-ESC are located at the start of the causality chain of Space Weather events starting from the Solar origin up to the effects on Geospace, Earth and, eventually, the impact on user infrastructure. Hence, the focus of the S-ESC is very much on the production and assessment of synoptic Solar data.

Solar Weather products can have direct relevance to the end-user (e.g. a general probabilistic forecast on the likelihood of solar events given the current Solar Weather conditions). In addition, Solar Weather ESC products are often digested by other Centres that, based on their subsequent expertise in the physical systems impacted, further specify and predict the more precise characteristics of these impacts [16-20].

Products

Below we give a list of products provided by the S-ESC, illustrated with the CACTus CME detection (Figure 2) and a full disk magnetogram with eligible active regions (Figure 3) produced by Athens Effective Solar Flare Forecasting (AEFFort).

a. Solar imaging

• White light continuum images from ground based stations (SIDC/USET, UniGraz/KSO, INAF/CAO).

• H-Alpha images from ground based stations (USET, KSO, CAO).

• PROBA2/SWAP EUV images.

• European mirror of SDO imagery.

b. Spectral radio observations

• Radio spectra from the Humain radio station, Belgium.

• Radio spectra from the eCallisto Global network.

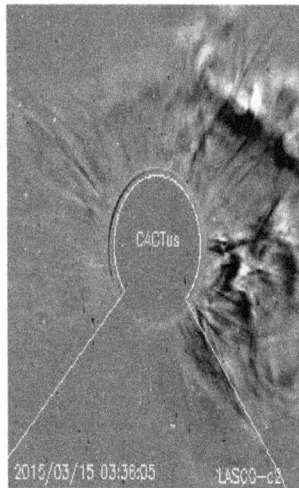

Figure 2: Cactus CME detection in SOHO/LASCO coronagraphic image [4,5].

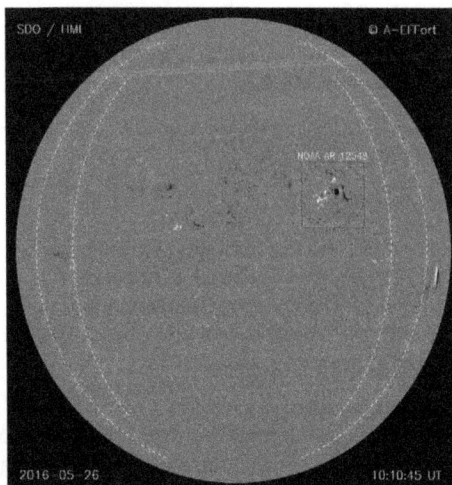

Figure 3: Eligible active regions identified on full-disk SDO/HMI line-of-sight magnetogram (from A-Effort) [6-8].

c. Solar irradiance

• PROBA2/LYRA 4channel (E) UV irradiance.

• Recalibrated LYRA data as proxy for X-ray flare magnitudes.

d. Solar feature and event detections/characterisation

Characterisation of Sunspot Groups: from USET and CAO.

Automated Radio Burst detections on SIDC/Humain radio observations.

• (CACTus) Automated CME detection in SoHO/LASCO coronagraphic images.

• Optical flare detections from KSO.

• X-ray flare detections based on NOAA/GOES flux measurements.

• Filament detections from KSO.

• Human operator (7d/7d, 16 h/24 h) moderation and expert annotation and alerting complementing the above automated event detections.

• Solar Activity indices.

• International Sunspot Number.

• Solar activity and indices forecasts.

• 3-day forecast of radio flux F10.7 index.

• Probabilistic Solar Flare predictions (2 distinct methods).

e. Service products

• Email alerts associated to the feature and event detections.

• Image browsing with Solar Active Region annotations.

• All quiet Alert: marking times with exceptionally low Space Weather risk. This product is targeted at Satellite operators.

• Daily bulletin on Solar (and Heliospheric and Geomagnetic) Weather (Human operator generated).

The Expert Service Centre – Heliospheric Weather

Mission statement

The mission of the Heliospheric Weather Expert Service Centre (H-ESC) is to provide and develop the functionalities, capabilities and expertise in the domain of Heliospheric Weather that are needed within the ESA SSA SWE network to achieve as a collaborative enterprise its mission of demonstrating and assessing the influences of Space Weather and informing and supporting end users through the provision of accurate, reliable and timely products and (pre-) operational services, tailored to their requirements.

The specific goal of the H-ESC is to fully exploit heliophysics assets and expertise from Europe, and further afield, to provide, improve and support the provision of alerts, forecasts and post event analysis of space weather conditions both near to the Earth and at other locations within the heliosphere.

This will be accomplished through the use of remote-sensing and in situ monitoring to identify and track key transient features (including coronal mass ejections, high-speed solar wind streams and solar energetic particles), combined with the use of advanced modelling and analysis techniques to predict their arrival at various points of interest in the heliosphere.

Expert groups

The H-ESC is being newly established as part of the ESA SSA SWE Period 2 ESC definition and development activities that kicked-off towards the end of September 2015. The initial H-ESC network consists of eight Expert Groups of which five will be providing products and the remaining three are consultants with specialised knowledge in the areas of CME characterisation, service assessment and development and modelling which are key areas for the initial development. The H-ESC network coordination is provided by STFC RAL Space, UK. The full list of H-ESC Expert Groups is:

• STFC RAL Space, UK (Coordination, scientific QA and service assessment).

• UK Met Office, UK (operational forecasting service, MHD modelling).

• University of Graz, Austria (solar wind and CME propagation).

• DTU, Denmark (near-Earth solar wind transient detection).

• IRAP, France (AMDA and propagation validation tools).

• KU Leuven, Belgium (European modelling assessment).

• DH Consultancy, Belgium (Existing service assessment).

• University of Göttingen, Germany (CME propagation).

Services and user types

The H-ESC has responsibility for three service areas falling within the general data, spacecraft operations and spacecraft design service domains:

• GEN/mod – End-to-end modelling capability for space weather.

• SCO/pla – Spacecraft operations for interplanetary missions.

• SCD/pla – Spacecraft design for interplanetary missions.

The SCO/pla and SCD/pla are two new service areas that are being defined as part of the current P2-SWE-I activity with a specific focus on the support of ESA interplanetary science missions.

Products

With the initial development of the H-ESC the focus is on the provision of products that address the heliospheric forecasting, nowcasting and alerting needs of the users and other ESCs with respect to background solar wind and solar wind transients and their arrival in the vicinity of the Earth. During the second part of the current project the products will be extended to provide broader support for heliospheric weather for other locations within the heliosphere.

The initial set of products to be released during the first half of 2016 consists of:

• Forecasts of near-Earth Solar wind based on MHD (WSA/ENLIL) and empirical modelling techniques and with lead times of 4 to 7 days depending on the model.

• Nowcasts of the near-Earth Solar wind measurements to provide information on current conditions and for verification and assessment of forecasts.

• Predictions of near-Earth CME arrival times based on MHD and empirical modelling techniques of the propagation from the Sun to the Earth.

• Near-Earth energetic particle measurements from GOES for SEP arrival assessment.

• Near-Earth space weather alerts based on forecaster and automated assessment of the near-Earth heliospheric weather conditions including shock, ICME and SIR arrival.

The second set of products that will be released during the autumn of 2016 are:

• Forecast of Solar wind at other locations in the heliosphere based on MHD modelling techniques.

• Tailored CME arrival predications for other locations within the heliosphere based on MHD and drag based modelling techniques.

• Tailored heliospheric weather alerts for other locations within the heliosphere.

Figure 4 shows an example of the WSA/ENLIL MHD heliospheric modelling. Empirical modelling products are shown in Figure 5 (left: coronal hole based near-Earth solar wind estimates, right: drag based model forecast of CME arrival.

In addition the H-ESC will provide a range of ancillary and support products and capabilities including:

• Case study and special event modelling.

• Assessment report on near-Earth forecast and alerting performance.

• Archive of H-ESC products.

• Integrated access to science archive (CDPP, UKSSDC) on near-Earth and interplanetary measurements in support of post event analysis.

• Long term statistical products on heliospheric weather conditions.

The Expert Service Centre – Space Radiation

Mission statement

The mission of the Space Radiation ESC (R-ESC) is to provide and develop the functionalities, capabilities and expertise in the domain of Space Radiation that are needed within the ESA SSA SWE Network to achieve as a collaborative enterprise its mission of demonstrating and assessing the influences of Space Weather and informing and

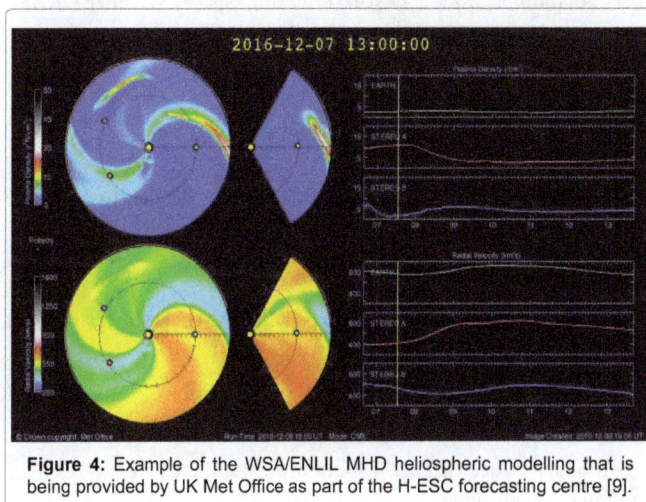

Figure 4: Example of the WSA/ENLIL MHD heliospheric modelling that is being provided by UK Met Office as part of the H-ESC forecasting centre [9].

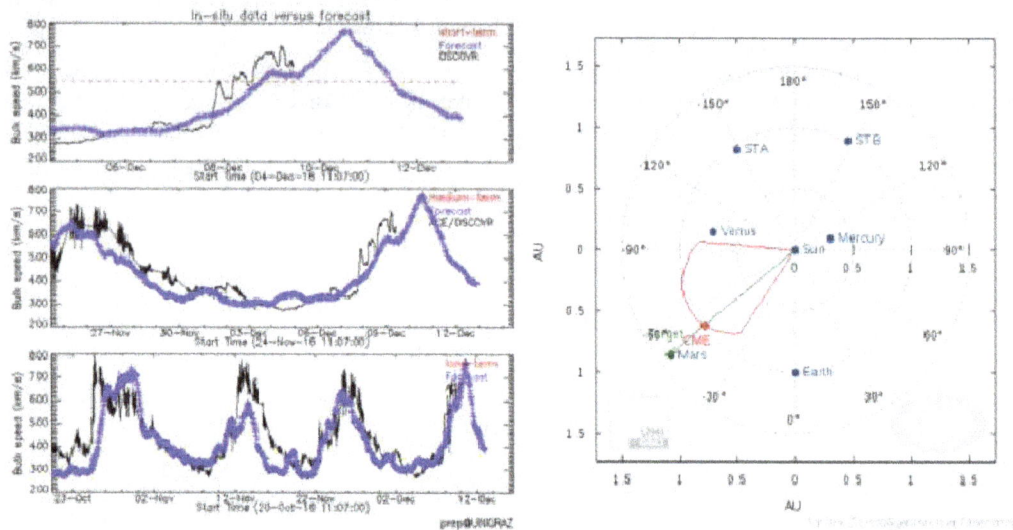

Figure 5: Example of the empirical modelling products provided by University of Graz. (left) coronal hole based near-Earth solar wind estimates [10], (right) drag based model forecast of CME arrival [11].

supporting end users through the provision of accurate, reliable and timely products and (pre-) operational services, tailored to their requirements.

The R-ESC ensures that the monitoring, modelling and forecasting of space particle radiation (ambient plasma, solar energetic particles, radiation belts, galactic cosmic rays), micron-size particulates (from meteoroids and space debris), as well as all types of phenomena induced effects on technologies and biological systems, are fully covered in regard to the near-Earth space environment.

Expert groups

The R-ESC is currently a network of eleven Expert Groups comprised of eight product providers and three consultants with expertise in charged particle radiation. Expert group names and their associated countries in alphabetical order are listed below. During P2-SWE-I the R-ESC is being coordinated by the Royal Belgian Institute for Space Aeronomy (BIRA-IASB).

a. Providers

• Radiation Hardness Assurance and Space Weather (RAS), Seibersdorf Laboratories Gmbh (Austria).

• Paul Buehler (Austria).

• Centre for Space Radiations (CSR), Université Catholique de Louvain (Belgium).

• Space Physics Division (BIRA-IASB), Royal Belgian Institute for Space Aeronomy (Belgium).

• Space Research Laboratory (SRL), University of Turku (Finland).

• Department Radiation Biology (DLR-IAM), DLR Institute of Aerospace Medicine (Germany).

• Athens Neutron Monitor Station (ANeMoS), National and Kapodistrian University of Athens (Greece).

• Mullard Space Science Laboratory (MSSL), University College London (United Kingdom).

b. Consultants

• Ondrej Santolik, Institute of Atmospheric Physics (IAP) (Czech Republic).

• Ilya Usoskin, Sodankylä Geophysical Observatory (SGO), University of Oulu (Finland).

• Clive Dyer, CSDRadConsultancy Ltd (United Kingdom).

C. Services and user types

Seventeen services from five of the service domains introduced in section I fall under the responsibility of the R-ESC. Domains and their respective services are listed here:

Spacecraft design

• SCD/arv – Data archive.

• SCD/orb – In orbit verification.

• SCD/pst – Post event analysis.

Spacecraft operation

• SCO/orb – In orbit environment and effects monitoring.

• SCO/pst – Post event analysis.

• SCO/for – In-orbit environment and effects forecast.

• SCO/ana – In-orbit mission risk analysis.

Human space flight

• SCH/orb – In-flight crew radiation exposure.

• SCH/pstb – Cumulative crew radiation exposure.

• SCH/for – Increased crew radiation exposure risk.

Launch operation

• LAU/orb – In-flight monitoring of radiation effects in sensitive electronics.

• LAU/pst – Estimate of radiation effects in sensitive electronics.

• LAU/for – Forecast of radiation storms.

• LAU/drg – Atmospheric density forecast.

• LAU/ios – Risk estimate of service disruption caused by ionospheric scintillations.

• LAU/mcp – Risk estimate of microparticle impacts.

Non-space systems operations

• NSO/air – Service to airlines

New products being provided during P2-SWE-I (see Section IV.D) were specifically chosen to improve these four services: SCD/arv, SCD/pst, SCO/orb, and NSO/air. During P2-SWE-I it is also being determined to what extent the seventeen R-ESC services as specified in the SSA SWE Customer Requirements document can be fully provided on the basis of existing European capabilities. This is done by reviewing existing capabilities (assets), analysing where there are gaps and when necessary identifying development needs. The Space Radiation ESC network expansion is thus directly linked to the latter.

d. Products

During P2-SWE-I the R-ESC continues to provide users with seven existing products that were delivered during the SSA Preparatory Programme. These products are available on the R-ESC homepage on the SSA SWE portal (http://swe.ssa.esa.int/web/) and are listed below:

• Radiation exposure estimation at aircraft altitude (AVIDOS).

• GLE alert service.

• Multi-station Neutron Monitor data.

• Space Environment Information System (SPENVIS).

• Space Environment Data System (SEDAT).

• Environment Information System for Operations (SEISOP).

• European Debris Impact Database (EDID).

Figure 6 displays two of these products, respectively the GLE alert service (upper panel) and AVIDOS that provides radiation exposure estimation at aircraft altitudes (lower panel).

During P2-SWE-I 28 new products will be delivered, implemented and provided to users, also via the SSA SWE portal (http://swe.ssa.esa.int/web/). These products can broadly be classified as follows:

• PROBA-V / EPT e, p, He flux (spectra time series, geographical maps).

• PROBA-V / EPT energy spectra (auroral electron, SAA p and He).

• SREM data on Proba-1, Integral, Rosetta, Herschel, Planck.

• Radiation and accumulated doses at ISS.

• SEP post-event analysis for aviation radiation exposure.

• Very high-energy SEP environment (proton fluence, worst-case proton flux).

• SEP event catalogue.

• Electron radiation belt models (GEO, MEO, LEO).

• Coronal Mass Ejections and Solar Energetic Particles (COMESEP) Alert System.

• Solar Energetic Particle Environment Modelling (SEPEM) Application Server.

• Plasmasphere electron density distribution model.

Existing as well as new products provide users a large spectrum of applications ranging from technologies to biological systems covering different orbits as well as airline altitudes (Figure 7).

The Expert Service Centre – Ionospheric Weather

Mission statement

The mission of the Expert Service Centre Ionospheric Weather (I-ESC) is to provide and develop the functionalities, capabilities and expertise in the domain of ionosphere and upper atmosphere weather that are needed within the ESA SSA SWE Network to achieve as a collaborative enterprise its mission of demonstrating and assessing the influences of Space Weather and informing and supporting end-users through the provision of accurate, reliable and timely products and (pre-) operational services, tailored to their requirements.

Expert groups

Currently, nine expert groups are contributing by product provision or development to the services for GNSS users in the I-ESC. An overview of these expert groups is shown in Figure 8.

The coordination of the I-ESC is performed by DLR. The currently nine expert groups are complemented during Period 2 through new expert groups, data and products. An I-ESC advisory board is constantly supporting the I-ESC coordinator in his duty to organize the development and operation of the I-ESC services. The vision of the ESC includes a continuous network extension and development. Therefore, the initial assets and expert groups involved in the ESC

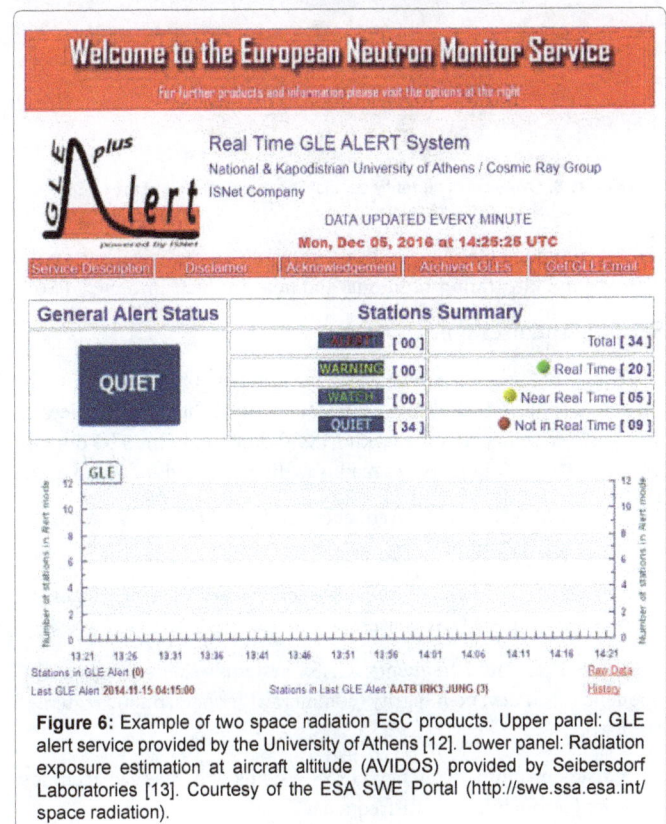

Figure 6: Example of two space radiation ESC products. Upper panel: GLE alert service provided by the University of Athens [12]. Lower panel: Radiation exposure estimation at aircraft altitude (AVIDOS) provided by Seibersdorf Laboratories [13]. Courtesy of the ESA SWE Portal (http://swe.ssa.esa.int/space radiation).

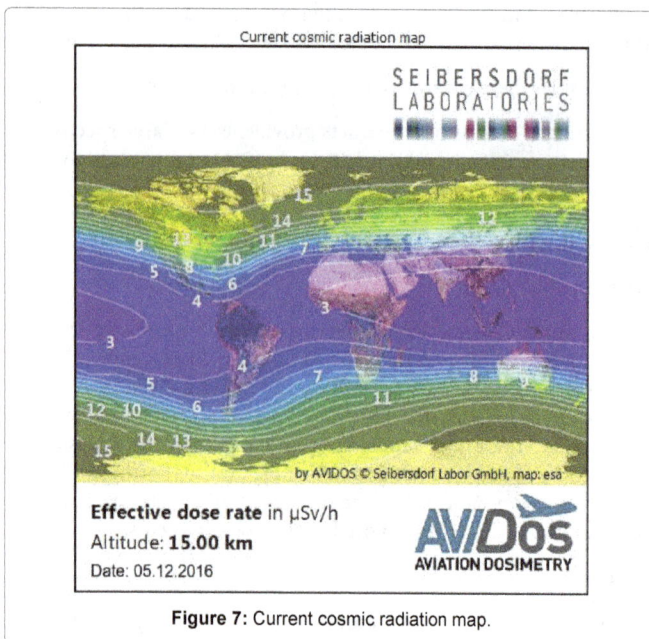

Figure 7: Current cosmic radiation map.

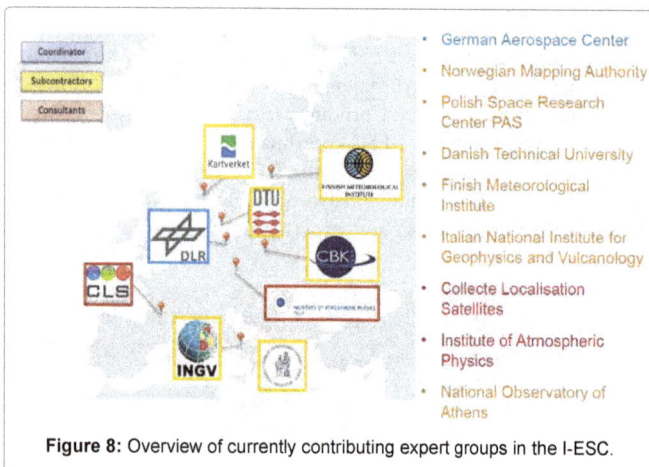

Figure 8: Overview of currently contributing expert groups in the I-ESC.

will be complemented by additional experts and assets extending the network and integrating additional and new products into the I-ESC.

Services and user types

The I-ESC is responsible for a number of services in the Transionospheric Radiolink (TIO) domain and the SWE services to the Space Surveillance and Tracking (SST) domain. In the TIO domain the I-ESC is providing services mainly to different kinds of GNSS users:

• Users of GNSS single frequency services with average accuracy, no integrity (e.g. typical GNSS mass market user).

• Users of GNSS single frequency services with average accuracy, using integrity (e.g. EGNOS user).

• Users of multi-frequency GNSS systems with average multi-frequency accuracy, no integrity (commercial services, public regulated services).

• Users of multi-frequency GNSS systems with average accuracy, integrity (aeronautical multi-frequency).

• Users or multi-frequency GNSS systems with very high accuracy (e.g. GNSS geodetic users, Real-Time Kinematics).

• Users of satellite data communications with high availability / continuity (e.g. Search-and-Rescue, Air Traffic Control/Management via Satellite, high availability/continuity data networks such as Galileo Ground Segment Data Network).

• Other space-based services/products users affected by the ionosphere (UHF - C-band radars, GNSS-R altimetry, UHF/low microwave radio astronomy and deep space communications). For these users, the I-ESC is providing the following services:

• TIO/tcr – Provide near real-time Total Electron Content (TEC) maps.

• TIO/tcf – Provide forecast TEC maps.

• TIO/qua – Provide information on whether standard corrections to GNSS signal are applicable (Quality assessment of ionospheric correction).

• TIO/sci – Provide near real-time estimate of ionospheric scintillation maps.

• TIO/for – Provide estimate of the occurrence risk of ionospheric disturbances (monitoring and forecast of ionospheric disturbances).

The SWE services targeting the the SST domain are mainly addressed to users in the following areas:

• Surveillance and tracking centre(s), stations and services.

• Spacecraft operators.

• Collision warning services.

• Re-entry risk assessment services.

For these users the I-ESC will provide the following services:

• SST/atm – Provide estimates of atmospheric density in the past years and predicted in near real-time.

• SST/arv – Provide a database of past values of solar and geomagnetic indices relevant to drag calculation.

• SST/for – Provide forecast of geomagnetic and solar induces for drag calculation

• SST/ion – Provide now cast of ionospheric group delay to estimate effects on radar signals.

Products

The I-ESC provides a large variety of products serving the needs of the addressed users and contributing to the provision of the mentioned services. The relation between the products and services is described in the documents [2] and [3] and will be presented in this way on the I-ESC homepage on the SSA SWE portal (http://swe.ssa.esa.int/web/). Here, currently diverse products in the following categories are provided:

• TEC maps.

• Scintillation maps.

• URSI parameters (ionosonde measurements).

• Databases of solar and geomagnetic indices.

• Ionospheric disturbances.

• Event based alarms for ionospheric disturbances.

A small subset of I-ESC products is shown in Figure 9.

The Expert Service Centre – Geomagnetic Conditions

Mission statement

The mission of the Geomagnetic Conditions Expert Service Centre (G-ESC) is to provide and develop the functionalities, capabilities and expertise in the domain of geomagnetism that are needed within the ESA SSA SWE Network to achieve as a collaborative enterprise its mission of demonstrating and assessing the influences of Space Weather and informing and supporting end-users through the provision of accurate, reliable and timely products and (pre-) operational services, tailored to their requirements.

The G-ESC thus provides implements and supports the geomagnetic products and capabilities of the ESA SSA SWE network for monitoring, interpreting, and forecasting variations of the geomagnetic field on various timescales, relying on in situ ground-based, as well as space-borne measurements and related derived quantities and models.

Expert groups

Currently the G-ESC comprises the following eight expert groups:

• Technical University of Denmark – DTU Space.

• Finnish Meteorological Institute (FMI).

• Swedish Institute for Space Physics (IRF).

• German Research Centre for Geosciences (GFZ).

• Polar Geophysical Institute (PGI).

• Solar Influences Data Analysis Center (SIDC).

• Norwegian Mapping Authority (NMA).

Services and user types

Four user domains have been identified that fall under the responsibility of the G-ESC within the Non-Space System Operators Services (NSO) domain:

• NSO/pow – Service to power system operators (high priority).

• NSO/ppl – Service to pipeline operators (low priority).

• NSO/res – Service to resource exploitation system operators (high priority).

• NSO/tou – Service to auroral tourism sector (medium priority).

Products

The list of existing and new (that will be delivered during P2-SWE-I) products provided by the G-ESC are displayed in Table 1. In Figures 10 and 11 respectively, a 27 days forecast of the daily global Ap index and 30-minutes dB/dt forecast map are displayed.

Tailoring of Space Weather Products

To address specific requests from key users, SSCC is tailoring products of the SWE network into dedicated space weather bulletins. In particular such bulletins have been generated in support to the following ESA missions:

• Launch of GAIA spacecraft (16–20 December 2013) and L2 insertion manoeuvre (6–14 January 2014).

• Last aerobraking campaign of Venus Express (May–July 2014).

• Rosetta mission (since October 2014) including the landing of the Rosetta's sounder Philae on comet 67P/Churyumov–Gerasimenko (12 November 2014).

• Launch of the experimental spaceplane IXV (11 February 2015).

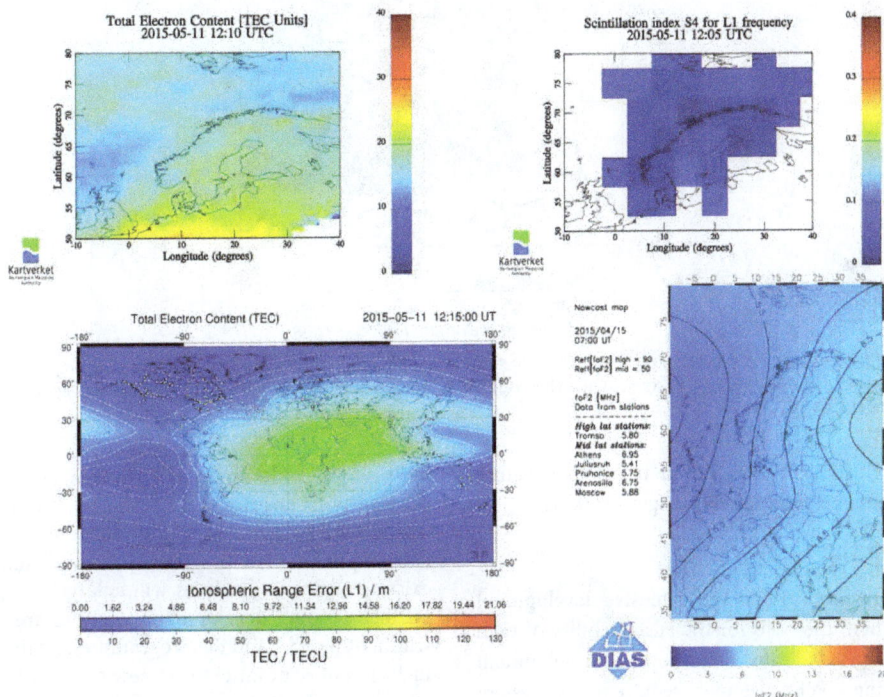

Figure 9: Sample subset for ionospheric products provided in the I-ESC. Upper left: DLR TEC map [14], Upper right: NMA S4 map [15], lower left: NMA TEC map [15], lower right: NOA Now-cast foF2 map [16-18].

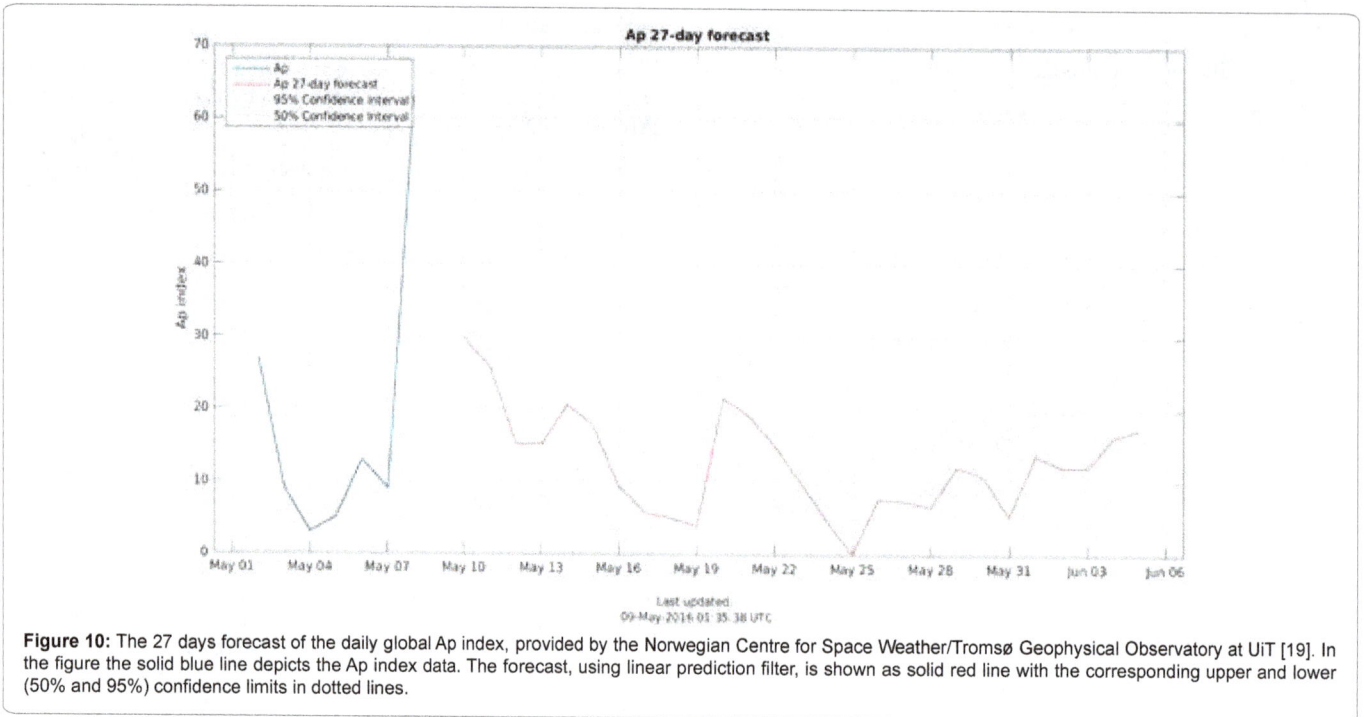

Figure 10: The 27 days forecast of the daily global Ap index, provided by the Norwegian Centre for Space Weather/Tromsø Geophysical Observatory at UiT [19]. In the figure the solid blue line depicts the Ap index data. The forecast, using linear prediction filter, is shown as solid red line with the corresponding upper and lower (50% and 95%) confidence limits in dotted lines.

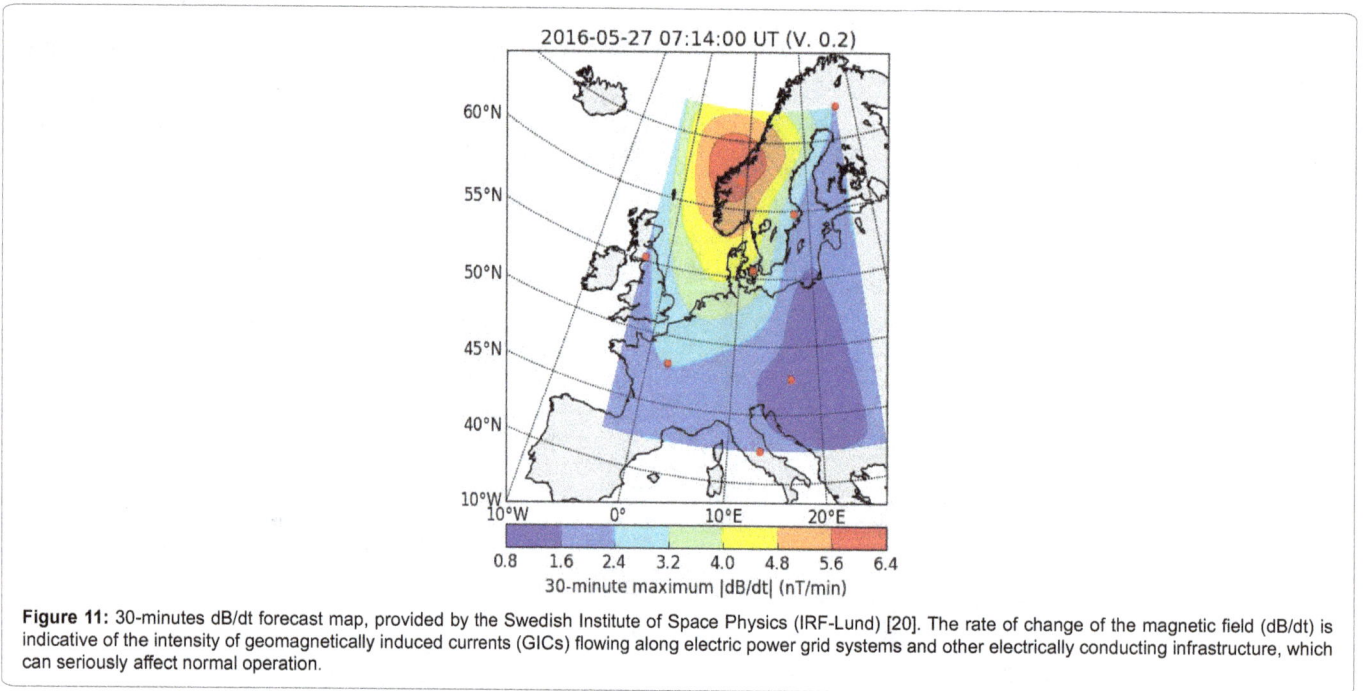

Figure 11: 30-minutes dB/dt forecast map, provided by the Swedish Institute of Space Physics (IRF-Lund) [20]. The rate of change of the magnetic field (dB/dt) is indicative of the intensity of geomagnetically induced currents (GICs) flowing along electric power grid systems and other electrically conducting infrastructure, which can seriously affect normal operation.

Figure 12 shows an illustration included in the space weather bulletin tailored for the Venus Express aero braking operations.

Future Perspectives of the Network

The SWE network is foreseen to undergo extensive development within the next years. At first, the basis for the sustainability of each ESC will be set by compiling a dedicated definition and development plan based on the existing SWE customer requirements, system requirements and service development roadmaps. This will ensure the consistent presentation of structured SWE services to the end users built on the wide range of European expertise and assets available, coupled with targeted developments. Constant user feedback and a programme of test campaigns will be used to highlight areas where user requirements are evolving. Second, substantial development of existing and new products will enlarge the capabilities of each ESC. Further product development activities are expected to be executed within other ESA and EU programmes. As these developments mature, inclusion of relevant existing new or updated products into each ESC will be investigated in terms of their possible further contribution to meeting SWE user requirements. Following this assessment, their inclusion into the SWE network may be foreseen.

Product	Description	Provider
Magnetogrammes from North (west) Europe and Greenland	Real-time ground based vector measurements (Include Greenland and Finland)	TGO/DTU FMI
Provisional K-indices from Northwest Europe	--	TGO/DTU
Geomagnetic Activity index for Aurora zone (AZ), last 33 days	--	TGO
Geomagnetic Activity index for auroral zone (AZ), long term variation	--	TGO
Provisional AA index	--	SIDC/ROB
Aurora alert and forecast service	Forecast of visible aurorae	FMI
Quick look Kp index	Near Real-time Kp index plot	GFZ
Most recent definitive Kp index	Kp musical diagram	GFZ
Kp and Ap index on tabular form	Latest quicklook Kp and Ap on tabular form	GFZ
Kp and Ap index archive	Historic Kp and Ap ftp database	GFZ
Maps for power and pipeline operators	Maps showing estimated GIC, E- fields dB/dt and B in European model grid as well as user specified grid	FMI, PGI
Table of modelled GIC	Tailored service for specific users providing a table of modelled GIC value for the users network in the last minute and peak GIC in the last 60 mins.	FMI
Forecasts of dB/dT	~30 min ahead	IRF
PSP difference	Tailored service for specific users providing pipe-to-soil potential difference (PSP) variations in the users pipe network	FMI
27 day local forecast issued every 24 hours	Forecast of local, 24 hour activity index using ARMA (CME override; ENLIL and/or Graz model may be used to improve the forecast in a later update)	TGO
Next 24 hours and next 24-48 hours local forecast issued every hour	Sliding forecast of local, 24 hour activity index using ARM (CME override; ENLIL and/or Graz model may be used to improve the forecast in a later update)	TGO
27 day global forecast issued every 24 hours	Forecast of Ap index using ARMA (CME override; ENLIL and/or Graz model may be used to improve the forecast in a later update)	TGO
Short term (1 hour) Kp forecast	Forecast using MAK method.	TGO
Short term (1 hour) local geomagnetic forecast	Forecast using GAFS method	TGO
Real-time and historic geomagnetic activity plots and data files for geomagnetic surveying (Total field)	Based on data from Norwegian magnetometer network	TGO
Real-time and historic geomagnetic activity plots and data files for directional drilling (Total field, declination and inclination)	Based on data from selected stations in Norwegian magnetometer network	TGO
E-mail alerts for geomagnetic disturbances	User receives e-mail when activity reaches certain threshold in a particular region.	TGO

Table 1: List of G-ESC products.

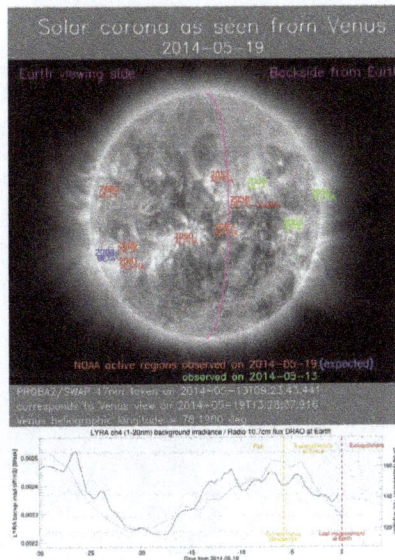

Figure 12: Illustration of a space weather bulletin for Venus Express operations showing an extrapolation of the solar disk image, EUV background irradiation and F10.7 index matching Venus viewpoint.

Conclusion

The ESA SSA SWE programme is driving the development of a system to monitor, predict and disseminate space weather information and alerts, including products and services available for use by both space an ground based user communities. This development process is guided by regular assessments of the requirements and of the product quality, with end users in the loop. The challenge of this process is to understand the user needs and to translate them into targeted services providing accurate and timely space weather information. The space community is therefore invited to contribute with feedback and recommendations.

Acknowledgment

This paper has been prepared in the frame of the ESA Contract No. 4000112444/14/D/MRP.

References

1. Colorado B (2015) ESA SSA space weather service system. Juha-Pekka Luntama Space Weather Workshop.

2. SSA-SWE (2011) Space weather customer requirements document. SSA-SWE-RS-CRD-1001.

3. SSA-SWE (2013) Space weather system requirements document. SSA-SWE-RS-RD-0001.

4. Robbrecht E, Berghmans D (2004) Automated recognition of coronal mass ejections (CMEs) in near-real-time data, Astronomy and Astrophysics 425: 1097-1106.

5. Robbrecht E, Berghmans D, Van Der Linden RAM (2009) Automated LASCO CME Catalog for solar cycle 23: are CMEs scale invariant? The Astrophysical Journal 691: 1222-1234.

6. Georgoulis P, Manolis K, Rust C, David M (2007) Quantitative forecasting of major solar flares, The Astrophysical Journal 661: L109-L112.

7. Georgoulis MK (2011) Pre-eruption magnetic configurations in the active-region solar photosphere. The Physics of Sun and Star Spots Proceedings of the International Astronomical Union IAU Symposium 273: 495-498.

8. Georgoulis MK (2010) Towards an efficient prediction of solar flares: Which parameters and how? Entropy 15: 5022-5052.

9. Odstrcil D, Pizzo VJ, Arge CN (2005) Propagation of the 12 May 1997 interplanetary coronal mass ejection in evolving solar wind structures, Geophysical Research: Space Physics.

10. Vrsnak B, Temmer M, Veronig AM (2007) Coronal holes and solar wind high-speed streams: I Forecasting the solar wind parameters. Solar Physics 240: 315-330.

11. Zic T, Vrsnak B, Temmer M (2015) Heliospheric propagation of coronal mass ejections: Drag-based model fitting. The Astrophysical Journal Supplement Series 218: 7.

12. Souvatzoglou G, Papaioannou A, Mavromichalaki H, Dimitroulakos J, Sarlanis C (2014) Optimizing the real-time ground level enhancement alert system based on neutron monitor measurements: Introducing GLE Alert Plus. Space Weather 12: 633-649.

13. Latocha M, Bek P, Rollet S (2009) AVIDOS a software package for European accredited aviation dosimetry Oxford Journals Science & Mathematics Radiation Protection Dosimetry 136: 286-290.

14. Jakowski N, Mayer C, Borries C, Wilken V (2009) Space weather monitoring by ground and space based GNSS measurements Proc. ION – International Technical Meeting Anaheim CA.

15. SWE (2011) Service Validation Report (SWE-DELI5-0001).

16. Tsagouri I, Belehaki A (2008) An upgrade of the solar wind driven empirical model for the middle latitude ionospheric storm time response. J Atmos Sol Terr Phys.

17. Tsagouri I, Zolesi B, Belehaki A, Cander LR (2005) Evaluation of the performance of the real-time updated Simplified Ionospheric Regional Model for the European area. J Atmos Sol-Terr Phys 67: 1137-1146.

18. Zolesi B, Belehaki A, Tsagouri I, Cander LR (2011) Real-time updating of the simplified ionospheric regional model for operational applications. Radio Science 39: 2.

19. Wintoft P, Wik M, Viljanen A (2015) Solar wind driven empirical forecast models of the time derivative of the ground magnetic field. J Space Weather Space Clim 5: 9.

20. Mc Pherron RL (1999) Predicting the Ap index from past behaviour and solar wind velocity. Phys Chem. Earth (C) 24: 45-56.

21

Buoyancy Explains Terminal Velocity in Skydiving

Landell-Mills N*

Edinburgh University, 75 Chemin Sous Mollards, Argentiere 74400, France

Abstract

Estimates show that skydivers in free-fall displace a mass of air downwards equal to their own mass every second, in order to maintain a constant terminal velocity. This is also demonstrated at indoor skydiving centers where air blown upwards can suspend skydivers in mid-air. Like a boat floating in water, the skydiver is floating on air. Consequently, Archimedes principle of buoyancy can be used to explain the physics of terminal velocity in skydiving. Conventional physics explains that drag, the force needed to push air out of a skydiver's path, sets a limit to a skydiver's velocity. Which is correct but incomplete. It is more accurate to add that according to buoyancy, the skydiver's velocity will increase until a mass of air equal to his own mass is displaced each second.

Drag on a skydiver is defined by the equation:

Drag = 0.5 (Velocity^2 × Air Density × Surface Area × Drag Coefficient)

This equation has severe limitations as It relies on a drag coefficient which must be already known in order to calculate terminal velocity. Worse, this drag coefficient cannot be directly measured and changes constantly. Why is this important? This demonstrates that buoyancy applies to objects that move and is measured over a one second time period. At present, buoyancy is only applied to stationary objects, such as boats or balloons. Also, buoyancy provides a simpler and more accurate method to estimate terminal velocity, without having to know the drag coefficient. This paper predicts that all objects falling at terminal velocity will displace a mass of fluid equal to their own mass each second to maintain buoyancy and a constant terminal velocity. An explanatory video: "Buoyancy explains terminal velocity in skydiving," is available on youtube, on channel of 'N Landell' (the author of this paper).

Keywords: Physics; Buoyancy; Skydiving; Archimedes; Terminal; Velocity; Airborne; Fly; Float

Method

This work was completed after extensive research, as well as numerous discussions with academics, engineer and skydivers. Experiments were conducted skydiving to test the validity of the assertions documented in this paper. The findings from the experiments were consistent with the assertions made in this paper (Figure 1).

Definitions

Terminal velocity

Terminal velocity is the highest velocity attainable by an object as it falls through a fluid (e.g. water or air) and there is no acceleration.

Free-fall

Free-fall is the downward movement of an object towards the ground under the force of gravity only.

Key Equations

Equation A

The equation for the drag on a skydiver at terminal velocity:

Drag = 0.5 (Velocity^2 × Air Density × Surface Area × Drag Coefficient)

Equation B

Weight = Mass × Gravity

Force = Mass × Acceleration

Equation C

Fluid Pressure = (Volume of Air × Air Density × Acceleration) / Area

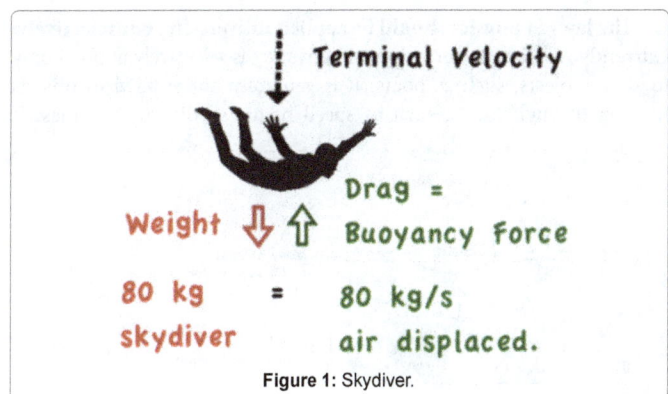

Figure 1: Skydiver.

Equation D

Upward Buoyancy Force = Net Air Pressure × Surface Area

= (Volume of Air × Air Density × Acceleration) / Area) × Surface Area

= Volume of Air × Air Density × Acceleration

= Mass of Air × Gravity

where the buoyancy force is measured every second.

***Corresponding author:** Landell-Mills N, Edinburgh University, 75 Chemin Sous Mollards, Argentiere 74400; France, E-mail: nicklm@gmx.com

Physics, Philosophy and Logic

Buoyancy is consistent with the laws of physics

This paper claims that Archimedes 2,300 years-old principle of buoyancy is a fundamental principle of physics that applies to all objects at all times. Therefore, buoyancy should be applied to moving objects, such as a skydiver in free-fall.

As a skydiver falls, energy is used to displace the air mass in his path, as well as indirectly displacing air. There is no net loss or gain of mass, energy or momentum. Energy is transferred from the falling skydiver to the air to generate a buoyancy force. In a vacuum the lack of air means that a buoyancy force is impossible; so, objects will fall at the same terminal velocity.

Philosophy and logic

This argument for buoyancy is consistent with what is observed in reality and the factors that affect the terminal velocity of a skydiver, as explained later. This paper also takes note of Occam's razor: The simplest explanation given the evidence is often true. Current explanations of drag tend to be complex and abstract.

The logic of applying buoyancy to moving objects is straight forward:

a. Gravity applies universally to all stationary and moving objects.

b. Buoyancy is a product of gravity.

c. Therefore, buoyancy should apply universally to all stationary and moving objects at all times, such as skydivers. Gravity and buoyancy are universal.

The laws of physics should be applied universally, not selectively: Currently Archimedes principle of buoyancy is selectively applied only to static objects, such as boats. It is generally not applied to objects moving through fluids, such as speed boats, skydivers or planes. It

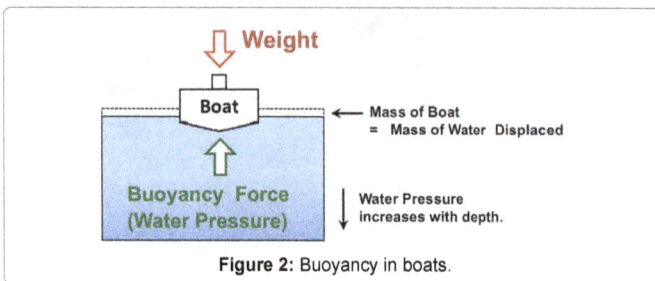
Figure 2: Buoyancy in boats.

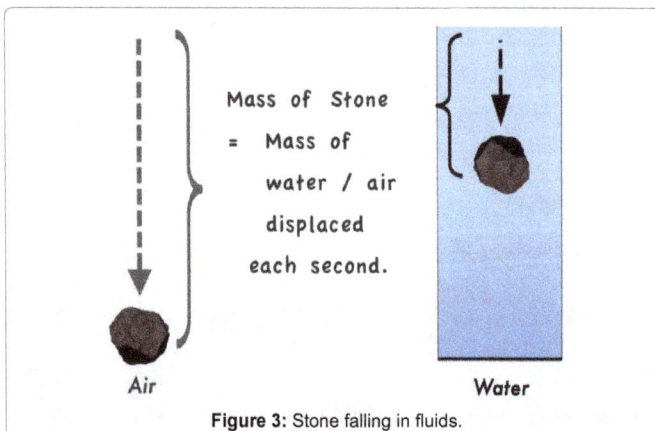
Figure 3: Stone falling in fluids.

has never been proved, or disproved, whether Archimedes' principle of buoyancy applies to objects that move. Boats and hot air balloons float due to buoyancy. There is no reason that moving objects should be exempt from the need to maintain buoyancy. Buoyancy doesn't stop acting on the object just because it's moving.

Archimedes principle of buoyancy

Archimedes principle of buoyancy [1,2] states that boats float due to an upward buoyancy force equalling the downward weight of the boat, as shown by Figure 2. To float, boats must displace a mass of water equal to their own mass. The buoyancy force is just the upward water pressure at the bottom of the boat; which is equal to the amount of water pushed up by the boat. Pressure and weight are both due to gravity.

Archimedes principle of buoyancy:

Weight of boat = Buoyancy Force (Water Pressure).

Mass of Boat × Gravity = Mass of Water Displaced × Gravity.

Mass of Boat = Mass of Water Displaced.

Example of a stone falling in different fluids

A stone falling at terminal through water and air demonstrates the principle of buoyancy applied to moving objects. This paper claims that at terminal velocity a stone displaces a mass of air or water downwards equal to its own mass each second, consistent with the principle of buoyancy (Figure 3).

The speed of the stone depends primarily on how far it must fall each second before it displaces a mass equal to its own mass. Water has a much greater density (mass per unit volume) than air. Consequently, the stone falls a much smaller distance in water than air each second. The mass of water or air displaced, includes both the mass directly in the path of the stone, as well as the mass that is indirectly displaced.

Note that water is more viscous than air, so the friction will be greater than in the air. But this remains a relatively minor consideration in this example.

Results

The difference speeds of stone falling in water and air at terminal velocity is consistent with the principle of buoyancy. Experimentation needs to be done to verify exactly what the difference is.

The Physics of Terminal Velocity in Skydiving

The principle of buoyancy is applied to skydiving to provide insight into how buoyancy affects objects falling through the air.

Buoyancy force vs. drag at terminal velocity

According to conventional physics, drag sets a limit to the skydiver's velocity. Drag is the force (energy) needed to physically push the air out of the path of the skydiver. At terminal velocity drag will equal the weight of the skydiver. Drag does not analyse how much air is displace nor in what direction it is displaced (up or down).

Whereas, according to buoyancy, the skydiver's velocity will increase until the mass of air displaced each second equals to his own mass. At terminal velocity, the buoyancy force will equal the weight of the skydiver. For example; At terminal velocity, a 80 kg skydiver falls at about 66.7 m³ (240 km/hr). Conventional physics explains that this is the maximum speed possible, where the skydiver has sufficient force to push air out of his way. Drag prevents his velocity from increasing. Which is correct but incomplete.

It is more accurate to add this statement that 80 kg/s of air is supporting the skydiver in the air, due to the buoyancy force from the air being displaced above and below the skydiver. This is demonstrated by the fact that if air is blown upwards at 66.7 m/s at an indoor skydiving centre; it will suspend the same 80 kg skydiver in mid-air. Then there is no drag on the skydiver (the skydiver is the drag on the air blown upwards). But the same buoyancy force exists. Here the skydiver is floating in the air, for the same principles of physics that explain how a boat floats in water: buoyancy.

The equation for buoyancy force

Buoyancy Force = Mass of air displaced (every second) × Gravity

The equation for the buoyancy force is consistent with Newton's 2nd law of motion (F = ma) [1].

Force = Mass × Acceleration

Applying Newtons 3rd law of motion to skydiving; In stable, free-fall descent at terminal velocity the forces are in balance. Therefore, every force will have an equal and opposite force. The weight of the skydiver will equal the buoyancy force.

Weight of skydiver = Mass of skydiver × Gravity Mass of skydiver

= Buoyancy Force

= Mass of Air displaced × Gravity

= Mass of Air displaced

Consequently, the skydiver will displace a mass of air equal to his own mass each second. For example, to maintain buoyancy an 80 kg

Figure 4: Water jet pack.

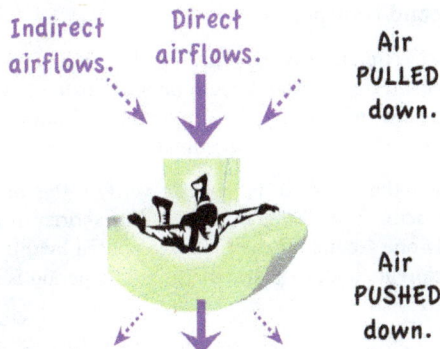

Figure 5: Direct and indirect airflows around the skydiver.

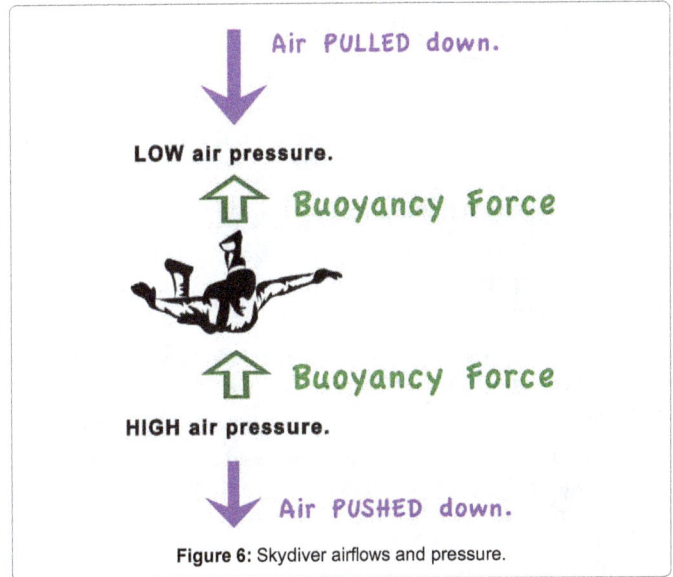

Figure 6: Skydiver airflows and pressure.

skydiver will displace 80 kg/s of air downwards; in order to create a sufficient equal & opposite force to push the skydiver upwards.

Jet packs example

To put it another way. If the skydiver could hypothetically (somehow) blow or displace 80 kg of air downwards each second, then he could remain suspended in mid-air. By comparison, water jet packs work on this principle (albeit, displacing water downwards, not air) (Figure 4).

The skydiver directly and indirectly displaces air

Critically, the total mass of air displaced downwards by the skydiver includes the air directly in his path and air indirectly displaced (Figure 5). Substantial energy is being transferred from the skydiver to the air. This energy has to go somewhere. It is used to directly and indirectly displace air.

Two airflows and two Newtonian forces

At terminal velocity, there are two separate forces that keep the skydiver's velocity from increasing. These force result from the skydiver pushing the air below him downwards, and pulling air above him down (Figures 6 and 7).

Applying Newtons 3rd law of motion [1] it is shown that there are two separate equal and opposite reactions to the forces acting on the skydiver. As the skydiver falls at terminal velocity:

1. A vacuum of air develops immediately above (behind) the skydiver. This causes the air above the skydiver to expand creating LOW air pressure. This PULLS air above the skydiver down. The equal & opposite force PULLS the skydiver UP.

2. The skydiver compresses the air below him as he falls; creating HIGH air pressure. This air is PUSHED down. The equal & opposite force PUSHES the skydiver UP.

At terminal velocity the forces are in balance. The buoyancy force (and net air pressure) equals the weight of the skydiver. Consequently, the mass of the skydiver will equal the mass of air displaced downwards.

Weight of skydiver = Mass of skydiver × Gravity Mass of skydiver

= Buoyancy Force

= Mass of Air displaced × Gravity

= Mass of Air displaced

Note that buoyancy is no more a "Newtonian Theory of skydiving" than walking or swimming are "Newtonian Theories of walking or swimming." Newtons laws of motion are applied universally.

How far down each kilogram of air is displaced

The farther that each 1 kg of air is displaced (pushed or pulled) by the skydiver, then the more air that is displaced; and the greater air mass displaced in total.

As 1 m³ of air has a mass of 1.2 kg (at a standard air density of 1.2 kg/m³); Then in order for it to displace 1 kg of air, this 1 m³ of air must be displaced down about 0.83 meters, (as: 0.83 m = 1 kg / 1.2 kg/m³).

Consequently, if each 1 m³ of air is displaced down over 1.66 meters by the skydiver. Then a total of 2 kg of air mass is displaced (as: 0.83 m × 2 = 1.66 m).

Figure 7: Tandem skydivers and mini parachutes.

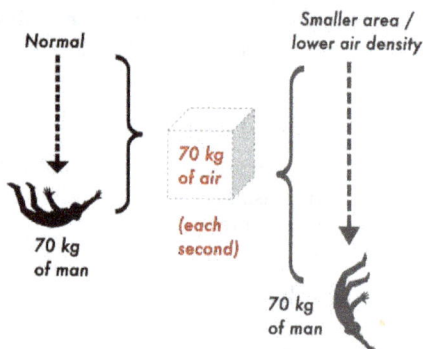

Figure 8: Skydiver; area and air density.

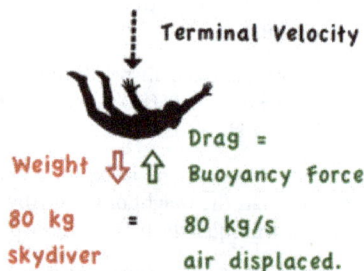

Figure 9: Skydiver at terminal velocity.

Key determinants of terminal velocity

This demonstrates that the key determinants of a skydiver's terminal velocity are consistent with buoyancy and empirical observations. These key determinants the skydiver's mass and the mass of air that he displaces. i.e., This is how far the skydiver falls each second before he displaces a mass of air equal to his own mass.

The skydiver's mass: Heavier skydivers fall faster: Heavier skydivers achieve a higher terminal velocities (assuming a constant surface area). A bigger mass of a skydiver will fall a bigger distance each second, in order to displace a correspondingly bigger air mass. Also, a bigger mass provides greater momentum and energy to displace more air each second. This is easiest shown by tandem skydivers, who fall a lot faster than solo skydivers (while having approximately the same surface area exposed to the direction of descent). Tandem skydivers typically deploy mini-parachutes to slow themselves down in free-fall (and to provide extra stability); so as to fall at the same velocity as solo skydivers (Figure 7).

Result: This observation and logic is consistent with buoyancy.

The mass of air displaced by the skydiver (surface area and air density): The amount of air mass displaced by the skydiver each second depends on his surface area exposed to the direction of travel (downwards); and the air density. The evidence for this is that skydivers go faster when head first and when the air is thinner. (Figure 8). In fact, the skydiver in the 2012 Red Bull Sky Jump reportedly broke the sound barrier when falling through the stratosphere, and slowed down as he came closer to the earth.

In a vacuum, no buoyancy force is possible. There is no air to provide any resistance to the object as it falls. So, objects in a vacuum will fall at the same velocity.

Result: This observation and logic is consistent with buoyancy.

Conclusions: Empirical observations of the factors that affect the terminal velocity of a skydiver, is consistent with buoyancy.

Consequently, a skydiver in free-fall will have a higher terminal velocity if:

1. The object is heavier (i.e., bigger mass); so, will travel farther in one second to displace more air. Heavier people or tandem skydivers will fall faster in free-fall; if the surface area is constant.

2. The surface area of the object exposed to the downward direction is smaller; or the lower the air density. In these cases, the object will go a farther distance down each second, before displacing a given mass of air each second.

The one-second time period

This paper estimates that Archimedes principle of buoyancy acts over a one second time period. i.e., An object must displace a mass of air equal to its own mass each second to maintain buoyancy. This one second time period is just an initial estimate.

Experimentation needs to be done to verify if this one second is accurate. The actual time frame may be slightly shorter or longer, and not be exactly one second. The core idea proposed by this paper does not alter if experimentation shows that the time period is different to one second.

A skydiver displaces air downwards as he falls. It is unclear over exactly what time period the skydiver will displace a mass of air downwards equal to his own mass.

The theoretical explanation for this one second time period is uncertain. Why one second; rather than 1.2 seconds or 0.8 seconds (for example), in unclear. There does not appear to be anything intrinsically or fundamentally special about this one second time period.

Conclusions: These arguments demonstrate that buoyancy applies to moving objects. That it is logically feasible for a skydiver to displace a mass of air downwards equal to his own mass each second (at terminal velocity, in stable free-fall descent).

Skydiving Calculation Example

Objective: This is to demonstrate via an example calculation, that it is theoretically feasible for a man to displace a mass of air equal to his own mass each second, at terminal velocity while in a free-fall descent (Figure 9). Experimentation needs to be done to confirm this is actually what happens.

Calculation of buoyancy

Assumptions: A skydiver is in a stable free-fall descent at terminal velocity.

a. An 80-kg skydiver descends at a terminal velocity of 66.7 m/s (or 240 km/hr).

b. This assumption is an estimate is based on anecdotal evidence from

c. Skydivers and skydiving organizations.

d. This assumption is also consistent with the terminal velocity provided by the conventional equation (A) for drag, based on a drag coefficient of 0.294. See

e. Section 6.4 (Example calculation of drag), for the detailed calculation of this.

f. The 80 kg includes his clothes and equipment.

g. Lower surface area of the man = 1.0 m².

This is about half the estimated total surface skin area of a 80 kg man with a height of 180 cm (5 ft. 10 inches); of about 2.0 m²; based on the body surface area (BSA) calculator; (using the Motseller formula). In respect to this assumption of surface area; Note that the skydiver's lower legs are partially obscured from the direction of descent, but this is compensated for by the skydiver's clothes and equipment adding to the total surface area [2,3].

i. Standard air density = 1.2 kg/m³.

ii. These calculations exclude any significant friction that would materially slow the skydiver's descent. The viscosity of air is low, so air friction is considered to be negligible. Evidence for this is the lack of any heating due to friction during a free-fall descent. However, note that as velocity increases, friction becomes increasingly significant.

iii. The skydiver only displaces air that is directly in his path.

iv. For simplicity purposes, there is no air that is indirectly displaced by the skydiver. This parameter is extremely difficult to estimate. Therefore, the estimate of the total air displaced by the skydiver is under-estimated by this assumption.

v. These estimates and assumptions are approximate, as they are only used to demonstrate the feasibility and reasonableness of the argument for buoyancy.

Calculations

First, the volume of air displaced by the skydiver is estimated:

Volume of air displaced by the skydiver = Velocity of Skydiver × Lower Surface Area of Skydiver

= 66.7 m/s × 1.0 m²

= 66.7 m³/s

As Mass = Volume × Density; then,

Mass of Air Directly Displaced by Skydiver = Volume of Air Displaced × Air Density

= 66.7 m³/s × 1.2 kg/m³

= 80 kg/s

Each 1 kg/s of air displaced must be displaced down by the skydiver about 0.83 meters; given that the density of the air is 1.2 kg/m³ (0.83 m = 1 kg / 1.2 kg/m³). So, 80 kg of air mush be displaced down about 0.83 meters, each second (by the 80 kg skydiver).

Note that these results are consistent with the data quoted by most skydiving organizations.

These estimate the terminal velocity of a 70-80 kg skydiver at 55-67 m/s (200–240 km/hr).

Conclusion

Based on the assumptions above, an 80 kg skydiver will directly displace 80 kg of air downwards each second, at terminal velocity. Therefore, at terminal velocity a skydiver will displace a mass of air equal to his own mass each second. Skydiving centres use large fans used to blow air upwards, to suspend skydivers in mid-air, to practice free-fall (Figure 10). This provides a reasonableness check that the estimate of terminal velocity for an 80-kg skydiver is about 66.7 m/s. In physics, all movement is relative. So to demonstrate buoyancy, it does not matter if the skydiver is falling through stationary air, or if a stationary skydiver is having air blown upwards. The physics is the same, assuming that air is Galilean invariant.

Results

Anecdotal evidence shows that the skydiver remains suspended mid-air if the fan blows the air up at the same speed as the man's terminal velocity when actually skydiving. So a fan blowing with a 80 kg/s force (with air blown up at 66.7 m/s), can suspend a 80 kg man mid-air. This speed of 66.7 m/s is well within the range that indoor skydiving centers advertise for the capacities of their fans. The skydiver is floating in the air based on the same principles of physics that explain how boats float on water.

Figure 10: Skydiving practice.

Sensitivity analysis

Using similar assumptions to those as above in Section 8.1:

a) Air Density = 1.2 kg/m³

b) Lower surface area of the skydiver is based on half the estimated the body surface area (BSA) for a ma 180 cm tall; using the Motseller formula.

c) Air friction is not included.

d) The skydiver only displaces air that is directly in his path.

Using similar methodology to that above in Section 5.1; Based on the principle of buoyancy, the terminal velocity of the skydiver is estimated based on his mass and the corresponding mass of air he is estimated to displace (Table 1).

Results

Terminal velocity for skydivers with a mass on 60-90 kg is 208-255 km/hr. Each additional 1 kg on the skydiver adds about 0.4 m/s (1.6 km/hr) to the skydiver's terminal velocity. The terminal velocities estimated are consistent with what is observed in reality, albeit slightly higher; based on evidence from skydiving institutions and skydivers [4].

Conclusion

Sensitivity analysis demonstrates it is feasible for buoyancy to explain the differences in terminal velocity of skydivers with a different mass. Buoyancy is consistent with what is observed in reality.

The Limitations of the Equation for Drag

The equation for drag

Conventional physics states that the for formula for drag on a falling object is (Equation A):

Drag = $0.5 \times Velocity^2 \times$ Air Density $\times$ Area $\times$ Drag Coefficient

Drag is defined in conventional physics as the physical force needed to push the air out of the way of the falling skydiver. This sounds a lot like the kinetic energy transferred from the skydiver to the air. Conventional physics claims that at terminal velocity, the weight of the skydiver will equal his drag (Figure 11). "The drag coefficient then expresses the ratio of the drag force to the force produced by the dynamic pressure times the area". The drag coefficient is typical a number between zero and one; but is rarely above two.

Problems with the equation for drag

This paper claims that the conventional equation for drag only partially describes what is observed in reality. That the equation is incomplete as it doesn't analyse how nor where the air is displaced.

The equation is merely a description of what is observed in reality. It does not answer why the equation has the form that it does.

Mass of Skydiver	kg	60.0	70.0	80.0	90.0
Surface Area	$_m2$	0.87	0.94	1.00	1.06
Terminal Velocity	m/s	57.8	62.4	66.7	70.8
Terminal Velocity	km/hr	208	225	240	255
Mass of Air Directly Displaced	kg	60.0	70.0	80.0	90.0

Table 1: The terminal velocity of the skydiver based on his mass and the corresponding mass of air he is estimated to displace.

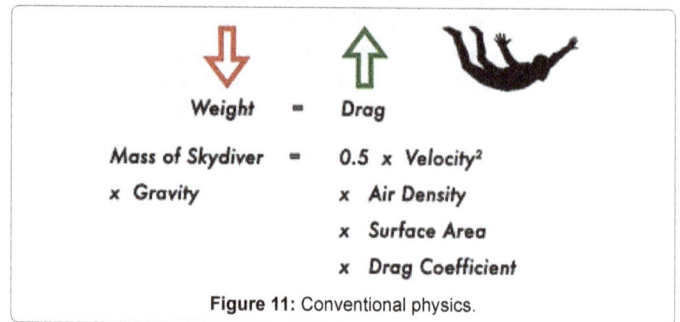

Figure 11: Conventional physics.

Problems with the drag coefficient

This paper also claims that the conventional equation for drag has limited usefulness due to the problems with the drag coefficient, as explained below. In short, the drag coefficient appears to be the residual factor that is used to make the drag equation work.

Problems with the drag coefficient include:

a. The drag coefficient can only be indirectly observed. It cannot be directly measured. Whereas parameters like velocity and area can be directly measured. This is a significant problem because it means that the terminal velocity of an object in free-fall cannot be estimated unless its drag coefficient is already known. But to accurately calculate the drag coefficient, the terminal velocity must be known.

b. The drag coefficient is not a fixed amount for an object, varies between objects and can change constantly with circumstances.

c. The drag coefficient varies with the speed and typed of airflow. Laminar airflow can produce a different drag coefficient to disrupted or turbulent airflow. The drag coefficient varies with the Reynolds number.

d. Similar objects (similar shapes) can have different drag coefficients. A heavier skydiver has a different drag coefficient to a lighter skydiver.

e. The drag coefficient can be different between water and air for the same object.

f. The drag coefficient changes with the amount of surface area. So the same object can have a different drag coefficient based on different amounts of surface area being exposed to the direction of travel. A skydiver falling head-first down, has a different drag coefficient to a skydiver in a standard flat alignment.

g. The appropriate drag coefficient for the skydiver is uncertain. But the calculation for terminal velocity is extremely sensitive to the drag coefficient.

Example calculation of drag

The conventional equation for drag is used to demonstrate that calculation for the terminal velocity of the skydiver above of 66.7 m/s is reasonable, but nothing more (Figure 9).

Conventional physics states that the for formula for drag on a falling object is (Equation A):

Drag = $0.5 \times Velocity^2 \times$ Air Density $\times$ Area $\times$ Drag Coefficient

At terminal velocity, all forces are in balance; so, Weight = Drag

Therefore,

Mass of Skydiver $\times$ Gravity = $0.5 \times$ Terminal Velocity$^2 \times$ Air Density $\times$ Surface Area $\times$ Drag Coefficient

Re-stating these equations:

Terminal Velocity^2 = (Mass of Skydiver × Gravity) / (0.5 × Air Density × Surface Area × Drag Coefficient)

Assumptions

- Air Density = 1.2 kg/m^3
- Gravity = 9.8 kg m/ s^2
- Surface area of skydiver = 1.0 m^2
- Mass of skydiver = 80 kg
- Drag Coefficient for the skydiver = 0.294

This is simply the drag coefficient that makes the equation work so that the mass of the skydiver will equal the mass of air displaced.

Terminal velocity can be calculated

Terminal Velocity^2 = (80 kg × 9.8 kg m/s^2) / (0.5 × 1.2 kg/m^3 × 1.0 m^2 × 0.294) = 784 / 0.176 = 4444.4

Terminal Velocity = 66.7 m/s (or 240 km/hr)

Results

Based on these assumptions, a 80 kg skydiver would have a terminal velocity of 66.7 m/s. This is consistent with the calculations above using buoyancy to estimate terminal velocity of the skydiver, in Section 5.

The assumption for the drag coefficient

Note that this calculation (using the conventional equation for drag to estimate terminal velocity), is based on selecting a drag coefficient that makes the equation work. A drag coefficient was deliberately selected so that the mass of the skydiver will equal the mass of air displaced. The question therefore is whether the drag coefficient used (of 0.294) is reasonable. As the drag coefficient cannot be directly measured, this is impossible to confirm.

For comparison purposes

1. A cube has a drag coefficient of 1.0;

2. A sphere has drag coefficient of 0.5, and

3. A streamlined object (e.g. aircraft wing) has a drag coefficient of about 0.04.

Therefore, a drag coefficient of about 0.3 for a skydiver is only reasonable if the skydiver is more streamlined than a sphere but less than a wing. The appropriate drag coefficient for the skydiver is uncertain. But the calculation for terminal velocity is extremely sensitive to the drag coefficient. However, a drag coefficient of over 0.5, provides estimates for terminal that are inconsistent with what is observed in reality. This would arise if the skydiver was less streamlined than a sphere. For example, a drag coefficient of 0.6 suggest that an 80-kg skydiver would fall at a terminal velocity of 46.7 m/s (168 km/hr); which is well below what is observed in reality [4].

Conclusion - Drag Coefficient

The drag coefficient is not a very reliable source it to make calculations for terminal velocity using the equation for drag. There is a substantial amount of variation possible when applying the drag coefficient in the conventional equation for drag to estimate terminal velocity. As the drag coefficient cannot be directly measured it is open to a lot of speculation when using it to estimate terminal velocity of an object such as a skydiver. Consequently, enormous care must be used when relying on the drag coefficient.

Acknowledgement

This paper and the related analysis is entirely the work of the author.

References

1. Newton I (1872) Philosophiae Naturalis Principia Mathematica. Newton's Laws of Motion.

2. NASA (2017) Glenn Research Centre. National Aeronautics and Space Administration, USA.

3. Cavcar M (2016) International Standard Atmosphere.

4. Publically available data from skydiving institutions.

22

Nanosatellites - The Tool for a New Economy of Space: Opening Space Frontiers to a Wider Audience

Garcia-Cuadrado G*

Celestia Aerospace, Av. Gran via Carles III 157, 08017 Barcelona, Spain

Abstract

Space is still a frontier. The advantages of research in microgravity conditions are still somehow a private niche of big aerospace contractors and main space agencies. But the landscape is changing and an incipient effort is being pursued to open space frontiers to small and medium-sized companies, universities, under-developed countries and non-profits. We will revise the advantages of microgravity research and a tool to conduct it at low-cost, rapid response and flexibility through the use of nanosatellites opening thus space frontiers to a wider audience. These highly capable satellites can support a wide range of mission objectives, from pure research to technology demonstrators and space qualification tests. The small satellites market is valued 600 M USD to 1.000 M USD yearly with an estimated 2.200 to 2.700 needed launches in the 2015-2020 timeframe. We will also introduce a new launcher under development to serve specifically the nanosatellite incipient market to help solve the scarce launching opportunities served today by conventional launchers.

Keywords: Space; Economy; Nanosatellite; Microgravity

Introduction

The global space economy in 2012 was valued at over £200Bn in commercial revenue and government budgets, and is expected to double by 2030 [1]. This growth is expected to come from the commercial sector, driven primarily by entrepreneurs and new business models. With historical average growth of 37% per year over the last four years, and estimated growth of 24% per year over the next six years, the small satellite market [2] is expected to be an important driver of growth in the overall space sector [2,3].

The global SmallSat market is valued 600 M USD to 1.000 M USD yearly [4] (most conservative and most optimistic estimations, respectively) with an estimated 2.200 to 2.700 needed launches in the 2015-2020 timeframe, and all these, under the current restricted circumstances, that is: without a disruptive solution as the one we are proposing that could boost the former figures.

The Small Satellites Market

In a recent study by the company Commercial Spaceworks, projections indicate strong growth in nano/microsatellite (1-50 kg) development, with an estimated range of 121 to 188 nano/microsatellites that will need launches globally in 2020 (versus 33 in 2012). Projections are summarized in Figure 1 below. Historical and announced future data sets suggest that the average number of nano/microsatellites launched per year triples with every five year period.

Nano/microsatellite (1-50 kg) development continues to be led by the civil sector, but the defense/intelligence community is showing increased interest and involvement. Applications for nano/microsatellites are diversifying, with increased use in the future for science, Earth observation, and reconnaissance missions.

The Space works Commercial study focuses on nano/microsatellites with masses between 1 kg and 50 kg. The study limits the upper end of microsatellite mass to 50 kg given the relative large amount of satellite development activity in the 1-50 kg range by comparison to the 50-100 kg range; Pico satellites (masses below 1 kg) are not within the scope of the afore mentioned study. The data source for the study is the SpaceWorks Satellite Launch Demand Database (LDDB), a database of all known historical and future satellite projects, including all known nano/microsatellites:

• Currently 377 known future nano/microsatellites (1-50 kg) in the LDDB.

• Currently 47 known future picosatellites in the LDDB (not included in this study).

Two projections were developed from "Announced" and "Optimistic" data sets:

• The "Announced" data set contains all publicly announced nano/microsatellite projects and programs from the SpaceWorks LDDB.

• The "Optimistic" data set consists of the announced plus an inflating factor for known unknowns plus assumed sustainment of certain current projects and programs (e.g. follow-on to NASA Ames EDSN, Cubesat Launch Initiative, DARPA SeeMe).

Nano/Microsatellite CAGR (Compound Annual Growth Rate) indicates an average growth of 8.6% per year over the last 12 years (2000-2012). According to the Announced Dataset, the average growth is of 16.8% per year over the next 7 years (2013-2020); according to the Optimistic Dataset (see below), the average growth is of 23.4% per year over the next 7 years (2012-2020).

Projections derived from the referred study are summarized in Figures 1 and 2 below. Concerning nano/microsatellite purpose trend, there is an evidence of adoption of small satellites for applications beyond technology demonstration, summarized in Figure 3.

*Corresponding author: Garcia-Cuadrado G, President and CEO - Celestia Aerospace, Av. Gran via Carles III 157, 08017 Barcelona, Spain
E-mail: gloria.garciacuadrado@celestiaaerospace.com

As an assessment background for the launch demands projections, there is evidence for an emerging and sustained launch, market for small satellites as historical data indicates (Figure 4). A previous study by the

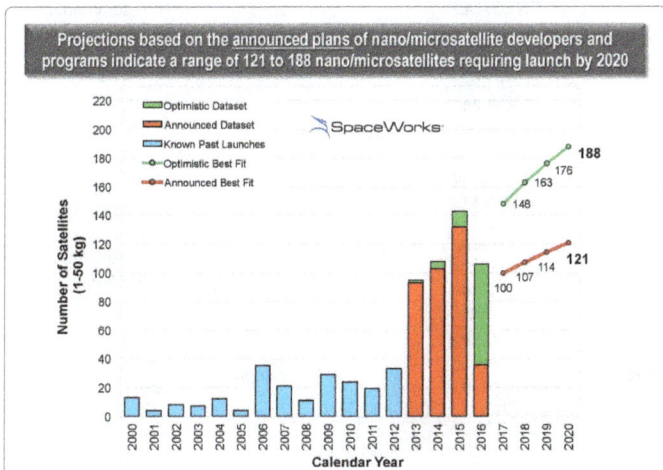

Figure 1: Nano/Microsatellite Launch History and Projections (2000-2020) [Source: Nano/Microsatellite Market Assessment in February 2013 - Spaceworks Commercial].

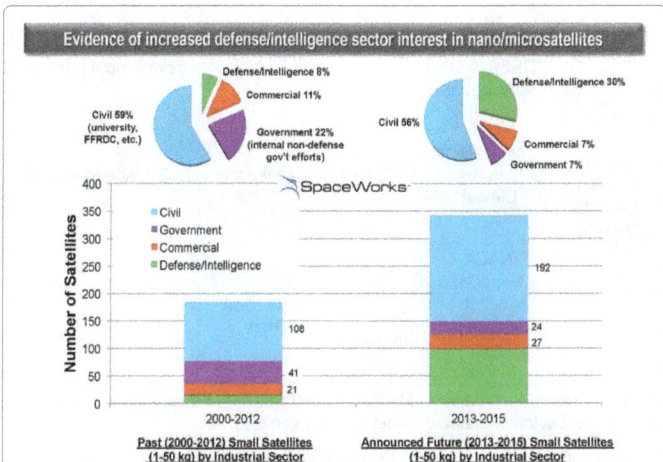

Figure 2: Nano/Microsatellite Launch Demand History and Projections (2000-2015) by target user [Source: Nano/Microsatellite Market Assessment in February 2013 - Spaceworks Commercial].

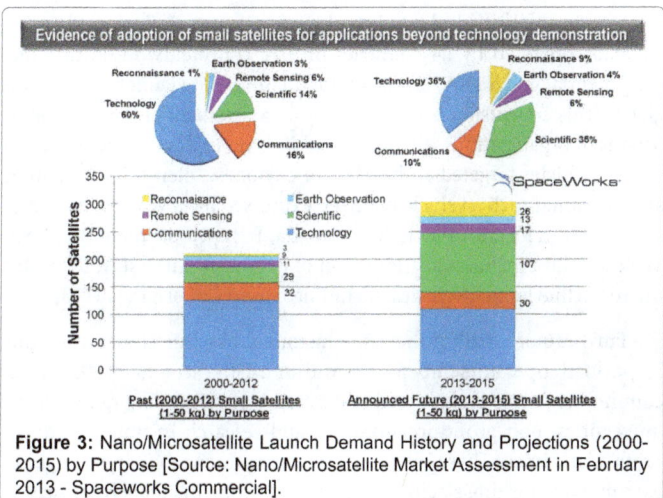

Figure 3: Nano/Microsatellite Launch Demand History and Projections (2000-2015) by Purpose [Source: Nano/Microsatellite Market Assessment in February 2013 - Spaceworks Commercial].

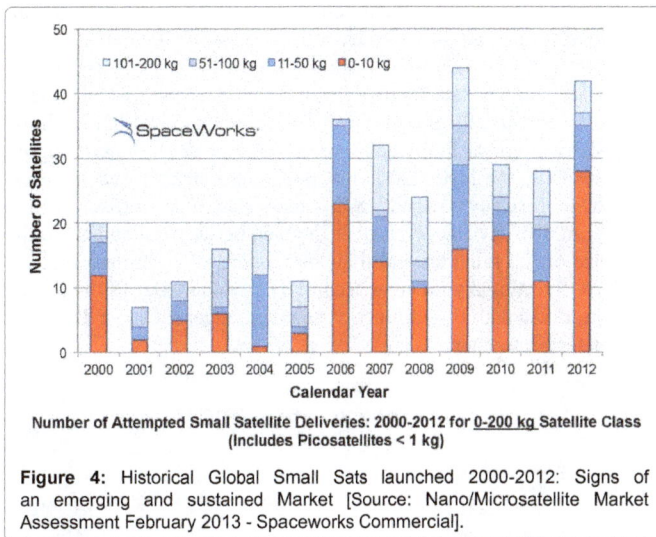

Number of Attempted Small Satellite Deliveries: 2000-2012 for 0-200 kg Satellite Class (Includes Picosatellites < 1 kg)

Figure 4: Historical Global Small Sats launched 2000-2012: Signs of an emerging and sustained Market [Source: Nano/Microsatellite Market Assessment February 2013 - Spaceworks Commercial].

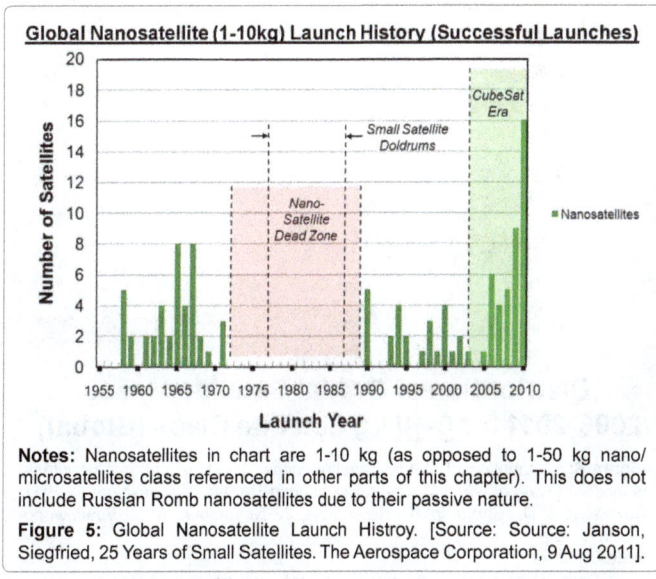

Notes: Nanosatellites in chart are 1-10 kg (as opposed to 1-50 kg nano/microsatellites class referenced in other parts of this chapter). This does not include Russian Romb nanosatellites due to their passive nature.

Figure 5: Global Nanosatellite Launch Histroy. [Source: Source: Janson, Siegfried, 25 Years of Small Satellites. The Aerospace Corporation, 9 Aug 2011].

same company, dating back to 2011, remarks also the impressive boost that the nano/microsatellite market has received with the adoption of the cubesat standard, technology developments, entrance of new developers, and furthering of applications, furthermore if one takes into account that these satellites have existed since the very beginning of the space age era - in fact, the first satellites were of these category - and the "sterile" era which followed, between 1972 and 1989, when no "active" nanosatellites were launched due to the focus on the big ones [5-10]. The last decade (2000-2011) has showed an impressive resurgence of the Nano/Microsatellite niche with evident cubesat popularity, throwing signs for an emerging and sustained market (Figures 5 and 6).

Futron's proprietary Electronic Library of Space Activities (ELSA) Database features over 20.000 interlocking records on all global past, current, and projected future space activity. ELSA contains comprehensive program and technical data on launch events, spacecraft, vehicles, launch sites, and space related organizations. Historical launch data, maintained in ELSA, covering the period from 2000 through all projected 2010 launches, for launches to all inclinations in LEO, indicates that an average of 12 spacecraft in the microsatellite mass class (10-100 kilogram launch mass range) have launched worldwide per year. This figure does not include launch of nanosatellites with a mass of less than 10 kilograms, at least 54 of which were successfully launched during the period from January 2005 through August 2010. Futron's analysis of historical launch data therefore indicates a baseline of 12 launches per year as estimate for the amount of launch demand which might be accessed by microsatellite launch vehicles under development. Figure 7 below shows the historical number of microsatellite-class launches per year. Historically payloads in this mass class have not represented a significant driver of launch vehicle demand because due in part to the small size of the satellites and their developers' often limited budgets they have traditionally flown as secondary payloads. But future trends can be modified with the entrance of a game changer such as a dedicated nanosatellites launcher.

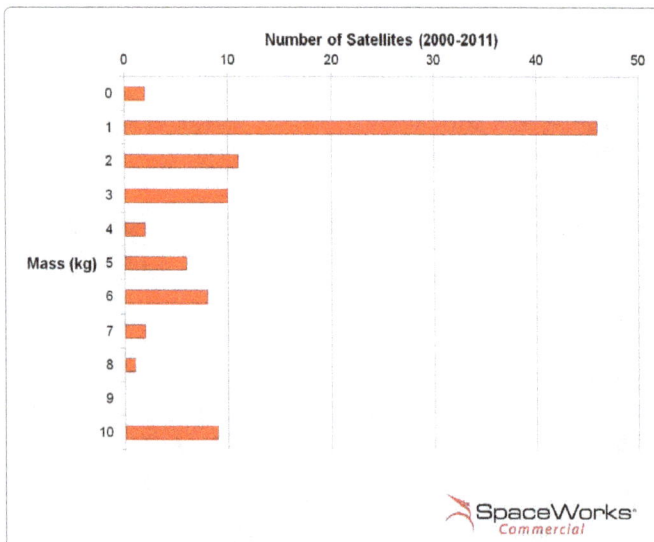

Distribution of Orbital Satellite Mass: 2000-2011 for 0-10 kg Satellite Class (Global)

Figure 6: Distribution of Orbital Satellite Mass: 1U (1Kg) cubesat popularity is evident [Source: Nano/Microsatellite Launch Demand Assessment in 2011, DePasquale, D., Charania, A., Revision A, 22 November 2011 - Spaceworks Commercial].

Notes:
- Data values refer to attempted launches, and may include failures.
- The number of satellites "launched" may not equal the number of launches since many satellites are multiple-manifested (i.e. more than one satellite on a particular launch).
- Many times in this presentation, the term "launch" or "launches" may refer to the number of satellites launched (even though they may be multiple-manifested).
- Data aggregated by Spaceworks Commercial from its Orbital Satellite Database.

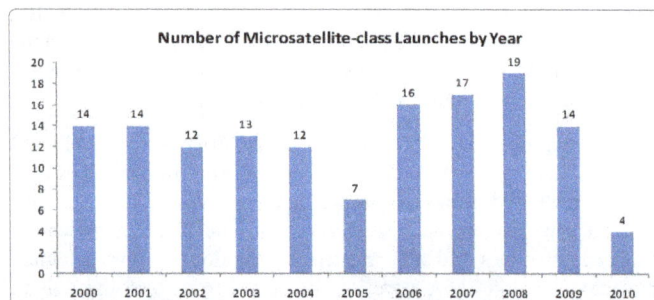

Figure 7: Historical Launch of Microsatellite-class Payloads (2000-2010) [Source: Futron Whitepapers "Market Characterisation: Launch of very-small and nano sized payloads enabled by new launch vehicles].

Vehicle Name	Organization (County)	Max Payload Capacity (LEO)	Development status	Target User/Buyer Markets
Aldebaran'	CNES, DLR, CDTI (France, Germany. Spain)	Up to 3004	Concept Study	Non-commercial, government science and tech. demo missions
Microsatellite Launch Vehicle [9]	Canadian Space Agency (Canada)	Up to 150 kg	Market Study	--
Multipurpose Nanomissile"	U.S Army Space and Missile Defense Command (United States)	Apx 23 kg	In design phases	U.S. Government military payloads, operationally responsive space
Neptune 30"	Interorbital Systems (United States)	30 kg	In-development	CubeSats. Universities and non-profits
Scorpius/ Mini- Sprite"	Microcosm Inc. (United States)	225 kg	Design	1) U.S Military and operationally responsive space 2) U.S Civil Government 3) Educational Organizations
Nano-Launcher [13]	IHI Aerospace, USEF, CSP Japan(Japan)	100 kg	Concept Study	Academia and government missions
Virgin Galactic Small Satellite Launch Vehicle	Virgin Galactic (United States)	100 kg	In-development	Science missions
NA	NASA NanoSat Launch Challenge (United States)	>1 kg and twice in one week	Innovation Prize	Cubesats

Source: Futron Whitepapers - Market Characterization: Launch of very-small and nano-sized payloads enabled by new launch vehicles.

Table 1: Dedicated microsatellite launch vehicles in development or planning.

Low cost piggy-back opportunities have historically attracted small satellite payloads to international launch vehicles as shown in Figure 8.

Within the historical baseline shown in Figure 7, Futron identifies payloads operated by 24 countries on five continents; indicating that the potential geographic extent of the microsatellite launches market is global. This conclusion is supported by an assessment of the geographic origins of organizations with expressed interest in developing dedicated launch vehicles targeted at this class of payloads. Interest in developing and operating such a vehicle has been expressed by governments and/or companies in locations including Canada, Europe, Japan and the USA. Table 1 below, includes a summation of expressed interest in dedicated microsatellite launch vehicles, including country of origin [11-13].

Furthermore, within the same historical baseline shown in Figure 7, payload operators are found within many sectors of the space community. These include civil and military government organizations, universities, non-profit organizations and research institutes, and for-profit companies. These operator groups represent the target buyer or user markets for those actors interested in supplying launch services in

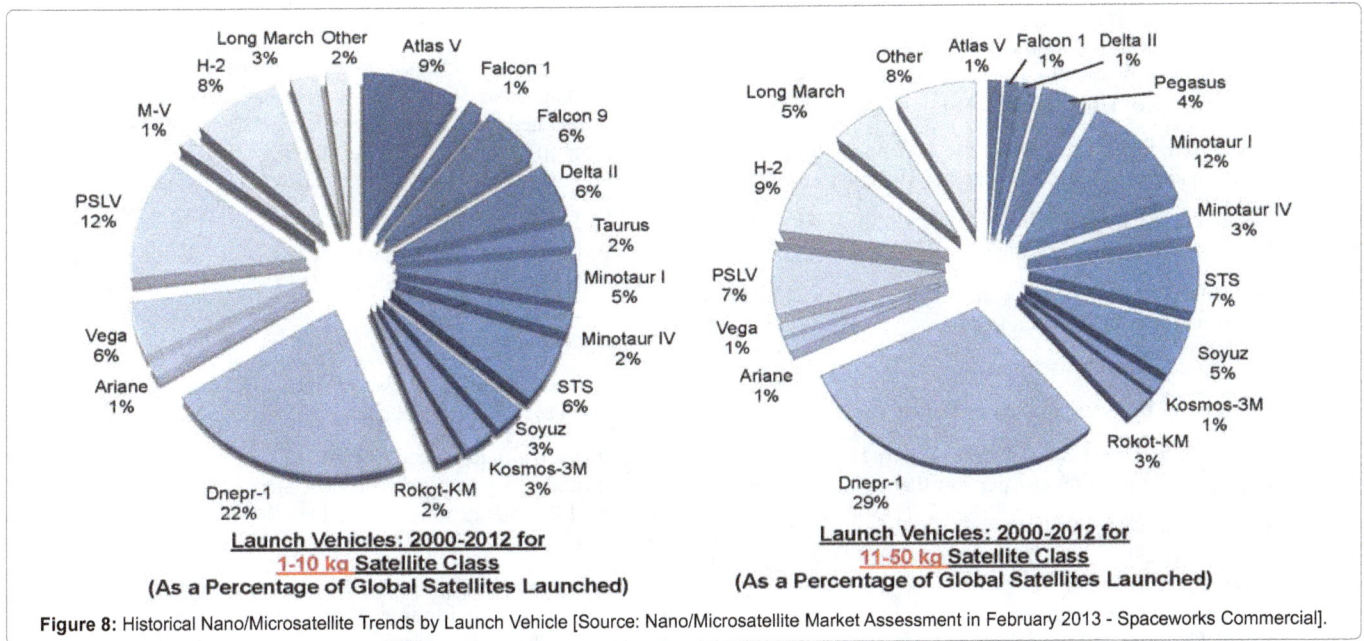

Figure 8: Historical Nano/Microsatellite Trends by Launch Vehicle [Source: Nano/Microsatellite Market Assessment in February 2013 - Spaceworks Commercial].

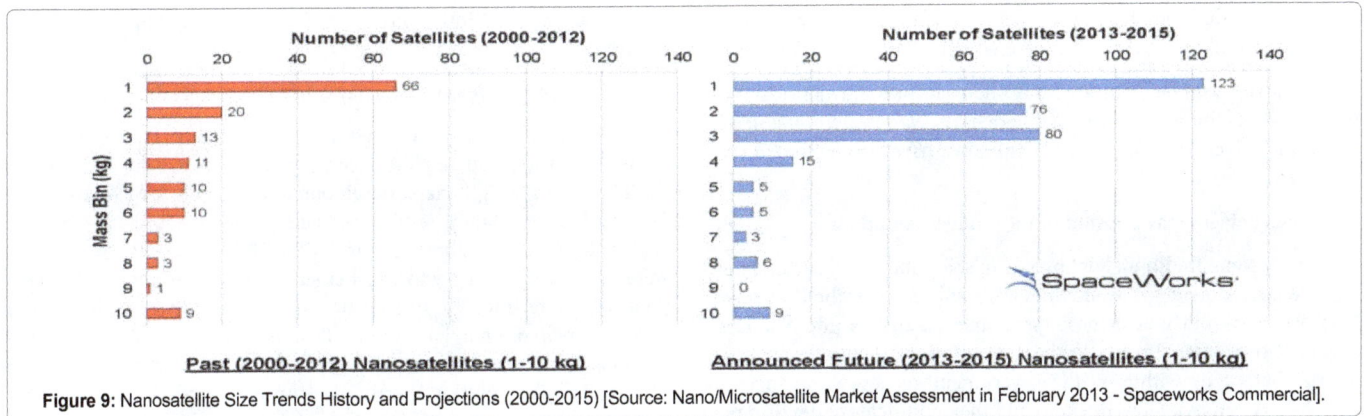

Figure 9: Nanosatellite Size Trends History and Projections (2000-2015) [Source: Nano/Microsatellite Market Assessment in February 2013 - Spaceworks Commercial].

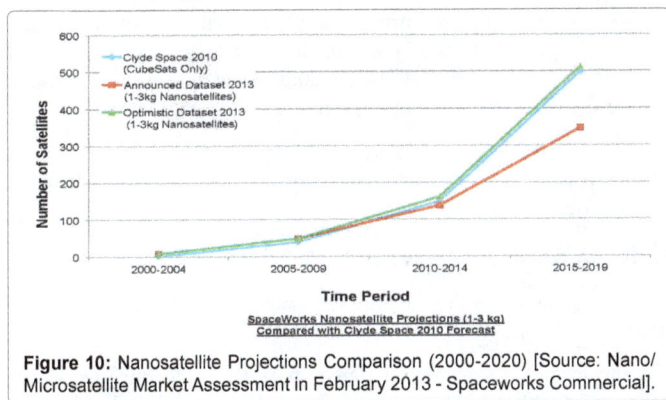

Figure 10: Nanosatellite Projections Comparison (2000-2020) [Source: Nano/ Microsatellite Market Assessment in February 2013 - Spaceworks Commercial].

this industry segment. Table 1, includes also a summation of targeted markets, where known, for dedicated microsatellite launch vehicles in development or planning (Figure 8).

The announced future nanosatellites suggest sustainment of the historically popular 1U (1 kg) Cubesat as well as the emerging 2U and 3U nanosatellites, showing also a future growth of the 6U (8 kg) class. The former is summarized in Figures 9 and 10 below.

Some ending notes are worthwhile taking into account following the Spaceworks Commercial study:

1. The number of satellites may not equal the number of launches since many small satellites are multiple-manifested (i.e., more than one satellite co-manifested on a particular launch vehicle). Data values refer to attempted launches and may include launch failures.

2. All data for nano/microsatellite projects and programs is from publically sourced information. This may not represent all global nano/microsatellite activities.

3. All NSF satellites thus far have launched through the NASA CSLI. In the table, these historical NSF satellites are included in both the count of number launched for NSF and the count for CSLI (double counted in this sense). The bar graph of future launches shows only those NSF satellites that expected, but currently not manifested (thus they are appropriately single counted for future launches).

4. The Announced data set includes some known nano/microsatellite programs for which a specific launch date has not been announced. The satellites belonging to these programs are distributed across the period (date range) for launches according to the announced program objectives

5. Future projections from 2016-2020 are determined by Gompertz logistic curve "best fit" regression with market saturation point (asymptote for number of satellites) set at 170 nano/micro satellites in a year for the Announced Dataset and, 250 for the Optimistic Dataset.

• The Gompertz logistic regression provides an accurate market growth prediction for many industries, particularly high tech.

• Regression curve based upon best fit to data while still accounting for a market saturation point.

• Market saturation point (launch demand asymptote in 2030) set at 170 nano/micro satellites in a year, limited by:

i. Realistic number of manufacturers.

ii. Limits imposed by requiring a shared launch with a larger satellite.

iii. Others within Cubesat industry have proposed that total market just for Cubesats is over 600 Cubesats per year that will need launches

6. The Optimistic data set contains all currently known past and future nano/microsatellites from the SpaceWorks LDDB, with the addition of an inflating factor for known unknowns plus assumed sustainment of certain current projects and programs (e.g. follow-on to NASA Ames EDSN, Cubesat Launch Initiative, DARPA SeeMe).

7. These graphs are based on the Announced data set only, and do not include additional satellites contained in the Optimistic data set.

8. The sum number of future nano/microsatellites shown in this chart may not equal the sum shown on other charts. Nano/microsatellites for which the subject data of interest is unknown have been excluded from this chart.

9. Percentages may not sum to 100% due to rounding.

10. By some traditional definitions of space industrial sectors, non-defense government space activities are a subsector of the civil sector. Here we break out non-defense government activities into a separate sector. "Government" refers to those nano/microsatellite development efforts that occur within/by the government agency or organization (e.g. NASA, JAXA). Civil refers to all other non-defense development activities (e.g. universities, federally funded research institutions), though the funding source may be a government agency.

11. Nanosatellites are binned by rounding mass to the nearest whole number. Picosatellites less than 1 kg are not included.

12. 70 percent of future satellites in the Announced Dataset are less than 3 kg. This percentage is applied to the projections for 2012-2019 to arrive at the estimated number of satellites under 3 kg for each data point in the projection.

13. Based on 2010 ClydeSpace analysis.

It is also worthwhile to point out that the methodology used to project the above figures has some risks implicit. First, it assumes launches will occur as planned, but launches are often delayed and projects may not launch due to lack of funding. Second, it is dependent on the market saturation assumption; and third, data sets contain an inflating factor based on unknown satellites that could potentially not exist.

Specifically concerning launch delays, Spaceworks Commercial analyzed in a case study published in 2013, the main causes for delay of launches during 2012, being the main one launcher delays/challenges as summarized in Figure 11 below.

Identified reasons for delay in anticipated launch date of nano/microsatellites include the following:

Launch: A slip in schedule of the launch vehicle, delay in schedule of the primary payload, delay in development of the launch vehicle, inability of the satellite developer to identify or contract with a suitable launch provider.

Development: Satellite development technical or management challenges, delays in funding, delays due to suppliers, testing and qualification challenges, delays in government approvals (e.g. ITAR).

Combination: Both launch and satellite development delays occurred.

Unknown: The reason for delay cannot be readily determined.

In another study performed by the company Futron in 2006 focused on the small satellites segment characterized by satellites in the 100-200 kg weight class, development cost from $5M to 10M and 1-2 year life expectancy, results showed a potential addressable market size of 40-75 such satellites/year and $290-570M/year. Over 30 markets were identified, six of which emerged as most promising: Military (Science and technology; Intelligence, surveillance, and reconnaissance (ISR)); Civil/commercial communications (Polling of unattended sensors; Remote site communications); and Civil/commercial remote sensing (High-resolution Earth observation; Landsat-class data for environmental monitoring).

As a sum-up, in the current paradigm, that is, without a game changer in play such as a new launcher as Sagitarius, the projected estimate for nanosatellite launch demand in the optimistic scenario is 400 for the 2015-2019 timeframe, that is, around 100/year.

To the former figure, we may add up the new market scenario arising from the entrance of new and dedicated nanosatellites launchers, as Sagitarius,

Celestia Aerospace Products and Services

The cost associated to the production and launch of a satellite is very high.

The former has so far restricted space access to huge aerospace contractors, either public, private or a mixed formula of both, such as Space Agencies (ESA, NASA....) and main industry contractors (Astrium, Lockheed Martin, Space Systems Loral...). Meanwhile, newcomers to the space industry are pushing to get introduced to the sector and demand a cost-effective solution. Launch opportunities are still scarce whilst demand is increasing, thus resulting in a bottle-neck for the small satellites market.

Celestia Aerospace proposes a solution. On the one hand, Celestia Aerospace satellites are less costly to build, launch, operate and maintain.

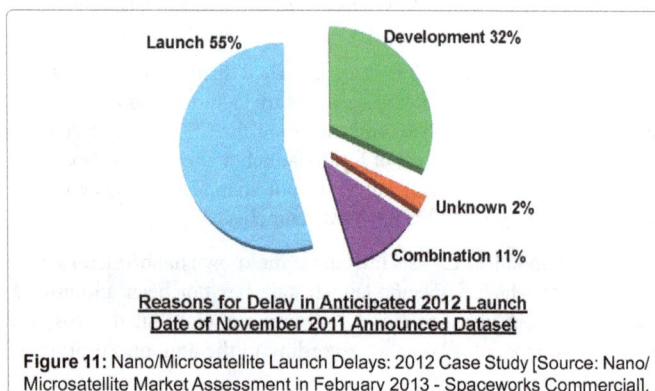

Launch 55% — Development 32% — Unknown 2% — Combination 11%

Reasons for Delay in Anticipated 2012 Launch
Date of November 2011 Announced Dataset

Figure 11: Nano/Microsatellite Launch Delays: 2012 Case Study [Source: Nano/Microsatellite Market Assessment in February 2013 - Spaceworks Commercial].

On the other, Celestia Aerospace provides a unique proprietary launch system specifically devoted to service the nanosatellite market: Sagitarius.

Celestia Aerospace is a start-up company headquartered in Barcelona (Spain) founded with the purpose of providing an integral, accessible, low-cost service for small satellite users, covering the whole value chain from design to manufacture, launch, operations and data delivery.

Celestia Aerospace goal is to build a low-cost, miniaturized and "user-friendly" small satellite, the "i-Sat", the i-phone of the satellites, both for the professional market as well as for the general public one, and deliver it to orbit with a dedicated nanosatellite launcher: Sagitarius.

Our solution focuses on meeting the needs of the small satellite users experiencing two main roadblocks: first one, the existence of a launch bottle-neck due to the lack of a launch system dedicated specifically to small satellites; second one, the lack of an integral service provider.

Celestia Aerospace will build cube-shaped low-weight satellites – ranging from 1 to 10 kg – and small dimensions – from 10 centimeters of edge – termed nanosatellites. The manufacturing will be undertaken from the company premises in Barcelona, and the integration of the satellite with the launcher will take place in the company's hangar, based at an airport in Spain.

The goal is for the client company to have a single interlocutor in the aerospace chain: client needs are determined and a customized solution is designed, which then is developed into a nanosatellite.

One of the keys for the low development cost is standardization and the use of off-the-shelf technologies, that is, widely used commercial applications, from sectors such as the multimedia or mobile communications.

For the development of systems and standards Celestia Aerospace has built up teams and established collaborations with Universities, such as the Technical University of Catalonia - Barcelona Tech.

The launch, orbital operation, and data management are developed by Celestia Aerospace as well. Satellite operations services covers in-orbit control; servicing in orbit including repairing and de-orbit; data gathering; and data handling to final user format. Therefore, the final aim is that the client has just to sit in front of the computer and wait for the desired data to download.

For the first time ever, the small satellite users will be able to cover all the value chain interacting with just one provider and have a dedicated launch system specifically addressed to this market, allowing for fast response times.

Celestia will develop, own and operate five specific satellites: BioPharmaSAT, SemicondSAT, TestSAT, SnapSAT and ClusterSAT aimed at the pharmaceutics/biotechnology sectors, semiconductor industry, aerospace industry and commercial services for Earth observation and Communications, respectively, to take full advantages of the microgravity research opportunities and offer them to small and medium enterprises, emerging economies, as well as educational and non-profit entetitis, opening space access to a wider audience for the benefit of all.

The Benefits of Microgravity Research

BioPharmaSAT

There is growing interest in biotech research in space, primarily on the International Space Station (ISS), but with challenges of access and cost. Small satellites can potentially provide a lower-cost (and more responsive means) of access to space for this option. Previous missions such as GeneSat have served as pathfinders for the former. The International Space Station can itself also be a target market for BioPharmaSAT as there exist small satellite synergies through the use of Nanoracks, cubesat adapters specifically designed for the ISS.

According to a study by Andrews Space and Technology [5], current and on-going research demonstrates the significant advantages of on-orbit research and manufacturing which has attracted the interest of pharmaceutical market leaders.

Tissue engineering is enabled by microgravity and may lead to treatments for many medical conditions. Liver tissue is the most likely early candidate for commercially viable space-based tissue engineering. Tissue engineering technologies, for example, have the potential to address diseases and disorders that account for about half of the USA's total healthcare costs. Tissue culture experiments performed on the space Shuttle and Russian Mir station have demonstrated the positive effects of microgravity on three-dimensional tissue growth and differentiation, and thus the potential for improved products. Liver disease in the United States resulted in 25.175 deaths in 1997, while only 4.000 people received a liver transplant (in 1996). Based on 1985 data, liver and gall bladder disease cost the US health care industry around US$17 billion. Space based tissue engineering could possibly save tens of thousands of lives and has the potential of saving care industry billions of dollars.

Small yield increases of recombinant drugs produced in microgravity may save millions in production costs. Space-based manufacture of recombinant drug could represent a substantial market. Recombinant protein drugs and diagnostic agents are one of the fastest growing segments of the pharmaceutical industry generating US$20 billion in annual revenues. Microgravity production of recombinant drugs offers the potential of improved quality and yield. An improvement in yield of only a few percent has the potential to save millions in production costs.

Access to microgravity laboratories is needed to drive market development. While biotechnology firms are aware of some of the advantages of microgravity, very few have performed microgravity experimentation for the manufacturing of biotechnology products. Like the semiconductor industry, biotechnology firms are in an extremely competitive and risky market arena. Also like the semiconductor industry, biotechnology firms spend between 10% and 15% of their annual revenues on R&D. In addition to the products mentioned before and their potential revenues, there is a significant demand for unique research and development facilities, which would likely include an orbital R&D laboratory in the future and small satellites at present. The biotechnology industry has US$365 billion in global annual revenues, which translates to between US$700 million and US$1.05 billion in weekly R&D expenditures.

SemicondSAT

According to a study by Andrews Space and Technology [6], the potential of space-based semiconductor manufacturing for the foreseeable future seems to be low. Industry leaders are continuing to scale down geometry features via wet processes and limited vacuum application. The potentially cleaner environment of space may not reduce defects and increase yield of semiconductor production because 95% of the contamination in today's processes are believed to come from process tools and are thus inherently internal to those processes.

Radical tool redesigns, aimed at eliminating those contaminants, are needed. Also, in two more generations microchips will have features less than 30 nm and semiconductors as we know them will not function due to quantum mechanical limits (electronic tunneling through CMOS gates). There are a number of alternate approaches in work and the availability of laboratories with microgravity and ultra-hard vacuum were definitely of interest.

Although it was not the focus of the Andrews Space and Technology study, interviews with "traditional" semiconductor manufacturers did uncover a significant interest for an On-Orbit Research Facility. The study highly recommends the investigation of an On-Orbit R&D facility as part of future studies. This stems from the fact that, within the next years, semiconductor companies will reach physical limits of material and present manufacturing processes, which they have refined over the last decade. Currently, they are searching for "revolutionary" methods of manufacturing follow-on generations of products. If an on-orbit research facility existed today, interviewees of the afore mentioned study would be willing to pay up to US$20 million for a single flight to conduct tests and build certain production elements that could lead to breakthrough material and manufacturing advancements. However, this market is only addressable if the companies are offered routine access: no less than once a month. Demand would significantly increase if the price for a week's research could be reduced to less than US$1 million.

The semiconductor market spends between US$20B and US$30B annually on R&D. This works out to between US$385M and US$577M per week. Based on the study's interviews feedback, if a new launch vehicle could provide weekly access, semiconductor companies could spend up to US$20M per week (3% of the world semi-conductor R&D funds) for the use of an Orbital R&D facility.

TestSAT and SnapSAT

TestSAT is a standardized cubesat aimed at offering technology validator and component testing services in space conditions. The payload compartment is highly configurable to meet the specific customer needs. Celestia Aerospace engineering service will constitute a team with its customer to identify the technology validator/testing setting and equipment, and data gathering to meet customer needs. The mission will be carried out during the in-orbit time, with data gathering and handling managed and provided by Celestia Aerospace, who will directly provide the data under the final user format needed by its customer.

SnapSAT is a standardized cubesat aimed at servicing current and future "big satellites" - satellites of more than 1.000 kg developed by main Space Agencies and main industry contractors, covering communications, navigation and scientific missions-. SnapSAT will have a standardized data communications port to dock to its standardized complimentary port aboard the target satellite. This communication port will have to become an industry standard so that it is adopted by main industry contractors and incorporated in their satellites. Once docked, SnapSAT will transfer data to update the target satellite systems. An extended version of SnapSAT will include extended satellite docking capabilities and component handling to broaden satellite services, comprising target satellite orbit correction, de-orbit, re-fuelling, repair and equipment replacing. All the former capabilities will be under Celestia Aerospace patent technology.

ClusterSAT for communications

ClusterSAT is a constellation of standardized cubesats that will jointly work to provide Earth Observation and Communication services.

As far as communications is concerned, the applications are varied. One of them concerns access to remote areas. Internet access worldwide is not yet guaranteed: either due to isolation and lack of terrestrial infrastructure in remote areas – currently 2/3rds of the world have no internet access at all-; either by Government control of the information flux in some countries (for example, China, North Korea). And even in those countries with internet access guaranteed, another problem adds to the previous ones: privacy and secure communications.

A Constellation of around 100 nanosatellites populating different orbital planes in a low-Earth orbit (LEO) of around 500 Km altitude orbit could provide private communication networks. For remote areas coverage, Wi-fi enabled devices and/or home radio antennas similar to those from satellite TV connected to a computer, or receivers plugged directly to a PC, would communicate with the satellites in their region, which in-turn communicate with other satellites and ground-based networks, thus forming the global network providing internet access in remote areas.

An intensive research area in Delay and Disruption Tolerant Networking (DTN) also known as "Ring Road network" is being applied to cubesat constellations [7]. Basically the DTN allows a cubesat to act as a data "mule" storing uploaded "bundles" of data until the cubesat's orbit takes it over a ground station connected to the Internet. The height and speed of the cubesat's orbit provides wide area coverage and speedy transit to its next download link. Thus, the Ring Road network aims to provide reliable epistolary data transfer and as such, it need not ensure uninterrupted end-to-end connectivity to assets on Earth. By tolerating episodic contacts, Ring Road may comprise solely nadir-pointing cubesats in LEO: during the time that a satellite is out of the view of any ground station, it retains in-transit data in a queue in its own local storage medium, awaiting its next ground station over flight.

While the innately high latency of this DTN communication rules out some kinds of network services such as Internet telephony, highly interactive web browsing, and massive multiplayer games, for example, supported applications would include:

1. Warnings of disaster events.

2. Requests for relief services.

3. Relief worker consultation, reporting, and direction.

4. Search and rescue support in remote areas.

5. Disease control information.

6. Weather forecasts.

7. Fish and game migration data.

8. Commodity pricing.

9. Distance learning.

10. Acquisition of data from remote sensors.

11. E-mail.

12. Research queries.

DTN offers also another advantage: incremental deployment. Because cross-links need not be maintained amongst the satellites, each satellite functions independently as a data transfer device. Ring Road could provide data communication service – albeit with extremely high latency and at very low data rates – throughout its intended coverage area even if only a single satellite were in operation.

Due to the low cost of building and launching each satellite, deploying a very large network is relatively inexpensive. Perhaps more importantly, deployment of such a network need not be accomplished all at once, so no large initial investment is needed and satellites can be deployed individually and opportunistically over a period of years. Also network maintenance will be significantly lower than with other satellite concepts as failed satellites could be replaced quickly and at low cost.

Main current restrictions concerning the use of cubesats for internet communications are basically three: limitations in band-width; rapid velocity of the satellite due to its LEO orbit; and inter-satellite communication. Nevertheless the former can be overcome.

First one concerning the available band-width can be overcome by increasing the power available to the satellite by means of larger solar cell panels, and/or by means of future technology improvements in this area. Band-width of up to 2Mbs are available right now with current state-of-the art technology and our team is working on reaching up to 5 Mbps, thus making already possible the use of narrow-band internet communications.

LEO orbit causes rapid transition times of the satellite over the ground network stations, thus posing restrictions in window-time for ground-link communications. This is overcome by increasing the number of cubesats in orbit. Our proposal is using up to 150 cubesats in a 500 Km altitude near-circular polar orbit. This restriction can be further diminished by increasing the number of "ground stations". Keeping in mind that ground stations for cubesats could be as simple as a radio receiver plugged-in a PC, potentially a huge number for cubesats ground stations could exist.

The last restriction, inter-satellite communications is really the one posing the major issue. Even for big satellites, this issue is still an under development one in part because up to now big satellites, with some exemptions, have been working alone instead of being integrated in constellations or clusters. Exemptions are mainly in research missions by NASA or ESA, where big satellites need to work in constellations such as the missions Darwin or the under development one LISA, where coordination is needed to perform interferometry. Or the big satellite constellation in LEO of IRIDIUM. NASA is working in inter-satellite optical communications. In the domain of constellations of cubesats, as there is none yet in orbit, nothing has been implemented, constituting this, therefore, another technology niche. Many research papers are currently available in the literature dealing with this issue. The DTN or Ring Road Network introduced here could be a solution to this problem as it does not require inter-satellite communications.

Many applications are innately delay-tolerant, such as e-mail, news feeds, weather advisories, archiving, backups, documentation of events, etc…i.e., any communication where reliability and security are more important than timeliness. These include also much electronic commerce and finance: merchandise used to be purchased by e-mail order, from catalogues received in the email; Banking – deposits, payments, statements – used to be done by e-mail. In every case, applications on a high-latency network can provide the same functionality but much more quickly and securely.

ClusterSAT for Earth Observations (EO)

Earth Observation has wide spread applications: use of near real time imaginary for social media, to hydrology, agriculture, resource monitoring, environmental monitoring, forestry, intelligence services, topography, traffic or urban development.

These services have been provided so far by big satellites but nanosatellites could serve them at more affordable prices and with resolutions starting at 5 meters.

From the plethora of cubesat missions designed in the last decade, only nine missions were identified that perform or would perform Earth Observation measurements other than space weather. These missions carry optical cameras, GNSS receivers for occultation measurements, photometers, and millimeter-wave sounders. The small number or missions, and the lack of variety in their payloads, are noticeable. The reasons for that are not much of a surprise: the stringent mass and dimension requirements of the cubesat bus translate into reduced mass, power, and data rate capabilities offered to payloads when compared to those of larger missions. Many of the currently used Earth Observation payload technologies (SAR, lidar, high-resolution optical imagers, hyperspectral imagers) are simply not compatible with these constraints.

However, many studies [8] have identified at least a few technologies that are compatible of these stringent constraints. These technologies include spectrometers with limited imaging capability, precise accelerometers, and broadband radiometers. These technologies would enable a broad variety of measurements with high societal and scientific return including ocean color, ocean mass distribution, glacier mass distribution, vegetation state, and Earth radiation budget amongst others.

A cubesat-based component having large constellations of cubesat carrying these technologies and taking these measurements would be also an extremely high value added asset for the so-called Earth Observing System-of-Systems (EOSS). First, such a system could take care of a fraction of the requirements for the EOSS, thus reducing the burden on larger satellites, and allowing them to focus on the highest performance missions, which cubesats are incapable of. Second, it would help close some of the expected data gaps in key measurements (e.g., gravity measurements). And finally, they would provide unprecedented data products with very high temporal resolution and relatively high spatial resolution, which could potentially create new opportunities for science. In particular, such measurements could be combined with data products from higher performance instruments using disaggregation schemes.

Within the Earth Observation domain, we can concentrate also on three main market opportunities: small satellite manufacture; EO data sale; and resulting downstream applications.

Upstream technology – Manufacturing: Nano/micro-satellite supply chain

The market for Nano/Micro-Satellites is growing quickly. In 2013, the number of small satellites (weighing less than 50 kg) launched was nearly as many as were launched in the previous three years. Indeed, the adoption of constellations by new businesses and corporations' intent on activities such as monitoring terrestrial assets is driving a forecast of 2000-2750 small satellites to be launched from 2014-2020, more than four times the number launched from 2000- 2012 [9].

While only 12% of small satellites were for EO and Remote Sensing purposes in 2009-2013, this proportion is expected to grow dramatically. Not only are the number of small satellites expected to treble over the next three years, the number of Nano/Micro-Satellites purposed for EO and Remote Sensing is expected to grow from 12% to 52% (i.e., from 24 to 338 spacecraft).

A recent market study by Euroconsult agrees with these growth estimates, forecasting the number of Earth Observation satellites (non-meteorology) launched by civil government and commercial entities to more than double over the next eight years to 290 satellites and £17.4 billion in manufacturing revenues over the period, an 84% increase over the previous decade [10]. The market for EO satellite manufacturing is estimated to reach £2.1 billion in 2020 alone. While Euroconsult does provide a comprehensive industry analysis and forecast in the growing EO sector, it does not distinguish between Nano/Micro-Satellites and their larger brethren. One can however extrapolate the value of the Nano/Micro-Satellite manufacturing market [11] based on the number of small satellites forecasted by SpaceWorks and a representative cost per satellite of £250.000 [12]. Therefore, 1.040 Nano/Micro EO satellites (52% of 2.000), at an average cost of £250.000 per satellite, give an estimated market value of £260 million over the next six years, or approximately £100 million in 2020 alone, for EO Nano/micro satellites manufacturing.

EO Data sale – Disruptive nano/micro-satellite constellations

Conventional imaging satellites are the size of a van, weigh tones, and cost hundreds of millions to build and launch. As a result, there are only a handful of operators, and the world fleet of commercial imaging satellites consists of less than 20 (excluding the 28 Nano- Satellite constellations that was just launched by Planet Labs).

Additionally, commercial satellite operators have not embraced a proactive approach to drive innovation in the sector, and tend to only take pictures when their operators receive orders from customers. Indeed, a limited number of large expensive satellites have resulted in a product that is expensive and untimely, and as a result, has not been adopted by the applications market.

However, it is expected that new entrants will challenge the existing commercial data pricing strategies, moving EO data sales opportunities from an end to a means for delivering commercial value-added downstream applications. Small satellite trends are shifting away from one-time stints and moving toward more widespread adoption, enabled by constellations.

The challenge is that space assets are traditionally extremely valuable, extremely expensive, and extremely risky. So the next technological frontier for quality imaging from space involves systems that can both capture high quality (resolution) data to show economic activity and be cost-effective enough to deploy in large numbers (timeliness). Resolution and timeliness are the two critical elements required to realize the growth potential of EO, reach a critical mass of adoption, and ultimately become self-sustaining.

Since most businesses are interested in economic activity of some kind, the most valuable data is that which can monitor at the economic scale (≤ 1 m) and temporal scale (daily). Existing EO systems have not met the needs of many business applications because they lack the critical combination of timeliness and sub-meter resolution. High-profile start-ups such as Planet Labs have taken a huge step forward in terms of improving temporal resolution, but still lack the economic resolution to drive mainstream adoption.

The biggest customers of conventional commercial imaging satellites are governments, in particular intelligence agencies and the military. These high cost, high capability products are priced far out of reach for many other potential users, including researchers, in areas

as diverse as farming, forest carbon management, regional and local planning, and environmental stewardship. By bringing the cost down and accessibility up, new entrants in this marketplace hope to spur a proliferation of innovative uses. Another innovative approach under discussion is to offer heavy discounts, or even make imagery free, to academics and non-governmental organizations, in order to increase usage.

Making EO data freely available in order to build services has been and will continue to deliver success. It is not expected that all data will lead to the development of revenue-generating services; however, the wider dissemination of EO data across applications, industries, and universities, etc., will increase adoption of EO as a new data set. In addition to creating more cost-effective services (by reducing the cost of the data product), it is felt that exposing more individuals and sectors to EO data will help increase the demand of paid- for commercial solutions if a higher resolution, high-accuracy product is required to support applications' development. The quality of service attached to the data is still recognized as carrying value: Co-existence of free and paying data is therefore still possible, provided that the paying data is delivered in a suitable approach.

Euroconsult estimates that the market for commercial EO data will reach £1.8 billion by 2020 (9% CAGR), with the greatest amount of growth coming from non-defense markets. According to their latest industry report, this is driven by demand to "support wider economic growth with applications spanning natural resources monitoring, engineering and infrastructure, and defense [13]".

The needs of the £1.8 billion market opportunity cannot be met entirely by future.

Nano/Micro-Satellite missions due to the following technical considerations:

a. Synthetic Aperture Radar (SAR): £340 million is attributed to SAR images, which are not currently feasible using a Nano/Micro-Satellite platform due to size and power constraints. While technology is improving, this is not likely feasible for another 5-10 years.

b. Very High Resolution (VHR): £1.1 billion is attributed to VHR optical images (ground resolution less than 1 meter) which are currently difficult to acquire from Nano/Micro- Satellite platforms, although new technologies are enabling greater capability. Additionally, £400 million of the VHR value is expected to come from US Defense, who demands resolution that is likely out of reach of Nano/Micro-Satellite platforms. Therefore, the VHR market captured by Nano/Micro-Satellite platforms in 2020 is expected to be £530 million, or 50% of the total.

c. High-Medium Resolution (H-MR): H-MR optical images (>1 meter ground resolution), an estimated £440 million market value in 2020, is fully within the reach of Nano/Micro- Satellite mission capabilities.

Therefore, the size of the commercial EO data market reachable by Nano/Micro-Satellites platforms is estimated to reach £970 million by 2020, or 54% of the total £1.8 billion market.

1. Downstream applications include increasing the profitability of business.

2. A strong upstream sector will enable the growth of the downstream sector (Figure 12). Indeed, the downstream sector will experience significant disruption with the arrival of Nano/Micro-Satellite constellations. For reference, the ratio of upstream turnover to downstream in other major space-faring nations such as Germany and

France (including satellite manufacturing, launch services, and ground systems) is approximately 1:12 (for example, for the United Kingdom this ratio is of 1:6).

The following examples have been identified as emerging opportunities already addressed by Nano/Micro-Satellite missions' precursors:

Agriculture health monitoring: Monitor crop health and forecast crop yields with timely sub-meter imagery. Identify pest infestation and plan irrigation levels to augment your precision agriculture techniques.

Humanitarian aid: Monitor refugee movements and infrastructure development in conflict areas to aid humanitarian efforts.

Insurance modelling: Inform risk exposure models to increase efficiency and profitability. Frequently monitor high value assets for change.

Oil storage monitoring: Monitor oil storage containers with sub-meter imagery for changes in volumes to inform commodity trading decisions.

Natural disaster response: Aid first responders in rescue coordination. Monitor long- term recovery and relief efforts.

Oil and gas infrastructure monitoring: Explore potential sites or monitor existing property and infrastructure for safety and security. Detect the intrusion of vehicles, new construction, or vegetation on pipeline corridors.

Financial trading intelligence: Access proprietary information to make more informed and competitive investment decisions. Identify changes in relevant metrics like the number of cars in a retailer's parking lot or size of stockpiles of natural resources in ports.

Mining operations monitoring: Explore new sites or monitor ongoing projects. Identify specific rock topologies and geological structures associated with mineralized areas. Obtain up to date imagery for evacuation planning.

Carbon monitoring: Create reliable carbon stock baselines and improve land cover maps.

Maritime monitoring: Monitor ships entering and exiting ports with HD video to inform supply chain optimization decisions. Validate AIS data and analyse container activity in ports.

Retail: Traffic gauging of parking lots to find out how many shoppers expected at every hour of every day.

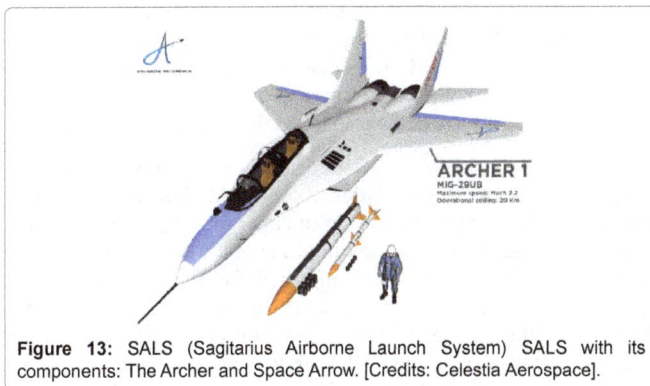

Figure 12: Nano/Micro-satellites downstream market nano/micro-satellites to deliver 54% of downstream market value in 2020 source: Satellite applications: Small is the new big – white papers.

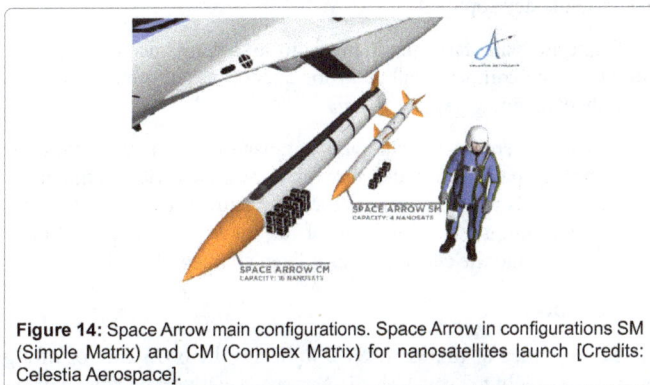

Figure 13: SALS (Sagitarius Airborne Launch System) SALS with its components: The Archer and Space Arrow. [Credits: Celestia Aerospace].

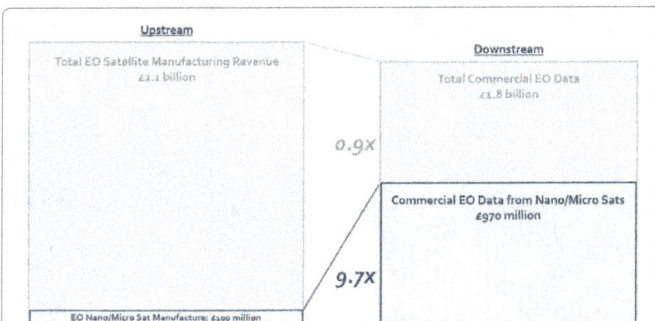

Figure 14: Space Arrow main configurations. Space Arrow in configurations SM (Simple Matrix) and CM (Complex Matrix) for nanosatellites launch [Credits: Celestia Aerospace].

Figure 15: Target Orbit for Sagitarius [Credits: Celestia Aerospace].

Celestia Aerospace unique launch solution: The Sagitarius Launch System (SALS)

The SALS is the first of its kind and it will service both Celestia Aerospace's own nanosatellites and other companies that require a fast, flexible, and affordable launch solution (Figure 12).

The Sagitarius launch system is an airborne platform capable of reaching orbits of 600 Km of altitude, and it is composed of two components: The Archer, a demilitarized MiG-29UB supersonic jet (Figure 13); and The Space Arrow, a launcher based on a modified missile, with two load configurations: simple matrix, with a load capacity of 4 nanosatellites; and complex matrix, with a load capacity of 16 (Figure 14). In a single flight, the Archer can deploy four simple matrix Space Arrows or a single one in the complex matrix set-up. Therefore, Celestia Aerospace can transport up to 16 nanosatellites to orbit in a single operation.

Sagitarius will operate from an airport in Spain with targets orbits reaching LEO (Figure 15). Celestia Aerospace will also offer its clients the possibility of actually taking part of the flight from the backseat of the MiG-29UB and deploy the Space Arrow carrying their own nanosatellite. This is a world-wide first. The Archer will also be used to offer space tourism flights to 20 Km of altitude, where the flier will be able to see the curvature of the planet and the darkness of space.

The advantages of this new launch system are varied: its low cost compared to normal launch platforms for this kind of satellites, which now have to fly as 'piggyback' cargo in larger rockets; the just-in-time service, allowing for a record waiting time of 2 weeks, radically more flexible than the average year-and-a-half for the available launch opportunities; total mission focus, as the launch is strictly prepared for the nanosatellites, compared to the limitations of flying as secondary cargo of a larger satellite; and total flexibility in terms of calendar, as the launch can be moved without problems to accommodate delays in the nanosatellite development.

Sagitarius will start operations from an airport located in Spain. Growth of the company will allow for operations from many airports not only in Spain.

Celestia Aerospace will also manage launch opportunities through third parties provided through agreements with main launch companies. This will also serve as the main launch service for Celestia during appropriate development of its own launching platform, Sagitarius, so that the customer can still get an integral service.

Conclusion

Performance of the MiG-29UB indicates the possibility of lifting even more weight to orbit. Celestia Aerospace will start first by securing the service for the nanosatellite community. Future upgrading of the systems can allow to place up to 50 kg satellites in orbit (microsatellites).

With SALS Celestia Aersopace can offer far reaching market opportunities and benefits:

1. Low cost services compared to normal launch platforms for this kind of satellites, which now have to fly as 'piggyback' cargo in larger rockets

2. Just-in-time service, allowing for a record waiting time of 2 weeks (1 week when operating with two Archers), radically more flexible than the average year-and-a-half for the available launch opportunities

3. Total mission focus, as the launch is strictly prepared for the nanosatellites, compared to the limitations of flying as secondary cargo of a larger satellite

4. Total flexibility in terms of calendar, as the launch can be moved without problems to accommodate delays in the nanosatellite development.

5. New experience for the customer who can take part of the launch flying aboard the launch platform.

6. Sagitarius is the final ingredient of a full-cycle service offered by Celestia Aerospace to help a wider audience to reach space.

References

1. Space Foundation (2013) The Space Report.

2. De-Pasquale D, Bradford J (2013) Nano/Microsatellite Market Assessment. Spaceworks.

3. New Space Global (2014) Smallsat report information packet.

4. Future Space Technology (2001) Future Space Transportation Study - Phase 1 - Executive Summary (FSTS-1) Andrews Space & Technology.

5. Burleigh SC (2011) Towards a communications satellite network for humanitarian relief. Proceedings of the 1st International Conference on Wireless Technologies for Humanitarian Relief. Kollam, Kerala, India pp: 219-224.

6. Selva D, Krejci D (2012) A survey and assessment of the capabilities of cubesats for earth observation. Elsevier, Acta Astronáutica.

7. Buchen E (2014) Spaceworks 2014 Nano/Microsatellite Market Assessment.

8. Euroconsult/ESA (2016) Satellite-based earth observation. Market Prospects to 2022, Euroconsult.

9. Satellite Applications (2014) Small is the new big: Nano/Micro-Satellite missions for earth observation and remote sensing. Catapult.

10. Euroconsult/ESA (2025) Satellite-based earth observation. Market Prospects for 2025.

11. Euroconsult/ESA (2007) Value added services in the European space sector. Euroconsult/ESA.

12. Planet (2016) Aerospace knows-how meets Silicon Valley ingenuity. Agile Aerospace, Australia.

13. Stock K (2014) The most powerful sales tool at Lowe's satellites. Bloomberg.

23

Variation in Argument of Perigee for Near-Earth Satellite Orbits Perturbed by Earth's Oblateness and Atmospheric Drag Interms of Ks Elements

Lila S Nair*

Department of Mathematics, HHMSPB NSS College for Women, Thiruvananthapuram, India

Abstract

Analytical solutions with the KS element equations of motion due to the combined effect of zonal harmonics J2,J3 and J4 and drag by considering an analytical oblate diurnal exponential density model when density scale height varies with altitude is obtained using series expansion method. Terms up to third terms in e, eccentricity, c, a small parameter depending on the ellipticity of the atmosphere and second order terms in μ, gradient of the scale height altitude are considered. The KS element equations are numerically integrated (NUM) through a fixed step size fourth-order Runge-Kutta-Gill method having a very small step-size of half degree in the eccentric anomaly for comparing analytically integrated (ANAL) values. After 100 revolutions, decrease in argument of perigee, ω, at perigee height = 400 kilometer, e = 0.1 and inclination i = 20 and 80 degrees, are found to be 7.42 and 39.8 degrees. At i =80 degree, the percentage error = (ANAL - NUM) / NUM after 1 and 100 revolutions are 0.61 and 2.09.

Keywords: Ks elements; Zonal harmonics; Atmospheric drag; Analytic integration

Introduction

The dynamical system of a satellite motion perturbed by both atmospheric drag and gravitational attraction is nonlinear, non conservative in form and the integration of the system, in general, is analytically intractable. To predict the motion precisely a mathematical representation for these forces must be selected for integrating the resulting differential equations of motion. Some of the early studies and analytical difficulties for the coupled problem were addressed by de Nike [1]. Hoots [2] used the gravitational and atmospheric models as used by Lane [3] and arrived at an improved analytical solution. Well known and commonly used models [4-9]. The KS total-energy elements equations [10] is a very powerful method for numerical solution with respect to any type of perturbing forces as the equations are less sensitive to round-off and truncation errors in the numerical integration algorithm. Sharma worked with these KS element equations to compute very accurate short-periodic terms due to J_2, even for very high eccentricity orbits [11,12]. Sharma [13,14] expanded analytic solutions by series expansion method using analytical models for oblate exponential atmospheric density model, and a model of the same with the effect of diurnal bulge.

In this paper my attempt is to get analytical solutions with the KS element equations of motion for long-term motion by considering the perturbations due to the combined effect of Earth's zonal harmonics J_2 to J_4 and atmospheric drag. The model used here is an oblate diurnally varying atmosphere with variation of the scale height depending on altitude almost similar to my work with Sharma [15]. Using series expansion method third-order terms in e, eccentricity, c, a small parameter depending on the ellipticity of the atmosphere and second order terms in μ, gradient of the scale height altitude are collected. Only one of the eight equations is solved analytically to obtain the state vector at the end of each revolution due to symmetry in KS equations. Numerical studies with test cases reveal that there is a good comparison between the analytical (ANAL) and numerically integrated (NUM) values of the position as well as velocity vectors $\bar{x}$ and $\bar{x}^*$.

Equations of Motion

The KS element equations of motion of a satellite under the effect of perturbing potential V and additional perturbing force $\vec{P}$ [10] are

$$\frac{d\omega}{dE} = -\frac{r}{8\omega^2}\frac{\partial V}{\partial t} - \frac{1}{2\omega}\left(\frac{d\bar{u}}{dE}, L^T(\bar{u})\vec{P}\right). \tag{1}$$

$$\frac{d\tau}{dE} = \frac{1}{8\omega^3}\left[K^2 - 2rV\right] - \frac{r}{16\omega^3}\left(\bar{u}, \frac{\partial V}{\partial \bar{u}} - 2L^T(\bar{u})\vec{P}\right) - \frac{2}{\omega^2}\frac{d\omega}{dE}\left(\bar{u}, \frac{d\bar{u}}{dE}\right). \tag{2}$$

$$\frac{d\bar{\alpha}}{dE} = \left\{\frac{1}{2\omega^2}\left[\frac{V}{2}\bar{u} + \frac{r}{4}\left(\frac{\partial V}{\partial \bar{u}} - 2L^T(\bar{u})\vec{P}\right)\right] + \frac{2}{\omega}\frac{d\omega}{dE}\frac{d\bar{u}}{dE}\right\}\sin\frac{E}{2}. \tag{3}$$

$$\frac{d\bar{\beta}}{dE} = -\left\{\frac{1}{2\omega^2}\left[\frac{V}{2}\bar{u} + \frac{r}{4}\left(\frac{\partial V}{\partial \bar{u}} - 2L^T(\bar{u})\vec{P}\right)\right] + \frac{2}{\omega}\frac{d\omega}{dE}\frac{d\bar{u}}{dE}\right\}\cos\frac{E}{2}. \tag{4}$$

$K^2 = k^2(M+m)$, specifying attraction between two masses M and m, E, ω, t, r and k^2 are, respectively, the eccentric anomaly, angular frequency, physical time, radial distance and the gravitational constant.

The perturbing potential V[10] and the aerodynamic drag force $\vec{P}$ [16] per unit mass acting on a satellite of mass m are respectively

$$V = \frac{K^2}{r}\sum_{n=2}^{\infty} J_n\left(\frac{R}{r}\right)^n P_n \cos(v) \text{ where } \cos(v) = \frac{x_3}{r}, \tag{5}$$

R, equatorial radius, J_n's, dimensionless constants known as zonal harmonics. Using Equation (5),

$$V_2 = \frac{K^2 J_2 R^2}{2r^3}\left[-1 + 3\frac{x_3^2}{r^2}\right]; \quad V_3 = \frac{K^2 J_3 R^3}{r^4}\left[-\frac{3}{2}\frac{x_3}{r} + \frac{5}{2}\left(\frac{x_3}{r}\right)^3\right];$$

$$V_4 = \frac{K^2 J_4 R^4}{r^5}\left[\frac{3}{8} - \frac{15}{4}\left(\frac{x_3}{r}\right)^2 + \frac{35}{8}\left(\frac{x_3}{r}\right)^4\right]; \quad \left(\bar{u}, \frac{\partial V}{\partial u}\right) = -2(n+1)V_n \text{ and } \frac{\partial V}{\partial t} = 0,$$

where

*Corresponding author: Lila S Nair, Department of Mathematics, HHMSPB NSS College for Women, Thiruvananthapuram, India
E-mail: lilasnair@yahoo.co.in

$$P_n(x) = \sum_{k=0}^{m} \frac{(-1)^k (2n-2k)! x^{(n-2k)}}{2^n k!(n-k)!(n-2k)!} \text{ where } m = \frac{n}{2} \text{ if n is even and } m = \frac{n-1}{2} \text{ if n is odd.}$$

As in [16], $\vec{P} = -\frac{1}{2}\rho \frac{SC_D}{m}|\vec{v}_r|\vec{v}_r$ the effective area of the satellite, C_D, the drag coefficient, ρ, the atmospheric density at the point of calculating atmospheric drag force and $\vec{v}_r$, the velocity of the satellite relative to the ambient air. If $\vec{v}$ is the velocity of the satellite relative to the Earth's centre, then $\vec{v}_r = \vec{v} - \vec{v}_\alpha$ where $\vec{v}_\alpha$ is the velocity of the air relative to the Earth's centre. $\vec{v}_\alpha$ is assumed to be west to east,

$$\vec{v}_r \approx \vec{v}\left(1 - \frac{r_{p_0}}{v_{p_0}}\Lambda \cos i_0\right), \quad \Lambda \text{, the rotational rate of the atmosphere}$$

about the Earth's axis and i_0, the initial inclination, $r_{p_0} = a_0(1-e_0)$ is the initial perigee radius, $\vec{v}_{p_0}$, the velocity at the initial perigee. Then the drag force per unit mass tangential to the orbit can be written as $\vec{P} = -\frac{1}{2}\rho\delta|\vec{v}|\vec{v}$ where $\delta = \frac{FSC_D}{m}$, and

$$F = \left(1 - \frac{r_{p_0}}{v_{p_0}}\Lambda \cos i_0\right)^2. \tag{6}$$

Following [17] the density function for an oblateness atmosphere together with day-to-night density variation is

$$\rho = \rho_{p_0}(1 + F\cos\phi)\exp\{-\beta(r-\sigma)\}. \tag{7}$$

To express $\cos\varphi$ in terms of the true anomaly θ and then in terms of the eccentric anomaly E let (α_s, δ_s) and (α_B, δ_B) are the right ascension and declination of the sun and the day time bulge respectively, then $\alpha_B = \alpha_s + h; \delta_B = \delta_s$. We can write

$$\cos\phi = A\cos\theta + B\sin\theta \tag{8}$$

$$A = \sin\delta_B \sin i \sin\omega + \cos\delta_B\{\cos\Omega - \alpha_B)\sin\omega + \cos i \sin(\Omega - \alpha_B)\cos\omega \tag{9}$$

$$B = \sin\delta_B \sin i \sin\omega + \cos\delta_B\{\cos\Omega - \alpha_B)\sin\omega + \cos i \sin(\Omega - \alpha_B)\cos\omega \tag{10}$$

The scale height H is known to increase with altitude and this variation of H will have an influence upon its motion. The value of H may be taken as $H = H_p + \mu(r - r_p)$ where $|\mu| < 0.2$ and for any particular value of r_p. To sum up, expression for the density, similar to that of Swinerd Boulton [17], is

$$\rho_v = \left[1 + \frac{1}{2}\mu z^2(1 - 2\cos E + \cos^2 E) + O(\mu c)\right]\rho, \rho \text{ from equation (7) and } z = \frac{ae}{H} \tag{11}$$

Analytical Integration

$$\vec{u} = (u_1, u_2, u_3, u_4) = \vec{\alpha}\cos\left(\frac{E}{2}\right) + \vec{\beta}\sin\left(\frac{E}{2}\right) \quad \vec{u}^* = \frac{d\vec{u}}{dE} = \frac{1}{2}\left[-\vec{\alpha}\sin\left(\frac{E}{2}\right) + \vec{\beta}\cos\left(\frac{E}{2}\right)\right]$$

$$\vec{x} = (x_1, x_2, x_3) = L(\vec{u})\vec{u}, \qquad r = (x_1^2 + x_2^2 + x_3^2)^{\frac{1}{2}} = u_1^2 + u_2^2 + u_3^2 + u_4^2,$$

$$\dot{\vec{x}} = (\dot{x}_1, \dot{x}_2, \dot{x}_3) = L(\vec{u})\vec{u}^* \quad w = \left[\frac{1}{2}\left(\frac{K^2}{r} - \frac{1}{2}|\dot{\vec{x}}|^2 - V\right)\right]^{\frac{1}{2}}, \quad t = \tau - \frac{1}{\omega}(\vec{u}, \vec{u}^*)$$

$$L(\vec{u}) = \begin{pmatrix} u_1 & -u_2 & -u_3 & u_4 \\ u_2 & u_1 & -u_4 & -u_3 \\ u_3 & u_4 & u_1 & u_2 \\ u_4 & -u_3 & u_2 & -u_1 \end{pmatrix}.$$

In terms of E,

$$r\cos\theta = a(\cos E - e), \quad r\sin\theta = a(1-e^2)^{\frac{1}{2}}\sin E \tag{12}$$

$$|\vec{v}| = \frac{K^2}{a^{\frac{1}{2}}}\left[1 + e\cos E + \frac{e^2}{2}\cos^2 E + \frac{e^3}{2}\cos^3 E\right] \tag{13}$$

$$\frac{|\vec{v}|}{r} = \frac{K^2}{a^{\frac{3}{2}}}\left[1 + 2e\cos E + \frac{5e^2}{2}\cos^2 E + 3e^3\cos^3 E\right] \tag{14}$$

The integrals available in the above theory are of the form

$$I(m,n) = \int_0^{2\pi}\cos^m E \sin^n E \, dE = 4\int_0^{\frac{\pi}{2}}\sin^m E \cos^n E \, dE \text{, if m and n are even}$$

and 0 if either m or n is odd.

$$= 2B\left(\frac{m+1}{2}, \frac{n+1}{2}\right) = 2\left(\frac{\Gamma\left(\frac{m+1}{2}\right)\Gamma\left(\frac{n+1}{2}\right)}{\Gamma\left(\frac{m+n+2}{2}\right)}\right), \quad \Gamma\left(\frac{1}{2}\right) = \sqrt{\pi} \text{ and } \Gamma(m) = (m-1)! \text{ if } m > 0.$$

Initial Conditions

Knowing the position and velocity vectors $\vec{x}$ and $\dot{\vec{x}}$ at the instant $t = 0$, the values of $r, \omega, t, \vec{u}_i$ and $\vec{u}_i^*$ can be computed [10], (pp. 91-92), and by adopting $E = 0$ as the initial value of the eccentric anomaly, we obtain $\vec{\alpha} = \vec{u}$, $\beta = 2\vec{u}^*$, $\tau = \frac{(\vec{u}, \vec{u}^*)}{w}$.

Numerical Results

In the entire test cases reported here, the values of ω, Right Ascension Node, Ω and mean anomaly, M are 60, 30 and 0 degrees respectively. The value of K^2, R, J_2, J_3 and J_4 utilized for numerical computations are $398600.8 km^3 s^{-2}$, 6378.135 kilometer and $1.0826157\text{Å} \times 10^{-3}, -2.53648D - 06$ and -1.52D-06 respectively. Jacchia (1977) atmospheric density model, which is relatively easier to use, is employed to compute the values of ρ_{p_0}, the density at the perigee and H, the density scale height at the end of each revolution. Arbitrarily 22 August 2002 is chosen as the initial epoch. The values of ε, $\wedge$ and $b_n = (m/c_D A)$, utilized during the computations are 0.00335, 1.2 and 50.0 respectively. In this model $c = \frac{1}{2}\beta r_{p_0}\sin^2 i$ approaches maximum value 0.2042 at e = 0.003 and i= 80° while minimum value 0.0232 at e = 0.005 and i= 20° respectively at the end of 100 revolutions (Figures 1 and 2). The values of 10.7cm solar flux ($F_{10.7}$) and averaged geomagnetic index (A_p) are taken as 150 and 10, respectively, which approximately represents an average density and results in exospheric temperatures between 1000 and 1100 K for the different cases we considered (Table 1).

We have transformed equations for $\frac{d\omega}{dE}, \frac{d\tau}{dE}$ and $\frac{d\vec{\alpha}_i}{dE}$, and using

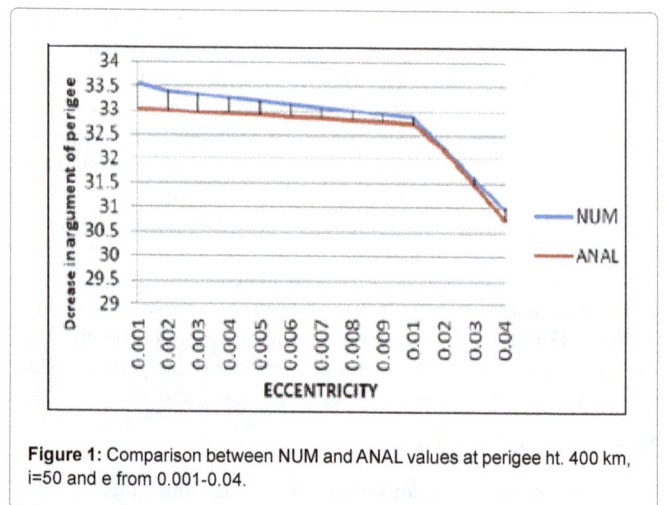

Figure 1: Comparison between NUM and ANAL values at perigee ht. 400 km, i=50 and e from 0.001-0.04.

e	Method	Perigee heights in Km					
		350			400		
		Inclination in Degrees					
		20	50	80	20	50	80
0.003	NUM	49.658	33.851	9.213	48.891	33.336	9.069
	ANAL	49.185	33.585	9.181	48.416	32.971	8.884
	% ERROR	0.95	0.79	0.35	0.97	1.1	2.04
0.004	NUM	49.561	33.783	9.193	48.796	33.269	9.05
	ANAL	49.104	33.584	9.156	48.327	32.942	8.866
	% ERROR	0.92	0.59	0.4	0.96	0.99	2.04
0.005	NUM	49.463	33.715	9.173	48.7	33.203	9.032
	ANAL	49.017	33.582	9.115	48.236	32.912	8.852
	% ERROR	0.9	0.39	0.64	0.95	0.88	1.99
0.007	NUM	49.2681	33.58	9.137	48.507	33.071	9
	ANAL	48.832	33.57	9.063	48.049	32.849	8.833
	% ERROR	0.89	0.03	0.8	0.94	0.67	1.81
0.008	NUM	49.171	33.513	9.118	48.411	33.005	8.978
	ANAL	48.734	33.558	9.039	47.953	32.814	8.825
	% ERROR	0.89	0.13	0.87	0.95	0.58	1.71
0.009	NUM	49.073	33.447	9.1	48.316	32.939	8.96
	ANAL	48.633	33.542	9.024	47.857	32.777	8.817
	% ERROR	0.9	0.28	0.84	0.945	0.49	1.6

Table 1: Decrease in Argument of Perigee.

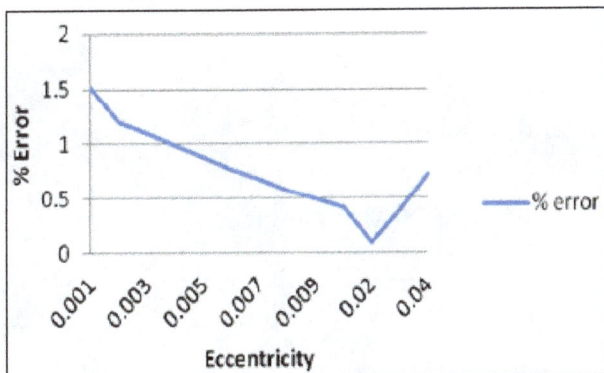

Figure 2: % error between NUM and ANAL values at perigee ht. 400 km, i = 50 and e from 0.001-0.04.

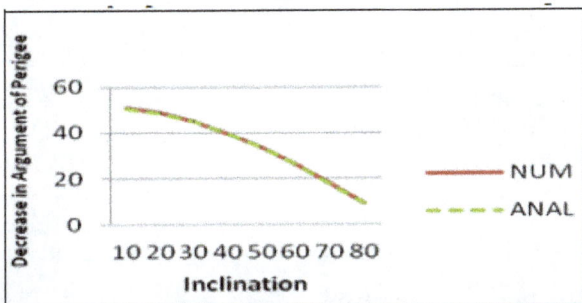

Figure 3: Comparison between NUM and ANAL values at perigee ht 350 km, e = 0.01 and i from 10-80 degree.

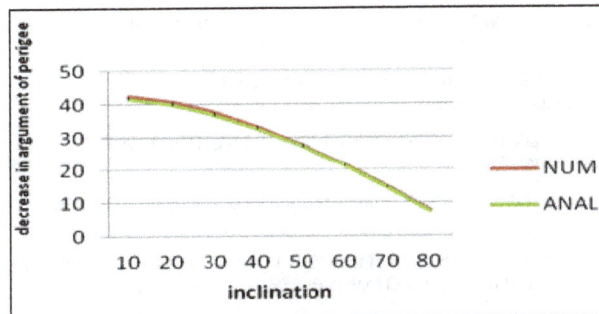

Figure 4: Comparison between NUM and ANAL values at perigee ht.400 km, e=0.1 and i from 10-80 degree.

e	After 1 revolution			After 100 revolutions		
	NUM	ANAL	% error	NUM	ANAL	% error
0.01	0.5029	0.5019	0.2	48.2202	47.7593	0.96
0.02	0.4931	0.4919	0.24	47.2798	46.7755	1.07
0.03	0.4835	0.4822	0.28	46.3662	45.8329	1.15
0.04	0.4743	0.4728	0.32	45.4785	44.9364	1.19
0.05	0.4653	0.4636	0.36	44.616	44.0643	1.24
0.06	0.4566	0.4548	0.4	43.7776	43.2031	1.31
0.07	0.4481	0.4461	0.44	42.9625	42.3474	1.43
0.08	0.4398	0.4377	0.49	42.1698	41.4952	1.6
0.09	0.4318	0.4295	0.55	41.3988	40.6458	1.82
0.10	0.424	0.4214	0.61	40.6485	39.7981	2.09

Table 2: Comparison of ANAL to NUM with % error at perigee ht 400 km, i = 80 after 1and 100 revolutions.

$\vec{\alpha}_i$ and $\vec{\beta}_i$ are known, they are converted into and $\bar{x}$ and $\dot{\bar{x}}$, which are further converted to the osculating orbital elements (Figures 3 and 4). Percentage Error = (ANAL – NUM)/NUM is calculated to check validity of the work. The algebraic computations are made with MAPLE12 mathematical software (Table 2).

Conslusion

The KS element equations are integrated analytically by a series expansion method by assuming an oblate diurnal atmosphere when density scale height varies with altitude and by including the terms corresponds to Earth's zonal harmonics J_2, J_3 and J_4. A wide range of eccentricity and inclination is considered for calculating the change in argument of perigee by present analytical theory and by numerical integration. Comparison between analytically and numerically integrated values for 1 and 100 revolutions shows that the analytically integrated values are reasonably accurate and thus highlights the usefulness of the analytical expressions. Graphical representation as well as the table presented here emphasizes the importance of developing the theory to find the decrease in argument of perigee.

References

1. De Nike J (1956) The Effect of the Earth's Oblateness and Atmosphere in a Satellite Orbit J of the Franklin Institute Monograph: 79-88.

2. Hoots FR (1981)Theory of the Motion of An Artificial Earth Satellite Celestial Mechanics 23: 307-363.

3. Lane MH (1965) The Development of an Artificial Satellite Theory Using a Power-Law Atmospheric Density Representation AIAA paper 65-35 AIAA 2 nd Aerospace Sciences Meeting New York.

4. Jacchia LG (1964) Static Diffusion Models of the Upper Atmosphere with

equations (11), (12), (13), (14) and programmed in double precision arithmetic to compute the KS elements $\bar{\alpha}_i$ and $\bar{\beta}_i$ (i = 1,2,3,4). Once

Empirical Temperature Profiles SAO SP: 170.

5. Jacchia LG (1970)New Static Models of the Thermosphere and Exosphere with Empirical Temperature Profile SAO SP: 313.

6. Jacchia LG (1971) Revised Static Models of the Thermosphere and Exosphere with Empirical Temperature Profiles SAO SP: 332.

7. Jacchia LG (1977) Thermospheric Temperature Density and Composition: New Models SAO SP: 375.

8. Hedin AE (1987) MSIS-86 Thermospheric Model J Geophys Res 92(A5): 4649-4662.

9. Hedin AE (1991) Extension of the MSIS Thermosphere Model into the Middle and Lower Atmosphere. J Geophys Res 96(A2): 1159-1172.

10. Stiefel EL, Scheifele G (1971) Linear and Regular Celestial Mechanics. Berlin/ Heidelberg/New York 1971. Springer-Verlag.

11. Sharma RK (1993) Analytical Short-Term Orbit Predictions with J3 and J4 in terms of KS Elements. Celes Mech and Dynam Astron 56: 503.

12. Sharma RK (1997)Analytical Integration of K-S Elements Equations with J2 for Short-Term Orbit Predictions. Planet Space Sci 45: 1481- 1486.

13. Sharma RK (1991) Analytical Approach using KS Elements to Near-Earth Orbit Predictions Including Drag. Proc Roy Soc Lond A 433: 121-130.

14. Sharma RK (1997) Contraction of Satellite Orbits using KS Elements in an Oblate Diurnally Varying Atmosphere. Proc Roy Soc Lond A 453: 2353- 2368.

15. Nair LS, Sharma RK (2003) Decay of Satellite Orbits Using KS Elements in an Oblate Diurnally Varying Atmosphere with Scale-height Dependent on Altitude. Adv Space Res 31: 2011-2017.

16. King-Hele DG (1987) Satellite Orbits in an Atmosphere. Theory and Applications Blackie Glasgow and London.

17. Swinerd GG, Boulton WJ (1982) Contraction of Satellite Orbits in an Oblate Atmosphere with a Diurnal Density Variation. Proc Roy Soc A 383: 127-145.

Permissions

List of Contributors

Ahmed Soliman M.Sherif
Novosibirsk State Technical University, Russia

Wang H, Guo L and Wu P
Key Laboratory of Space Utilization, Technology and Engineering Center for Space Utilization, Chinese Academy of Sciences, PR China

Luiz Sampaio Athayde Junior
Professor at School of Accounting Sciences, Federal University of Bahia, Brazil

Xuan H, Cheng S and Fang L
China Academy of Aerospace Aerodynamics, Yungang West Road, Beijing, 100074, P. R. China

Rathina Kumar V, Nanda M and Jayanthi J
Department of Aerospace, Electronics and System Division, CSIR-National Aerospace Laboratories, Bangalore, India

Mazzola L, Bruno G, Galasso B, Quaranta V, Albano F and Auletta A
CIRA - Italian Aerospace Research Centre, Via Maiorise 1, 81043 Capua, Italy

Coriand L
Fraunhofer Institute for Applied Optics and Precision Engineering, Albert-Einstein-Str. 7, 07745 Jena, Germany

Gangadharan S
Department of Mechanical Engineering, Embry-Riddle Aeronautical University, 600 South Clyde Morris Blvd, Daytona Beach, Florida 32114, USA

Baliga SV, Sonawane NH and Sathya-narayan P
Department of Aerospace Engineering, Embry-Riddle Aeronautical University, 600 South Clyde Morris Blvd, Daytona Beach, Florida 32114, USA

Shruti Kamdar
Department of Design Engineering, EGA, Honda Engineering North America, 640 Colemans Crossing Blvd, Marysville, OH 43040, USA

Dinesh Kumar B, Shishira Nayana B and Shravya Shree D
Department of Aerospace engineering from M.L.R Institute of Technology, Dundigal, Hyderabad

Behbahani-Pour MJ and Radice G
Division of Aerospace Sciences, School of Engineering, University of Glasgow, Glasgow G12 8QQ, UK

Reddy MVR
Department of Aeronautical Engineering, Nitte Meenakshi Institute of Technology, India

Kis KI
Geophysics and Space Sciences Department, Loránd Eötvös University, 1117 Budapest Pázmány Péter Sétány 1/c. Hungary

Taylor PT
Planetary Geodynamics Laboratory, NASA/GSFC, Greenbelt, MD 20771, USA

Wittmann G
MOL Hungarian Gas and Oil Company, 1117 Budapest, Október Huszonharmadika u. 18. Hungary

Diaa AM, El-Dosoky MF and Ahmed MA
Department of Mechanical Engineering, Assiut University, Egypt

Alsarheed M and Sedaghat A
Department of Mechanical Engineering, Australian College of Kuwait Mishref, Kuwait

Clarence T Chang and Chi-Ming Lee
NASA Glenn Research Center, Cleveland, OH 44135, USA

John T Herbon
General Electric Company, Cincinnati, OH 45215, USA

Stephen K Kramer
Pratt and Whitney, East Hartford, CT 06108, USA

Abdelghany ES
Institute of Aviation Engineering, Cairo, Egypt

Abdellatif OE
Department of Mechanical Engineering, Benha University, Egypt

Elhariry G
Department of Mechanical Engineering, Cairo University, Egypt

Khalil EE
Department of Mechanical Engineering, Cairo University, Egypt

László Bacsárdi, Árpád Huszák and Aliz Szeile
Budapest University of Technology and Economics, Budapest, Hungary

Tamás Haidegger
Óbuda University, Budapest, Hungary

Shreeharsha HV, Shivakumar SG and Mallikarjun SG
Department of Aeronautical Engineering, MVJ College of Engineering, Bangalore, India

Hennecke C, Von der Haar H and Dinkelacker F
Institute of Technical Combustion (ITV), Leibniz University Hanover, Welfengarten 1A, 30167 Hanover, Germany

Donder ED1 Crosby N1 and Kruglanski M
Royal Belgian Institute for Space Aeronomy, Uccle, Belgium

Andries J and Devos A
Royal Observatory of Belgium, Uccle, Belgium

Perry C
STFC RAL Space, UK

Borries C
Institute of Communications and Navigation, German Aerospace Centre, Neustrelitz, Germany

Martini D
Norwegian Center for Space Weather/Tromsø Geophysical Observatory, Norway

Glover A and Luntama JP
ESA SSA Programme Office European Space Operation Centre Darmstadt, Germany

Landell-Mills N
Edinburgh University, 75 Chemin Sous Mollards, Argentiere 74400, France

Garcia-Cuadrado G
Celestia Aerospace, Av. Gran via Carles III 157, 08017 Barcelona, Spain

Lila S Nair
Department of Mathematics, HHMSPB NSS College for Women, Thiruvananthapuram, India

Index

www.ingramcontent.com/pod-product-compliance
Lightning Source LLC
Chambersburg PA
CBHW080657200326

41458CB00013B/4894